INTRODUCTION TO STOCHASTIC SEARCH AND OPTIMIZATION

A complete list of titles in this series appears at the end of this volume.

INTRODUCTION TO STOCHASTIC SEARCH AND OPTIMIZATION

Estimation, Simulation, and Control

JAMES C. SPALL

The Johns Hopkins University
Applied Physics Laboratory

A JOHN WILEY & SONS, INC., PUBLICATION

Published by John Wiley & Sons, Inc., Hoboken, New Jersey.
Published simultaneously in Canada.

For general information on our other products and services please contact our Customer Care Department within the U.S. at 877-762-2974, outside the U.S. at 317-572-3993 or fax 317-572-4002.

Wiley also publishes its books in a variety of electronic formats. Some content that appears in print, however, may not be available in electronic format.

Library of Congress Cataloging-in-Publication Data:

Spall, James C.
Introduction to stochastic search and optimization : estimation, simulation, and control / James C. Spall.
p. cm. — (Wiley-Interscience series in discrete mathematics)
Includes bibliographical references and index.
ISBN 978-0-471-33052-3 (cloth : acid-free paper)
1. Stochastic processes. 2. Search theory. 3. Mathematical optimization. I. Title. II. Series

QA274 .S63 2003
519.2—dc21 2002038049

10 9 8 7 6 5 4 3 2

CONTENTS

PREFACE

THE SUBJECT

Individuals and organizations are often faced with making tradeoffs in order to achieve desirable outcomes. Choosing these tradeoffs in the "best" way is the essence of the search and optimization problem. Mathematical algorithms for search and optimization play a large role in finding the best options in many problems in engineering, business, medicine, and the physical and social sciences. *Stochastic* search and optimization pertains to problems where there is random noise in the measurements provided to the algorithm and/or there is injected (Monte Carlo) randomness in the algorithm itself. A countless number of stochastic search and optimization problems arise in industry, academia, and government.

SUMMARY OF THE BOOK

Introduction to Stochastic Search and Optimization provides a broad survey of many of the most important methods in stochastic search and optimization. These include random search, recursive least squares, stochastic approximation, simulated annealing, evolutionary computation (including genetic algorithms), and reinforcement learning. Also included is a discussion of closely related subjects such as multiple statistical comparisons, model selection, simulation-based optimization, Markov chain Monte Carlo, and experimental design. These subjects are covered in 17 chapters. Each chapter ends with some concluding remarks, including comments on the historical perspective and on the nexus with other topics in the book. Five appendices review essential background information in multivariate analysis, matrix theory, statistical testing, probability theory, pseudorandom number generation, and Markov chains. All chapters and appendices include exercises. The book concludes with an extensive list of references.

Broadly speaking, the first part of the book (Chapters 1–12) is devoted to a description of core algorithms for stochastic search and optimization. The second part (Chapters 13–17) discusses some closely related subjects in modeling, simulation, and estimation, including some important applications of the algorithms in Chapters 1–12. The two courses summarized in "Some Educational Experiences" below are divided roughly along these two lines.

PREREQUISITES AND INTENDED AUDIENCE

Readers should have a working knowledge of basic probability and statistics, together with knowledge of multivariate calculus and basic matrix algebra. Previous optimization experience is helpful, but not required. With the exception of some of the material on probabilistic convergence (Section C.2 in Appendix C), random number generation (Appendix D), and possibly Markov processes (Appendix E), the appendices are largely a review of the prerequisite subject matter.

The book may serve as either a reference book for researchers and practitioners or as a textbook, the latter use being supported by exercises at the end of every chapter and appendix. The relatively modest prerequisites are intended to make the book broadly accessible, including to those who will actually *use* the methods in practical settings.

PHILOSOPHY OF THE BOOK

As any author must do, I had to make many choices about overall philosophy and topics to include or exclude. Aside from the inevitable personal biases, my choices were guided by the need to keep the book from reaching biblical proportions and the desire to make the book accessible to industrial practitioners and master's level students in engineering and other areas.

This book focuses on methods that have a solid theoretical foundation and that have a track record of effectiveness in a *broad range* of practical applications. Without such restrictions, there are an almost limitless number of methods that could be discussed (in particular, there are a huge number of highly specialized and/or ad hoc methods discussed in the literature). Even with these restrictions, there are a very large number of possible algorithms. Although the methods here are among the most popular in the field, the coverage is not encyclopedic.

Rather than simply present various stochastic search and optimization algorithms as a collection of distinct techniques, the book compares and contrasts the algorithms within a broader context of stochastic methods. From this perspective, the book demonstrates that relatively simple algorithms may perform as well as—or better than—other "exotic" algorithms that may have received a disproportionate amount of attention. A correct choice of algorithm may lead to significant cost savings through the production of a better solution to the problem and through savings in the effort required for implementation.

Because of the diverse areas in which this material applies, the book is not directed toward any specific application area. Nevertheless, it does include *some* applications-oriented examples, but these are designed to be easily understandable and not to require specialized knowledge. These simple examples are drawn from areas such as control engineering, transportation, music, chemistry, economics, medicine, and artificial intelligence. Further, the book includes an extensive list of references, many of which provide detail on specific applications.

TRUTH IN ADVERTISING

There are perfect books and there are finished books. As such, allow me to mention a few items and topics the book does *not* include. First, as mentioned above, this book does not include detailed discussions of many "serious" applications, although there are countless real-world uses of the methods here. Rather, the focus is on generally useful principles for a broad audience without devoting significant attention to the inevitable application-specific details of a serious implementation. Related to this, this book bucks the trend (particularly among statistics books) of including analyses of numerous real-world data sets. The aim is to encourage the reader to grasp basic principles and not be distracted by the nuances and scale of many such data sets. In an educational context, an instructor can introduce such data as desired or have students pursue one or more of the many listed references (and associated Web sites) that analyze such data.

A related distinction between this work and many others on algorithms is that this book is not filled with a large number of Monte Carlo simulation (numerical) studies. Certainly, the book includes *some* such studies, but the overall philosophy is to emphasize approaches that have a solid theoretical foundation and to present this foundation together with a *limited* number of simulation studies that augment the theory. The rationale for such an approach is that a strong reliance on numerical studies alone can provide misleading indications about the effectiveness (or lack thereof) of an approach. All simulation studies are problem specific, with results that may not transfer to other problems. Further, it is difficult to provide a fully objective numerical comparison when the algorithms must be "tuned" to the application. The temptation exists to spend more time tuning favored algorithms than tuning other algorithms.

Although this book presents the theoretical basis of the algorithms, it does not include lengthy proofs of such results. Rather, the reader is directed to the literature for detailed derivations and proofs. (Many of the proofs in the literature are quite intricate and use mathematics beyond the level of this book.) Nevertheless, results are presented rigorously, with the aim of having the reader gain an appreciation for the rationale and limits of various algorithms.

There is no standard list of topics to be covered in a book on stochastic search and optimization. Some experienced researchers and practitioners may find their favorite algorithms excluded here. Please accept my apologies for such omissions. *Some* of the specific technical topics not covered—which could well qualify as stochastic search and optimization—are stochastic programming, dynamic programming, Markov decision processes, and many of the optimization metaheuristics (such as tabu search and ant colony optimization). Some of these are omitted in the interest of providing balanced coverage of different general approaches (gradient-based versus nongradient-based, single path versus population-based, etc.) while keeping the book between only two covers. Others are omitted because there is little evidence of greater general efficiency or ease of use when compared to some of the basic methods that *are* discussed in the book. Although some may challenge the choice of topics, it is incontrovertible that the topics included are important, are relevant to the book's theme, and are widely used in industrial and other practice.

Finally, this book does not review, endorse, or seriously discuss available commercial software in stochastic search and optimization. (I am referring to specialized software—e.g., a commercial package implementing a genetic algorithm—rather than generic packages such as MATLAB or Microsoft EXCEL.) Many "serious" implementations of the methods here rely on such specialized software. An Internet search or perusal of the advertisements appearing in many technical magazines or related publications will provide a number of possible vendors.

COMPUTATIONAL EXPERIMENTS

The simulation studies are performed using Microsoft EXCEL and/or MATLAB (version 6). These familiar packages are powerful—and relatively easy—means of constructing and testing stochastic algorithms of the type in this book. Selected MATLAB M-files are available at the book's Web site (address below). The use of EXCEL in some statistical applications has caused controversy, mainly because of documented failures in certain cases.[1] The controversy has been exacerbated by an unfortunate tendency of Microsoft Corp. to stonewall in providing information to external researchers that would help in understanding the limits of EXCEL.[2] Nevertheless, the uses of EXCEL in this text are relatively benign applications of regression and basic descriptive statistics (we especially avoid using the EXCEL pseudorandom number generator, which appears to be one of EXCEL's weakest points). The ubiquity and relative ease of use of EXCEL make it a compelling tool for many of the experiments in this text. While EXCEL and MATLAB are powerful tools for carrying out the numerical exercises, the book is not explicitly tied to any specific computing environment.

WEB SITE

The Web site below contains supplementary information such as MATLAB M-files for many of the algorithms, instructional material, links to related sites, data files, and errata:

www.jhuapl.edu/ISSO

The book's site is also accessible through the John Wiley & Sons, Inc. site: *www.wiley.com/mathematics*.

---

[1]McCullough, B. D. and Wilson, B. (2002), "On the Accuracy of Statistical Procedures in Microsoft Excel 2000 and Excel XP," *Computational Statistics and Data Analysis*, vol. 40, pp. 713–721.

[2]McCullough, B. D. and Wilson, B. (1999), "On the Accuracy of Statistical Procedures in Microsoft Excel 1997," *Computational Statistics and Data Analysis*, vol. 31, pp. 27–37.

SOME EDUCATIONAL EXPERIENCES

Draft versions of this book have been used in courses at the Johns Hopkins University, Morgan State University, and the University of Maryland. At Johns Hopkins, I taught two graduate courses from this book, one in stochastic optimization and one in simulation and Monte Carlo methods. These were offered in the Department of Mathematical Sciences as part of the full-time graduate program and in the Applied and Computational Mathematics and Electrical and Computer Engineering Programs, which are divisions of the Johns Hopkins Part-Time Engineering Program (catering largely to people employed full-time and pursuing a master's degree by taking courses in the evening).

As a potential guide to faculty at other universities considering similar courses, the table on the following page lists the subjects in the above-mentioned two courses. The prerequisites for these courses are the same as the prerequisites for the book, as mentioned above. The courses cover a blend of theory and practical algorithm descriptions. The aim is to appeal to a diverse student audience, some of whom have "academic" (research-oriented) interests and some of whom have very practical needs for solving problems in industry and government. Given the nature of the subject matter, it is not possible—nor desirable—to fully divorce the theory and the applications.

Each chapter and appendix contains examples and exercises for the reader to use in practicing with the methods being described. A partial set of solutions to the exercises is included at the back of the book and at the book's Web site. Faculty using this book as a text for a course may request a more complete set of solutions by sending a letter to the author on institutional letterhead. The exercises are a blend of conceptual ("pencil and paper") problems and numerical experiments.

For the most part, the two courses in the table cover distinct material. However, because neither course is a prerequisite for the other, there is a small amount of overlap. Appendices A and B review material that most students should have encountered prior to these classes, but since "encountered" and "retained" are different things, I have found it useful to spend some time on these subjects. Appendices C, D, and E, on the other hand, include material that may not be familiar to some students, suggesting that these appendices might be covered more slowly. In the simulation course, some material is drawn from supplementary sources, as indicated directly below the table.

Aside from the graduate courses above, the material here has also been used in short courses at conferences sponsored by the Institute of Electrical and Electronics Engineers (IEEE), the American Statistical Association, the U. S. Department of Defense, and the Society for Computer Simulation.

Two one-semester courses taught at the Johns Hopkins University using material from this text. Table shows approximate order of presentation.

Stochastic Optimization	Simulation and Monte Carlo Methods*
Appendix A—Multivariate analysis Chapter 1—Issues in stochastic optimization	Handout—General issues in simulation Appendix A—Multivariate analysis Appendix B—Statistics
Appendix B—Statistics Appendix C—Probability theory Chapter 2—Direct search	Appendix C—Probability theory Chapter 13 (Sect. 13.1)—Modeling and bias–variance tradeoff
Chapter 3—Recursive linear estimation	Chapter 13 (Sects. 13.2 & 13.3)—Model selection and Fisher information matrix
Chapter 4—Stochastic approximation for root-finding	Appendix D and supplementary handout—Pseudorandom number generation
Chapter 5—Stochastic gradient methods	Chapter 7 (Sects. 7.1, 7.2, & 7.5) and Chapter 14 (Sects. 14.1 & 14.2)—SPSA and regenerative simulation systems
Chapter 6—Nongradient stochastic approximation (finite differences)	Chapter 14 (Sects. 14.3 & 14.4)—Common random numbers in simulation-based optimization
Chapter 7—Simultaneous perturbation stochastic approximation (SPSA)	Chapter 15—Gradient-based methods for simulation-based optimization
Appendix E—Markov processes Chapter 8—Annealing-type algorithms	Appendix E—Markov processes Chapter 16—Markov chain Monte Carlo
Chapter 9—Evolutionary computation I	Chapter 17 (Sects. 17.1 & 17.2)—Optimal experimental design for linear models
Chapter 10—Evolutionary computation II	Chapter 17 (Sects. 17.3–17.6)—Response surfaces and optimal design for nonlinear and dynamic models
Chapter 11—Reinforcement learning by temporal differences	Chapter 12 (Sects. 12.1, 12.2, & 12.5) and Chapter 14 (Sect. 14.5)—Statistical methods for selection using simulations
Chapter 12—Statistical methods for selection	

*In addition to the indicated material, several sessions are included in the simulation and Monte Carlo class on subjects such as variance reduction (importance sampling, common random numbers, etc.), model validation, discrete-event systems, and sensitivity analysis. Material for these subjects is drawn from sources such as the tutorial articles in recent volumes of the *Proceedings of the Winter Simulation Conference*, or the texts *Modern Simulation and Modeling* (1998) by R. Y. Rubinstein and B. Melamed (Wiley, New York) and *Simulation Modeling and Analysis* (2000) by A. M. Law and W. D. Kelton (McGraw-Hill, New York).

ACKNOWLEDGEMENTS

I greatly appreciate the support I received from my employer, the Johns Hopkins University, Applied Physics Laboratory (JHU/APL), and from my line supervisors. A project like this is well out of the mainstream at my workplace, but I had an environment and sustained support that allowed it to reach fruition. I also appreciate the support that I received from the JHU/APL Janney Fellowship Program, the U. S. Navy (contract N00024-98-D-8124), the JHU/APL Independent Research and Development (IRAD) Program, DARPA (contract MDA972-96-D-002), and the Maryland Applied Information Technology Initiative (MAITI).

Many individuals contributed to various aspects of this book. I thank the students in classes that I taught in the period 1996–2002, many of whom offered valuable comments on the suboptimal handouts that served as a temporary textbook. I apologize to them for having to suffer from material that was often more sub than optimal. Lieutenant Colonel (ret.) David Hutchison provided significant assistance in the design and solution for numerous exercises and examples. Dr. I.-Jeng Wang helped in writing Chapter 15 and Appendix D.

I am most grateful to the following people for their comments as reviewers. Their suggestions greatly improved the book's correctness and clarity.

Hossein Arsham	University of Baltimore
John A. Cristion	JHU/APL
Marco Dorigo	Université Libre de Bruxelles
Michael C. Fu	University of Maryland
Gu Ming Gao	Chinese University of Hong Kong
Laszlo Gerencsér	Hungarian Academy of Sciences
David Goldsman	Georgia Institute of Technology
Shannon M. Hall	JHU/APL
Stacy D. Hill	JHU/APL
Tito Homem-de-Mello	Ohio State University
David W. Hutchison	U. S. Army (ret.)
Victor A. Ilenda	JHU/APL
Matthew I. Koch	JHU/APL
Larry J. Levy	JHU/APL
William R. Martin	JHU/APL
John L. Maryak	JHU/APL
Allan D. McQuarrie	JHU/APL
J. Barry Nelson	JHU/APL
Fernando Pineda	JHU School of Public Health
Arlene C. Rhodes	Morgan State University
Payman Sadegh	United Technologies Corp.
John W. Sheppard	ARINC Engineering Services, LLC
Richard H. Smith	JHU/APL
I-Jeng Wang	JHU/APL
George Yin	Wayne State University
Thomas R. Young	JHU/APL

I also appreciate the comments from several anonymous reviewers that I received via John Wiley & Sons, Inc.

The JHU/APL in-house editorial and production team of Christine Fatz, Marion Sparks, Dawna Swartz, and George Travez helped immensely in taking my raw material and turning it into camera-ready form. Angioline Loredo of Wiley also provided many useful editorial comments. On a personal note, I appreciate—and extensively used in this project—the "stick-to-it-ness" that was instilled in me by my parents, Raymond and Dorothy Spall. And last, but not least, I would like to thank my long-suffering wife, Kathy Ceasar-Spall. Over the course of this six-year project, she "cleared the decks" in countless instances to allow me the needed time and freedom to work on the book. I dedicate this book to her and to our epsilons, Daniel and Sarah. They share with me the joy in finishing this book.

Disclaimer: The views expressed in the book do not necessarily reflect those of JHU/APL or any of the organizations that provided funding for work that is reported here. The author takes full responsibility for the contents, including the inevitable shortcomings. Readers may feel free to provide comments and corrections via the book's Web site.

CHAPTER 1

STOCHASTIC SEARCH AND OPTIMIZATION: MOTIVATION AND SUPPORTING RESULTS

Preparation is required before starting any journey. This chapter lays the groundwork for our study of stochastic algorithms. Section 1.1 introduces the two basic—and closely related—problems that are the focus of this book. Section 1.2 summarizes some of the fundamental issues one faces in tackling a problem in stochastic search and optimization. This includes some of the immutable limits to what can be expected in any practical problem. Although classical deterministic optimization is not a prerequisite to this book, a casual understanding of some deterministic methods will be useful here, especially where the methods are analogous to certain stochastic approaches. Hence, Sections 1.3 and 1.4 provide a brief review of some important results related to deterministic search and optimization. Section 1.5 offers some concluding remarks.

Although the majority of the results in this book do not require prerequisites beyond multivariate calculus, basic matrix theory, and introductory probability and statistics, it is not possible to fully analyze and understand some of the stochastic algorithms here without some more advanced mathematics. In that sense, Appendices A, B, and C may be considered companions to this introductory chapter. These appendices review some of the most important prerequisites and introduce a limited amount of material that goes slightly beyond the prerequisites. Appendix A is a summary of some results from multivariate analysis and matrix theory. Appendix B summarizes some of the standard one- and two-sample statistical tests that will be used here. Appendix C summarizes some essential concepts from probability theory.

And with that, let the journey begin!

1.1 INTRODUCTION

1.1.1 General Background

- Managers at a firm are faced with making short- and long-term investment decisions in order to increase profit.
- An aerospace engineer runs computer simulations and conducts wind tunnel tests to refine the design of a missile or aircraft.

- Researchers at a pharmaceuticals firm design laboratory experiments to extract the maximum information about the efficacy of a new drug.
- Traffic engineers in a municipality set the timing strategy for the signals in a traffic network to minimize the delay incurred by vehicles in the network.

The list above is a tiny sample of problems for which stochastic search and optimization have been used. All such problems involve tradeoffs. Choosing among the tradeoffs in the best way is the essence of multivariate search and optimization problems.

Mathematical techniques of search and optimization are aimed at providing a formal means for making the best decisions in problems of the type above. Given the difficulties in many real-world problems and the inherent uncertainty in information that may be available for carrying out the task, *stochastic* search and optimization methods—the theme of this book—have been playing a rapidly growing role.

Much of this book focuses on two general—and closely related—mathematical problems. Let Θ be the domain of allowable values for a vector $\boldsymbol{\theta}$.[1] The two main problems of interest are:

Problem 1	Find the value(s) of a vector $\boldsymbol{\theta} \in \Theta$ that minimize a scalar-valued *loss function* $L(\boldsymbol{\theta})$
	— or —
Problem 2	Find the value(s) of $\boldsymbol{\theta} \in \Theta$ that solve the equation $\boldsymbol{g}(\boldsymbol{\theta}) = \mathbf{0}$ for some vector-valued function $\boldsymbol{g}(\boldsymbol{\theta})$.

The vector $\boldsymbol{\theta}$ represents a collection of "adjustables" that one is aiming to pick in the best way. The loss function $L(\boldsymbol{\theta})$ is a scalar measure that summarizes the performance of the system for a given value of the adjustables. The domain Θ reflects allowable values (constraints) on the elements of $\boldsymbol{\theta}$. Other common names for the loss function are *performance measure*, *objective function*, *fitness function*, or *criterion*. While Problem 1 above refers to *minimizing* a loss function, a maximization problem (e.g., maximizing profit) can be trivially converted to a minimization problem by changing the sign of the criterion. This book focuses on the problem of minimization. The root-finding function $\boldsymbol{g}(\boldsymbol{\theta})$ (which is generally a vector) often arises via calculating the gradient (derivative) of the loss function (i.e., $\boldsymbol{g}(\boldsymbol{\theta}) = \partial L(\boldsymbol{\theta})/\partial \boldsymbol{\theta}$). More generally, $\boldsymbol{g}(\boldsymbol{\theta})$ may represent a collection of functions that are derived from physical principles related to the system under study.

---

[1] One digression before proceeding to mention a notational convention for this and the remaining chapters: As discussed in the "Frequently Used Notation" section near the back of the book, a vector will be treated as a column vector and a superscript T on a matrix or vector denotes the transpose operation. So, $\boldsymbol{\theta}^T$ is a row vector for a vector $\boldsymbol{\theta}$. All matrices and vectors are in boldface.

Versions of the two problems above arise in countless areas of practice. A huge amount of effort in society is devoted to "doing the most with the least," which is the essence of the optimization objective. More specifically, one faces problems 1 and/or 2 in virtually all areas of engineering, the physical sciences, business, and medicine. In many problems of practical interest, mathematical search *algorithms*—step-by-step procedures usually implemented on a computer—are used to produce a solution. The focus on algorithms, of course, is central to this book.

In many applications, the two problems above are effectively equivalent and one may choose a formulation oriented directly at optimization (Problem 1) or at root-finding (Problem 2). The choice of formulation typically depends on issues such as the basic applied problem structure and data format, the available and relevant search algorithms, the proclivities of the analyst and traditions of his/her field, and the available software. For example, if one starts with a problem of minimizing $L = L(\boldsymbol{\theta})$ (Problem 1) for a differentiable loss function, standard methods of calculus can sometimes convert this to a problem of finding a solution (root) of the vector equation $\boldsymbol{g}(\boldsymbol{\theta}) = \partial L/\partial\boldsymbol{\theta} = \boldsymbol{0}$ (Problem 2). There may be advantages to such a conversion if the appropriate mathematical assumptions are satisfied and if the information required to compute the gradient is available.

Conversely, Problem 2 can be converted directly into an optimization problem by noting that a $\boldsymbol{\theta}$ such that $\boldsymbol{g}(\boldsymbol{\theta}) = \boldsymbol{0}$ is equivalent to a $\boldsymbol{\theta}$ such that $\|\boldsymbol{g}(\boldsymbol{\theta})\|$ is minimized for any vector norm $\|\cdot\|$ (the most common norm is the Euclidean norm $\|\boldsymbol{g}\| = \sqrt{\boldsymbol{g}^T\boldsymbol{g}}$). This conversion from Problem 2 to Problem 1 is used to advantage in a wastewater treatment example in Section 2.3 and in a second-order stochastic approximation method discussed in Section 4.5. We clearly see that there is a close connection between optimization and root-finding, even in root-finding problems that are not motivated by optimization problems per se. Hence, to avoid the cumbersome requirement of separately discussing both optimization and root-finding in the text, most (but not all!) of the discussion in this book will be on search algorithms for optimization problems, with the expectation that the appropriate interpretation as a root-finding problem will be transparent. There will, however, sometimes be cases where a point needs to be highlighted relevant to a unique structure for root-finding. Chapter 4, in particular, discusses such cases.

The two remaining subsections in this section define some basic quantities and demonstrate their role in stochastic search and optimization.

1.1.2 Formal Problem Statement; General Types of Problems and Solutions; Global versus Local Search

As mentioned above, much of this book focuses on search and optimization problems associated with minimizing a loss function $L = L(\boldsymbol{\theta})$. This optimization problem can be formally represented as finding the set:

$$\Theta^* \equiv \arg\min_{\boldsymbol{\theta}\in\Theta} L(\boldsymbol{\theta}) = \{\boldsymbol{\theta}^* \in \Theta : L(\boldsymbol{\theta}^*) \le L(\boldsymbol{\theta}) \text{ for all } \boldsymbol{\theta}\in\Theta\}, \tag{1.1}$$

where $\boldsymbol{\theta}$ is a p-dimensional vector of parameters that are being adjusted and $\Theta \subseteq \mathbb{R}^p$ is the domain for $\boldsymbol{\theta}$ representing constraints on allowable values for $\boldsymbol{\theta}$. ($\mathbb{R}^p$ is Euclidean space of dimension p; see Appendix A.) The "$\arg\min_{\boldsymbol{\theta}\in\Theta}$" statement in (1.1) may equivalently be read as: Θ^* is the set of values $\boldsymbol{\theta} = \boldsymbol{\theta}^*$ ($\boldsymbol{\theta}$ the "argument" in "arg min") that minimize $L(\boldsymbol{\theta})$ subject to $\boldsymbol{\theta}^*$ satisfying the constraints represented in the set Θ. The elements $\boldsymbol{\theta}^* \in \Theta^* \subseteq \Theta$ are equivalent solutions in the sense that they yield identical values of the loss function.

The solution set Θ^* in (1.1) may be a unique point, a countable (finite or infinite) collection of points, or a set containing an uncountable number of points. The three examples below show simple cases illustrating these types of solution sets Θ^*.

Example 1.1—Θ^* contains unique solution. Suppose that $L(\boldsymbol{\theta}) = \boldsymbol{\theta}^T\boldsymbol{\theta}$ and $\Theta = \mathbb{R}^p$ (i.e., $\boldsymbol{\theta}$ is unconstrained). The value $\boldsymbol{\theta} = \mathbf{0}$ uniquely minimizes L. Hence, Θ^* is the single point $\boldsymbol{\theta}^* = \mathbf{0}$. ❑

Example 1.2—Θ^* has countable (finite or infinite) number of points. Let θ be a scalar and $L(\theta) = \sin(\theta)$. If $\Theta = [0, 4\pi]$, then $\sin(\theta) = -1$ (its minimum) at the points $\Theta^* = \{3\pi/2,\ 7\pi/2\}$, a countable set with a finite number (two) of elements. On the other hand, if $\Theta = \mathbb{R}^1$, then $\Theta^* = \{\ldots, -5\pi/2, -\pi/2, 3\pi/2, 7\pi/2, \ldots\}$, a countable set with an infinite number of elements. ❑

Example 1.3—Θ^* has uncountable number of points. Suppose that $L(\boldsymbol{\theta}) = (\boldsymbol{\theta}^T\boldsymbol{\theta} - 1)^2$ and $\Theta = \mathbb{R}^p$. This loss function is minimized when $\boldsymbol{\theta}^T\boldsymbol{\theta} = 1$, which is the set of points lying on the surface of a p-dimensional sphere having radius 1. When $p \ge 2$, Θ^* is an uncountable (but bounded) set. ❑

Unlike the simple motivational examples above, this book focuses on problems where L is sufficiently complex so that it is not possible to obtain a closed-form analytical solution to (1.1). This is far and away the most common setting for large-scale optimization problems encountered in practice. Hence to solve for at least one $\boldsymbol{\theta}^*$ in (1.1), an iterative algorithm is used—a step-by-step procedure for moving from an initial guess (or set of guesses) about likely values of $\boldsymbol{\theta}^*$ to a final value that is expected to be closer to the true $\boldsymbol{\theta}^*$ than the initial guess(es).

Because any one value in Θ^* is—with respect to L—as good as any other, it is common in practice to seek only one $\boldsymbol{\theta}^*$ in the algorithmic search process. That is, there is often no reason to use scarce resources to determine the full set Θ^*. Further, there may be "side considerations" that cannot be quantifiably represented in a loss function that may make it desirable to seek only one solution. For example, one solution may be more consistent with historical precedent or physical intuition and may therefore be preferable. Finally, in many problems, Θ^* may, in fact, contain only one $\boldsymbol{\theta}^*$ (i.e., L may have a *unique* minimum).

This book considers both continuous and discrete problems, reflecting the available methods in stochastic search and optimization. Figure 1.1 depicts three loss functions with $p = 2$. These loss functions illustrate the generic forms of most practical interest—a continuous function, a hybrid function (discrete in one element of $\boldsymbol{\theta}$; continuous in the other element), and a purely discrete function. In the continuous case, it is often assumed that L is a "smooth" (perhaps several times differentiable) function of $\boldsymbol{\theta}$. Continuous problems arise frequently in applications such as model fitting (parameter estimation), adaptive control, neural network training, signal processing, and experimental design. In the discrete case, $\boldsymbol{\theta}$ can take on only a countable (finite or infinite) number of values (e.g., $\Theta = \{0, 1, 2,\ldots\}$ for a scalar θ). A typical discrete problem is: For a fixed

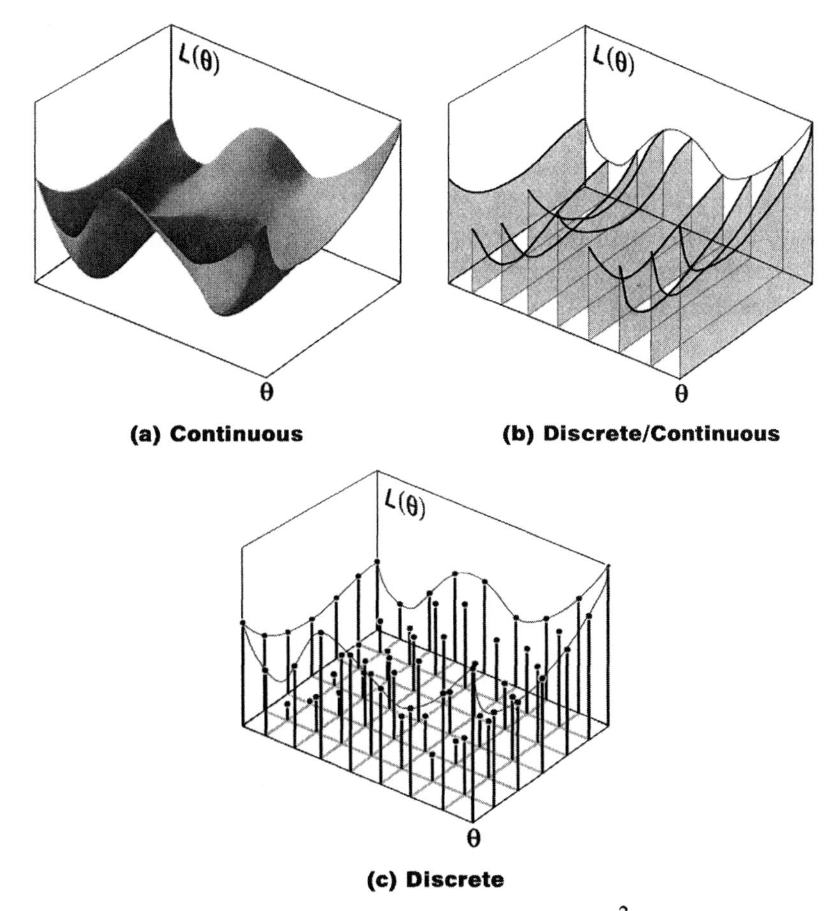

Figure 1.1. Three common types of loss functions for $\boldsymbol{\theta} \in \mathbb{R}^2$. Case (a) is continuous; case (b) is hybrid discrete/continuous (discrete in one element of $\boldsymbol{\theta}$; continuous in the other element); and case (c) is purely discrete in both elements of $\boldsymbol{\theta}$.

number (p) of categories of items, how many of each category are needed to optimize performance? Here, each element of $\boldsymbol{\theta}$ represents one category of items. Discrete optimization—sometimes called *combinatorial optimization*—is a large subject unto itself (resource allocation, network routing, policy planning, etc.).

Together with continuous search and optimization, discrete optimization appears as part of Chapters 2 and 7–10 on random search, stochastic approximation, simulated annealing, and evolutionary computation. Moreover, Chapter 12 and Section 14.5 are devoted exclusively to discrete problems via a discussion of statistical selection methods for comparing a finite number of options.

One of the major distinctions in optimization is between global and local optimization. All other factors being equal, one would always want a globally optimal solution to the optimization problem (i.e., at least one $\boldsymbol{\theta}^*$ in the set of values Θ^*, with each $\boldsymbol{\theta}^* \in \Theta^*$ providing a lower value of L than any $\boldsymbol{\theta} \notin \Theta^*$). In practice, however, a global solution may not be available and one must be satisfied with obtaining a *local* solution. A local solution is better than any in its vicinity (i.e., has a lower value of L), but it will not necessarily correspond to a best (global) solution $\boldsymbol{\theta}^*$ in Θ^*. For example, L may be shaped such that there is a clearly defined minimum point over a broad region of the domain Θ, while there is a very narrow spike at a distant point. If the trough of this spike is lower than any point in the broad region, the local optimal solution $\boldsymbol{\theta}_{\text{local}}$ is better than any nearby $\boldsymbol{\theta}$, but it is not be the best possible $\boldsymbol{\theta}$.

For a case where the problem dimension $p = 1$ (i.e., a scalar θ), Figure 1.2 depicts two problems with distinct local and global minima. Figure 1.2(a) shows a relatively benign setting where, despite the presence of the dip containing the local minimum, many algorithms will be able to find θ^* due to the broad indications of its presence in the nearby values of L. Figure 1.2(b), on the other hand, is an extreme form of the narrow spike idea. Here θ^* is one isolated point. This example illustrates that, *in general*, one cannot be guaranteed of ever

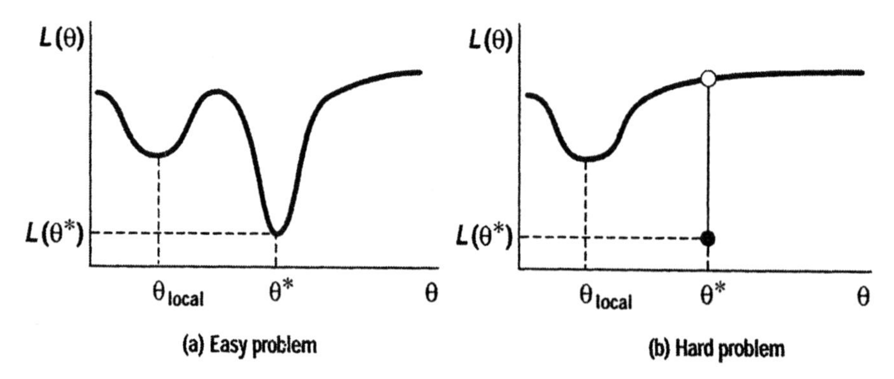

Figure 1.2. Examples of easy and hard problems for global optimization. In (a), a global algorithm will easily avoid θ_{local} and find θ^*; in (b), a global algorithm will only find θ_{local} since it is effectively impossible to find θ^*.

obtaining a global solution. Without prior knowledge about the possible existence of such a point, no typical algorithm would be able to find this solution because there is nothing near $\boldsymbol{\theta}^*$ to indicate the presence of a minimum (i.e., no "dip" in L).

Because of the inherent limitations of the vast majority of optimization algorithms, it is usually only possible to ensure that an algorithm will approach a local minimum with a finite amount of resources being put into the optimization process. However, since the local minimum may still yield a significantly improved solution (relative to no formal optimization process at all), the local minimum may be a fully acceptable solution for the resources available (human time, money, computer time, etc.) to be spent on the optimization. Much of this book will focus on algorithms that are only guaranteed to yield a local optimum. However, we will also consider some algorithms (random search, stochastic approximation, simulated annealing, genetic algorithms, etc.) that are sometimes able to find global solutions from among multiple local solutions.

1.1.3 Meaning of "Stochastic" in Stochastic Search and Optimization

The focus in this book is *stochastic* search and optimization. The problems and algorithms considered here apply where:

Property A	There is random noise in the measurements of $L(\boldsymbol{\theta})$ or $\boldsymbol{g}(\boldsymbol{\theta})$
	—and/or —
Property B	There is a random choice made in the search direction as the algorithm iterates toward a solution.

The above two properties contrast with classical deterministic search and optimization, where it is assumed that one has perfect information about the loss function (and derivatives, if relevant) and that this information is used to determine the search direction in a deterministic manner at every step of the algorithm. In many practical problems, such information is not available, indicating that deterministic algorithms are inappropriate.

Let $\hat{\boldsymbol{\theta}}_k$ be the generic notation for the estimate for $\boldsymbol{\theta}$ at the kth iteration of whatever algorithm is being considered, $k = 0, 1, 2, \ldots$. Throughout this book, the *specific* mathematical form of $\hat{\boldsymbol{\theta}}_k$ will change as the algorithm being considered changes. With the exception of the deterministic optimization algorithms considered briefly in Section 1.4, $\hat{\boldsymbol{\theta}}_k$ will always be a random vector since it is derived from input under stochastic Properties A and/or B above.

The following notation will be used throughout the book to represent noisy measurements of L and $\boldsymbol{g}$ at a specific $\boldsymbol{\theta}$:

$$y(\boldsymbol{\theta}) \equiv L(\boldsymbol{\theta}) + \varepsilon(\boldsymbol{\theta}), \tag{1.2a}$$

$$\boldsymbol{Y}(\boldsymbol{\theta}) \equiv \boldsymbol{g}(\boldsymbol{\theta}) + \boldsymbol{e}(\boldsymbol{\theta}), \tag{1.2b}$$

where ε and $\boldsymbol{e}$ represent the noise terms. Note that the noise terms show dependence on $\boldsymbol{\theta}$. This dependence is relevant for many applications. It indicates that the common statistical assumption of independent, identically distributed (i.i.d.) noise does not necessarily apply since $\boldsymbol{\theta}$ will be changing as the search process proceeds.

It will sometimes be convenient to use notation slightly different from that in (1.2a, b). Reflecting the fact that the measurements of L and/or $\boldsymbol{g}$ will frequently be at a current estimate $\boldsymbol{\theta} = \hat{\boldsymbol{\theta}}_k$, and that the noise-generating process may change with k (perhaps as a real-time system evolves in concert with the algorithm evolution in k), let

$$y_k \equiv y_k(\hat{\boldsymbol{\theta}}_k) = L(\hat{\boldsymbol{\theta}}_k) + \varepsilon_k(\hat{\boldsymbol{\theta}}_k), \tag{1.3a}$$

$$\boldsymbol{Y}_k \equiv \boldsymbol{Y}_k(\hat{\boldsymbol{\theta}}_k) = \boldsymbol{g}(\hat{\boldsymbol{\theta}}_k) + \boldsymbol{e}_k(\hat{\boldsymbol{\theta}}_k). \tag{1.3b}$$

Depending on whether it is important to emphasize the dependence on $\boldsymbol{\theta}$ or the dependence on k, the notation $y(\boldsymbol{\theta})$ and y_k will be used interchangeably in the text as a measurement of L; likewise for $\boldsymbol{Y}(\boldsymbol{\theta})$ and $\boldsymbol{Y}_k$ as a measurement of $\boldsymbol{g}$. Slight variations on (1.3a, b) are also seen in applications. For instance, if the noise is not dependent on $\boldsymbol{\theta}$ (e.g., is statistically independent), then, in the case of the loss measurement, $y_k = y_k(\hat{\boldsymbol{\theta}}_k) = L(\hat{\boldsymbol{\theta}}_k) + \varepsilon_k$, where ε_k is an independent noise process.

For stochastic Property A above, the noise is relative to the measurements of L (and/or $\boldsymbol{g}$). This is different from some other uses of the term *noise* in the technical literature. In particular, in many estimation problems, one may have noisy data. This does not, however, necessarily mean that the values of L and/or $\boldsymbol{g}$ used in the optimization for forming the estimate will be noisy. The fundamental distinction comes in how the data are treated in the optimization process. In many estimation problems, a full set of data is collected and an L or $\boldsymbol{g}$ is chosen that is *conditioned on this set of data*. This conditioning removes the randomness from the problem as far as Property A is concerned. For example, in classical (batch) least-squares or maximum likelihood estimation methods, one finds the best parameter estimates *conditioned* on an existing set of noisy data. The estimation criterion (the sum of squares or the likelihood function) therefore becomes deterministic. Such estimation is often associated with so-called "off-line" methods of estimation. This deterministic setting for least-squares estimation is not to be confused with the (stochastic) *recursive* least-squares methods of Chapter 3.

In addition to such recursive methods, there are many problems where the noise associated with Property A *is* relevant. Many examples are sprinkled throughout this book. Noise fundamentally alters the search and optimization process because the algorithm is getting misleading information throughout the search process. The following example illustrates the dangers of a naïve treatment of noise in an optimization context.

Example 1.4—Effect of noisy data on solution to optimization problem. Consider the following loss function with a scalar θ: $L(\theta) = e^{-0.1\theta}\sin(2\theta)$. If the domain for optimization is Θ = [0, 7], the (unique) minimum occurs at $\theta^* = 3\pi/4 \approx 2.36$, as shown in Figure 1.3. Suppose that the analyst carrying out the optimization is not able to calculate $L(\theta)$, obtaining instead only *noisy* measurements $y(\theta) = L(\theta) + \varepsilon$, where the noises ε are i.i.d. with distribution $N(0, 0.5^2)$ (a normal distribution with mean zero and variance 0.5^2). The analyst uses the $y(\theta)$ measurements in conjunction with an algorithm to attempt to find θ^*.

Suppose that the algorithm here is the simple method of collecting one measurement at each increment of 0.1 over the interval defined by Θ (including the endpoints 0 and 7). The estimate of the minimum is the value of θ on this sampling grid [0, 0.1, 0.2,..., 6.9, 7] corresponding to the lowest *measured* loss. Using the MATLAB normal (pseudo) random number generator **randn**, Figure 1.3 shows the set of 71 measurements obtained for the optimization process. These noisy measurements are superimposed on the true loss function. Based on these measurements, the analyst would falsely conclude that the minimum is at θ = 5.9. As shown, this false minimum is far from the actual θ^*. ❑

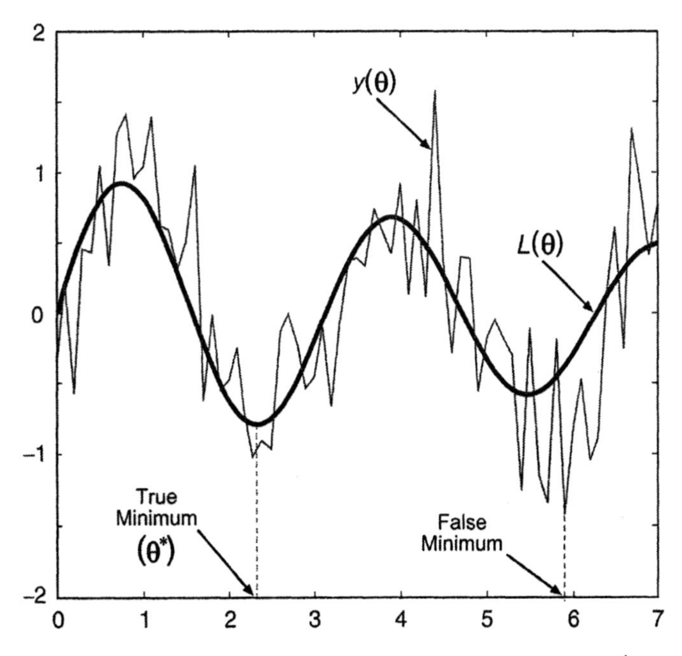

Figure 1.3. Simple loss function $L(\theta)$ with indicated minimum θ^*. Note how noise causes the algorithm to be deceived into sensing that the minimum is at the indicated false minimum.

Noise arises in almost any case where physical system measurements or computer simulations are used to approximate a steady-state criterion. Some specific areas of relevance include:

(i) Real-time estimation and control problems where data are collected "on the fly" as a system is operating. Property A (noisy loss measurements) is relevant because the data arrive in concert with the estimation of $\boldsymbol{\theta}$, with each increment of data being used to *approximate* some long-run (average) criterion such as a mean-squared error (MSE). Each such real-time approximation may be considered a noisy "measurement" of the long-run criterion (Chapters 3–7).

(ii) Problems where estimates of a criterion are formed by computer-based Monte Carlo sampling according to a statistical distribution (e.g., certain quantile estimation problems or Markov chain sampling schemes—see Chapters 4 and 16).

(iii) Problems where large-scale simulations are run as estimates of actual system behavior (Chapters 14 and 15).

(iv) Problems where physical data are processed sequentially, with each sequential data point being used to estimate some overall (average) criterion, such as the MSE. Unlike item (i), the data are not necessarily processed in real time. The sequential processing may be done to reduce the computational burden (versus the batch processing of all data) or to expose the unique impact of each data point.

Examples 1.5 and 1.6 provide conceptual descriptions of two distinct problems involving noise in the loss function measurements. Example 1.5 illustrates setting (i) above, while Example 1.6 depicts setting (iii).

Example 1.5—Noisy loss measurements in a tracking problem. As one specific example of a setting where Property A is relevant, consider the popular MSE criterion mentioned above. An analyst's goal might be to optimize the performance of some system (e.g., the tracking error in a robot control problem) while it is operating. The MSE criterion is

$$L(\boldsymbol{\theta}) = E\left(\left\|\text{actual output} - \text{desired output}\right\|^2\right),$$

where $E(\cdot)$ denotes the expectation (Appendix C) and the actual output depends on the design parameter vector $\boldsymbol{\theta}$ and other factors, including randomness. Only rarely will one be able to compute the MSE (or its derivatives!), as required in deterministic algorithms. In general, computation of the MSE requires complete knowledge of: (i) the probability distribution of all the randomness in the system, (ii) the mechanism by which the randomness manifests itself in the output, and (iii) the ability to carry out the multivariate numerical integration associated with the expected value in the MSE calculation. Rather, one might be able to get a *specific observation* of the squared error $\|\cdot\|^2$, but this differs from the *mean-squared* error. For instance, an engineer may have some spatial and temporal target values for a robot arm to meet in an operation; in lieu of computing the (hopeless) MSE, the engineer measures the deviation from these target values in a set of experiments.

More generally, one can write the observation as a noisy measurement of the mean-squared error,

$$y(\boldsymbol{\theta}) = \|\cdot\|^2 = E\left(\|\cdot\|^2\right) + \varepsilon(\boldsymbol{\theta}),$$

using the notation of (1.2a). The noise $\varepsilon(\boldsymbol{\theta})$ in this case is simply the arithmetic difference between the observed squared error and mean-squared error. Note that the noise in the measurements of L and/or $\boldsymbol{g}$ will often *not* be statistically independent of other randomness in the problem. In particular, the noise often depends on the current estimate for $\boldsymbol{\theta}$, which sometimes complicates the theoretical and practical analysis of an algorithm. ❑

Example 1.6—Optimization via Monte Carlo simulations. Suppose that an analyst is trying to optimize some complex system. As usual, let $\boldsymbol{\theta}$ represent the vector of terms being optimized and $L(\boldsymbol{\theta})$ be the loss function representing some type of "average" performance for the system. Computer simulation models are widely used to analyze and improve systems for which simple analytical analysis is inadequate. If the simulation includes randomness—a.k.a. the simulation is a *Monte Carlo* process—the simulation will, in general, produce results that change randomly with each run. Suppose that the analyst is going to use the computer simulation as a proxy for the true system in the process of optimizing $\boldsymbol{\theta}$. The analyst picks a value for $\boldsymbol{\theta}$, runs the simulation one or more times, evaluates the performance, and then updates $\boldsymbol{\theta}$. The process is then repeated at the new value of $\boldsymbol{\theta}$. The final value of $\boldsymbol{\theta}$ is used in the *real* system.

The evaluation of performance at a given $\boldsymbol{\theta}$, however, is not straightforward because of the randomness in the simulation (which should be acting like the randomness in the real system if the simulation is properly designed). While the goal is to optimize the *average* measure $L(\boldsymbol{\theta})$, the information available at any $\boldsymbol{\theta}$ is only a particular result from one or a small number of simulation runs. The analyst can, in principle, run the simulation many times at a given $\boldsymbol{\theta}$ to effectively average out the randomness. Such averaging, however, is usually impractical, especially if many $\boldsymbol{\theta}$ values must be examined and the simulation is expensive to run. So, for the optimization process, the information available to the analyst is in the general form associated with Property A above:

$$\text{simulation output at } \boldsymbol{\theta} = L(\boldsymbol{\theta}) + \text{noise at } \boldsymbol{\theta}.$$

Hence, an algorithm that formally accommodates noisy loss measurements in attempting to find an optimum $\boldsymbol{\theta}^*$ should be used. Chapters 14 and 15 delve into many issues of optimization that are specific to Monte Carlo simulations. Many of the generic principles in the chapters preceding Chapters 14 and 15 also apply to simulations. ❑

Relative to the second defining property of a stochastic algorithm, Property B above, it is sometimes beneficial to deliberately introduce randomness into the search process as a means of speeding convergence and making the algorithm less sensitive to modeling errors. This injected (Monte Carlo) randomness is usually created via computer-based pseudorandom number generators of the type discussed in Appendix D. Although the introduction of randomness may seem at first thought counterproductive, it is well known to have beneficial effects in some settings. One of the roles of injected randomness in stochastic search is to allow for "spontaneous" movements to unexplored areas of the search space that may contain an unexpectedly good $\boldsymbol{\theta}$ value. The randomness may provide the necessary kick. This is especially relevant in seeking out a global optimum when the search is stalled near a local solution.

Related to this application is the use of Monte Carlo randomness in the important class of algorithms that emulate evolutionary principles of optimization (Chapters 9 and 10); randomness is a central part of both physical and simulated evolution through the introduction of mutations and through the choice of parents. These mutations may sometimes have a beneficial effect by allowing unexpected solutions to be evaluated.

Injected randomness may also be used for the creation of simple random quantities that act like their deterministic counterparts, but which are much easier to obtain and more efficient to compute. An example of this is the simultaneous perturbation approximation of a gradient vector in Chapter 7. This gradient approximation can be used in place of the true gradient in certain optimization schemes, yielding similar general performance.

Yet another area where injected randomness is useful is in numerical integration. Often, Monte Carlo methods are implementable when analytical methods are impractical or impossible. Monte Carlo methods are usually more efficient in high-dimensional problems than deterministic quadrature approaches provided that one is willing to tolerate a small probability of achieving a poor estimate. Chapter 16, on Markov chain Monte Carlo, discusses one very popular general approach for numerical integration via such injected randomness.

1.2 SOME PRINCIPLES OF STOCHASTIC SEARCH AND OPTIMIZATION

We discussed above the problem formulation and some of the basic issues that surround the field of stochastic search and optimization. The two subsections in this section summarize some fundamental underpinnings and limits. The aim here is to provide the reader with some principles for interpreting the results in the remainder of this book. Some of these principles will seem self-evident and some will become apparent as the reader absorbs the results in the book. The first subsection lists some important facts that are relevant to understanding the results of this book and the second subsection is a summary of the immutable limits implied by the "no free lunch" theorems for search and optimization.

1.2.1 Some Key Points

Below is a list of facts that the reader should keep in mind while reading this book or implementing any of the algorithms described here.

Relative efficiency via function evaluations. One of the key issues in comparing algorithms is "efficiency" in solving problems of interest. In essence, this is some representation of the cost of finding an acceptable solution. There are many ways of measuring efficiency, including computer run time, number of algorithm iterations, and number of L and/or $\boldsymbol{g}$ function evaluations. This book will emphasize the latter as generally the most objective measure. Computer times tend to be machine and software dependent and hence do not readily transfer from one setting (or person) to another. Iterations can also be misleading, as one algorithm may have much more demanding per-iteration requirements than another; hence an algorithm producing a solution in a lower number of iterations does not generally do so at lower cost when compared to another algorithm taking more iterations but with lower per-iteration demands. In addition, the algorithm using fewer iterations may have required more "off-line" information in the form of research and analysis to provide information not used or required in the other algorithm (e.g., loss function gradient information in the first algorithm but not in the second).

Comparing algorithms based on number of L and/or $\boldsymbol{g}$ evaluations puts things on a common basis for contrast provided that the algorithms being compared use the same type of information (e.g., the algorithms use only loss evaluations L). The prime motivation is that in practice the L and/or $\boldsymbol{g}$ measurements are usually the dominant cost in the optimization process; the other calculations in the algorithms are often relatively unimportant (and becoming more unimportant as computing power grows). This philosophy is consistent with most complex stochastic optimization problems, where each loss function or gradient measurement may represent a large-scale simulation or a physical experiment. Although we use relatively simple loss functions in the numerical comparisons of this book, these are merely proxies for the more complex functions encountered in practice.

Implications of noisy function measurements. There are fundamental limits of search and optimization with noisy information about the L or $\boldsymbol{g}$ functions (fundamental Property A above). Foremost, perhaps, is that the statistical error of the information fed into the algorithm—and the resulting error of the output of the algorithm—can only be reduced by incurring a significant cost in number of function evaluations. (Ideally, of course, one would like to use perfect information in the algorithm as a way of enhancing the speed of the search.) For the simple case of independent noise, the error decreases at the rate $1/\sqrt{N}$, where N represents the number of L or $\mathbf{g}$ measurements fed into the algorithm. This is a classical result in statistics, indicating that a 100-fold increase in function evaluations reduces the error by a factor of 10. More generally, it takes an increase of $2K$ orders of magnitude in the number of measurements to reduce the error by K orders of magnitude. In fact, this $2K$ versus K barrier is frequently optimistic in practical problems, as the noise is often dependent in a way that increases the amount of averaging required to effectively reduce the statistical error.

Curse of dimensionality. A further limit for multivariate ($p > 1$) searches is that the volume of the search region generally grows geometrically with dimension. This implies that a "naïve" search in a high-dimensional problem will generally be hopeless. The famous control theorist and mathematician Richard Bellman (1920–1984) coined the piquant phrase "curse of dimensionality" to capture this phenomenon. Ho (1997), for example, gives an illustration in a discrete problem where $p = 10$ (which is relatively small by "industrial" standards): if each of the 10 elements of $\boldsymbol{\theta}$ can take on 10 values, there are 10^{10} possible outcomes. If we randomly sample 10,000 values of L uniformly in the domain Θ, the probability of finding one of the best 500 values for $\boldsymbol{\theta}$ (a much easier problem than finding a unique optimum $\boldsymbol{\theta}^*$) is a minuscule 0.0005. With noisy measurements of L, the problem is even worse because it is not known which $\boldsymbol{\theta}$ value corresponds to the lowest loss value from among the sampled $\boldsymbol{\theta}$ points. The lesson here is that, in practice, one will generally have to exploit some problem knowledge to have any hope of carrying out search and optimization in most real-world problems. Exercise 1.6 explores this point further.

Ordinal versus cardinal rankings. An area of stochastic optimization has developed over the last 10–15 years based on the idea of ordinal optimization (e.g., Ho et al., 1992). The basic philosophy here is to seek an answer that is "good enough" rather than to seek the optimal answer. The basis of the approach is ordinal rankings (i.e., is $A > B$?) rather than cardinal rankings (i.e., what is the precise value of $A - B$?). It is customarily much easier to determine relative rankings than it is to determine exact differences. The former usually requires only a minimal amount of averaging, while the latter needs a much more demanding averaging process. Some of the philosophy of ordinal optimization is illustrated in the discussion related to the curse of dimensionality above. In fact, relaxing the goals from finding the optimal value can yield orders of magnitude increases in efficiency in large-dimensional problems with noisy function evaluations.

Constraints. Constraints on the allowable values of $\boldsymbol{\theta}$ are a fact of life. All practical problems involve some restrictions on the relevant variables, although in some applications it may be possible to effectively ignore the constraints. Constraints can be encountered in many different ways, as motivated by the specific application. Note that the constraint set Θ does not necessarily correspond to the set of allowable values for $\boldsymbol{\theta}$ *in the search* since some problems allow for the "trial" values of the search to be outside the set of allowable final estimates. Constraints are usually handled in practice on an ad hoc basis, especially tuned to the problem at hand. There are few general, practical methods that apply broadly in stochastic search and optimization. Michalewicz and Fogel (2000, Chap. 9), for example, discuss some of the practical methods by which constraints are handled in evolutionary computation. Similar methods apply in other stochastic algorithms. Let us summarize the principal types of constraints seen in search problems:

> ***Hard vs. soft constraints.*** In hard constraints, no value of $\boldsymbol{\theta}$ can ever be taken outside of the allowable set. With soft constraints, values of $\boldsymbol{\theta}$ outside the constrained set are allowed during the search process, but it is required that the final estimate for $\boldsymbol{\theta}$ lie inside the constraint set. The former typically arises when the search is performed using an actual physical system, while the latter often arises in simulation-based search (so in the latter, sampling for $\boldsymbol{\theta}$ in the search algorithm might be done over some set Θ' such that $\Theta \subset \Theta'$, which can be of potential interest due to the additional information about the nature of L that this "extended sampling" can sometimes provide).
>
> ***Explicit vs. implicit constraints.*** For explicit constraints, limits on the values of $\boldsymbol{\theta}$ can be specified directly (e.g., one constraint might be $\|\boldsymbol{\theta}\| \leq c$ for some vector norm $\|\cdot\|$ and constant c; another constraint might be that each component of $\boldsymbol{\theta}$ lies in the interval $[-c, c]$). For implicit constraints, the limits are specified in terms of some complicated function of $\boldsymbol{\theta}$, possibly in terms of allowable values for L. Given the complicated function, there is no method for explicitly representing the constraints on $\boldsymbol{\theta}$. For example, in a control system, if $\boldsymbol{\theta}$ represents some parameters inside the controller software and the real system output must stay within some bounds, then there are implied constraints on what values of $\boldsymbol{\theta}$ are allowed in order to maintain the system inside its allowable region (and if the $\boldsymbol{\theta}$ estimation is done in real time on the physical control system, this may also represent one of the hard constraint cases cited above).

Stopping criteria. In search and optimization, there is traditionally a concern with developing a good *stopping criterion* (i.e., a means of indicating when the algorithm is close enough to the solution that it can be stopped). Unfortunately, the quest for an automatic means of stopping an algorithm with a guaranteed level of accuracy seems doomed to failure in general stochastic search problems. The fundamental reason for this pessimistic message is that in nontrivial problems, there will always be a significant region within Θ that will remain unexplored in any finite number of iterations. Without prior knowledge, there is always the possibility that $\boldsymbol{\theta}^*$ could lie in this unexplored region. This applies even when the functions involved are relatively benign; see Solis and Wets (1981) for mention of this in the context of twice-differentiable convex L (convexity is discussed in Appendix A). Difficulties are compounded when the function measurements include noise. For this reason, most of the comparisons in this book will be done under the constraint that the competing algorithms use the same number of function evaluations. This provides an objective comparison for the same "cost" of search.

Time-varying problems. In many practical settings, the environment changes over time. Hence, the "best" solution to a problem now may not be the best (or even a good) solution to the corresponding problem in the future. Such a formulation is obvious in dynamic problems such as building control systems, where the optimum may change continuously. Although less obvious, the time-varying problem also arises in settings that may appear at first glance to be static. For example, an optimum financial plan for a business or family depends on the external environment, which, of course, changes over time. In some search and optimization problems, the algorithm will be explicitly designed to adapt to a changing environment and "automatically" provide a new estimate at the optimal value (e.g., a control system). In other cases, one needs to restart the process and find a new solution. In either sense, the problem solving may never stop!

Limits of numerical comparisons by Monte Carlo. A common means of comparing algorithms is to run Monte Carlo studies on some test cases (e.g., with particular forms for L). This can be a sound scientific method of gaining insight and can be a useful supplement to theory, much of which is based on asymptotic (infinite sample) analysis. In fact, it is especially popular in certain branches of optimization to create "test suites" of problems, where various algorithms face-off in a numerical showdown.

A danger arises, however, in making *broad* claims about the performance of an algorithm based on the results of numerical studies. Performance can vary tremendously under even small changes in the form of the functions involved or the coefficient settings within the algorithms themselves. One must be careful about drawing conclusions beyond those directly supported by the specific numerical studies performed. This, in fact, is a manifestation of the no free lunch theorems discussed below—outstanding performance on some types of functions is consistent with poor performance on certain other types of functions.

For purposes of drawing objective conclusions about the relative performance of algorithms, it is preferable to use *both* theory and numerical studies. The theory—despite its limitations (e.g., asymptotically based derivations)—provides a basis for drawing broad formal conclusions and pointing to possible limitations in the range of appropriate problems, while the numerical studies give the user a direct feel for algorithm behavior in specific problems in finite samples. Unfortunately, in some popular algorithms, the theory is incomplete or even virtually nonexistent. This has led to some overly broad claims in the literature about algorithm performance on the basis of limited numerical studies. The use of test suites, as mentioned above, offers some enlightenment, but the reader is cautioned to interpret all numerical studies—including those in this book!—with the proverbial grain of salt.

Special note for numerical comparisons with noisy measurements. With the foregoing caveats in mind, Monte Carlo simulations are frequently used to evaluate different algorithms based on noisy function measurements $y = y(\boldsymbol{\theta})$ or $\boldsymbol{Y} = \boldsymbol{Y}(\boldsymbol{\theta})$. In such evaluations, the simulation designer usually knows "truth"; that is, L or $\boldsymbol{g}$ may usually be calculated exactly. In contrast, the algorithms being tested use only the noisy measurements y or $\boldsymbol{Y}$ in their simulated processing to emulate the operations of a real application. In comparing algorithms in such a Monte Carlo fashion, the *final evaluation* should be done based on the exact L or $\boldsymbol{g}$ values, not the noisy values y or $\boldsymbol{Y}$. So, for example, in comparing several optimization algorithms on a specific problem, one may select as "best" the algorithm that produces the lowest value of L at the terminal estimate for $\boldsymbol{\theta}$. It wastes information that is available to the simulation designer to evaluate the algorithms with only the noisy measurements that are used by the algorithms.

Uniqueness vs. nonuniqueness of $\boldsymbol{\theta}^*$. A number of results appear in the literature pertaining to analysis and convergence when there may be multiple global solutions $\boldsymbol{\theta}^*$. In practice, however, this is often not a concern. Many (but certainly not all) real systems have one (unique) globally "best" operating point ($\boldsymbol{\theta}^*$) in the domain Θ, although, of course, there could be many *locally* optimal solutions. So, in order to avoid excessively cumbersome discussion of algorithms and supporting implementation issues and theory, we will often refer to "the" solution $\boldsymbol{\theta}^*$ (versus "a" solution $\boldsymbol{\theta}^*$). In cases where there are multiple $\boldsymbol{\theta}^*$, the results presented here (such as convergence) will typically apply to any one of the multiple solutions (guaranteeing, e.g., convergence into the *set* Θ^*). In practice, an analyst may be quite satisfied to reach a solution at or close to *any* one $\boldsymbol{\theta}^* \in \Theta^*$, so the potential issue of nonuniqueness of $\boldsymbol{\theta}^*$ is sometimes of limited practical concern.

Notational convention. A slight conflict arises when trying to avoid excessively cumbersome notation to indicate a particular element of a vector and to indicate a particular iteration of an estimate for the vector. We will attempt to live with that conflict by establishing the following notational convention. Let $\hat{\boldsymbol{\theta}}_k$ represent the estimate for $\boldsymbol{\theta}$ at the kth iteration of an algorithm and $\hat{\theta}_{ki}$ represent the ith element of the estimate at the kth iteration. On the other hand, the generic representation of the elements of $\boldsymbol{\theta}$ is according to $\boldsymbol{\theta} = [t_1, t_2,\ldots, t_p]^T$. We represent the ith value of the vector $\boldsymbol{\theta}$ by $\boldsymbol{\theta}_i$. This notation is useful in time-varying $\boldsymbol{\theta}$ cases to indicate the value of $\boldsymbol{\theta}$ at time i, and useful when Θ is composed of a discrete number of elements to denote the ith possible value of $\boldsymbol{\theta}$. In cases where a vector quantity is not associated with an iteration process, a simple subscript will denote the indicated element of the vector. For example, x_i denotes the ith element of the vector $\boldsymbol{x}$. This should not create any significant difficulties, as the meaning of the subscript should always be clear from the context.

1.2.2 Limits of Performance: No Free Lunch Theorems

There is a fundamental tradeoff between algorithm efficiency and algorithm robustness (reliability and stability in a broad range of problems). In essence, algorithms that are designed to be very efficient on one type of problem tend to be "brittle" in the sense that they do not reliably transfer to problems of a different type. Hence, there can never be a universally best search algorithm just as there is rarely (never?) a universally best solution to any general problem of society. For example, an airframe design for a plane requiring maneuverability and speed in a battle setting is a very poor design for commercial aviation. One must consider the characteristics of the problem together with the goals of the search and the resources available (computing power, human analysis time, etc.) in choosing an approach.

This lack of a universal best algorithm is a manifestation of the *no free lunch* (NFL) theorems in Wolpert and Macready (1997), which say, in essence, that an algorithm that is effective on one class of problems is *guaranteed* to be ineffective on another class. One may get a feel for the NFL theorems by considering the famous "needle in a haystack" problem. In the absence of clues about the needle's location, it is clear that no search approach can, on average, beat blind random search. The NFL theorems are discussed in more detail in Chapter 10 on evolutionary computation, reflecting the area of optimization from which they arose. The discussion here is intended to provide a flavor of the implications.

While the NFL theorems are established for discrete optimization with a finite (but arbitrarily large) number of options, their applicability includes most practical continuous problems because virtually all optimization is carried out on 32- or 64-bit digital computers. The theorems apply to the cases of both noise-free and noisy loss measurements. Because of the restriction to a finite number of discrete options, there is a corresponding finite number of possible mappings of the input space (values of $\boldsymbol{\theta}$) to the output space (values of L or y). This number may be huge (Exercise 1.7). In particular, in the noise-free case, if there are N_θ possible values for $\boldsymbol{\theta}$ and N_L possible values for the loss, then by direct enumeration there are $(N_L)^{N_\theta}$ possible mappings of $\boldsymbol{\theta}$ to possible loss values. The NFL theorems indicate that:

> When averaging over all $(N_L)^{N_\theta}$ possible mappings from Θ to the output space (loss values), all algorithms work the same (i.e., none can work better than a blind random search).

The *mappings* in the statement above are sometimes called *problems*. Each mapping $L(\boldsymbol{\theta})$ corresponds to one problem that an analyst may be solving.

The averaging mentioned at the beginning of the NFL statement above is a simple arithmetic mean over the $(N_L)^{N_\theta}$ mappings. The following small-scale example with noise-free loss measurements is intended to provide a flavor of the theorems.

Example 1.7—NFL implications in a small problem. Suppose that $\Theta = \{\boldsymbol{\theta}_1, \boldsymbol{\theta}_2, \boldsymbol{\theta}_3\}$ and that there are two possible outcomes for the noise-free loss measurements, $\{L_1, L_2\}$. Hence, $(N_L)^{N_\theta} = 2^3 = 8$. Table 1.1 summarizes the eight possible mappings.

Note that all rows in Table 1.1 have the same number of L_1 and L_2 values. Hence, the mean loss value across all eight possible problems is the same regardless of which $\boldsymbol{\theta}_i$ is chosen. Because the algorithm chooses the $\boldsymbol{\theta}_i$, all algorithms produce the same mean loss value when averaging across all possible problems. As a specific instance of the implication of the NFL theorems, suppose that $L_1 < L_2$. Then, an algorithm that puts priority on picking $\boldsymbol{\theta}_1$ will work well on problems 1, 2, 3, and 7, but poorly on problems 4, 5, 6, and 8. The *average* performance is the same as an algorithm that puts priority on picking $\boldsymbol{\theta}_2$ or $\boldsymbol{\theta}_3$. ❑

Because the NFL theorems seem to paint a discouraging picture that "all algorithms work the same," one might wonder about the value of studying various algorithms, as we do in this book. A book devoted to blind search alone would be a short book indeed! A key qualifier in the NFL theorems is the "averaging over all possible mappings" statement. If a problem has some known structure—and all conceivable practical problems do—*and* the algorithm uses that structure, it is certainly possible that one algorithm will work better than another on the given problem. If the needle is more likely to be located near the surface of the haystack, one can save a lot of time that would otherwise be devoted to burrowing into the haystack.

Table 1.1. The eight possible mappings (problems), representing all possible combinations of inputs ($\boldsymbol{\theta}$) and outputs (loss values). Values shown are subscripts of loss outcome (i.e., 1 represents L_1; 2 represents L_2). For example, under mapping 4, $L(\boldsymbol{\theta}_1) = L_2$ and $L(\boldsymbol{\theta}_2) = L_1$.

Mapping / $\boldsymbol{\theta}$	1	2	3	4	5	6	7	8
$\boldsymbol{\theta}_1$	1	1	1	2	2	2	1	2
$\boldsymbol{\theta}_2$	1	1	2	1	1	2	2	2
$\boldsymbol{\theta}_3$	1	2	2	1	2	1	1	2

Given the above intuition associated with the value of problem structure, an informal mathematical representation of the NFL theorems is

Size of domain of applicability
× Efficiency on domain of applicability = Universal constant.

That is, an algorithm cannot have *both* wide applicability and uniformly high efficiency.

In the spirit of the NFL theorems, much of the discussion in this book focuses on algorithms that are appropriate in restricted domains. For example, Chapter 3 considers methods that are especially appropriate in linear models; these methods are relatively ineffective in nonlinear models. Cumulatively, the algorithms discussed in this book span a broad range of problems, but any one algorithm will be applicable in a relatively restricted range of problems.

1.3 GRADIENTS, HESSIANS, AND THEIR CONNECTION TO OPTIMIZATION OF "SMOOTH" FUNCTIONS

1.3.1 Definition of Gradient and Hessian in the Context of Loss Functions

As mentioned, many of the algorithms in this book apply to problems where L is continuous, or, more strongly, once or several times differentiable in $\boldsymbol{\theta}$ (some of the same algorithms—e.g., simulated annealing, evolutionary computation, etc.—may also apply to discrete problems). Hence, certain results in multivariate calculus are useful. In particular, some of the methods discussed in this book rely—at least indirectly—on the gradient and (possibly) Hessian matrix of the loss function (possibly embedded in noise).

When considering problems with such "smooth" $L(\boldsymbol{\theta})$, it is of interest to consider some calculus-based aspects of analysis. Recall that $\boldsymbol{\theta} = [t_1, t_2, \ldots, t_p]^T$. The gradient of the loss function is defined as the p-dimensional column vector

$$\boldsymbol{g}(\boldsymbol{\theta}) \equiv \frac{\partial L}{\partial \boldsymbol{\theta}} = \begin{bmatrix} \partial L / \partial t_1 \\ \partial L / \partial t_2 \\ \vdots \\ \partial L / \partial t_p \end{bmatrix}.$$

The Hessian is the matrix of second partial derivatives

$$H(\boldsymbol{\theta}) \equiv \frac{\partial^2 L}{\partial\boldsymbol{\theta}\partial\boldsymbol{\theta}^T} = \begin{bmatrix} \partial^2 L/\partial t_1^2 & \partial^2 L/\partial t_1\partial t_2 & \cdots & \partial^2 L/\partial t_1\partial t_p \\ \partial^2 L/\partial t_2\partial t_1 & \partial^2 L/\partial t_2^2 & & \vdots \\ \vdots & & \ddots & \\ \partial^2 L/\partial t_p\partial t_1 & \cdots & \cdots & \partial^2 L/\partial t_p^2 \end{bmatrix},$$

where $\partial^2 L/\partial t_i\partial t_j = \partial^2 L/\partial t_j\partial t_i$ under the assumption of continuity of the terms in the Hessian matrix (so then $\boldsymbol{H}$ is a symmetric matrix). Note that the term "$\partial\boldsymbol{\theta}\partial\boldsymbol{\theta}^T$" appearing in the denominator of the derivative expression is merely a notational convenience for suggesting a matrix outcome in the differentiation process; it is not to be interpreted as referring to a product of $\boldsymbol{\theta}$ and $\boldsymbol{\theta}^T$ or a derivative with respect to such a product.

The optimization setting being emphasized in this book motivates the gradient and Hessian terminology above. In problems of root-finding, the terminology is slightly different. The terms *function* and *Jacobian matrix* typically replace the terms *gradient* and *Hessian matrix*, respectively (so the Jacobian matrix is a collection of first derivatives of the vector-valued function). Specifically, in root-finding, $\boldsymbol{g}(\boldsymbol{\theta})$ is the *function* for which a zero is to be found (a root to $\boldsymbol{g}(\boldsymbol{\theta}) = \boldsymbol{0}$) and $\boldsymbol{H}(\boldsymbol{\theta})$ is the *Jacobian matrix*. The author can attest from experience that it is important to recognize this semantic difference when communicating optimization results in fields traditionally based in root-finding!

The Taylor theorems from multivariate analysis are important in the construction and formal analysis of both deterministic and stochastic algorithms. Under appropriate conditions, these theorems allow one to convert a possibly messy nonlinear function into a relatively benign low-order polynomial approximation. Appendix A summarizes some of the important results associated with Taylor theorems.

1.3.2 First- and Second-Order Conditions for Optimization

Algorithms for solving unconstrained local minimization problems (i.e., (1.1) with $\Theta = \mathbb{R}^p$) with smooth loss functions are frequently based upon the first- and second-order derivative conditions. These derivative conditions also apply in some constrained problems and in some of the stochastic search techniques of interest here.

The first-order condition states that at $\boldsymbol{\theta}^*$ it is necessary that

$$\boldsymbol{g}(\boldsymbol{\theta}^*) = \boldsymbol{0} \tag{1.4}$$

when L is a continuously differentiable function. The proof of this result follows from simple Taylor series arguments showing that if (1.4) is not true, then one can move $\boldsymbol{\theta}$ in some direction that reduces the value of $L(\boldsymbol{\theta})$ (use Taylor's theorem A.1 in Appendix A). In particular, if $\boldsymbol{d} \in \mathbb{R}^p$ denotes a vector such that $\boldsymbol{g}(\boldsymbol{\theta})^T\boldsymbol{d} < 0$, then moving $\boldsymbol{\theta}$ to a new value $\boldsymbol{\theta} + \lambda \boldsymbol{d}$ for any $\lambda \in (0, u)$ and some $u > 0$ will guarantee a lower value of L. Directions $\boldsymbol{d}$ such that $\boldsymbol{g}(\boldsymbol{\theta})^T\boldsymbol{d} < 0$ are called *descent directions* for L at $\boldsymbol{\theta}$. However, since one can repeat the same line of reasoning to show that $\boldsymbol{g}(\boldsymbol{\theta}^*) = \boldsymbol{0}$ is a necessary condition for $\boldsymbol{\theta}^*$ to be a local *maximizer* of L, we need some information to distinguish the solution $\boldsymbol{\theta}^*$ as being a local minimum versus a local maximum (or saddlepoint).

The Hessian matrix $\boldsymbol{H} = \boldsymbol{H}(\boldsymbol{\theta})$ plays an important role in making the distinction between local minima and maxima for loss functions that are twice continuously differentiable. In particular, if $\boldsymbol{\theta}'$ is a root of the equation $\boldsymbol{g}(\boldsymbol{\theta}) = \boldsymbol{0}$, then $\boldsymbol{H}(\boldsymbol{\theta}')$ being positive definite (see Appendix A) implies that $\boldsymbol{\theta}'$ is a local minimizer of L. Further, it is necessary that $\boldsymbol{H}(\boldsymbol{\theta}')$ be at least positive semidefinite for $\boldsymbol{\theta}'$ to be a minimizing point $\boldsymbol{\theta}^*$. If $\boldsymbol{H}(\boldsymbol{\theta}')$ is positive semidefinite but not positive definite, then one can sometimes evaluate higher-order derivatives to determine if $\boldsymbol{\theta}'$ is a local mimimizer (e.g., Bazaraa et al., 1993, pp. 134–138). Conversely, if $-\boldsymbol{H}(\boldsymbol{\theta}')$ is positive definite (i.e., $\boldsymbol{H}(\boldsymbol{\theta}')$ is negative definite), then $\boldsymbol{\theta}'$ is a local maximizer of L. If $\boldsymbol{H}(\boldsymbol{\theta}')$ is indefinite (i.e., has both positive and negative eigenvalues), then $\boldsymbol{\theta}'$ is a saddlepoint, implying that $\boldsymbol{\theta}'$ is a local minimizer in some components of $\boldsymbol{\theta}$ and a local maximizer in other components. The connection of the Hessian to convexity is discussed in Appendix A.

Exercise 1.10 shows an example where a root of $\boldsymbol{g}(\boldsymbol{\theta}) = \boldsymbol{0}$ with a positive semidefinite Hessian is a unique (global) minimum. Two examples of functions where roots of $\boldsymbol{g}(\boldsymbol{\theta}) = \boldsymbol{0}$ with positive semidefinite Hessians correspond to saddlepoints are given in Bazaraa et al. (1993, pp. 135–138).

1.4 DETERMINISTIC SEARCH AND OPTIMIZATION: STEEPEST DESCENT AND NEWTON–RAPHSON SEARCH

There are many deterministic optimization algorithms for the problem in (1.1) and it is not the purpose here to survey these algorithms. However, two algorithms in particular are worth discussing since many stochastic search algorithms for "smooth" (differentiable) $L(\boldsymbol{\theta})$ or $\boldsymbol{g}(\boldsymbol{\theta})$ can be motivated by connections to these deterministic algorithms. These are the methods of steepest descent and Newton–Raphson, considered in Subsections 1.4.1 and 1.4.2.

1.4.1 Steepest Descent Method

The method of steepest descent is one of the oldest formal optimization techniques. Nevertheless, it remains one of the more popular deterministic approaches. For example, it corresponds to the widely used backpropagation algorithm for neural networks when one is working with a fixed set of input–output data—see Chapter 5. Steepest descent is based on the simple principle that from a given value $\boldsymbol{\theta}$ the best direction to go is the one that produces the largest local change in the loss function (the steepest descent). The gradient vector at the given $\boldsymbol{\theta}$ defines this direction. Hence the algorithm is

$$\hat{\boldsymbol{\theta}}_{k+1} = \hat{\boldsymbol{\theta}}_k - a_k \boldsymbol{g}(\hat{\boldsymbol{\theta}}_k), \; k = 0, 1, 2, \ldots, \tag{1.5}$$

where k is the iteration count, $\hat{\boldsymbol{\theta}}_0$ is the initial "guess" at $\boldsymbol{\theta}^*$, and $a_k > 0$ is the *step size*, which may be specified a priori (often as a constant $a_k = a$) or picked on an iteration-to-iteration basis as a solution to $\min_{a \geq 0}\{L(\hat{\boldsymbol{\theta}}_k - a\boldsymbol{g}(\hat{\boldsymbol{\theta}}_k))\}$ (this secondary optimization problem is called a *line search*). So, (1.5) states that the new estimate of the best value of $\boldsymbol{\theta}$ is equal to the previous value minus a term proportional to the gradient at the current value.

Dennis and Schnabel (1989, p. 38) discuss implementations of steepest descent when the secondary (line search) optimization problem of solving for a_k at each iteration is eliminated and a simplified (possibly predetermined) form of a_k employed. In fact, in the stochastic approximation and neural network methods discussed in Chapters 4 and 5, the stochastic analogues of steepest descent typically have a_k as a predetermined decaying sequence. The scale factors a_k play a critical role in the algorithm, often determining whether the algorithm converges or diverges. Aside from being called step sizes, these factors are sometimes referred to as *gains* or *learning coefficients*, depending on the field of application. Conditions guaranteeing that the steepest descent iterate converges to $\boldsymbol{\theta}^*$ as $k \to \infty$ are presented in many places (e.g., Bazaraa et al., 1993, pp. 300–308).

Figure 1.4 illustrates why the simple steepest descent recursion in (1.5) can lead to reductions in the L value. Given a multivariate functional relationship of $L(\boldsymbol{\theta})$ versus $\boldsymbol{\theta}$, consider the plane showing L as a function of the ith component of $\boldsymbol{\theta}$ with the other components of $\boldsymbol{\theta}$ fixed. For a relatively simple convex ("bowl-shaped") function, Figure 1.4 shows that when the ith component of the current estimate is on either side of the current minimum for that component, the updated estimate will move towards the minimum in a direction *opposite* the sign of the corresponding element of the gradient vector. In particular, when the current value of $\hat{\theta}_{ki}$ is left of the minimum, (1.5) implies that the new estimate will move to the right according to the search direction given by the negative of the negative sign of the gradient component (= positive direction). Conversely, when the current value is to the right of the optimum, the new estimate will move left since the negative of the positive sign of the ith gradient component leads to a negative direction.

The role of a_k is to regulate how large a step the algorithm takes. Clearly, steps that are too large or too small may prevent the algorithm from ever converging to $\boldsymbol{\theta}^*$, even if the steps are in the correct directions. Steps that are neither too large nor too small will move the algorithm toward $\boldsymbol{\theta}^*$ in a controlled

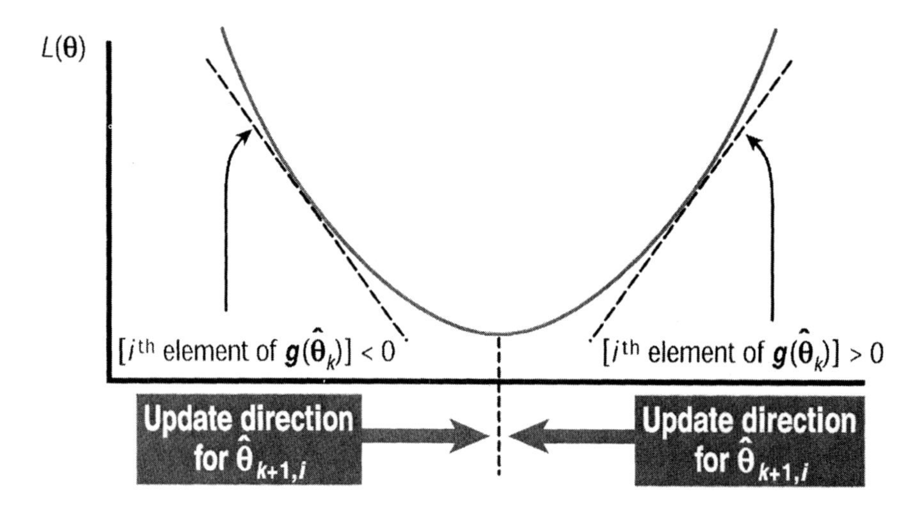

Figure 1.4. Motivation for steepest descent: Update directions for ith component of $\boldsymbol{\theta}$ in iterating from $k \to k+1$.

manner. This involves interplay with the other components of the $\boldsymbol{\theta}$ vector since, in general, the ith element of $\boldsymbol{\theta}^*$ is not the same as the *current* minimum in the plane of L versus the ith element of $\boldsymbol{\theta}$. These two minima will correspond when all other components of $\boldsymbol{\theta}$ are at their $\boldsymbol{\theta}^*$ values.

Despite steepest descent's virtues of formal convergence and widespread use, it is a relatively inefficient algorithm in contrast to some other approaches that use gradient information. Further, it is sensitive to transformations and scaling. That is, a linear transformation of the underlying $\boldsymbol{\theta}$ vector can dramatically alter the algorithm performance even though the two vectors contain equivalent information about the underlying physical process if the transformation is invertible. Related to this sensitivity to transformations, the algorithm will have difficulty in optimizing some components if the magnitudes of the components differ dramatically (see the discussion in Subsection 1.4.2 and Exercise 1.11 for contrasting behavior with the Newton–Raphson method).

The steepest descent algorithm, however, provides a useful starting point for deriving more powerful algorithms. For example, other algorithms may have a generic steepest descent-type form with some differing details or may use steepest descent as part of the per-iteration calculations, as in trust region and conjugate gradient methods (Dennis and Schnabel, 1989, Sects. 4.2 and 5.2). The steepest descent algorithm also has the virtues of relative simplicity, of guaranteeing at least a local descent in the loss function, and (as we will see with the Robbins-Monro stochastic approximation algorithm of Chapters 4 and 5) of having a ready extension to stochastic problems where the gradient is known only to within some random noise. Example 1.8 illustrates the steepest descent algorithm on a simple $p = 2$ problem, comparing the results of a deterministic search and a search with noisy measurements of the gradient $\boldsymbol{g}(\boldsymbol{\theta})$. Calculations for this example (as for others in the book) are carried out in MATLAB.

Example 1.8—Steepest descent with and without noise. Let $\boldsymbol{\theta} = [t_1, t_2]^T$ and consider the simple loss function and associated (unconstrained) domain:

$$L(\boldsymbol{\theta}) = t_1^4 + t_1^2 + t_1 t_2 + t_2^2 \text{ with } \Theta = \mathbb{R}^2.$$

By solving for $\boldsymbol{\theta}$ in the equation $\boldsymbol{g}(\boldsymbol{\theta}) = \boldsymbol{0}$ and checking for the positive definiteness of the Hessian $\boldsymbol{H}(\boldsymbol{\theta})$ for all $\boldsymbol{\theta}$, it is easily seen that there is a unique solution $\boldsymbol{\theta}^* = [0, 0]^T$. For our purpose, however, let us assume that this solution is not known and that a steepest descent algorithm is to be used. In standard deterministic (noise-free) steepest descent, the gradient

$$\boldsymbol{g}(\boldsymbol{\theta}) = \begin{bmatrix} 4t_1^3 + 2t_1 + t_2 \\ t_1 + 2t_2 \end{bmatrix}$$

is assumed directly available.

In addition to the above-mentioned noise-free implementation, we also consider a case where only noisy measurements $\boldsymbol{Y}(\boldsymbol{\theta}) = \boldsymbol{g}(\boldsymbol{\theta}) + \boldsymbol{e}$ are available. Here, the noise $\boldsymbol{e}$ is i.i.d. $N(\boldsymbol{0}, \boldsymbol{I}_2)$, where $\boldsymbol{I}_2$ denotes the 2×2 identity matrix (so the variance of each element of $\boldsymbol{e}$ is 1). With the exception of this noisy measurement $\boldsymbol{Y}(\boldsymbol{\theta})$ being substituted for the true gradient, we use the basic steepest descent recursion in (1.5) with a constant gain $a_k = a$. Hence, the modified steepest descent algorithm (actually, a simple example of a Robbins–Monro root-finding stochastic approximation algorithm, discussed in Chapters 4 and 5) is $\hat{\boldsymbol{\theta}}_{k+1} = \hat{\boldsymbol{\theta}}_k - a\boldsymbol{Y}(\hat{\boldsymbol{\theta}}_k)$.

From an initial condition $\hat{\boldsymbol{\theta}}_0 = [1, 1]^T$, the noisy and noise-free versions of steepest descent are run for 50 iterations. Based on some limited numerical experimentation with sample runs to $k = 50$ in the noisy case, it is found that $a = 0.05$ tends to produce a mean final loss value (i.e., a mean of $L(\hat{\boldsymbol{\theta}}_{50})$ across several independent replications) that is approximately the lowest feasible under the noise. We use the same a in the noise-free case. Figures 1.5 and 1.6 show the relative performance for the search with and without noise. Figure 1.5 contrasts the search path for the two elements of $\boldsymbol{\theta}$. The two curves in Figure 1.6 show the mean loss over 50 independent replications (all starting at $\hat{\boldsymbol{\theta}}_0$) and the deterministic result. A single noisy run, of course, is much more erratic than the mean of the noisy runs. As motivated in Subsection 1.2.1, the plot in Figure 1.6 showing the mean of 50 runs is based on the exact (noise-free) L values even though the algorithm uses only noisy values $\boldsymbol{Y}$ in its iterations.

The noisy run in Figure 1.5 is slightly better than the typical run in the sense that the terminal loss is slightly below the mean terminal loss value over

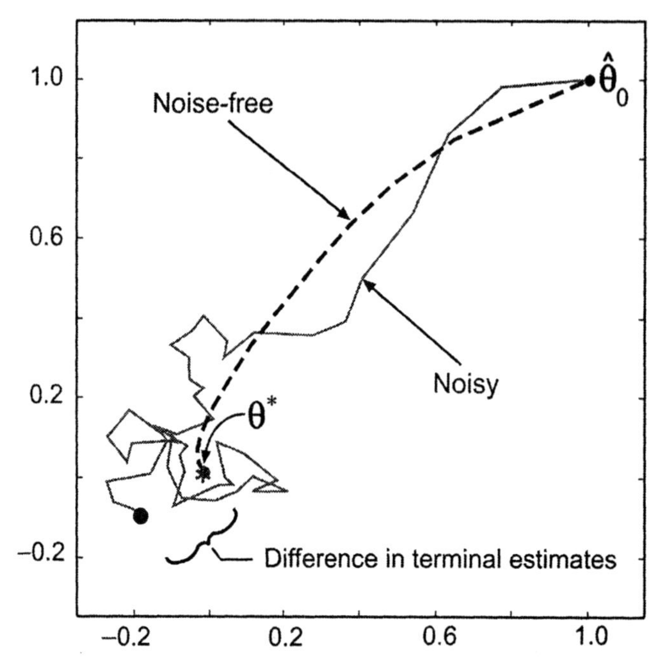

Figure 1.5. Comparison of steepest descent runs with noisy and noise-free gradient input. The indicated search paths are in the plane of t_2 (vertical axis) versus t_1 (horizontal axis). The indicated difference in terminal estimates shows the difference between the final noisy and noise-free estimates for $\boldsymbol{\theta}$.

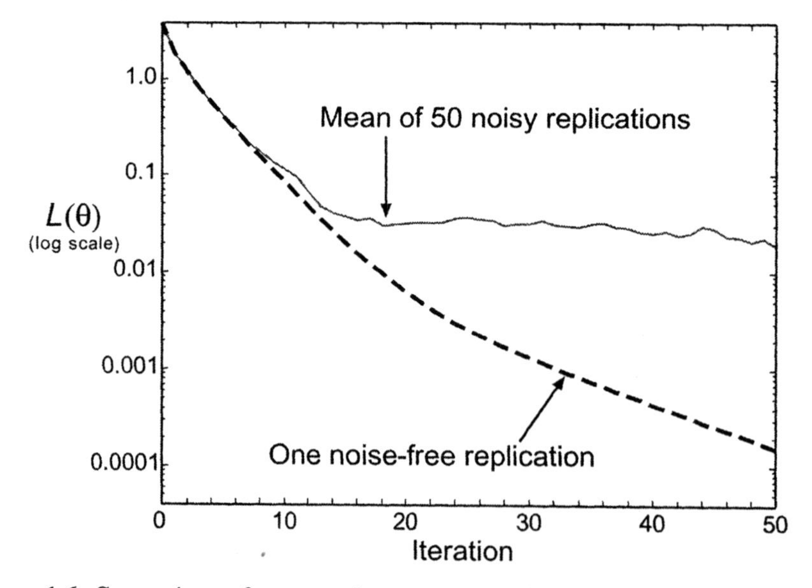

Figure 1.6. Comparison of steepest descent runs with noisy and noise-free gradient input: relative value of L as the iteration proceeds.

50 runs that is shown in Figure 1.6. Nevertheless, the terminal error in the noisy $\hat{\boldsymbol{\theta}}_{50}$ (relative to $\boldsymbol{\theta}^*$) represents about 16 percent of the total distance between $\hat{\boldsymbol{\theta}}_0$ and $\boldsymbol{\theta}^*$; in contrast, the terminal error for the noise-free case is approximately 1.0 percent. Figure 1.5 shows the relatively large difference between the terminal noisy and noise-free estimates of $\boldsymbol{\theta}$. Because the gain a is optimized for the noisy case, the deterministic steepest descent algorithm is capable of performing even better with a modification to the gain.

Figure 1.5 illustrates the common phenomenon—especially in stochastic problems—of the terminal iterate not being the best of the iterates, either in terms of distance between $\hat{\boldsymbol{\theta}}_k$ and the unknown $\boldsymbol{\theta}^*$ or in terms of the value of the loss function. This is not surprising in light of the "misleading" information being provided to the algorithm as a consequence of the noise. Figure 1.6 shows that the mean loss value in the noisy case decreases, although much more slowly than in the noise-free case. ❑

1.4.2 Newton–Raphson Method and Deterministic Convergence Rates

The Newton–Raphson algorithm (sometimes simply called *Newton's method*) builds on the basic recursion in (1.5) by introducing a scaling via the inverse Hessian matrix. This has the advantage of potentially speeding the convergence significantly, but has the possible disadvantage of making the algorithm more unstable. In particular, if it is assumed that L is at least twice continuously differentiable and that $\boldsymbol{H}$ is invertible at all $\boldsymbol{\theta}$ encountered in the search, the algorithm has the form

$$\hat{\boldsymbol{\theta}}_{k+1} = \hat{\boldsymbol{\theta}}_k - \boldsymbol{H}(\hat{\boldsymbol{\theta}}_k)^{-1}\boldsymbol{g}(\hat{\boldsymbol{\theta}}_k),\ k = 0, 1, 2, \ldots. \tag{1.6}$$

It is simple to informally derive this algorithm by expanding $\boldsymbol{g}(\boldsymbol{\theta})$ to first order around an old $\boldsymbol{\theta}$ value to get its approximate value at a new $\boldsymbol{\theta}$:

$$\boldsymbol{g}(\boldsymbol{\theta}_{\text{new}}) \approx \boldsymbol{g}(\boldsymbol{\theta}_{\text{old}}) + \boldsymbol{H}(\boldsymbol{\theta}_{\text{old}})(\boldsymbol{\theta}_{\text{new}} - \boldsymbol{\theta}_{\text{old}}) \tag{1.7}$$

Setting $\boldsymbol{g}(\boldsymbol{\theta}_{\text{new}}) = \boldsymbol{0}$ and taking "$\approx$" as an exact equality yields the Newton–Raphson recursion, where the subscripts k and k+1 represent old and new, respectively, and provided that $\boldsymbol{H}(\boldsymbol{\theta}_{\text{old}}) = \boldsymbol{H}(\hat{\boldsymbol{\theta}}_k)$ is invertible at each k. Note that if L is a quadratic function, the expansion in (1.7) is exact and the Newton–Raphson algorithm will converge in one step from any starting point—no algorithm can converge faster! This is obviously not a common situation in practice since few practical loss functions will be quadratic. However, near the solution $\boldsymbol{\theta}^*$, L will be nearly quadratic for any twice continuously differentiable function (by Taylor's theorem in Appendix A) and thus one can expect the Newton–Raphson algorithm to be fast once it is near $\boldsymbol{\theta}^*$.

In nonlinear deterministic optimization, "there is a strong suspicion that if any iterative method for any problem in any field is exceptionally effective, then it is a Newton [–Raphson] method in some appropriate context" (Dennis and Schnabel, 1989). For general nonlinear (nonquadratic) loss functions, $\boldsymbol{H}$ may not be positive definite at $\boldsymbol{\theta}$ away from $\boldsymbol{\theta}^*$, which can cause stalling or divergence. For this reason, many modifications of the basic Newton–Raphson method have been introduced with the goal of preserving most of the very fast local convergence while stabilizing the algorithm when it is not close to $\boldsymbol{\theta}^*$ (e.g., Dennis and Schnabel, 1989; Jang et al., 1997, Sect. 6.4). One simple way of helping control possible poor behavior of Newton–Raphson is to introduce a coefficient a_k in front of the $\boldsymbol{H}^{-1}\boldsymbol{g}$ term in (1.6), analogous to the step size in steepest descent. If $a_k < 1$ for all or most k, this may help stabilize the algorithm at the possible expense of slowing its convergence.

In contrast to steepest descent, Newton–Raphson has the desirable property of being transform invariant and unaffected by large scaling differences in the $\boldsymbol{\theta}$ elements (see Exercise 1.13). In particular, if $\boldsymbol{\theta}$ is transformed to $\boldsymbol{\theta}' = \boldsymbol{A\theta}$ for some invertible matrix $\boldsymbol{A}$, then optimization based on $\boldsymbol{\theta}'$ yields the same value of $\boldsymbol{\theta}$ after transforming back from $\boldsymbol{\theta}'$ to $\boldsymbol{\theta}$. For steepest descent, this is *not* true since the underlying Hessian in the $\boldsymbol{\theta}'$-based search becomes $(\boldsymbol{A}^{-1})^T\boldsymbol{H}(\boldsymbol{\theta})\boldsymbol{A}^{-1}$, affecting the convergence properties via a fundamental change in the loss function curvature. Related to this scaling issue is the issue of when a steepest descent and Newton–Raphson algorithm are equivalent. For a quadratic loss function, if $\boldsymbol{H} = \boldsymbol{H}(\boldsymbol{\theta}) = c\boldsymbol{I}_p$ for all $\boldsymbol{\theta}$ and some $c > 0$, c not a function of $\boldsymbol{\theta}$, then the algorithms (1.5) and (1.6) are identical when $a_k = 1/c$ in (1.5) ($\boldsymbol{I}_p$ denotes the $p \times p$ identity matrix).

Figure 1.7 illustrates the relative behavior of the steepest descent (first-order) and Newton–Raphson (second-order) algorithms on three different loss functions with $p = 2$. The level curves in the three cases indicate points in $\boldsymbol{\theta}$-space having the same level of L value. A steepest descent algorithm always moves in a direction perpendicular to the level curve at the current point. Cases (a) and (b) are quadratic surfaces. In (a), the circular level curves correspond to $\boldsymbol{H} = c\boldsymbol{I}_p$ as above. Thus, the two algorithms move in the same direction from the indicated starting point. In (b), the Hessian is nondiagonal, implying a difference in the search directions of the algorithms. While steepest descent moves perpendicular to the level curve, Newton–Raphson in (b) moves directly toward $\boldsymbol{\theta}^*$ because the function is quadratic (the one-step solution mentioned above).

Case (c) in Figure 1.7 illustrates the danger of moving in the second-order direction with a nonquadratic loss function. Here, Newton–Raphson moves in a direction *worse* than steepest descent. In fact, in some such cases, Newton–Raphson can diverge from the solution wildly. This figure is an illustration of the principle mentioned in Subsection 1.2.2: algorithms well suited to particular settings (e.g., quadratic loss functions) tend to be "brittle," performing poorly in other settings. (Newton–Raphson may, however, sometimes perform very well with nonquadratic loss functions; one must consider the algorithm on a case-by-case basis.)

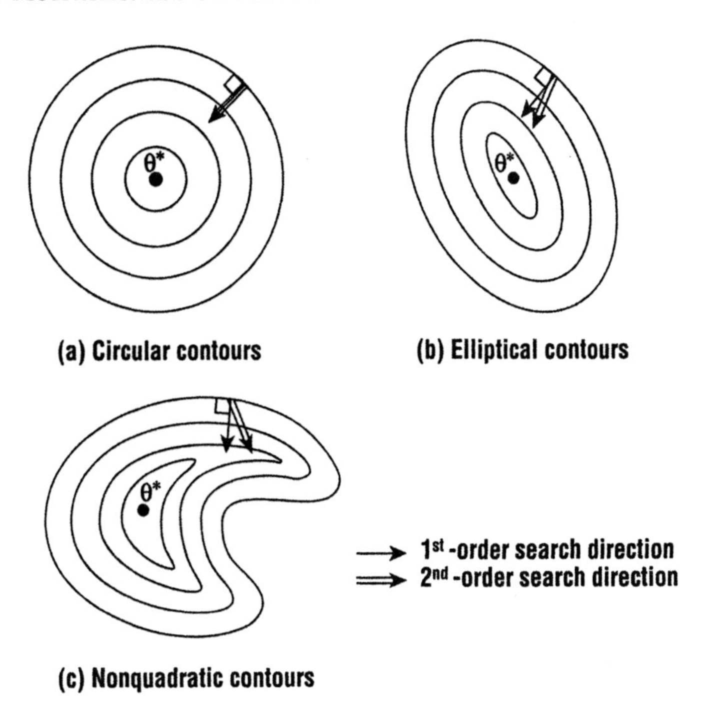

Figure 1.7. Contrast of first-order (steepest descent) and second-order (Newton–Raphson) search directions on $p = 2$ problems. In (a), both search directions are the same. In the elliptical (quadratic) surface (b), second-order direction points directly at $\boldsymbol{\theta}^*$ while first-order direction is not toward $\boldsymbol{\theta}^*$. Example (c) shows that the second-order direction can be *worse* than the first-order direction in nonquadratic regions of search space (as when starting far from $\boldsymbol{\theta}^*$).

Under the conditions of $\boldsymbol{H}$ being positive definite in a neighborhood of $\boldsymbol{\theta}^*$ and assuming that $\hat{\boldsymbol{\theta}}_k$ stays in this neighborhood, then the deterministic rate of convergence for Newton–Raphson is quadratic in the sense that

$$\left\|\hat{\boldsymbol{\theta}}_{k+1} - \boldsymbol{\theta}^*\right\| = O\left(\left\|\hat{\boldsymbol{\theta}}_k - \boldsymbol{\theta}^*\right\|^2\right),$$

where $O(x)$ is the standard big-O order notation implying $O(x)/x \le c$ for some constant $c > 0$ as $x \to 0$. (In the *stochastic* algorithms of this book, different types of rate of convergence measures will be used to accommodate the additional inherent variability.) The above quadratic rate result is proved in many places (e.g., Dennis and Schnabel, 1989; Kelly, 1999a, pp. 13–17). For $\hat{\boldsymbol{\theta}}_k$ sufficiently close to $\boldsymbol{\theta}^*$, this result implies very fast convergence: $\hat{\boldsymbol{\theta}}_{k+1}$ has an error that is proportional to the *square* of the (assumed small) error in the previous iterate $\hat{\boldsymbol{\theta}}_k$. This rate of convergence contrasts with the slower linear rate for the method of steepest descent:

$$\left\| \hat{\boldsymbol{\theta}}_{k+1} - \boldsymbol{\theta}^* \right\| = O\left(\left\| \hat{\boldsymbol{\theta}}_k - \boldsymbol{\theta}^* \right\| \right),$$

where the implied constant (c) in the order bound is less than unity (Kelly, 1999a, pp. 45–46). We emphasize that the above rates for both Newton–Raphson and steepest descent depend on the iterates being "sufficiently close" to $\boldsymbol{\theta}^*$.

Many algorithms aim to capture most of the enhanced speed of Newton–Raphson, but aim to do so with greater stability and without the need to explicitly calculate the Hessian matrix (which is difficult to obtain in many practical problems). These algorithms are designed to avoid the potential divergence illustrated in Figure 1.7(c). Included in this class of algorithms are the popular conjugate gradient or quasi-Newton methods (e.g., Dennis and Schnabel, 1989; Bazaraa et al., 1993, pp. 312–345; Rustagi, 1994, pp. 72–75; Kelly, 1999a, Chap. 4). Most of these algorithms have a deterministic convergence rate between linear and quadratic.

All of the above rates are critically dependent on the deterministic structure of the problem. If one has only gradient and Hessian information in the presence of random noise, these rates and/or implied constants no longer apply. As part of the stochastic approximation discussion in Chapters 4–7, we see analogous algorithms, where modifications are made to a basic first-order algorithm to capture the speedup due to second-order search directions while preserving stability.

1.5 CONCLUDING REMARKS

This chapter sets the stage for the remainder of the book. We discussed the meaning of *stochastic* in the context of search and optimization, noting that it pertains to random noise in measurements of the loss function or root-finding function and/or to randomness that is injected in a Monte Carlo fashion. Although much of this book focuses on standard algorithms for stochastic search and optimization, we also consider some closely related aspects of stochastic modeling. These include model selection and experimental design.

Real-world problem solving is often difficult. We discussed some of the reasons why this is so. These include: (i) the presence of noise in the function evaluations; (ii) the difficulties in distinguishing a globally optimal solution from locally optimal solutions; (iii) the "curse of dimensionality," which causes the size of the search space to grow exponentially with problem dimension; (iv) the difficulties associated with constraints and the need for problem-specific methods; and (v) the lack of stationarity in the solution as a result of the conditions of the problem changing over time. A good treatment of both technical and philosophical challenges in real-world problem solving is Michalewicz and Fogel (2000).

The "no free lunch" theorems are a fundamental barrier to exaggerated claims of the power and efficiency of any specific algorithm. These theorems indicate that if an algorithm is especially efficient on one set of problems, it is *guaranteed* to be inefficient on another set of problems. The way to cope with the negative implication of these theorems in practice is to restrict an algorithm to a particular class of problems *and* to have the algorithm exploit the structure in this class. Most of the algorithms in this book are based on this principle of exploiting available structure, although in many cases the available structure is an incomplete representation of the process.

One related general area not emphasized in this book is *stochastic programming* (e.g., Birge and Louveaux, 1997). Although not universally defined as such, stochastic programming tends to emphasize methods that are direct applications of deterministic linear and nonlinear programming techniques. Often, the random aspects of the problem are removed by replacing the relevant random quantities with corresponding (fixed) mean values. Although such an approach is powerful in the right context, it requires information that is often not available in practical applications. For example, in the framework of this chapter, many methods in stochastic programming require direct knowledge of the loss function $L(\boldsymbol{\theta})$, rather than using only the noisy measurements $y(\boldsymbol{\theta})$. One of the appealing features of stochastic programming is the ability to deal with nontrivial constraints in the same way as deterministic methods. This benefit accrues because of the transformation of the stochastic problem to an effectively deterministic problem.

Many of the methods in this book are designed to handle the challenges outlined above, having applicability to broad areas such as estimation, Monte Carlo simulation, and control. It will also be clear that there are many exciting areas—in both algorithm development and applications—that have yet to be fully explored. In applying methods from this book, it is important to keep in mind the fundamental goals of a scientific or other investigation: "Better a rough answer to the right question than an exact answer to the wrong one" (aphorism possibly due to Lord Kelvin). In a wide range of problems, the extensive set of tools here will include the right tools for the right question.

EXERCISES[2]

1.1 Suppose that the noise ε has a symmetric triangular probability density function over the interval $[-1, 1]$ (i.e., with x the dummy variable for the density function, the density is $1 - |x|$ for $x \in [-1, 1]$ and 0 for $x \notin [-1, 1]$). What is $\text{var}(\varepsilon)$ (the variance of ε)?

1.2 For a scalar θ, identify the maximum and minimum of the function $\theta/2+\sqrt{1-\theta^2}$ on the domain $[-1, 1]$.

---

[2]Appendices A–C also contain exercises relevant to this chapter. Answers to selected exercises for this and other chapters/appendices are at the back of the book and at the book's Web site.

1.3 Based on investments represented by $\boldsymbol{\theta}$, suppose that each component of some deterministic vector function $s(\boldsymbol{\theta})$ represents the amount of output in one segment of a firm's business and that each component of the random vector $\boldsymbol{\pi}$ represents the profit resulting from 1 unit of output in the corresponding segment (so $\dim(s(\boldsymbol{\theta})) = \dim(\boldsymbol{\pi})$ with all segments of the business represented in these vectors). Suppose that the total profit of the firm can be measured and that $\text{cov}(\boldsymbol{\pi}) = \boldsymbol{\Sigma}$. Formulate the stochastic optimization problem of finding the investments to maximize mean profit subject to having the variance of the profit be no greater than some bound $C > 0$. In particular, write down the *general forms* for $L(\boldsymbol{\theta})$, $y(\boldsymbol{\theta})$, and Θ, and show that ε depends on $\boldsymbol{\theta}$. (In practice, a simulation or real data on the firm could be used to produce the $y(\boldsymbol{\theta})$ values needed in the optimization.)

1.4 Suppose that Θ contains only two options, $\Theta = \{\boldsymbol{\theta}_1, \boldsymbol{\theta}_2\}$, and that $L(\boldsymbol{\theta}_1) = 0$ and $L(\boldsymbol{\theta}_2) = 1$. Suppose that the analyst's budget allows only one (noisy) measurement $y(\boldsymbol{\theta}_i)$ at each $\boldsymbol{\theta}_i$ and that the noise terms (ε) are independently $N(0, 1)$ distributed for both measurements. What is the probability of incorrectly selecting $\boldsymbol{\theta}_2$ as the optimal $\boldsymbol{\theta}$ when the choice is based on the $\boldsymbol{\theta}_i$ having the lowest $y(\boldsymbol{\theta}_i)$?[3]

1.5 Suppose that $\Theta = \{\boldsymbol{\theta}_1, \boldsymbol{\theta}_2\}$ and that only noisy measurements of L are available, where the noise process ε is i.i.d. over all measurements (i.e., does not depend on $\boldsymbol{\theta}$). Suppose that the sample means of 20 measurements at each of $\boldsymbol{\theta}_1$ and $\boldsymbol{\theta}_2$ are 5.4 and 6.6, respectively, with corresponding sample variances 2.2 and 3.2 (using the standard definition in Appendix B). Based on the appropriate two-sample test (Appendix B), discuss whether such data provide sufficient evidence to claim that there is a significant difference in $L(\boldsymbol{\theta}_1)$ and $L(\boldsymbol{\theta}_2)$ and discuss at least one caveat associated with the conclusion in light of what is *not* stated in the assumptions.[4]

1.6 Suppose that each of the p components in $\boldsymbol{\theta}$ can take on one of N values and suppose that we sample uniformly and independently (with replacement) in Θ in search of an optimum.

(a) What is the general expression for the probability of finding at least one of the V best points in Θ when we generate K (noise-free) values of $L(\boldsymbol{\theta})$?

(b) With $p = 10$ and $N = 15$, what is the probability of finding at least one of the 10,000 best points in Θ when we generate 100,000 values of $L(\boldsymbol{\theta})$?

1.7 Consider a two-dimensional $\boldsymbol{\theta}$ where $\Theta = [1.0, 2.1) \times [1.0, 2.1)$ (i.e., a Cartesian product of intervals with endpoints 2.1 not in Θ) and $L(\boldsymbol{\theta}) \in \{0, 1, 2, 3, 4\}$. Suppose the computer accuracy is such that each component of $\boldsymbol{\theta}$ is evaluated at increments of 2.2×10^{-16} in Θ. (Using IEEE standards, this is the approximate accuracy of floating-point numbers on a 32-bit

---

[3] A normal distribution table is needed for this and several other exercises in the book. Such tables are available in almost any introductory statistics textbook or in various forms in MS EXCEL (e.g., the NORMDIST function).

[4] Analogous to Exercise 1.4, this exercise requires a t-distribution table, available in almost any introductory statistics textbook or in MS EXCEL (e.g., the TDIST function).

computer operating in double precision near the value 1.0.) What is the number of mappings in the average that forms the basis for the NFL theorems? Express the answer in the general form 10^x by solving for x.

1.8 Let $\boldsymbol{\theta} = [t_1, t_2, t_3]^T$ and

$$L(\boldsymbol{\theta}) = t_1^2 + 2t_2^2 + 3t_3^2 + 3t_1t_2 + 4t_1t_3 - 3t_2t_3.$$

Verify that $\boldsymbol{\theta} = \mathbf{0}$ is a solution to $\boldsymbol{g}(\boldsymbol{\theta}) = \mathbf{0}$. Determine whether this solution is a minimum, maximum, or saddlepoint.

1.9 Using Taylor's theorem A.2 in Appendix A, prove that L is a convex function on Θ if $\boldsymbol{H}$ is positive semidefinite at all points in Θ.

1.10 Let $\boldsymbol{\theta} = [t_1, t_2]^T$. Apply the steepest descent method to the function $L(\boldsymbol{\theta}) = (t_1 - 2)^4 + (t_1 - 2t_2)^2$ beginning with the initial guess $\hat{\boldsymbol{\theta}}_0 = [0, 3]^T$ and using $a_k = 0.062, 0.24, 0.11, 0.31, 0.12$, and 0.36 for $k = 0, 1, 2, 3, 4$, and 5, respectively (yielding $\hat{\boldsymbol{\theta}}_1$ to $\hat{\boldsymbol{\theta}}_6$). Report values for L and $\boldsymbol{\theta}$ after each of the six iterations. (For comparison purposes, note that $L(\hat{\boldsymbol{\theta}}_0) = 52.0$, $\boldsymbol{\theta}^* = [2, 1]^T$ and $L(\boldsymbol{\theta}^*) = 0$.)

1.11 Using the same function and initial guess as Exercise 1.10, apply the Newton–Raphson method in eqn. (1.6) and report the stated values at the indicated number of iterations.

1.12 For the function in Exercise 1.10, do the following:

(a) Run the standard deterministic steepest descent algorithm for 1000 iterations with $a_k = 0.0275$ for all k.

(b) Suppose that we only observe $\boldsymbol{Y}(\boldsymbol{\theta}) = \boldsymbol{g}(\boldsymbol{\theta}) + \boldsymbol{e}$, where the noise $\boldsymbol{e}$ is independently $N(\mathbf{0}, 0.1^2\boldsymbol{I}_2)$ distributed. Use the steepest descent algorithm with $\boldsymbol{Y}(\boldsymbol{\theta})$ replacing $\boldsymbol{g}(\boldsymbol{\theta})$ and $a_k = 0.0275$ (as in part (a)). Determine the sample mean of the final loss values from 100 realizations of 1000 iterations per realization, i.e., form a sample mean of the 100 values of $L(\hat{\boldsymbol{\theta}}_{1000})$. Compare the mean final loss function value here with the deterministic value from part (a). (This is an example of a root-finding stochastic approximation algorithm, as discussed in Chapters 4 and 5.)

1.13 Fill in the missing details in the arguments in Subsection 1.4.2 regarding the invariance of the Newton–Raphson method to transformations and different scaling.

1.14 Suppose that $\|\hat{\boldsymbol{\theta}}_k - \boldsymbol{\theta}^*\| = f(k)$ in the steepest descent or Newton–Raphson algorithm.

(a) Give an example $f(k)$ that satisfies the big-O result of Subsection 1.4.2 for the steepest descent algorithm.

(b) Give an analogous example $f(k)$ for the Newton–Raphson algorithm.

CHAPTER 2

DIRECT METHODS FOR STOCHASTIC SEARCH

This chapter examines several direct search methods for optimization, "direct" in the sense that the algorithms use minimal information about the loss function. These direct methods have the virtues of being simple to implement and having broad applicability (e.g., allowing for either continuous- or discrete-valued $\boldsymbol{\theta}$). Further, it is sometimes possible to make rigorous statements about the convergence properties of the algorithms, which is not always possible in more complex procedures.

Prior to presenting the algorithms, we discuss in Section 2.1 some motivating issues in direct random search methods. The algorithms discussed in Sections 2.2 to 2.4 represent two distinct types of the most popular direct search methods. Sections 2.2 and 2.3 consider the first type—direct search methods involving random selections of $\boldsymbol{\theta}$ within the search domain Θ. These are among the most popular and versatile search methods available. Section 2.2 considers noise-free loss measurements and Section 2.3 considers noisy loss measurements. Section 2.4 considers the second type of direct search method—a nonlinear algorithm inspired by the geometric notion of a simplex (and not to be confused with the well-known simplex method of linear programming). Although this technique is frequently presented as a deterministic search method, its origins are in stochastic search involving noisy loss measurements.

In support of this and other chapters, Appendix D discusses some issues in computer-based random number generation; such *pseudorandom* numbers play a critical role in the random search algorithms of this chapter as well as in some of the algorithms to be seen later.

2.1 INTRODUCTION

This chapter focuses on relatively simple *direct search* methods for solving the optimization problem of minimizing a loss function $L = L(\boldsymbol{\theta})$ subject to the parameter vector $\boldsymbol{\theta}$ lying in some set Θ. The information required to implement these methods is essentially only input–output data where $\boldsymbol{\theta}$ is the input and $L(\boldsymbol{\theta})$ (noise-free) or $y(\boldsymbol{\theta})$ (noisy) is the output. Further, the underlying assumptions about L are relatively minimal. In particular, there are no requirements that the gradient of L be computable (or even that the gradient *exist*) or that L be unimodal (i.e., that L have only one local optimum, corresponding to the global optimum). This chapter focuses on two distinct types of popular direct search methods. The first is random search and the second is based on the geometric notion of a simplex.

Note that when the random search or nonlinear simplex methods are applied to problems with differentiable L, these techniques make no use of the gradient that exists. Avoiding the use of the gradient can often significantly ease implementation. Additional—and often more powerful—gradient-free methods will be discussed later in this book (principally, in Chapters 6–10, covering simultaneous perturbation stochastic approximation, simulated annealing, and evolutionary computation methods).

With the exception of *some* of the discussion on the nonlinear simplex method, the algorithms of this chapter have at least one of the two characteristics of stochastic search and optimization mentioned in Chapter 1—noisy input information (Property A in Section 1.1) or injected algorithm randomness (Property B). For purposes of algorithm *evaluation* by Monte Carlo, pseudorandom numbers generated via a computer algorithm are used to produce both the noise and the injected randomness. For purposes of *actual application* to a physical system producing its own randomness, pseudorandom numbers are used only for the injected randomness. All pseudorandom numbers in this (and other chapters) are generated via the default generators **`rand`** (uniform) and **`randn`** (normal) in MATLAB (see Appendix D for a brief discussion of these and other generators).

Unless noted otherwise, we restrict the random sampling of the algorithms here to be in the constraint domain Θ (i.e., there is no sampling over a larger set Θ', where $\Theta \subset \Theta'$, as mentioned in Subsection 1.2.1 as being of potential interest in some applications; see also Exercise 2.11). This appears to be the most common form of implementation.

Recall from Subsection 1.2.1 that there are never likely to be fully acceptable automated stopping criteria for stochastic search algorithms. For that reason, this book will generally emphasize algorithm comparisons and stopping based on "budgets" of function evaluations. Nonetheless, sometimes one may wish to augment function budgets or analyst "intuition" via some automated approach.

Two common stopping criteria are given below. Let η represent a small positive number, N be the number of iterations for which the algorithm should be "stable," and $\hat{\boldsymbol{\theta}}_k$ be the estimate for $\boldsymbol{\theta}$ at iteration k. An algorithm may be stopped at iteration $n \geq N$ when, for all $1 \leq j \leq N$, at least one of the two criteria below hold:

$$\left|L(\hat{\boldsymbol{\theta}}_n) - L(\hat{\boldsymbol{\theta}}_{n-j})\right| \leq \eta \quad \text{or} \quad \left\|\hat{\boldsymbol{\theta}}_n - \hat{\boldsymbol{\theta}}_{n-j}\right\| \leq \eta.$$

These criteria express the fact that the algorithm has "settled down" in some sense. The first criterion above is useful when noise-free loss measurements are available, while the second criterion is useful in noise-free *and* noisy cases. Although the criteria have some intuitive appeal, they provide no guarantee of the terminal iterate $\hat{\boldsymbol{\theta}}_n$ being close to an optimum $\boldsymbol{\theta}^*$. Some additional stopping criteria that are closely related to the above are given in Section 2.4.

The next three sections present some simple direct search algorithms, focusing on random search and the nonlinear simplex method (a simplex is the region contained within a collection of $p+1$ points in $\mathbb{R}^p$, as defined more formally in Section 2.4). Although the simplex algorithm of Section 2.4 is not as simple as the random search methods of Sections 2.2 and 2.3, it is popular and often powerful. As an illustration of its popularity, the simplex method is the main MATLAB gradient-free optimization algorithm (**`fminsearch`**).

2.2 RANDOM SEARCH WITH NOISE-FREE LOSS MEASUREMENTS

2.2.1 Some Attributes of Direct Random Search

Random search methods for optimization are based on exploring the domain Θ in a random manner to find a point that minimizes $L = L(\boldsymbol{\theta})$. These are the simplest methods of stochastic optimization and can be quite effective in some problems. Their relative simplicity is an appealing feature to both practitioners and theoreticians. These direct random search methods have all or some of the following advantages relative to most other search methods. Some of these attributes were mentioned in the prescient paper of Karnopp (1963):

(i) **Ease of programming.** The methods described below are relatively easy to code in software and can thereby significantly reduce the *human* cost of an optimization process (not to be ignored in practice, but often ignored in published results comparing the efficiency of one algorithm against another!).

(ii) **Use of only *L* measurements.** The reliance on L measurements alone can significantly reduce the incentive to pick a loss function largely for analytical convenience—perhaps at a sacrifice to the true optimization goals—so that gradients or other ancillary information may be computed. Karnopp (1963) aptly calls this a reduction in "artful contrivance."

(iii) **Reasonable computational efficiency.** Although not generally the most computationally efficient algorithms in practical problems, the algorithms can often provide *reasonable* solutions fairly quickly, especially if the problem dimension p ($=\dim(\boldsymbol{\theta})$) is not too large. This is especially true in those direct search algorithms that make use of some local information in their search (e.g., random search algorithms B and C below). For example, as demonstrated in Anderssen and Bloomfield (1975), random search may be more efficient than corresponding deterministic algorithms based on searches over multidimensional grids when p is large. The solution from a random search method can usually be augmented with some more powerful—but perhaps more complex—search algorithm if a more accurate solution is required. In fact, the random search algorithms may provide a means of finding "good" initial conditions for some of the more sophisticated algorithms to be presented later.

(iv) **Generality.** The algorithms can apply to virtually any function. The user simply needs to specify the nature of the sampling randomness to allow an adequate search in Θ. Thus, if $\boldsymbol{\theta}$ is continuous valued, the sampling distribution should be continuous (e.g., Gaussian or continuously uniform on Θ); likewise, a discrete-valued $\boldsymbol{\theta}$ calls for a discrete sampling distribution with nonzero probability of hitting the candidate points and a hybrid $\boldsymbol{\theta}$ calls for the appropriate mix of continuous and discrete sampling distributions.

(v) **Theoretical foundation.** Unlike some algorithms, supporting theory is often available to provide guarantees of performance and guidance on the expected accuracy of the solution. In fact, the theory may even be exact in finite samples, which is virtually unheard of in the analysis of other stochastic algorithms.

2.2.2 Three Algorithms for Random Search

Beginning with the most basic algorithm, this subsection describes three direct random search techniques. Because direct random search is a large subject unto itself, only a small selection of algorithms is being presented here in order to keep this subsection of manageable length (see, e.g., Solis and Wets, 1981; Zhigljavsky, 1991; and some references discussed below for additional algorithms). The three algorithms here are intended to convey the essential flavor of most available direct random search algorithms. The methods of this subsection assume perfect (noise-free) values of L.

In the noise-free case, it can be shown that many random search methods converge to an optimum $\boldsymbol{\theta}^*$ in one of the probabilistic senses discussed in Appendix C—almost surely (a.s.), in probability (pr.), or in mean square (m.s.)—as the number of L evaluations gets large (see below). Realistically, however, these convergence results may have limited utility in practice since the algorithms may take a prohibitively large number of function evaluations to reach a value close to $\boldsymbol{\theta}^*$, especially if p is large. This is illustrated below. Nevertheless, the formal convergence provides a guarantee not always available in other approaches.

The simplest random search method is one where we repeatedly sample over Θ such that the current sampling for $\boldsymbol{\theta}$ does not take into account the previous samples. This "blind search" approach does not adapt the current sampling strategy to information that has been garnered in the search up to the present time. The approach can be implemented in batch (nonrecursive) form simply by laying down a number of points in Θ and taking the value of $\boldsymbol{\theta}$ yielding the lowest L value as our estimate of the optimum. The approach can also be implemented in recursive form as we illustrate below.

The simplest setting for conducting the random sampling of new (candidate) values of $\boldsymbol{\theta}$ is when Θ is a hypercube (a p-fold Cartesian product of intervals on the real line) and we are using uniformly generated values of $\boldsymbol{\theta}$. The uniform distribution is continuous or discrete for the elements of $\boldsymbol{\theta}$ depending on the definitions for these elements. *In fact, this particular (blind search) form of the algorithm is unique among all general stochastic search and optimization algorithms in this book: It is the only one without any adjustable algorithm coefficients that need to be "tuned" to the problem at hand.*

For a domain Θ that is not a hypercube or for other sampling distributions, one may use transformations, rejection methods, or Markov chain Monte Carlo to generate the sample $\boldsymbol{\theta}$ values. For example, if Θ is an irregular shape, one can generate a sample on a hypercube superset containing Θ and then reject the sample point if it lies outside of Θ. Appendix D and Chapter 16 provide a discussion of techniques for random sampling in nontrivial cases (see also Exercise 2.4).

A recursive implementation of the simple random search idea is as follows. This algorithm, called algorithm A here, applies when $\boldsymbol{\theta}$ has continuous, discrete, or hybrid elements.

Algorithm A: Simple Random ("Blind") Search

Step 0 **(Initialization)** Choose an initial value of $\boldsymbol{\theta}$, say $\hat{\boldsymbol{\theta}}_0 \in \Theta$, either randomly or deterministically. (If random, usually a uniform distribution on Θ is used.) Calculate $L(\hat{\boldsymbol{\theta}}_0)$. Set $k = 0$.

Step 1 Generate a new independent value $\boldsymbol{\theta}_{\text{new}}(k+1) \in \Theta$, according to the chosen probability distribution. If $L(\boldsymbol{\theta}_{\text{new}}(k+1)) < L(\hat{\boldsymbol{\theta}}_k)$, set $\hat{\boldsymbol{\theta}}_{k+1} = \boldsymbol{\theta}_{\text{new}}(k+1)$. Else, take $\hat{\boldsymbol{\theta}}_{k+1} = \hat{\boldsymbol{\theta}}_k$.

Step 2 Stop if the maximum number of L evaluations has been reached or the user is otherwise satisfied with the current estimate for $\boldsymbol{\theta}$ via appropriate stopping criteria; else, return to step 1 with the new k set to the former $k + 1$.

Example 2.1 demonstrates algorithm A on a simple loss function. At the possible expense of belaboring the obvious, we present more details here than in other examples because this is the first stochastic optimization algorithm being covered.

Example 2.1—Illustration of algorithm A on a simple function. With $p = 2$, consider the simple quadratic loss function $L(\boldsymbol{\theta}) = \boldsymbol{\theta}^T\boldsymbol{\theta}$ on the domain $\Theta = [1, 3]\times[1, 3]$. The unique minimizing solution to this constrained problem is $\boldsymbol{\theta}^*$

$= [1, 1]^T$. Suppose that an analyst has no knowledge of the functional form of $L(\boldsymbol{\theta})$; rather, the analyst is only able to obtain output L values for given input $\boldsymbol{\theta}$ values. Consider the use of algorithm A with a uniform distribution on Θ for generating the candidate points $\boldsymbol{\theta}_{\text{new}}$. Because of the lack of information about L, the analyst lets $\hat{\boldsymbol{\theta}}_0 = [2, 2]^T$, the center point of Θ. Each of the uniformly distributed candidate points in Θ is generated according to the formula $2\boldsymbol{U} + [1, 1]^T$, where $\boldsymbol{U}$ is a two-dimensional random vector in $[0, 1]\times[0, 1]$ generated via the MATLAB pseudorandom generator **rand**.

Table 2.1 shows the details from the first six iterations of one run. Even though only a few iterations are shown, there is ample evidence of the algorithm converging to $\boldsymbol{\theta}^*$. The $\boldsymbol{\theta}$ estimate moves from $[2, 2]^T$ to $[1.34, 1.76]^T$ and the loss value drops from 8.00 to 4.89. Note that in three of the six iterations (at $k = 2$, 4, and 5), the candidate point $\boldsymbol{\theta}_{\text{new}}$ is rejected because it does not offer an improvement over the current estimate. The fraction of candidate points rejected will increase as the number of iterations increase because the portion of Θ offering a lower loss value decreases. Equivalently, the *rate* of decay in the loss value (as shown in the last column in Table 2.1) will be reduced. ❑

Theorem 2.1 shows that algorithm A converges a.s. to $\boldsymbol{\theta}^*$ under reasonable conditions. Informally, Theorem 2.1 says that if values of L at or near $L(\boldsymbol{\theta}^*)$ can be reached with nonzero probability based on the sampling probability chosen for generating $\boldsymbol{\theta}_{\text{new}}(k)$ (the sampling probability is allowed to vary with k), then $\hat{\boldsymbol{\theta}}_k$ will converge to $\boldsymbol{\theta}^*$ as $k \to \infty$. This theorem uses the concept of the infimum (inf) of a function, which corresponds to the *greatest lower bound* to the function on the specified domain (see Appendix A).

Table 2.1. Details of the first several iterations of algorithm A for the constrained problem with quadratic loss function in Example 2.1. Note decreasing $L(\hat{\boldsymbol{\theta}}_k)$.

k	$\boldsymbol{\theta}_{\text{new}}(k)^T$	$L(\boldsymbol{\theta}_{\text{new}}(k))$	$\hat{\boldsymbol{\theta}}_k^T$	$L(\hat{\boldsymbol{\theta}}_k)$
0	—	—	[2.00, 2.00]	8.00
1	[2.25, 1.62]	7.69	[2.25, 1.62]	7.69
2	[2.81, 2.58]	14.55	[2.25, 1.62]	7.69
3	[1.93, 1.19]	5.14	[1.93, 1.19]	5.14
4	[2.60, 1.92]	10.45	[1.93, 1.19]	5.14
5	[2.23, 2.58]	11.63	[1.93, 1.19]	5.14
6	[1.34, 1.76]	4.89	[1.34, 1.76]	4.89

Unlike most of the other algorithms in this book, we present the proof of convergence here. Although not trivial, this proof is *relatively* simple as a consequence of the simplicity of the algorithm. Nevertheless, despite the relative simplicity, the proof provides a flavor of some of the arguments that are made in the more difficult proofs not included in this book.

Theorem 2.1. Suppose that $\boldsymbol{\theta}^*$ is the unique minimizer of L on the domain Θ (i.e., Θ^* contains only one point), where $L(\boldsymbol{\theta}^*) > -\infty$ and

$$\inf_{\boldsymbol{\theta}\in\Theta,\|\boldsymbol{\theta}-\boldsymbol{\theta}^*\|\geq\eta} L(\boldsymbol{\theta}) > L(\boldsymbol{\theta}^*) \qquad (2.1)^1$$

for all $\eta > 0$. Suppose further that for any $\eta > 0$ and for all k, there exists a function $\delta(\eta) > 0$ such that

$$P[\boldsymbol{\theta}_{\text{new}}(k): L(\boldsymbol{\theta}_{\text{new}}(k)) < L(\boldsymbol{\theta}^*) + \eta] \geq \delta(\eta). \qquad (2.2)$$

Then, for algorithm A with noise-free loss measurements, $\hat{\boldsymbol{\theta}}_k \to \boldsymbol{\theta}^*$ a.s. as $k \to \infty$.

Proof. (The mathematical level of this proof is slightly beyond the prerequisites of this book, but should be largely accessible to those who understand the material of Appendix C. The proof may be skipped without significantly affecting the understanding of subsequent material.) Because $L(\boldsymbol{\theta}^*) > -\infty$ and $L(\hat{\boldsymbol{\theta}}_k)$ is monotonically nonincreasing due to having noise-free loss measurements, $\lim_{k\to\infty} L(\hat{\boldsymbol{\theta}}_k(\omega))$ exists for each underlying sample point ω, as in Appendix C (e.g., Apostol, 1974, p. 185). Let $S_\eta = \{\boldsymbol{\theta}: L(\boldsymbol{\theta}) < L(\boldsymbol{\theta}^*) + \eta\}$. By condition (2.2), $P(\boldsymbol{\theta}_{\text{new}}(k) \in S_\eta) \geq \delta(\eta)$ for any $\eta > 0$. Hence, by the independence of the sampling distribution, $P(\boldsymbol{\theta}_{\text{new}}(k) \notin S_\eta \; \forall\, k) \leq [1 - \delta(\eta)]^k \to 0$ as $k \to \infty$. So, $\lim_{k\to\infty} L(\hat{\boldsymbol{\theta}}_k) < L(\boldsymbol{\theta}^*) + \eta$ in pr. Because $\eta > 0$ is arbitrarily small, we know that $L(\hat{\boldsymbol{\theta}}_k) \to L(\boldsymbol{\theta}^*)$ in pr. as $k \to \infty$. Then by a standard result in probability theory (e.g., Laha and Rohatgi, 1979, Proposition 1.3.4 and problem 25(b) on p. 63), this implies that $L(\hat{\boldsymbol{\theta}}_{k_j}) \to L(\boldsymbol{\theta}^*)$ a.s. as $j \to \infty$ for some subsequence $\{k_j\}$.

Given the convergence of $L(\hat{\boldsymbol{\theta}}_k)$ to *some* limit for each sample point ω mentioned above, and the fact that the subsequence has to converge to the same point as the full sequence, we know that the a.s. limit of $L(\hat{\boldsymbol{\theta}}_k)$ must also be

---

[1] The left-hand side of inequality (2.1) should be read as: "The infimum of $L(\boldsymbol{\theta})$ over the set of $\boldsymbol{\theta}$ such that $\boldsymbol{\theta} \in \Theta$ and $\|\boldsymbol{\theta}-\boldsymbol{\theta}^*\| \geq \eta$." Obviously, η must be such that at least some $\boldsymbol{\theta}$ satisfying $\|\boldsymbol{\theta}-\boldsymbol{\theta}^*\| \geq \eta$ lie in Θ.

$L(\boldsymbol{\theta}^*)$. The uniqueness of $\boldsymbol{\theta}^*$ as given in (2.1) then implies that $\hat{\boldsymbol{\theta}}_k \to \boldsymbol{\theta}^*$ a.s., as we set out to prove. ❑

Figure 2.1 presents three example functions and sampling distributions, two of which (examples (a) and (b)) satisfy the important theorem condition (2.2) and one of which (example (c)) does not. In Figure 2.1(a), there is always some nonzero probability of producing a loss value arbitrarily close to $L(\theta^*)$, while in Figure 2.1(b), there is nonzero probability of landing directly on θ^*; in either case, clearly condition (2.2) is satisfied. On the other hand, for the discontinuous function of Figure 2.1(c), the probability is zero of generating $\theta_{\text{new}} = \theta_{\text{new}}(k)$ such that $|L(\theta_{\text{new}}) - L(\theta^*)| < \eta$ for all η sufficiently small because of the gap between $L(\theta)$ and $L(\theta^*)$ for all $\theta \neq \theta^*$ (recall that $P(\theta_{\text{new}} = \theta^*) = 0$ because the sampling is from a continuous distribution). Clearly, condition (2.2) is violated because $P(\theta_{\text{new}}: L(\theta_{\text{new}}) < L(\theta^*) + \eta) = 0$ for any $0 < \eta < |L(0) - L(\theta^*)|$ since $\theta^* \neq 0$. (Exercise 2.1 shows that Theorem 2.1 applies to the problem of Example 2.1.)

While Theorem 2.1 establishes convergence of the simple random search algorithm, it is also of interest to examine the *rate* of convergence. The rate is intended to tell the analyst how close $\hat{\boldsymbol{\theta}}_k$ is likely to be to $\boldsymbol{\theta}^*$ for a given cost of search. As motivated in Subsection 1.2.1, the cost of search here (and through

(a) Continuous $L(\theta)$; sampling probability density for θ_{new} that is > 0 on $\Theta = [0, \infty)$

(b) Discrete $L(\theta)$ defined on $\Theta = \{0, 1, 2,\ldots\}$; discrete sampling for θ_{new} with $P(\theta_{\text{new}} = i) > 0$ for $i = 0, 1, 2,\ldots$

(c) Noncontinuous $L(\theta)$; sampling probability density for θ_{new} that is > 0 on $\Theta = [0, \infty)$

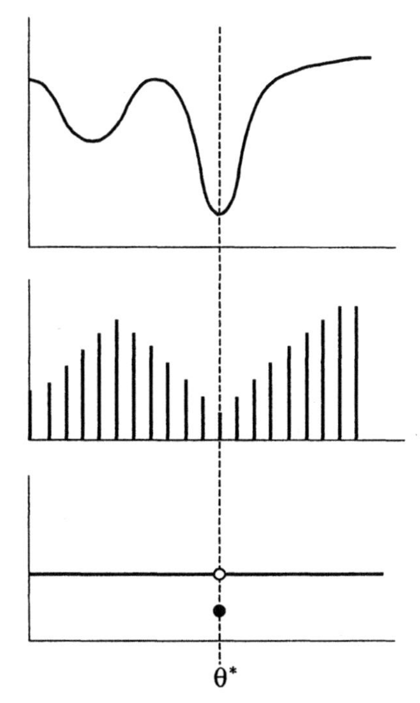

Figure 2.1. Applicability/nonapplicability of Theorem 2.1. Three example loss functions ($L(\theta)$ versus θ for scalar $\theta \geq 0$) and sampling distributions for $\theta_{\text{new}} = \theta_{\text{new}}(k)$. Condition (2.2) of Theorem 2.1 holds in settings (a) and (b), but does not hold in (c).

much of this book) is expressed in terms of number of loss function evaluations. Knowledge of the rate of convergence is critical in practical applications. Simply knowing that an algorithm converges begs the question of whether the algorithm will yield a practically acceptable solution in any reasonable period. To evaluate the rate, let us specify a *satisfactory region* $S(\boldsymbol{\theta}^*)$ representing some neighborhood of $\boldsymbol{\theta}^*$ providing acceptable accuracy in our solution (e.g., $S(\boldsymbol{\theta}^*)$ might represent a hypercube about $\boldsymbol{\theta}^*$ with the length of each side representing a tolerable error in each coordinate of $\boldsymbol{\theta}$). Suppose that the algorithm is terminated once an iterate lands in $S(\boldsymbol{\theta}^*)$ (perhaps because the L value for any $\boldsymbol{\theta}$ in $S(\boldsymbol{\theta}^*)$ is fully acceptable for the application of interest). So, if $\hat{\boldsymbol{\theta}}_k$ represents the iterate that first lands in $S(\boldsymbol{\theta}^*)$, then $\hat{\boldsymbol{\theta}}_k = \hat{\boldsymbol{\theta}}_{k+1} = \ldots = \hat{\boldsymbol{\theta}}_n$, where n is some prior value of maximum number of iterations.

Given the independence of the samples at each iteration, the probability of never landing in $S(\boldsymbol{\theta}^*)$ in n iterations for algorithm A is $\prod_{k=1}^{n}[1-P(\boldsymbol{\theta}_{\text{new}}(k)\in S(\boldsymbol{\theta}^*))]$. Hence, the probability of having a $\boldsymbol{\theta}_{\text{new}}(k)$ land in $S(\boldsymbol{\theta}^*)$ within $k = 1, 2,\ldots, n$ iterations (effectively stopping the algorithm) is

$$P(\hat{\boldsymbol{\theta}}_n \in S(\boldsymbol{\theta}^*)) = 1 - \prod_{k=1}^{n}[1-P(\boldsymbol{\theta}_{\text{new}}(k)\in S(\boldsymbol{\theta}^*))]. \tag{2.3}$$

One can use (2.3) to answer the question: How many iterations (with one loss function evaluation per iteration) will it take before the $\boldsymbol{\theta}$ estimate is guaranteed to land in $S(\boldsymbol{\theta}^*)$ with probability $1-\rho$ (ρ a small number)? Setting the left-hand side of (2.3) to $1-\rho$ and supposing that there is a constant sampling probability $P^* = P(\boldsymbol{\theta}_{\text{new}}(k)\in S(\boldsymbol{\theta}^*))$ for all k, we have

$$n = \frac{\log \rho}{\log(1-P^*)} \tag{2.4}$$

(see also Exercise 2.3). Here, and elsewhere in the book, log(x) or $\log x$ represents the *natural* (base $e = 2.71828\ldots$) logarithm of x.

Although (2.4) may appear benign at first glance, this expression grows rapidly as p gets large because of the rate at which P^* approaches 0 (since the relative volume of $S(\boldsymbol{\theta}^*)$ to volume of Θ decays geometrically as p increases). Hence, (2.4) shows the extreme inefficiency of algorithm A in high-dimensional problems, as illustrated in Example 2.2. While (2.3) is in terms of the iterate $\hat{\boldsymbol{\theta}}_n$, a result related to the rate of convergence for $L(\hat{\boldsymbol{\theta}}_n)$ is given in Pflug (1996, p. 24). This result is in terms of extreme-value distributions and also confirms the inefficiency of algorithm A in high-dimensional problems. Despite this inefficiency, the no free lunch theorems of Subsection 1.2.2 indicate that *when averaged across all possible problems*, algorithm A is as efficient as any other method. In practice, however, other algorithms usually have greater *practical* efficiency because of the restrictions to *specific* problems having specific characteristics.

Example 2.2—Varying efficiency of simple random search (algorithm A) as p changes. Let $\Theta = [0, 1]^p$ (the p-dimensional hypercube with minimum and maximum values of 0 and 1 for each component of $\boldsymbol{\theta}$) and suppose that uniform sampling on Θ is used to generate $\boldsymbol{\theta}_{\text{new}}(k)$ for all k. We want to guarantee with probability at least 0.90 that each element of $\boldsymbol{\theta}$ is within 0.04 units of the optimal value. Let the (unknown) optimal $\boldsymbol{\theta}$, $\boldsymbol{\theta}^*$, lie in a p-fold Cartesian product of open intervals, $(0.04, 0.96)^p$, and $\boldsymbol{\theta}^* = [t_1^*, t_2^*, \ldots, t_p^*]^T$. Hence,

$$S(\boldsymbol{\theta}^*) = [t_1^* - 0.04,\ t_1^* + 0.04] \times [t_2^* - 0.04,\ t_2^* + 0.04] \times \ldots \times [t_p^* - 0.04,\ t_p^* + 0.04]$$

and $P^* = 0.08^p$. How many iterations are required to ensure that we land in $S(\boldsymbol{\theta}^*)$ with a probability of 0.90 (i.e., $\rho = 0.10$)? Table 2.2 provides the answer for some values $1 \leq p \leq 10$.

Because each iteration requires one loss evaluation, Table 2.2 illustrates the explosive growth in the number of loss evaluations needed as p increases. This demonstrates the inefficiency of algorithm A in practical high-dimensional problems. ❑

Algorithm A is the simplest random search in that the sampling generating the new $\boldsymbol{\theta}$ value at each iteration is over the entire domain of interest. The sampling does not take account of where the previous estimates of $\boldsymbol{\theta}$ have been. The two algorithms below, although still simple, are slightly more sophisticated in that the random sampling is a function of the position of the current best estimate for $\boldsymbol{\theta}$. In this way, the search is more localized in the neighborhood of that estimate, allowing for a better exploitation of information that has previously been obtained about the shape of the loss function.

Such algorithms are sometimes referred to as *localized algorithms* to emphasize their dependence on the local environment near the current estimate for $\boldsymbol{\theta}$. This terminology is not to be confused with the global versus local algorithms discussed in Chapter 1, where the emphasis is on searching for a global or local solution to the optimization problem. In fact, sometimes a localized algorithm is guaranteed to provide a global solution, as discussed below following the presentation of the first localized algorithm.

Algorithm B is the first of the two localized algorithms we consider. This algorithm was described in Matyas (1965) and Jang et al. (1997, pp. 186–189).

Table 2.2. Number of iterations n to ensure iterate lands in $S(\boldsymbol{\theta}^*)$ with probability 0.90 for varying problem dimension p.

p	1	2	5	10
n	28	359	7.0×10^5	2.1×10^{11}

Algorithm B: Localized Random Search

Step 0 **(Initialization)** Pick an initial guess $\hat{\boldsymbol{\theta}}_0 \in \Theta$, either randomly or with prior information. Set $k = 0$.

Step 1 Generate an independent random vector $\boldsymbol{d}_k \in \mathbb{R}^p$ and add it to the current $\boldsymbol{\theta}$ value, $\hat{\boldsymbol{\theta}}_k$. Check if $\hat{\boldsymbol{\theta}}_k + \boldsymbol{d}_k \in \Theta$. If $\hat{\boldsymbol{\theta}}_k + \boldsymbol{d}_k \notin \Theta$, generate a new $\boldsymbol{d}_k$ and repeat or, alternatively, move $\hat{\boldsymbol{\theta}}_k + \boldsymbol{d}_k$ to the nearest valid point within Θ. Let $\boldsymbol{\theta}_{\text{new}}(k+1)$ equal $\hat{\boldsymbol{\theta}}_k + \boldsymbol{d}_k \in \Theta$ or the aforementioned nearest valid point in Θ.

Step 2 If $L(\boldsymbol{\theta}_{\text{new}}(k+1)) < L(\hat{\boldsymbol{\theta}}_k)$, set $\hat{\boldsymbol{\theta}}_{k+1} = \boldsymbol{\theta}_{\text{new}}(k+1)$; else, set $\hat{\boldsymbol{\theta}}_{k+1} = \hat{\boldsymbol{\theta}}_k$.

Step 3 Stop if the maximum number of L evaluations has been reached or the user is otherwise satisfied with the current estimate for $\boldsymbol{\theta}$ via appropriate stopping criteria; else, return to step 1 with the new k set to the former $k + 1$.

Although Matyas (1965) and others have used the (multivariate) normal distribution for generating $\boldsymbol{d}_k$, the user is free to set the distribution of the deviation vector $\boldsymbol{d}_k$. The distribution should have mean zero and each component should have a variation (e.g., standard deviation) consistent with the magnitudes of the corresponding $\boldsymbol{\theta}$ elements. So, for example, if the magnitude of the first component in $\boldsymbol{\theta}$ lies between 0 and 0.05 while the magnitude of the second component is between 0 and 500, the corresponding standard deviations in the components of $\boldsymbol{d}_k$ might also vary by a magnitude of 10,000). This allows the algorithm to assign roughly equal weight to each of the components of $\boldsymbol{\theta}$ as it moves through the search space.

Although not formally allowed in the convergence theorem below, it is often advantageous in practice if the variability in $\boldsymbol{d}_k$ is reduced as k increases. This allows one to focus the search more tightly as evidence is accrued on the location of the solution (as expressed by the location of our current estimate $\hat{\boldsymbol{\theta}}_k$). A simple implementation of this idea would be to reduce the variances by a factor such as k when the normal distribution is used in generating the $\boldsymbol{d}_k$. For the numerical studies below, we use the simple (constant variance) sampling, $\boldsymbol{d}_k \sim N(\boldsymbol{0}, \rho^2 \boldsymbol{I}_p)$ for all k where ρ^2 represents the (common) variance of each of the components in $\boldsymbol{d}_k$.

The convergence theory for the localized algorithms tends to be more restrictive than the theory for algorithm A. Solis and Wets (1981) provide a theorem for global convergence of localized algorithms, but the theorem conditions may not be verifiable in many practical applications. Their theorem would, in principle, cover both algorithm B and the enhanced algorithm C below. Other results related to formal convergence to global optima of various localized random search algorithms appear, for example, in Yakowitz and Fisher (1973) and Zhigljavsky (1991, Chap. 3). An earlier theorem from Matyas (1965) (with proof corrected in Baba et al., 1977) provides for global convergence of algorithm B if L is a continuous function. The convergence is in the "in probability" (pr.) sense (Appendix C). The theorem allows for more than one global minimum to exist in Θ. Therefore, in general, the result provides no guarantee of $\hat{\boldsymbol{\theta}}_k$ ever settling near any one value $\boldsymbol{\theta}^*$. We present the theorem below.

Theorem 2.2. Let Θ^* represent the set of global minima for L (see Section 1.1). Suppose that L is continuous on a bounded domain Θ and that if $\hat{\boldsymbol{\theta}}_k + \boldsymbol{d}_k \notin \Theta$ at a given iteration, a new $\boldsymbol{d}_k$ is randomly generated (versus the other option of bringing $\hat{\boldsymbol{\theta}}_k + \boldsymbol{d}_k$ back to the nearest point within Θ). For any $\eta > 0$, let $R_\eta = \bigcup_{\boldsymbol{\theta}^* \in \Theta^*} \{\boldsymbol{\theta} : |L(\boldsymbol{\theta}) - L(\boldsymbol{\theta}^*)| < \eta\}$ (i.e., the union over all $\boldsymbol{\theta}^* \in \Theta^*$ of the sets of $\boldsymbol{\theta}$ such that $L(\boldsymbol{\theta})$ is near each $L(\boldsymbol{\theta}^*)$). Then, for algorithm B with the $\boldsymbol{d}_k$ having an i.i.d. $N(\boldsymbol{0}, \boldsymbol{I}_p)$ distribution, $\lim_{k \to \infty} P(\hat{\boldsymbol{\theta}}_k \in R_\eta) = 1$.

Algorithm B above might be considered the most naïve of the localized random search algorithms. More sophisticated approaches are also easy to implement. For instance, if a search in one direction increases L, then it is likely to be beneficial to move in the opposite direction. Further, successive iterations in a direction that tend to consistently reduce L should encourage further iterations in the same direction. Many algorithms exploiting these simple properties exist (e.g., Solis and Wets, 1981; Zhigljavsky, 1991, Chap. 3; Li and Rhinehart, 1998; and the nonlinear simplex algorithm of Section 2.4). An extensive survey emphasizing such algorithms developed prior to 1980 is given in Schwefel (1995, pp. 87–89).

An example algorithm is given below (from Jang et al., 1997, pp. 187–188), which is a slight simplification of an algorithm in Solis and Wets (1981). The full Solis and Wets algorithm includes an even greater degree of adaptivity to the current environment, but this comes at the expense of more complex implementation.

Algorithm C: Enhanced Localized Random Search

Step 0 **(Initialization)** Pick an initial guess $\hat{\boldsymbol{\theta}}_0 \in \Theta$, either randomly or with prior information. Set $k = 0$. Set bias vector $\boldsymbol{b}_0 = 0$.

Step 1 Generate an independent random vector $\boldsymbol{d}_k$ and add it and the bias term $\boldsymbol{b}_k$ to the current value $\hat{\boldsymbol{\theta}}_k$ (as in algorithm B, a standard distribution for $\boldsymbol{d}_k$ is $N(\boldsymbol{0}, \rho^2 \boldsymbol{I}_p)$). Check if $\hat{\boldsymbol{\theta}}_k + \boldsymbol{b}_k + \boldsymbol{d}_k \in \Theta$. If $\hat{\boldsymbol{\theta}}_k + \boldsymbol{b}_k + \boldsymbol{d}_k \notin \Theta$, generate a new $\boldsymbol{d}_k$ and repeat; alternatively, move $\hat{\boldsymbol{\theta}}_k + \boldsymbol{b}_k + \boldsymbol{d}_k$ to the nearest valid point within Θ. Let $\boldsymbol{\theta}_{\text{new}}(k+1)$ equal $\hat{\boldsymbol{\theta}}_k + \boldsymbol{b}_k + \boldsymbol{d}_k \in \Theta$ or the above-mentioned nearest valid point in Θ.

Step 2 If $L(\boldsymbol{\theta}_{\text{new}}(k+1)) < L(\hat{\boldsymbol{\theta}}_k)$, set $\hat{\boldsymbol{\theta}}_{k+1} = \boldsymbol{\theta}_{\text{new}}(k+1)$ and $\boldsymbol{b}_{k+1} = 0.2\boldsymbol{b}_k + 0.4\boldsymbol{d}_k$; go to step 5. Otherwise, go to step 3.

Step 3 Analogous to step 1, let $\boldsymbol{\theta}'_{\text{new}}(k+1) = \hat{\boldsymbol{\theta}}_k + \boldsymbol{b}_k - \boldsymbol{d}_k \in \Theta$ or its altered version within Θ. If $L(\boldsymbol{\theta}'_{\text{new}}(k+1)) < L(\hat{\boldsymbol{\theta}}_k)$, set $\hat{\boldsymbol{\theta}}_{k+1} = \boldsymbol{\theta}'_{\text{new}}(k+1)$ and $\boldsymbol{b}_{k+1} = \boldsymbol{b}_k - 0.4\boldsymbol{d}_k$; go to step 5.[2] Otherwise, go to step 4.

Step 4 Set $\hat{\boldsymbol{\theta}}_{k+1} = \hat{\boldsymbol{\theta}}_k$ and $\boldsymbol{b}_{k+1} = 0.5\boldsymbol{b}_k$. Go to step 5.

Step 5 Stop if the maximum number of L evaluations has been reached or the user is otherwise satisfied with the current estimate for $\boldsymbol{\theta}$ via appropriate stopping criteria; else, return to step 1 with the new k set to the former $k+1$.

2.2.3 Example Implementations

Examples 2.3 and 2.4 report on a comparison of the three random search techniques above for a relatively simple $p = 2$ quartic (fourth-order) polynomial loss function. This function was used in Styblinski and Tang (1990, Example 1) to test simulated annealing algorithms. While the examples suggest the superiority of algorithm C, one should keep in mind the cautionary note on the limits of numerical studies in Subsection 1.2.1 together with the limits implied by the no free lunch theorems of Subsection 1.2.2. In particular, the conclusion of these examples may not apply to another problem of interest to the reader. Further, the low dimensionality of this example allows algorithm A to be more competitive than it would be in many realistic problems where $p > 2$. A larger-dimensional ($p = 10$) comparison is given in Example 2.5. Consistent with the discussion of Subsection 1.2.1, the algorithms are compared based on a common number of loss evaluations.

Example 2.3—Comparison of random search techniques A, B, and C (part 1). Let $\boldsymbol{\theta} = [t_1, t_2]^T$ and consider the loss function

$$L(\boldsymbol{\theta}) = \tfrac{1}{2}\left[(t_1^4 - 16t_1^2 + 5t_1) + (t_2^4 - 16t_2^2 + 5t_2)\right]$$

as given in Styblinski and Tang (1990, Example 1). Suppose that $\Theta = [-8, 8]^2$. Figure 2.2 shows that this function has four local minima, one of which is the unique global minimum at $\boldsymbol{\theta}^* = [-2.9035, -2.9035]^T$ (see Exercise 2.8).

The three algorithms are initialized at $\hat{\boldsymbol{\theta}}_0 = [4.0, 6.4]^T$ as in Styblinski and Tang (1990). Note that $L(\hat{\boldsymbol{\theta}}_0) = 537.2$ and $L(\boldsymbol{\theta}^*) = -78.33$. Each algorithm is run using 500 loss measurements. The random perturbations $\boldsymbol{d}_k$ in algorithms B and C are generated according to a $N(\boldsymbol{0}, \rho^2 \boldsymbol{I}_2)$ distribution; ρ is set to 3.0 for both algorithms, found to be approximately optimal after some preliminary numerical testing. Table 2.3 presents the mean values of the loss function at the final $\boldsymbol{\theta}$ estimates over 40 independent runs. Below the mean values are approximate 95

---

[2]There is no typographical error in step 3; unlike step 2, $\boldsymbol{b}_{k+1}$ in step 3 does not include the multiplier 0.2 on $\boldsymbol{b}_k$.

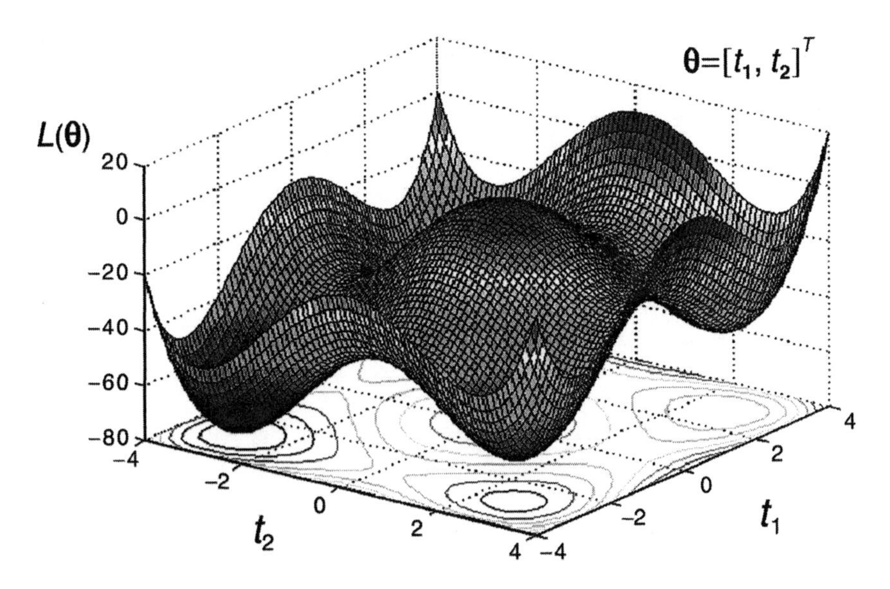

Figure 2.2. Multimodal loss function with unique global minimum at $\boldsymbol{\theta}^*$ near $[-2.9, -2.9]^T$ (Examples 2.3 and 2.4). Plot is on subset of Θ.

percent confidence intervals constructed according to a t-distribution with $40 - 1 = 39$ degrees of freedom (see Appendix B). For each algorithm, these intervals are derived from the sample variance s^2 of the terminal loss values over the sample of 40 terminal values. Using basic statistical principles discussed in Appendix B, each confidence interval is constructed according to

$$\left[\text{sample mean} - 2.023\sqrt{s^2/40}\,,\ \ \text{sample mean} + 2.023\sqrt{s^2/40}\right],$$

where 2.023 is the t-value for 0.025 probability (for each of the two tails) with 39 degrees of freedom.

Table 2.3 shows slight evidence that algorithm C is superior in this problem, although the upper value of the confidence interval for algorithm C

Table 2.3. Sample means and approximate 95 percent confidence intervals for terminal values $L(\hat{\boldsymbol{\theta}}_k) - L(\boldsymbol{\theta}^*)$ from algorithms A, B, and C.

Algorithm A	Algorithm B	Algorithm C
2.51 [1.94, 3.08]	0.78 [0.51, 1.04]	0.49 [0.32, 0.67]

overlaps the lower value of the interval for algorithm B (even with shorter 90 percent—versus 95 percent—confidence intervals, there remains some small overlap; see Exercise 2.10). Using the information here (also from Exercise 2.10), a more direct comparison of algorithms B and C is available by constructing the two-sample t-statistic in (B.4) from Appendix B. This yields $t =$ 1.80 (for the loss from algorithm B minus the loss from algorithm C). Given that we do not know that the variances of the loss measurements are identical, the nonidentical variance form in (B.3c) of Appendix B is relevant, leading to a degrees of freedom of approximately $68.9 \approx 69$. From the TDIST function in the MS EXCEL spreadsheet, the two-sided P-value is 0.076 (P-values are discussed in Appendix B). This P-value provides *some*—but not overwhelming—evidence of a significant difference between algorithms B and C in this problem. ❑

Example 2.4—Typical loss values and $\boldsymbol{\theta}$ estimates for random search algorithms A, B, and C (part 2). Let us continue with the problem of Example 2.3. To evaluate what a "typical" (single) run of algorithm A, B, or C would produce, let us compare the values of $\boldsymbol{\theta}$ after 500 loss evaluations. For algorithms A and B, this corresponds to the value of $\hat{\boldsymbol{\theta}}_{499}$; for algorithm C, the number of iterations is generally less than 499 because some of the iterations may take two loss measurements. Table 2.4 shows typical standardized loss values $L(\hat{\boldsymbol{\theta}}_k) - L(\boldsymbol{\theta}^*)$ together with the terminal $\boldsymbol{\theta}$ estimate that produced those loss values. The loss values are chosen to be the ones closest to the mean loss value in Example 2.3. Each $\boldsymbol{\theta}$ estimate is the value for the final iterate $\hat{\boldsymbol{\theta}}_k$ on the run producing the typical loss value.

The order of the loss values in Table 2.4 is consistent with the relative distances from $\hat{\boldsymbol{\theta}}_k$ to $\boldsymbol{\theta}^*$ (as given by $\|\hat{\boldsymbol{\theta}}_k - \boldsymbol{\theta}^*\|$) of 0.404, 0.224, and 0.173 for algorithms A, B, and C. Such consistent ordering does not always occur in practice. ❑

Table 2.4. Typical terminal $L(\hat{\boldsymbol{\theta}}_k) - L(\boldsymbol{\theta}^*)$ and $\hat{\boldsymbol{\theta}}_k$ values for algorithms A, B, and C (recall that $\boldsymbol{\theta}^* = [-2.9035, -2.9035]^T$).

	Algorithm A	Algorithm B	Algorithm C
$L(\hat{\boldsymbol{\theta}}_k) - L(\boldsymbol{\theta}^*)$ (one of the 40 runs in Example 2.3)	2.60	0.80	0.49
$\hat{\boldsymbol{\theta}}_k$ for above L value	$\begin{bmatrix}-2.547\\-3.093\end{bmatrix}$	$\begin{bmatrix}-2.680\\-2.898\end{bmatrix}$	$\begin{bmatrix}-2.740\\-2.959\end{bmatrix}$

Recall from the discussion of Subsection 1.2.1 that the tunable algorithm coefficients can have a dramatic effect on algorithm performance. For algorithms B and C, the adjustable coefficients pertained to the distribution of the perturbation vector $\boldsymbol{d}_k$. Since we took $\boldsymbol{d}_k$ as normally distributed with covariance matrix of the form $\rho^2\boldsymbol{I}_2$, this reduced to picking the value of the standard deviation ρ. A value of $\rho = 1.0$ (versus 3.0 in the example) significantly degraded the performance of algorithms B and C (increasing the relevant loss differences from below 1.0 in Tables 2.3 and 2.4 to over 20).

We now present a larger dimensional example related to a well-known test function.

Example 2.5—Random search algorithms applied to Rosenbrock function. Consider the well-known Rosenbrock test function in the optimization literature. This test function was first presented in Rosenbrock (1960) for the $p = 2$ setting, with higher-dimensional extensions given in, among other places, Moré et al. (1981) (note the typographical error in the function presentation on line 21(b) of p. 26 of the Moré et al. reference: an $i - 1$ should be $2i - 1$). This test function has an interesting shape in that the solution lies in a curved valley when considered in two-dimensional space. For general p, the function has the fourth-order polynomial form

$$L(\boldsymbol{\theta}) = \sum_{i=1}^{p/2}\left[100(t_{2i} - t_{2i-1}^2)^2 + (1 - t_{2i-1})^2\right], \tag{2.5}$$

where $\boldsymbol{\theta} = [t_1, t_2, \ldots, t_p]^T$ and p is divisible by 2.

Let us consider a problem where $p = 10$. Note that $L(\boldsymbol{\theta}^*) = 0$ at $\boldsymbol{\theta}^* = [1, 1, \ldots, 1]^T$. Let the constraint set be the Cartesian product of intervals $[-4, 4]$: $\Theta = [-4, 4]^{10}$. Table 2.5 presents the results of the study based on 40 independent runs for each algorithm. Each run of algorithms A and B uses 1000 loss evaluations; algorithm C is terminated at the 1000th loss evaluation or the first possible loss after the 1000th (it is not possible to specify a priori the exact number of measurements needed by algorithm C). Each run was started at the

Table 2.5. Sample means and approximate 95 percent confidence intervals for terminal Rosenbrock loss values $L(\hat{\boldsymbol{\theta}}_k)$ from algorithms A, B, and C (loss values relative to $L(\hat{\boldsymbol{\theta}}_0) = 121$; $L(\boldsymbol{\theta}^*) = 0$).

Algorithm A	Algorithm B	Algorithm C
121 [121, 121]	20.07 [19.78, 20.36]	19.80 [19.23, 20.37]

initial condition $\hat{\boldsymbol{\theta}}_0 = [-1.2, 1, -1.2, 1, \ldots, 1]^T$ (so $L(\hat{\boldsymbol{\theta}}_0) = 121$; this is a common initial condition in numerical studies with this function). The confidence intervals are computed in the manner of Example 2.3. In algorithm A, standard uniform sampling over Θ is used. For algorithms B and C, the perturbations are in the standard form $\boldsymbol{d}_k \sim N(\boldsymbol{0}, \rho^2 \boldsymbol{I}_{10})$, with $\rho = 0.05$.

Algorithm A does not move away from the initial condition in any of the 40 runs due to the large size of the sampling region (relative to regions of Θ that produce improvements in the loss function). That is, not one of the 40,000 total loss evaluations from uniform random sampling in Θ produced a loss value lower than the initial value (recall the "curse of dimensionality" in Subsection 1.2.1). Algorithms B and C perform better that algorithm A, with both B and C showing significant improvement relative to the initial condition. It is clear, however, that more loss evaluations are required to bring the solution close to $L(\boldsymbol{\theta}^*) = 0$. ❑

2.3 RANDOM SEARCH WITH NOISY LOSS MEASUREMENTS

The description and analysis of the random search algorithms above assume perfect (noise-free) measurements of the loss function. This is usually considered a critical part of such algorithms (Pflug, 1996, p. 25). In contrast to the noise-free case, random search methods with noisy loss evaluations of the form $y(\boldsymbol{\theta}) = L(\boldsymbol{\theta}) + \varepsilon(\boldsymbol{\theta})$ generally do not formally converge. However, there are means by which the random search techniques can be modified to accommodate noisy measurements, at least on a heuristic basis. Some of the limited formal convergence theory for random search as applied to the noisy measurement case includes Yakowitz and Fisher (1973, Sect. 4) and Zhigljavsky (1991, Chap. 3).

The most obvious way to attempt to cope with noise is to collect several (possibly many) function evaluations $y(\boldsymbol{\theta})$ at each value of $\boldsymbol{\theta}$ generated in the search process and then average these values. If the number of function evaluations at each $\boldsymbol{\theta}$ is sufficiently large (relative to the noise magnitude), then the averaged value can be effectively treated as one perfect measurement of the loss. As mentioned in Subsection 1.2.1, however, this can be costly since the error decreases only at the rate of $1/\sqrt{N}$ when averaging N function evaluations with independent noise.

Another approach is to alter the algorithm's key decision criterion to build in some robustness to the noise. In particular, the algorithm resists changing the current best estimate for $\boldsymbol{\theta}$ unless there is "significant" probabilistic evidence that the new value will lead to an improved value of the loss. As an example, the acceptance threshold in step 2 of algorithm B can be altered to:

Step 2–modified If $y(\boldsymbol{\theta}_{\text{new}}(k+1)) < y(\hat{\boldsymbol{\theta}}_k) - \tau_k$, set $\hat{\boldsymbol{\theta}}_{k+1} = \boldsymbol{\theta}_{\text{new}}(k+1)$; else set $\hat{\boldsymbol{\theta}}_{k+1} = \hat{\boldsymbol{\theta}}_k$, where $\tau_k \geq 0$. (2.6)

The same idea applies to step 1 of algorithm A or steps 2 and 3 of algorithm C.

It may be convenient to view the τ_k in terms of the number of standard deviations in the measurement noise that the new measurement must improve on the old measurement before $\boldsymbol{\theta}$ will be changed. So, if one had estimated the standard deviation of the measurement noise to be 0.1 for all k, then $\tau_k = 0.2$ indicates that a new $\boldsymbol{\theta}$ is accepted only if it shows a "two-sigma" improvement in the *measured* loss value (using the standard vernacular where *sigma* represents a standard deviation). The sigma interpretation is often convenient if the noise is at least approximately normally distributed. A simple table lookup reveals the approximate probability of making a right or wrong decision (in terms of whether the new $\boldsymbol{\theta}$ value really is better or worse than the current value).

Obviously, both of the averaging and altered threshold approaches for coping with noise have significant shortcomings. Averaging may greatly increase the number of loss evaluations required to obtain sufficient accuracy. The altered threshold may suffer the same shortcoming by rejecting too many changes in $\boldsymbol{\theta}$ due to the conservative criterion. Nevertheless, the presence of noise in the loss evaluations—even a small amount of noise—makes the optimization problem so much harder than deterministic problems that there is little choice but to accept these penalties if one wants to use a simple random search. We will see in later chapters that other general approaches, such as stochastic approximation, tend to be more adept at coping with noise at the price of a more restrictive problem setting.

Let us consider an example of algorithm B applied in a problem with noise. This example uses both averaging of the loss measurements and the modification of the acceptance threshold as in (2.6). The example is motivated by a problem in wastewater treatment, as described in Spall and Cristion (1997) (based on an earlier model of Dochain and Bastin, 1984). The aim is to find an open-loop controller. More sophisticated closed-loop controllers that take account of the time-varying current state of a system are discussed in Chapters 3, 6, and 7. As part of this example, we illustrate the process of converting a root-finding problem to a minimization problem. The general conversion approach outlined below would apply in other problems as well.

The three related examples below consider this open-loop control problem. Example 2.6 goes through the general process of converting the root-finding problem to an optimization problem. Example 2.7 presents numerical results. This numerical example compares a naïve implementation (ignoring the noise) of algorithm B with a "proper" implementation based on averaging of the loss function measurements. Example 2.8 continues with the wastewater example as an illustration of the general degrading effects of having noise in the optimization process.

Example 2.6—Conversion of root-finding to optimization in the context of two-dimensional problem. Without getting into most of the specifics of the wastewater treatment problem, let us outline how root-finding can be converted to optimization for which algorithms such as the random search methods A, B, and C readily apply. The original root-finding problem is one of determining a $\boldsymbol{\theta}$ such that $\boldsymbol{g}(\boldsymbol{\theta}) = \boldsymbol{0} \in \mathbb{R}^2$ based on noisy measurements $\boldsymbol{Y}(\boldsymbol{\theta}) = \boldsymbol{g}(\boldsymbol{\theta}) + \boldsymbol{e}$, with $\boldsymbol{e}$ being a normally distributed vector. Suppose that the noise $\boldsymbol{e}$ has mean zero and covariance matrix $\text{diag}[\sigma_1^2, \sigma_2^2]$ (so the components of $\boldsymbol{e}$ are independent because of the normality).

Let $\boldsymbol{W}$ be a matrix that assigns relative weights to the two elements of $\boldsymbol{g}(\boldsymbol{\theta})$. In particular, assume that

$$\boldsymbol{W} = \begin{bmatrix} w & 0 \\ 0 & 1-w \end{bmatrix}$$

with $0 < w < 1$. Clearly, finding a $\boldsymbol{\theta}$ such that $\boldsymbol{g}(\boldsymbol{\theta}) = \boldsymbol{0}$ is equivalent to finding a $\boldsymbol{\theta}$ such that $\boldsymbol{g}(\boldsymbol{\theta})^T\boldsymbol{W}\boldsymbol{g}(\boldsymbol{\theta}) = 0$. To convert this relationship into one directly usable with measurements $\boldsymbol{Y}(\boldsymbol{\theta})$, the root-finding problem is converted to an optimization problem $\min_{\boldsymbol{\theta}\in\Theta} L(\boldsymbol{\theta})$ according to the loss function

$$L(\boldsymbol{\theta}) = E\left[\boldsymbol{Y}(\boldsymbol{\theta})^T \boldsymbol{W}\boldsymbol{Y}(\boldsymbol{\theta})\right] = \boldsymbol{g}(\boldsymbol{\theta})^T \boldsymbol{W}\boldsymbol{g}(\boldsymbol{\theta}) + [\sigma_1, \sigma_2]\boldsymbol{W}\begin{bmatrix} \sigma_1 \\ \sigma_2 \end{bmatrix}. \tag{2.7}$$

(Differentiating (2.7) with respect to $\boldsymbol{\theta}$ illustrates that $\boldsymbol{g}(\boldsymbol{\theta})$ is a general root-finding quantity not necessarily corresponding to $\partial L/\partial\boldsymbol{\theta}$.)

The additive constant (the term after the "+" sign) in (2.7) does not affect the solution for $\boldsymbol{\theta}$. If $\boldsymbol{\theta}$ is such that $\boldsymbol{g}(\boldsymbol{\theta}) = \boldsymbol{0}$, then $\boldsymbol{\theta}$ minimizes $L(\boldsymbol{\theta})$. Including the additive constant in the loss function, as in (2.7), is convenient because the readily available noisy observation

$$\begin{aligned} y(\boldsymbol{\theta}) &= \boldsymbol{Y}(\boldsymbol{\theta})^T \boldsymbol{W}\boldsymbol{Y}(\boldsymbol{\theta}) \\ &= L(\boldsymbol{\theta}) + \varepsilon(\boldsymbol{\theta}) \end{aligned} \tag{2.8}$$

is an unbiased measurement of $L(\boldsymbol{\theta})$ at any $\boldsymbol{\theta}$ (i.e., $E[y(\boldsymbol{\theta})] = L(\boldsymbol{\theta})$ for all $\boldsymbol{\theta}$). Combining (2.7) and (2.8) indicates that the noise $\varepsilon(\boldsymbol{\theta})$ in the loss function measurement is

$$\varepsilon(\boldsymbol{\theta}) = 2\boldsymbol{g}(\boldsymbol{\theta})^T \boldsymbol{W}\boldsymbol{e}(\boldsymbol{\theta}) + \boldsymbol{e}(\boldsymbol{\theta})^T \boldsymbol{W}\boldsymbol{e}(\boldsymbol{\theta}) - [\sigma_1, \sigma_2]\boldsymbol{W}\begin{bmatrix} \sigma_1 \\ \sigma_2 \end{bmatrix}. \tag{2.9}$$

From (2.9), note that the noise in the loss measurements collected during a search process depends on the current estimate for $\boldsymbol{\theta}$. ❑

Example 2.7—Numerical results on optimization with noisy loss measurements in wastewater treatment model. With conventional notation that u represents a control variable, let $\boldsymbol{\theta} = [u_1, u_2]^T$ be the two control inputs for the physical model for the wastewater treatment process. Let the two variances satisfy $\sigma_1^2 = \sigma_2^2 = 1$ and the constraint region $\Theta = [0, 5] \times [0, 5]$; the measurement noise $\boldsymbol{e}$ is normally distributed (so the measurement noise $\varepsilon(\boldsymbol{\theta})$ in (2.9) is *not* normal). From Spall and Cristion (1997), suppose that "nature" provides measurements of the form

$$\boldsymbol{Y}(\boldsymbol{\theta}) = \begin{bmatrix} 1-u_1 \\ 1-0.5u_1u_2-u_2 \end{bmatrix} + \boldsymbol{e}, \tag{2.10}$$

where the top expression on the right-hand side is associated with the methane gas byproduct state and the bottom expression is associated with the water cleanliness state. It is assumed that the analyst does not know the analytical structure on the right-hand side of (2.10), but does know that the two control inputs u_1 and u_2 are somehow used to regulate $\boldsymbol{Y}$. Note that the control u_1 affects both states. Following the discussion in Example 2.6, the noisy measurement $\boldsymbol{Y}$ above can be converted into the loss function measurement in an optimization setting. Suppose that the analyst puts 10 percent weighting ($w = 0.10$) on the methane and 90 percent weighting on the water cleanliness.

From (2.7), an exact solution is available by minimizing the quadratic expression

$$L(\boldsymbol{\theta}) = 0.10(1 - u_1)^2 + 0.90(1 - 0.5u_1u_2 - u_2)^2 + 1 \tag{2.11}$$

(the additive "1" contribution from the variances σ_1^2 and σ_2^2 is, of course, irrelevant here since the variances do not depend on $\boldsymbol{\theta}$). The above L yields the unique solution $\boldsymbol{\theta}^* = [1, 2/3]^T$ and $L(\boldsymbol{\theta}^*) = 1.0$ by solving $\partial L/\partial\boldsymbol{\theta} = [\partial L/\partial u_1, \partial L/\partial u_2]^T = \mathbf{0}$ and verifying that the root is a minimum of L.

Table 2.6 contrasts the mean values of $L(\hat{\boldsymbol{\theta}}_k) - L(\boldsymbol{\theta}^*)$ at the terminal iteration over 50 independent runs (the loss value $L(\boldsymbol{\theta}^*)$ has been subtracted out

Table 2.6. Sample means and approximate 95 percent confidence intervals for terminal values of $L(\hat{\boldsymbol{\theta}}_k) - L(\boldsymbol{\theta}^*)$ from algorithm B without and with averaging of noisy loss measurements.

Number of loss measurements y	Algorithm B without averaging	Algorithm B with averaging
100	0.80 [0.59, 1.01]	0.59 [0.47, 0.70]
2000	0.78 [0.64, 0.92]	0.38 [0.28, 0.48]

to more clearly show the relative performance). Below the means are approximate 95 percent confidence intervals derived as in Appendix B ("approximate" because the loss measurements based on eqns. (2.7)−(2.10) are not normally distributed). The table shows four settings. Two are for the small-sample case where only 100 noisy loss measurements are used; the other two are when 2000 measurements are used.

Note that the *true* loss function (2.11) (not the noisy measurements in (2.8)) is used to produce the numbers in Table 2.6. Hence, while the algorithm produces a solution using only noisy measurements, the *evaluation* of the algorithm is with the true loss function. Of course, the true loss would not be available in real-world applications where only noisy measurements are available. The same approach of evaluation based on true loss functions is used throughout this book when Monte Carlo simulations are being used for testing algorithms with noisy function measurements (this was discussed in Subsection 1.2.1).

For each of the sample sizes, the table shows results for algorithm B where no averaging is used (i.e., the noisy loss measurements are used directly in the algorithm in place of the L values) and where an average of four noisy measurements is used at each iteration. To preserve a common number of measurements being used for both the no-averaging and averaging cases, the no-averaging case ran more iterations than the averaging case. All results use the simple sampling, $\boldsymbol{d}_k \sim N(\boldsymbol{0}, \rho^2 \boldsymbol{I}_2)$, for the perturbations with $\rho = 2$. Constraints are imposed by mapping any component of $\boldsymbol{\theta}$ that lies outside the interval [0, 5] to its nearest end point (0 or 5). An initial condition of $\hat{\boldsymbol{\theta}}_0 = [3, 3]^T$ is used for each of the 50 runs (so $L(\hat{\boldsymbol{\theta}}_0) - L(\boldsymbol{\theta}^*) = 38.43$).

Table 2.6 shows strong evidence of the superiority of algorithm B with averaging in the large-sample case and modest evidence in the small-sample case. The altered acceptance criterion mentioned in (2.6) was also tested in this application, but the averaging method produced lower loss values (of course, this is one specific case and results here may not transfer to another setting). ❑

Example 2.8—Illustration of degrading effects of noise on the optimization process using wastewater treatment model. As a final study in this sequence of examples related to the wastewater model, let us compare the results in Table 2.6 with results based on noise-free loss measurements. The purpose here is to illustrate the detrimental effects of noise in the optimization process, although noise-free loss measurements would *not* be available in a real-world implementation for this application. That is, we are simply using the loss function above as a convenient example while dropping the noise shown in eqns. (2.7) − (2.9) (i.e., $\sigma_1^2 = \sigma_2^2 = 0$ and $\boldsymbol{e} = \boldsymbol{0}$).

With the algorithm coefficient settings remaining as in Example 2.7—which is not optimal for the noise-free case!—the sample mean loss values are 0.041 and 0.0019 for 100 and 2000 loss measurements, respectively. The best corresponding numbers in Table 2.6 are at least 14 and 200 times larger, respectively, indicating the significantly greater difficulty of reaching an optimum in the presence of noisy function measurements. The greater distance between $\boldsymbol{\theta}^*$ and the terminal $\hat{\boldsymbol{\theta}}_k$ is also apparent. Based on the standard distance measure (Euclidean norm $\|\cdot\|$), the mean of the distances between $\boldsymbol{\theta}^*$ and the terminal iterate $\hat{\boldsymbol{\theta}}_k$ for 2000 measurements is 11 times greater (0.93 versus 0.084) in the noisy case than the noise-free case. These comparisons become even more dramatic if the coefficient ρ in algorithm B is tuned for noise-free measurements. ❑

2.4 NONLINEAR SIMPLEX (NELDER–MEAD) ALGORITHM

2.4.1 Basic Method

Another popular direct search method is the nonlinear simplex algorithm of Nelder and Mead (1965). This algorithm for nonlinear optimization is based on the concept of a simplex, a geometric object that is the *convex hull* of $p+1$ points in $\mathbb{R}^p$ not lying in the same hyperplane. (The convex hull is the smallest set enclosing the $p+1$ points such that a line segment connecting any two points in the set lies in the set.) Note that the nonlinear simplex algorithm is not directly related to the popular simplex method for linear programming. With $p = 2$, a simplex is a triangle; with $p = 3$, a simplex is a pyramid.

The nonlinear simplex algorithm is the baseline gradient-free multivariate optimization technique in MATLAB (version 6) (the function **`fminsearch`**) and has arguably been called the most popular of the optimization techniques based on comparing loss function values only (Barton and Ivey, 1996). Based on extensive practical experience, the method often produces significant decreases in the loss function in the first few iterations. Interestingly, however, despite its popularity and practical effectiveness in a range of problems, there is no *general* convergence theory in either the deterministic or stochastic case (Barton and Ivey, 1996). In fact, many demonstrations of nonconvergence for specific functions or classes of functions have been given (e.g., Lagarias et al., 1998; McKinnon, 1998; Kelly, 1999b).

Although the nonlinear simplex algorithm was proposed in the Nelder and Mead (1965) paper as a deterministic method, it has frequently been used in a stochastic setting with noisy loss function measurements. Further, the algorithm is built on the simplex ideas in Spendley et al. (1962), which *do* explicitly account for noise in the function measurements.

The basic idea in the nonlinear simplex algorithm is that at each iteration a new point is generated in or near the current simplex. Usually, this new point replaces one of the current simplex vertices, yielding a new simplex. To determine this new point, a reflection step is introduced where the new point is chosen as the reflection (across the hyperplane spanned by the other vertices) of the vertex that currently has the worst (highest) value of $L(\boldsymbol{\theta})$. The aim is to move the simplex in a direction away from the high values of the loss function and toward the lowest values of the loss. This process is continued until the size of the simplex is sufficiently small, in which case the solution is taken as a point inside the simplex.

The basic nonlinear simplex steps are given below for unconstrained optimization ($\Theta = \mathbb{R}^p$). Figure 2.3 illustrates these steps for the $p = 2$ case. If the problem involves explicit constraints, then the steps below can be modified so that any vertex outside of Θ is moved to the nearest point in Θ.

The algorithm steps below are identical to the steps in the MATLAB **fminsearch** routine (also, e.g., Kelly, 1999b). These steps differ from the original Nelder–Mead simplex algorithm in two small ways: (i) Certain strict inequalities here (< or >) are nonstrict inequalities (≤ or ≥) in the original algorithm (or vice versa) and (ii) the condition $L(\boldsymbol{\theta}_{\text{exp}}) < L(\boldsymbol{\theta}_{\text{refl}})$ in step 2b here is $L(\boldsymbol{\theta}_{\text{exp}}) < L(\boldsymbol{\theta}_{\text{min}})$ in the original algorithm. These changes from the original algorithm have become standard practice (Lagarias et al., 1998). Modification (i) is useful in tie breaking, while modification (ii) has been adopted based on accumulated experience of generally better performance. To avoid excessive subscript notation, the iteration counter k is suppressed in the algorithm presentation below. Note that one or more function evaluations may be required in each iteration.

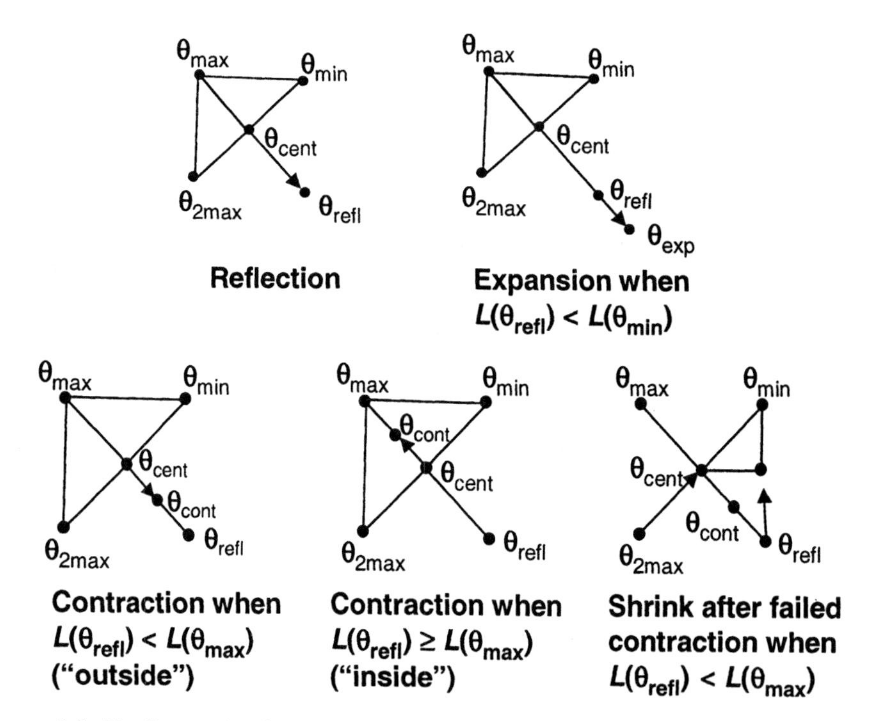

Figure 2.3. Nonlinear simplex algorithm for $p = 2$ with noise-free loss measurements (adapted from Barton and Ivey, 1996).

While the steps below are given for noise-free loss measurements, the same basic algorithm is often used with noisy measurements. Then, $y(\boldsymbol{\theta})$ values replace the indicated $L(\boldsymbol{\theta})$ evaluations. In Subsection 2.4.2, we comment on some fine tuning of these steps to accommodate noisy measurements of the loss function.

Nonlinear Simplex (Nelder–Mead) Algorithm (Unconstrained)

Step 0 **(Initialization)** Generate an initial set of $p+1$ extreme points in $\mathbb{R}^p$, say $\boldsymbol{\theta}^i$ ($i = 1, 2,\ldots, p+1$), representing the vertices of the initial simplex. Calculate $L(\boldsymbol{\theta}^i)$ for each extreme point. Set values for the algorithm coefficients α, β, γ, and δ. Default settings are specified by Nelder and Mead (1965) as 1.0, 0.5, 2.0, and 0.5, respectively (also used in MATLAB `fminsearch`).

Step 1 **(Reflection)** Identify the vertices where the maximum, second highest, and minimum L values occur. Let $\boldsymbol{\theta}_{\text{max}}$, $\boldsymbol{\theta}_{\text{2max}}$, and $\boldsymbol{\theta}_{\text{min}}$ represent these points, respectively. Let $\boldsymbol{\theta}_{\text{cent}}$ represent the centroid (mean) of all $\boldsymbol{\theta}^i$ except for $\boldsymbol{\theta}_{\text{max}}$. Generate a new candidate vertex $\boldsymbol{\theta}_{\text{refl}}$ by reflecting $\boldsymbol{\theta}_{\text{max}}$ through $\boldsymbol{\theta}_{\text{cent}}$ according to $\boldsymbol{\theta}_{\text{refl}} = (1+\alpha)\boldsymbol{\theta}_{\text{cent}} - \alpha\boldsymbol{\theta}_{\text{max}}$ ($\alpha > 0$).

Step 2a **(Accept reflection)** If $L(\boldsymbol{\theta}_{\text{min}}) \leq L(\boldsymbol{\theta}_{\text{refl}}) < L(\boldsymbol{\theta}_{\text{2max}})$, then $\boldsymbol{\theta}_{\text{refl}}$ replaces $\boldsymbol{\theta}_{\text{max}}$ in the simplex and proceed to step 3; else, go to step 2b.

Step 2b **(Expansion)** If $L(\boldsymbol{\theta}_{\text{refl}}) < L(\boldsymbol{\theta}_{\text{min}})$, then the reflection is expanded according to $\boldsymbol{\theta}_{\text{exp}} = \gamma\boldsymbol{\theta}_{\text{refl}} + (1-\gamma)\boldsymbol{\theta}_{\text{cent}}$, where the expansion coefficient $\gamma > 1$; else, go to step 2c. If $L(\boldsymbol{\theta}_{\text{exp}}) < L(\boldsymbol{\theta}_{\text{refl}})$, then $\boldsymbol{\theta}_{\text{exp}}$ replaces $\boldsymbol{\theta}_{\text{max}}$ in the simplex; otherwise, the expansion is rejected and $\boldsymbol{\theta}_{\text{refl}}$ replaces $\boldsymbol{\theta}_{\text{max}}$. Go to step 3.

Step 2c **(Contraction)** If $L(\boldsymbol{\theta}_{\text{refl}}) \geq L(\boldsymbol{\theta}_{\text{2max}})$, then the simplex contracts to reflect the poor $\boldsymbol{\theta}_{\text{refl}}$. Consider two cases: (i) $L(\boldsymbol{\theta}_{\text{2max}}) \leq L(\boldsymbol{\theta}_{\text{refl}}) < L(\boldsymbol{\theta}_{\text{max}})$ (sometimes called *outside contraction*) or (ii) $L(\boldsymbol{\theta}_{\text{refl}}) \geq L(\boldsymbol{\theta}_{\text{max}})$ (*inside contraction*). The contraction point is determined by $\boldsymbol{\theta}_{\text{cont}} = \beta\boldsymbol{\theta}_{\text{max/refl}} + (1-\beta)\boldsymbol{\theta}_{\text{cent}}$, $0 \leq \beta \leq 1$, where $\boldsymbol{\theta}_{\text{max/refl}} = \boldsymbol{\theta}_{\text{refl}}$ in case (i) or $\boldsymbol{\theta}_{\text{max/refl}} = \boldsymbol{\theta}_{\text{max}}$ in case (ii). In case (i), if $L(\boldsymbol{\theta}_{\text{cont}}) \leq L(\boldsymbol{\theta}_{\text{refl}})$, the contraction is accepted. In case (ii), if $L(\boldsymbol{\theta}_{\text{cont}}) < L(\boldsymbol{\theta}_{\text{max}})$, the contraction is accepted. If the contraction is accepted, replace $\boldsymbol{\theta}_{\text{max}}$ with $\boldsymbol{\theta}_{\text{cont}}$ and go to step 3. If the contraction is not accepted, go to step 2d.

Step 2d **(Shrink)** The contraction has failed and the entire simplex shrinks according to a factor of $0 < \delta < 1$, retaining only $\boldsymbol{\theta}_{\text{min}}$. This is done by replacing each vertex $\boldsymbol{\theta}^i$ (except $\boldsymbol{\theta}_{\text{min}}$) by $\delta\boldsymbol{\theta}^i + (1-\delta)\boldsymbol{\theta}_{\text{min}}$. Go to step 3.

Step 3 **(Termination)** Stop if a convergence criterion is satisfied or if the maximum number of function evaluations has been reached; else, return to step 1.

2.4.2 Adaptation for Noisy Loss Measurements

It is sometimes acceptable to replace the exact L evaluations shown in the algorithm steps above with measurements of the loss function that include noise (so values of $y(\boldsymbol{\theta}) = L(\boldsymbol{\theta}) + \varepsilon(\boldsymbol{\theta})$ replace the indicated $L(\boldsymbol{\theta})$ values). As discussed in Barton and Ivey (1996), the justification for making this direct substitution of possibly noisy function evaluations is the relative insensitivity of the algorithm to noise by virtue of its using only *ranks* of the loss function values, not the function values themselves (reminiscent of the ordinal versus cardinal discussion of Subsection 1.2.1). Small noise effects have little impact on ranks and therefore have little effect on the path of the algorithm. On the other hand, large noise effects *will* frequently change the relative ranks of the loss values, leading to changes in the algorithm convergence properties. In such a case, the nonlinear simplex algorithm may perform poorly, and one may do better with one of the stochastic approximation algorithms in Chapters 4–7, which are explicitly designed to cope with noise.

Several authors have provided modifications of the nonlinear simplex algorithm to adapt to noisy loss measurements. The basic structure of the algorithm remains the same. The modifications tend to be small—but crucial—refinements. The most obvious modification is to average several loss measurements at a given $\boldsymbol{\theta}$, as discussed in Section 2.3 for the random search methods. If there is an adequate amount of averaging given the level of noise, the algorithm described above can be used verbatim with the averaged loss values representing the loss measurements. Another change involves the choice of the convergence criterion. (This, of course, is irrelevant if one simply uses a "budget" of loss evaluations as discussed in Subsection 1.2.1.) A common criterion for the deterministic nonlinear simplex algorithm is to stop when

$$\left\{\sum_{i=1}^{p+1}\left[L(\boldsymbol{\theta}^i)-L(\boldsymbol{\theta}_{\text{cent}})\right]^2\right\}^{1/2}$$

is sufficiently small.

The above criterion may be inappropriate with noisy loss values. For convergence, the differences in the loss values in the simplex are supposed to be small. The algorithm may never converge in practice because the noises may dominate the underlying differences in the true (noise-less) loss values. Hence, a criterion based on simplex size may be more appropriate. Dennis and Woods (1987) suggest the stopping criterion

$$\frac{\max_i \left\|\boldsymbol{\theta}^i - \boldsymbol{\theta}_{\min}\right\|}{\max\left\{1,\ \left\|\boldsymbol{\theta}_{\min}\right\|\right\}},$$

where the maximization in the numerator is over all extreme points in the simplex (the denominator is simply a normalization quantity). Because one always has exact knowledge of the $\boldsymbol{\theta}$ values, this criterion is less affected by noise than criteria involving loss measurements.

Noise introduces an unwelcome feedback process when the simplex is small. Recall that the noises tend to dominate the differences in loss values at the extreme points. Tomick et al. (1995) and Barton and Ivey (1996) show that noise can lead to premature reductions of simplex size due to an increase in the probabilities of invoking the contraction or shrink steps of the algorithm (steps 2c or 2d). Certainly, *some* tendency to reduce the size of the simplex is desirable to enhance the likelihood of the algorithm clustering around $\boldsymbol{\theta}^*$, but an excessive shrinkage tendency leads to the algorithm stopping before it has approached $\boldsymbol{\theta}^*$.

Ernst (1968), Deming and Parker (1978), Tomick et al. (1995), Barton and Ivey (1996), and Kelly (1999b) are among the references that present methods for coping with this premature convergence. Probably the easiest means by which this tendency can be arrested with noisy loss measurements is to alter the algorithm coefficients. Barton and Ivey (1996) recommend an increase of δ from the nominal value of 0.5 above to 0.9; this will cause a shrinkage of only 10 percent instead of the standard 50 percent.

The above modifications once again illustrate the important role in stochastic optimization of the user-specified algorithm coefficients. Two other possible modifications suggested by Barton and Ivey (1996) pertain to resampling loss measurements with the aim of providing stronger statistical evidence of the need for a reduction in the simplex size; this resampling involves discarding the previous value at a given $\boldsymbol{\theta}$ (vs. simply averaging multiple values at a given $\boldsymbol{\theta}$). Barton and Ivey (1996) provide statistical evidence over a range of 18 test problems that the increase in δ leads to improved behavior (also see Example 2.9). The evidence for their other recommended changes is more mixed. In any specific application, it is likely that tuning the coefficients α, β, γ, and δ would lead to improved behavior in the presence of noise (and perhaps even without noise). Unfortunately, this is sometimes difficult, especially if the loss measurements are expensive.

Example 2.9—Nonlinear simplex algorithm applied to Rosenbrock function with and without noise. Using the **`fminsearch`** function in MATLAB (version 6), this study evaluates the nonlinear simplex method in a problem with the Rosenbrock function in (2.5) with $p = 10$. The evaluation considers noise-free and noisy function measurements. We also compare the performance with random search algorithm B. In the noisy case, $y(\boldsymbol{\theta}) = L(\boldsymbol{\theta}) + \varepsilon$, where ε represents an i.i.d. $N(0, \sigma^2)$ noise term. As in Example 2.5, let $\hat{\boldsymbol{\theta}}_0 = [-1.2, 1, -1.2, 1, \ldots, 1]^T$. Suppose that 4000 measurements of the loss function are used in each replication and that the problem is unconstrained (i.e., $\Theta = \mathbb{R}^{10}$).

Two sets of algorithm coefficient values are used for the simplex method. The first is the default set, as given in step 0 above, and the second follows the recommendation above for coping with noisy loss measurements (i.e., δ is manually changed from 0.5 to 0.9 in **`fminsearch`** while other coefficients remain at the default levels). The algorithm B results are based on perturbations $\boldsymbol{d}_k \sim N(\boldsymbol{0}, 0.1^2\boldsymbol{I}_{10})$. Algorithm B is implemented in a "baseline" form as given in the steps in Subsection 2.2.2 (i.e., y values are directly substituted for L values). Improved results are possible in the noisy case by using "thresholding" (modified step 2 in (2.6)); see comments below.

For values of $\sigma = 0$ and $\sigma = 5$, Table 2.7 shows the mean terminal loss values together with the approximate 95 percent confidence intervals computed using a t-distribution as in Example 2.3. The confidence intervals are computed from 40 independent runs, each using the 4000 measurements mentioned above.

For this problem, the simplex method outperforms algorithm B in the noise-free case, while the converse is true in the noisy case. Note, however, that the change in δ produces a significant improvement in the simplex method (nonoverlapping confidence intervals), as anticipated by Barton and Ivey (1996). (Although not shown in Table 2.7, thresholding likewise improves the performance of algorithm B by dropping the sample mean loss value to below 20 when $\sigma = 5$.) While the results of this study do not necessarily transfer to other noisy problems, the relatively poor performance of an "industry standard" search method such as **`fminsearch`** serves as a cautionary demonstration of the effects of noise. ❑

2.5 CONCLUDING REMARKS

The focus in this chapter has been some of the popular search methods for optimization based only on direct measurements of the loss function. We considered two broad types of such algorithms. The first is random search, based on searching through the domain Θ using randomly generated steps, usually generated in a Monte Carlo fashion via a pseudorandom number generator. The second is a geometrically motivated search based on the idea of a simplex. The specific nonlinear simplex algorithm emphasized here is sometimes called the

Table 2.7. Sample means and approximate 95 percent confidence intervals for terminal Rosenbrock loss values from the nonlinear simplex algorithm and algorithm B. Simplex method is implemented with default parameter settings and with δ changed. (For comparison purposes, note that $L(\hat{\boldsymbol{\theta}}_k) = 121$ and $L(\boldsymbol{\theta}^*) = 0$).

Noise level	Simplex (Default)	Simplex ($\delta = 0.90$)	Algorithm B
$\sigma = 0$	10.11 (no interval)	10.11 (no interval)	17.97 [16.90, 19.04]
$\sigma = 5$	28.31 [26.80, 29.82]	25.28 [24.41, 26.15]	22.95 [22.03, 23.86]

Nelder–Mead algorithm. The simplex approach is based on moving a shrinking region (a simplex) through Θ. When the algorithm works properly, the simplex collapses onto $\boldsymbol{\theta}^*$. As a testament to the popularity of this algorithm, the baseline gradient-free multivariate optimization technique in MATLAB (version 6) (the function **`fminsearch`**) is an implementation of the nonlinear simplex algorithm. This method is not to be confused with the simplex method of linear programming.

The methods of this chapter are versatile and broadly applicable. Because they require only loss measurements, they are often easier to implement than other methods that require relatively detailed information about the loss function (e.g., gradient information). In particular, a user might find the localized random search algorithm B to be useful in many problems, especially if only modest precision is required in the solution. This algorithm is relatively easy to implement and has a long record of reasonable practical efficiency.

In their standard forms, the algorithms of this chapter are used with noise-free measurements of the loss function. We also discussed some modifications to accommodate noisy measurements. Nonetheless, we saw, as predicted in Chapter 1, that noise can significantly degrade the results of the search process. The stochastic approximation methods in Chapters 4–7 are explicitly designed to cope with noise.

EXERCISES

2.1 Show that the conditions of Theorem 2.1 apply to $L(\boldsymbol{\theta}) = \boldsymbol{\theta}^T\boldsymbol{\theta}$ when using a uniform distribution for $\boldsymbol{\theta}_{\text{new}}$ on the domain $\Theta = [-1, 1] \times [0, 2]$. Given the knowledge of $\boldsymbol{\theta}^*$, present an explicit form for the lower bound $\delta(\eta)$ that appears in (2.2).

2.2 Let $\boldsymbol{\theta} = [a, b, c]^T$ and suppose that

$$L(\boldsymbol{\theta}) = -\left\{\det\begin{bmatrix} 1 & a & a^2 \\ 1 & b & b^2 \\ 1 & c & c^2 \end{bmatrix}\right\}^2 .$$

(a) Write $L(\boldsymbol{\theta})$ as an algebraic function of a, b, c.

(b) Using algorithm A, search for $\boldsymbol{\theta}^*$ from an initial condition $\hat{\boldsymbol{\theta}}_0 = [0, 0, 0]^T$ subject to $\Theta = [-1, 1]^3$ (i.e., each of a, b, c are between -1 and 1). For one realization of the search process, show how the estimates of $\boldsymbol{\theta}$ and the corresponding values of $L(\boldsymbol{\theta})$ change as the number of loss evaluations range over 100, 1000, 10,000, 100,000, and 1,000,000.

(c) Comment on the nonuniqueness of the solution $\boldsymbol{\theta}^*$ and its observed or likely effect on the search process. (The general form of $L(\boldsymbol{\theta})$ in this problem arises in the experimental design discussion of Chapter 17; see, in particular, Example 17.6 and some of the accompanying exercises.)

2.3 Formula (2.4) may be numerically unreliable when P^* is very small. A more numerically stable form that is accurate for small P^* is $n \approx -\log \rho / P^*$. Derive this approximation and compare its accuracy with the exact values from (2.4) for the problem of Example 2.2 with the corresponding values of $1 \le p \le 10$ in Table 2.2. Show three significant digits of accuracy for both the exact and approximate values of n (this is greater accuracy than shown in Table 2.2).

2.4 Suppose that $p = 3$ and $\Theta = \{\boldsymbol{\theta}: \boldsymbol{\theta}^T\boldsymbol{\theta} \le 1\}$. Consider algorithm A where uniform sampling on a superset of Θ, $\Theta' = [-1, 1]^3$, is used (with the sample rejected if it lies outside of Θ) to generate each of the candidate $\boldsymbol{\theta}_{\text{new}} = \boldsymbol{\theta}_{\text{new}}(k)$ values. What is the expected number of sample values of $\boldsymbol{\theta} \in \Theta'$ needed to produce the $\boldsymbol{\theta}_{\text{new}} \in \Theta$ values used in 100 iterations?

2.5 Sketch or describe a loss function and domain Θ where condition (2.1) in Theorem 2.1 does *not* hold, but where Θ^* still has only one (unique) element $\boldsymbol{\theta}^*$. Discuss or numerically demonstrate how algorithm A may not converge to this unique $\boldsymbol{\theta}^*$ for this problem.

2.6 Let $\Theta = [-1, 1]^p$ and suppose that uniform sampling on Θ is used to generate $\boldsymbol{\theta}_{\text{new}}(k)$ for all k. We want to guarantee with probability 0.90 that the first $p/2$ elements of $\boldsymbol{\theta}$ are within 0.04 units of the optimal and the next $p/2$ elements are within 0.10 units (p an even number). Let the (unknown) $\boldsymbol{\theta}^*$ lie in $[-0.90, 0.90]^p$. How many loss measurements in algorithm A are needed if $p = 2$? $p = 10$?

2.7 Consider a modified form of algorithm A where the sampling in step 1 is replaced with sampling over a domain that changes with k. In particular, suppose that $\boldsymbol{\theta}_{\text{new}}(k+1)$ is generated uniformly on a sampling domain $\Theta_{k+1} \equiv \Theta_k \cap \{\boldsymbol{\theta}: L(\boldsymbol{\theta}) < L(\hat{\boldsymbol{\theta}}_k)\}$, where Θ_0 is the problem domain Θ (so $\Theta_{k+1} \subseteq \Theta_k \subseteq \ldots \subseteq \Theta_0 = \Theta$). Let V_k denote the volume of Θ_k. Show that the expected volume is reduced by half at every iteration. That is, show that $E(V_k/V_{k-1}) = 1/2$. (Hint: First show that $P(V_k/V_{k-1} \le r) = r$ for all $r \in [0, 1]$. Note that this algorithm is an idealized adaptive search algorithm that is not usually implementable due to the unknown shape of the Θ_k domains; the reduction of candidate volume by half at every iteration is very fast.)

2.8 By any numerical or analytical means, identify the minima, maxima, and saddlepoints of the loss function in Example 2.3.

2.9 Consider the function $L(\boldsymbol{\theta}) = \sum_{i=1}^{p/2} t_i^4 + \boldsymbol{\theta}^T\boldsymbol{B}\boldsymbol{\theta}$, where p is an even number, $\boldsymbol{\theta} = [t_1, t_2, \ldots, t_p]^T$, and $\boldsymbol{B}$ is a symmetric matrix with 1's on the diagonals and 0.5's elsewhere. Apply algorithm B with $\boldsymbol{d}_k \sim N(\boldsymbol{0}, \rho^2\boldsymbol{I}_p)$, $\Theta = \mathbb{R}^p$, and $\hat{\boldsymbol{\theta}}_0 = [1, 1, \ldots, 1]^T$. Use 100 noise-free loss measurements per replication as follows:

(a) Using $p = 20$, compute the mean terminal loss values based on 40 replications for $\rho = 0.125$, 0.25, 0.5, and 1.

(b) Using $p = 2$, compute the mean terminal loss values based on 40 replications and the same four values of ρ as in part (a).

(c) Normalize the mean loss values in part (a) and the mean loss values in part (b) by their respective $L(\hat{\boldsymbol{\theta}}_0)$ (i.e., divide the mean loss values by $L(\hat{\boldsymbol{\theta}}_0)$, which is the only normalization required since $L(\boldsymbol{\theta}^*) = 0$). How do the normalized loss values with $p = 20$ compare with the normalized values for $p = 2$? Comment on the effects of increased dimension.

2.10 From the information provided in Example 2.3, determine the approximate sample standard deviations (s) for the terminal loss values from algorithms A, B, and C, and compute 90 percent confidence intervals for the mean values shown in the example. Show that there is still some overlap in these intervals.

2.11 If the problem setting allowed it, determine on a conceptual basis which, if any, of the algorithms A, B, or C would potentially benefit by allowing sampling over a region Θ' that contains Θ in the sense that Θ is a strict subset of Θ' (i.e., $\Theta \subset \Theta'$). Here, *benefit* refers to greater accuracy of the $\boldsymbol{\theta}$ estimate or a lower final value of the loss function for a given number of loss evaluations. If any algorithm shows a potential benefit, give a conceptual or numerical example where the benefit is realized (if numerical, do repeated runs until there is strong statistical evidence of a benefit).

2.12 In contrast to general results with noisy loss measurements, provide a conceptual example (with justification) of a problem with noise where algorithm A is guaranteed to converge. (Hint: Discrete or continuous $\boldsymbol{\theta}$ may be considered.)

2.13 Suppose that only noisy measurements of the loss function $L(\boldsymbol{\theta}) = t_1^4 + t_1^2 + t_1 t_2 + t_2^2$ (introduced in Subsection 1.4.1) are available ($\boldsymbol{\theta} = [t_1, t_2]^T$). The noise is additive independent $N(0, \sigma^2)$. Suppose that algorithm B is going to be used with $\hat{\boldsymbol{\theta}}_0 = [1, 1]^T$, $\boldsymbol{d}_k \sim N(\boldsymbol{0}, 0.125^2 \boldsymbol{I}_2)$ for all k, and $\Theta = \mathbb{R}^2$. Carry out numerical analysis to determine whether the loss averaging or altered threshold (2.6) should be used to cope with the noise. In particular:

(a) For σ = 0.001, 0.01, 0.1, and 1.0 with 10,000 loss measurements, determine approximately optimal values for the amount of averaging and for the altered threshold. For each of these values compute the sample mean of the terminal loss value in 40 independent runs. Which method (averaging or altered threshold) seems to work better for this problem? (You may base your analysis on only the sample means—it is not necessary to do a formal statistical analysis.)

(b) Compute the sample mean of the terminal loss value in the noise-free case (40 runs). Show that this is significantly lower than the best of the values in part (a) for the smallest noise level.

2.14 Consider the wastewater treatment problem in Example 2.7. For the initial condition, noise levels, and domain Θ in the example, use algorithms A and C to produce tables analogous to Table 2.6 in Example 2.7 (choose ρ appropriately for algorithm C). In particular, calculate the means of 50 terminal loss values for 100 and 2000 measurements in the algorithms, including 95 percent confidence intervals, for algorithms A and C. For each algorithm, use the averaged and unaveraged implementations discussed in Example 2.7 (so the table you produce should have eight numerical entries—

averaged and unaveraged implementations for the two algorithms for the two sample sizes). Comment on how the results of these two algorithms compare with those of algorithm B in Example 2.7.

2.15 Using the $p = 20$ version of the loss function in Exercise 2.9 together with $\hat{\theta}_0 = [1, 1, \ldots, 1]^T$ and 1000 loss measurements:

(a) Implement the nonlinear simplex algorithm using the default coefficient settings and noise-free measurements (you may use **`fminsearch`**).

(b) Repeat part (a) with the exception of having the algorithm use loss measurements with additive, independent $N(0, 0.2^2)$ noise. Generate 40 replications and report the sample mean and approximate 95 percent confidence interval for the terminal loss value.

(c) Repeat part (b) with the exception of setting $\delta = 0.9$. Contrast the results of parts (b) and (c).

2.16 For the nonlinear simplex algorithm, elaborate on why noise causes an increased tendency to reduce the size of the simplex relative to a deterministic case with the same values of algorithm coefficients α, β, γ, and δ.

CHAPTER 3

RECURSIVE ESTIMATION FOR LINEAR MODELS

This chapter describes recursive optimization algorithms for linear models that have static or time-varying parameters $\boldsymbol{\theta}$. Consistent with the discussion of Section 1.1, the use of *recursive* here is meant to suggest that the data arrive over time, and that the algorithm processes the data approximately as they arrive. In fact, however, the algorithms can also be applied as general iterative methods for "off-line" analysis. Linear models are considered here for two reasons: (i) the models are popular and the associated recursive algorithms are widely used and (ii) the algorithms provide motivation for the more general nonlinear stochastic approximation framework to be introduced in Chapter 4. Broadly speaking, the linear models here lead to loss functions $L = L(\boldsymbol{\theta})$ that are quadratic in $\boldsymbol{\theta}$. So, the algorithms of this chapter are oriented to finding effective means for solving quadratic optimization problems with possibly noisy loss function or associated gradient information. This chapter marks the beginning of an emphasis on continuous-valued $\boldsymbol{\theta}$ problems that will last through Chapter 7.

Section 3.1 summarizes some essential properties of the linear model and main estimation criterion, introducing the estimation problem via consideration of a fixed (static) $\boldsymbol{\theta}$ vector. Section 3.2 describes some popular recursive methods for estimating a static $\boldsymbol{\theta}$. Section 3.3 extends the ideas in Sections 3.1 and 3.2 to the problem of estimating time-varying $\boldsymbol{\theta}$. This discussion includes the celebrated Kalman filter. Section 3.4 is a case study devoted to demonstrating the recursive methods on a set of data associated with a musical instrument—the oboe. Section 3.5 offers some concluding remarks. Chapter 13 considers some general issues associated with building and analyzing mathematical models, including models of the type here.

3.1 FORMULATION FOR ESTIMATION WITH LINEAR MODELS

This section focuses on the parameter estimation problem when it is assumed that there is no change over time in the underlying true values of the parameters $\boldsymbol{\theta}$. The first subsection in this section summarizes the principal linear model of this chapter and some of its applications. (Some extensions of this model are considered in Section 3.3.) The second subsection discusses the least-squares estimation criterion of interest, showing the connection of this criterion to an idealized (but infeasible) mean-squared error criterion. This subsection discusses classical least squares (essentially a deterministic optimization approach) and then makes connections to the stochastic optimization setting of prime interest in this book.

3.1.1 Linear Model

The model of interest here has the classical linear form

$$z_k = \boldsymbol{h}_k^T \boldsymbol{\theta} + v_k, \tag{3.1}$$

where z_k is the kth measured output (assumed for now to be a scalar), $\boldsymbol{h}_k$ is the corresponding *design vector*, v_k is the unknown noise value, and $\boldsymbol{\theta}$ represents the parameter vector to be estimated. Because we will be interested in squared-error-type loss functions, the linearity in $\boldsymbol{\theta}$ in (3.1) implies a loss function that is quadratic in $\boldsymbol{\theta}$. This form of criterion has a single global (= local) minimum, eliminating the concern about finding a global minimum among many local minima.

Of course, few *real* systems truly produce responses in the clean linear form of (3.1), but as a reasonable approximation of reality, model (3.1) has been used in countless applications. In fact, it is more general than it may first appear in that it also accommodates a certain type of nonlinearity—namely, the so-called curvilinear models where the model contribution $\boldsymbol{h}_k^T \boldsymbol{\theta}$ represents a sum of *nonlinear* functions of input variables (appearing within $\boldsymbol{h}_k$). The nonlinear summands are weighted by the elements of $\boldsymbol{\theta}$. Let us illustrate both pure linear and curvilinear models in the examples below. Another example of both a pure linear and curvilinear model is given in the case study of Section 3.4. The texts by Widrow and Stearns (1985), Haykin (1996), Neter et al. (1996), Ljung (1999), and Moon and Stirling (2000) are several of *many* books in a variety of fields that present detailed analysis and excellent examples of model (3.1). Chapters 13 and 17 here also consider aspects of model (3.1) as related to model selection and experimental design.

Example 3.1—Time series model as an example of (3.1). A large number of systems have been successfully analyzed with models described by a linear (or affine) relationship between input variables and output variables. Linear time series models describing the state of a system as a function of previous states are one important class of linear models. (Here, *state* refers to a vector summarizing the essential characteristics of the system *relative to the needs of the analysis.*) For example, for a macroeconomist, a state vector of interest might include—among other quantities—the gross national product and inflation rate, whereas to a physicist, a state vector might be the position, velocity, and acceleration of a moving object. More specifically, autoregressive (AR) models are an important subset of time series models. Scalar AR models have the form

$$x_{k+1} = \beta_0 x_k + \beta_1 x_{k-1} + \ldots + \beta_m x_{k-m} + w_k ,$$

where x_k represents the (scalar) state of the system at time point k, β_i is the coefficient describing the effect of the state at an earlier time $k - i$ (i.e., x_{k-i}) on the next state x_{k+1}, and w_k is the input noise. This AR form fits into the framework of the linear model in (3.1) by letting $z_k = x_k$, $v_k = w_{k-1}$, $\boldsymbol{\theta} = [\beta_0, \beta_1, \ldots, \beta_m]^T$, and $\boldsymbol{h}_k = [x_{k-1}, x_{k-2}, \ldots, x_{k-m-1}]^T$. (Note that there is no dependence of $\boldsymbol{h}_k$ on the unknown variables $\boldsymbol{\theta}$, which would make model (3.1) *nonlinear* in $\boldsymbol{\theta}$. The lagged values x_{k-i} in $\boldsymbol{h}_k$ represent *actual* values of the physical system. We are only *modeling* those values as depending on $\boldsymbol{\theta}$. Even if the AR model form were precisely correct, the physical values x_{k-i} would be generated by some *fixed* value $\boldsymbol{\theta} = \boldsymbol{\theta}^*$.) ❑

Example 3.2—Nonlinear model transformed to linear model (3.1). Let us now present an example of a model that is nonlinear in both the parameters to be estimated and the input variables, but which can be converted to a linear model by a simple transformation. Suppose that the scalar output, say q_k, for the kth set of m inputs $x_{k1}, x_{k2}, \ldots, x_{km}$ is given by the multiplicative model

$$q_k = \beta_0 x_{k1}^{\beta_1} x_{k2}^{\beta_2} \cdots x_{km}^{\beta_m} \xi_k ,$$

where the β_i, $i = 0, 1, \ldots, m$, represent the parameters of the model to be estimated and ξ_k is a multiplicative (positive) noise term. Taking logarithms of both sides of this expression yields

$$\log q_k = \log \beta_0 + \beta_1 \log x_{k1} + \beta_2 \log x_{k2} + \ldots + \beta_m \log x_{km} + \log \xi_k ,$$

which is *linear* in a redefined set of parameters and input variables. In particular, setting $z_k = \log q_k$, $v_k = \log \xi_k$, $\boldsymbol{\theta} = [\log \beta_0, \beta_1, \beta_2, \ldots, \beta_m]^T$, and $\boldsymbol{h}_k = [1, \log x_{k1}, \log x_{k2}, \ldots, \log x_{km}]^T$ leads to a model of the form (3.1). There is no loss of information about the original parameters and variables in this logarithmic transformation.

The multiplicative model, for example, appears in microeconomics as the Cobb–Douglas production function (e.g., Section 4.2 here; Kmenta, 1997, pp. 252–253). This function measures the output (q_k) of a production facility as a function of relevant inputs where the parameters $\beta_1, \beta_2, \ldots, \beta_m$ reflect the relative importance of the various inputs. Here k is either an index across time for a given production facility or an index across multiple facilities of the same type. One special case is when $m = 2$ with x_{k1} representing a summary measure of labor input and x_{k2} representing a summary measure of capital equipment. In some applications of the Cobb–Douglas function, the effective dimension of $\boldsymbol{\theta}$ is reduced from $p = m+1$ to $p = m$ because there is a constraint $1 = \beta_1 + \beta_2 + \ldots + \beta_m$ (i.e., once an estimate for $m-1$ of the m parameters $\beta_1, \beta_2, \ldots, \beta_m$ is available, the remaining estimate is directly available from the constraint). ❑

Example 3.3—Polynomial class of curvilinear models for use in (3.1). Suppose that the observed output of a process is modeled as a sum of additive noise and an mth-order polynomial of a scalar input variable x. Let the mth-order polynomial be $\beta_0 + \beta_1 x + \beta_2 x^2 + \ldots + \beta_m x^m$ where the β_i, $i = 0, 1, \ldots, m$, represent the parameters to be estimated. This fits into the framework of the linear model in (3.1), where repeated values of x, indexed as x_k, are used to generate outputs for estimating the β_i. In particular, the connection to (3.1) is apparent by letting $\boldsymbol{\theta} = [\beta_0, \beta_1, \beta_2, \ldots, \beta_m]^T$ and $\boldsymbol{h}_k = [1, x_k, x_k^2, \ldots, x_k^m]^T$. The additive noise for the kth measurement is represented by v_k.

A simple specific example of this curvilinear (polynomial) model is the motion of a particle along a line where time τ represents the fundamental input variable. From elementary physics, the position, say $\pi(\tau)$, of a particle undergoing constant acceleration is given by the formula $\pi(\tau) = \pi_0 + V_0\tau + A_0\tau^2/2$, where π_0 is the initial ($\tau = 0$) position, V_0 is the initial velocity, and A_0 is the constant acceleration. Suppose there is instrumentation that provides noisy measurements $z_1, z_2, \ldots$ of the position $\pi(\tau)$ at discrete times $\tau_1, \tau_2, \ldots$ (so $z_k = \pi(\tau_k) + v_k$). One may estimate the initial position and velocity and the constant acceleration from these noisy measurements by letting $\boldsymbol{\theta} = [\pi_0, V_0, A_0]^T$ and $\boldsymbol{h}_k = [1, \tau_k, \tau_k^2/2]^T$. ❑

3.1.2 Mean-Squared and Least-Squares Estimation

This subsection presents the mean-squared and least-squares estimation criteria and discusses the batch least-squares estimate for model (3.1). The aim is to estimate $\boldsymbol{\theta}$ from a sequence of input–output pairs $\{\boldsymbol{h}_k, z_k\}$. In deriving the mean-squared criterion, suppose that $z_k = \boldsymbol{h}_k^T\boldsymbol{\theta}^* + v_k$, where $\boldsymbol{\theta}^*$ is the unknown true value of $\boldsymbol{\theta}$ (so that the model and the true system have the same functional form). Under the assumption that the noises v_k have mean zero, a natural conceptual criterion for estimation is the mean-squared error

$$L(\boldsymbol{\theta}) = \frac{1}{n} E\left[\tfrac{1}{2} \sum_{k=1}^{n} (z_k - \boldsymbol{h}_k^T\boldsymbol{\theta})^2 \right]$$

$$= \frac{1}{2n} \sum_{k=1}^{n} E\left[(\boldsymbol{h}_k^T\boldsymbol{\theta}^* + v_k - \boldsymbol{h}_k^T\boldsymbol{\theta})^2 \right], \tag{3.2}$$

where the immaterial factor 1/2 is included to facilitate later derivative calculations and the division by n is a normalization so that (3.2) reflects the *average* error over the n measurements. The expectations shown in (3.2) are with respect to randomness in v_k; the expectation may also be with respect to $\boldsymbol{h}_k$ if the inputs $\boldsymbol{h}_k$ are random. (Estimation based on the inclusion of weighted summands—i.e., the implied multiplier of unity on $(z_k - \boldsymbol{h}_k^T\boldsymbol{\theta})^2$ is replaced by a scale factor changing with k—is discussed in Subsection 3.2.6 and Section 3.3.)

Because $L(\boldsymbol{\theta})$ is a quadratic function of $\boldsymbol{\theta}$ (i.e., there is one unique minimum), the solution is available by setting the gradient $\boldsymbol{g}(\boldsymbol{\theta}) = \partial L/\partial\boldsymbol{\theta} = \boldsymbol{0}$ when the problem is unconstrained (i.e., $\Theta = \mathbb{R}^p$). If var(v_k) is independent of $\boldsymbol{\theta}$ and the $\boldsymbol{h}_k$ are deterministic, the derivative of the kth summand above (ignoring the immaterial division by n for the moment) is

$$\frac{\partial E\left[(\boldsymbol{h}_k^T\boldsymbol{\theta}^* + v_k - \boldsymbol{h}_k^T\boldsymbol{\theta})^2\right]/2}{\partial\boldsymbol{\theta}} = \frac{\partial(\boldsymbol{h}_k^T\boldsymbol{\theta}^* - \boldsymbol{h}_k^T\boldsymbol{\theta})^2/2}{\partial\boldsymbol{\theta}} = -\boldsymbol{h}_k(\boldsymbol{h}_k^T\boldsymbol{\theta}^* - \boldsymbol{h}_k^T\boldsymbol{\theta}).$$

Hence, an estimate, in principle, is available by solving for $\boldsymbol{\theta}$ such that

$$\sum_{k=1}^{n}\boldsymbol{h}_k(\boldsymbol{h}_k^T\boldsymbol{\theta}^* - \boldsymbol{h}_k^T\boldsymbol{\theta}) = \boldsymbol{0}.$$

This, however, is not useful for implementation since it depends on the very quantity that one is trying to find ($\boldsymbol{\theta}^*$). In other words, one has to know $\boldsymbol{\theta}^*$ to find $\boldsymbol{\theta}^*$! So, although the mean-squared error criterion is a desirable *conceptual* criterion, it does not lead to a usable solution strategy.

The method of least squares is a practical alternative to the infeasible method based on the mean-squared error criterion. This method works with a criterion, say $\hat{L}(\boldsymbol{\theta})$, that is an unbiased estimate of $L(\boldsymbol{\theta})$ in (3.2). The least-squares loss function is

$$\begin{aligned}\hat{L}(\boldsymbol{\theta}) &= \frac{1}{2n}\sum_{k=1}^{n}(z_k - \boldsymbol{h}_k^T\boldsymbol{\theta})^2 \\ &= \frac{1}{2n}(\boldsymbol{Z}_n - \boldsymbol{H}_n\boldsymbol{\theta})^T(\boldsymbol{Z}_n - \boldsymbol{H}_n\boldsymbol{\theta}),\end{aligned} \tag{3.3}$$

where $\boldsymbol{Z}_n = [z_1, z_2, \ldots, z_n]^T$ and $\boldsymbol{H}_n$ is the $n \times p$ concatenated matrix of $\boldsymbol{h}_k^T$ row vectors. This function is a relatively simple criterion that is quadratic in the elements of $\boldsymbol{\theta}$ but, unlike the mean-squared criterion (3.2), does not depend on the unknown $\boldsymbol{\theta}^*$. Further, the minimum of $\hat{L}(\boldsymbol{\theta})$ is not generally $\boldsymbol{\theta}^*$, but is a point tied to the specific observed data $\boldsymbol{Z}_n$. As n gets large, the minimum of $\hat{L}(\boldsymbol{\theta})$ approaches (in a stochastic sense) $\boldsymbol{\theta}^*$ under standard conditions.

Criterion (3.3) has some desirable connections to the mean-squared criterion. In particular, for any $\boldsymbol{\theta}$, the criterion and its gradient are unbiased estimators of the mean-squared criterion and its gradient. That is,

$$E[\hat{L}(\boldsymbol{\theta})] = L(\boldsymbol{\theta}), \tag{3.4a}$$

$$E\left[\frac{\partial \hat{L}(\boldsymbol{\theta})}{\partial \boldsymbol{\theta}}\right] = \frac{\partial L(\boldsymbol{\theta})}{\partial \boldsymbol{\theta}}. \tag{3.4b}$$

Result (3.4b) follows since $\frac{1}{2}E[\partial(z_k - \boldsymbol{h}_k^T\boldsymbol{\theta})^2/\partial\boldsymbol{\theta}] = -\boldsymbol{h}_k(\boldsymbol{h}_k^T\boldsymbol{\theta}^* - \boldsymbol{h}_k^T\boldsymbol{\theta})$, which corresponds to the derivatives of the summands in (3.2) that form $\partial L/\partial\boldsymbol{\theta}$. This is an example of an interchange of derivative and integral since $\partial L/\partial\boldsymbol{\theta}$ on the right-hand side of (3.4b) can be written as $\partial E[\hat{L}(\boldsymbol{\theta})]/\partial\boldsymbol{\theta}$. Therefore, unlike some general nonlinear problems considered in Chapter 5 and elsewhere, the relatively simple quadratic structure here does not require the general machinery given in Appendix A to justify the interchange.

By (3.4a, b), optimizing based on the least-squares criterion is equivalent to optimizing an unbiased estimate of the mean-squared criterion. Further, optimizing based on setting $\partial\hat{L}/\partial\boldsymbol{\theta} = \boldsymbol{0}$ is equivalent to root finding with an unbiased estimate of $\partial L/\partial\boldsymbol{\theta}$. These connections of least-squares estimation to minimum mean-squared estimation can be used to make formal statements about the convergence of the recursive estimates to $\boldsymbol{\theta}^*$ via stochastic approximation theory (see, especially, Sections 4.3 and 5.1).

The close connection of the desirable (but infeasible) minimum mean-squared error estimation to the feasible least-squares estimation provides the rationale for focusing on the least-squares criterion (3.3). A unique solution to the problem of estimating $\boldsymbol{\theta}$ is readily available if $n \geq p$ (so that there are at least as many data points as parameters to be estimated in $\boldsymbol{\theta}$). The form (3.3) is most appropriate when one assumes that $E(v_k) = 0$ and the var(v_k) terms are identical for all k (but not necessarily known). Criterion (3.3) leads to the ordinary least-squares solution. This solution follows by setting $\partial\hat{L}/\partial\boldsymbol{\theta} = \boldsymbol{0}$ and solving the resulting "normal equations" to yield the batch least-squares estimate for $\boldsymbol{\theta}$:

$$\hat{\boldsymbol{\theta}}^{(n)} = (\boldsymbol{H}_n^T\boldsymbol{H}_n)^{-1}\boldsymbol{H}_n^T\boldsymbol{Z}_n, \tag{3.5}$$

assuming, of course, that the indicated matrix inverse exists. (The notation $\hat{\boldsymbol{\theta}}^{(n)}$ is used here and elsewhere to denote a batch estimate based on n data points. In contrast, the notation $\hat{\boldsymbol{\theta}}_n$ is generally used to denote a recursive estimate based on n iterations.) Suppose that var(v_k) = σ^2 for all k and that, as with the mean-squared criterion (3.2), the model form (3.1) is accurate in the sense that the physical data are generated by the "true process" $z_k = \boldsymbol{h}_k^T\boldsymbol{\theta}^* + v_k$. Then, two important properties of the batch solution are that it is unbiased (i.e., $E(\hat{\boldsymbol{\theta}}^{(n)}) = \boldsymbol{\theta}^*$) and has covariance matrix $\text{cov}(\hat{\boldsymbol{\theta}}^{(n)}) = (\boldsymbol{H}_n^T\boldsymbol{H}_n)^{-1}\sigma^2$ (Exercise 3.1).

If there is a symmetric weighting matrix $\boldsymbol{W}_n$ in the inner product of (3.3) (i.e., the second line of (3.3) becomes $(\boldsymbol{Z}_n - \boldsymbol{H}_n\boldsymbol{\theta})^T\boldsymbol{W}_n(\boldsymbol{Z}_n - \boldsymbol{H}_n\boldsymbol{\theta})/(2n)$), then the solution in (3.5) for ordinary least squares is replaced by the weighted least-squares solution. Such a weighting matrix is typically used when the variances of the v_k terms are not identical for all k. In such a solution, the two $\boldsymbol{H}_n^T$ terms in (3.5) are replaced by $\boldsymbol{H}_n^T\boldsymbol{W}_n$. We give a general form for the recursive implementation of least squares under a weighting matrix with multivariate z_k in Subsection 3.2.6.

Figure 3.1 provides a geometric interpretation of the least-squares estimate. In particular, the famous orthogonal projection principle (e.g., Sorenson, 1980, pp. 35–36; Jang et al., 1997, pp. 110–112; Moon and Stirling, 2000, pp. 114–115) states that the error in the predicted $\boldsymbol{Z}_n$, as given by the predictor $\boldsymbol{H}_n\hat{\boldsymbol{\theta}}^{(n)}$ (i.e., the error $\boldsymbol{Z}_n - \boldsymbol{H}_n\hat{\boldsymbol{\theta}}^{(n)}$), is orthogonal to each of the $n \times 1$ columns in $\boldsymbol{H}_n$. For convenience in Figure 3.1 and the following discussion, let us suppress the sub/superscript n. Formally, the orthogonality is then expressed as

$$\boldsymbol{H}^T(\boldsymbol{Z} - \boldsymbol{H}\hat{\boldsymbol{\theta}}) = \boldsymbol{0}.$$

One can see the origins of this relationship by observing that the prediction $\boldsymbol{H}\hat{\boldsymbol{\theta}}$ must lie in the hyperplane spanned by the p columns, say $\boldsymbol{h}_{\bullet j}$, in $\boldsymbol{H} \equiv [\boldsymbol{h}_{\bullet 1}, \boldsymbol{h}_{\bullet 2},\ldots, \boldsymbol{h}_{\bullet p}]$ (the notation $\boldsymbol{h}_{\bullet j}$ is chosen to avoid confusion with the jth design vector $\boldsymbol{h}_j$ from (3.1)). In particular, with $\boldsymbol{\theta} = [t_1, t_2,\ldots, t_p]^T$:

$$\boldsymbol{H}\boldsymbol{\theta} = [\boldsymbol{h}_{\bullet 1}, \boldsymbol{h}_{\bullet 2}, \ldots, \boldsymbol{h}_{\bullet p}]\begin{bmatrix} t_1 \\ t_2 \\ \vdots \\ t_p \end{bmatrix} = t_1[\boldsymbol{h}_{\bullet 1}] + t_2[\boldsymbol{h}_{\bullet 2}] + \ldots + t_p[\boldsymbol{h}_{\bullet p}],$$

indicating that the prediction $\boldsymbol{H}\hat{\boldsymbol{\theta}}$ is a linear combination of the basis vectors $[\boldsymbol{h}_{\bullet 1}, \boldsymbol{h}_{\bullet 2},\ldots, \boldsymbol{h}_{\bullet p}]$.

Figure 3.1 depicts the case $p = 2$ and $n = 3$. The prediction $\boldsymbol{H}\hat{\boldsymbol{\theta}}$ is constrained to lie in the lower plane since that is the plane containing the two basis vectors. Clearly, the prediction $\boldsymbol{H}\hat{\boldsymbol{\theta}}$ is made as close as possible to the observed $\boldsymbol{Z}$ by picking $\hat{\boldsymbol{\theta}}$ such that $\boldsymbol{H}\hat{\boldsymbol{\theta}}$ represents the projection of $\boldsymbol{Z}$ into the plane containing the basis vectors. Geometrically, the error $\boldsymbol{Z} - \boldsymbol{H}\hat{\boldsymbol{\theta}}$ is orthogonal to the prediction $\boldsymbol{H}\hat{\boldsymbol{\theta}}$.

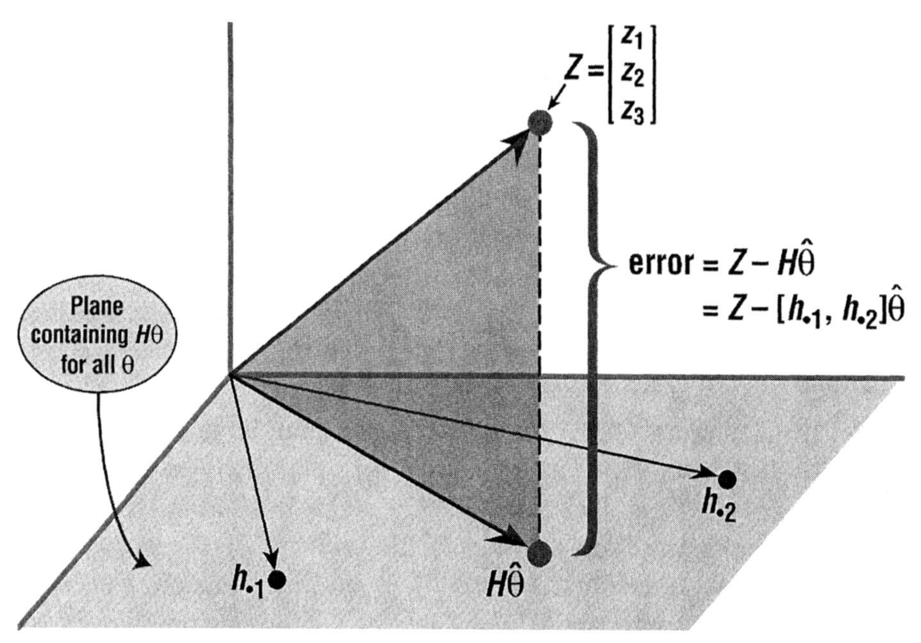

Figure 3.1. Geometric interpretation of least-squares estimate when $p = 2$ and $n = 3$. All possible predictions $\boldsymbol{H}\hat{\boldsymbol{\theta}}$ lie in indicated horizontal plane.

3.2 LEAST-MEAN-SQUARES AND RECURSIVE-LEAST-SQUARES FOR STATIC θ

3.2.1 Introduction

Suppose that $k = 1, 2,\ldots, n$ is a time index and suppose that the input–output data pairs $\{\boldsymbol{h}_k, z_k\}$ arrive sequentially: $\{\boldsymbol{h}_1, z_1\}$, $\{\boldsymbol{h}_2, z_2\},\ldots,$ $\{\boldsymbol{h}_n, z_n\}$. (Viewing k as a time index is for ease of discussion; it could as easily be viewed as another type of index—spatial, cross-sectional, and so on.) The focus in this and the next section is the development of *recursive* estimates $\hat{\boldsymbol{\theta}}_k$ that are intended to approximate corresponding batch estimates $\hat{\boldsymbol{\theta}}^{(k)}$. If $\hat{\boldsymbol{\theta}}_k$ represents an estimate for $\boldsymbol{\theta}$ based on k data pairs, we seek a way of computing $\hat{\boldsymbol{\theta}}_{k+1}$ with a minimum of effort when the $(k+1)$st data pair arrives. Such recursive solutions are based on combining the currently available estimate with the $(k+1)$st data pair $\{\boldsymbol{h}_{k+1}, z_{k+1}\}$ in an efficient way to obtain the new estimate.

Aside from potential computational advantages, the recursive form clearly exhibits the value of a new data point. In fact, with advances in computing power, the computational advantages alone are becoming less important in many applications. However, it is likely that there will continue to be some real-time applications with massive data quantities where performing the required matrix inversion for a new batch solution at every time point may be difficult.

In the remaining five subsections, we present the least-mean-squares (LMS) and recursive-least-squares (RLS) solutions to this recursive estimation problem. We will see that these algorithms have, respectively, an interpretation as stochastic versions of the deterministic steepest descent and Newton–Raphson algorithms discussed in Section 1.4. The focus here is on cases where the optimal $\boldsymbol{\theta}$ is fixed over time; Section 3.3 considers the time-varying $\boldsymbol{\theta}$ setting.

3.2.2 Basic LMS Algorithm

A popular algorithm for parameter estimation in linear models is the LMS algorithm. LMS is widely used in signal processing and control applications, with one of the classic references being Widrow and Stearns (1985). The LMS algorithm has an interpretation as a stochastic version of the steepest descent algorithm of Section 1.4. LMS for linear models represents a special case of the nonlinear stochastic gradient algorithms to be considered in Chapter 5. (Additional detail on LMS is given in Sections 3.3 and 5.1.)

Before presenting the LMS algorithm, let us introduce the notion of the *instantaneous gradient*. When the least-squares criterion (3.3) is summed through the $(k+1)$st time point, the gradient of the latest summand at any $\boldsymbol{\theta}$ is

$$\frac{1}{2}\frac{\partial(z_{k+1}-\boldsymbol{h}_{k+1}^T\boldsymbol{\theta})^2}{\partial\boldsymbol{\theta}} = \boldsymbol{h}_{k+1}(\boldsymbol{h}_{k+1}^T\boldsymbol{\theta}-z_{k+1}). \tag{3.6}$$

The expression above is sometimes called the instantaneous gradient because it corresponds to the gradient of only the latest part of the least-squares loss function (3.3). Note that the expression in (3.6) is an unbiased, noisy measurement of the gradient of the $(k+1)$st summand in (3.2) (i.e., a noisy measurement of $\frac{1}{2}\partial E[(z_{k+1}-\boldsymbol{h}_{k+1}^T\boldsymbol{\theta})^2]/\partial\boldsymbol{\theta}$). As discussed in Section 3.3, the instantaneous gradient is especially important in time-varying systems where the aim is to estimate a dynamically changing value of $\boldsymbol{\theta}$. In such systems, it is important to put emphasis on the most recent information, corresponding to the last term in the sum of (3.2).

LMS is based on using the (noisy) instantaneous gradient expression in (3.6) as the gradient input in a steepest descent-type search. The qualifier "steepest descent-*type*" is used here because the gradient input is not deterministic, as pure steepest descent requires. In most practical applications, the general gain coefficient a_k in the steepest descent algorithm of Section 1.4 is set to a constant $a > 0$ that regulates the speed and stability of the algorithm. (A large a often helps the algorithm converge more quickly to the vicinity of $\boldsymbol{\theta}^*$, but a small a helps the algorithm avoid instability and divergence; what is meant by large and small in this context is highly problem dependent.) The iterative update for LMS is given below.

LMS Algorithm

$$\hat{\boldsymbol{\theta}}_{k+1} = \hat{\boldsymbol{\theta}}_k - a\boldsymbol{h}_{k+1}(\boldsymbol{h}_{k+1}^T\hat{\boldsymbol{\theta}}_k - z_{k+1}). \tag{3.7}$$

Interestingly, the convergence theory for such a simple algorithm is not so simple! Informal arguments that $\hat{\boldsymbol{\theta}}_k$ is close to $\boldsymbol{\theta}^*$ in the sense that $E\left(\|\hat{\boldsymbol{\theta}}_k - \boldsymbol{\theta}^*\|^2\right)$ is small are given in, for example, Widrow and Stearns (1985, pp. 101–103), Haykin (1996, pp. 392–399), and Moon and Stirling (2000, pp. 646–647). These arguments do not constitute a formal proof of convergence for LMS, but they are often used in guiding the implementation for many practical applications. Let us now outline the conditions given by Widrow and Stearns (1985, pp. 101–103).

Suppose that the $\boldsymbol{h}_k$ are random and mutually independent across k, having a common mean and second moment matrix $\boldsymbol{S}_h \equiv E(\boldsymbol{h}_k\boldsymbol{h}_k^T)$ (i.e., $E(\boldsymbol{h}_j\boldsymbol{h}_j^T) = E(\boldsymbol{h}_k\boldsymbol{h}_k^T)$ for all j, k). Further, suppose that $\boldsymbol{h}_{k+1}$ is independent of $\hat{\boldsymbol{\theta}}_k$ and

$$0 < a < \frac{2}{\lambda_{\max}(\boldsymbol{S}_h)},$$

where $\lambda_{\max}(\cdot)$ denotes the largest eigenvalue of the argument matrix. Then, it is claimed that $E\left(\|\hat{\boldsymbol{\theta}}_k - \boldsymbol{\theta}^*\|^2\right) \approx 0$ for large k. However, one should be careful in interpreting this result. Aside from some "casual" portions in the derivation, another potentially misleading aspect is that it ignores the magnitude of the noise (v_k). Noise typically has a dramatic affect on practical convergence.

A more rigorous analysis for fixed a (as in (3.7)) is given in Macchi and Eweda (1983) and Gerencsér (1995), where it is shown that $E\left(\|\hat{\boldsymbol{\theta}}_k - \boldsymbol{\theta}^*\|^2\right) = O(a)$ for small a and $k \to \infty$. So, for small a and large k, the magnitude of the error in $\hat{\boldsymbol{\theta}}_k$ is proportional (in a stochastic sense) to $\sqrt{a}$, where the constant of proportionality depends on the noise level. Wang et al. (2000) consider a different setting where there is a fixed amount of data and LMS makes multiple passes through the fixed data. Although a type of convergence is possible in this batch setting for a constant a, the convergence is not to $\boldsymbol{\theta}^*$ but to a value that depends on the given data (similar ideas are considered in Sections 5.1, 5.2, and 15.4).

Conditions for formal convergence of recursive LMS to $\boldsymbol{\theta}^*$ are given, for example, in Ljung et al. (1992, Part III) and Guo and Ljung (1995). We also present a convergence result for LMS in Chapter 5 (Proposition 5.1). These analyses use stochastic approximation theory (Chapter 4) and require that the gain a be indexed by k and approach zero as $k \to \infty$. Under appropriate conditions, $\hat{\boldsymbol{\theta}}_k$ converges to $\boldsymbol{\theta}^*$ in the mean-squared or almost sure sense (as defined in Appendix C). Further, in a general setting that includes the linear model here as a special case, it is shown in Gerencsér (1993) that the difference between the LMS solution with decaying gain and the batch solution is proportional in a stochastic sense to $\log k/k$ for large k. That is, $\|\hat{\boldsymbol{\theta}}_k - \hat{\boldsymbol{\theta}}^{(k)}\| = O(\log k/k)$, where the big-$O$ term has an appropriate stochastic interpretation.

3.2.3 LMS Algorithm in Adaptive Signal Processing and Control

This subsection discusses the use of LMS in adaptive signal processing and control. We begin by introducing a slight generalization of the AR model of Section 3.1. We then discuss several ways that this model is used in adaptive control.

Suppose that a process is modeled as evolving according to

$$x_{k+1} = \beta_0 x_k + \beta_1 x_{k-1} + \ldots + \beta_m x_{k-m} + \gamma u_k + w_k, \tag{3.8}$$

where x_k represents the (scalar) state of the system at time point k, β_i is an unknown parameter describing the effect of the state at an earlier time $k - i$ (i.e., x_{k-i}) on the next state x_{k+1}, u_k is a deterministic or stochastic input, γ is an unknown scale factor that applies to the input, and w_k is the state noise accounting for random fluctuations in the process not captured via the other part of the linear model. The AR model of Example 3.1 is a special case of this model (corresponding to $\gamma = 0$). This form fits into the framework of the linear model in (3.1) by letting $\boldsymbol{\theta} = [\beta_0, \beta_1, \ldots, \beta_m, \gamma]^T$ and $\boldsymbol{h}_k = [x_{k-1}, x_{k-2}, \ldots, x_{k-m-1}, u_{k-1}]^T$. Hence, the LMS algorithm in (3.7) can be used to estimate $\boldsymbol{\theta}$ based on the noise-free measurements $z_k = x_k$. The model above is widely used in areas such as adaptive signal processing (e.g., noise cancellation) and adaptive control.

Note that this model violates one of the conditions discussed above from Widrow and Stearns (1985, pp. 101–103) and others. (Recall that if these conditions apply, there is informal evidence that $E\left(\|\hat{\boldsymbol{\theta}}_k - \boldsymbol{\theta}^*\|^2\right) \approx 0$ for large k.) Namely, the $\boldsymbol{h}_k$ are *not* independent because they share common elements across k. Nevertheless, despite violating the independence condition, LMS is widely used in such models because the above conditions are merely *sufficient* conditions (and nonrigorous sufficient conditions at that!).

Two popular areas for such problems are noise cancellation and adaptive control. In noise cancellation, there exists an external sequence representing the sum of noise plus a useful signal. The aim is to try to recover the useful signal in this external sequence. Suppose that u_k represents a reference input signal that has frequency characteristics similar to the noise. LMS attempts to adapt $\boldsymbol{\theta}$ so that a predicted state value x_k can be differenced from the corresponding external value to recover the useful signal (e.g., Haykin, 1996, pp. 377–385).

For the remainder of this subsection, let us focus on the application of the above model in the area of adaptive control. There are many ways that adaptive controllers may be constructed. Let us outline one relatively simple (but useful!) *indirect* adaptive control method for solving a tracking problem. The method is indirect in that it relies on estimating the parameters of model (3.8) and then using the estimates to form the controller. (In contrast, a *direct* method bypasses the model estimation step while estimating unknown parameters in the control function u_k; see, e.g., Landau et al., 1998, pp. 12–20, for a more detailed discussion of direct versus indirect adaptive control.)

Consider a tracking problem, where the aim is to have the state x_k track a target ("desired") value d_k. This target value may be a deterministic sequence or a random sequence that is independent of the randomness in the system being controlled. Suppose further that there are noise-free measurements (i.e., $z_k = x_k$) and that the state noise w_k is independent with mean zero and variance q^2. (If only noisy measurements are available, the Kalman filter [see Section 3.3] can be used to estimate x_k.) Assuming for the moment that the model parameters $\boldsymbol{\theta}$ are known, then the control function u_k that forces x_{k+1} to be as close to d_{k+1} as possible is

$$u_k = \frac{d_{k+1} - \beta_0 x_k - \beta_1 x_{k-1} - \ldots - \beta_m x_{k-m}}{\gamma}, \quad k \geq m. \tag{3.9}$$

That is, substituting the above u_k into (3.8) produces a tracking error $x_{k+1} - d_{k+1} = w_k$ that has the smallest possible mean-squared magnitude given the presence of the state noise (i.e., $E[(x_{k+1} - d_{k+1})^2] = q^2$ for all $k \geq m$).

Of course, $\boldsymbol{\theta}$ in (3.9) is not known and must be estimated. There are both open- and closed-loop methods for estimating $\boldsymbol{\theta}$ using LMS. In the *open-loop* approach, $\boldsymbol{\theta}$ is estimated via some experimentation with predetermined inputs u_k *before* putting the controller "online" with u_k as given in (3.9). In such open-loop training, it is desirable to pick the inputs with the general goal of enhancing the estimation of $\boldsymbol{\theta}$. This ties directly to Chapter 17 on optimal input (experimental) design and to the related idea of *persistency of excitation* (e.g., Ljung, 1999, Chap. 13). A different suboptimal method that is sometimes used is to randomly generate a set of inputs via a pseudorandom generator as in Appendix D. After the open-loop training is completed, the terminal estimate, say $\hat{\boldsymbol{\theta}}_n$, is substituted for $\boldsymbol{\theta}$ in (3.9). The system can now be run in *closed-loop* mode based on the controller (3.9) (closed-loop because now the controller u_k depends on feedback via the previous state values as opposed to the predetermined values used in the open-loop phase).

Alternatively, the training can be done in closed-loop mode based on using (3.9) as input to the system, with the most recent $\boldsymbol{\theta}$ estimate used in (3.9) at each time point. That is, the system is started with the controller (3.9) relying on

$\boldsymbol{\theta} = \hat{\boldsymbol{\theta}}_0$. The state x_1 is observed, and then LMS is used to update $\boldsymbol{\theta}$ to $\hat{\boldsymbol{\theta}}_1$ based on the difference between x_1 and d_1. This procedure is repeated as long as desired. This has the advantage of not requiring a separate "off-line" open-loop set of experiments, but has the disadvantage that the inputs u_k are not chosen to enhance the estimation of $\boldsymbol{\theta}$. Among others, Ljung (1999, Sects. 13.4 and 13.5) discusses some of the issues associated with such closed-loop model estimation.

Sometimes open-loop training, as mentioned above, is combined with closed-loop training. The open-loop phase is used to get $\boldsymbol{\theta}$ estimates that are reasonable, with training in closed-loop being used to refine the values or adapt to changes in the optimal values.

The LMS recursion in (3.7) can be used to estimate $\boldsymbol{\theta}$ regardless of the means for determining the inputs u_k for the open-loop model estimation. The quality of estimate depends on the number of measurements used in the training and the way the inputs are chosen for the given number of measurements. In open- or closed-loop estimation, the justification for the substitution of an estimate for $\boldsymbol{\theta}$ in formula (3.9) rests on the *certainty equivalence principle* of adaptive control (e.g., Landau et al., 1998, p. 504). The example below considers an implementation of LMS in closed-loop estimation.

Example 3.4—LMS in closed-loop estimation. Suppose that the state evolution is modeled according to $x_{k+1} = \beta x_k + \gamma u_k + w_k$ with noise-free state measurements $z_k = x_k$. Then $\boldsymbol{\theta} = [\beta, \gamma]^T$ and $\boldsymbol{h}_k = [x_{k-1}, u_{k-1}]^T$. Given an initial state x_0, parameter vector $\hat{\boldsymbol{\theta}}_0$, and target value d_1, the initial control u_0 is constructed based on (3.9). The system is operated with this control, producing a new state value x_1. LMS is then used to update $\boldsymbol{\theta}$ according to

$$\hat{\boldsymbol{\theta}}_1 = \hat{\boldsymbol{\theta}}_0 - a\begin{bmatrix} x_0 \\ u_0 \end{bmatrix}\left([x_0 \;\; u_0]\hat{\boldsymbol{\theta}}_0 - x_1\right)$$

(note that $[x_0, u_0]\hat{\boldsymbol{\theta}}_0 = d_1$). This updating process continues for as long as desired. The general process is illustrated in the feedback diagram of Figure 3.2. (Exercise 3.4 considers a numerical implementation.) ❑

3.2.4 Basic RLS Algorithm

Another standard method for processing data in a sequential manner is RLS. While the LMS method is connected to stochastic steepest descent-type methods, RLS is a type of stochastic Newton–Raphson method, as shown below. Further, the RLS solution approaches the batch solution for large n. Large in this context may actually be quite small in practice, say $n = 5$ (see Exercise 3.8). That is, for the same set of n input–output pairs $\{\boldsymbol{h}_k, z_k\}$, $k = 1, 2, \ldots, n$, the RLS estimate of $\boldsymbol{\theta}$ is nearly the same as the batch estimate of $\boldsymbol{\theta}$ found by (3.5).

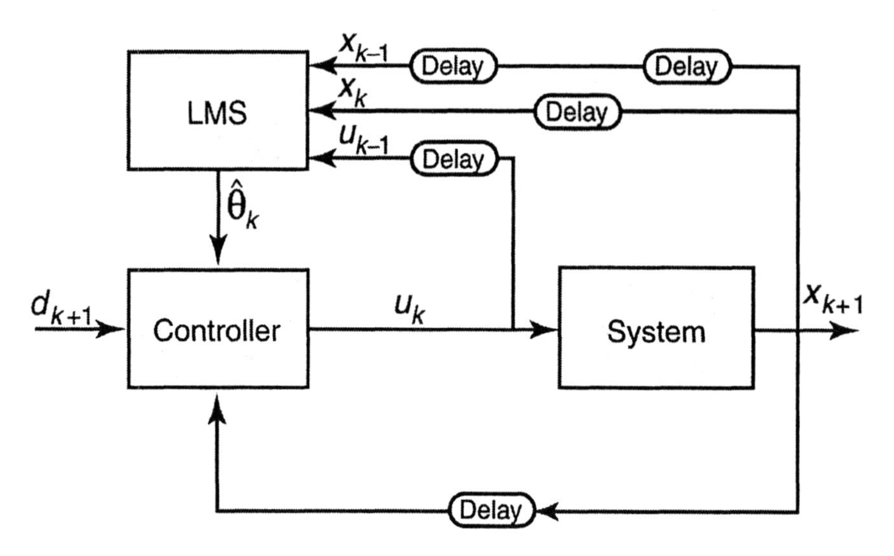

Figure 3.2. Estimation via LMS for first-order process with noise-free state measurements. Parameter values are updated while system operates in closed-loop mode.

Derivation of the RLS solution follows using the batch solution as a starting point. It is assumed that all indicated matrix inverses exist. In deriving the RLS algorithm, first note by (3.5) that the batch estimator satisfies

$$\hat{\theta}^{(k+1)} = \left(\begin{bmatrix} H_k \\ h_{k+1}^T \end{bmatrix}^T \begin{bmatrix} H_k \\ h_{k+1}^T \end{bmatrix}\right)^{-1} \begin{bmatrix} H_k \\ h_{k+1}^T \end{bmatrix}^T \begin{bmatrix} Z_k \\ z_{k+1} \end{bmatrix} \tag{3.10}$$

(recall that H_k is a $k \times p$ matrix). Because we have assumed that the necessary inverse exists (see (3.5)), let us define

$$P^{(k)} \equiv (H_k^T H_k)^{-1}.$$

From (3.10),

$$\hat{\theta}^{(k+1)} = P^{(k+1)}(H_k^T Z_k + h_{k+1} z_{k+1}). \tag{3.11}$$

Note that (3.5) implies that $[P^{(k)}]^{-1}\hat{\theta}^{(k)} = H_k^T Z_k$ for $k \geq p$.

We are now in a position to convert from the above batch forms for the θ estimate ((3.10) and (3.11)) and $P^{(k)}$ value to corresponding recursive forms. Consequently, we switch the notation from $\hat{\theta}^{(k)}$ to $\hat{\theta}_k$ and from $P^{(k)}$ to P_k. Although $\hat{\theta}_k$ and P_k are derived directly from the corresponding batch quantities, $\hat{\theta}^{(k)} \neq \hat{\theta}_k$ and $P^{(k)} \neq P_k$ in general. The discrepancies result from the need for user-specified initial conditions $\hat{\theta}_0$ and P_0 to initiate the recursive estimates. The effects of the initial conditions decay over time (see Exercise 3.8).

From the implied form for $P^{(k+1)}$ in (3.10) (i.e., the $(\cdot)^{-1}$ part; see Exercise 3.5), the corresponding recursive form is

$$P_{k+1}^{-1} = P_k^{-1} + h_{k+1}h_{k+1}^T. \tag{3.12}$$

Likewise, (3.11) can be rewritten in recursive form as

$$\begin{aligned}\hat{\theta}_{k+1} &= P_{k+1}(P_k^{-1}\hat{\theta}_k + h_{k+1}z_{k+1})\\ &= P_{k+1}[(P_{k+1}^{-1} - h_{k+1}h_{k+1}^T)\hat{\theta}_k + h_{k+1}z_{k+1}]\\ &= \hat{\theta}_k - P_{k+1}h_{k+1}(h_{k+1}^T\hat{\theta}_k - z_{k+1}),\end{aligned} \tag{3.13}$$

where (3.12) is used in the second line of (3.13).

Note that (3.13) nicely exposes the relationship between the new and old estimates for θ. The new estimate is equal to the old estimate plus an adjustment due to the new data $\{h_{k+1}, z_{k+1}\}$. The correction term is proportional to the prediction error $h_{k+1}^T\hat{\theta}_k - z_{k+1}$, where from (3.1) it is apparent that $h_{k+1}^T\hat{\theta}_k$ represents a prediction of z_{k+1}. In the idealized case of a perfect prediction, there would be no change in the subsequent θ estimate.

Although the above recursive formulas for $\hat{\theta}_k$ and P_k are sufficient to yield an RLS solution, the form in (3.12) for P_k is computationally undesirable since it involves the inverse of a $p\times p$ matrix when used in (3.13). A form based on the *matrix inversion lemma* (Appendix A, matrix relationship (xxii)) eliminates this requirement. Applying this lemma to (3.12) yields the following final recursions for the RLS algorithm based on model (3.1) (Exercise 3.6):

RLS Algorithm

$$P_{k+1} = P_k - \frac{P_k h_{k+1}h_{k+1}^T P_k}{1 + h_{k+1}^T P_k h_{k+1}}, \tag{3.14a}$$

$$\hat{\theta}_{k+1} = \hat{\theta}_k - P_{k+1}h_{k+1}(h_{k+1}^T\hat{\theta}_k - z_{k+1}). \tag{3.14b}$$

Initial conditions $\hat{\theta}_0$ and P_0 need to be specified to start the RLS recursion in (3.14a, b). A common means for initialization is to set $\hat{\theta}_0 = \mathbf{0}$ or some other vector reflecting prior knowledge about θ. For P_0, a common choice is cI_p for some $c > 0$. In particular, in the absence of prior knowledge suggesting another choice, $c >> 0$ is recommended in order to have the RLS solution and the batch solution be close to each other for small k (Young, 1984, p. 27; Jang et al., 1997, p. 115). Choosing some other form for P_0 is associated with cases where one has knowledge about θ that might suggest a value different from the "prior-free" batch solution. One alternative might arise if there is available a batch solution based on some data available *prior* to $\{\boldsymbol{h}_1, z_1\}$; $\hat{\boldsymbol{\theta}}_0$ and $\boldsymbol{P}_0$ may then be chosen based on this prior batch estimate. Another setting where an alternative prior might be useful is when the underlying true $\boldsymbol{\theta}$ varies in time; this is discussed in Section 3.3.

Expression (3.14b) has an intuitively appealing form. The recursion indicates that the new estimate $\hat{\boldsymbol{\theta}}_{k+1}$ is a linear combination of the old estimate $\hat{\boldsymbol{\theta}}_k$ and a correction term that is a weighting of the error in predicting the new measurement from the old $\boldsymbol{\theta}$ estimate (the error is $\boldsymbol{h}_{k+1}^T\hat{\boldsymbol{\theta}}_k - z_{k+1}$). This prediction error is sometimes called a *residual*.

In the case study of Section 3.4, which is based on real data for a complex process, we will see that the RLS and batch solutions also effectively reach the same points over the range of data available. In contrast, in some applications, there will be a slow drift in the data that produces a discrepancy between the true data-generating mechanism and the assumed model for batch estimation. In particular, the batch assumption of the noise being mean zero with a common variance may be violated. In cases with such a drift, it is likely that the RLS solution will tend to be better (for predictive purposes) since it puts more weight on recent measurements. Section 3.3 considers this setting in more detail.

It was mentioned in Subsection 3.1.2 that there is a close connection between the least-squares and minimum-mean-squared error approaches. Although RLS is usually presented (as above) as a recursive implementation of a least-squares algorithm (basically a deterministic algorithm), it also has an interpretation as *stochastic* optimization algorithms in the sense of using noisy input information to optimize the *mean-squared* error. The loss function can be changed to a mean-squared error criterion by taking the expectation of the right-hand side of (3.3). In some sense, this is what it *should* be since then we are trying to find the $\boldsymbol{\theta}$ that minimizes the cost over all possible outcomes (not just those that have actually been seen in the n observations). The input $\boldsymbol{h}_{k+1}(\boldsymbol{h}_{k+1}^T\hat{\boldsymbol{\theta}}_k - z_{k+1})$ on the right-hand side of (3.14b) represents a noisy measurement of the instantaneous gradient $\frac{1}{2}\partial E[(z_{k+1} - \boldsymbol{h}_{k+1}^T\boldsymbol{\theta})^2]/\partial\boldsymbol{\theta}$ (see (3.6)) weighted by the gain matrix $\boldsymbol{P}_{k+1}$.

Using stochastic approximation theory (see Chapter 4), conditions can be established such that the RLS algorithm converges to the minimum ($\boldsymbol{\theta}^*$) of the mean-squared error loss function (3.2) as the sample size $n \to \infty$. This may or may not change the actual mechanics of the algorithm, depending on whether the input–output pairs $\{\boldsymbol{h}_k, z_k\}$ satisfy the stochastic approximation conditions, but it does change the interpretation of the algorithm and provides for the invocation of asymptotic convergence and distribution theory available in stochastic approximation. Note also that this theory allows the design vectors (sometimes called *regressors*) $\boldsymbol{h}_k$ to be random. Macchi and Eweda (1983) and Ljung et al.

(1992, Part III) are two references that discuss the connection of RLS (and LMS) algorithms to stochastic approximation, establishing formal convergence in the process.

3.2.5 Connection of RLS to the Newton–Raphson Method

The RLS algorithm above has a close connection to the Newton–Raphson algorithm discussed in Section 1.4. As with LMS, this connection is via the instantaneous gradient (3.6). For convenience in the recursions of this subsection, let us index the least-squares criterion by k, writing $\hat{L}(\boldsymbol{\theta}, k)$ for $\hat{L}(\boldsymbol{\theta})$ $= (2k)^{-1}\sum_{j=1}^{k}(z_j - \boldsymbol{h}_j^T\boldsymbol{\theta})^2$.

Suppose that the RLS estimate $\hat{\boldsymbol{\theta}}_k$ approximately minimizes $\hat{L}(\boldsymbol{\theta}, k)$. (Because of the usually small discrepancy between the RLS solution and the batch solution, it cannot generally be assumed that $\hat{\boldsymbol{\theta}}_k$ exactly minimizes $\hat{L}(\boldsymbol{\theta}, k)$.) Then

$$\left.\frac{\partial \hat{L}(\boldsymbol{\theta}, k)}{\partial \boldsymbol{\theta}}\right|_{\boldsymbol{\theta}=\hat{\boldsymbol{\theta}}_k} \approx \mathbf{0}.$$

Because $\hat{L}(\boldsymbol{\theta}, k+1) = k\hat{L}(\boldsymbol{\theta}, k)/(k+1) + (z_{k+1} - \boldsymbol{h}_{k+1}^T\boldsymbol{\theta})^2/[2(k+1)]$, this approximation implies from (3.6),

$$\begin{aligned}\left.\frac{\partial \hat{L}(\boldsymbol{\theta}, k+1)}{\partial \boldsymbol{\theta}}\right|_{\boldsymbol{\theta}=\hat{\boldsymbol{\theta}}_k} &= \frac{k}{k+1}\left.\frac{\partial \hat{L}(\boldsymbol{\theta}, k)}{\partial \boldsymbol{\theta}}\right|_{\boldsymbol{\theta}=\hat{\boldsymbol{\theta}}_k} + \frac{\boldsymbol{h}_{k+1}(\boldsymbol{h}_{k+1}^T\hat{\boldsymbol{\theta}}_k - z_{k+1})}{k+1} \\ &\approx \frac{\boldsymbol{h}_{k+1}(\boldsymbol{h}_{k+1}^T\hat{\boldsymbol{\theta}}_k - z_{k+1})}{k+1}.\end{aligned}$$

Here, the gradient of the *instantaneous* part serves as a proxy for the gradient of the *overall* loss function in (3.3) (which is a sum over all measurements). Further, by (3.12), the Hessian of $\hat{L}(\boldsymbol{\theta}, k+1)$ is given by

$$\begin{aligned}\text{Hessian}_{k+1} &= \frac{\partial^2 \hat{L}(\boldsymbol{\theta}, k+1)}{\partial \boldsymbol{\theta}\partial \boldsymbol{\theta}^T} = \frac{1}{k+1}\sum_{j=1}^{k+1}\boldsymbol{h}_j\boldsymbol{h}_j^T \\ &= \frac{1}{k+1}(\boldsymbol{P}_{k+1}^{-1} - \boldsymbol{P}_0^{-1}),\end{aligned}$$

where the second line above follows by solving the difference equation in (3.12) in terms of $\boldsymbol{P}_0^{-1}$ and the sum of $\boldsymbol{h}_j\boldsymbol{h}_j^T$ over $j = 1$ to $k+1$. Hence, if we assume that $\boldsymbol{P}_0^{-1}$ is small relative to $\boldsymbol{P}_{k+1}^{-1}$ (usually reasonable given (3.12)), then, from the RLS recursion in (3.13),

$$\begin{aligned}\hat{\boldsymbol{\theta}}_{k+1} &= \hat{\boldsymbol{\theta}}_k - \boldsymbol{P}_{k+1}\boldsymbol{h}_{k+1}(\boldsymbol{h}_{k+1}^T\hat{\boldsymbol{\theta}}_k - z_{k+1}) \\ &\approx \hat{\boldsymbol{\theta}}_k - [\text{Hessian}_{k+1}]^{-1}\left.\frac{\partial \hat{L}(\boldsymbol{\theta}, k+1)}{\partial \boldsymbol{\theta}}\right|_{\boldsymbol{\theta}=\hat{\boldsymbol{\theta}}_k}.\end{aligned}$$

Hence, one step of RLS is closely related to one step of the Newton–Raphson algorithm. The discrepancy (hence the $\approx$) is due to the approximations involving the gradient of $\hat{L}(\boldsymbol{\theta}, k)$ being nearly $\boldsymbol{0}$ and $\boldsymbol{P}_0^{-1}$ being relatively small. This close connection between RLS and Newton–Raphson helps explain the widespread observation of fast convergence in practice for RLS.

3.2.6 Extensions to Multivariate RLS and Weighted Summands in Least-Squares Criterion

The RLS form in (3.14a, b) readily extends to multivariate z_k (say, m-dimensional) with a least-squares criterion that is possibly weighted differently across k. In particular, suppose that the multivariate model for output z_k is

$$z_k = \overline{\boldsymbol{H}}_k\boldsymbol{\theta} + \boldsymbol{v}_k, \tag{3.15}$$

where $\overline{\boldsymbol{H}}_k$ is an $m \times p$ design matrix and $\boldsymbol{v}_k$ is a mean-zero noise vector with (possibly varying with k) full-rank covariance matrix $\boldsymbol{\Sigma}_k$. Consider the weighted least-squares criterion

$$\hat{L}(\boldsymbol{\theta}) = \frac{1}{2n}\sum_{k=1}^{n}(z_k - \overline{\boldsymbol{H}}_k\boldsymbol{\theta})^T\boldsymbol{\Sigma}_k^{-1}(z_k - \overline{\boldsymbol{H}}_k\boldsymbol{\theta}), \tag{3.16}$$

where the weighting $\boldsymbol{\Sigma}_k^{-1}$ makes the expected value of each summand in (3.16) identical (= 1) when the model (3.15) is an accurate description of the output z_k. Criterion (3.16) retains the $\hat{L}(\boldsymbol{\theta})$ notation of (3.3) (versus $L(\boldsymbol{\theta})$) to emphasize that this least-squares criterion is a noisy representation of the corresponding mean-squared criterion.

Multivariate analogues of the arguments yielding (3.14a, b) can be used to derive the RLS algorithm. In converting the batch solution to a recursive solution, the matrix inversion lemma (Appendix A, matrix relationship (xxii)) is used (e.g., Sorenson, 1980, pp. 56–57; Ljung, 1999, pp. 366–367). The resulting RLS algorithm is:

RLS Algorithm—Multivariate Output with Varying Noise Covariance Matrix

$$P_{k+1} = P_k - P_k \bar{H}_{k+1}^T (\Sigma_{k+1} + \bar{H}_{k+1} P_k \bar{H}_{k+1}^T)^{-1} \bar{H}_{k+1} P_k, \tag{3.17a}$$

$$\hat{\theta}_{k+1} = \hat{\theta}_k - P_{k+1} \bar{H}_{k+1}^T \Sigma_{k+1}^{-1} (\bar{H}_{k+1} \hat{\theta}_k - z_{k+1}). \tag{3.17b}$$

A comparison of the above with the scalar version in (3.14a, b) shows a strong resemblance. The primary differences are the substitution of the matrix $\bar{H}_k$ for the vector h_k^T and the inclusion of the covariance matrix Σ_k to account for the weighting in the criterion (3.16). If one has scalar measurements z_k, but time-varying noise variances, say σ_k^2, then, analogous to (3.16), the criterion in (3.3) may be changed to

$$\begin{aligned}\hat{L}(\theta) &= \frac{1}{2n} \sum_{k=1}^{n} \sigma_k^{-2} (z_k - h_k^T \theta)^2 \\ &= \frac{1}{2n} (Z_n - H_n \theta)^T \begin{bmatrix} \sigma_1^{-2} & 0 & 0 & 0 \\ 0 & \sigma_2^{-2} & 0 & 0 \\ 0 & 0 & \ddots & 0 \\ 0 & 0 & 0 & \sigma_n^{-2} \end{bmatrix} (Z_n - H_n \theta).\end{aligned}$$

Consequently, the recursions in (3.17a, b) apply directly with σ_k^2 replacing Σ_k and h_k^T replacing $\bar{H}_k$. As before, initial conditions $\hat{\theta}_0$ and P_0 need to be specified to start the RLS recursion in (3.17a, b). Also, as before, the recursion for θ has an intuitively appealing form by being a linear combination of the old estimate and the weighted residual error.

3.3 LMS, RLS, AND KALMAN FILTER FOR TIME-VARYING θ

3.3.1 Introduction

In many applications of linear models, the underlying θ evolves in time. Typical applications of time-varying θ include target tracking, adaptive control, time-series forecasting, sequential design and response surface methods (Chapter 17), and signal processing. This section discusses some popular approaches to this estimation problem.

The model considered here is based on scalar measurements $z_1, z_2, \ldots$ obtained according to the following form:

$$z_k = h_k^T \theta_k + v_k, \quad k = 1, 2, \ldots, n, \tag{3.18}$$

where $\boldsymbol{h}_k$ is the $p \times 1$ design vector, v_k is the random noise term, and $\boldsymbol{\theta}_k$ is the *time-varying* true (unknown) value of the parameter vector of interest. In general, the variation in $\boldsymbol{\theta}_k$ will be stochastic in nature, but this is not necessarily the case. Similar to Section 3.2, our interest is in taking the input–output pairs $\{\boldsymbol{h}_k, z_k\}$ and estimating $\boldsymbol{\theta}_k$.[1] Of course, the problem differs fundamentally from that in Sections 3.1 and 3.2 in that $\boldsymbol{\theta}_k$ may be constantly changing. Hence, in contrast to the discussion of Sections 3.1 and 3.2, the standard batch regression solution (eqn. (3.5)) for a fixed $\boldsymbol{\theta}$ is an inappropriate starting place for developing recursive forms.

Given the time variation in $\boldsymbol{\theta}_k$, the optimization problem is also time varying in that the aim is to minimize some time-varying loss L_k or associated least-squares estimate $\hat{L}_k$ that expresses, in some sense, the desire to pick an estimate $\hat{\boldsymbol{\theta}}_k$ that is close to the true value of $\boldsymbol{\theta}_k$. We discuss below three common approaches to such estimation: LMS, RLS, and the Kalman filter. Many books and articles consider one or more of these approaches in the context of time-varying estimation. Among such references are Anderson and Moore (1979), Sorenson (1980), Spall (1988a), Haykin (1996), Ljung (1999), and Moon and Stirling (2000).

The prototype recursive form for time-varying parameter estimation is

$$\hat{\boldsymbol{\theta}}_{k+1} = \boldsymbol{A}_k\hat{\boldsymbol{\theta}}_k - \boldsymbol{\kappa}_{k+1}(\boldsymbol{h}_{k+1}^T\boldsymbol{A}_k\hat{\boldsymbol{\theta}}_k - z_{k+1}), \tag{3.19}$$

where $\boldsymbol{A}_k$ is a potentially time-varying $p \times p$ matrix and $\boldsymbol{\kappa}_k$ ("kappa") is some appropriately chosen $p \times 1$ gain vector. The specific forms for $\boldsymbol{A}_k$ and $\boldsymbol{\kappa}_k$ depend on the choice of LMS, RLS, or the Kalman Filter. Each of the three subsections below discusses the essential assumptions of one of these methods, the algorithmic form vis-à-vis (3.19), and the core theoretical results in support of the method.

Unlike the fixed parameter case emphasized in Sections 3.1 and 3.2, it is usually impossible for an estimation algorithm to formally converge to a time-varying solution in the sense that $\hat{\boldsymbol{\theta}}_k - \boldsymbol{\theta}_k$ goes to zero in a probabilistic manner, as discussed in Appendix C.[2] In essence, the algorithm is not able to acquire enough information about any one $\boldsymbol{\theta}_k$. Hence, the theory discussed below pertains to *other* aspects of the algorithm performance, such as bounds on the mean-squared error of the parameter estimate.

---

[1]Following the precedent of using $\boldsymbol{\theta}^*$ for an optimal value, we could also write $\boldsymbol{\theta}_k^*$ for the true parameter $\boldsymbol{\theta}_k$ of interest. However, we write $\boldsymbol{\theta}_k$ for ease of notation and because the time-varying solution is often a random quantity, unlike the *deterministic* quantity $\boldsymbol{\theta}^*$.

[2]The qualifier "usually" is included because there are rare cases where extensive prior information is available on the evolution of the parameter, with the algorithm able to encompass this information in a manner that allows formal convergence.

3.3.2 LMS

Model Assumptions

There are minimal statistical assumptions made about the evolution of the process $\boldsymbol{\theta}_k$ (which may be random or deterministic). For example, a typical assumption is that the increments $\boldsymbol{\theta}_{k+1} - \boldsymbol{\theta}_k$ are uncorrelated across k and have finite-magnitude covariance matrix. The aim with LMS is to minimize the instantaneous criterion $L_k(\boldsymbol{\theta}) = \frac{1}{2}E[(z_k - \boldsymbol{h}_k^T\boldsymbol{\theta})^2]$ as k evolves (minimizing L_k yields the ideal solution $\boldsymbol{\theta} = \boldsymbol{\theta}_k$). LMS allows for $\boldsymbol{h}_k$ to be random with the expectation in $L_k(\boldsymbol{\theta})$ being an average over the noise together with the randomness in $\boldsymbol{h}_k$.

Algorithm Terms for Use in (3.19)

In LMS, A_k in (3.19) is set to $\boldsymbol{I}_p$ for all k. The standard unnormalized LMS algorithm has a gain vector of

$$\boldsymbol{\kappa}_{k+1} = a\boldsymbol{h}_{k+1}, \tag{3.20a}$$

where $a > 0$ is a constant (nondecaying) scale factor useful in tracking problems. An often-used normalized variation on the basic form above is

$$\boldsymbol{\kappa}_{k+1} = \frac{a\boldsymbol{h}_{k+1}}{1 + a\|\boldsymbol{h}_{k+1}\|}, \tag{3.20b}$$

which sometimes helps to adapt to time variation in the system by scaling for different magnitudes in the design vector $\boldsymbol{h}_{k+1}$ (as usual, $\|\cdot\|$ denotes the standard Euclidean norm).

Discussion and Summary of Theory

Because LMS is an analogue of first-order steepest descent, it is generally easier to implement (in either the unnormalized or normalized form) than RLS described below (an analogue of Newton-Raphson). Further, because LMS only pertains to the instantaneous loss $L_k(\boldsymbol{\theta}) = \frac{1}{2}E[(z_k - \boldsymbol{h}_k^T\boldsymbol{\theta})^2]$, it tends to have less inertia than RLS. (As shown below, RLS works with a loss function representing a weighted sum over all time points up to the current time.) Some theory for this time-varying $\boldsymbol{\theta}_k$ case is presented in Guo and Ljung (1995) and Delyon and Juditsky (1995). These results pertain to the error $E\left(\|\hat{\boldsymbol{\theta}}_k - \boldsymbol{\theta}_k\|^2\right)$ and to the approximate distribution of $\hat{\boldsymbol{\theta}}_k - \boldsymbol{\theta}_k$ for large k and small a.

3.3.3 RLS

Model Assumptions

As with LMS, there are minimal statistical assumptions made about the evolution of the process $\boldsymbol{\theta}_k$. However, to reflect the greater importance of current measurements z_k relative to older measurements, the standard least-squares criterion in (3.3) is altered to assign greater weight to the new information. In particular, the RLS algorithm for time-varying $\boldsymbol{\theta}$ is tied to a criterion of the form

$$\hat{L}(\boldsymbol{\theta}, k) = \frac{1}{2k}\sum_{j=1}^{k} \gamma^{k-j}(z_j - \boldsymbol{h}_j^T\boldsymbol{\theta})^2 \tag{3.21}$$

for $0 < \gamma \leq 1$. Note that (3.21) exponentially de-weights past measurements.

Algorithm Terms for Use in (3.19)

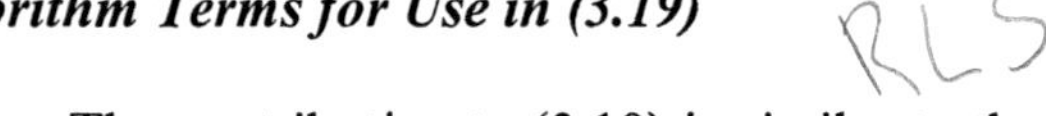

The contribution to (3.19) is similar to the constant-parameter solution in Subsection 3.2.4. In particular, $\boldsymbol{A}_k$ in (3.19) is set to $\boldsymbol{I}_p$ for all k and

$$\boldsymbol{\kappa}_{k+1} = \boldsymbol{P}_{k+1}\boldsymbol{h}_{k+1}, \tag{3.22a}$$

$$\boldsymbol{P}_{k+1}^{-1} = \gamma\boldsymbol{P}_k^{-1} + \boldsymbol{h}_{k+1}\boldsymbol{h}_{k+1}^T, \tag{3.22b}$$

where $\boldsymbol{P}_0 > 0$. The efficiency of the above form can be improved by avoiding the required matrix inverse to obtain the $\boldsymbol{P}_{k+1}$ appearing in $\boldsymbol{\kappa}_{k+1}$. A preferred (equivalent) form for implementation based on the matrix inversion lemma (Appendix A, matrix relationship (xxii)) is

$$\boldsymbol{P}_{k+1} = \frac{1}{\gamma}\left[\boldsymbol{P}_k - \frac{\boldsymbol{P}_k\boldsymbol{h}_{k+1}\boldsymbol{h}_{k+1}^T\boldsymbol{P}_k}{\gamma + \boldsymbol{h}_{k+1}^T\boldsymbol{P}_k\boldsymbol{h}_{k+1}}\right]. \tag{3.22c}$$

Discussion and Summary of Theory

The weighting of past input information in the RLS loss function tends to create an "inertia" that will slow down the variations in the $\boldsymbol{\theta}_k$ estimate from time point to time point; as γ increases from 0 to 1, this weighting of earlier points increases. When $\gamma = 1$, the solution reduces to the "standard" (no damping) RLS solution in (3.14a, b) where all previous points are weighted equally (i.e., the greatest inertia). Note that no matrix inverse is required to compute $\boldsymbol{\kappa}_{k+1}$ when using the form in (3.22c). So, (3.22a) and (3.22c) together (versus (3.22a) and (3.22b)) constitute the generally preferred RLS algorithm to accommodate time-varying $\boldsymbol{\theta}_k$. While the standard RLS algorithm assumes deterministic $\boldsymbol{h}_k$, some theory exists for random $\boldsymbol{h}_k$ in this time-varying $\boldsymbol{\theta}_k$ setting. In particular, Ljung et al. (1992, pp. 95–113) and Guo and Ljung (1995) provide bounds on the mean-squared tracking error $E\left(\|\hat{\boldsymbol{\theta}}_k - \boldsymbol{\theta}_k\|^2\right)$ when the inputs $\boldsymbol{h}_k$ are random and satisfy certain conditions, such as independence and bounded second moments across k.

3.3.4 Kalman filter

Model Assumptions

Unlike the LMS and RLS approaches above, the Kalman filter (Kalman, 1960) is based on a precise statistical representation for the evolution of $\boldsymbol{\theta}_k$. This representation, called a *state equation*, is given by the expression

$$\boldsymbol{\theta}_{k+1} = \boldsymbol{\Phi}_k \boldsymbol{\theta}_k + \boldsymbol{w}_k, \tag{3.23}$$

where $\boldsymbol{\Phi}_k$ is the transition matrix and $\boldsymbol{w}_k$ is a mean-zero noise vector having covariance matrix $\boldsymbol{Q}_k$. Hence, from (3.23), the evolution of $\boldsymbol{\theta}_k$ is via a first-order vector difference equation with random input. In the Kalman filter literature, (3.23) and (3.18) together constitute the *state-space model* for the process. In general, the state-space model and Kalman filter allows for multivariate measurements z_k (as in (3.15)), but for consistency with the discussion of RLS and LMS, we focus on the scalar measurement case.

Algorithm Terms for Use in (3.19)

The algorithm makes use of the state-space model terms, including σ_k^2 = $\text{var}(v_k)$ and $\boldsymbol{Q}_k = \text{cov}(\boldsymbol{w}_k)$. In particular, $\boldsymbol{A}_k = \boldsymbol{\Phi}_k$ for all k and

$$\boldsymbol{\kappa}_{k+1} = \frac{(\boldsymbol{\Phi}_k \boldsymbol{P}_k \boldsymbol{\Phi}_k^T + \boldsymbol{Q}_k)\boldsymbol{h}_{k+1}}{\sigma_{k+1}^2 + \boldsymbol{h}_{k+1}^T(\boldsymbol{\Phi}_k \boldsymbol{P}_k \boldsymbol{\Phi}_k^T + \boldsymbol{Q}_k)\boldsymbol{h}_{k+1}}, \tag{3.24a}$$

$$\boldsymbol{P}_{k+1} = (\boldsymbol{I}_p - \boldsymbol{\kappa}_{k+1}\boldsymbol{h}_{k+1}^T)(\boldsymbol{\Phi}_k \boldsymbol{P}_k \boldsymbol{\Phi}_k^T + \boldsymbol{Q}_k), \tag{3.24b}$$

where $\sigma_k^2 > 0$, $\boldsymbol{P}_0 \geq \boldsymbol{0}$, and $\boldsymbol{Q}_k \geq \boldsymbol{0}$. (With multivariate measurements z_k, the gain $\boldsymbol{\kappa}_{k+1}$ becomes a matrix and the denominator in (3.24a) is replaced by an analogous matrix inverse; see any reference on Kalman filtering.)

Discussion and Summary of Theory

The Kalman filter form above is a special case of several more general forms. Aside from multivariate measurements, the more general forms encompass correlation between the process and measurement noises. The form above *does* allow for one significant extension, in that time-varying state-space parameters are allowed ($\boldsymbol{\Phi}_k$, $\boldsymbol{Q}_k$, σ_k^2). The Kalman filter also has important statistical properties. One is unbiasedness in the sense that $E(\hat{\boldsymbol{\theta}}_k - \boldsymbol{\theta}_k) = \boldsymbol{0}$; another is that $\boldsymbol{P}_k$ in (3.24b) is equal to $E\left[(\hat{\boldsymbol{\theta}}_k - \boldsymbol{\theta}_k)(\hat{\boldsymbol{\theta}}_k - \boldsymbol{\theta}_k)^T\right]$ (hence $\boldsymbol{P}_k$ is called the error-covariance matrix). Note that $\boldsymbol{P}_k$ is an essential quantity in assessing the filter accuracy. Both the unbiasedness and error-covariance interpretation of $\boldsymbol{P}_k$ depend critically on the validity of the state-space model (including parameter values $\hat{\boldsymbol{\theta}}_0$, $\boldsymbol{P}_0$, $\boldsymbol{\Phi}_k$, $\boldsymbol{Q}_k$, σ_k^2).

As with RLS and LMS, the Kalman filter is a *linear* estimator in the sense that (3.19) yields a $\hat{\boldsymbol{\theta}}_k$ that is a linear combination of all previous measurements z_j, $j \le k$. The Kalman filter minimizes the time-varying MSE $E\left(\left\|\hat{\boldsymbol{\theta}}_k - \boldsymbol{\theta}_k\right\|^2\right)$ at each k among all possible linear estimators (i.e., among all estimators that are linear combinations of the measurements). It is *not* generally the minimum MSE estimator when one allows for estimators that are nonlinear functions of the measurements (see, e.g., Anderson and Moore, 1979, pp. 46–49; Spall, 1988a, pp. xx–xxi). However, under the *additional* assumption that the initial state, measurement noises, and process noises are independent normally distributed processes, the Kalman filter minimizes the time-varying MSE $E\left(\left\|\hat{\boldsymbol{\theta}}_k - \boldsymbol{\theta}_k\right\|^2\right)$ among *all* estimators (linear or nonlinear).

Because RLS and LMS do not require the assumption of an underlying state equation for $\boldsymbol{\theta}_k$, it would not be expected that those methods would track as well as the Kalman filter *if* a state-space model were available. The LMS and RLS algorithms require a gain a or weight γ to be user specified, while the Kalman filter has all parameters other than $\hat{\boldsymbol{\theta}}_0$ and $\boldsymbol{P}_0$ uniquely specified by the assumptions of the state-space model. Hence, one cannot say that any one of these tracking algorithms is uniformly better than the others since the best choice depends on the assumptions governing the parameter variation (see, e.g., Ljung et al., 1992, pp. 96–101, and Guo and Ljung, 1995, for further comparative discussion).

3.4 CASE STUDY: ANALYSIS OF OBOE REED DATA

...an ill wind that nobody blows good.

—Comedian Danny Kaye, paraphrasing a proverb of English playwright John Heywood (approx. 1497–1580), in speaking of the oboe in "The Secret Life of Walter Mitty" (1947).

This section applies some of the recursive estimation algorithms above to the estimation of models associated with a musical instrument. The oboe is a medium-to-high-pitched wind instrument that has long played a central role in symphony orchestras. Its nasal tone and piercing quality make it one of the most distinctive of the orchestral sounds. Like the lower-pitched bassoon, the oboe is a double-reed instrument, so called because the musician blows into a reed separated into two parts that vibrate against one another to produce a sound. The quality of the reed is absolutely critical to the sound quality produced by the oboe. Figure 3.3 shows an oboe with an expanded display of the reed.

A fact unknown to many concertgoers is that almost all professional oboists make their own reeds. This can be a tedious and lengthy process of shaping a treated piece of cane (*Arundo donax*) into a reed that attaches to the oboe and meets stringent requirements for playing quality (e.g., easy articulation of notes and the ability to play in tune with other instruments). It is common for an oboist to spend several hours making a reed that lasts through only one concert or rehearsal, or, even worse, is ultimately unacceptable and must be discarded before any use in a concert or rehearsal.

Ceasar-Spall and Spall (1997) (CS&S) report on a study involving the application of statistical regression analysis to predicting the ultimate quality of a reed. The essential idea is to evaluate characteristics of a reed early in the reed-making process and make a prediction as to whether the reed will ultimately be successful. If the prediction indicates that the reed is likely to be unacceptable, the oboist may wish to stop working on that reed. Six input variables and one output variable are described in CS&S. Five of the input variables pertain to inherent qualities of the cane. These are termed "top close" (T), "appearance"

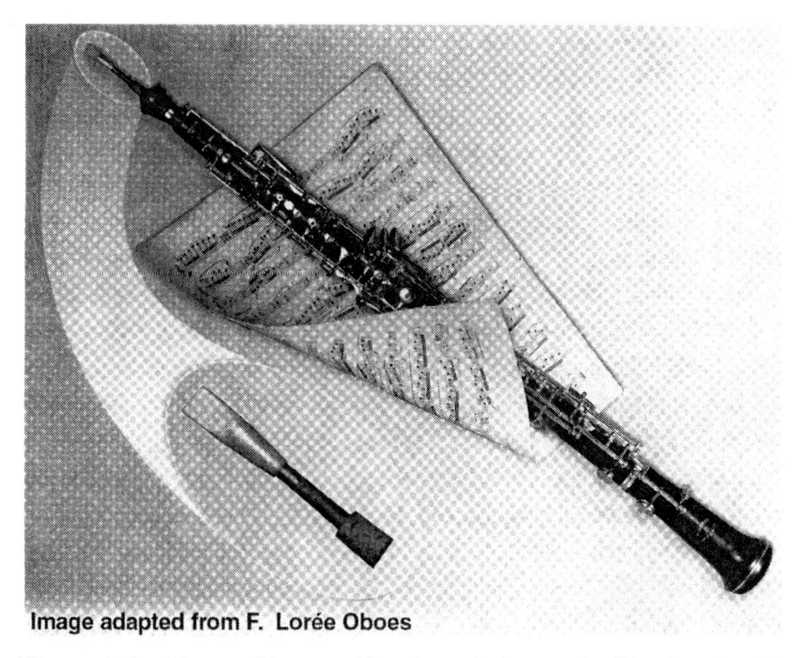
Image adapted from F. Lorée Oboes

Figure 3.3. Oboe with magnification of the reed. (Reprinted with permission from the *Journal of Testing and Evaluation*, July 1997, copyright American Society for Testing and Materials.)

(A), "ease of gouge" (E), "vasculur bundles" (V), and "shininess" (S). The sixth input variable, "first blow" (F), pertains to the oboist's impression of the reed after an initial tryout about one-fourth of the way through the total reed-making process. The output variable (z_k) is a measure of quality of a finished reed.

Of course, the ratings assigned for each of the input and output variables represent a subjective impression of the oboist; this is unavoidable in such a process. The input and output variables are rated on a scale of 0 to 2, with 2 representing a greater degree of the attribute in question (so a rating of $z_k = 2$ indicates that the kth reed is of top quality). The physical interpretation of the variables and other aspects of the analysis are thoroughly described in CS&S, but the summary above is sufficient for our purposes here.

Let us focus on the use of RLS for two of the four models considered in CS&S (Exercises 3.16 and 3.17 consider LMS). The first is a standard linear model

$$z = \theta_{\text{const}} + \theta_T T + \theta_A A + \theta_E E + \theta_V V + \theta_S S + \theta_F F + v, \tag{3.25}$$

and the second is a curvilinear model

$$z = \theta_F F + \theta_{AE} AE + \theta_{TF} TF + \theta_{VS} VS + v, \tag{3.26}$$

where for convenience we have suppressed the subscript k on the input, output, and noise variables. Each of the $\boldsymbol{\theta}$ elements is labeled according to the input variable to which it is attached, with θ_{const} representing the additive constant. The corresponding parameter vector $\boldsymbol{\theta}$ for model (3.25) has seven elements (θ_{const}, θ_T, etc.), while for model (3.26) it has four elements (θ_F, θ_{AE}, etc.). For reasons of parsimony (fewer elements in $\boldsymbol{\theta}$), predictive ability, and connections to the underlying aspects of the physical process, CS&S recommend the model form in (3.26) as the preferred form from the four models considered.

Given the highly subjective nature of the process, it cannot be expected that any mathematical model will fully (or even largely) capture the input–output relationship. Using the traditional statistical measure of the amount of variability explained by a linear regression model (the coefficient of multiple determination, R^2 in the vernacular), CS&S find that about *one-third* of the variation in outputs can be explained by the models, leaving about two-thirds to unexplained phenomena or inherent randomness. (R^2 is the square of a measure of linear association between z and the input variables; see, e.g., Montgomery and Runger, 1999, pp. 521–522.) This is an example of a hard-to-model process that is typical of many in the humanities, business, and social sciences, where the human element is a significant part of the process. Tests are also performed in CS&S on an independent data set (different from that used to fit the models) and evidence is found of reasonable predictive ability for the models, but, of course, there are nontrivial uncertainties in the predictions.

Tables 3.1 and 3.2 present results comparing the batch and RLS solutions for 160 input–output points available from the fitting set, **reeddata-fit** (available at the book's Web site). These 160 data points are different from the fitting data used in CS&S. Initial values $\hat{\boldsymbol{\theta}}_0 = \mathbf{0}$ and $\boldsymbol{P}_0 = \boldsymbol{I}_4$ or $\boldsymbol{I}_7$ (as appropriate) are used for the RLS solutions in the tables.

Let us now compare the predictive ability of the four models in Tables 3.1 and 3.2. An independent data set of 80 input-output values, **reeddata-test**, is used for comparing the sample mean of the absolute prediction deviations (the sample mean of the 80 values $|\boldsymbol{h}_k^T \hat{\boldsymbol{\theta}}_{160} - z_k|$, where $\hat{\boldsymbol{\theta}}_{160}$ represents one of the four sets of parameter estimates from Tables 3.1 and 3.2; this is sometimes called the mean absolute deviation [MAD]). Table 3.3 is a summary of the mean prediction error for the four models.

A standard statistical test for comparing two means with matched samples (Appendix B, expression (B.3a)) can be used to determine whether the differences in the mean prediction errors in Table 3.3 are significant. This test is based on comparing the sample means of the absolute prediction errors for each of the 80 samples (the means of $|\boldsymbol{h}_k^T \hat{\boldsymbol{\theta}}_{160} - z_k|$). In each comparison of models

Table 3.1. Comparison of batch and RLS estimates for $\boldsymbol{\theta}$ parameters in basic linear model (3.25).

$\boldsymbol{\theta}$ parameters	Batch Model (3.25)	RLS Model (3.25)
θ_{const}	−0.156	−0.079
θ_T	0.102	0.101
θ_A	0.055	0.046
θ_E	0.175	0.171
θ_V	0.044	0.043
θ_S	0.056	0.056
θ_F	0.579	0.540

Table 3.2. Comparison of batch and RLS estimates for $\boldsymbol{\theta}$ parameters in curvilinear model (3.26).

$\boldsymbol{\theta}$ parameters	Batch Model (3.26)	RLS Model (3.26)
θ_F	0.584	0.557
θ_{AE}	0.101	0.106
θ_{TF}	0.078	0.086
θ_{VS}	0.034	0.036

Table 3.3. Mean and median absolute prediction errors for the four models from Tables 3.1 and 3.2.

	Batch Model (3.25)	RLS Model (3.25)	Batch Model (3.26)	RLS Model (3.26)
Mean	0.242	0.242	0.236	0.235
Median	0.243	0.250	0.227	0.224

from Table 3.3, it is enlightening to compute the *P*-value of the test statistic (the probability of seeing an outcome at least as extreme as that observed if there is, in fact, no underlying difference between the predictive abilities of the models). For example, the one-sided *P*-value in comparing the RLS/(3.26) model (fifth column in Table 3.3) with the RLS/(3.25) model (third column in Table 3.3) is 0.103. This value is moderately small, but is not typically considered small enough to conclude that there is a statistically significant difference between those two models (see Exercise 3.18).

3.5 CONCLUDING REMARKS

This has been the first of several chapters devoted to the study of search algorithms that are stochastic analogues of the deterministic steepest descent and Newton–Raphson algorithms. This chapter focused on estimation where the underlying system model has a linear form in $\boldsymbol{\theta}$, leading to loss functions that are quadratic in the parameters $\boldsymbol{\theta}$.

Methods of the type in this chapter are among the most widely used stochastic search and optimization techniques. As any quick Internet search will reveal, countless industrial and other systems rely on LMS, RLS, or the Kalman filter. These methods typically represent a reasonable balance between generality and ease of implementation. While the underlying linear models may be restrictive in some sense, they are general enough to apply to many practical systems, especially when care is used in defining the key variables (see Sections 13.1 and 13.2) and when the range of operation is restricted to values reasonably close to some nominal operating point. Further, the linearity provides significant advantages in implementation with respect to both derivations (e.g., easy gradient calculations) and numerical stability. Results in matrix theory (such as matrix factorizations) can be used to create numerically stable versions of methods such as the Kalman filter (e.g., Moon and Stirling, 2000, pp. 604–605).

Nonetheless, there are also a large number of problems where nonlinearity is significant to the point where the linear-based methods of this chapter do not apply. Chapters 4–10 discuss general stochastic search and optimization methods that apply in nonlinear systems.

EXERCISES

3.1 Given that the unknown true process has a form identical to the model in (3.1), show that the batch least-squares solution (3.5) is an unbiased estimator and derive its covariance matrix. Assume that the v_k are independent, identically distributed (i.i.d.) with mean 0 and variance σ^2.

3.2 Suppose that data are generated according to an AR process $x_{k+1} = -0.99x_k + w_k$, where w_k is i.i.d. $N(0, 1)$ and $x_0 = 0$. Based on a model of the same AR form as the true process with an unknown β replacing the constant -0.99, use 500 iterations of LMS to estimate β when $z_k = x_k$ for all k. Use $\hat{\theta}_0 = 0$ and $a = 0.005$.

(a) Produce a plot showing one realization of estimates for β.

(b) Give the sample mean and approximate 95 percent confidence interval for the terminal estimate for β over 50 independent realizations (each of 500 iterations).

3.3 Suppose that random $\boldsymbol{h}_k$ are i.i.d. with mean $[2, 1]^T$ and covariance matrix $\begin{bmatrix} 3 & 1 \\ 1 & 1 \end{bmatrix}$. Based on the informal bound in Subsection 3.2.2, give an upper bound to the gain a to ensure that $E\left(\left\|\hat{\boldsymbol{\theta}}_k - \boldsymbol{\theta}^*\right\|^2\right) \approx 0$ in LMS for sufficiently large k.

3.4 Consider the adaptive control problem of Example 3.4, where the model and true process have the same AR form, the unknown true values of β and γ are 0.9 and 0.5, respectively, and w_k is i.i.d. $N(0, 0.25^2)$. Given $x_0 = 0$, $\hat{\boldsymbol{\theta}}_0 = [1, 1]^T$, $d_k = \sin(\pi k/10)$, and a value of a of your choosing:

(a) Plot one realization of estimates for $\boldsymbol{\theta}$ as a function of $k = 0, 1, \ldots, 200$.

(b) Give the sample means for the terminal estimate for β and γ over 50 independent replications (each of 200 iterations).

3.5 Fill in the steps in going from (3.10) to (3.12) in the derivation of the basic RLS algorithm.

3.6 Use the matrix inversion lemma (Appendix A) to derive (3.14a) from (3.12).

3.7 Fill in the steps in deriving the multivariate form of the RLS algorithm in (3.17a, b).

3.8 Using (3.1) and assuming a true model having $p = 2$, $\boldsymbol{h}_k = [k, 1]^T$, $\boldsymbol{\theta}^* = [1, 2]^T$, and $v_k \sim N(0, 1)$, run some numerical cases comparing batch regression solutions to RLS solutions as a function of n. What is the sensitivity to $\hat{\boldsymbol{\theta}}_0$, $\boldsymbol{P}_0$, and n?

3.9 Suppose that $\boldsymbol{h}_k = [1, 1]^T$ for all k. Is RLS in (3.14a, b) implementable? If so, discuss reasons to doubt the validity of the solution. (Note that the batch least-squares solution does not exist.)

3.10 Assume the true model in Exercise 3.8, except that $\text{var}(v_k) = 1$ for k even and $\text{var}(v_k) = c$ for k odd, $c >> 1$. Compare the estimation accuracy of the unweighted RLS in (3.14a, b) with the weighted version of RLS in (3.17a, b). In particular, how do the algorithms compare in estimating $\boldsymbol{\theta}$? Try varying n, c, and the initial conditions (try $c = 100$ and 1000).

3.11 Derive the efficient matrix RLS form (3.22c) from the "basic" form (3.22b).

3.12 Explain in intuitive terms the types of problems for which each of RLS, LMS, and the Kalman filter are best suited. What types of applications are most appropriate for each of the approaches?

3.13 For a system with constant state vector and scalar measurements, comment on the connection between the RLS solution in (3.14a, b) and the Kalman filter. How do the terms in (3.14a, b) relate to the Kalman gain and error covariance matrix?

3.14 Consider a state-space model with $\boldsymbol{P}_0 = \boldsymbol{0}$, $\boldsymbol{Q}_0 = \begin{bmatrix} 1 & \rho \\ \rho & 1 \end{bmatrix}$, and $\boldsymbol{h}_1 = [0, 1]^T$. Based on the diagonal elements of $\boldsymbol{P}_1$, comment in qualitative terms on the accuracy of the Kalman filter estimate for both elements of $\boldsymbol{\theta}_1$ as ρ approaches 0 or 1 and as σ_1^2 approaches 0 or ∞.

3.15 Reproduce the numbers in Table 3.1 using the data set **`reeddata-fit`** (from the book's Web site).

3.16 Estimate the four parameters in the curvilinear model (3.26) using basic LMS (eqn. (3.7)) with the data set **`reeddata-fit`**. Use $\hat{\boldsymbol{\theta}}_0 = \boldsymbol{0}$ as in the RLS studies and let $a = 0.005$, 0.05, and 0.10 (so you produce three $\boldsymbol{\theta}$ estimates, one for each a). Comment on the differences in the values produced by LMS and the values produced by RLS as given in Table 3.2.

3.17 In the setting of Exercise 3.16 (same $\hat{\boldsymbol{\theta}}_0$ and a), use the normalized LMS algorithm given in (3.19) and (3.20b). Comment on differences in the results from the basic LMS algorithm if you performed Exercise 3.16.

3.18 Calculate the one-sided P-value (using the matched-pairs t-test in Appendix B) for comparing the mean absolute prediction error of models batch/(3.25) and batch/(3.26) as given in the second and fourth columns of Table 3.3. Use the data set **`reeddata-test`** (from the book's Web site).

CHAPTER 4

STOCHASTIC APPROXIMATION FOR NONLINEAR ROOT-FINDING

We now broaden our horizons considerably from the linear setting of Chapter 3 and consider nonlinear systems. This is the first of four chapters that focus on a core approach in nonlinear stochastic search and optimization—*stochastic approximation* (SA). As might be expected, the problem of nonlinear estimation is usually more challenging than estimation in linear models (where the loss function is quadratic—or equivalently, the loss gradient is linear—in the elements of $\boldsymbol{\theta}$). Stochastic search and optimization for nonlinear models is used in a large number of areas. These include (to name just a few) the estimation of connection weights in artificial neural networks, system design for discrete-event dynamic (queuing) systems, image restoration from blurred image data, dose response analysis for drugs, and stochastic adaptive control.

Section 4.1 introduces the basic root-finding (Robbins–Monro) SA algorithm form and Section 4.2 presents several motivating examples. Sections 4.3 and 4.4 discuss some of the theoretical properties related to convergence and asymptotic distributions. Section 4.5 summarizes four extensions to the basic algorithm form of this chapter and Section 4.6 provides some concluding remarks.

4.1 INTRODUCTION

This chapter introduces the method of *stochastic approximation* (SA) for solving nonlinear root-finding problems in the presence of noisy measurements. The basic approach is sometimes referred to as the Robbins–Monro algorithm in honor of the two people who introduced the modern general setting (Robbins and Monro, 1951). Root-finding SA is a cornerstone of stochastic search and optimization as a generalization of the well-known deterministic algorithms in Section 1.4 (steepest descent and Newton–Raphson).

Root-finding SA provides a general framework for convergence analysis of many algorithms that may not appear directly as root-finding methods. These include the recursive-least-squares (RLS) and least-mean-squares (LMS) algorithms of Chapter 3, nonlinear parameter estimation methods of Chapter 5 (e.g., neural network backpropagation), gradient-free SA algorithms of Chapters

6 and 7, simulated annealing and related algorithms of Chapter 8, reinforcement (temporal difference) learning in Chapter 11, simulation-based optimization of Chapters 14 and 15, and some optimal experimental design methods in Chapter 17.

In the notation of Section 1.1, the focus in this chapter is to find at least one root $\boldsymbol{\theta}^* \in \Theta^* \subseteq \Theta \subseteq \mathbb{R}^p$ to

$$\boldsymbol{g}(\boldsymbol{\theta}) = \boldsymbol{0} \tag{4.1}$$

based on noisy measurements of $\boldsymbol{g}(\boldsymbol{\theta})$. This root-finding problem was introduced in Section 1.1. Note that $\boldsymbol{g}(\boldsymbol{\theta}) \in \mathbb{R}^p$. So the problem is the classic "p equations and p unknowns" with, in general, the significant complications of nonlinear $\boldsymbol{g}(\boldsymbol{\theta})$ and noisy input information. A very important special case is finding a root to $\boldsymbol{g}(\boldsymbol{\theta}) = \partial L/\partial\boldsymbol{\theta} = \boldsymbol{0}$ when faced with a problem of minimizing $L(\boldsymbol{\theta})$. Given the importance of this special case, a full chapter—Chapter 5—is devoted to the stochastic gradient problem of optimization when only noisy measurements of $\partial L/\partial\boldsymbol{\theta}$ are available.

This chapter focuses on root-finding per se, without dwelling on the special case of optimization ($\partial L/\partial\boldsymbol{\theta} = \boldsymbol{0}$). Root-finding via SA was introduced in modern form in Robbins and Monro (1951), with important generalizations and extensions following close behind as given in Kiefer and Wolfowitz (1952), Chung (1954), and Blum (1954a, b). Some classic books covering SA are Albert and Gardner (1967), Nevel'son and Has'minskii (1973), and Kushner and Clark (1978). A more recent thorough mathematical treatment is Kushner and Yin (1997).

A central aspect of SA is the allowance for noisy input information in the algorithm. In fact, as we will see, the SA methods in this and the next three chapters are often better at coping with noisy input information than other search methods in this book. Moreover, the theoretical foundation for SA is deeper than the theory for other stochastic search methods with noisy measurements. In the case of root-finding SA, the noise manifests itself in the measurements of $\boldsymbol{g}(\boldsymbol{\theta})$ used in the search as $\boldsymbol{\theta}$ varies. More specifically, as in Section 1.1, suppose that measurements of $\boldsymbol{g}(\boldsymbol{\theta})$ at any $\boldsymbol{\theta}$ are available as

$$\boldsymbol{Y}_k(\boldsymbol{\theta}) = \boldsymbol{g}(\boldsymbol{\theta}) + \boldsymbol{e}_k(\boldsymbol{\theta}),\ \ k = 0, 1, 2, \ldots, \tag{4.2}$$

where $\boldsymbol{e}_k(\boldsymbol{\theta})$ is assumed to be some noise term of dimension p.

Closely related to (4.2) is the case where there are input measurements $\boldsymbol{x}_k$ as part of $\boldsymbol{Y}_k(\boldsymbol{\theta})$. These inputs may represent random or deterministic terms (e.g., randomly generated or user-specified target values in a target-tracking problem). So, for a specified $\boldsymbol{\theta}$ *and* $\boldsymbol{x}_k$, a noisy measurement $\boldsymbol{Y}_k(\boldsymbol{\theta})$ is returned. The measurements in this setting are assumed to come from

$$\boldsymbol{Y}_k(\boldsymbol{\theta}) = \tilde{\boldsymbol{g}}_k(\boldsymbol{\theta}, \boldsymbol{x}_k) + \tilde{\boldsymbol{e}}_k(\boldsymbol{\theta}, \boldsymbol{x}_k), \tag{4.3}$$

where $\tilde{g}_k(\boldsymbol{\theta}, \boldsymbol{x}_k)$ and $\tilde{e}_k(\boldsymbol{\theta}, \boldsymbol{x}_k)$ represent function and noise terms analogous to $\boldsymbol{g}(\boldsymbol{\theta})$ and $\boldsymbol{e}_k(\boldsymbol{\theta})$ in (4.2). We can reexpress (4.3) in the conventional form of (4.2) by defining

$$\boldsymbol{e}_k(\boldsymbol{\theta}) \equiv \tilde{\boldsymbol{g}}_k(\boldsymbol{\theta}, \boldsymbol{x}_k) - \boldsymbol{g}(\boldsymbol{\theta}) + \tilde{\boldsymbol{e}}_k(\boldsymbol{\theta}, \boldsymbol{x}_k). \tag{4.4}$$

Substituting this noise term in (4.2) yields a measurement the same as (4.3). The introduction of inputs $\boldsymbol{x}_k$ does not fundamentally alter the basic root-finding problem of (4.1). That is, there exists a $\boldsymbol{g}(\boldsymbol{\theta})$ and $\boldsymbol{e}_k(\boldsymbol{\theta})$ as in (4.4) such that measurements of form (4.3) yield a solution to (4.1). For example, if the inputs $\boldsymbol{x}_k$ are random and $E[\boldsymbol{e}_k(\boldsymbol{\theta})] = E[\tilde{\boldsymbol{e}}_k(\boldsymbol{\theta}, \boldsymbol{x}_k)] = \boldsymbol{0}$, then $\boldsymbol{g}(\boldsymbol{\theta}) = E[\tilde{\boldsymbol{g}}_k(\boldsymbol{\theta}, \boldsymbol{x}_k)]$.

The mechanics of the algorithm are essentially the same in either measurement setting, (4.2) or (4.3). In fact, in much of the literature on SA, the concept of inputs as in (4.3) is suppressed. That is, the problem is usually stated as in (4.1) and (4.2) while assuming (implicitly, at least) a noise structure as in (4.4). In a like manner, to avoid having to continuously distinguish the two settings, we generally just discuss problems in the context of $\boldsymbol{g}(\boldsymbol{\theta})$ without specific reference to inputs $\boldsymbol{x}_k$.

In cases where an average $\boldsymbol{g}(\boldsymbol{\theta})$ does not exist (i.e., $\boldsymbol{g}(\boldsymbol{\theta}) \neq E[\tilde{\boldsymbol{g}}_k(\boldsymbol{\theta}, \boldsymbol{x}_k)]$), there may be an inherent time-varying root-finding problem where $\boldsymbol{g}(\boldsymbol{\theta})$ is replaced by $\boldsymbol{g}_k(\boldsymbol{\theta}) \equiv \tilde{\boldsymbol{g}}_k(\boldsymbol{\theta}, \boldsymbol{x}_k)$. Some of the theory and methodology of SA extends to the time-varying $\boldsymbol{g}_k(\boldsymbol{\theta})$ case, as discussed in Subsections 4.5.1 and 4.5.4.

In root-finding SA, the measurements at $\boldsymbol{\theta} = \hat{\boldsymbol{\theta}}_k$ are:

$$\boldsymbol{Y}_k(\hat{\boldsymbol{\theta}}_k) = \boldsymbol{g}(\hat{\boldsymbol{\theta}}_k) + \boldsymbol{e}_k(\hat{\boldsymbol{\theta}}_k), \tag{4.5}$$

where $\boldsymbol{e}_k = \boldsymbol{e}_k(\hat{\boldsymbol{\theta}}_k)$ is the general error term in (4.2) or (4.4). Although $\boldsymbol{e}_k$ may have general statistical properties (e.g., dependence across k and/or nonidentical distributions), an important special case is where $\{\boldsymbol{e}_k\}$ is an independent, identically distributed (i.i.d.) sequence of mean-zero random vectors.

Recalling the basic steepest descent algorithm in Section 1.4 (i.e., $\hat{\boldsymbol{\theta}}_{k+1} = \hat{\boldsymbol{\theta}}_k - a_k \boldsymbol{g}(\hat{\boldsymbol{\theta}}_k)$), an obvious implementation with noisy measurements of $\boldsymbol{g}(\boldsymbol{\theta})$ is to average $\boldsymbol{Y}_k(\boldsymbol{\theta})$ values at $\boldsymbol{\theta} = \hat{\boldsymbol{\theta}}_k$. Such averaging is used to approximate $\boldsymbol{g}(\hat{\boldsymbol{\theta}}_k)$ from multiple values of $\boldsymbol{Y}_k(\hat{\boldsymbol{\theta}}_k)$. A significant innovation of Robbins and Monro (1951) was the recognition that this is a wasteful use of the measurements. Recall that $\boldsymbol{g}(\hat{\boldsymbol{\theta}}_k)$ is merely an intermediate calculation towards the ultimate goal of trying to find a root $\boldsymbol{\theta}^*$. There is little interest in $\boldsymbol{g}(\hat{\boldsymbol{\theta}}_k)$ per se. So, the main innovation in SA is to do a form of averaging *across iterations*. At first thought, this type of averaging may seem dubious, since the underlying evaluation point $\boldsymbol{\theta}$ is changing across iterations. But, as suggested by Robbins and Monro (1951), this across-iteration averaging can lead to a more effective use of the input information than expending a large amount of resources in getting accurate estimates for $\boldsymbol{g}(\boldsymbol{\theta})$ at each iteration.

In implementing the across-iteration averaging, the core root-finding SA algorithms are given below (unconstrained and constrained versions). Let the scalar a_k be a nonnegative "gain" value, $\hat{\theta}_0$ be the initial condition, and $\Psi_\Theta[\cdot]$ be a user-defined mapping that projects any point not in the constraint domain Θ to a new point inside Θ.

Basic Root-Finding (Robbins–Monro) SA Algorithms

Unconstrained:

$$\hat{\theta}_{k+1} = \hat{\theta}_k - a_k Y_k(\hat{\theta}_k) \tag{4.6}$$

Constrained:

$$\hat{\theta}_{k+1} = \Psi_\Theta[\hat{\theta}_k - a_k Y_k(\hat{\theta}_k)] \tag{4.7}$$

The above root-finding algorithms are clearly motivated by the steepest descent algorithm with the noisy measurement $Y_k(\hat{\theta}_k)$ replacing the exact root-finding function $g(\hat{\theta}_k)$. A major innovation in the algorithm is the specification of precise conditions on the gain coefficients a_k to ensure that the process in (4.6) or (4.7) properly invokes the across-iteration averaging and converges to a root θ^*. As expected, these conditions generally differ from those in the easier deterministic steepest descent setting. (Of course, these conditions also apply in steepest descent because that is a special case of SA.) We discuss these conditions in Sections 4.3 and 4.4. Because of the minus sign on the right-hand side of (4.6) and (4.7), each component of $g(\theta)$ should be positive when the corresponding component of θ is greater than the corresponding component of θ^* and negative for the opposite case (this applies in the typical case where e_k has mean zero) (why?). For example, a trivial scalar problem of finding θ such that $1 - \theta = 0$ should be changed to the equivalent $g(\theta) = \theta - 1 = 0$. More formal statements of this sign requirement are given in Section 4.3.

While (4.7) provides a nice conceptual framework for constrained search, the vast majority of practical theoretical results pertain to the unconstrained version (4.6). In applications, it is often difficult to implement (4.7) unless the constraints are fairly benign (such as a hypercube constraint on θ where each component of θ has a distinct lower and upper bound). As discussed in Subsection 1.2.1, this issue of difficult analysis and implementation for constrained algorithms is not unique to SA. It affects *all* search and optimization methods, and is especially challenging in stochastic methods as considered in this book.

4.2 POTPOURRI OF STOCHASTIC APPROXIMATION EXAMPLES

Four examples of root-finding SA are given below. The first shows that the classical sample mean of a sequence of random vectors is a special case of SA. The second is a problem of estimating a *quantile*, a point on the real line such that a process will have an outcome below this point with a specified probability.

This problem appears in many contexts. For example, in pharmacology there is interest in determining the required dose such that (say) 90 percent of a population achieves a desired therapeutic response to a treatment at that dose. A bivariate extension of the quantile idea appears in the third example. The goal is to find the radius of a circle about a target such that a projectile directed towards the target is likely to land in this circle with specified probability. If the probability is set to 0.5, this radius is referred to as the *circular error probable* (CEP), which is the single most important measure of accuracy for many military weapon systems. The fourth example is from microeconomics, the aim being to estimate a *production function* relating labor and capital inputs to production outputs for firms in a sector of the economy. This example compares SA with the method of maximum likelihood.

Example 4.1—Sample mean as an SA algorithm. Suppose that independent measurements $\boldsymbol{X}_i$ are available, where the measurements share a common mean $\boldsymbol{\mu}$ (i.e., $E(\boldsymbol{X}_i) = \boldsymbol{\mu}$ for all i). The goal is to estimate $\boldsymbol{\mu}$. The sample mean of the $\boldsymbol{X}_i$ represents the most common estimator:

$$\begin{aligned}\bar{\boldsymbol{X}}_{k+1} &\equiv \frac{1}{k+1}\sum_{i=1}^{k+1}\boldsymbol{X}_i \\ &= \frac{k}{k+1}\bar{\boldsymbol{X}}_k + \frac{1}{k+1}\boldsymbol{X}_{k+1} \\ &= \bar{\boldsymbol{X}}_k - \frac{1}{k+1}\left(\bar{\boldsymbol{X}}_k - \boldsymbol{X}_{k+1}\right),\end{aligned}$$

where $\bar{\boldsymbol{X}}_0 = \boldsymbol{0}$ in the recursive representation in the second and third lines. Letting $\hat{\boldsymbol{\theta}}_k = \bar{\boldsymbol{X}}_k$, $a_k = 1/(k+1)$, and $\boldsymbol{Y}_k(\hat{\boldsymbol{\theta}}_k) = \bar{\boldsymbol{X}}_k - \boldsymbol{X}_{k+1} = \hat{\boldsymbol{\theta}}_k - \boldsymbol{X}_{k+1}$ puts this recursion for the sample mean in the framework of SA recursion (4.6). Further, connections to (4.2) (and thus (4.5)) are apparent by letting $\boldsymbol{g}(\boldsymbol{\theta}) = \boldsymbol{\theta} - \boldsymbol{\mu}$ and $\boldsymbol{e}_k = \boldsymbol{\mu} - \boldsymbol{X}_{k+1}$ (i.e., $\boldsymbol{e}_k$ is independent of $\boldsymbol{\theta}$ and has mean zero). Hence, the simple sample mean calculation for a sequence of random vectors represents a special case of an SA algorithm. (Exercise 4.3 considers the problem of estimating the mean using a different gain a_k.) ❑

Example 4.2—$\mathbf{LD_{50}}$ quantile. Consider the following problem in quantile estimation, similar to that described in Robbins and Monro (1951). Let $F_X(x) = P(X \leq x)$ be an unknown distribution function for a scalar random variable X. Suppose that a researcher wants to estimate the scalar quantile θ such that $F_X(\theta) = 0.5$, i.e., find the root of the equation $g(\theta) = F_X(\theta) - 0.5 = 0$. In clinical trials, the indicated probability 0.5 has special importance in determining a quantity called LD_{50}—*lethal dosage 50*—the dosage that is lethal to 50 percent of a population of organisms.

Suppose that the researcher is not allowed to know the value of X in an experiment, but is allowed to know whether X is above or below a specified threshold. In particular, in running a sequence of experiments generating outcomes $X_0, X_1, \ldots$ according to the distribution $F_X(x)$, the researcher is free to specify a value $\hat{\theta}_k$ such that information about whether X_k is smaller or larger than $\hat{\theta}_k$ (a "success" or "nonsuccess," respectively) is returned. Formally, let us introduce the success variable:

$$s_k(\hat{\theta}_k) = \begin{cases} 1 \text{ if } X_k \leq \hat{\theta}_k \text{ (success)}, \\ 0 \text{ otherwise (nonsuccess)}. \end{cases}$$

With $\hat{\theta}_0$ as the best guess of the quantile θ such that $F_X(\theta) = 0.5$, the root-finding SA recursion has the form

$$\hat{\theta}_{k+1} = \hat{\theta}_k - a_k\left(s_k(\hat{\theta}_k) - 0.5\right).$$

So, $Y_k(\hat{\theta}_k) = s_k(\hat{\theta}_k) - 0.5$. Note that $E[Y_k(\hat{\theta}_k) | \hat{\theta}_k] = F_X(\hat{\theta}_k) - 0.5 = g(\hat{\theta}_k)$, so that the noise e_k has mean zero. Intuitively, it is apparent that if $\hat{\theta}_k$ is too small, then $s_k(\hat{\theta}_k) = 0$ is more likely than $s_k(\hat{\theta}_k) = 1$, leading to $Y_k(\hat{\theta}_k) = 0 - 0.5 < 0$. This, in turn, tends to make $\hat{\theta}_{k+1}$ larger than $\hat{\theta}_k$ (as desired) according to the update $\hat{\theta}_{k+1} = \hat{\theta}_k + 0.5a_k$. The other possible outcome, $Y_k(\hat{\theta}_k) = 1 - 0.5 > 0$, which is less likely to occur, leads to an update with the same magnitude of change in θ but in the wrong direction of decreasing θ (i.e., $\hat{\theta}_{k+1} = \hat{\theta}_k - 0.5a_k$). Overall, therefore, there is a tendency to make θ larger, as desired. Conversely, if $\hat{\theta}_k$ is too large, the next value $\hat{\theta}_{k+1}$ is likely to be smaller than $\hat{\theta}_k$.

This example points to the important role of a_k. If a_k is too small, the progress of the algorithm to the optimal quantile θ^* will be sluggish because the increment $0.5a_k$ (the magnitude of change in θ value) will be too small. Conversely, too large a value for a_k may cause the new estimate to vastly overshoot the LD_{50} quantile. Exercise 4.4 demonstrates some numerical performance associated with this quantile example. ❑

Example 4.3—Circular error probable (CEP). Let us now consider a numerical example for the CEP problem mentioned above. More detail on this problem is given in Grubbs (1964) and Spall and Maryak (1992). The algorithm processes projectile impact measurements $\boldsymbol{X}_k \in \mathbb{R}^2$ having a (generally unknown) bivariate probability distribution. The first coordinate in $\boldsymbol{X}_k$ represents the crossrange direction and the second coordinate represents the downrange direction. The mean of the bivariate distribution represents the impact bias relative to the target and the covariance matrix represents the dispersion of the impact points in the two coordinates.

Recall that the Euclidean norm $\|\cdot\|$ has the geometric interpretation of distance from the origin to the argument point. Then, following Example 4.2, let

$$s_k(\hat{\theta}_k) = \begin{cases} 1 \text{ if } \|\boldsymbol{X}_k\| \leq \hat{\theta}_k & \text{(success)}, \\ 0 \text{ otherwise (nonsuccess)}. \end{cases}$$

The SA recursion has a form identical to that in Example 4.2: $\hat{\theta}_{k+1} = \hat{\theta}_k - a_k(s_k(\hat{\theta}_k) - 0.5)$ (implying that $Y_k(\hat{\theta}_k) = s_k(\hat{\theta}_k) - 0.5$). Arguments similar to those in Example 4.2 can be used to show the rationale of this form.

Suppose that the impacts $\boldsymbol{X}_k$ arrive according to a $N([1, 1]^T, \text{diag}[4, 1])$ distribution (the target is the point $[0, 0]^T$). For this statistical model, the true CEP equals 2.17. (This value is found from a combination of deterministic numerical optimization and numerical quadrature applied to the integral of the bivariate normal density function; an alternative approach is the highly accurate closed-form approximation in Grubbs, 1964.) In practice, of course, the data-generating mechanism (true model) is not known exactly and so the exact solution is unavailable for comparison.

Based on simulated data from the assumed model, one replication of the root-finding SA algorithm in (4.7) finds estimates of 1.60, 1.82, and 2.16, respectively, after n = 20, 100, and 1000 experimental impact points $\boldsymbol{X}_k$ (each generating one $Y_k(\cdot)$). These estimates are based on $\hat{\theta}_0 = 0.5$ and $a_k = 1/(k+1)$. A better initial condition or a tuned a_k sequence produces a better SA estimate at a specified n. Figure 4.1 shows a plot of the 1000 impacts and the associated true CEP and (indistinguishable) CEP estimate. A visual inspection shows that the circle contains approximately half the impact points, as expected by the

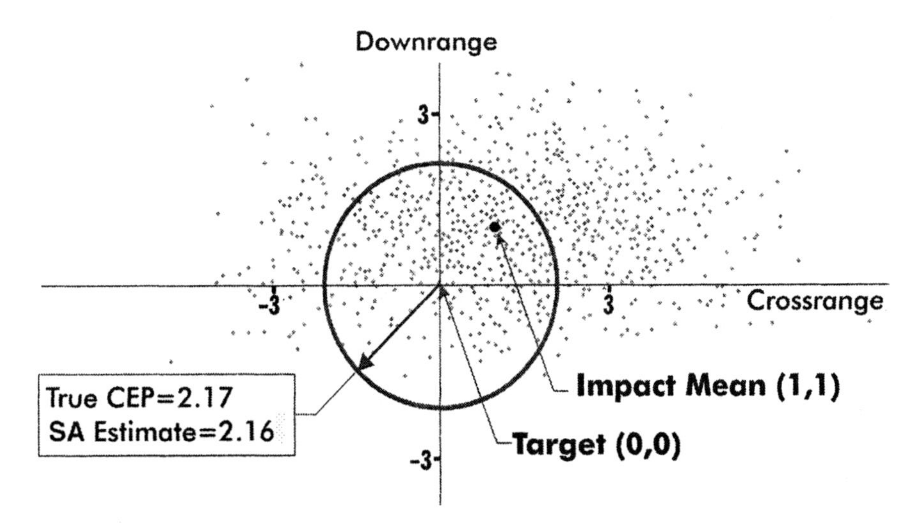

Figure 4.1. 1000 impact points with impact mean differing from target point. The indicated circle is centered at the target with radius equal to the CEP.

definition of CEP. Note the bias in the impacts toward the upper right quadrant and greater spread in the crossrange direction (the standard deviation for crossrange is double that of downrange). Because the CEP radius is relative to the target, not the mean of impacts, the CEP is larger with a nonzero mean than with a zero mean given that the covariance matrix is the same. ❑

Example 4.4—Production function in microeconomics. An important concern in economics is the tradeoff between different types of inputs that can be used in making a product. At its simplest level, this tradeoff may be between work hours (W) and capital equipment (C) in the manufacturing of a product. A common means of analyzing the tradeoff is through the use of a production function $h(\cdot)$ that relates the quantities of labor and capital to the amount of output z_k:

$$z_k = h(\boldsymbol{\theta}, \boldsymbol{x}_k) + v_k,$$

where $\boldsymbol{\theta}$ is a vector of parameters of the function, $\boldsymbol{x}_k$ represents the inputs used in the production of the product for index k (e.g., $\boldsymbol{x}_k = [W_k, C_k]^T$ represent the work hours and capital inputs at the kth time period), and v_k is the random noise. In the numerical results below, W_k and C_k are generated randomly (uniformly) in [1, 10] and [11, 100] at each k, respectively.

An important special case of a production function for two inputs is the Cobb–Douglas form.[1] Let $\boldsymbol{\theta} = [\lambda, \beta]^T$, where $\lambda > 0$ and $0 \le \beta \le 1$ are parameters having economic significance relative to total production capability and the degree of tradeoff possible between the two inputs. The functional form for this production function together with additive noise is

$$z_k = h(\boldsymbol{\theta}, \boldsymbol{x}_k) + v_k = \lambda C_k^{\beta} W_k^{1-\beta} + v_k \tag{4.8}$$

(Kmenta, 1997, pp. 252–253). One well-known approach to estimating $\boldsymbol{\theta}$ from a sequence of n input–output data pairs $\{(\boldsymbol{x}_1, z_1), (\boldsymbol{x}_2, z_2), \ldots, (\boldsymbol{x}_n, z_n)\}$ is the method of maximum likelihood (ML) (e.g., Kmenta, 1997, pp. 582–583). An alternative approach is to apply the SA algorithm to estimate $\boldsymbol{\theta}$ based on the above form for $h(\boldsymbol{x}_k, \boldsymbol{\theta})$. Note that a model similar to (4.8) was considered in Section 3.1 (Example 3.2), but unlike the earlier model, the form in (4.8) cannot be converted to an equivalent linear form via a logarithmic (or other) transformation because the noise is additive (versus multiplicative in Example 3.2).

One of the points illustrated in this example is that there is a fundamental difference between SA and standard approaches such as ML in that SA operates under weaker conditions. In particular, SA does not require assumptions about

---

[1] This production function was introduced in the 1930s by economists Charles W. Cobb and Paul H. Douglas; Douglas served as a U. S. senator from 1948 to 1966.

the distributional form of the outputs z_k. Although ML has certain optimality properties, it may yield a poor estimate when the actual data have a distribution significantly different from that assumed in forming the likelihood criterion.

The ML estimator is derived by maximizing the joint probability density function of the n data points when viewed as a function of $\boldsymbol{\theta}$. It is common to assume that the data are independent, normally distributed, which leads to an ML criterion based on the normal density function. Reflecting common practice in ML estimation, the results below are based on this criterion, even when the true data are not normally distributed.

The loss function for the SA recursion is the expected squared error

$$L(\boldsymbol{\theta}) = \frac{1}{2}E\left\{[z_{k+1} - h(\boldsymbol{\theta}, \boldsymbol{x}_{k+1})]^2\right\},$$

where the expectation is taken with respect to the noise v_k and the randomness in $\boldsymbol{x}_k$. The stochastic gradient (i.e., the derivative of the argument in the above expectation) is

$$\boldsymbol{Y}_k(\boldsymbol{\theta}) = \frac{1}{2}\frac{\partial[z_{k+1} - h(\boldsymbol{\theta}, \boldsymbol{x}_{k+1})]^2}{\partial\boldsymbol{\theta}}.$$

So, the input at step k is calculated as

$$\boldsymbol{Y}_k(\hat{\boldsymbol{\theta}}_k) = [h(\hat{\boldsymbol{\theta}}_k, \boldsymbol{x}_{k+1}) - z_{k+1}]\left.\frac{\partial h(\boldsymbol{\theta}, \boldsymbol{x}_{k+1})}{\partial\boldsymbol{\theta}}\right|_{\boldsymbol{\theta}=\hat{\boldsymbol{\theta}}_k}.$$

We use this gradient in the root-finding algorithm (4.7). Note that SA assumes no knowledge of the distributions generating the data (i.e., the expectation associated with $L(\boldsymbol{\theta})$ is never actually computed).

The results of the analysis are shown on Table 4.1. Data are generated according to the form in (4.8) with $\boldsymbol{\theta}$ taken as $\boldsymbol{\theta}^* = [2.5,\ 0.70]^T$. The table contrasts the ML estimates with those of SA. The first row of results in Table 4.1 shows the parameter estimates when the added noise matches the assumptions for ML. The second row shows the results when the noise is not normally distributed, but is distributed according to $v_k = h(\boldsymbol{\theta}^*, \boldsymbol{x}_k)(\xi_k - 1)$, where ξ_k is generated from an exponential distribution with a mean of 1.0 (see Appendix D for a brief discussion of the exponential distribution). To test the estimates, 100 new data points z_k are generated for each of the normal and nonnormal cases using $\boldsymbol{\theta} = \boldsymbol{\theta}^*$. Then, for each $\boldsymbol{\theta}$ (corresponding to one of the four estimates in the table), the sample root-mean-squared (RMS) errors are computed according to

$$\text{RMS} = \sqrt{\frac{1}{100}\sum_{k=1}^{100}[z_k - h(\boldsymbol{\theta}, \boldsymbol{x}_k)]^2}.$$

Table 4.1. ML and SA estimates for production function parameters $\boldsymbol{\theta} = [\lambda, \beta]^T$ from one realization of 1000 measurements ($\boldsymbol{\theta}^* = [2.5, 0.70]^T$). RMS errors are determined from prediction errors using measurements independent of the measurements used for estimating $\boldsymbol{\theta}$.

Noise distribution	ML estimate	RMS error	SA estimate	RMS error
Normal	$[2.54, 0.71]^T$	0.129	$[2.47, 0.70]^T$	0.268
Nonnormal	$[2.76, 0.70]^T$	2.822	$[2.48, 0.67]^T$	1.725

(So each summand involves the difference between an actual output z_k and a prediction $h(\boldsymbol{\theta}, \boldsymbol{x}_k)$.)

While the ML estimate does better than the SA estimate when the data are normally distributed (consistent with the assumptions), the opposite is true when the data are not normally distributed. Relative to SA, the ML estimate significantly degrades when the true noise distribution deviates from the assumptions. (Both of the SA and ML estimates degrade in the nonnormal case as a reflection of greater noise variability.) This example is an introduction to the notion of stochastic gradient, covered in more detail in Chapter 5. ❑

4.3 CONVERGENCE OF STOCHASTIC APPROXIMATION

4.3.1 Background

As with any search algorithm, it is of interest to know whether the iterate $\hat{\boldsymbol{\theta}}_k$ converges to a solution $\boldsymbol{\theta}^* \in \Theta^*$ as k gets large. In fact, one of the strongest aspects of SA is the rich convergence theory that has developed over many years. The versatility of SA theory allows it to be used to show convergence of stochastic algorithms that, on the surface, may not look like SA. For example, SA is used to analyze the convergence of neural network backpropagation (see Section 5.2), simulated annealing (Section 8.6), evolutionary computation (Section 10.5), temporal difference learning (Section 11.6), simulation-based optimization (Chapters 14 and 15), Markov chain Monte Carlo (Section 16.6), and sequential experimental design (Section 17.4). This allows researchers and analysts in many fields to establish formal convergence where otherwise that may have remained an open question.

Most of the stated convergence results for SA are in the almost sure (a.s.) sense. Of historical note is that Robbins and Monro (1951) gave conditions for *mean-squared* (m.s.) convergence, which implies "in probability" (pr.) convergence. Blum (1954a, b) was the first to give conditions for a.s.

convergence. As discussed in Appendix C, neither a.s. nor m.s. convergence is implied by the other, but pr. convergence is implied by a.s. convergence.

Many sufficient conditions have been given over the years for a.s. convergence of the SA recursions in (4.6) and (4.7). The convergence results have largely evolved out of two general settings. One focuses on the imposition of statistical conditions on the function $\boldsymbol{g}(\boldsymbol{\theta})$ and noise $\boldsymbol{e}_k(\boldsymbol{\theta})$. Young (1984, pp. 33–41), Ruppert (1991), and Rustagi (1994, Chap. 9), for example, discuss such results, which have largely evolved out of a statistics perspective. The other setting is based on defining an underlying ordinary differential equation (ODE) that roughly emulates the SA algorithm in (4.6) for large k and as the random effects disappear. This approach has been particularly popular in the applied mathematics and engineering literature. Ljung (1977), Kushner and Clark (1978, Chap. 2), Benveniste et al. (1990, Part I, Chap. 2), and Kushner and Yin (1997, Chaps. 5 and 6) discuss convergence results from the ODE perspective. It turns out that the convergence properties of this *deterministic* differential equation are closely related to the *stochastic* convergence properties of (4.6).

Neither of the "statistics" or "engineering" conditions mentioned above is a special case of the other, so neither set is necessarily weaker. One must consider the application in determining which set of conditions is easier to verify. Note that the conditions given here are not the weakest possible. Rather, they have been chosen to convey the essential flavor of the conditions commonly used to guarantee convergence of SA algorithms. Some of the references cited above (and elsewhere) present conditions more general than those here. As is typical in convergence results for stochastic algorithms, some of these conditions involve aspects of the problem that may be unknown to the analyst (e.g., conditions requiring full knowledge of $\boldsymbol{g}(\boldsymbol{\theta})$). This conundrum seems unavoidable. Meaningful convergence results naturally require assumptions about essential aspects of the problem structure.

These conditions apply when there is a unique root $\boldsymbol{\theta}^*$. Hence, when used for optimization (à la $\partial L/\partial\boldsymbol{\theta} = \boldsymbol{0}$), they apply when there are no local minima different from the (unique) global minimum. Sections 7.7 and (especially) 8.4 discuss the use of SA for global optimization in the face of multiple local minima.

Also note that while these conditions—together with other similar conditions in the literature—play a central role in the theoretical analysis of SA, they are *sufficient* conditions. Many practical applications of SA produce satisfactory results when one or more of the conditions are not satisfied.

4.3.2 Convergence Conditions

This subsection presents the "statistics" and "engineering" conditions for strong (a.s.) convergence of the SA iterate $\hat{\boldsymbol{\theta}}_k$. Some conditions of the first (statistics) type are given below. These are drawn from Blum (1954a, b) and Nevel'son and Has'minskii (1973, Sect. 4.4).

A.1 **(Gain sequence)** $a_k > 0$, $a_k \to 0$, $\sum_{k=0}^{\infty} a_k = \infty$, and $\sum_{k=0}^{\infty} a_k^2 < \infty$.

A.2 **(Search direction)** For some symmetric, positive definite matrix $\boldsymbol{B}$ and every $0 < \eta < 1$,

$$\inf_{\eta < \|\boldsymbol{\theta}-\boldsymbol{\theta}^*\| < 1/\eta} (\boldsymbol{\theta}-\boldsymbol{\theta}^*)^T \boldsymbol{B}\, \boldsymbol{g}(\boldsymbol{\theta}) > 0 .$$

(The "inf" statement pertains to the infimum [see Appendix A] of the expression $(\boldsymbol{\theta}-\boldsymbol{\theta}^*)^T \boldsymbol{B}\, \boldsymbol{g}(\boldsymbol{\theta})$ over the set of $\boldsymbol{\theta}$ such that $\eta < \|\boldsymbol{\theta}-\boldsymbol{\theta}^*\| < 1/\eta$. One may choose any convenient $\boldsymbol{B}$; often, $\boldsymbol{B} = \boldsymbol{I}_p$.)

A.3 **(Mean-zero noise)** $E[\boldsymbol{e}_k(\boldsymbol{\theta})] = \boldsymbol{0}$ for all $\boldsymbol{\theta}$ and k.

A.4 **(Growth and variance bounds)** $\|\boldsymbol{g}(\boldsymbol{\theta})\|^2 + E\left(\|\boldsymbol{e}_k(\boldsymbol{\theta})\|^2\right) \le c\left(1+\|\boldsymbol{\theta}\|^2\right)$ for all $\boldsymbol{\theta}$ and k and some $c > 0$.

Let us offer a few comments about the above conditions. From the point of view of the user's input, condition A.1 is the most relevant. This condition provides a careful balance in having the gain a_k decay neither too fast nor too slow. In particular, the gain should approach zero sufficiently fast ($a_k \to 0$, $\sum_k a_k^2 < \infty$) to damp out the noise effects as the iterate gets near the solution $\boldsymbol{\theta}^*$ but should approach zero at a sufficiently slow rate ($\sum_k a_k = \infty$) to avoid premature (false) convergence of the algorithm. Condition A.2 is a fairly stringent condition on the shape of $\boldsymbol{g}(\boldsymbol{\theta})$. The analyst need only find one valid $\boldsymbol{B}$ that satisfies the indicated inequality. For example, in the linear case, if $\boldsymbol{g}(\boldsymbol{\theta}) = \boldsymbol{A}(\boldsymbol{\theta} - \boldsymbol{\theta}^*)$ for some matrix $\boldsymbol{A}$, then one must identify a $\boldsymbol{B}$ such that $\boldsymbol{BA}$ is positive definite. Condition A.3 is the standard mean-zero noise condition and A.4 provides restrictions on the magnitude of $\boldsymbol{g}(\boldsymbol{\theta})$. In particular, this condition says that $\|\boldsymbol{g}(\boldsymbol{\theta})\|^2$ and the variance elements of $\boldsymbol{e}_k$ cannot grow faster than a quadratic function of $\boldsymbol{\theta}$ (note that $\boldsymbol{e}_k$ is allowed to be a function of $\boldsymbol{\theta}$, as discussed in Section 4.1). By the mean-zero condition in A.3, the left-hand side of the inequality in A.4 can be written as $E\left(\|\boldsymbol{Y}_k(\boldsymbol{\theta})\|^2\right) = \|\boldsymbol{g}(\boldsymbol{\theta})\|^2 + E\left(\|\boldsymbol{e}_k(\boldsymbol{\theta})\|^2\right)$ (why?). Note that there are no conditions on the smoothness of $\boldsymbol{g}(\boldsymbol{\theta})$, such as a requirement that $\boldsymbol{g}(\boldsymbol{\theta})$ be differentiable.

Using the ODE approach mentioned above, conditions of the second (engineering) type are given below. These conditions are special cases of more general conditions in Kushner and Clark (1978, Theorem 2.3.1), Metivier and Priouret (1984), and Kushner and Yin (1997, Theorem 5.2.1). While these conditions are included here because of the importance the ODE approach plays in the analysis of SA algorithms, it is recognized that ODEs and some other concepts used in the conditions (such as "infinitely often") are not within the prerequisites of this book. A reader may simply skim these conditions to get the flavor of the ODE approach. With the exception of Examples 4.5 and 4.6 below, most other aspects of SA to follow do not rest critically on the details here. The broader ODE-based approach, as discussed throughout Kushner and Yin (1997) and in many references cited therein, lends itself to significant generalizations beyond the conditions here. These generalizations include cases involving some types of discontinuities in $\boldsymbol{g}(\boldsymbol{\theta})$ (which are useful, e.g., in areas such as manufacturing and signal processing; see Kushner and Yin, 1997, Chap. 9).

B.1 **(Gain sequence)** $a_k > 0$, $a_k \to 0$, $\sum_{k=0}^{\infty} a_k = \infty$.

B.2 **(Relationship to ODE)** Let $\boldsymbol{g}(\boldsymbol{\theta})$ be continuous on $\mathbb{R}^p$. With $\boldsymbol{Z}(\tau) \in \mathbb{R}^p$ representing a time-varying function (τ denoting time), suppose that the differential equation given by $d\boldsymbol{Z}(\tau)/d\tau = -\boldsymbol{g}(\boldsymbol{Z}(\tau))$ has an asymptotically stable equilibrium point at $\boldsymbol{\theta}^*$ (we use τ, rather than t, to denote time to avoid potential confusion with the elements of $\boldsymbol{\theta}$: $t_1, t_2, \ldots, t_p$). (An asymptotically stable equilibrium has the following two requirements: (i) For every $\eta > 0$, there exists a $\delta(\eta)$ such that $\|\boldsymbol{Z}(\tau) - \boldsymbol{\theta}^*\| \le \eta$ for all $\tau > 0$ whenever $\|\boldsymbol{Z}(0) - \boldsymbol{\theta}^*\| \le \delta(\eta)$, and (ii) there exists a δ_0 such that $\boldsymbol{Z}(\tau) \to \boldsymbol{\theta}^*$ as $\tau \to \infty$ whenever $\|\boldsymbol{Z}(0) - \boldsymbol{\theta}^*\| \le \delta_0$.)

B.3 **(Iterate boundedness)** $\sup_{k \ge 0} \|\hat{\boldsymbol{\theta}}_k\| < \infty$ a.s. Further, $\hat{\boldsymbol{\theta}}_k$ lies in a compact (i.e., closed and bounded) subset of the "domain of attraction" for the differential equation in B.2 infinitely often. (The domain of attraction is that set such that $\boldsymbol{Z}(\tau)$ will converge to $\boldsymbol{\theta}^*$ for any starting point in the domain; "infinitely often" is largely self-descriptive, but is defined more formally, e.g., in Laha and Rohatgi, 1979, p. 73.)

B.4 **(Bounded variance property of measurement error)** Let $\Im_k \equiv \{\hat{\boldsymbol{\theta}}_0, \hat{\boldsymbol{\theta}}_1, \ldots, \hat{\boldsymbol{\theta}}_k\}$ (for $k \ge 1$, the information equivalent to $\hat{\boldsymbol{\theta}}_0$ plus the cumulative inputs $\boldsymbol{Y}_i = \boldsymbol{Y}_i(\hat{\boldsymbol{\theta}}_i)$, $0 \le i \le k-1$). Let $\boldsymbol{b}_k = E[\boldsymbol{e}_k(\hat{\boldsymbol{\theta}}_k) | \Im_k]$ ($\boldsymbol{b}_k$ for "bias"). Then $E\left[\left\|\sum_{k=0}^{\infty} a_k(\boldsymbol{e}_k - \boldsymbol{b}_k)\right\|^2\right] < \infty$.

B.5 **(Disappearing bias)** $\sup_{k \ge 0} \|\boldsymbol{b}_k\| < \infty$ a.s. and $\boldsymbol{b}_k \to \boldsymbol{0}$ a.s. as $k \to \infty$.

Let us comment briefly on the above conditions. Although B.1 appears slightly weaker than its companion condition, A.1, it is for many practical purposes equivalent. This follows from the practical implications of B.4. In particular, if $\boldsymbol{b}_k = \boldsymbol{0}$ for all k, and $\boldsymbol{e}_i$ and $\boldsymbol{e}_j$ are uncorrelated for all $i \ne j$, then the more general boundedness condition in B.4 can be replaced by $E\left[\sum_{k=0}^{\infty} a_k^2 \|\boldsymbol{e}_k\|^2\right] < \infty$. In addition, if $\text{cov}(\boldsymbol{e}_k) \ge \eta \boldsymbol{I}_p$ for all k and some $\eta > 0$ (implying that $\inf_{k \ge 0} E\left(\|\boldsymbol{e}_k\|^2\right) \ge \eta p$; see Exercise 4.6), then condition B.4 requires that $\sum_k a_k^2 < \infty$ (i.e., condition A.1 and B.1 are then effectively the same). There are, however, other convergence results based on ODEs (see, e.g., Kushner and Yin, 1997, Chaps. 5 and 6) where B.1 is sufficient and the condition $\sum_k a_k^2 < \infty$ is not required. Condition B.2 pertains to the above-mentioned ODE analysis, relating the SA recursion to a deterministic path for a related ODE. Unlike the previous "statistics" conditions, it is assumed that $\boldsymbol{g}(\boldsymbol{\theta})$ is smooth in the sense that it is continuous.

The boundedness condition B.3 is somewhat controversial (e.g., Benveniste et al., 1990, p. 46) since it imposes a requirement on the very iterate that one is trying to analyze. Kushner and Clark (1978, p. 40) point out that this condition is, in fact, not strong since one typically imposes bounds on $\boldsymbol{\theta}$ in practice. Borkar and Meyn (2000) present sufficient conditions for the boundedness to hold in terms of the associated ODE while Chen (2002, Chaps. 2 and 3) focuses on a method of expanding (iterate-varying) truncations to eliminate the condition. Condition B.4 is a special case of the so-called convergence systems (e.g., Lai, 1985) and ensures that the important martingale convergence theorem from probability theory (e.g., Laha and Rohatgi, 1979, pp. 396–400) can be used to cope with the noise effects in the SA recursion. (This book does not delve into martingales, but the subject plays an important role in the convergence theory for SA.) Finally, condition B.5 is a generalization of the mean-zero noise condition in A.3; here the noise is only required to have a mean that *converges to* (versus being *equal to*) zero. Among other uses, this extension is useful in proving convergence for some of the SA methods for optimization without direct gradient measurements (Chapters 6 and 7).

Theorem 4.1. Consider the unconstrained algorithm (4.6) (i.e., $\boldsymbol{g}(\boldsymbol{\theta})$ has the domain $\Theta = \mathbb{R}^p$). Suppose that either conditions A.1–A.4 hold or conditions B.1–B.5 hold. Further, suppose that $\boldsymbol{\theta}^*$ is a unique solution to $\boldsymbol{g}(\boldsymbol{\theta}) = \boldsymbol{0}$ (i.e., the set of solutions Θ^* is the singleton $\boldsymbol{\theta}^*$). Then $\hat{\boldsymbol{\theta}}_k \to \boldsymbol{\theta}^*$ a.s. as $k \to \infty$.

Comments on proof of Theorem 4.1. The proof under either conditions A.1–A.4 or B.1–B.5 uses mathematical machinery beyond the level of this book. The proof based on A.1–A.4 follows as in the proof of Corollary 4.1 in Nevel'son and Has'minskii (1973, p. 93), which is closely related to the proof in Blum (1954b). For the result based on B.1–B.5, the result follows from the proof of Theorem 2.3.1 in Kushner and Clark (1978, pp. 39–43). Note that B.4 is a sufficient condition for Kushner and Clark's more general sufficient condition A2.2.4 by the martingale convergence result mentioned in Kushner and Clark (1978, p. 27). ❑

4.3.3 On the Gain Sequence and Connection to ODEs

The choice of the gain sequence a_k is critical to the performance of SA. The scaled harmonic sequence $a_k = a/(k+1)$, $a > 0$, $k \geq 0$, is the best-known example of a gain sequence that satisfies condition A.1 (and, of course, B.1). As discussed in Section 4.4, this harmonic decay rate of $O(1/k)$ is optimal with respect to the limiting rate of convergence of $\hat{\boldsymbol{\theta}}_k$, although slower decay rates may be superior in practical (finite-sample) problems. Note that the sequences $a_k = a/(k+1)^2$ and $a_k = a/\sqrt{k+1}$ do *not* satisfy condition A.1. Usually, some numerical experimentation is required to choose the best value of the coefficient a that appears in the gain.

A common generalization of the harmonic sequence is $a_k = a/(k+1)^\alpha$ for strictly positive a and α. From basic calculus, picking $1/2 < \alpha \leq 1$ yields an a_k satisfying the conditions $\sum_{k=0}^{\infty} a_k = \infty$ and $\sum_{k=0}^{\infty} a_k^2 < \infty$ appearing in A.1. Section 4.4 includes more discussion on the choice of the gain sequences.

A key aspect of the "engineering" conditions B.1−B.5, is the connection of the SA recursion to the underlying differential equation

$$\frac{d\boldsymbol{Z}}{d\tau} = -\boldsymbol{g}(\boldsymbol{Z}), \quad \boldsymbol{Z} = \boldsymbol{Z}(\tau). \tag{4.9}$$

Let us now provide some intuitive basis for this connection. Because $\boldsymbol{g}(\boldsymbol{\theta}^*) = \boldsymbol{0}$, the constant solution $\boldsymbol{Z}(\tau) = \boldsymbol{\theta}^*$ represents an equilibrium point of the above differential equation. That is, because $d\boldsymbol{Z}(\tau)/d\tau = \boldsymbol{0}$ at $\boldsymbol{Z}(\tau) = \boldsymbol{\theta}^*$, the system $\boldsymbol{Z}(\tau)$ is not going to move from $\boldsymbol{\theta}^*$ unless an external disturbance is introduced.

To motivate the connection of SA to the ODE, note that a *deterministic* version of (4.6) (equivalent to the steepest descent algorithm in Section 1.4) can be written as

$$\frac{\hat{\boldsymbol{\theta}}_{k+1} - \hat{\boldsymbol{\theta}}_k}{a_k} = -\boldsymbol{g}(\hat{\boldsymbol{\theta}}_k). \tag{4.10}$$

Suppose that a_k is viewed as an increment in time, say $a_k = \tau_{k+1} - \tau_k$, where τ_k represents the kth time point. Equivalently, $\tau_{k+1} = \sum_{i=0}^{k} a_i$. Because $\hat{\boldsymbol{\theta}}_k$ is now a deterministic process, we can write $\hat{\boldsymbol{\theta}}_k = \boldsymbol{Z}(\tau_k)$ for some deterministic function $\boldsymbol{Z}(\cdot)$. Then, (4.10) can be reexpressed as

$$\frac{\boldsymbol{Z}(\tau_{k+1}) - \boldsymbol{Z}(\tau_k)}{\tau_{k+1} - \tau_k} = -\boldsymbol{g}(\boldsymbol{Z}(\tau_k)). \tag{4.11}$$

Because $a_k = \tau_{k+1} - \tau_k \to 0$ as $k \to \infty$ (assumption B.1), the ODE in (4.9) can be regarded as a limiting form of the difference equation (4.11). Hence, for sufficiently large k, the behavior of (4.10) and (4.11) bear close resemblance to the ODE in (4.9).

Of course, the deterministic recursion in (4.10) is *not* identical to the SA recursion of interest, (4.6). The addition of the noise (i.e., $\boldsymbol{Y}(\boldsymbol{\theta})$ measurements instead of $\boldsymbol{g}(\boldsymbol{\theta})$ measurements) represents a fundamental distinction between these two recursions. The non-ODE-related conditions for convergence (B.1 on the gains, B.4 on the bounded variance, etc.) bridge the gap between the deterministic ODE and the stochastic algorithm of interest.

Let us now present two examples of the construction and analysis of the associated ODE. The first is a simple linear root-finding problem and the second is a nonlinear problem. The ability to verify the ODE conditions in detail (i.e., to write down and solve the ODEs) is generally not possible in practice. In both of the examples below, there is complete knowledge of $\boldsymbol{g}(\boldsymbol{\theta})$, which will not be the case when noise is present. Nevertheless, this idealized structure provides some insight into the role of ODEs in the convergence of SA algorithms.

Example 4.5—ODE for a linear problem. Suppose that $p = 2$ and

$$\boldsymbol{g}(\boldsymbol{\theta}) = \begin{bmatrix} 2 & 1 \\ 1 & 2 \end{bmatrix} \boldsymbol{\theta}.$$

Expressed as an ODE as in (4.9), the above is

$$\frac{d\boldsymbol{Z}}{d\tau} = -\begin{bmatrix} 2 & 1 \\ 1 & 2 \end{bmatrix} \boldsymbol{Z},$$

leading to the solution

$$\boldsymbol{Z}(\tau) = -\begin{bmatrix} C_0 e^{-3\tau} + C_1 e^{-\tau} \\ C_0 e^{-3\tau} - C_1 e^{-\tau} \end{bmatrix},$$

where C_0 and C_1 are constants (Exercise 4.9). The constants are determined from the initial condition $\boldsymbol{Z}(0)$; in particular $-[C_0 + C_1,\ C_0 - C_1]^T = \boldsymbol{Z}(0)$. We see that $\boldsymbol{\theta}^* = \boldsymbol{0}$ is an asymptotically stable equilibrium since all initial conditions $\boldsymbol{Z}(0)$ near $\boldsymbol{\theta}^*$ produce solutions $\boldsymbol{Z}(\tau)$ that stay near $\boldsymbol{\theta}^*$ and that converge to $\boldsymbol{\theta}^*$. The domain of attraction is all of $\mathbb{R}^2$ because for any $\boldsymbol{Z}(0) \in \mathbb{R}^2$, $\boldsymbol{Z}(\tau) \to \boldsymbol{0}$ as $\tau \to \infty$. Hence, the ODE aspects of conditions B.2 and B.3 are satisfied for this problem. ❑

Example 4.6—ODE for a nonlinear problem. Again, suppose that $p = 2$. Consider the nonlinear function

$$\boldsymbol{g}(\boldsymbol{\theta}) = \begin{bmatrix} 2 + 2t_1 + t_2 \exp(t_1 t_2) \\ 6t_2 + t_1 \exp(t_1 t_2) \end{bmatrix},$$

where $\boldsymbol{\theta} = [t_1, t_2]^T$. Let us check the ODE-based convergence conditions by solving the associated ODE for its critical points and inspecting the domain of attraction. Let $\boldsymbol{Z}(\tau) = [Z_1(\tau), Z_2(\tau)]^T$. Analogous to Example 4.5, the system of ODEs based on the form of $\boldsymbol{g}(\boldsymbol{\theta})$ is

$$\frac{d\boldsymbol{Z}}{d\tau} = -\begin{bmatrix} 2 + 2Z_1 + Z_2 \exp(Z_1 Z_2) \\ 6Z_2 + Z_1 \exp(Z_1 Z_2) \end{bmatrix}.$$

This system has no analytical solution, but it can be solved using standard numerical methods for ODEs. An isolated critical point (corresponding to $d\boldsymbol{Z}/d\tau = \boldsymbol{0}$) is found at $\boldsymbol{Z} = [-1.0643, 0.1510]^T$. This critical point is the unique solution $\boldsymbol{\theta}^*$. The domain of attraction for this point is all of $\mathbb{R}^2$, so the ODE aspects of conditions B.2 and B.3 are all satisfied. A phase plot for this system is shown in Figure 4.2. ❑

4.4 ASYMPTOTIC NORMALITY AND CHOICE OF GAIN SEQUENCE

Of central importance in SA is knowledge that the iterate $\hat{\boldsymbol{\theta}}_k$ will converge to a solution $\boldsymbol{\theta}^*$ as $k \to \infty$. Such knowledge ensures that $\hat{\boldsymbol{\theta}}_k$ gets to within a small neighborhood of $\boldsymbol{\theta}^*$ with enough computational and/or experimental resources devoted to the search process. However, convergence by itself gives no information about the speed with which the iterate approaches $\boldsymbol{\theta}^*$. To address the issue of convergence rate, this section discusses the probability distribution of the iterate. Knowledge of the distribution also provides insight into two related aspects of the algorithm: (i) error bounds for the iterate and (ii) guidance into the choice of the gain a_k so as to minimize the likely deviation of $\hat{\boldsymbol{\theta}}_k$ from $\boldsymbol{\theta}^*$.

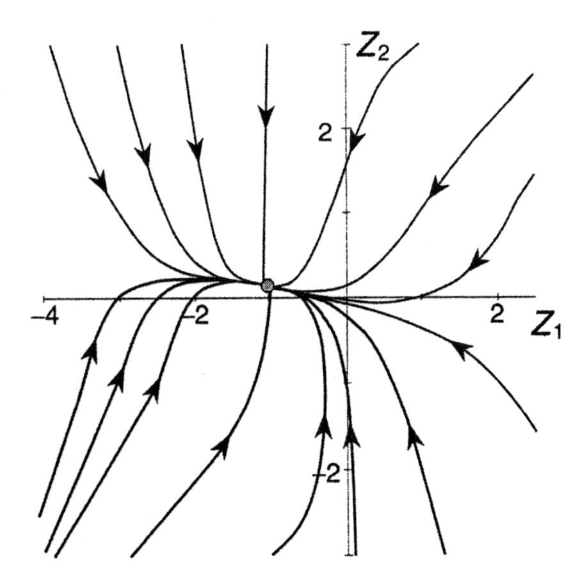

Figure 4.2. ODE convergence paths for Example 4.6. Each line depicts a path that $\boldsymbol{Z}(\tau) = [Z_1(\tau), Z_2(\tau)]^T$ follows over time from a particular initial condition $\boldsymbol{Z}(0)$. There is global convergence to $\boldsymbol{\theta}^* = [-1.0643, 0.1510]^T$ from any $\boldsymbol{Z}(0)$.

Unfortunately, for general nonlinear problems, there is no known finite-sample ($k < \infty$) distribution for the SA iterate. Further, the theory governing the asymptotic ($k \to \infty$) distribution is rather difficult. This is to be expected given the nonlinear transformations of the underlying random effects $\boldsymbol{e}_k$ that enter at each iteration. The nonlinear transformation is apparent by examining the forms in (4.2), (4.6), and (4.7): A value $\boldsymbol{e}_k$ at one iteration is transformed via the generally nonlinear mapping $\boldsymbol{g}(\cdot)$ in obtaining the next measurement $\boldsymbol{Y}_{k+1}$, and this measurement forms the basis for $\hat{\boldsymbol{\theta}}_{k+1}$, which, in addition to the new random effect $\boldsymbol{e}_{k+1}$, is the point of evaluation for $\boldsymbol{g}(\cdot)$ in the next iteration. Additional complications ensue when the noise itself is statistically dependent on the previous $\boldsymbol{\theta}$ estimates. Since this process is repeated for as many iterations as the algorithm takes, the nonlinear transformations of $\boldsymbol{e}_0, \boldsymbol{e}_1, \boldsymbol{e}_2, \ldots$ are usually very complex.

General results on the asymptotic distribution of the SA iterate are given in Fabian (1968). His work is a generalization of the first asymptotic distribution results for SA in Chung (1954) and Sacks (1958). Fabian shows that, under appropriate regularity conditions,

$$k^{\alpha/2}(\hat{\boldsymbol{\theta}}_k - \boldsymbol{\theta}^*) \xrightarrow{\text{dist.}} N(\boldsymbol{0}, \boldsymbol{\Sigma}) \qquad (4.12)$$

as $k \to \infty$, where $\xrightarrow{\text{dist.}}$ denotes "converges in distribution" (as discussed in Appendix C), $\boldsymbol{\Sigma}$ is some covariance matrix that depends on the coefficients in the gain sequence a_k and on the Jacobian matrix of $\boldsymbol{g}(\boldsymbol{\theta})$ (i.e., the derivative $\boldsymbol{H}(\boldsymbol{\theta}) = \partial \boldsymbol{g}/\partial \boldsymbol{\theta}^T$), and α governs the decay rate for the SA gain a_k (e.g., $a_k = a/(k+1)^{\alpha}$). The intuitive interpretation of (4.12) is that $\hat{\boldsymbol{\theta}}_k$ is approximately normally distributed with mean $\boldsymbol{\theta}^*$ and covariance matrix $\boldsymbol{\Sigma}/k^{\alpha}$ for k reasonably large. Ruppert (1991) also discusses this result. Various special cases of this result dealing with the situation $\alpha = 1$ are presented in Rustagi (1994, pp. 258–259), Ljung et al. (1992, pp. 71–78), and Ruppert (1991).

Expression (4.12) implies that the rate at which the iterate $\hat{\boldsymbol{\theta}}_k$ approaches $\boldsymbol{\theta}^*$ is proportional in a stochastic sense to $k^{-\alpha/2}$ for large k. That is, allowing for random variation at each k, $\hat{\boldsymbol{\theta}}_k - \boldsymbol{\theta}^*$ decays at a rate proportional to $k^{-\alpha/2}$ to balance the $k^{\alpha/2}$ "blow-up" factor on the left-hand side of (4.12) and yield a well-behaved random vector with the distribution $N(\boldsymbol{0}, \boldsymbol{\Sigma})$ on the right-hand side of (4.12) (i.e., "well-behaved" in the sense of being neither degenerate 0 nor ∞ in magnitude). Under condition A.1 or B.1 on the gain a_k for convergence of the iterate (Theorem 4.1), the rate of convergence of $\hat{\boldsymbol{\theta}}_k$ to $\boldsymbol{\theta}^*$ is maximized at $\alpha = 1$ when the gain has the standard form $a_k = a/(k+1)^{\alpha}$. That is, the maximum rate of convergence for the root-finding SA algorithm under the general conditions above is $O(1/\sqrt{k})$ in an appropriate stochastic sense.

Further, through minimizing a norm of the matrix $\boldsymbol{\Sigma}$ appearing in (4.12), it is known (see, e.g., Benveniste et al., 1990, pp. 110–116) that the asymptotically optimal gain for root-finding SA is a *matrix* gain given by the scaled inverse Jacobian matrix

$$\boldsymbol{a}_k = \frac{\boldsymbol{H}(\boldsymbol{\theta}^*)^{-1}}{k+1}, \quad k \geq 0. \tag{4.13}$$

Recall that this Jacobian matrix corresponds to the Hessian matrix of $L(\boldsymbol{\theta})$ if $\boldsymbol{g}(\boldsymbol{\theta})$ represents a gradient for an underlying problem of minimizing $L(\boldsymbol{\theta})$. With the exception of the decay factor $k+1$ included to damp out the noise effects, the gain in (4.13) is identical to the multiplier of $-\boldsymbol{g}(\boldsymbol{\theta})$ in the deterministic Newton–Raphson search of Section 1.4. With the gain given in (4.13), the limiting covariance matrix $\boldsymbol{\Sigma}$ is equal to the inverse of the average Fisher information matrix across the measurements (see Section 13.3 for a discussion of this matrix). The inverse Fisher information matrix is the smallest possible covariance matrix in the matrix sense discussed in Appendix A (so the optimal gain achieves this smallest possible covariance matrix).

Unfortunately, the gain in (4.13) is largely of theoretical interest only since in practice one does not know either $\boldsymbol{\theta}^*$ or the Jacobian matrix as a function of $\boldsymbol{\theta}$. It is also an asymptotic result, and, as discussed below, optimality for practical finite-sample analysis may impose other requirements. Nevertheless, this asymptotic result provides a type of ideal in designing adaptive SA algorithms (see Subsection 4.5.2 for a discussion of some practical implementations of (4.13)).

In practical problems, it may not be best to choose $\alpha = 1$. Most practitioners find that a lower value of α yields superior finite-sample behavior. This fact is also mentioned occasionally in the more-theoretical literature (e.g., Ruppert, 1991; Kushner and Yin, 1997, p. 328). The intuitive reason for the desirability of $\alpha < 1$ is that a slower decay provides a larger step size in the iterations with large k, allowing the algorithm to move in bigger steps toward the solution. This observation is a practical *finite-sample* result, as the asymptotic theory showing optimality of $\alpha = 1$ is unassailable.

Given the desirability for a gain sequence that balances algorithm stability in the early iterations with nonnegligible step sizes in the later iterations, a recommended gain form is

$$a_k = \frac{a}{(k+1+A)^{\alpha}}, \tag{4.14}$$

where the *stability constant* $A \geq 0$. Choosing $A = 0$ (i.e., no stability constant) is the most widely discussed form of gain sequence in the literature. However, there is a potential problem with $A = 0$. Choosing a large numerator a in the hopes of producing nonnegligible step sizes after the algorithm has been running awhile may cause unstable behavior in the early iterations (when the denominator is still small). Choosing a small a leads to stable behavior in the early iterations but sluggish performance in later iterations. For this reason, picking $A > 0$ is usually recommended.

A strictly positive A allows for a larger a without risking unstable behavior in the early iterations. Then, in the later iterations, the A in the denominator becomes negligible relative to the k, while the relatively large a in the numerator helps maintain a nonnegligible step size. These larger step sizes often enhance practical convergence. For values of $1/2 < \alpha \leq 1$, a reasonable choice for the stability constant is to pick A such that it is approximately 5 to 10 percent of the total number of expected (or allowed) iterations in the search process. This defines the sense in which A is "negligible" for large k. Ruppert (1991) and Okamura et al. (1995) also consider the gain form including the stability constant; the latter reference demonstrates the improvement possible in a signal processing application of estimation in adaptive filters.

In fact, in many applications, a *constant* step size ($\alpha = 0$) is used as a way of avoiding gains that are too small for large k. Typical applications involve adaptive tracking or control problems where $\boldsymbol{\theta}^*$ is changing in time. The constant gain provides enough impetus to the algorithm to keep up with the variation in $\boldsymbol{\theta}^*$. In contrast, a decaying gain provides too little weight to the current input information to allow for the algorithm to track the solution. Such constant-gain algorithms are also frequently used in neural network training even when there is no variation in the underlying $\boldsymbol{\theta}^*$ (White, 1989; Kuan and Hornik, 1991). As discussed with LMS (Section 3.2), algorithms with constant step sizes will generally not formally converge.[2] Also, note that the limiting distribution for the standardized SA iterate (analogous to the left-hand side of (4.12)) is *not* generally normal with constant step sizes (Mukherjee and Fine, 1996; Pflug, 1986).

Another common ad hoc "trick" to avoid potentially sluggish behavior is to periodically restart the algorithm. That is, after starting the search with a bona fide initial condition $\hat{\boldsymbol{\theta}}_0$, periodically reset the current estimate $\hat{\boldsymbol{\theta}}_k$ to $\hat{\boldsymbol{\theta}}_0$ while restarting the gain sequence at a_0. Ruppert (1991) provides an interpretation of the stability constant in (4.14) in the context of such a restarting method.

---

[2]A form of convergence theory is possible for constant gains. This is typically based on limiting arguments as the gain magnitude gets small. Essentially, one is able to show that the iterate from a constant-gain algorithm will approach the optimal $\boldsymbol{\theta}$ to within some error that decreases as the gain magnitude is made smaller (see, e.g., Macchi and Eweda, 1983; Kushner and Huang, 1983; Pflug, 1986). This was also discussed in the context of LMS in Subsection 3.2.2. Another form of convergence that can cope with nondecaying gains is the *weak convergence* approach. This form of convergence is a generalization of convergence in distribution. As discussed in Kushner and Yin (1997, Chaps. 7 and 8) and Yin and Yin (1996), this form allows one to say that with high probability, the iteration process will emulate the random behavior of the solution to a stochastic differential equation with Wiener noise input. A detailed discussion of weak convergence is beyond the scope of this book.

4.5 EXTENSIONS TO BASIC STOCHASTIC APPROXIMATION

The four subsections here discuss some extensions to the basic SA framework presented above. Subsection 4.5.1 considers the setting where the observation $\boldsymbol{Y}_k = \boldsymbol{Y}_k(\hat{\boldsymbol{\theta}}_k)$ includes a state vector that evolves as $\boldsymbol{\theta}$ is being updated. Subsection 4.5.2 discusses acceleration methods through intelligent gain selection and/or adaptive choices of whether to accept a new iteration value. Subsection 4.5.3 introduces iterate averaging for SA as a means for accelerating convergence and Subsection 4.5.4 considers the setting where the function $\boldsymbol{g}(\cdot)$ may change with time.

4.5.1 Joint Parameter and State Evolution

Consider an important special case of (4.3), where the input $\boldsymbol{x}_k$ represents a state vector related to the system being optimized. There is a statistical model governing the evolution of $\boldsymbol{x}_k$ in this special case, and this model is dependent on previous values of the estimated $\boldsymbol{\theta}$. Much of the book by Benveniste et al. (1990) is devoted to this framework, which was also considered by Ljung (1977) and Metivier and Priouret (1984). We may view $\boldsymbol{x}_k$ as a system observation evolving according to certain restrictions on the conditional probability $P(\boldsymbol{x}_{k+1} \in S \,|\, \hat{\boldsymbol{\theta}}_0, \hat{\boldsymbol{\theta}}_1, \ldots, \hat{\boldsymbol{\theta}}_k; \boldsymbol{x}_0, \boldsymbol{x}_1, \ldots, \boldsymbol{x}_k)$ for some set S. The convergence of $\hat{\boldsymbol{\theta}}_k$ (similar to that discussed in Section 4.3) then depends on the properties of this conditional probability. An important special case is when this conditional probability is a Markov process (Appendix E). Then, the convergence of $\hat{\boldsymbol{\theta}}_k$ is tied to the stationarity of the Markov process. Benveniste et al. (1990, Part I, Chaps. 1 and 4) discuss several applications of this framework in the context of telecommunications, fault detection, and signal processing.

One of the common representations for the evolution of $\boldsymbol{x}_k$ is a state equation that is linear in $\boldsymbol{x}_k$ (as in Subsection 3.3.4):

$$\boldsymbol{x}_{k+1} = \boldsymbol{A}(\hat{\boldsymbol{\theta}}_k)\boldsymbol{x}_k + \boldsymbol{B}(\hat{\boldsymbol{\theta}}_k)\boldsymbol{w}_k, \tag{4.15}$$

where $\boldsymbol{A}(\cdot)$ and $\boldsymbol{B}(\cdot)$ are appropriately dimensioned matrices and $\boldsymbol{w}_k$ is a sequence of independent random vectors (see, e.g., Ljung, 1977). In this case, the full SA algorithm is the recursion for $\boldsymbol{\theta}$ in (4.6) or (4.7) together with the state equation (4.15). As noted by Benveniste et al. (1990, p. 27), the coupling of a recursion for $\boldsymbol{\theta}$ together with (4.15) is a common form in the identification of linear systems. Then the evolution of $\boldsymbol{x}_k$ (and hence behavior of the SA recursion for $\boldsymbol{\theta}$) can be tied directly to the stability of the state equation. In particular, in the special constant-coefficient case where $\boldsymbol{A}$ is independent of $\boldsymbol{\theta}$, the well-known requirement that the eigenvalues of $\boldsymbol{A}$ lie inside the unit circle (the circle of radius 1 about the origin) guarantees the existence of the required stationary transition probabilities for the above-mentioned Markov process.

4.5.2 Adaptive Estimation and Higher-Order Algorithms

There are a large number of methods for adaptively estimating the gain a_k (or multivariate analogue of the gain). The aim is to enhance the convergence rate of the SA algorithm. Some of the results rely on choosing the gains adaptively depending on recent information acquired by the algorithm in its search process. Other results are built on SA analogues of the Newton–Raphson search in Section 1.4. In particular, some of the results are aimed at adaptively estimating the Jacobian (or Hessian) matrix in the asymptotically optimal gain $\boldsymbol{a}_k = \boldsymbol{H}(\boldsymbol{\theta}^*)^{-1}/(k+1)$, $k \geq 0$, discussed in Section 4.4 (eqn. (4.13)).

One of the first adaptive techniques is given in Kesten (1958), which is based on the signs (±) of the differences $\hat{\theta}_{k+1} - \hat{\theta}_k$ in a scalar θ process as a means of designing an adaptive gain sequence a_k. This approach does not explicitly use the connection to the Jacobian matrix as mentioned above. If there are frequent sign changes, this is an indication that the iterate is near θ^*; if the signs are not changing, this is an indication that the iterate is far from θ^*. This forms the basis for an adaptive choice of the gain a_k, where a larger gain is used if there are no sign changes and a smaller gain is used if the signs change frequently. Kesten (1958) established a.s. convergence to θ^* with such a scheme. A multivariate extension (including theoretical justification) of the Kesten idea is given in Delyon and Juditsky (1993).

There exist a number of stochastic analogues of the Newton–Raphson search in the context of parameter estimation for *particular* (possibly linear) models. The scalar gain a_k is then replaced by a matrix that approximates the (unknown) true inverse of the Jacobian (Hessian) matrix. Ljung and Soderstrom (1983, Sects. 2.4 and 3.4, Chap. 4), Macchi and Eweda (1983), Benveniste et al. (1990, Part I, Chaps. 3 and 4), Ljung et al. (1992, Part III), Yin and Zhu (1992), and Ljung (1999, Sect. 11.6) discuss some of the philosophy and mechanics of such adaptive approaches in various special cases.

While the above methods for special cases are effective ways to increase convergence speed, they are restricted in their range of application. More general approaches are described, for example, in Nevel'son and Khas'minskii (1973), Ruppert (1985), and Wei (1987) (note: due to inconsistent Russian translations, Khas'minskii here is the same as Has'minskii in the 1973 book cited in Sections 4.1 and 4.3). The problem formulation used by Ruppert (1985) differs slightly from the standard root-finding form in that he converts the basic problem from one of finding the root to $\boldsymbol{g}(\boldsymbol{\theta}) = \boldsymbol{0}$ to one of minimizing $\|\boldsymbol{g}(\boldsymbol{\theta})\|^2$ (as discussed in Section 1.1, this conversion yields the same $\boldsymbol{\theta}^*$ when there is a unique root).

In both Ruppert (1985) and Wei (1987), each column of the Jacobian matrix is approximated at the current value of $\boldsymbol{\theta}$ by perturbing one of the components of $\boldsymbol{\theta}$ in a positive and negative direction and evaluating a (noisy) $\boldsymbol{g}(\boldsymbol{\theta})$ at each of those two perturbed $\boldsymbol{\theta}$ values. This finite-difference process is repeated for all elements of $\boldsymbol{\theta}$ to get a full Jacobian matrix. In the process of producing each column, some averaging can be used to smooth out the noise effects. With such a scheme, Ruppert (1985) and Wei (1987) give conditions for a.s. convergence and asymptotic normality of the iterate (analogous to results in Sections 4.3 and 4.4). The author is unaware of any numerical evaluation of this type of approach.

Recent results in Spall (2000) suggest an easier and more efficient way of estimating the Jacobian using the ideas of simultaneous perturbation discussed in Chapter 7. At each iteration, only *two* noisy measurements of $\boldsymbol{g}(\boldsymbol{\theta})$ are required to estimate the Jacobian matrix, irrespective of the dimension p. This contrasts with a number of measurements of order of p in the finite-difference-based approaches above. Section 7.8 discusses this adaptive SA approach in detail. This use of only two measurements can lead to a large cost savings (i.e., fewer $\boldsymbol{g}(\boldsymbol{\theta})$ measurements) when p is large.

4.5.3 Iterate Averaging

Iterate averaging is an important and relatively recent development in SA. Like many good ideas in science and engineering, this idea is simple, and, in some problems, can be very effective. Ruppert (1991, based on an earlier 1988 internal technical report) and Polyak and Juditsky (1992) jointly introduced iterate averaging, together with the key supporting theory. There are several variations, but the basic idea is to replace $\hat{\boldsymbol{\theta}}_k$ as the final best estimate of $\boldsymbol{\theta}$ with the average

$$\overline{\boldsymbol{\theta}}_k \equiv (k+1)^{-1} \sum_{j=0}^{k} \hat{\boldsymbol{\theta}}_j \tag{4.16}$$

where each of the $\hat{\boldsymbol{\theta}}_j$ summands in (4.16) is computed as in (4.6) or (4.7). The basic SA recursion (4.6) or (4.7) does not change (i.e., $\overline{\boldsymbol{\theta}}_k$ is not used in (4.6) or (4.7)); $\overline{\boldsymbol{\theta}}_k$ is only used as the final estimate in place of $\hat{\boldsymbol{\theta}}_k$. Ruppert (1991) and Polyak and Juditsky (1992) show that $\sqrt{k}(\overline{\boldsymbol{\theta}}_k - \boldsymbol{\theta}^*)$ is asymptotically normally distributed with mean zero and a covariance matrix as small as possible in an appropriate statistical sense. That is, the limiting covariance matrix is the inverse of the average Fisher information matrix across the measurements (see Section 13.3); this is identical to the covariance matrix based on the optimal gain in (4.13).

The optimality of (4.16) holds for gain sequences satisfying the standard conditions mentioned in Section 4.3 plus the condition $a_{k+1}/a_k = 1 - o(a_k)$ (with "little-o" implying a term that goes to zero faster than the argument; note that $o(a_k)$ is a positive term in the standard case of a monotonically decaying gain). This important additional condition implies that a_k must decay at a rate *slower* than the asymptotically optimal rate of $O(1/k)$ for the individual iterates $\hat{\boldsymbol{\theta}}_k$ (as discussed in Section 4.4; see Exercise 4.17).

The implications of the above are quite impressive. Namely, one can use the standard algorithm in (4.6) together with the simple averaging in (4.16) to achieve the same optimal rate of convergence that otherwise is possible only with exact knowledge (or a convergent estimate) of the Jacobian matrix $\boldsymbol{H}(\boldsymbol{\theta}^*)$. This asymptotic optimality is available for *any* gain satisfying the above-mentioned general conditions. Hence, in principle, iterate averaging greatly reduces one of the traditional banes of SA—that of choosing the gain sequence in some "good" way.

Variations on the basic iterate averaging approach are readily available. One obvious one is to not include the first few iterations in the average, instead starting the average at some $N > 0$ or else using only a sliding window of the last $k - N$ (say) measurements. In practical implementations, such modifications are likely to help since the first few iterations tend to produce the poorest estimates. In the sliding window approach, formal asymptotic optimality can be shown if the window length grows with time (see, e.g., Kushner and Yang, 1993; Kushner and Yin, 1997, Chap. 11). A further modification to the basic approach is to use the averaged value $\overline{\boldsymbol{\theta}}_k$ (together with $\hat{\boldsymbol{\theta}}_k$) in a modified form of the SA iteration (instead of $\hat{\boldsymbol{\theta}}_k$ alone on the right-hand side of the basic form (4.6) or (4.7)). This is the feedback approach in Kushner and Yang (1995) (see also Kushner and Yin, 1997, Chap. 11), which can be shown to yield further improvement in certain situations.

In practice, however, the results on iterate averaging are more mixed than the above would suggest. While some numerical results have shown considerable promise (e.g., Yin and Zhu, 1992; Kushner and Yin, 1997, Chap. 11), other numerical studies (e.g., Spall and Cristion, 1998; Spall, 2000) have shown that a reasonable tuned a_k sequence often yields results superior to those possible by averaging, including averaging using the same tuned gains (see also Wang, 1996, p. 57 for some cautionary notes). This is not surprising upon reflection as a consequence of the *finite-sample* properties of practical problems.

For iterate averaging to be successful, it is necessary that a large proportion of the individual iterates hover in some balanced way around $\boldsymbol{\theta}^*$, leading to the sample mean of the iterates being nearer $\boldsymbol{\theta}^*$ than the bulk of the individual iterates. A well-designed (stable) SA algorithm will not be jumping approximately uniformly around $\boldsymbol{\theta}^*$ when the iterates are far from the solution (else it is likely to diverge). The only way for the bulk of the iterates to be distributed uniformly around the solution is for the individual iterates to be near the solution, where *near* here is relative to the level of noise in the problem. That is, other things being equal, the domain in which the iterates begin bouncing roughly uniformly around $\boldsymbol{\theta}^*$ gets smaller as the level of noise gets smaller.

With a well-designed algorithm, $\hat{\boldsymbol{\theta}}_k$ in most practical settings moves in a way that *approximately* decreases the distance to $\boldsymbol{\theta}^*$ in a monotonic manner. The "approximately" clause follows partly from the inherent stochastic variability. A user terminates the algorithm when either the budget of iterations has been exceeded or when $\hat{\boldsymbol{\theta}}_k$ begins to move very slowly near (one hopes!) $\boldsymbol{\theta}^*$. The latter situation is precisely when iterate averaging *starts* to work well. In fact, while the algorithm is in its monotonic phase, iterate averaging tends to *hurt* the accuracy of those components in $\hat{\boldsymbol{\theta}}_k$ that have not yet settled near $\boldsymbol{\theta}^*$!

Figure 4.3 illustrates this dichotomy on a search path for a typical $p = 2$ problem. Figure 4.3(a) shows a standard case of termination without the iterate bouncing around $\boldsymbol{\theta}^*$. Figure 4.3(b) depicts a case where the algorithm is not terminated until the iterate oscillates around the solution for some time. Case (b) is favorable to iterate averaging.

A further contrast of iterate averaging with an optimal algorithm based on the Jacobian matrix or an approximation (Subsection 4.5.2) is that the transform invariance property of Newton–Raphson-type algorithms (Section 1.4) may enhance convergence by improving the search direction when the magnitudes of the $\boldsymbol{\theta}$ components differ significantly. This improvement may occur without the iterate having to bounce around $\boldsymbol{\theta}^*$ (i.e., when the iteration process is in the monotonic phase mentioned above). That is, the optimal algorithm based on the Jacobian matrix does not require the algorithm to run long enough so that it is bouncing around $\boldsymbol{\theta}^*$. (However, the algorithm must run long enough to obtain a good estimate when the Jacobian matrix is being estimated.) This suggests that despite the simplicity and asymptotic justification, iterate averaging in practical finite-sample problems *may* not achieve the efficiency of the optimal algorithm based on the Jacobian matrix.

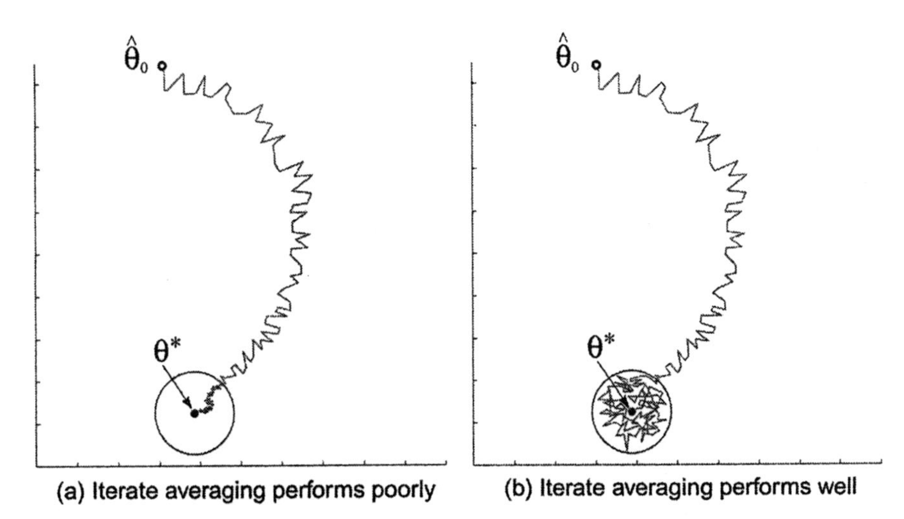

Figure 4.3. Two search paths for a $p = 2$ problem. Note *approximate* monotonic improvement in $\hat{\boldsymbol{\theta}}_k$ until search reaches circled area. Iterate average $\overline{\boldsymbol{\theta}}_k$ is usually poorer estimate than $\hat{\boldsymbol{\theta}}_k$ before search process enters circled area. If process does not oscillate about $\boldsymbol{\theta}^*$ (case (a)), terminal iterate average $\overline{\boldsymbol{\theta}}_k$ will be poorer estimate than $\hat{\boldsymbol{\theta}}_k$. If process oscillates around $\boldsymbol{\theta}^*$ as in case (b), then terminal $\overline{\boldsymbol{\theta}}_k$ is likely to provide better estimate than $\hat{\boldsymbol{\theta}}_k$ if process is allowed to run long enough inside circled area. Case (a) is commonly associated with low noise, while case (b) may be seen with high noise.

4.5.4 Time-Varying Functions

A further generalization is one where the root-finding function varies with k. Among other references, this problem is formally treated in Goodsell and Hanson (1976) and Evans and Weber (1986). The basic idea is that, while the function for which one is finding a root may change shape with k, it is assumed that the underlying root of $\boldsymbol{g}_k(\boldsymbol{\theta}) = \boldsymbol{0}$, say $\boldsymbol{\theta}_k^*$, is either constant for all k or reaches a limiting value as $k \to \infty$ (even though $\boldsymbol{g}_k(\boldsymbol{\theta})$ may change shape indefinitely). The two references mentioned above show a.s. convergence in these cases for the scalar θ setting; the proofs have to be changed from those commonly seen in SA to accommodate the time-varying loss. A multivariate extension is in Spall and Cristion (1998) in the context of nonlinear adaptive control. As an example of time-varying functions, Section 4.1 discussed the common situation where there are inputs $\boldsymbol{x}_k$ at each noisy evaluation of the function. In particular, $\boldsymbol{g}_k(\boldsymbol{\theta}) \equiv \tilde{\boldsymbol{g}}_k(\boldsymbol{\theta}, \boldsymbol{x}_k)$ for some function $\tilde{\boldsymbol{g}}_k$.

Figure 4.4 depicts the concept for the scalar ($p = 1$) case where $g_k(\theta)$ represents the gradient in a problem of minimizing time-varying loss functions $L_k(\theta)$. This figure depicts a problem where the time-varying loss (and corresponding gradient) functions never settle down, but where the time-varying solution θ_k^* does approach a limiting value θ_∞^*. The above-mentioned theory applies in such a setting, leading to an SA algorithm with an a.s. convergent $\hat{\theta}_k$ (i.e., $\hat{\theta}_k \to \theta_\infty^*$ a.s. as $k \to \infty$).

The time-varying setting in Figure 4.4 arises frequently in control problems. The time-varying loss follows when the target for the system performance may be perpetually varying, even when the underlying system dynamics remain essentially unchanged. Let us sketch an example of an application of this setting to a control problem.

Example 4.7—Sketch of implementation in adaptive control. Suppose that a system output vector z_k is modeled as $z_k = \boldsymbol{h}(\boldsymbol{\theta}, \boldsymbol{x}_k) + \boldsymbol{v}_k$, where $\boldsymbol{h}(\cdot)$ is a representation of the process under study and $\boldsymbol{v}_k$ is the noise (a nonlinear analogue of the linear setting of Chapter 3). The input $\boldsymbol{x}_k$ represents some input (control) value that is being set to try to make z_k perform in a desired way.

A common specific case of a control problem is target tracking, where the aim is to have z_k follow a time-varying desired value $\boldsymbol{d}_k$. As part of determining the best input values $\boldsymbol{x}_k$, it is common to have to estimate $\boldsymbol{\theta}$. (As mentioned in Subsection 3.2.3, *indirect adaptive control* refers to this process of estimating model parameters as a vehicle to obtain optimal control inputs; in contrast, *direct adaptive control* is where the control inputs are determined directly without estimating the model.) In target tracking, one faces an estimation problem associated with minimizing a time-varying error

$$L_k(\boldsymbol{\theta}) = E\left(\|z_k - \boldsymbol{d}_k\|^2\right).$$

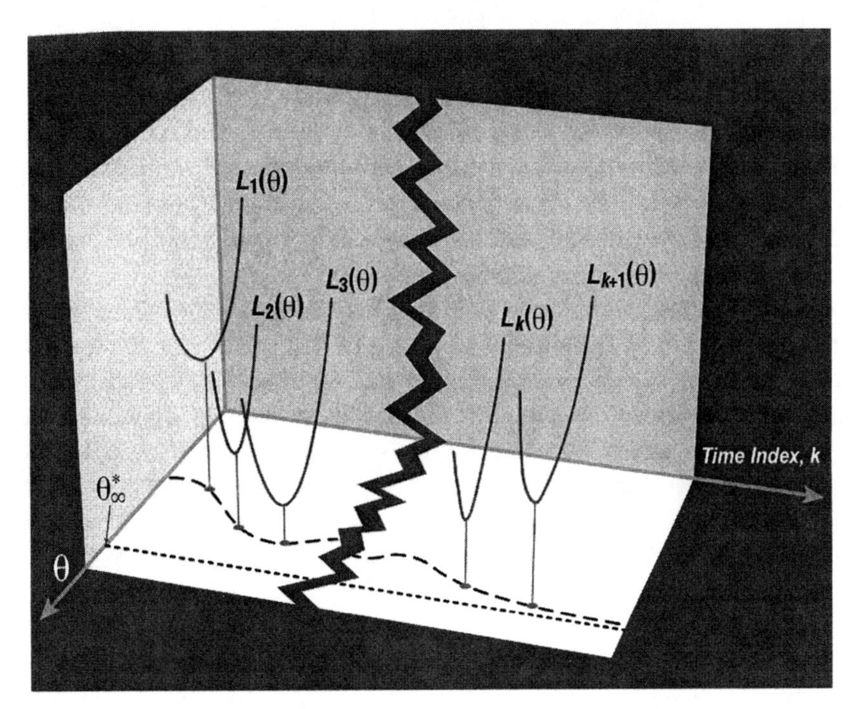

Figure 4.4. Time-varying $L_k(\theta)$ (leading to time-varying $g_k(\theta) = \partial L_k(\theta)/\partial\theta$) for minimization problem with limiting minimum at θ^*_∞.

The function $\boldsymbol{g}_k(\boldsymbol{\theta})$ represents a gradient of the above time-varying loss. Based on noisy measurements of $\boldsymbol{g}_k(\boldsymbol{\theta})$ at each k, SA may be used to find the estimate for $\boldsymbol{\theta}$ according to $\boldsymbol{g}_k(\boldsymbol{\theta}) = \boldsymbol{0}$. If it is assumed that $\boldsymbol{h}(\boldsymbol{\theta}_{\text{true}}, \boldsymbol{x})$ represents the true process, then, under modest conditions, $\boldsymbol{\theta}^*_k = \boldsymbol{\theta}_{\text{true}}$ for all k. That is, the best value for $\boldsymbol{\theta}$ in the *model* (i.e., the value minimizing $L_k(\boldsymbol{\theta})$) is the true value for $\boldsymbol{\theta}$. As a simple example, suppose that both the model and the true system are represented by the scalar polynomial $h(\boldsymbol{\theta}, x) = b_0 + b_1 x + b_2 x^2$, where $\boldsymbol{\theta} = [b_0, b_1, b_2]^T$. Then, $\boldsymbol{\theta}^*_k = \boldsymbol{\theta}_{\text{true}}$ for all k since there is one set of true b_0, b_1, and b_2. This constant solution for all k applies even though $L_k(\boldsymbol{\theta})$ and $\boldsymbol{g}_k(\boldsymbol{\theta})$ may be perpetually changing. ❑

4.6 CONCLUDING REMARKS

The root-finding stochastic approximation framework is a cornerstone of stochastic search and optimization. Many popular search methods are special cases of root-finding SA. Moreover, there is a deep theory associated with the convergence of SA algorithms. Because of the web of connections to many other methods, the SA theory has broad implications for the performance of general stochastic search and optimization algorithms, especially in cases with only noisy measurements for use in the search process.

This chapter discussed two general approaches to convergence analysis of SA—the "statistics" and "engineering" approaches. The former relies on classical assumptions about the shape of the function $\boldsymbol{g}(\boldsymbol{\theta})$ and the noise. The latter is built on connections to the stability and convergence of an associated ordinary differential equation. Both approaches include conditions on the all-important gain sequence (a_k). In fact, the choice of the gain sequence remains one of the key challenges in many practical problems.

Some of the conditions for convergence may be difficult to check and/or difficult to satisfy in practical problems (e.g., A.2 or B.2 and B.3 in Section 4.3). Nonetheless, the theory provides general guidance on the expected performance of SA, and, with the appropriate qualifications, may provide guidance in cases that do not entirely satisfy the conditions (e.g., guidance in multiple root problems provided that the search is restricted to certain neighborhoods of the search space Θ). The results of this chapter do not *directly* apply with multiple local minima when used in an optimization context. Extensions of SA to the global optimization problem (introduced in Section 1.1) are discussed in Sections 7.7 and 8.4.

There have been countless applications of SA in the greater than half century since the seminal publication of Robbins and Monro (1951). This chapter touched on only a few of these (univariate and bivariate quantile estimation, model estimation, and adaptive control). Some areas not discussed in detail in this chapter include neural network training, simulation-based optimization, evolutionary algorithms, machine learning, experimental design, and signal processing applications such as noise cancellation and pattern recognition. Some of these are discussed elsewhere in this book. Benveniste et al. (1990, Part I), Kushner and Yin (1997, Chaps. 2 and 3), Yin (2002), and some of the other references cited are among the publications that summarize these and other applications.

While all of the remaining chapters in this book involve at least *some* connection to SA, the next three chapters focus *directly* on SA, especially as related to optimization problems. The range of applications for SA now extends well beyond those envisioned by some of the pioneers in the field. This blossoming is expected to continue as technical challenges exceed the reach of classical deterministic search and optimization methods.

EXERCISES

4.1 Based on (4.3) and $p = 1$, suppose that $Y_k(\theta) = \tilde{g}_k(\theta, \boldsymbol{x}_k) = x_{k1}\theta - \sin(x_{k1} + x_{k2})$, where $\boldsymbol{x}_k = [x_{k1}, x_{k2}]^T$ is uniformly distributed over $[0, \pi/2] \times [0, \pi/2]$ for all k (note that $\tilde{e}_k(\theta, \boldsymbol{x}_k) = 0$). Further, suppose that $E[e_k(\theta)] = 0$ for all k. What are $g(\theta)$ and θ^*?

4.2 Consider the problem of minimizing $L(\boldsymbol{\theta}) = t_1^4 + t_1^2 + t_1 t_2 + t_2^2$, $\boldsymbol{\theta} = [t_1, t_2]^T$. Suppose that the loss gradient $\boldsymbol{g}(\boldsymbol{\theta})$ can only be measured in the presence of independent $N(\mathbf{0}, \sigma^2 \boldsymbol{I}_2)$ noise (the $\boldsymbol{e}$ term). In particular, using the two gain sequences, $a_k = 0.1/(k+1)$ and $a_k = 0.1/(k+1)^{0.501}$, $k = 0, 1, \ldots$, compare the mean terminal loss function value after 1000 iterations using 200 realizations at each gain sequence (it is not necessary to perform a statistical test in this comparison). Take $\sigma = 0.1$ and $\hat{\boldsymbol{\theta}}_0 = [1, 1]^T$. Repeat the above for the case where $\sigma = 1.0$. What can you say about the relative performance of the two gain sequences for each of the two noise levels?

4.3 For the problem of estimating μ from a sample of independent measurements, consider the recursion in Example 4.1, $\hat{\theta}_{k+1} = \hat{\theta}_k - a_k Y_k(\hat{\theta}_k)$, where $Y_k(\hat{\theta}_k) = \hat{\theta}_k - X_{k+1}$. In contrast to Example 4.1, however, now suppose that $a_k = 0.25/(k+1)$. Calculate 20 replications of n = 100, 10,000, and 1,000,000 iterations, where X_k is uniformly distributed over (0, 2) (i.e., $X_k \sim U(0, 2)$) for all k. For each n, report the sample mean of the terminal estimate for μ from the 20 replications together with an approximate 95 percent confidence interval (using the t-distribution). Comment on the observed results relative to those expected from $a_k = 1/(k+1)$, as in Example 4.1.

4.4 In Example 4.2 on quantile estimation, assume that experimental process X is governed by a $U(-1, 1)$ distribution. Suppose that $\hat{\theta}_0 = 0.6$.

(a) Estimate the LD_{50} point using $a_k = 1/(k+1)$ and $a_k = 1/(k+1)^{0.501}$. For each of the two gains, use 100 Monte Carlo replications with n = 200 experiments (number of X values) for each replication. Use a different random number seed to initialize each of the sets of 100 Monte Carlo replications. Applying the appropriate unmatched pairs two-sample test in Appendix B, can you draw any valid statistical conclusion about the relative performance of the SA algorithm with the two gain sequences?

(b) After some tuning experiments, choose a new gain sequence of the form in (4.14). Generate another 100 replications with n = 200. Does this new gain sequence yield a statistically significant improvement?

4.5 Implement the root-finding procedure in the stochastic gradient mode of Example 4.4 to estimate the parameters $\boldsymbol{\theta} = [\lambda, \beta]^T$. For each replication as requested below, generate data on a grid defined by the integer pairs (C, W) where $C \in [1, 10]$ and $W \in [11, 110]$ (e.g., (5, 15), (4, 87), (8, 33), etc.). Perform 1000 iterations by randomly sampling (uniformly) the 1000 grid points without replacement until all 1000 points have been used. Let $\boldsymbol{\theta}^* = [2.5, 0.7]^T$ and use (4.8) to compute the output z_k with random error v_k defined below. Use $\hat{\boldsymbol{\theta}}_0 = [1.0, 0.5]^T$ as the starting point and $a_k = 0.0015/(k+100)^{0.501}$ as the gain. Consider two distributions for the noise in the measurements z_k: (i) normal error where $v_k \sim N(0, 5^2)$ and (ii) dependent error where $v_k = 0.2h(\boldsymbol{\theta}^*, \boldsymbol{x}_k)w_k$ with $w_k \sim N(0, 1)$ and $\boldsymbol{x}_k = [W_k, C_k]^T$. Do five replications (each of 1000 iterations as described above) for each of the two noise distributions. Compute the distance of the final estimate to $\boldsymbol{\theta}^* = [2.5, 0.7]^T$ for each replication. Report the results for each replication.

4.6 It was stated in Section 4.3 that if $\text{cov}(\boldsymbol{e}_k) \geq \eta \boldsymbol{I}_p$ for some $\eta > 0$, then $E(\|\boldsymbol{e}_k\|^2) \geq \eta p$ when $\boldsymbol{b}_k = \boldsymbol{0}$. Prove this result by using the definition of positive semidefiniteness in matrix relationship (xiii) of Appendix A.

4.7 Show that the condition $\sup_{k\geq 0}\|\hat{\boldsymbol{\theta}}_k\| < \infty$ a.s. in B.3 of Section 4.3 does *not* imply that $\|\hat{\boldsymbol{\theta}}_k\|$ is *uniformly* bounded in magnitude (a.s.) for all k. (This exercise requires a knowledge of measure-theoretic probability at the level of Appendix C.)

4.8 Suppose $p = 2$ and $\boldsymbol{g}(\boldsymbol{\theta}) = \begin{bmatrix} 4 & 0 \\ 0 & 2 \end{bmatrix}\boldsymbol{\theta}$. Based on the associated ODE (4.9), determine whether $\boldsymbol{\theta}^*$ is an asymptotically stable equilibrium and determine the domain of attraction.

4.9 Prove that the indicated solution $\boldsymbol{Z}(\tau)$ in Example 4.5 is the solution to the associated ODE.

4.10 Consider the function $\boldsymbol{g}(\boldsymbol{\theta}) = \left[t_1^2 - t_1 - t_1t_2 + t_2^2/2,\ t_2^2 - 2t_2 + t_1t_2 - t_1^2/2\right]^T$, $\boldsymbol{\theta} = [t_1, t_2]^T$. Numerically or analytically identify the candidate points $\boldsymbol{\theta}^*$ where $\boldsymbol{g}(\boldsymbol{\theta}) = \boldsymbol{0}$. For each of these points, show that $\boldsymbol{g}(\boldsymbol{\theta})$ fails convergence condition A.2 when $\boldsymbol{B} = \boldsymbol{I}_2$.

4.11 Consider the function $\boldsymbol{g}(\boldsymbol{\theta})$ in Exercise 4.10. Write the associated ODE, and identify the points where $d\boldsymbol{Z}/d\tau = \boldsymbol{0}$. Show that only one of these points is a (locally) stable equilibrium for the ODE (if $\boldsymbol{Z}(\tau)$ is sufficiently close to this equilibrium for any τ, then $\boldsymbol{Z}(\tau)$ converges to this point as $\tau \to \infty$). Graphically identify the approximate domain of attraction for this equilibrium. Explain why $\boldsymbol{g}(\boldsymbol{\theta})$ violates convergence condition B.2, but intuitively why the ODE analysis indicates that convergence may still be possible with appropriate additional restrictions.

4.12 Verify convergence condition A.2 for the $\boldsymbol{g}(\boldsymbol{\theta})$ in Example 4.6.

4.13 Identify at least one gain form *other* than (4.14) that satisfies condition A.1 together with the range of coefficient values for this form.

4.14 Let $\boldsymbol{\theta} = [t_1, t_2, \ldots, t_{20}]^T$ and $\boldsymbol{g}(\boldsymbol{\theta}) = [4t_1^3, \ldots, 4t_{10}^3; 0, \ldots, 0]^T + 2\boldsymbol{B}\boldsymbol{\theta}$, where $\boldsymbol{B}$ is a symmetric matrix with 1's on the diagonals and 0.5's elsewhere. Measurements $\boldsymbol{Y}(\boldsymbol{\theta}) = \boldsymbol{g}(\boldsymbol{\theta}) + \boldsymbol{e}$ are available to the algorithm of interest, where $\boldsymbol{e}$ is i.i.d. $N(\boldsymbol{0}, \sigma^2\boldsymbol{I}_{20})$ distributed. Let us compare the unconstrained and constrained algorithms (4.6) and (4.7). Use a gain sequence $a_k = a/(k+1+A)^{0.501}$, $a = 0.01$, and $A = 100$, with $\hat{\boldsymbol{\theta}}_0 = 0.1\times[1, 1, \ldots, 1]^T$. For $\sigma = 20$ and $\sigma = 1$, compare the unconstrained and constrained results. For (4.7), assume that $\Theta = [-0.1, 0.1]^{20}$ and that the projection Ψ_Θ simply moves any component of $\boldsymbol{\theta}$ lying outside of $[-0.1, 0.1]$ back to the nearest endpoint of the interval. Based on 25 replications and 1000 iterations/replication, use the observed RMS error in the terminal $\boldsymbol{\theta}$ estimate (relative to $\boldsymbol{\theta}^*$) as the basis for comparison (i.e., the square root of the average terminal MSE across the 25 replications).

4.15 For the setting of Exercise 4.2 with $\sigma = 0.1$, implement the iterate averaging form of SA discussed in Subsection 4.5.3 with $a_k = 0.1/(k+1)^{0.501}$. Use both the basic averaging approach in (4.16) together with a sliding window of the most recent 100 iterates. Compare these two iterate averaging approaches with each other and with the results from a standard SA (no iterate averaging) using the available loss function (not always available in root-finding problems!). In particular, calculate the sample mean of the terminal loss values from 50 realizations of 1000 iterations per realization for each of the three SA implementations.

4.16 Consider the function of Exercise 4.14 with $\sigma = 0.1$. Implement the iterate averaging form of SA discussed in Subsection 4.5.3 based on $a_k = a/(k+1+A)^{0.501}$, $a = 0.2$, and $A = 100$, with $\hat{\boldsymbol{\theta}}_0 = [1, 1,\ldots, 1]^T$. Use both the basic averaging approach in (4.16) together with a sliding window of the most recent 100 iterates. Compare these two iterate averaging approaches with each other and with the results from a standard SA (no iterate averaging). In particular, based on 25 replications and 1000 iterations/replication, calculate the sample RMS error in the terminal $\boldsymbol{\theta}$ estimate (relative to $\boldsymbol{\theta}^*$) for each of the three SA implementations (i.e., the square root of the sample mean of the squared errors across the 25 replications).

4.17 For gains of the form $a_k = a/(k+1)^{\alpha}$, $a > 0$, $\alpha > 0$, show that $a_{k+1}/a_k = 1 - o(a_k)$ implies that $\alpha < 1$ (relevant for iterate averaging in Subsection 4.5.3).

CHAPTER 5

STOCHASTIC GRADIENT FORM OF STOCHASTIC APPROXIMATION

Chapter 4 introduced the root-finding (Robbins-Monro) stochastic approximation (SA) algorithm as a general method for nonlinear problems. The aim is to find one or more zeros of the function $\boldsymbol{g}(\boldsymbol{\theta})$ (i.e., roots of $\boldsymbol{g}(\boldsymbol{\theta}) = \boldsymbol{0}$) when only noisy measurements of the function $\boldsymbol{g}(\boldsymbol{\theta})$ are available. Now, let us return to the essential optimization problem of minimizing a loss function $L = L(\boldsymbol{\theta})$. The root-finding SA algorithm discussed in Chapter 4 plays a central role in optimization through the classical problem of finding a $\boldsymbol{\theta}$ such that $\boldsymbol{g}(\boldsymbol{\theta}) = \partial L/\partial\boldsymbol{\theta} = \boldsymbol{0}$. That is, noisy measurements of the gradient $\boldsymbol{g}(\boldsymbol{\theta})$ are used in the SA algorithm. For this reason, the SA algorithm in this application is often referred to as a *stochastic gradient* algorithm. Many practical problems have been solved via the stochastic gradient formulation, a few of which are reviewed in this chapter.

Section 5.1 discusses the fundamental principles in the stochastic gradient interpretation of SA. The subsequent three sections are brief discussions of applications of the stochastic gradient-based SA algorithm. Section 5.2 considers neural network training and the interpretation of the well-known backpropagation algorithm as an SA algorithm. Section 5.3 treats discrete-event dynamic systems (e.g., queuing networks), focusing on the "infinitesimal perturbation analysis" method of gradient estimation. Section 5.4 considers image restoration. The discussions of Sections 5.2–5.4 merely touch on the subjects, the aim being to provide a sense of how the stochastic gradient method is used in applications. Section 5.5 offers some concluding remarks.

5.1 ROOT-FINDING STOCHASTIC APPROXIMATION AS A STOCHASTIC GRADIENT METHOD

This relatively long section is in four subsections. These subsections introduce the stochastic gradient algorithm for both recursive and batch processing, and give a brief discussion of the relationship of the least-mean-squares (LMS) algorithm of Chapter 3 to the stochastic gradient method.

5.1.1 Basic Principles

With a differentiable loss function $L(\boldsymbol{\theta})$, recall the familiar set of p equations and p unknowns for use in finding a minimum $\boldsymbol{\theta}^*$:

$$g(\boldsymbol{\theta}) = \frac{\partial L}{\partial \boldsymbol{\theta}} = \mathbf{0}\,. \tag{5.1}$$

Consider the setting where only noisy information is available about the loss function and gradient. We may then express the loss function as

$$L(\boldsymbol{\theta}) = E[Q(\boldsymbol{\theta}, V)]\,, \tag{5.2}$$

where V represents the random effects in the process generating the system output and $Q(\boldsymbol{\theta}, V)$ represents some "observed" cost as a function of the chosen $\boldsymbol{\theta}$ and random effects V (V for *variability* is perhaps a useful mnemonic). The variable V may represent the amalgamation of many individual random effects. The expectation in (5.2) is with respect to all randomness embodied in V. So $L(\boldsymbol{\theta})$ represents an average cost over all possible values of V at the specified $\boldsymbol{\theta}$.

For the discussion of this subsection, Q will represent either an instantaneous (recursive) input or a batch input. In the instantaneous case, as discussed in Subsections 5.1.2 and 5.1.3, Q and V will generally be indexed by k, as they reflect an iteration (k)-varying cost function and associated random input. So, $Q(\boldsymbol{\theta}, V)$ is replaced by $Q_k(\boldsymbol{\theta}, V_k)$. In the batch case, we write $\overline{Q}(\boldsymbol{\theta}, V)$ to denote an average observed cost function, averaged over the number of measurements as in the least-squares criterion of Chapter 3. To avoid having to distinguish between the instantaneous and batch cases in the discussion of this subsection, we simply let $Q(\boldsymbol{\theta}, V)$ represent $Q_k(\boldsymbol{\theta}, V_k)$ or $\overline{Q}(\boldsymbol{\theta}, V)$ as appropriate. It should be understood, for example, that for recursive applications below requiring an unbiased gradient measurement, expression (5.2) is interpreted as $L(\boldsymbol{\theta}) = E[Q_k(\boldsymbol{\theta}, V_k)]$.

Because of the nonlinearity (and possible lack of knowledge about the probability distribution of V), it is almost never the case that L can be computed. However, $Q(\boldsymbol{\theta}, V)$ is typically available since that just represents the outcome for a particular experiment (no expectation involved). Further, it is assumed that $Q(\boldsymbol{\theta}, V)$ is a differentiable function (in $\boldsymbol{\theta}$) for *almost all* V (i.e., for all values of V except possibly a set of values having probability zero; see Appendix C) and that we have knowledge of $\partial Q/\partial\boldsymbol{\theta}$ for any reasonable $\boldsymbol{\theta}$. The expression $\partial Q/\partial\boldsymbol{\theta}$ is called a *stochastic gradient* because it depends on the random term V.

The ability to compute the derivative $\partial Q/\partial\boldsymbol{\theta}$ is, of course, fundamental to the stochastic gradient approach. In many practical problems, it is not possible to compute $\partial Q/\partial\boldsymbol{\theta}$, motivating the gradient-free SA approaches of Chapters 6 and 7. Nonetheless, there is a large set of problems for which it *is* possible to compute the gradient. This set is the focus of this chapter.

Unlike the linear case in Chapter 3 (where the loss function is quadratic in $\boldsymbol{\theta}$), a solution to (5.1) is not necessarily a global minimum to (5.2). Nevertheless, this root-finding approach is adopted in the stochastic gradient SA framework since there is usually a guarantee of at least a local minimum. Obviously, this root-finding approach differs from direct search techniques for optimization, such as those described in Chapter 2. In those, one solves the optimization problem directly rather than through the proxy problem of root finding. Significant advantages of the SA-based root-finding approach over the earlier direct methods are the greater efficiency and the strong theory for guaranteeing performance with noisy measurements of the root-finding function.

While this chapter focuses on basic forms of the stochastic gradient algorithm, Section 8.4 discusses a modification to the basic algorithm to provide for *global* convergence. This modification involves the addition of a Monte Carlo-generated noise term to the right-hand side of the fundamental SA recursions (4.6) or (4.7) when implemented in a stochastic gradient context. This additional noise gives the algorithm enough "jumpiness" to ensure that it will not get prematurely trapped in local minima. The user controls the degree of Monte Carlo noise using principles of annealing (analogous to simulated annealing).

If we assume for the moment that V has a probability density function, then the most general representation of the expectation in (5.2) is

$$E[Q(\boldsymbol{\theta},V)] = \int_{\Lambda} Q(\boldsymbol{\theta},\upsilon)p_V(\upsilon\,|\,\boldsymbol{\theta})\,d\upsilon, \tag{5.3}$$

where $p_V(\upsilon|\boldsymbol{\theta})$ is the density function for V given $\boldsymbol{\theta}$ and the integral is over the domain Λ for V (υ is the dummy variable of integration). In the stochastic gradient formulation, however, the usual form is a special case of that shown in (5.3). In particular, V is usually viewed as some "fundamental" underlying random process that is generated *independently* of $\boldsymbol{\theta}$. Then, (5.3) is replaced by

$$E[Q(\boldsymbol{\theta},V)] = \int_{\Lambda} Q(\boldsymbol{\theta},\upsilon)p_V(\upsilon)\,d\upsilon, \tag{5.4}$$

where $p_V(\upsilon)$ is the density function for V. Although the formulations in (5.3) and (5.4) and the discussion to follow are given in terms of density functions for V, precisely the same line of reasoning applies to distributions for which density functions do not exist. In particular, if V involves discrete (or hybrid discrete/continuous) random variables, the arguments apply with probability mass functions replacing probability density functions and sums replacing integrals.

The distinction between (5.3) and (5.4)—with their density functions dependent and independent of $\boldsymbol{\theta}$—is fundamental in the stochastic gradient implementation. It is shown below that (5.4) is the preferred form for the stochastic gradient algorithm. Let us present an example showing how a simple transformation can *sometimes* convert the general setting of (5.3) to the simpler (5.4).

Example 5.1—Conversion to preferred stochastic gradient form. Suppose that $Q(\theta, V) = f(\theta, Z) + W$, where the scalar $W = W(\theta) \sim N(0, \theta^2\sigma^2)$ (with θ and σ both strictly positive), Z is some scalar random variable independent of θ, and $f(\cdot)$ is some function. Then V represents the combined random effects $\{Z, W(\theta)\}$, which we assume to have a joint probability density function. Clearly, with this definition, the density function for V must be represented by $p_V(\upsilon|\theta)$, not $p_V(\upsilon)$, since the distribution of W depends on θ. Suppose, on the other hand, that $Q(\theta, V)$ is expressed in the equivalent form $Q(\theta, V) = f(\theta, Z) + \theta W'$, where $W' \sim N(0, \sigma^2)$. Then, V represents the combined random effects $\{Z, W'\}$, where V has a density *independent* of θ. Hence, the problem is now in the realm of (5.4), as desired. ❑

Our interest is in unbiased measurements of the gradient $\boldsymbol{g}(\boldsymbol{\theta})$ at any $\boldsymbol{\theta}$. That is, $E[\boldsymbol{Y}(\boldsymbol{\theta})] = \boldsymbol{g}(\boldsymbol{\theta})$. Hence, the mean-zero noise conditions for convergence of root-finding SA discussed in Section 4.3 hold (condition A.3 or the special case of condition B.5 where the bias $\boldsymbol{b}_k = \boldsymbol{0}$). Suppose for the moment that it is valid to interchange the derivative $\partial(\cdot)/\partial\boldsymbol{\theta}$ and integrals in (5.3) or (5.4). We present formal conditions for the interchange in Theorem 5.1 below. Then, in case (5.3) where the density function depends on $\boldsymbol{\theta}$,

$$\begin{aligned} \boldsymbol{g}(\boldsymbol{\theta}) &= \frac{\partial}{\partial\boldsymbol{\theta}}\int_\Lambda Q(\boldsymbol{\theta},\boldsymbol{\upsilon})p_V(\boldsymbol{\upsilon}\,|\,\boldsymbol{\theta})\,d\boldsymbol{\upsilon} \\ &= \int_\Lambda\left[Q(\boldsymbol{\theta},\boldsymbol{\upsilon})\frac{\partial p_V(\boldsymbol{\upsilon}\,|\,\boldsymbol{\theta})}{\partial\boldsymbol{\theta}} + \frac{\partial Q(\boldsymbol{\theta},\boldsymbol{\upsilon})}{\partial\boldsymbol{\theta}}p_V(\boldsymbol{\upsilon}|\boldsymbol{\theta})\right]d\boldsymbol{\upsilon}. \end{aligned} \tag{5.5}$$

When (5.4) applies, the following simpler form results:

$$\boldsymbol{g}(\boldsymbol{\theta}) = \frac{\partial}{\partial\boldsymbol{\theta}}\int_\Lambda Q(\boldsymbol{\theta},\boldsymbol{\upsilon})p_V(\boldsymbol{\upsilon})\,d\boldsymbol{\upsilon} = \int_\Lambda \frac{\partial Q(\boldsymbol{\theta},\boldsymbol{\upsilon})}{\partial\boldsymbol{\theta}}p_V(\boldsymbol{\upsilon})\,d\boldsymbol{\upsilon}. \tag{5.6}$$

The expression appearing on the right-hand side of (5.6) forms the basis for the stochastic gradient algorithm. Let us now use Theorem A.3 in Appendix A to formally state conditions under which the derivative–integral interchange operation in (5.6) is valid.

Theorem 5.1 (Validity of interchange of derivative and integral in (5.6)). Let Λ be the domain for V. Suppose that Θ is an open set. Let $Q(\boldsymbol{\theta}, \boldsymbol{\upsilon})p_V(\boldsymbol{\upsilon})$ and $p_V(\boldsymbol{\upsilon})\partial Q(\boldsymbol{\theta},\boldsymbol{\upsilon})/\partial\boldsymbol{\theta}$ be continuous on $\Theta \times \Lambda$. Suppose that there exist nonnegative functions $q_0(\boldsymbol{\upsilon})$ and $q_1(\boldsymbol{\upsilon})$ such that

$$\left|Q(\boldsymbol{\theta},\boldsymbol{\upsilon})p_V(\boldsymbol{\upsilon})\right| \le q_0(\boldsymbol{\upsilon}),\ \left\|p_V(\boldsymbol{\upsilon})\frac{\partial Q(\boldsymbol{\theta},\boldsymbol{\upsilon})}{\partial\boldsymbol{\theta}}\right\| \le q_1(\boldsymbol{\upsilon}) \text{ for all } (\boldsymbol{\theta},\boldsymbol{\upsilon}) \in \Theta\times\Lambda,$$

where $\int_\Lambda q_0(\boldsymbol{\upsilon})\,d\boldsymbol{\upsilon} < \infty$ and $\int_\Lambda q_1(\boldsymbol{\upsilon})\,d\boldsymbol{\upsilon} < \infty$. Then

$$\frac{\partial}{\partial\boldsymbol{\theta}}\int_\Lambda Q(\boldsymbol{\theta},\boldsymbol{\upsilon})p_V(\boldsymbol{\upsilon})\,d\boldsymbol{\upsilon} = \int_\Lambda \frac{\partial Q(\boldsymbol{\theta},\boldsymbol{\upsilon})}{\partial\boldsymbol{\theta}}p_V(\boldsymbol{\upsilon})\,d\boldsymbol{\upsilon}.$$

That is, (5.6) holds.

If (5.6) and the above interchange hold, a realization $\partial Q/\partial\boldsymbol{\theta}$ yields an unbiased estimate of the true gradient, à la

$$E\left[\frac{\partial Q(\boldsymbol{\theta},V)}{\partial\boldsymbol{\theta}}\right] = \boldsymbol{g}(\boldsymbol{\theta}). \tag{5.7}$$

Expression (5.5), on the other hand, does not yield such an "automatic" means of getting an unbiased gradient approximation. Unlike the right-hand side of (5.6), the integrand in (5.5) (specifically, the part before the "+" in the second line) is not in the form (*function*) × (*density*). Hence, there is no natural way of generating a random outcome that is an unbiased estimator of the gradient.

Chapter 15 discusses a "trick" that gets around the problem of an integrand not being in the form (*function*) × (*density*). In particular, one can rewrite the integrand by multiplying through by unity expressed as a ratio of identical density functions $p_V(\boldsymbol{\upsilon}\,|\,\boldsymbol{\theta}')/p_V(\boldsymbol{\upsilon}\,|\,\boldsymbol{\theta}')$, where $\boldsymbol{\theta}'$ is a fixed value of $\boldsymbol{\theta}$. The new (equivalent) integrand can be expressed in the form (*function*) × (*density*), where the density is $p_V(\boldsymbol{\upsilon}|\boldsymbol{\theta}')$ (see Chapter 15 on the likelihood ratio and sample path methods for gradient estimation in simulation-based optimization). However, in contrast to simulation-based optimization, this trick does not work in *general* SA applications since it requires that "nature" generate V from a density with the known value $\boldsymbol{\theta}'$. In general SA applications, the analyst has no control over how the underlying randomness in the problem is produced. (The "trick" *does* work in the simulation-based optimization setting of Chapter 15 because the analyst controls "nature" by controlling the Monte Carlo randomness produced in the simulation.) A further difficulty with (5.5) relative to general stochastic gradient problems is that the analyst needs complete knowledge of $p_V(\boldsymbol{\upsilon}|\boldsymbol{\theta})$ and $\partial p_V(\boldsymbol{\upsilon}\,|\,\boldsymbol{\theta})/\partial\boldsymbol{\theta}$ for varying $\boldsymbol{\upsilon}$ and $\boldsymbol{\theta}$.

In contrast to the problems in using (5.5) for gradient estimation, (5.4) as it manifests itself in (5.6) yields the fundamental gradient estimate:

$$\boldsymbol{Y}(\boldsymbol{\theta}) = \frac{\partial Q(\boldsymbol{\theta},V)}{\partial\boldsymbol{\theta}}. \tag{5.8}$$

SA with input (5.8) is the *stochastic gradient algorithm* of interest here (Chapter 15 considers the more general form using (5.5)). From (5.6), $\boldsymbol{Y}(\boldsymbol{\theta})$ is an unbiased estimate of $\boldsymbol{g}(\boldsymbol{\theta})$ (i.e., A.3 and B.5 in Section 4.3 hold). This input is available without knowing the distribution of V because $\boldsymbol{Y}$ is an *observed* outcome.

Algebraically, one can always write the manifestation of V as an error in the L measurement, where the error generally depends on $\boldsymbol{\theta}$. So, $Q(\boldsymbol{\theta}, V) = L(\boldsymbol{\theta}) + \varepsilon(\boldsymbol{\theta})$, although the analytical form of $\varepsilon(\boldsymbol{\theta})$ is usually unknown. Note that $y(\boldsymbol{\theta}) = Q(\boldsymbol{\theta}, V)$ in the notation introduced in Chapter 1. In some practical problems, V manifests itself as an additive *independent* random noise term. Then, *assuming that the gradient in (5.8) is available* is equivalent to assuming that the deterministic gradient $\boldsymbol{g}(\boldsymbol{\theta})$ is available. The examples below illustrate the distinction between an error dependent on $\boldsymbol{\theta}$ and not dependent on $\boldsymbol{\theta}$.

Example 5.2—Case where the stochastic gradient method is appropriate. Suppose that $Q(\boldsymbol{\theta}, V) = f(\boldsymbol{\theta}) + \boldsymbol{\theta}^T V$, where $f(\boldsymbol{\theta})$ is a differentiable function and $V \in \mathbb{R}^p$ has an unknown distribution independent of $\boldsymbol{\theta}$. The minimum of $L(\boldsymbol{\theta}) = E[Q(\boldsymbol{\theta}, V)] = f(\boldsymbol{\theta}) + \boldsymbol{\theta}^T E(V)$ is clearly dependent on the (unknown) mean of V (and $\varepsilon(\boldsymbol{\theta}) = Q(\boldsymbol{\theta}, V) - L(\boldsymbol{\theta}) = \boldsymbol{\theta}^T[V - E(V)]$). Given that one knows the form for $Q(\boldsymbol{\theta}, V)$, the stochastic gradient method would be appropriate in this problem. The random input according to (5.8) is $\boldsymbol{Y}(\boldsymbol{\theta}) = \partial f(\boldsymbol{\theta})/\partial \boldsymbol{\theta} + V$. ❑

Example 5.3—Stochastic problem solvable by deterministic methods. Now, suppose that the form for Q changes from that in Example 5.2 to $Q(\boldsymbol{\theta}, V) = f(\boldsymbol{\theta}) + V$, where the distribution of the scalar V does not depend on $\boldsymbol{\theta}$. If the analyst has enough information to invoke (5.8), then $\boldsymbol{Y}(\boldsymbol{\theta}) = \partial f(\boldsymbol{\theta})/\partial \boldsymbol{\theta}$. But this is clearly independent of the random effect V. Hence, standard deterministic methods can be used on $f(\boldsymbol{\theta}) = L(\boldsymbol{\theta}) - E(V)$ (i.e., minimizing $f(\boldsymbol{\theta})$ is the same as minimizing $L(\boldsymbol{\theta})$ since they only differ by the constant corresponding to the mean of V). The key qualifier here is the above statement: "…assuming that the gradient in (5.8) is available…." If this qualifier is not true, then one is generally unable to use deterministic methods to solve the problem because of the *random* form of $Q(\boldsymbol{\theta}, V)$. In that case, the SA methods for optimization with noisy loss measurements in Chapters 6 and 7 are more appropriate. ❑

5.1.2 Stochastic Gradient Algorithm

Let us now combine the basic SA algorithm in (4.6) or (4.7) with the definition of $\boldsymbol{Y}(\boldsymbol{\theta})$ in (5.8) to obtain a stochastic gradient algorithm. To reflect the common recursive situation where a Q may result from random effects together with a deterministic input at each iteration (as in the instantaneous gradient of Section 3.2 or the external input setting of Section 4.1), we index Q and V by the iteration counter k. Hence, $Q_k(\boldsymbol{\theta}, V_k)$ may represent the instantaneous observation of an overall loss $L(\boldsymbol{\theta})$ or an observation of an instantaneous loss $L_k(\boldsymbol{\theta})$, the latter case usually associated with time-varying tracking problems. A related implementation is associated with *batch processing*, where the recursive input $Q_k(\boldsymbol{\theta}, \boldsymbol{V}_k)$ is replaced by the overall (sample mean) response $\bar{Q}(\boldsymbol{\theta}, \boldsymbol{V})$ at any $\boldsymbol{\theta}$. In batch processing, the algorithm repeatedly passes through a given fixed set of data (so $\boldsymbol{V}$ represents the randomness in the given set of data). A special case of $\bar{Q}(\boldsymbol{\theta}, \boldsymbol{V})$ is the least-squares criterion for linear systems, $\hat{L}(\boldsymbol{\theta})$, used in Chapter 3.

For the general recursive unconstrained case (SA algorithm (4.6)), we have

$$\begin{aligned}\hat{\boldsymbol{\theta}}_{k+1} &= \hat{\boldsymbol{\theta}}_k - a_k \left.\frac{\partial Q_k(\boldsymbol{\theta}, \boldsymbol{V}_k)}{\partial \boldsymbol{\theta}}\right|_{\boldsymbol{\theta}=\hat{\boldsymbol{\theta}}_k} \\ &= \hat{\boldsymbol{\theta}}_k - a_k \boldsymbol{Y}_k(\hat{\boldsymbol{\theta}}_k), \end{aligned} \tag{5.9}$$

where $\boldsymbol{V}_k$ represents the random input at the kth iteration. Eqn. (5.9) is reflective of recursive processing given that Q_k and $\boldsymbol{V}_k$ may change with k.

If the SA recursion is being used in the above-mentioned batch processing mode, $\boldsymbol{V}$ is the overall set of random inputs, leading to an iterative process of the form

$$\begin{aligned}\hat{\boldsymbol{\theta}}_{k+1} &= \hat{\boldsymbol{\theta}}_k - a_k \left.\frac{\partial \bar{Q}(\boldsymbol{\theta}, \boldsymbol{V})}{\partial \boldsymbol{\theta}}\right|_{\boldsymbol{\theta}=\hat{\boldsymbol{\theta}}_k} \\ &= \hat{\boldsymbol{\theta}}_k - a_k \boldsymbol{Y}(\hat{\boldsymbol{\theta}}_k), \end{aligned} \tag{5.10}$$

where $\boldsymbol{Y}$ is not indexed by k in the second line to reflect the reprocessing of the same gradient $\partial \bar{Q}/\partial \boldsymbol{\theta}$ with only $\boldsymbol{\theta} = \hat{\boldsymbol{\theta}}_k$ changing.

Because the stochastic gradient quantity $\partial Q_k/\partial \boldsymbol{\theta}$ in (5.9) does not, in general, equal the deterministic quantity $\partial L/\partial \boldsymbol{\theta}$, the SA algorithm is fundamentally different from the deterministic steepest descent algorithm in Section 1.4. There is, however, an intuitive connection between (5.9) and steepest descent because $E(\partial Q_k/\partial \boldsymbol{\theta}) = \partial L/\partial \boldsymbol{\theta}$ (or $\partial L_k/\partial \boldsymbol{\theta}$ in the time-varying loss case) under the regularity conditions mentioned above justifying the interchange of a derivative and an (expectation) integral. (In fact, there is some theory that allows for biased $E(\partial Q_k/\partial \boldsymbol{\theta}) \neq \partial L/\partial \boldsymbol{\theta}$ cases. See, for example, condition B.5 of Section 4.3, where the bias $\boldsymbol{b}_k$ can be nonzero but where $\boldsymbol{b}_k \to 0$ as $k \to \infty$. This theory is most useful in the setting of Chapters 6 and 7, where gradients are approximated from loss function measurements.) Figure 5.1 depicts a simple scalar case, suggesting why the recursion in (5.9) yields convergence to $\boldsymbol{\theta}^*$. Note that the a_k sequence is critical to controlling the magnitude of the steps to ensure an algorithm that is neither too timid nor too aggressive.

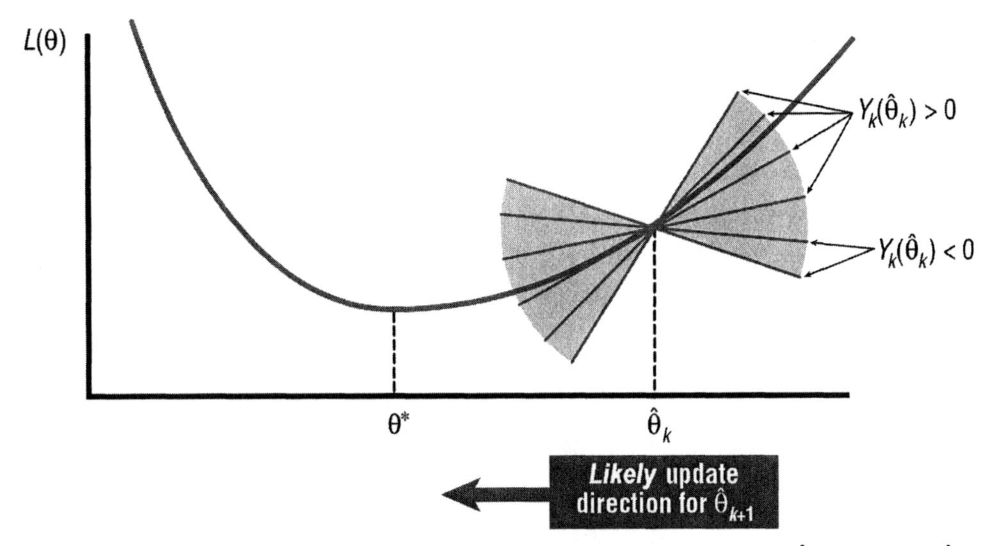

Figure 5.1. Intuitive basis for stochastic gradient search when $p = 1$. For $\hat{\theta}_k > \theta^*$, $Y_k(\hat{\theta}_k) = \partial Q_k/\partial\theta$ at $\theta = \hat{\theta}_k$ *tends* to be positive, leading to updated $\hat{\theta}_{k+1} < \hat{\theta}_k$, as desired. The bow-tie-shaped shaded area is the region likely to contain the line with measured slope $Y_k(\hat{\theta}_k)$. Four of the six measured derivatives shown have the correct (positive) sign. When $\hat{\theta}_k < \theta^*$, the complementary relationship holds, leading to $\hat{\theta}_{k+1} > \hat{\theta}_k$.

There exists formal convergence theory for $\hat{\boldsymbol{\theta}}_k$ in both the recursive and batch settings. The theory for the recursive setting is a special case of the conditions for general root-finding SA in Section 4.3 (a.s. convergence) and Section 4.4 (asymptotic normality). In particular, convergence relies on properly chosen decaying gains a_k. The theory for the batch case is tied to deterministic optimization (see Section 1.4). Batch convergence may *sometimes* occur with fixed gains $a_k = a$ for all k (e.g., in linear models; see Wang et al., 2000). As discussed below, convergence in the batch case is not generally to $\boldsymbol{\theta}^*$, but to a minimum tied to the specific set of data.

Note the connection of the general recursive algorithms in Chapter 3, particularly LMS (Subsection 3.2.2) and recursive least squares (RLS) (Subsection 3.2.4), to the SA form in (5.9). The instantaneous gradient in Section 3.2 corresponds to $\partial Q_k/\partial\boldsymbol{\theta}$ in (5.9). This gradient, of course, is tied to the special linear structure of the model in Chapter 3. We discuss the connection of LMS to the stochastic gradient recursion in more detail in Subsection 5.1.4.

RLS and the stochastic gradient algorithm of (5.9) also have the same generic form except that the specially chosen matrix $\boldsymbol{P}_{k+1}$ replaces the gain coefficient a_k in (5.9). General SA theory allows for choosing a matrix gain for a_k, but in most nonlinear problems (unlike RLS) there is usually not enough information to know how to choose such a matrix. Hence, analysts *typically* use the simpler scalar gain or use an adaptive scheme that seeks to estimate the matrix gain as the algorithm iterates (see Subsection 4.5.2).

While the recursive and batch modes represent, in some sense, the two extremes in implementation of the stochastic gradient algorithm, another hybrid form is popular. We refer to this hybrid form as a *multiple-pass implementation*. This form operates on the data using the instantaneous gradient, as in the recursive form, yet (like batch processing) makes multiple passes through the full set of data. In particular, after the algorithm has passed through the data one time, it returns to the first data point with the initial $\boldsymbol{\theta}$ equal to the terminal estimate from the first pass (the first pass may be carried out concurrent with the arrival of the data). The user may choose to reset the gain value at a_0 as in starting the first pass or may continue with the current gain sequence, processing the first data point in the second pass with gain a_n (producing $\hat{\boldsymbol{\theta}}_{n+1}$) if the terminal estimate for $\boldsymbol{\theta}$ from the first pass was $\hat{\boldsymbol{\theta}}_n$. This process is repeated for as many passes through the data as desired.

The multiple-pass method is particularly popular in neural network estimation, where each pass through the data is referred to as an *epoch*. Exercise 5.8 considers the multiple-pass method.

5.1.3 Implementation in General Nonlinear Regression Problems

An important example of the nonlinear setting of this chapter is nonlinear regression with scalar output. Suppose that the kth output of some process is modeled by

$$z_k = h_k(\boldsymbol{\theta}, \boldsymbol{x}_k) + v_k, \tag{5.11}$$

where $h_k(\cdot)$ is some nonlinear regression function dependent on the parameters being estimated and some input variables $\boldsymbol{x}_k$, and v_k is a mean-zero noise term. In a neural network problem, for example, $h_k(\boldsymbol{\theta}, \boldsymbol{x}_k)$ represents the network output for the given weight values $\boldsymbol{\theta}$ and particular input $\boldsymbol{x}_k$ (see Section 5.2).

Within the context of the nonlinear regression problem associated with (5.11), the algorithm can be implemented in a recursive mode as one collects outputs z_k for each $\boldsymbol{x}_k$ input or in a batch mode where all data have been collected and we work with an overall noisy loss measurement.

In the recursive form based on instantaneous input, (5.9) is implemented with input

$$\begin{aligned} \boldsymbol{Y}_k(\hat{\boldsymbol{\theta}}_k) &= \left.\frac{\partial Q_k(\boldsymbol{\theta}, \boldsymbol{V}_k)}{\partial \boldsymbol{\theta}}\right|_{\boldsymbol{\theta}=\hat{\boldsymbol{\theta}}_k} = \left.\frac{1}{2}\frac{\partial (z_{k+1} - \hat{z}_{k+1})^2}{\partial \boldsymbol{\theta}}\right|_{\boldsymbol{\theta}=\hat{\boldsymbol{\theta}}_k} \\ &= \left[\left(h_{k+1}(\boldsymbol{\theta}, \boldsymbol{x}_{k+1}) - z_{k+1}\right)\frac{\partial h_{k+1}}{\partial \boldsymbol{\theta}}\right]_{\boldsymbol{\theta}=\hat{\boldsymbol{\theta}}_k}, \end{aligned} \tag{5.12}$$

where $\hat{z}_k = h_k(\boldsymbol{\theta}, \boldsymbol{x}_k)$ is the prediction at each iteration for the given $\boldsymbol{\theta}$ and $\boldsymbol{x}_k$. The above implies that $Q_k(\boldsymbol{\theta}, \boldsymbol{V}_k) = \frac{1}{2}(z_{k+1} - \hat{z}_{k+1})^2$. This form clearly exhibits the role of the gradient $\partial h_k/\partial\boldsymbol{\theta}$ in the gradient-based algorithms of interest here. Note that (5.12) is a noisy observation of the corresponding instantaneous gradient $\frac{1}{2}\partial E[(z_{k+1}-\hat{z}_{k+1})^2]/\partial\boldsymbol{\theta}$ at $\boldsymbol{\theta} = \hat{\boldsymbol{\theta}}_k$. In a neural network context, for example, a backpropagation type of algorithm would typically be used to calculate $\partial h_{k+1}/\partial\boldsymbol{\theta}$ at the particular $\boldsymbol{x}_{k+1}$ and $\hat{\boldsymbol{\theta}}_k$, as appears in (5.12) (see Section 5.2). If the problem were such that a gradient of this type was unavailable, then we could not use the SA algorithm in (5.9) (but see Chapters 6 and 7).

The randomness V_k is reflected in the real data z_{k+1}, although it does not show explicitly in the bottom line of (5.12). For example, suppose that the process generating the data z_k has the same form as the assumed model (i.e., the actual measurements come from $z_k = h_k(\boldsymbol{\theta}^*, \boldsymbol{x}_k) + v_k$). If the $\boldsymbol{x}_k$ are chosen randomly—as in the LMS discussion of Subsection 3.2.2 and the discussion of Subsection 4.5.1, where the inputs are treated as Markov state variables—then $V_k \equiv \{\boldsymbol{x}_{k+1}, v_{k+1}\}$. Alternatively, if the inputs are chosen deterministically, $V_k = v_{k+1}$. In more general cases, where the true process and assumed model have different forms, V_k represents all randomness in the true process.

In the batch mode, we iterate on $\boldsymbol{\theta}$ for a *fixed* set of data. The batch mode recursion uses (5.10) with input based on n data points:

$$Y(\hat{\boldsymbol{\theta}}_k) = \left.\frac{\partial \bar{Q}}{\partial\boldsymbol{\theta}}\right|_{\boldsymbol{\theta}=\hat{\boldsymbol{\theta}}_k} = \frac{1}{2n}\sum_{i=1}^{n}\left.\frac{\partial(z_i-\hat{z}_i)^2}{\partial\boldsymbol{\theta}}\right|_{\boldsymbol{\theta}=\hat{\boldsymbol{\theta}}_k}$$

$$= \frac{1}{n}\sum_{i=1}^{n}[h_i(\hat{\boldsymbol{\theta}}_k,\boldsymbol{x}_i)-z_i]\left.\frac{\partial h_i(\boldsymbol{\theta},\boldsymbol{x}_i)}{\partial\boldsymbol{\theta}}\right|_{\boldsymbol{\theta}=\hat{\boldsymbol{\theta}}_k}. \tag{5.13}$$

Because n is fixed, $\hat{\boldsymbol{\theta}}_k$ will not generally converge to $\boldsymbol{\theta}^*$, but will converge to some minimum that is tied to the fixed n (and associated data z_1, $z_2, \ldots, z_n$). In the linear case, this minimum corresponds to the batch solution $\hat{\boldsymbol{\theta}}^{(n)}$ (see Subsection 3.1.2). More generally, for linear and nonlinear problems, this minimum based on n data is generally close to the true minimum $\boldsymbol{\theta}^*$ if n is large. To implement (5.13), it is necessary to have the full set of data $z_1, z_2, \ldots, z_n$ available at the start of the search process.

5.1.4 Connection of LMS to Stochastic Gradient SA

This subsection shows how the recursive framework associated with input (5.12) applies in the special case of estimating $\boldsymbol{\theta}$ in linear models with LMS as in Chapter 3. Let us begin by considering the following example.

Example 5.4—LMS as an example of the stochastic gradient method. Suppose that $z_k = \boldsymbol{h}_k^T\boldsymbol{\theta} + v_k$, as introduced in Section 3.1, where, in general, $\boldsymbol{h}_k$

depends on some inputs $\boldsymbol{x}_k$ (i.e., $\boldsymbol{h}_k = \boldsymbol{h}_k(\boldsymbol{x}_k)$). In the context of (5.11), $h_k(\boldsymbol{\theta}, \boldsymbol{x}_k)$ is equivalent to $\boldsymbol{h}_k^T\boldsymbol{\theta} = \boldsymbol{h}_k(\boldsymbol{x}_k)^T\boldsymbol{\theta}$. This represents a model linear in $\boldsymbol{\theta}$, which is a special case of the nonlinear models of interest here. Subsection 3.2.2 introduced the basic LMS recursion:

$$\hat{\boldsymbol{\theta}}_{k+1} = \hat{\boldsymbol{\theta}}_k - a_k \boldsymbol{h}_{k+1}(\boldsymbol{h}_{k+1}^T\hat{\boldsymbol{\theta}}_k - z_{k+1}), \tag{5.14}$$

with $a_k = a > 0$ for all k. Because $\partial \boldsymbol{h}_{k+1}^T\boldsymbol{\theta}/\partial\boldsymbol{\theta} = \boldsymbol{h}_{k+1}$ (representing $\partial h_{k+1}(\boldsymbol{\theta}, \boldsymbol{x}_{k+1})/\partial\boldsymbol{\theta}$ in (5.12)), the LMS algorithm above corresponds to the recursive algorithm (5.9) with

$$\boldsymbol{Y}_k(\hat{\boldsymbol{\theta}}_k) = \boldsymbol{h}_{k+1}(\boldsymbol{h}_{k+1}^T\hat{\boldsymbol{\theta}}_k - z_{k+1}).$$

Similar arguments can be used for RLS in Subsection 3.2.4 provided that one allows for a matrix gain in place of the scalar a_k. ❑

Let us now continue with LMS as an example of the application of the general SA convergence conditions in Section 4.3. Recall that the two sets of conditions (the "statistics" conditions "A" and the "engineering conditions" "B") are sufficient to guarantee strong (a.s.) convergence of the root-finding SA algorithm to a solution $\boldsymbol{\theta}^*$. The engineering conditions, in particular, are especially powerful for analyzing the LMS algorithm (e.g., Ljung, 1984; Gerencsér, 1995). Proposition 5.1 below gives one set of convergence conditions for LMS, derived from the engineering conditions of Section 4.3. While this illustrates the connection between SA and LMS, the conditions are not the most general possible (see Ljung, 1984, Gerencsér, 1995, or Sections 3.2 and 3.3 for alternative references). The proposition statement uses the Kronecker product ($\otimes$) discussed in Appendix A.

Proposition 5.1 (Convergence of LMS). Consider the LMS recursion in (5.14). Suppose that the real system produces data according to $z_k = \boldsymbol{h}_k^T\boldsymbol{\theta}^* + v_k$, where the v_k are independent, identically distributed (i.i.d.) with mean zero and finite variance and the $\boldsymbol{h}_k$ are random and i.i.d. (independent of the v_k) such that all components of $E(\boldsymbol{h}_k \otimes \boldsymbol{h}_k \otimes \boldsymbol{h}_k \otimes \boldsymbol{h}_k)$ are finite in magnitude (i.e., all possible mixed and unmixed fourth moments of $\boldsymbol{h}_k$ are finite). Suppose that conditions B.1–B.3 (Section 4.3) hold, $\sum_{k=0}^{\infty} a_k^2 < \infty$, and $E\left(\|\hat{\boldsymbol{\theta}}_k\|^2\right)$ is uniformly bounded above for all k. Then, $\hat{\boldsymbol{\theta}}_k \to \boldsymbol{\theta}^*$ a.s. as $k \to \infty$.

Proof. First, note that $\boldsymbol{\theta}^*$ is the minimum of $L(\boldsymbol{\theta}) \equiv \frac{1}{2}E[(z_k - \boldsymbol{h}_k^T\boldsymbol{\theta})^2]$, where $L(\boldsymbol{\theta})$ does not depend on k because of the assumptions about v_k and $\boldsymbol{h}_k$. From Section 4.3, it is sufficient to have conditions B.1 – B.5 hold. Because B.1–B.3

are assumed to hold, we need only to show that B.4 and B.5 are implied by the other assumptions or the problem structure.

Let us first show that B.5 holds. From Example 5.4, $\boldsymbol{Y}_k(\boldsymbol{\theta}) = \boldsymbol{h}_{k+1}(\boldsymbol{h}_{k+1}^T\boldsymbol{\theta} - z_{k+1})$. This can be expressed as $\boldsymbol{Y}_k(\boldsymbol{\theta}) = \boldsymbol{g}(\boldsymbol{\theta}) + \boldsymbol{e}_k(\boldsymbol{\theta})$, where

$$\boldsymbol{g}(\boldsymbol{\theta}) = [E(\boldsymbol{h}_{k+1}\boldsymbol{h}_{k+1}^T)](\boldsymbol{\theta} - \boldsymbol{\theta}^*) \ ,$$

$$\boldsymbol{e}_k(\boldsymbol{\theta}) = [\boldsymbol{h}_{k+1}\boldsymbol{h}_{k+1}^T - E(\boldsymbol{h}_{k+1}\boldsymbol{h}_{k+1}^T)](\boldsymbol{\theta} - \boldsymbol{\theta}^*) - \boldsymbol{h}_{k+1}v_{k+1},$$

and z_{k+1} is generated by the true process $\boldsymbol{h}_{k+1}^T\boldsymbol{\theta}^* + v_{k+1}$. Recall that $\Im_k = \{\hat{\boldsymbol{\theta}}_0, \hat{\boldsymbol{\theta}}_1, \ldots, \hat{\boldsymbol{\theta}}_k\}$, as defined in condition B.4. The form for $\boldsymbol{e}_k$ above implies that

$$\boldsymbol{b}_k = E[\, \boldsymbol{e}_k(\hat{\boldsymbol{\theta}}_k) \,|\, \Im_k \,] = E[\, \boldsymbol{e}_k(\hat{\boldsymbol{\theta}}_k) \,|\, \hat{\boldsymbol{\theta}}_k \,] = \boldsymbol{0}$$

by the mutual independence of $\{\boldsymbol{h}_1, \boldsymbol{h}_2, \ldots; v_1, v_2, \ldots\}$ and the fact that $E(v_k) = 0$ (the substitution of $\hat{\boldsymbol{\theta}}_k$ for $\Im_k$ in the conditioning is allowed because the independence assumptions on $\{\boldsymbol{h}_k, v_k\}$ imply that conditioning on $\hat{\boldsymbol{\theta}}_k$ alone is equivalent to conditioning on $\hat{\boldsymbol{\theta}}_k$ plus the earlier terms in $\Im_k$).[1] Hence, condition B.5 holds.

Let us now show B.4. Let $\boldsymbol{e}_k = \boldsymbol{e}_k(\hat{\boldsymbol{\theta}}_k)$. Because $\boldsymbol{b}_k = \boldsymbol{0}$, B.4 may be expressed as

$$E\left[\left\|\sum_{k=0}^{\infty} a_k(\boldsymbol{e}_k - \boldsymbol{b}_k)\right\|^2\right] = E\left[\left\|\sum_{k=0}^{\infty} a_k\boldsymbol{e}_k\right\|^2\right] < \infty. \tag{5.15}$$

Note that for all $i < j$, $E(\boldsymbol{e}_i^T\boldsymbol{e}_j) = E[E(\boldsymbol{e}_i^T\boldsymbol{e}_j|\Im_j)] = E[\boldsymbol{e}_i^T E(\boldsymbol{e}_j|\Im_j)] = 0$ (Exercise 5.6). Because $E(\boldsymbol{e}_j^T\boldsymbol{e}_i) = E(\boldsymbol{e}_i^T\boldsymbol{e}_j)$, we have $E(\boldsymbol{e}_i^T\boldsymbol{e}_j) = 0$ for all $i \neq j$. Hence, the cross-product terms in (5.15) disappear, indicating that (5.15) is equivalent to

$$\sum_{k=0}^{\infty} a_k^2 \boldsymbol{E}\left(\|\boldsymbol{e}_k\|^2\right) < \infty. \tag{5.16}$$

It was assumed that $\boldsymbol{E}\left(\|\hat{\boldsymbol{\theta}}_k\|^2\right)$ is uniformly bounded for all k; further, the v_k are i.i.d. with finite variance and the $\boldsymbol{h}_k$ are i.i.d. with finite fourth moments. These facts imply that $\boldsymbol{E}\left(\|\boldsymbol{e}_k\|^2\right)$ is uniformly bounded for all k. Hence, (5.16) holds

---

[1] Strictly, all conditional expectations are true a.s.; we suppress that clause in the interest of a streamlined discussion.

when $\sum_{k=0}^{\infty} a_k^2 < \infty$. Because this corresponds to one of the assumptions, we have shown B.4, completing the proof. ❑

The sections to follow are *brief* discussions of the application of the stochastic gradient method to the areas of neural networks, discrete-event dynamic systems, and image restoration. These vignettes are not intended to be thorough discussions of the indicated subjects. The reader should consult the indicated references—or other source material—for more complete background on these subjects.

5.2 NEURAL NETWORK TRAINING

Neural networks (NNs) are mathematical models that have become popular over the last two decades as representations of complex nonlinear systems. Among many other areas, applications include forecasting, signal processing, and control. Consider the problem of training a feedforward NN based on input–output data using the well-known backpropagation algorithm (e.g., Jang et al., 1997, pp. 205–210). This has long been recognized as an application of the stochastic gradient version of the SA algorithm, with two of the earlier publications discussing this connection being White (1989) and Kosko (1992, pp. 190–199). One of the main virtues of feedforward (and certain other) NNs is that they are *universal approximators*. That is, they have the ability to approximate broad classes of functions to within any level of accuracy (Haykin, 1999, Sect. 4.13). This property is one of the reasons that NNs have attracted so much attention.

For our purposes here, suppose that the system output z_k can be represented by a NN according to the standard model $z_k = h(\boldsymbol{\theta}, \boldsymbol{x}_k) + v_k$, as introduced in Subsection 5.1.3. (More generally, the output can be a vector $\boldsymbol{z}_k$, but we restrict ourselves to a scalar in this brief discussion.) Here, $h(\boldsymbol{\theta}, \boldsymbol{x}_k)$ represents the NN output, $\boldsymbol{\theta}$ represents the NN weights to be estimated, and, as usual, $\boldsymbol{x}_k$ represents inputs (deterministic or random). The noise v_k captures the difference between a NN output and an actual measurement z_k. Unlike the more general context of Subsection 5.1.3, note that the output $h(\cdot)$ is without a subscript k since we are assuming a fixed NN structure across input–output pairs. Figure 5.2 shows a simple feedforward NN with two hidden layers. This diagram shows two inputs (i.e., $\dim(\boldsymbol{x}_k) = 2$). The output corresponds to a prediction of the scalar z_k.

The NN training problem can be posed as a nonlinear regression problem with $\boldsymbol{\theta}$ representing the NN weights to be estimated. In the nonlinear regression context above, the prediction at the kth input is $\hat{z}_k = h(\boldsymbol{\theta}, \boldsymbol{x}_k)$ for the given input $\boldsymbol{x}_k$ and weights $\boldsymbol{\theta}$. Here the loss function is the normalized sum of mean-squared errors:

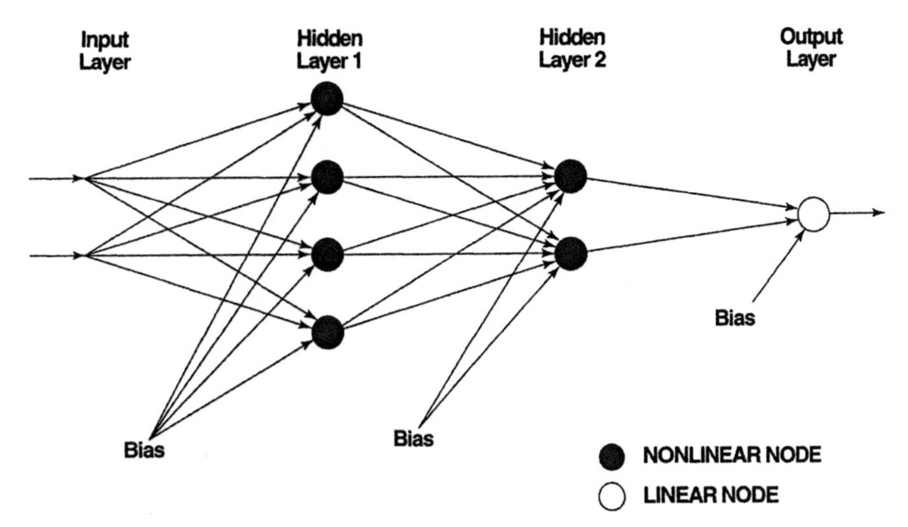

Figure 5.2. Simple feedforward neural network with two inputs in $\boldsymbol{x}_k$. Each of the p connections is weighted by an element of the vector $\boldsymbol{\theta}$ ($p = 25$ here). The inputs to each node are usually summed prior to application of the relevant linear or nonlinear function. The nonlinear nodes are frequently bounded (*sigmoidal*) functions that monotonically increase and have single inflection points. The bias terms represent constant inputs. The rightmost arrow represents the NN output $h(\boldsymbol{\theta}, \boldsymbol{x}_k)$.

$$L(\boldsymbol{\theta}) = \frac{1}{2n}\sum_{k=1}^{n} E\left[(z_k - h(\boldsymbol{\theta}, \boldsymbol{x}_k))^2\right], \tag{5.17}$$

where z_k represents the true system observation. In light of the practical situation where data arrive sequentially (possibly at a high rate), White (1989) emphasizes the backpropagation algorithm based on recursively processing one input–output pair $\boldsymbol{x}_k$, z_k at a time. Using (5.9), the stochastic gradient algorithm then becomes

$$\begin{aligned}\hat{\boldsymbol{\theta}}_{k+1} &= \hat{\boldsymbol{\theta}}_k - a_k \frac{\partial Q_k(\boldsymbol{\theta}, V_k)}{\partial \boldsymbol{\theta}}\bigg|_{\boldsymbol{\theta}=\hat{\boldsymbol{\theta}}_k} \\ &= \hat{\boldsymbol{\theta}}_k - a_k\left[(h(\boldsymbol{\theta}, \boldsymbol{x}_{k+1}) - z_{k+1})\frac{\partial h}{\partial \boldsymbol{\theta}}\right]_{\boldsymbol{\theta}=\hat{\boldsymbol{\theta}}_k},\end{aligned} \tag{5.18}$$

where the term inside the square brackets represents the noisy observation of the instantaneous gradient of the $(k+1)$st summand in (5.17). In practice, one can take multiple passes through the input–output pairs using (5.18) (e.g., White, 1989, does this in his examples). This only costs computer time (as opposed to the cost of real data) and is generally effective at improving the estimation of $\boldsymbol{\theta}$. It is also possible to run an algorithm with the batch gradient $\partial\bar{Q}/\partial\boldsymbol{\theta}$ as in (5.13) instead of the instantaneous gradient appearing in (5.18). Here, each iteration requires a set of backpropagation gradients, one for each input–output pair in the data set.

The essential innovation in NN backpropagation is the systematic implementation of the chain rule for computing the $\partial h/\partial \boldsymbol{\theta}$ term that appears in (5.18). Some historical perspective on backpropagation is found, for example, in Kosko (1992, pp. 196–199).

White (1989) also suggests using the backpropagation solution as the initial value for one or more iterations of a Newton–Raphson type of algorithm. This requires the Hessian matrix of the NN (see, e.g., Bishop, 1992). As discussed in Section 1.4, Newton–Raphson is fast, but may be unstable if the initial value is not close to the optimal value. The role of backpropagation in this case is to help avoid the instability.

The asymptotic distribution theory for SA discussed in Section 4.4 can be used to test hypotheses about the effect of particular inputs on the weight identification process. In particular, White (1989) shows how a test statistic can be formed to determine whether particular weights contribute toward the ability of the NN to approximate a function.

White (1989) conducted a numerical study on having a NN emulate a Henon map, which is a nonlinear time series of the form

$$U_t = 1 - 1.4U_{t-1}^2 + 0.3U_{t-2}$$

for scalar U_t. He generated 4000 data points and then sampled randomly (with replacement) from these to get his sequential input–output data for training the NN via (5.18). With output U_t, the inputs used by White were one of the three combinations: U_{t-1} and U_{t-2}, U_{t-1} only, or U_{t-2} only. He used a single hidden-layer NN with nonlinear nodes, each of which had a sigmoidal function. Both a decaying gain, $a_k = a/(k+1)$, and constant gain were used. As might be expected, when both of the inputs were used, the NN provided the best fit. In these studies, decaying gains a_k produced results slightly better than those where constant gains were used. White (1989) also demonstrates the idea of using Newton–Raphson steps on the full data set after achieving effective convergence with the recursion (5.18). This yields further improvement.

There are two other issues strongly related to the stochastic gradient form of SA that arise in the training of NNs. One of these is the momentum form of the gradient input and the other is overfitting, discussed briefly below. *Momentum* is used to smooth out the variations in the individual gradient inputs, especially when using the recursive (one measurement at a time) form of SA as in (5.18). With momentum, the individual gradient inputs are replaced by a weighted combination of the current instantaneous gradient $\partial Q_k/\partial \boldsymbol{\theta}$ and past instantaneous gradients. The modified version of (5.18) then becomes

$$\hat{\boldsymbol{\theta}}_{k+1} = \hat{\boldsymbol{\theta}}_k - a_k \tilde{\boldsymbol{Y}}_k ,$$

where

$$\tilde{Y}_0 = \left.\frac{\partial Q_0(\boldsymbol{\theta}, V_0)}{\partial \boldsymbol{\theta}}\right|_{\boldsymbol{\theta}=\hat{\boldsymbol{\theta}}_0},$$

$$\tilde{Y}_k = \gamma_k \tilde{Y}_{k-1} + (1-\gamma_k)\left.\frac{\partial Q_k(\boldsymbol{\theta}, V_k)}{\partial \boldsymbol{\theta}}\right|_{\boldsymbol{\theta}=\hat{\boldsymbol{\theta}}_k}, \quad k \geq 1,$$

and where $0 \leq \gamma_k < 1$. Momentum is sometimes useful in accelerating the algorithm, but the usefulness is at the expense of an additional algorithm sequence that must be specified by the analyst (the weighting coefficient γ_k). Typically, the weighting is greater on the current gradient and less on the previous gradients (Jang et al., 1997, pp. 158–159 and 236–237). Thus, the current input to the recursive algorithm has an exponentially decaying contribution from all past individual gradients when $\gamma_k = \gamma$ for all k, $0 < \gamma < 1$. By letting $\gamma_k \to 0$, the current gradient estimate is given full weight as $k \to \infty$; this is used to ensure that the stochastic gradient algorithm converges to $\boldsymbol{\theta}^*$ as $k \to \infty$ (Spall and Cristion, 1994).

Overtraining (or *overfitting*) is a well-known issue in NNs, regression, and time series. It pertains to tuning the parameter estimates too much to the noise in the data. In fact, n data points can often be used to estimate (uniquely) n parameters, but such a model is likely to be very poor at prediction for a new data set since the original estimate will have adjusted the n parameters to all of the nuances in the particular data set, including the noise. Hence, the general principle is that a parsimonious ("lean") model is preferable to a model with many free parameters. This lets the model parameter estimation uncover overall trends in the data without fitting too closely to the specific noise characteristics of the training data set. This is sometimes referred to as the principle of *Occam's Razor*. However, one must be careful to avoid making the model too parsimonious since there is a risk of providing too little flexibility in the NN to capture the important nonlinearities. Techniques such as cross-validation, the Vapnik–Chervonenkis dimension, and the Akaike information criterion are methods for expressing this tradeoff and choosing the appropriate dimension of the model. Sections 13.1 and 13.2 consider the issues of overtraining and model selection in much greater detail.

Overtraining is a greater threat in batch processing than recursive processing. In batch processing, it is relatively easy to fit the parameters to the noise in the data because there is a *fixed* amount of data and a potentially large number of parameters. For a given NN structure, one practical method for coping with overtraining is to restrict the number of batch iterations to avoid hewing too closely to the specific data set (e.g., Haykin, 1999, Sects. 4.12 and 4.13). There is less of a threat of overtraining in *recursive* processing with a fixed NN structure because there is no hard limit to the amount of data that is to be used in the estimation of the fixed number of weights.

5.3 DISCRETE-EVENT DYNAMIC SYSTEMS

Discrete-event dynamic systems (DEDS) are continuous time systems that evolve by changing states at random points in time. The changes in state correspond to a discrete event. The occurrence of a discrete event, which is one outcome from a discrete set of possible outcomes, triggers a change from one system state to another. Once in a given state, the system will stay in that state until the next discrete event occurs, triggering a jump to a new state. A comprehensive introduction to DEDS is Cassandras and Lafortune (1999).

A large number of practical systems fit within the framework of DEDS, including communications networks, manufacturing systems, computer networks, and transportation systems. Thus, DEDS are models of continuous-time systems whose possible states lie in a discrete set of possible outcomes. Equivalently, the trajectories of the process over time are piecewise constant. For example, if the system state represents the number of users on a computer network, the state remains constant until a user logs off or on, corresponding to a (random) discrete event. From an engineering perspective, the jumps associated with DEDS require fundamentally different tools than traditional models based on ordinary or partial differential equations. Often, DEDS are thought of in the context of queuing networks, where the discrete events might denote a customer arrival or a service completion. In this context, $\boldsymbol{\theta}$ represents parameters that the system designer has some control over, such as mean service rates or other parameters associated with service times.

A typical goal in DEDS design is to come up with the optimal choice of $\boldsymbol{\theta}$ in light of the inherent stochastic aspects of the problem. In DEDS, $\boldsymbol{\theta}$ might represent a collection of direct physical parameters for the system and/or parameters entering the probability distributions of various random processes in the system. Most of the discussion below is drawn from Suri (1989), Glasserman (1991a), L'Ecuyer and Glynn (1994), and Fu and Hu (1997). It is common to optimize DEDS using a Monte Carlo simulation of the process. Chapters 14 and 15 consider simulation-based optimization in detail, including methods relevant to DEDS.

A brute-force approach to optimizing DEDS might be to change various elements of $\boldsymbol{\theta}$, run the simulation several times to get average performance for that chosen value of $\boldsymbol{\theta}$, and then repeat this process for a new value of $\boldsymbol{\theta}$. Obviously, this may be a seriously flawed approach relative to obtaining the true optimal value of $\boldsymbol{\theta}$. If $\boldsymbol{\theta}$ is of even moderate dimension, the number of changes required to ensure reasonable confidence in obtaining a good solution is huge, and, furthermore, each simulation may be very costly to run.

A more systematic approach has been adapted under the rubric of *perturbation analysis* (PA). In this approach, one looks for ways to get a gradient estimate at any $\boldsymbol{\theta}$ value based on only one or a small number of simulation runs. Given the stochastic nature of the simulation, this gradient estimate is only a stochastic estimate of the true gradient $\boldsymbol{g}(\boldsymbol{\theta}) = \partial L/\partial\boldsymbol{\theta}$. A specific form of PA is the infinitesimal perturbation analysis (IPA) approach to generating a gradient estimate. IPA requires that the probability distribution generating V be independent of $\boldsymbol{\theta}$, consistent with the standard formulation for stochastic gradient methods discussed in Section 5.1. Thus, the IPA gradient estimate has the generic form $\partial Q_k(\boldsymbol{\theta}, V_k)/\partial\boldsymbol{\theta}$ or $\partial\bar{Q}(\boldsymbol{\theta}, V)/\partial\boldsymbol{\theta}$, as appropriate (recursive or batch, respectively).

The IPA method has a significant potential advantage in efficiency over traditional methods based on approximating the gradient using finite-difference methods (see Chapter 6). The finite-difference methods typically require between $p+1$ and $2p$ simulation runs to form a gradient approximation, in contrast to the one run for IPA. Most discussion of IPA heavily exploits structure that is unique to DEDS. Aside from the above-mentioned assumption regarding the distribution of V being independent of $\boldsymbol{\theta}$ (as in other stochastic gradient methods), IPA requires two problem characteristics to be true:

(i) Small perturbations in $\boldsymbol{\theta}$ do not cause a dramatic change in the output or a change in the order of occurrence of events in the simulation.

(ii) Enough information about the inner workings of the simulation is available to calculate $\partial\bar{Q}(\boldsymbol{\theta}, V)/\partial\boldsymbol{\theta}$ or $\partial Q_k(\boldsymbol{\theta}, V_k)/\partial\boldsymbol{\theta}$, as appropriate, using one run of the simulation.

Neither of (i) or (ii) is automatically true in most practical DEDS, but if (i) and (ii) do hold, then IPA can be used to provide the required input for the gradient-based SA recursion (5.9). Condition (i) is closely tied to the general requirements for interchange of derivative and integral (Theorem 5.1 in Subsection 5.1.1), so that $E[\partial Q_k(\boldsymbol{\theta}, V_k)/\partial\boldsymbol{\theta}] = \boldsymbol{g}(\boldsymbol{\theta})$ at all $\boldsymbol{\theta}$ of interest. A large fraction of Glasserman (1991a), for example, is devoted to establishing formal conditions in DEDS under which (i) holds. (Suri, 1989, presents a realistic example where the interchange of derivative and integral is *invalid*.) Nevertheless, even when (i) holds, the failure of condition (ii) will make it impossible to implement IPA. Monte Carlo simulations are often used for optimization of DEDS. In many complex simulations, it is difficult or impossible to calculate $\partial Q_k(\boldsymbol{\theta}, V_k)/\partial\boldsymbol{\theta}$, as that requires detailed analytical knowledge of the effect of each element of $\boldsymbol{\theta}$ on the simulation output.

Because DEDS are frequently associated with queuing systems, we present a simple *conceptual* example of a loss function and associated Q for a queuing-based DEDS and IPA implementation below.

Example 5.5—Simple queuing system. Suppose that a simulation output $Q(\boldsymbol{\theta}, V)$ represents the total waiting time in a system for a group of customers, where $\boldsymbol{\theta}$ is a vector of design parameters that can be adjusted to govern "typical" waiting times and V represents random effects whose distribution does not depend on $\boldsymbol{\theta}$. For example, V might represent random arrival times for customers into the system, which are unaffected by $\boldsymbol{\theta}$. The loss function is $L(\boldsymbol{\theta}) = E[Q(\boldsymbol{\theta}, V)]$, representing the typical waiting time given by the simulation, averaged over all possible values of V.

Suppose that a small change in $\boldsymbol{\theta}$ causes only small changes in simulation output Q. For example, we do *not* allow a case where, say, a slight change in the position of a piece of equipment—corresponding to one or more components of $\boldsymbol{\theta}$—causes the system to be blocked and forces the times in the queue to become indefinitely long (violating condition (i)). Further, suppose that enough is known about the simulation so that $\partial Q/\partial\boldsymbol{\theta}$ can be calculated (satisfying condition (ii) above). Then the stochastic gradient algorithm (5.9) can be used to estimate $\boldsymbol{\theta}$ based on the principles of IPA. Adding the subscript k to Q, the measurement $\boldsymbol{Y}_k(\boldsymbol{\theta}) = \partial Q_k(\boldsymbol{\theta}, V_k)/\partial\boldsymbol{\theta}$ represents the gradient associated with the kth simulation run. For example, each simulation output $Q = Q_k$ may represent the total system wait time for one day of operation. If the conditions for convergence apply (as in Section 4.3), the IPA-based SA algorithm converges to $\boldsymbol{\theta}^*$. ❑

5.4 IMAGE RESTORATION

In image restoration, the task is to recover a true image of a scene from a recorded image of the scene. The recorded image is typically corrupted by noise and/or otherwise degraded from the original image. The image restoration problem may be viewed as one of taking data $\boldsymbol{Z}$ and recovering an image $\boldsymbol{s}$ ("scene") that is observed through some nonlinear and noisy mapping. That is:

Estimate $\boldsymbol{s}$ such that $\boldsymbol{Z}$ is modeled according to $\boldsymbol{Z} = \boldsymbol{F}(\boldsymbol{s}, \boldsymbol{x}, \boldsymbol{V})$,

where $\boldsymbol{Z}$ is a vector of the observed pixel images, $\boldsymbol{F}$ is the relevant nonlinear mapping, $\boldsymbol{x}$ is some known input vector, and $\boldsymbol{V}$ is the random term, often taking the form of noise. For the problem as given above, the vector of parameters $\boldsymbol{\theta}$ corresponds to $\boldsymbol{s}$. We also consider a case below where $\boldsymbol{\theta}$ represents the parameters (weights) of a NN associated with an image processing problem.

Following Gonzalez and Woods (1992, Chap. 5) (hereafter G&W), suppose that a digitized form of an image is available. The digitized form is often a pixel-by-pixel collection of gray levels of some typically degraded image, as might have been obtained, say, through the quantification of an image obtained through cloud cover. G&W consider the case where the degradation is approximated by linear, position-invariant processes. However, they note that nonlinear mappings often make the restoration more accurate at the expense of greater complexity.

In the linear case, G&W construct a simple least-squares problem:

$$\min_{\boldsymbol{s}} \|\boldsymbol{Z} - \boldsymbol{H} \bullet \boldsymbol{s}\|^2, \tag{5.19}$$

where $\boldsymbol{H} \bullet \boldsymbol{s}$ represents a convolution of the measurement process ($\boldsymbol{H}$) and the true pixel-by-pixel image ($\boldsymbol{s}$) (the convolution represents a smoothing process, but the details are not important for the discussion here—see G&W). This can be solved by either batch linear regression methods or the LMS/RLS methods discussed in Chapter 3. As mentioned in Subsection 5.1.4, the LMS/RLS methods are special cases of the root-finding SA algorithm. G&W also show how constraints can be included through the well-known Lagrange multiplier method.

If we adopt the nonlinear approach mentioned in G&W, then the criterion analogous to that in (5.19) does not lend itself to the customary linear regression-type solution. In this case, the more general SA framework can be applied if the appropriate gradient is available. Suppose that $\boldsymbol{F}(\boldsymbol{s}, \boldsymbol{x}, \boldsymbol{V}) = \boldsymbol{h}(\boldsymbol{s}, \boldsymbol{x}) + \boldsymbol{V}$, where $\boldsymbol{h}(\cdot)$ is some nonlinear function and $\boldsymbol{V}$ is an independent, mean-zero noise. Then, (5.19) may be changed to the following optimization problem based on a stochastic criterion:

$$\min_{\boldsymbol{s}} E\left(\left\|\boldsymbol{Z} - \boldsymbol{h}(\boldsymbol{s}, \boldsymbol{x})\right\|^2\right). \tag{5.20}$$

Expression (5.20) is then appropriate for the root-finding SA algorithm. The implementation may be in a complete-data (batch) mode, where the algorithm repeatedly passes through all the pixel data at once, leading to each iteration relying on $\partial\left(\left\|\boldsymbol{Z} - \boldsymbol{h}(\boldsymbol{s}, \boldsymbol{x})\right\|^2\right)/\partial \boldsymbol{s}$. Alternatively, the implementation may be in a pixel-by-pixel mode, where the algorithm passes through the data using $\partial\left(\left\|z_k - h_k(\boldsymbol{s}, \boldsymbol{x}_k)\right\|^2\right)/\partial \boldsymbol{s}$ at each iteration, where z_k represents the gray level for the kth pixel, $h_k(\cdot)$ represents the relevant nonlinear mapping for the kth pixel, and $\boldsymbol{x}_k$ represents the relevant user input for that pixel.

To reflect correlation in the pixels and the relative quality of the image data, it is, in general, preferable to include some type of weighting for the various contributions in (5.19) or (5.20). The form shown above does not include such weighting since it is built from a simple Euclidean norm (i.e., sum of squared errors). Obviously, the generalization to weighting of the squared errors does not affect the overall optimization approach.

Specific implementations of the ideas discussed above are given, for example, in Abreu et al. (1996) and Cha and Kassam (1996). Abreu et al. (1996) is concerned with removing impulse noise that may be due, for example, to noisy sensor or transmission errors in collecting the image data. They note that nonlinear methods have often proven superior for this problem relative to linear methods. Their approach is based on a classification algorithm where the current pixel is replaced by a smoothed combination of neighboring pixels if the classifier detects that the current pixel is corrupted by noise. Although this involves a linear combining process, the *overall* process is nonlinear because of the ranking and classification aspects of the filter.

The essential stochastic gradient form of the SA recursion arises in this approach through an adaptive calculation of the weighting coefficients in the linear combining mentioned above; these weights express the emphasis to put on the current pixel value versus the rank-ordered mean. The number of pairs of weighting coefficients is equal to the number of states that the current set of neighboring pixels can occupy; each state has a separate SA recursion.

Cha and Kassam (1996) describe an approach to image restoration based on a type of NN called the *radial basis function network* (RBFN) (e.g., Jang et al 1997, Sect. 9.5). An N-node RBFN with scalar output has the generic form

$$\text{output}(\boldsymbol{x}) = \sum_{i=1}^{N} w_i \phi(\|\boldsymbol{x} - \boldsymbol{c}_i\|), \tag{5.21}$$

where $\boldsymbol{x}$ is some network input vector, w_i is the ith weight parameter of the RBFN, $\phi(\cdot)$ is some scalar-valued nonlinear basis function, and $\boldsymbol{c}_i$ is a centering vector. RBFNs have a single-layer structure as shown in (5.21), in contrast to the general multilayer structure of the feedforward NNs discussed in Section 5.2. The most common form for $\phi(\cdot)$ is the Gaussian function associated with the normal probability density function. RBFNs have the advantage of being generally faster to train than feedforward NNs (note the lack of weights to be estimated between the input and single hidden layer) and relative ease of interpretation due to their single-layer structure (Cha and Kassam, 1996; Jang et al, 1997, Sect. 9.5). As with feedforward NNs, RBFNs share the property of being universal approximators of continuous functions.

Cha and Kassam (1996) take the recorded gray scale images of a window of pixels surrounding the pixel of interest as the $\boldsymbol{x}$ vector in (5.21) (so $\boldsymbol{x}$ corresponds to part of the $\boldsymbol{Z}$ vector above). The output of the RBFN is the estimate of the pixel of interest. Through the use of the RBFN, it is not necessary to build up an explicit model for the measurement process (the RBFN acts essentially as an inverse model for recovering $\boldsymbol{s}$).

The stochastic gradient SA approach arises in the training of the weights w_i and centering vectors $\boldsymbol{c}_i$ shown in (5.21) (they also train an additional spread parameter analogous to the standard deviation in the Gaussian density function). The algorithm is aimed at minimizing a mean-squared error loss function measuring the deviation between the actual and estimated gray-scale level. As in Sections 5.1 and 5.2, Cha and Kassam (1996) use the instantaneous (pixel-by-pixel) error in the implementation of the algorithm. Although the estimation for the weights, centering parameters, and scale factors is conducted simultaneously, each set of parameters is assigned its own gain sequence (corresponding to the a_k in (5.9)).

Much of the attention in image restoration is on simulated annealing algorithms, which may be thought of as generalizations of stochastic gradient SA in (5.9) to the global optimization context (see Section 8.6). This is a large subject in image processing (see, e.g., Olsson, 1993). Simulated annealing is considered in Chapter 8.

5.5 CONCLUDING REMARKS

This chapter has focused on what is perhaps the most popular use of the root-finding SA class of algorithms—stochastic gradient algorithms for optimization. These algorithms are associated with the fundamental optimization equation $\boldsymbol{g}(\boldsymbol{\theta}) = \partial L/\partial\boldsymbol{\theta} = \boldsymbol{0}$. Relative to standard optimization, the main distinction is that only noisy measurements of the gradient $\boldsymbol{g}(\boldsymbol{\theta})$ are used. There are several general implementations in the stochastic gradient setting. These include recursive processing, batch processing, and hybrid forms involving multiple passes through a given set of data (epochs).

Many well-known algorithms are special cases of the stochastic gradient setting. These include the LMS and RLS algorithms of Chapter 3, backpropagation for neural networks, infinitesimal perturbation analysis for discrete-event systems, and many algorithms for signal processing and adaptive control (e.g., in noise cancellation, image restoration, target tracking).

While the stochastic gradient formulation enjoys wide use, its very name also points to its major restriction. The requirement for a gradient—even a noisy gradient—means that one must have available the details of the underlying model. In problems involving complex processes, it is frequently difficult or impossible to get such information. Associated with the fundamental issue of gradient availability, the stochastic gradient method requires that the probability distribution generating the randomness (V) be independent of $\boldsymbol{\theta}$ and that the derivative and integral (expectation) be interchangeable in $\partial E[Q(\boldsymbol{\theta},V)]/\partial\boldsymbol{\theta}$. There are numerous realistic applications where one or both of these requirements break down.

To address some of the difficulties in implementing stochastic gradient methods, Chapters 6 and 7 consider some gradient-free SA methods. These chapters show how the rich theory of SA can be used in optimization problems where only noisy measurements of the loss function are available.

EXERCISES

5.1 Consider the setting of Example 5.1, except that the scalar $W = W(\theta)$ now has a distribution governed by the density function $\exp(-w/\theta)/\theta$ (w is the dummy variable). Create an equivalent random process and associated replacement for W, say W', such that the joint distribution for $\{Z, W'\}$ (representing the random effects V) does not depend on θ.

5.2 Suppose that V is generated according to a density $p_V(\upsilon|\boldsymbol{\theta})$. Write down an unbiased estimate of $\boldsymbol{g}(\boldsymbol{\theta})$ (the estimate does not necessarily have to be implementable in all applications).

5.3 Paraphrase the arguments in Example A.3 (Appendix A) regarding the invalidity of interchange between the derivative and integral. Plot the function and comment on how this illustrates the violation of at least one condition for the interchange.

5.4 Suppose that $Q(\boldsymbol{\theta}, \boldsymbol{V}) = L(\boldsymbol{\theta}) + \boldsymbol{\theta}^T\boldsymbol{W}\boldsymbol{\theta} + \boldsymbol{\theta}^T\boldsymbol{w}$, where $L(\boldsymbol{\theta})$ is a differentiable function, $\boldsymbol{W}$ is a symmetric $p \times p$ random matrix, $\boldsymbol{w}$ is $p \times 1$ random vector, and $\boldsymbol{V} = [\boldsymbol{W}, \boldsymbol{w}]$ has an unknown distribution independent of $\boldsymbol{\theta}$ (it is known that the elements of $E(\boldsymbol{V})$ formally exist in the sense discussed in Appendix C). Determine $\boldsymbol{e}(\boldsymbol{\theta})$ and the conditions on $\boldsymbol{V}$ to ensure that condition A.3 in Section 4.3 holds when $\Theta = \mathbb{R}^p$ (i.e., $E[\boldsymbol{e}(\boldsymbol{\theta})] = \boldsymbol{0}$ for all $\boldsymbol{\theta}$).

5.5 Consider the function $Q(\boldsymbol{\theta}, \boldsymbol{V}) = \sum_{i=1}^{p/2} t_i^4 + \boldsymbol{\theta}^T\boldsymbol{B}\boldsymbol{\theta} + \boldsymbol{\theta}^T\boldsymbol{V}$, where p is an even number, $\boldsymbol{\theta} = [t_1, t_2, \ldots, t_p]^T$, $\boldsymbol{B}$ is a symmetric matrix, and $\boldsymbol{V}$ is a vector of random errors. Derive $\boldsymbol{Y}(\boldsymbol{\theta}) = \partial Q(\boldsymbol{\theta}, \boldsymbol{V})/\partial\boldsymbol{\theta}$ when $\boldsymbol{B}$ and $\boldsymbol{V}$ are independent of $\boldsymbol{\theta}$ (this function is considered in several numerical exercises throughout this book).

5.6 In the proof of Proposition 5.1 on convergence for LMS, show that for $i \neq j$, $E(\boldsymbol{e}_i^T\boldsymbol{e}_j) = 0$, as required in going from (5.15) to (5.16).

5.7 Consider a model of the form $z = \beta_0 + \boldsymbol{\beta}^T\boldsymbol{x} + \boldsymbol{x}^T\boldsymbol{B}\boldsymbol{x} + v$ (suppressing the subscript k), where β_0 is an additive constant, $\boldsymbol{\beta}$ is a vector, and $\boldsymbol{B}$ is a symmetric matrix. Let $\boldsymbol{\theta}$ represent the unique elements in $\{\beta_0, \boldsymbol{\beta}, \boldsymbol{B}\}$. If $\dim(\boldsymbol{x}) = r$, determine $\dim(\boldsymbol{\theta}) = p$ in terms of r.

5.8 Using the production function from Example 4.4, conduct an analysis of the improvement in the accuracy of estimates obtained by using the stochastic gradient implementation in batch and multiple-pass modes. Generate data on the grid defined by the integer pairs (C, W), where $C \in [1, 2, \ldots, 10]$ and $W \in [11, 12, \ldots, 110]$. Perform 1000 iterations by randomly sampling (uniformly) the 1000 grid points without replacement until all 1000 points have been used. Let $\boldsymbol{\theta}^* = [2.5, 0.7]^T$ and use the production function in Example 4.4 to compute the output z_k, with random error $v_k \sim N(0, 5^2)$. Compare your results with those obtained in Example 4.4. Pick an initial condition of $\hat{\boldsymbol{\theta}}_0 = [1.0, 0.5]^T$ for all runs. Produce a table showing the distance of the final estimate to $\boldsymbol{\theta}^*$ for each of the three implementations according to the following three settings: (i) The basic recursive form (5.9) with $a_k = 0.00125/(k+100)^{0.501}$ as the step size. (ii) The batch mode based on 500 iterations of (5.10) as applied to full data set of 1000 points. Let $a_k = 0.0075/(k+100)^{0.501}$. (iii) The multiple-pass implementation (Subsection 5.1.2) by cycling 10 times through the data (using the final parameter estimate of one cycle as the starting value for the next cycle). Use $a_k = 0.00015/(k+100)^{0.501}$ and do not reset to a_0 at each cycle (so a_k runs from a_0 to a_{9999} over the 10 passes through the data).

5.9 With the 160 input–output measurements in **`reeddata-fit`** (discussed in Section 3.4) (available at the book's Web site), perform a *batch* implementation of SA as in (5.10). Use a simplified form of the quadratic model in Exercise 5.7, where $\boldsymbol{B}$ is a diagonal matrix. (Here $\boldsymbol{x} \in \mathbb{R}^6$ represents the input variables defined in Section 3.4—T, A, E, V, S, and F). Let $\boldsymbol{\theta}$

represent β_0, the elements of $\boldsymbol{\beta}$, and the diagonal elements of $\boldsymbol{B}$. Use the standard sum of squared-error loss function. Calculate three estimates of $\boldsymbol{\theta}$ using a gain sequence of the form $a/(k+10)^{0.501}$: One estimate with a = 0.005, one estimate with a = 0.02, and one estimate with another value of a. Use the initial conditions $\beta_0 = 0$, $\boldsymbol{\beta} = \boldsymbol{0}$, and $\boldsymbol{B} = \boldsymbol{I}_6$. With the 80 independent measurements in **reeddata-test** (available at the book's Web site), report the mean absolute prediction errors (analogous to Table 3.3) for the three estimates of $\boldsymbol{\theta}$.

5.10 For the model form and problem setting of Exercise 5.9, estimate $\boldsymbol{\theta}$ using a *recursive* implementation of the SA algorithm ((5.9) and (5.12)) and *one* pass through the data **reeddata-fit**. Carry out this process three times with the gain $a/(k+10)^{0.501}$: Use a = 0.005, 0.02, and one other value of your choosing. Then, using the 80 independent measurements in **reeddata-test**, report the mean absolute prediction errors (i.e., the mean absolute deviation [MAD], analogous to Table 3.3) for the three different values of a. If you also did Exercise 5.9, how do the final recursive and iterative batch estimates compare for $\boldsymbol{\theta}$ (based on the analysis with **reeddata-test**)?

5.11 With the data in **reeddata-fit**, estimate a single hidden-layer feedforward NN using backpropagation in a batch implementation (pick eight to 10 nodes in the hidden layer, each node being a common form of sigmoidal function or hyperbolic tangent function). Use a gain sequence of the form $a/(k+10)^{0.501}$ and initial weights generated uniformly over (−0.5, 0.5). (It is not necessary to code backpropagation from scratch; use, e.g., the version in the MATLAB NNs toolbox or any reliable version from the Web where the gain sequence can be user-specified.) After completing the above, use the 80 measurements in **reeddata-test** to determine the mean absolute prediction error (a.k.a. the MAD) from the trained NN. Compare with the numbers in Table 3.3.

5.12 Perform a recursive implementation with a single hidden-layer feedforward NN of the form in Exercise 5.11. This recursive implementation should be run for one pass (160 iterations) through the data based on initial weights as in Exercise 5.11. Compare the MAD using data in **reeddata-test** for several values of a in a gain sequence of the form $a/(k+10)^{0.501}$ and with the values in Table 3.3. If you also did Exercise 5.11, how do the batch and recursive results compare?

CHAPTER 6

STOCHASTIC APPROXIMATION AND THE FINITE-DIFFERENCE METHOD

This and the next chapter continue in the spirit of Chapter 5 in focusing on the loss minimization problem with stochastic approximation (SA) (versus general root-finding as in Chapter 4). The approaches in this and the next chapter, however, differ in a fundamental way from the approach in Chapter 5. Namely, it is assumed that the stochastic gradient ($\partial Q/\partial \boldsymbol{\theta}$) is *not* available. Rather, it is assumed that only (generally noisy) measurements of the loss function are available. This gradient-free approach has significant value in many problems since the stochastic gradient is often difficult or impossible to obtain in practice.

Section 6.1 introduces some of the fundamental issues in gradient-free optimization and Section 6.2 discusses several problem areas where a gradient-free method is more appropriate than the gradient-based methods discussed in Chapter 5. Section 6.3 describes a core gradient-free SA method, which uses a classical finite-difference (FD) approximation to the gradient. Sections 6.4 and 6.5 summarize some of the theory available on convergence and convergence rates for the FD-based SA algorithm. Section 6.6 discusses the recurring issue of choosing the coefficients for the algorithm gain sequences. Section 6.7 presents some numerical results. Section 6.8 summarizes some extensions and provides a bridge to Chapter 7 on simultaneous perturbation SA. Section 6.9 offers some final remarks.

6.1 INTRODUCTION AND CONTRAST OF GRADIENT-BASED AND GRADIENT-FREE ALGORITHMS

As discussed in Chapters 4 and 5, stochastic approximation is a powerful tool for root-finding and optimization when only noisy measurements are available to solve the problem. Let us continue with the familiar problem of minimizing a differentiable loss function $L = L(\boldsymbol{\theta})$ via finding a root to $\partial L/\partial \boldsymbol{\theta} = \mathbf{0}$. Chapter 5 discussed the application of the root-finding (Robbins–Monro) SA methods (Chapter 4) to the minimization problem. This led to a stochastic gradient algorithm requiring knowledge of $\boldsymbol{Y}(\boldsymbol{\theta}) = \partial Q/\partial \boldsymbol{\theta}$ for Q in the representation $L(\boldsymbol{\theta}) = E[Q(\boldsymbol{\theta}, V)]$, where V corresponds to the cumulative randomness in the problem. Here $\partial Q/\partial \boldsymbol{\theta}$ represents a *direct* noisy measurement of the (unknown) true gradient $\boldsymbol{g}(\boldsymbol{\theta}) = \partial L/\partial \boldsymbol{\theta}$.

There are, however, a large number of problems where the direct measurement, $\partial Q/\partial \boldsymbol{\theta}$, is difficult or impossible to obtain. For this reason, there is considerable interest in SA algorithms that do not depend on direct gradient measurements. Rather, these algorithms are based on an *approximation* to the gradient formed from (generally noisy) measurements of L. This interest has been motivated, for example, by problems in the adaptive control and statistical identification of complex systems, the optimization of processes by Monte Carlo simulations, the training of recurrent neural networks, the design of complex queuing and discrete-event systems, and the recovery of images from noisy sensor data. To contrast with the stochastic gradient algorithms of Chapter 5, the methods of this chapter are *gradient-free*. This use of "gradient-free" is associated with the *implementation* of the algorithms. It does not refer to conditions imposed to guarantee convergence, where the gradient $\boldsymbol{g}(\boldsymbol{\theta}) = \partial L/\partial \boldsymbol{\theta}$ is assumed to exist.

The focus in this chapter is the SA algorithm based on the oldest method for gradient approximation—the finite-difference (FD) approximation (e.g., Dennis and Schnabel, 1989). The FD approximation relies on small one-at-a-time changes to each of the individual elements of $\boldsymbol{\theta}$. After each change, the (possibly noisy) value of L is measured. When measurements of L have been collected for perturbations in each of the elements of $\boldsymbol{\theta}$, the gradient approximation may be formed. The FD approach is motivated directly from the definition of the gradient as a collection of derivatives for each of the components in $\boldsymbol{\theta}$, holding all other components fixed. Unfortunately, the method can be very costly if the dimension p is high since one must collect at least one L measurement for each of the elements in $\boldsymbol{\theta}$. This cost motivates the Monte Carlo-based approaches of Section 6.8 and Chapter 7. Nevertheless, the FD method is fundamental in both stochastic and deterministic optimization.

Building on the seminal Robbins and Monro (1951) paper, an SA algorithm based on the FD gradient approximation was introduced for scalar θ in Kiefer and Wolfowitz (1952) and multivariate $\boldsymbol{\theta}$ in Blum (1954b). The FD-based SA (FDSA) algorithm is the oldest SA method using gradient approximations built from only loss measurements. Because of its relative ease of use, FDSA is much more widely used in practice than the stochastic gradient-based methods in at least one important area—simulation-based optimization (Fu and Hu, 1997, p. 5). It is probably more widely used in other areas as well, although this author can offer no proof of that.

Although they require only loss function measurements, gradient-free SA algorithms such as FDSA exhibit convergence properties similar to the properties of the stochastic gradient method of Chapter 5. The indirect connection to the gradient usually enhances the convergence when there is not a great danger of converging to an unacceptable local minimum. That is, basic FDSA—as with most other SA algorithms—is fundamentally a local optimizer. (Section 8.4 discusses global versions of SA.)

Recall that the random search methods of Chapter 2 also use only loss measurements (no gradients). Further, the random search methods may be *global* optimizers under appropriate conditions. However, one of the major restrictions in both the convergence theory for random search (such as Theorems 2.1 and 2.2) and in many practical implementations is the assumption of perfect *noise-free* loss measurements. This contrasts with FDSA, which is fundamentally based on noisy loss measurements (although a special case is noise-free measurements).

Direct measurements of the gradient $\partial Q/\partial \boldsymbol{\theta}$ do not typically arise naturally in the course of operating or simulating a system. Hence, one must have detailed knowledge of the underlying system input–output relationships in order to calculate the $\partial Q/\partial \boldsymbol{\theta}$ from basic output measurements such as the z_k in the nonlinear settings of Section 5.1. In contrast, the FDSA approach requires only conversion of the basic output measurements to sample values of the loss function, which does *not* require full knowledge of the system input–output relationships.

Because of the fundamentally different information needed in implementing the stochastic gradient-based and gradient-free algorithms, it is difficult to construct meaningful methods of comparison. As a general rule, however, the stochastic gradient algorithms converge faster *when speed is measured in number of iterations*. This is not surprising given the additional information required for the gradient-based algorithms. In particular, based on asymptotic distribution theory, one can determine the optimal rate of convergence of the iterate $\hat{\boldsymbol{\theta}}_k$ to an optimum $\boldsymbol{\theta}^*$. In a stochastic sense and for a large number of iterations k, this rate is proportional to $1/\sqrt{k}$ for the stochastic gradient (i.e., root-finding) algorithms (see Section 4.4) and proportional to the more slowly decaying $1/k^{1/3}$ for the algorithms based on gradient approximations (Fabian, 1971). (Exceptions to this maximum rate of convergence for the gradient-free algorithms are discussed in Fabian, 1971; Glasserman and Yao, 1992; L'Ecuyer and Yin, 1998; and Kleinman et al, 1999, where special cases are presented that achieve a rate arbitrarily close, or equal, to $1/\sqrt{k}$. See also Chapter 14 and the discussion on common random numbers.)

In practice, many factors besides the asymptotic rate of convergence must be considered in determining which algorithm is most appropriate for a given circumstance. The stochastic gradient algorithms may be either infeasible (if no system model is available) or undependable (if a poor system model is used). Further, the total cost to achieve effective convergence depends not only on the number of iterations required, but also on the cost per iteration, which is typically greater in gradient-based algorithms. This cost may include greater computational burden, additional human effort required for determining and writing software for gradients, and experimental costs for model building such as labor, materials, and fuel. Finally, the rates of convergence are based on asymptotic theory, and may not represent practical convergence rates in finite-samples. As a general rule, however, if direct (stochastic) gradient information is *conveniently and reliably* available, it is generally to one's advantage to use this information in the optimization process. This chapter focuses on the setting where such information is *not* readily available.

6.2 SOME MOTIVATING EXAMPLES FOR GRADIENT-FREE STOCHASTIC APPROXIMATION

This section summarizes several problems where the stochastic gradient $\partial Q/\partial\boldsymbol{\theta}$ is difficult or impossible to obtain, motivating gradient-free SA methods such as FDSA or the simultaneous perturbation SA algorithm of the next chapter. The gradient-free methods use only (generally noisy) measurements of L. Three examples are given below, one in generic parameter estimation, one in feedback control, and one in simulation-based optimization.

Example 6.1—Generic parameter estimation for complex loss functions. Many general problems of statistical parameter estimation involve a complex relationship between $\boldsymbol{\theta}$ and $L(\boldsymbol{\theta})$, or, in the stochastic context, between $\boldsymbol{\theta}$ and $Q(\boldsymbol{\theta}, V)$. In such problems, difficulties in the calculation of the exact gradient $\boldsymbol{g}(\boldsymbol{\theta})$ or stochastic gradient $\partial Q/\partial\boldsymbol{\theta}$ may arise in several ways:

- It may be too costly in time and/or money to carry out the calculations required to obtain the gradient.
- There is the potential for human error in the derivations when it is possible—at least in principle—to carry out the calculations.
- For complex calculations, there is the possibility of software coding errors in implementing the algorithm even if the gradient derivation is correct.
- Although powerful for many low-dimensional problems of moderate difficulty, symbolic software packages such as MAPLE or MATHEMATICA have difficulty in handling the complex multivariate calculations of many serious gradient derivations.
- When trying to differentiate a function represented in software (e.g., taking the stochastic gradient of the output of a Monte Carlo simulation—see the example below), so-called *automatic differentiation* methods provide a means for calculating gradients (e.g., Griewank and Corliss, 1991). However, these methods require extensive knowledge of the "inner workings" of the software. While automatic differentiation provides a systematic means for gradient calculation, the term *automatic* is somewhat of a misnomer.

There are countless examples in the literature and/or in practice where it is difficult to derive the gradient. The classical method for avoiding unwieldy gradient calculations in optimization is to use the FD approximation to the gradient as mentioned in Section 6.1 (and discussed in detail in Section 6.3). The FD approximation can be used in approximating either the deterministic gradient $\boldsymbol{g}(\boldsymbol{\theta})$ or the stochastic gradient $\partial Q/\partial\boldsymbol{\theta}$ as appropriate. ❑

Example 6.2—Model-free feedback control system. The problem of parameter estimation above is one where it is *difficult* to obtain the gradient, but where—with enough care, patience, and resources—it *may* be possible to carry out the derivation. In contrast, we now present an example where it is *inherently impossible* to obtain the gradient.

Consider building a mathematical function that allows one to control and regulate a system when there is no mathematical model representing the open-loop dynamics of the system. The control function takes information about the current and past state of the system and produces the control signals to govern future behavior of the system. For example, in an urban vehicle traffic system with real-time control, the control function takes measurements of the traffic patterns throughout the network and produces the control signal governing the amount of green, yellow, and red time allotted to the intersections in the network.

The setup here differs from the customary control framework in that no attempt is made to model the system itself (the open-loop dynamics) when constructing the control function. Various forms of this model-free control approach are considered in Saridis (1977, pp. 375–376), Bayard (1991), and Spall and Cristion (1998). Two examples of available applications include Spall and Chin (1997) (vehicle traffic control) and Vorontsov et al. (2000) (adaptive optics). Applications in many other areas, including human–machine interface control, wastewater treatment, macroeconomic policy making, and chemical process control are given at *www.jhuapl.edu/SPSA* (also available through the book's Web site).

Let us consider the following special case for motivation. Suppose that $\boldsymbol{x}_k$ represents a state vector at time k; this vector is a summary of the relevant characteristics of the system. Let the process be governed by the following discrete-time recursion:

$$\boldsymbol{x}_{k+1} = \phi(\boldsymbol{x}_k, \boldsymbol{u}_k, \boldsymbol{w}_k)\,, \tag{6.1}$$

where $\phi(\cdot)$ is an *unknown* input–output relationship (the open-loop system), $\boldsymbol{u}_k$ is the control signal, and $\boldsymbol{w}_k$ is random noise. One example of an automatic controller $\boldsymbol{u}_k$ is some function mapping that takes information about the current state and produces a signal to help the next state get close to its target vector. There exists some unknown best function for the controller depending on the unknown $\phi(\cdot)$. The aim here is to construct a control function *without* building a separate model for the unknown process $\phi(\cdot)$. Hence, in the traffic control example, no model is built for the traffic flow and queues (which, given the highly nonlinear and uncertain aspects of human behavior, is a virtually hopeless task in complex multiple-intersection networks).

The goal in the specific problem here is to have the state be close to a desired vector $\boldsymbol{d}_k$. In particular, one might aim to build a controller that minimizes the time-varying loss function

$$L_k(\boldsymbol{\theta}) = E[Q_k(\boldsymbol{\theta}, V_k)] = E\left(\tfrac{1}{2}\|\boldsymbol{x}_{k+1} - \boldsymbol{d}_{k+1}\|^2\right) \tag{6.2}$$

at each k, where V_k represents the cumulative randomness in the state $\boldsymbol{x}_{k+1}$ (e.g., that due to the cumulative effects of the noises $\boldsymbol{w}_0, \boldsymbol{w}_1, \ldots, \boldsymbol{w}_k$ plus any other external random inputs). The need for the time index k on L_k and Q_k might be due, among other things, to the time variation in the target vector $\boldsymbol{d}_{k+1}$ (the dependence of the right-hand side of (6.2) on $\boldsymbol{\theta}$ is discussed below). Clearly, the dimensions of the vectors $\boldsymbol{x}_k$ and $\boldsymbol{d}_k$ must be identical.

Because the system relationship $\boldsymbol{\phi}(\cdot)$ is unknown, it is not possible to know the analytical form of the true optimal control function. Therefore, we introduce an *approximation*

$$\boldsymbol{u}_k = \boldsymbol{\mu}_k(\boldsymbol{x}_k, \boldsymbol{d}_{k+1} | \boldsymbol{\theta}), \tag{6.3}$$

where $\boldsymbol{\mu}_k(\cdot\,|\boldsymbol{\theta})$ is a user-specified function approximator that depends on some parameters $\boldsymbol{\theta}$. (This building of a control function *without* the building of a separate model for the system $\boldsymbol{\phi}(\cdot)$ is in the spirit of the principle of parsimony—Chapter 13—which favors a minimalist approach to modeling necessary to carry out the task at hand.)

As discussed in Section 5.2, one important class of "universal" function approximators is artificial neural networks. If a NN is used to represent the controller in (6.3), then the input to the NN at time k will be $\boldsymbol{x}_k$ and $\boldsymbol{d}_{k+1}$, the output will be the control signal, and $\boldsymbol{\theta}$ will represent the connection weights to be estimated. Note that $\boldsymbol{\theta}$ enters the right-hand side of (6.2) via the dependence of $\boldsymbol{x}_{k+1}$ on $\boldsymbol{u}_k$ and the dependence of $\boldsymbol{u}_k$ on $\boldsymbol{\theta}$. Figure 6.1 depicts a process associated with the setting of (6.1)–(6.3). Note the absence of any model for the system in the overall feedback process (hence the label *model free*). The details of the parameter estimation method are not shown in Figure 6.1, as they may vary depending on the nature of the system being controlled.

Let us now show why a gradient-based approach does not apply in the model-free control setting. For convenience, suppose that x_k and u_k are both scalars (but see Exercise 6.1). Returning to (6.2) and assuming that the necessary interchange of derivative and integral applies (as discussed in Section 5.1),

$$\frac{\partial Q_k}{\partial \boldsymbol{\theta}} = \frac{1}{2}\frac{\partial (x_{k+1} - d_{k+1})^2}{\partial \boldsymbol{\theta}} = (x_{k+1} - d_{k+1})\frac{\partial x_{k+1}}{\partial \boldsymbol{\theta}} = (x_{k+1} - d_{k+1})\frac{\partial \phi}{\partial u_k}\frac{\partial u_k}{\partial \boldsymbol{\theta}}. \tag{6.4}$$

The gradient $\partial \phi / \partial u_k$ is, however, unavailable since $\phi(\cdot)$ represents the *unknown* dynamics of the process. This breaks the chain in the chain rule on the right-hand side of (6.4). Hence, the overall stochastic gradient $\partial Q_k / \partial \boldsymbol{\theta}$ is unavailable even though $\partial u_k / \partial \boldsymbol{\theta}$ and $x_{k+1} - d_{k+1}$ are typically available (e.g., for a feedforward NN, $\partial u_k / \partial \boldsymbol{\theta}$ is readily available as part of the backpropagation calculations). So, as long as the dynamics are unknown, (6.4) implies that the gradient $\partial Q_k / \partial \boldsymbol{\theta}$ is unavailable. ❑

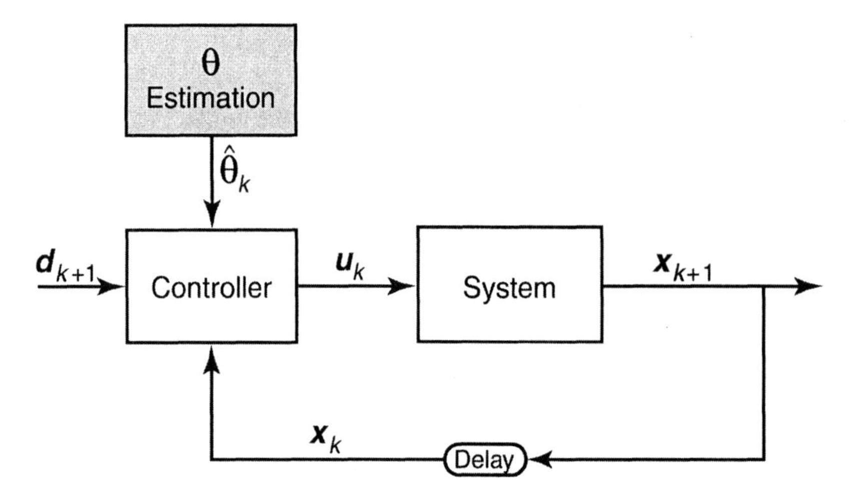

Figure 6.1. Model-free control setup with $\boldsymbol{u}_k$ representing the control input.

Example 6.3—Simulation-based optimization. Many practical systems are too complex to be analyzed via traditional analytical methods. Computer-based simulations are popular methods for investigating such systems. Suppose that one is going to use a simulation to optimize some real system. For example, in a public health scenario, there may be interest in determining some "best" combination of strategies related to vaccination, medical treatment, and quarantining, to stave off an epidemic. Here, $\boldsymbol{\theta}$ represents various terms associated with the available strategies. The simulation output represents some measure of public health (e.g., fraction of population that is ill), reflecting the value of $\boldsymbol{\theta}$ and the inherent randomness of the spread of disease.

For general problems, optimization of a *physical system* may be carried out by varying $\boldsymbol{\theta}$ in some intelligent way, running a *credible simulation* of the system, evaluating the outcome, and then repeating these steps until the "best" value of $\boldsymbol{\theta}$ has been obtained. In the notation of Chapter 5, $Q(\boldsymbol{\theta}, V)$ represents the simulation output and V represents the amalgamation of the Monte Carlo (pseudo) random effects in the simulation. The interest is in minimizing $L(\boldsymbol{\theta}) = E[Q(\boldsymbol{\theta}, V)]$, representing the mean outcome for a particular $\boldsymbol{\theta}$. So, each simulation output represents a noisy measurement of $L(\boldsymbol{\theta})$ (i.e., $y(\boldsymbol{\theta}) = Q(\boldsymbol{\theta}, V)$).

For the very reason that a simulation is being used (versus analytical analysis), the functional relationship between $\boldsymbol{\theta}$ and $Q(\boldsymbol{\theta}, V)$ is likely to be very complex. Hence, it is unlikely that $\partial Q/\partial\boldsymbol{\theta}$ is available; knowledge of the "inner workings" of the simulation is required to obtain $\partial Q/\partial\boldsymbol{\theta}$. Rather, one may simply specify $\boldsymbol{\theta}$ and run the simulation to produce an output $Q(\boldsymbol{\theta}, V)$. This is an example of gradient-free optimization based on noisy measurements of the loss function. Chapters 14 and 15 consider simulation-based optimization in detail. ❑

6.3 FINITE-DIFFERENCE ALGORITHM

The above discussion indicates that many problems do not allow for the computation of $\partial Q/\partial \boldsymbol{\theta}$, as needed in the stochastic gradient form of the root-finding (Robbins–Monro) algorithm. This section introduces an alternative to the stochastic gradient algorithm for the case where only the measurements $y(\boldsymbol{\theta}) = L(\boldsymbol{\theta}) + \varepsilon(\boldsymbol{\theta})$ are available at various values of $\boldsymbol{\theta}$. In fact, the algorithm has the even weaker requirement of only requiring measurements of the *difference* of two values of the loss function, as opposed to measuring the loss functions themselves.

The recursive procedure here is in the general SA form

$$\hat{\boldsymbol{\theta}}_{k+1} = \hat{\boldsymbol{\theta}}_k - a_k \hat{\boldsymbol{g}}_k(\hat{\boldsymbol{\theta}}_k), \tag{6.5}$$

where $\hat{\boldsymbol{g}}_k(\hat{\boldsymbol{\theta}}_k)$ is the estimate of the gradient $\partial L/\partial \boldsymbol{\theta}$ at the iterate $\hat{\boldsymbol{\theta}}_k$ based on measurements of the loss function. Hence, (6.5) is analogous to the stochastic gradient algorithm, with the gradient estimate $\hat{\boldsymbol{g}}_k(\boldsymbol{\theta})$ replacing the direct gradient measurement $\boldsymbol{Y}(\boldsymbol{\theta}) = \partial Q/\partial \boldsymbol{\theta}$ at $\boldsymbol{\theta} = \hat{\boldsymbol{\theta}}_k$. The gain $a_k > 0$ here acts in a way similar to its role in the stochastic gradient form. Under appropriate conditions, the iteration in (6.5) converges to $\boldsymbol{\theta}^*$ in some stochastic sense (usually almost surely, a.s.). Typical convergence conditions are very similar to those in Section 4.3 here for the root-finding SA algorithm. Convergence is discussed in Section 6.4.

The essential part of (6.5) is the gradient approximation $\hat{\boldsymbol{g}}_k(\hat{\boldsymbol{\theta}}_k)$. We discuss below the oldest and best-known means of forming the approximation—the FD method. Expression (6.5) with this approximation represents the FDSA algorithm. One-sided gradient approximations involve measurements $y(\hat{\boldsymbol{\theta}}_k)$ and $y(\hat{\boldsymbol{\theta}}_k + \text{perturbation})$, while two-sided approximations involve measurements of the form $y(\hat{\boldsymbol{\theta}}_k \pm \text{perturbation})$. The two-sided FD approximation for use with (6.5) is

$$\hat{\boldsymbol{g}}_k(\hat{\boldsymbol{\theta}}_k) = \begin{bmatrix} \dfrac{y(\hat{\boldsymbol{\theta}}_k + c_k \boldsymbol{\xi}_1) - y(\hat{\boldsymbol{\theta}}_k - c_k \boldsymbol{\xi}_1)}{2c_k} \\ \vdots \\ \dfrac{y(\hat{\boldsymbol{\theta}}_k + c_k \boldsymbol{\xi}_p) - y(\hat{\boldsymbol{\theta}}_k - c_k \boldsymbol{\xi}_p)}{2c_k} \end{bmatrix}, \tag{6.6}$$

where $\boldsymbol{\xi}_i$ denotes a vector with a 1 in the ith place and 0's elsewhere and $c_k > 0$ defines the difference magnitude. The pair $\{a_k, c_k\}$ are the gains (or gain sequences) for the FDSA algorithm. An obvious analogue to (6.6) holds for the one-sided FD approximation: The ith component of the gradient approximation is $[y(\hat{\boldsymbol{\theta}}_k + c_k\boldsymbol{\xi}_i) - y(\hat{\boldsymbol{\theta}}_k)]/c_k$. The two-sided form in (6.6) is the obvious multivariate extension of the scalar two-sided form in Kiefer and Wolfowitz (1952). The initial multivariate method in Blum (1954b) used a one-sided approximation.

6.4 CONVERGENCE THEORY

6.4.1 Bias in Gradient Estimate

The convergence theory for the FDSA algorithm is similar to the convergence theory for the root-finding SA algorithm as presented in Section 4.3. Additional difficulties, however, arise due to a bias in $\hat{\boldsymbol{g}}_k(\hat{\boldsymbol{\theta}}_k)$ as an estimator of $\boldsymbol{g}(\hat{\boldsymbol{\theta}}_k)$ and the need to control the extra gain sequence c_k. Recall from Section 4.3 that the standard "statistics" conditions for convergence of the iterate (A.1−A.4) required unbiased estimates of $\boldsymbol{g}(\cdot)$ at all k. (At the expense of other complications, the "engineering" conditions in Section 4.3 [B.1−B.5] required only an asymptotically unbiased estimate of $\boldsymbol{g}(\cdot)$.)

Let us begin by analyzing the bias in $\hat{\boldsymbol{g}}_k(\hat{\boldsymbol{\theta}}_k)$ as an estimator of $\boldsymbol{g}(\hat{\boldsymbol{\theta}}_k)$. Let $L_i'(\boldsymbol{\theta})$, $L_{ii}''(\boldsymbol{\theta})$, and $L_{iii}'''(\boldsymbol{\theta})$ represent the first, second, and third derivatives of L with respect to the ith component of $\boldsymbol{\theta}$. Suppose that the $L_{iii}'''(\boldsymbol{\theta})$ are continuous for all i and $\boldsymbol{\theta} \in \mathbb{R}^p$. Then by a third-order form of Taylor's theorem (Appendix A), we have

$$L(\hat{\boldsymbol{\theta}}_k + c_k\boldsymbol{\xi}_i) = L(\hat{\boldsymbol{\theta}}_k) + c_k L_i'(\hat{\boldsymbol{\theta}}_k) + \tfrac{1}{2}c_k^2 L_{ii}''(\hat{\boldsymbol{\theta}}_k) + \tfrac{1}{6}c_k^3 L_{iii}'''(\overline{\boldsymbol{\theta}}_k^{(i+)}), \quad (6.7a)$$

$$L(\hat{\boldsymbol{\theta}}_k - c_k\boldsymbol{\xi}_i) = L(\hat{\boldsymbol{\theta}}_k) - c_k L_i'(\hat{\boldsymbol{\theta}}_k) + \tfrac{1}{2}c_k^2 L_{ii}''(\hat{\boldsymbol{\theta}}_k) - \tfrac{1}{6}c_k^3 L_{iii}'''(\overline{\boldsymbol{\theta}}_k^{(i-)}), \quad (6.7b)$$

where $\overline{\boldsymbol{\theta}}_k^{(i+)}$ and $\overline{\boldsymbol{\theta}}_k^{(i-)}$ denote points on the line segments between $\hat{\boldsymbol{\theta}}_k$ and $\hat{\boldsymbol{\theta}}_k \pm c_k\boldsymbol{\xi}_i$ (so $\overline{\boldsymbol{\theta}}_k^{(i+)} = \hat{\boldsymbol{\theta}}_k + \lambda c_k\boldsymbol{\xi}_i$ for some $0 \le \lambda \le 1$; analogous for $\overline{\boldsymbol{\theta}}_k^{(i-)}$ with a generally different λ). Let $\mathfrak{I}_k = \{\hat{\boldsymbol{\theta}}_0, \hat{\boldsymbol{\theta}}_1, \ldots, \hat{\boldsymbol{\theta}}_k\}$. Suppose that the difference of the noise terms at any i, $\varepsilon(\hat{\boldsymbol{\theta}}_k + c_k\boldsymbol{\xi}_i) - \varepsilon(\hat{\boldsymbol{\theta}}_k - c_k\boldsymbol{\xi}_i)$, has conditional (on $\mathfrak{I}_k$) mean zero.

Combining (6.7a, b), the ith component of the FD gradient estimate (6.6) satisfies

$$
\begin{aligned}
E\left[\hat{g}_{ki}(\hat{\boldsymbol{\theta}}_k)|\Im_k\right] &= E\left[\frac{y(\hat{\boldsymbol{\theta}}_k + c_k\boldsymbol{\xi}_i) - y(\hat{\boldsymbol{\theta}}_k - c_k\boldsymbol{\xi}_i)}{2c_k}\bigg|\Im_k\right] \\
&= \frac{L(\hat{\boldsymbol{\theta}}_k + c_k\boldsymbol{\xi}_i) - L(\hat{\boldsymbol{\theta}}_k - c_k\boldsymbol{\xi}_i)}{2c_k} \\
&= L_i'(\hat{\boldsymbol{\theta}}_k) + \frac{1}{6}\frac{c_k^3 L_{iii}'''(\overline{\boldsymbol{\theta}}_k^{(i+)}) + c_k^3 L_{iii}'''(\overline{\boldsymbol{\theta}}_k^{(i-)})}{2c_k} \\
&\equiv L_i'(\hat{\boldsymbol{\theta}}_k) + b_{ki},
\end{aligned}
\qquad (6.8)^{1}
$$

where $L_i'(\cdot)$ is the ith term of $\boldsymbol{g}(\cdot)$ and b_{ki} is the ith term of the bias

$$\boldsymbol{b}_k \equiv E\left\{[\hat{\boldsymbol{g}}_k(\hat{\boldsymbol{\theta}}_k) - \boldsymbol{g}(\hat{\boldsymbol{\theta}}_k)]|\Im_k\right\}.$$

Assuming that the $L'''(\overline{\boldsymbol{\theta}}_k^{(i\pm)})$ terms are bounded, (6.8) implies that $\hat{\boldsymbol{g}}_k(\hat{\boldsymbol{\theta}}_k)$ has a bias of $O(c_k^2)$ (i.e., each component of $\hat{\boldsymbol{g}}_k(\hat{\boldsymbol{\theta}}_k)$ has an $O(c_k^2)$ bias):

$$E\left[\hat{\boldsymbol{g}}_k(\hat{\boldsymbol{\theta}}_k)|\Im_k\right] = \boldsymbol{g}(\hat{\boldsymbol{\theta}}_k) + \boldsymbol{b}_k;\ \boldsymbol{b}_k = O(c_k^2) \ \text{ a.s.}$$

Expression (6.8) also indicates that $\hat{\boldsymbol{g}}_k(\hat{\boldsymbol{\theta}}_k)$ is an unbiased estimator of the gradient (i.e., $O(c_k^2) = \boldsymbol{0}$) in the special case where L is a quadratic loss function since $L_{iii}'''(\boldsymbol{\theta}) = 0$ for all i and $\boldsymbol{\theta}$. (See Exercise 6.2 for a discussion of the bias in the one-sided FD approximation.)

6.4.2 Convergence

Given the analysis of the bias above, we are now in a position to present conditions for the formal (a.s.) convergence of the FDSA algorithm. As with the root-finding SA algorithm, there are a large number of sets of sufficient conditions in the literature (see, e.g., Fabian, 1971; Kushner and Yin, 1997, Sects. 5.3, 8.3, and 10.3). The two sets of conditions below (analogues of the "statistics" and "engineering" conditions in Section 4.3) are representative of those in the literature, but are not the most general.

The conditions below are a special case of those in Fabian (1971, Theorem 2.4) (see also Ruppert, 1991). This set is presented in a manner roughly corresponding to the "statistics" conditions A.1–A.4 in Section 4.3 for the root-finding algorithm (i.e., A.j' is an extension or modification of A.j, j = 1, 2, 3, and 4). For convenience, let $\varepsilon_k^{(i\pm)} = \varepsilon(\hat{\boldsymbol{\theta}}_k \pm c_k\boldsymbol{\xi}_i)$.

---

[1] Strictly, all conditional expectations only hold a.s. For convenience, we drop the a.s. qualifier.

A.1′ **(Gain sequences)** $a_k > 0$, $c_k > 0$, $a_k \to 0$, $c_k \to 0$, $\sum_{k=0}^{\infty} a_k = \infty$, $\sum_{k=0}^{\infty} a_k c_k < \infty$, and $\sum_{k=0}^{\infty} a_k^2 / c_k^2 < \infty$.

A.2′ **(Unique minimum)** There is a unique minimum $\boldsymbol{\theta}^*$ such that for every $\eta > 0$, $\inf_{\|\boldsymbol{\theta}-\boldsymbol{\theta}^*\| > \eta} \|\boldsymbol{g}(\boldsymbol{\theta})\| > 0$ and $\inf_{\|\boldsymbol{\theta}-\boldsymbol{\theta}^*\| > \eta} [L(\boldsymbol{\theta}) - L(\boldsymbol{\theta}^*)] > 0$ (the "inf" statements here represent the infimum—see Appendix A—with respect to all $\boldsymbol{\theta}$ such that $\|\boldsymbol{\theta} - \boldsymbol{\theta}^*\| > \eta$).

A.3′ **(Mean-zero and finite variance noise)** For all i and k, $E\left[\left(\varepsilon_k^{(i+)} - \varepsilon_k^{(i-)}\right) \middle| \Im_k\right] = 0$ a.s. and $E\left[\left(\varepsilon_k^{(i\pm)}\right)^2 \middle| \Im_k\right] \le C$ a.s. for some $C > 0$ that is independent of k and $\boldsymbol{\theta}$.

A.4′ **(Bounded Hessian matrix)** The Hessian matrix $\boldsymbol{H}(\boldsymbol{\theta}) = \partial^2 L / \partial\boldsymbol{\theta}\partial\boldsymbol{\theta}^T$ exists and is uniformly bounded in norm for all $\boldsymbol{\theta} \in \mathbb{R}^p$ (i.e., all components of $\boldsymbol{H}(\boldsymbol{\theta})$ are uniformly bounded in magnitude).

As with the root-finding SA algorithm, condition A.1′ is the most relevant from the point of view of the user's input. The condition includes restrictions on c_k as well as on a_k. It is apparent that $c_k \to 0$ slower than a_k. Condition A.2′ states that $\boldsymbol{\theta}^*$ is a unique global minimum in the sense that $L(\boldsymbol{\theta})$ always increases away from $\boldsymbol{\theta}^*$. Further, this condition rules out any local minima of L. A.3′ is the standard mean-zero and bounded variance noise condition. Although A.4′ places a bound on the Hessian, there is no need to assume the existence of third derivatives (as used in the bias discussion above); the third derivative assumption is used in the alternative conditions below.

There are also FDSA analogues of the "engineering" conditions B.1–B.5 for the root-finding approach (Section 4.3). Recall that the engineering conditions rely on connections of the SA recursion to an underlying ordinary differential equation (ODE). The FDSA algorithm can be expressed as a root-finding algorithm with a nonzero-mean noise (as allowed in condition B.5 of Section 4.3). Following the pattern of conditions A.1′–A.4′ above, we now present conditions B.1′–B.5′, which have a direct correspondence to B.1–B.5.

B.1′ **(Gain sequences)** $a_k > 0$, $c_k > 0$, $a_k \to 0$, $c_k \to 0$, $\sum_{k=0}^{\infty} a_k = \infty$, and $\sum_{k=0}^{\infty} a_k^2 / c_k^2 < \infty$.

B.2′ **(Relationship to ODE)** Same as condition B.2 in Section 4.3.

B.3′ **(Iterate boundedness)** Same as condition B.3 in Section 4.3.

B.4′ **(Mean and bounded variance of measurement error)** Same as condition A.3′ above.

B.5′ **(Smoothness of L)** The third derivatives $L_{iii}'''(\boldsymbol{\theta})$ are continuous and uniformly bounded for all $i = 1, 2, \ldots, p$ and $\boldsymbol{\theta} \in \mathbb{R}^p$.

The conditions above have largely been discussed in Section 4.3 or as part of the discussion following A.1′–A.4′. Because B.1′ does not include the condition $\sum_{k=0}^{\infty} a_k c_k < \infty$ appearing in A.1′, there is additional flexibility in picking the gains. In particular, the condition $\sum_{k=0}^{\infty} a_k c_k < \infty$ requires that a_k and c_k decay faster than sometimes recommended for practical applications (e.g., Section 6.6 suggests that a_k and c_k decay at rates proportional to $1/(k+1)^{0.602}$ and $1/(k+1)^{0.101}$ respectively; these gains satisfy B.1′ but not A.1′). On the other hand, B.2′ and B.3′ will generally be more formidable to verify than their closest counterpart A.2′. B.5′ is included to ensure that $\boldsymbol{b}_k = O(c_k^2)$ a.s., along the lines of the arguments surrounding (6.8).

The convergence conditions above provide an abstract ideal. In practice, one will rarely be able to check all of conditions A.1′–A.4′ or B.1′–B.5′ due to a lack of knowledge about L. In fact, the conditions may not be verifiable for the very reason that one is using the gradient-free FDSA algorithm! Nonetheless, the conditions are important in identifying the types of problems for which there are guarantees of algorithm convergence. Also, conditions on L that may be formally unverifiable may be at least intuitively plausible, providing some sense that the algorithm is appropriate for the problem.

We now present a convergence theorem for FDSA, together with a proof of the theorem. Although we omit most proofs in this book, this proof is included because of its relative accessibility (by connecting to the conditions of Theorem 4.1 for root-finding SA) and to convey some of the ideas that support SA. A reader may skip or skim the proof without great harm to subsequent understanding of FDSA.

Theorem 6.1. Consider the unconstrained problem (i.e., $\Theta = \mathbb{R}^p$). Suppose that conditions A.1′–A.4′ or conditions B.1′–B.5′ hold. Further, suppose that $\boldsymbol{\theta}^*$ is a unique minimum of L (i.e., Θ^* is the singleton $\boldsymbol{\theta}^*$). Then, for FDSA according to (6.5) and (6.6), $\hat{\boldsymbol{\theta}}_k \to \boldsymbol{\theta}^*$ a.s. as $k \to \infty$.

Proof. The proof under conditions A.1′–A.4′ is in Fabian (1971, Theorem 2.4). Let us work with conditions B.1′–B.5′. It is sufficient to show that these conditions imply or are equivalent to B.1–B.5 in Section 4.3 since B.1–B.5 allow for a biased gradient estimate. First, B.1′ implies B.1 through the addition of the decaying sequence c_k. There is nothing to show regarding B.2′ and B.3′ as these conditions directly assume the validity of the ODE (and related) conditions B.2 and B.3.

Let us now show the validity of B.4, which is satisfied if $E\left[\left\|\sum_{k=0}^{\infty} a_k(\boldsymbol{e}_k - \boldsymbol{b}_k)\right\|^2\right] < \infty$, where $\boldsymbol{b}_k = E[\boldsymbol{e}_k(\hat{\boldsymbol{\theta}}_k) \mid \Im_k]$. For the FDSA setting, $\boldsymbol{e}_k \equiv \hat{\boldsymbol{g}}_k(\hat{\boldsymbol{\theta}}_k) - \boldsymbol{g}(\hat{\boldsymbol{\theta}}_k)$ since, in the notation of Chapter 4, $\boldsymbol{Y}_k(\hat{\boldsymbol{\theta}}_k) = \hat{\boldsymbol{g}}_k(\hat{\boldsymbol{\theta}}_k)$. Then

$$\lim_{n\to\infty} E\left[\left\|\sum_{k=0}^{n} a_k(\boldsymbol{e}_k - \boldsymbol{b}_k)\right\|^2\right] = \lim_{n\to\infty} \sum_{k=0}^{n} a_k^2 E\left(\|\boldsymbol{e}_k - \boldsymbol{b}_k\|^2\right), \tag{6.9}$$

where the cross products disappear in producing the right-hand side of (6.9) because $E[(\boldsymbol{e}_i - \boldsymbol{b}_i)^T(\boldsymbol{e}_j - \boldsymbol{b}_j)] = E\{E[(\boldsymbol{e}_i - \boldsymbol{b}_i)^T(\boldsymbol{e}_j - \boldsymbol{b}_j)|\Im_j]\} = E\{(\boldsymbol{e}_i - \boldsymbol{b}_i)^T E[(\boldsymbol{e}_j - \boldsymbol{b}_j)|\Im_j]\} = 0$ for all $i < j$ (the last equality follows since $\boldsymbol{e}_i - \boldsymbol{b}_i$ is uniquely determined by $\hat{\boldsymbol{\theta}}_i$ and $\hat{\boldsymbol{\theta}}_{i+1}$, both of which are contained in $\Im_j$). Note that

$$\boldsymbol{e}_k - \boldsymbol{b}_k = \begin{bmatrix} \dfrac{\varepsilon_k^{(1+)} - \varepsilon_k^{(1-)}}{2c_k} \\ \vdots \\ \dfrac{\varepsilon_k^{(p+)} - \varepsilon_k^{(p-)}}{2c_k} \end{bmatrix}. \tag{6.10}$$

Given the bounded variance from B.4′, (6.9) and (6.10) imply that $E\left[\left\|\sum_{k=0}^{\infty} a_k(\boldsymbol{e}_k - \boldsymbol{b}_k)\right\|^2\right]$ is proportional to $\sum_{k=0}^{\infty} a_k^2 / c_k^2$, which, by B.1′, is bounded. This shows that B.4 in Section 4.3 is true. Finally, B.5′ implies, via the arguments surrounding (6.8), that $\boldsymbol{b}_k$ is uniformly bounded and has bias $O(c_k^2)$ a.s. So the bias satisfies the conditions $\sup_{k\geq 0}\|\boldsymbol{b}_k\| < \infty$ a.s. and $\boldsymbol{b}_k \to \boldsymbol{0}$ a.s. in B.5. Hence, all of B.1–B.5 hold under B.1′–B.5′, implying the a.s. convergence of the FDSA algorithm. ❑

6.5 ASYMPTOTIC NORMALITY

As with the root-finding SA algorithm, the FDSA iterates (appropriately standardized) have an asymptotic normal distribution under appropriate conditions. The asymptotic normality applies under more specific choices of the gain sequences than the general guidelines provided with the conditions for Theorem 6.1 in Section 6.4. In particular,

$$a_k = \frac{a}{(k+1+A)^{\alpha}} \text{ and } c_k = \frac{c}{(k+1)^{\gamma}}, \tag{6.11}$$

where a, c, α, and γ are strictly positive and the stability constant $A \geq 0$ plays the same role here that it did in Section 4.4. The results below on the large-sample distribution of $\hat{\theta}_k$ provide the asymptotically optimal values for the coefficients in (6.11).

Sacks (1958) was the first to establish the asymptotic normality of FDSA. Later, the more general result of Fabian (1968) on the asymptotic distribution of SA iterates was used to establish the asymptotic normality of FDSA under broader conditions than those of Sacks (1958). A simplified proof is given in Fabian (1971) for the $\alpha = 1$ special case. Generalized forms of asymptotic distributions are available via the *weak convergence* ideas in Kushner and Yin (1997, Chaps. 7 and 8); weak convergence is beyond the scope of this book.

Although we do not present a formal theorem here, the basic idea is to begin with an assumption that the algorithm converges (as in Theorem 6.1) and then add several conditions to guarantee asymptotic normality. Among the added conditions are

$$\beta \equiv \alpha - 2\gamma > 0 \text{ and } 3\gamma - \alpha/2 \geq 0. \tag{6.12}$$

Then, for the FDSA algorithm in (6.5) and (6.6) with gains as in (6.11),

$$k^{\beta/2}(\hat{\theta}_k - \theta^*) \xrightarrow{\text{dist.}} N(\mu_{\text{FD}}, \Sigma_{\text{FD}}) \tag{6.13}$$

as $k \to \infty$, where μ_{FD} is a mean vector that depends on the Hessian $H(\theta^*)$ and the third derivative $L'''(\theta^*)$, Σ_{FD} is some covariance matrix that depends on $H(\theta^*)$, and both μ_{FD} and Σ_{FD} depend on the coefficients a, α, c, and γ of a_k and c_k (A does not affect the asymptotic properties). While the forms for μ_{FD} and Σ_{FD} are somewhat unwieldy (see the references cited above), they are useful for shedding insight into the efficiency of FDSA. Chapter 7 uses these forms together with corresponding forms for another gradient-free SA algorithm to draw some conclusions about relative efficiency.

The presence of a generally nonzero mean (μ_{FD}) in the limiting distribution of FDSA is an interesting distinction from the limiting distributions commonly seen in estimation theory (including, e.g., the limiting distribution of root-finding SA as shown in Section 4.4). From the proofs of (6.13) in the references cited above, it is evident that this bias in the limiting distribution is a consequence of the bias in the gradient estimate. Note that the presence of $\mu_{\text{FD}} \neq \mathbf{0}$ does *not* necessarily imply that $\hat{\theta}_k$ has an asymptotic bias as an estimator of θ^*. In particular, under the additional conditions needed for the mean of the asymptotic distribution to correspond to the mean of the random process itself (e.g., Laha and Rohatgi, 1979, pp. 138–141), (6.13) implies that for large k, $E(\hat{\theta}_k)$ is approximately equal to $\theta^* + k^{-\beta/2}\mu_{\text{FD}}$. Hence, $E(\hat{\theta}_k)$ has a limiting value of θ^*. (Note that the conditions for asymptotic normality alone—as for (6.13)—do *not* guarantee that $\lim_{k\to\infty} E[k^{\beta/2}(\hat{\theta}_k - \theta^*)] = \mu_{\text{FD}}$; see Exercise 6.4.)

Akin to the discussion in Section 4.4 for the root-finding SA algorithm, expression (6.13) implies that the stochastic rate at which $\hat{\boldsymbol{\theta}}_k$ approaches $\boldsymbol{\theta}^*$ is proportional to $k^{-\beta/2}$ for large k. That is, $\hat{\boldsymbol{\theta}}_k - \boldsymbol{\theta}^*$ must be decaying in a stochastic sense at a rate proportional to $k^{-\beta/2}$ to balance the increasing $k^{\beta/2}$ term on the left-hand side of (6.13), yielding a well-behaved random vector with the distribution $N(\boldsymbol{\mu}_{\text{FD}}, \boldsymbol{\Sigma}_{\text{FD}})$.

With the gain forms in (6.11), and under condition B.1′ (weaker than A.1′) for convergence of the iterate, we know that $\alpha > 1/2$ and $\gamma > 0$. The conditions in (6.12) put further constraints on α and γ, implying that

$$0.6 < \alpha \leq 1,\ \ 0.1 < \gamma < 1/2,\ \text{and}\ \alpha - \gamma > 1/2 \tag{6.14}$$

(see Exercise 6.5). Note that the reverse implication does not hold: an arbitrary α and γ satisfying (6.14) may not satisfy the gain conditions. For example, $\alpha = 1$ and $\gamma = 0.15$ satisfy (6.14) but violate $3\gamma - \alpha/2 \geq 0$ in (6.12). Given B.1′ and (6.12), we find that β is maximized at $\alpha = 1$ and $\gamma = 1/6$, leading to a maximum rate of stochastic convergence of $\hat{\boldsymbol{\theta}}_k$ to $\boldsymbol{\theta}^*$ that is proportional to $k^{-\beta/2} = 1/k^{1/3}$ for large k. This contrasts with a maximum rate proportional to $1/\sqrt{k}$ in the stochastic gradient algorithms of root-finding type (see Section 4.4). Hence, the direct gradient information of the stochastic gradient implementation "buys" an increase in the maximum rate of convergence from $O\left(1/k^{1/3}\right)$ to $O\left(1/\sqrt{k}\right)$.

Glasserman and Yao (1992) and L'Ecuyer and Yin (1998) show that the maximum rate of convergence in FDSA can be the same as in root-finding SA (i.e., $O\left(1/\sqrt{k}\right)$) in the special case of simulation-based optimization with common random numbers. This contrasts with the standard maximum rate of $O\left(1/k^{1/3}\right)$. This issue is considered in detail in Chapter 14.

6.6 PRACTICAL SELECTION OF GAIN SEQUENCES

This section focuses on the selection of the coefficients a, A, c, α, and γ in the gains a_k and c_k appearing in (6.11). As with root-finding SA, the asymptotic analysis above may not point to the best choice of gains in practical *finite-sample* problems. It is often (but not always!) preferable in practice to have a slower decay rate than the asymptotically optimal $\alpha = 1$ and $\gamma = 1/6$. This provides more power to the algorithm through larger step sizes when k is large. Practical values of α and γ that are effectively as low as possible while satisfying B.1′ and (6.12) are 0.602 and 0.101, respectively.

In practical applications, the gains are usually chosen by trial and error on some small-scale (e.g., reduced number of iterations) version of the full problem. Obviously, one typically wants a minimal number of the overall "budget" of loss measurements to be devoted to the gain tuning process so that most of the effort in collecting loss measurements is directed to the optimization process per se.

It is sometimes possible to partially automate the gain selection process, as we now outline in a method referred to as the *semiautomatic* method for gain selection. The aim is to pick the coefficients a, A, c, α, and γ in the gains of (6.11). First, picking $\alpha = 0.602$ and $\gamma = 0.101$, as mentioned above, provide the generally more desirable slowly decaying gains. To cope with noise effects, it is effective to set c at a level approximately equal to the standard deviation of the measurement noise in $y(\boldsymbol{\theta})$ (see Exercise 6.9). This helps keep the p elements of $\hat{g}_k(\hat{\boldsymbol{\theta}}_k)$ from getting excessively large in magnitude before a_k has decreased enough to compensate during the search process (the standard deviation can be estimated by collecting several $y(\boldsymbol{\theta})$ values at the initial guess $\hat{\boldsymbol{\theta}}_0$). If the standard deviation changes dramatically with $\boldsymbol{\theta}$, this approach might not be useful. In the case where one has perfect (noise-free) measurements of $L(\boldsymbol{\theta})$, then c should be chosen as some small positive number.

The values of a and A can be chosen together to ensure effective practical performance of the algorithm. As discussed in Section 4.4, a stability constant $A > 0$ allows for a more aggressive search (larger a) through avoiding instabilities in the early iterations. A larger a often enhances performance in the later iterations by producing a larger step size when the effect of A is negligible compared to the iteration count k. A rule of thumb is to choose A such that it is much less than the maximum number of iterations allowed or expected (e.g., take it to be 10 percent or less of the maximum number of expected/allowed iterations).

After determining A, one must choose the most important coefficient, a. In making this choice, first pick $a_{\text{temp},i}$, $i = 1, 2, \ldots, p$, such that $\left[a_{\text{temp},i}/(A+1)^{0.602}\right]\hat{g}_{0i}(\hat{\boldsymbol{\theta}}_0)$ is approximately equal to the desired change magnitude in the ith element of $\boldsymbol{\theta}$ in the early iterations, where $\hat{g}_{0i}(\hat{\boldsymbol{\theta}}_0)$ represents the ith element of $\hat{g}_0(\hat{\boldsymbol{\theta}}_0)$. To do this reliably in the face of the noise in the loss measurements may require several replications of $\hat{g}_0(\hat{\boldsymbol{\theta}}_0)$; the typical magnitude for the ith element of the gradient estimate is then taken to be an average of $|\hat{g}_{0i}(\hat{\boldsymbol{\theta}}_0)|$ from several replications. Finally, $a = \min\{a_{\text{temp},1}, a_{\text{temp},2}, \ldots, a_{\text{temp},p}\}$, where the *minimum* is chosen to preserve the stability of the algorithm.[2] (At the expense of a more cumbersome implementation, it is also possible to use a diagonal *matrix* gain $\boldsymbol{a}_k$ with the ith diagonal element being the gain in (6.11) with numerator $a = a_{\text{temp},i}$.)

Note that this semiautomatic means of picking the gain coefficients is generally useful in getting *reasonable* values. In most applications, refinement of the values will enhance the performance of the algorithm. Unfortunately, this refinement is usually possible only after observing the search process or by

---

[2]This is an illustration of the *lack* of transform invariance of a first-order algorithm. Because (6.11) does not scale for each element of $\boldsymbol{\theta}$ separately, the gain is chosen conservatively so that no "wild" behavior is seen for any one element. A matrix gain—such as a second-order algorithm in Subsection 4.5.2—scales each element differently.

doing "trial runs" where the gains are tuned to provide better results. Usually (not always!), the greatest benefit follows by focusing this refinement on the single coefficient a, as the search tends to be more robust to changes in the other coefficients (A, c, α, and γ). We demonstrate the semiautomatic method of gain selection in the example below; the method is used in the studies in Section 6.7.

Example 6.4—Choosing gain coefficients using the semiautomatic method. Suppose that a maximum of 6000 loss measurements may be used during the search process for a $p = 3$ problem. Further, based on several measurements of the loss function at the initial condition, $y(\hat{\boldsymbol{\theta}}_0)$, it has been determined that the noise has an approximate standard deviation of 0.5 and that the standard deviation is not expected to change significantly as $\boldsymbol{\theta}$ changes. Then, together with $\alpha = 0.602$ and $\gamma = 0.101$, one chooses $c = 0.5$. Using a threshold of 10 percent of the maximum number of iterations implies that $A = 0.10\times 6000/(2\times 3) = 100$. Suppose further that it is felt that the elements of $\boldsymbol{\theta}$ should typically move by a magnitude 0.1 in the early iterations. Based on the c chosen above, one forms several $\hat{g}_0(\hat{\boldsymbol{\theta}}_0)$, finding that the magnitudes of $\hat{g}_{0,1}(\hat{\boldsymbol{\theta}}_0)$, $\hat{g}_{0,2}(\hat{\boldsymbol{\theta}}_0)$, and $\hat{g}_{0,3}(\hat{\boldsymbol{\theta}}_0)$ are around 10, 20, and 10, respectively. This leads to $a_{\text{temp},1} = a_{\text{temp},3} = 0.16$ and $a_{\text{temp},2} = 0.08$ (i.e., for the first and third elements, $\left[0.16/(100+1)^{0.602}\right]\times 10 = 0.1$). Hence, the choice is $a = 0.08$. All five of the gain coefficients have now been chosen. It is likely that refinements—especially of a—will be helpful after one has observed the optimization process. ❑

6.7 SEVERAL FINITE-DIFFERENCE EXAMPLES

Let us present three examples of FDSA: one for a simple $p = 2$ problem and two for a more challenging $p = 10$ problem. In comparing the performance of FDSA with the random search methods of Chapter 2, these examples illustrate the benefits of the assumptions about L that are required in FDSA. In particular, while the no free lunch theorems (Subsection 1.2.2) indicate that all algorithms perform the same when averaging over all possible problems, we have prior information stating that L is differentiable and unimodal. This prior information provides enough structure for an algorithm such as FDSA (which uses the *existence* of the gradient) to have an apparent advantage over more general algorithms such as the random search algorithms of Chapter 2. (One must be cautious about the connections to the NFL theorems because of NFL's restriction to discrete problems; nevertheless, since the implementation of FDSA is on a digital machine, FDSA is strictly a discrete algorithm.) This advantage applies even if, as with FDSA, the *mechanics* of the algorithm do not depend explicitly on this prior information. In particular, both FDSA and random search use the same information—noisy or noise-free loss measurements—in the operations of the algorithm.

Example 6.5—Optimization in wastewater treatment problem (redux). Let us revisit Examples 2.6–2.8, which consider a $p = 2$ problem related to wastewater treatment. The aim is to run FDSA on this problem and compare results with the random search results in Example 2.7 using the same number of loss measurements n (so each run of FDSA is for $n/(2p) = n/4$ iterations). As in Examples 2.6–2.8, the random search method used here is algorithm B, described in Sections 2.2 and 2.3 for noise-free and noisy problems. The problem setup of Example 2.7 is used here, including measurement process $y(\boldsymbol{\theta})$, noise level, initial condition ($\hat{\boldsymbol{\theta}}_0$), priority of weighting on methane gas and water cleanliness, and definition of Θ ($= [0, 5]\times[0, 5]$). To avoid confusion in reviewing Examples 2.6–2.8, note that because of the current focus on optimization via finding a solution to $\partial L/\partial\boldsymbol{\theta} = \mathbf{0}$, the mathematical form of $\boldsymbol{g}(\boldsymbol{\theta}) = \partial L/\partial\boldsymbol{\theta}$ here generally differs slightly from $\boldsymbol{g}(\boldsymbol{\theta})$ as used in Examples 2.6–2.8 (where $\boldsymbol{g}(\boldsymbol{\theta})$ is a generic root-finding function from which L is built).

If an iterate falls outside of Θ, each violating component of $\boldsymbol{\theta}$ is mapped to it nearest valid point in Θ. The subsequent FD gradient estimate is formed at the modified (valid) $\boldsymbol{\theta}$ value. For example, an iterate achieving the value $\boldsymbol{\theta} = [2.2, 6.2]^T$ via the update step in (6.5) is mapped to $\boldsymbol{\theta} = [2.2, 5.0]^T$; the next FD approximation is computed at $[2.2, 5.0]^T$. (The perturbed values $\hat{\boldsymbol{\theta}}_k \pm c_k\boldsymbol{\xi}_i$ for FDSA are allowed to go outside of Θ.)

Table 6.1 contrasts the mean values of $L(\hat{\boldsymbol{\theta}}_k) - L(\boldsymbol{\theta}^*)$ at the terminal iteration over 50 independent replications. The numbers in the table can be directly compared with the numbers in the table of Example 2.7. Below the means are 95 percent confidence intervals calculated from a t-distribution as in Appendix B. The first column of results is based on naïve gains $a_k = 1/(k+1)$ and $c_k = 1/(k+1)^{1/6}$, corresponding to the asymptotically optimal α and γ according to the discussion in Section 6.5. The second column uses gains of form (6.11) that have been partially tuned based on trial and error to adjust the numerator a in a_k to provide optimal results while maintaining the same $\alpha = 1$, $\gamma = 1/6$, $c = 1$, and $A = 0$ ($a_k = a/(k+1)$ with $a = 0.4$ for $n = 100$ loss measurements; $a = 1.5$ for $n = 2000$; $c_k = 1/(k+1)^{1/6}$ in both cases). Hence, the

Table 6.1. Sample means (50 replications) of $L(\hat{\boldsymbol{\theta}}_k) - L(\boldsymbol{\theta}^*)$ for FDSA under two a_k sequences; values contrast with $L(\hat{\boldsymbol{\theta}}_0) - L(\boldsymbol{\theta}^*) = 38.43$. Approximate 95 confidence intervals are also shown.

	FDSA with "naïve" gains	FDSA with partially tuned gains
$n = 100$ ($k = 25$ iterations)	0.11 [0.087, 0.140]	0.083 [0.057, 0.108]
$n = 2000$ ($k = 500$ iterations)	0.023 [0.017, 0.028]	0.021 [0.016, 0.026]

gain tuning did not involve the full complement of flexibility for the gains in (6.11).

Based on numerical experimentation with $n = 2000$, the asymptotically optimal $\alpha = 1$ and $\gamma = 1/6$ seemed to provide performance superior to that with the recommended finite-sample values of $\alpha = 0.602$ and $\gamma = 0.101$. This appears to be a consequence of asymptotic theory when using 2000 measurements to estimate only two parameters. (The larger-dimensional Example 6.6 shows a case where $\alpha = 0.602$ and $\gamma = 0.101$ provide superior performance.)

The four sample means in this table are significantly below the corresponding values in Example 2.7 based on random search algorithm B. For example, at $n = 2000$, the best value in Example 2.7 is 0.38, significantly greater than 0.021 or 0.023 as shown above. Further, the confidence intervals here and in Example 2.7 are not close to overlapping. This illustrates the benefit of the partial information about the gradient $\boldsymbol{g}(\boldsymbol{\theta})$ that is available from the FD approximation.

In comparing the numbers *within* Table 6.1, note that the confidence intervals for the naïve and tuned cases overlap for each n value. Although this may suggest nonsignificant differences between the naïve and tuned cases, one should dig a little further because the same random number seed is used to initialize all runs, providing a degree of shared randomness. The matched-pairs test (Appendix B) is the proper test in such a case. Based on the 50 terminal loss values, this t-test reveals a nonsignificant difference between the values for $n =$ 2000 (two-sided P-value of 0.18) and a significant difference between the values for $n = 100$ (two-sided P-value of 0.009). ❑

The next two examples are for a larger $p = 10$ problem. The primary aims are to illustrate the role of "gain tuning" and its effect on the quality of the solution, to compare FDSA against random search algorithm B from Chapter 2, and to illustrate that the closeness of $L(\hat{\boldsymbol{\theta}}_k)$ and $L(\boldsymbol{\theta}^*)$ does not necessarily translate into $\hat{\boldsymbol{\theta}}_k$ being close to $\boldsymbol{\theta}^*$. The loss function considered here is a fourth-order (i.e., quartic) polynomial with significant variable interaction and highly skewed level surfaces. By *highly skewed*, we mean that a given change in one of the $\boldsymbol{\theta}$ elements may affect L in a way very different than the same change in another element of $\boldsymbol{\theta}$. The first of the two examples, Example 6.6, compares algorithms based on their performance relative to the true loss function. The second example, Example 6.7, compares algorithms based on the accuracy of the final $\boldsymbol{\theta}$ estimate.

Example 6.6—Skewed quartic loss function: comparison of loss values in FDSA and random search. Consider the loss function

$$L(\boldsymbol{\theta}) = \boldsymbol{\theta}^T \boldsymbol{B}^T \boldsymbol{B}\boldsymbol{\theta} + 0.1\sum_{i=1}^{p} (\boldsymbol{B}\boldsymbol{\theta})_i^3 + 0.01\sum_{i=1}^{p} (\boldsymbol{B}\boldsymbol{\theta})_i^4 , \tag{6.15}$$

where $(\cdot)_i$ represents the ith component of the argument vector $\boldsymbol{B\theta}$, and $\boldsymbol{B}$ is such that $p\boldsymbol{B}$ is an upper triangular matrix of 1's (so elements below the diagonal are zero). Let $p = 10$. The minimum occurs at $\boldsymbol{\theta}^* = \boldsymbol{0}$ with $L(\boldsymbol{\theta}^*) = 0$; all runs are initialized at $\hat{\boldsymbol{\theta}}_0 = [1, 1,\ldots, 1]^T$ (so $L(\hat{\boldsymbol{\theta}}_0) = 4.178$). Unlike Example 6.5, the measurement noise ε is independent, identically distributed (i.i.d.); the distribution of the noise is $N(0, 1)$. All iterates are constrained to be in $\Theta = [-5, 5]^{10}$ (the 10-dimensional hypercube with minimum and maximum values of -5 and 5). The ratio of maximum to minimum eigenvalue of the Hessian $\boldsymbol{H}(\boldsymbol{\theta}^*)$ is approximately 65, indicating the high degree of skewness in (6.15).

Using $n = 1000$ loss measurements for all algorithms, the aim here is to compare estimation performance of: (i) FDSA ((6.5) and (6.6)) using the semiautomatic gain selection method outlined in Section 6.6, (ii) FDSA using the gains determined by manual trial and error following the semiautomatic gain selection, and (iii) random search algorithm B from Sections 2.2 and 2.3. For FDSA, the gains are of the form (6.11), while for algorithm B, trial and error is used to choose the approximate best value of the perturbation distribution and the amount of averaging needed to cope with the noise (Section 2.3). For both random search and FDSA, a point $\boldsymbol{\theta}$ lying outside of Θ is moved back to Θ on a component-wise basis in the same manner as Example 6.5 (so, in algorithm B, it is not necessary to generate extra perturbations $\boldsymbol{d}_k$, and associated loss measurements, to meet the constraints, as was mentioned in step 1 of the algorithm in Subsection 2.2.2).

The coefficient values used for the three algorithm implementations are shown in Exercises 6.10 and 6.11; for the manual tuning, a, A, α, and γ (but not c) are varied and experimentally tested, using the semiautomatic values as a starting point in the experiments. Note that both FDSA with manual tuning and algorithm B required a number of trial runs over *all* 1000 measurements. (In practice, it would be possible to use truncated runs in the hopes of getting good coefficient values for the later full-length runs.) FDSA-semiautomatic, on the other hand, required only a small number of trial loss measurements around $\hat{\boldsymbol{\theta}}_0$. We used 100 trial measurements here, producing five gradient approximations to be averaged in determining the "typical" magnitudes of the gradient elements. This is a major advantage in practical applications, where it is not typically possible to collect the required *multiple sets* of the full complement of measurements (e.g., 1000 here) to tune for *one* later run using the same number of measurements.

Figure 6.2 summarizes the results. Each curve represents the sample mean of 50 independent replications. An individual replication of any one of the three algorithms has much more wiggle than the corresponding smoothed curve in the figure. As indicated in Exercise 6.11, algorithm B has 20 measurements being averaged for each loss evaluation. Hence, the 1000 measurements translates into $1000/(2\times10) = 50$ iterations for FDSA and 49 iterations for algorithm B (including the initial loss evaluation).

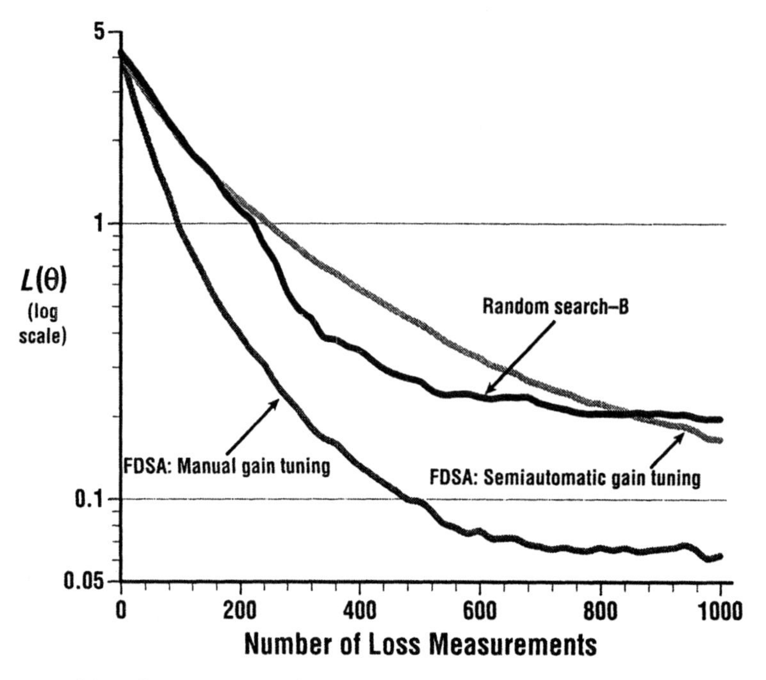

Figure 6.2. Comparison of two implementations of FDSA (with semiautomatic gains and tuned gains) and random search algorithm B. Each curve represents sample mean of 50 independent replications.

Figure 6.2 shows that all algorithm implementations produce an overall reduction in the loss function as the number of measurements approach 1000 (the true loss values $L(\hat{\boldsymbol{\theta}}_k)$ are used in constructing the figure; these values are not available to the algorithms, which use only noisy measurements $y(\boldsymbol{\theta})$ at the various values of $\boldsymbol{\theta}$). The curves illustrate that the FDSA algorithm with manual tuning performs (as expected) the best, but that the FDSA algorithm with semiautomatic tuning performs reasonably well. The manual tuning experiments reveal that $\alpha = 0.602$ and $\gamma = 0.101$ work better here than the asymptotically optimal $\alpha = 1$ and $\gamma = 1/6$, suggesting that 1000 measurements for $p = 10$ parameters is not yet asymptotic-like in its behavior for this problem (contrast with Example 6.5). Random search algorithm B outperforms the FDSA-semiautomatic algorithm for most of the search period, but the situation reverses after approximately 840 loss measurements. ❑

Example 6.7—Skewed quartic loss function: comparison of the accuracy of $\boldsymbol{\theta}$ estimates in FDSA and random search. Let us continue with the setting of Example 6.6. Interesting insight comes by looking at the accuracy of $\hat{\boldsymbol{\theta}}_k$ relative to $\boldsymbol{\theta}^*$, in contrast to Example 6.6 based on comparing $L(\hat{\boldsymbol{\theta}}_k)$ with $L(\boldsymbol{\theta}^*)$. Although, the loss function is formally the fundamental criterion of interest, it is often of interest in practice to be concerned with the distance from $\hat{\boldsymbol{\theta}}_k$ to $\boldsymbol{\theta}^*$

(i.e., $\|\hat{\boldsymbol{\theta}}_k - \boldsymbol{\theta}^*\|$). This is especially true in cases—such as maximum likelihood estimation—where the criterion L may not have an easy physical interpretation.

For each of the three algorithms considered in Figure 6.2, Table 6.2 shows the sample mean of the distances from $\hat{\boldsymbol{\theta}}_k$ to $\boldsymbol{\theta}^*$ for the 50 final estimates formed from the 1000 measurements. (Note: The table values are *not* the root-mean-squared values corresponding to the square root of the sample mean of the mean-squared errors.) The $\hat{\boldsymbol{\theta}}_k$ values producing the sample means in Table 6.2 are identical to the terminal estimates generating Figure 6.2. The distances are normalized by $\|\hat{\boldsymbol{\theta}}_0 - \boldsymbol{\theta}^*\| = 3.162$. The approximate 95 percent confidence intervals are calculated in the usual manner from a t-distribution.

There are two notable points in Table 6.2. One is that the FDSA-semiautomatic algorithm provides the best performance, unlike the results of Figure 6.2 (note the nonoverlap in the confidence intervals). This confirms a point that is easy to forget: A "better" $\boldsymbol{\theta}$ value with respect to L is not necessarily closer to $\boldsymbol{\theta}^*$ than a value with a poorer L value. In fact, the table shows that with algorithm B, the final solutions are typically *farther* from $\boldsymbol{\theta}^*$ than the initial condition, although the mean of the corresponding loss values is reduced by approximately 80 percent. Such a disparity between loss and distance measures is typical for highly skewed loss functions, where the $\boldsymbol{\theta}$ sensitivities vary considerably by element. An element with little impact on L may be poorly estimated, barely affecting the loss value at the final estimate, but having a devastating effect in the distance from the final estimate to $\boldsymbol{\theta}^*$.

The other interesting point in Table 6.2 is that the magnitudes of the numbers are much larger than the corresponding numbers from Figure 6.2 when normalized by the initial loss value $L(\hat{\boldsymbol{\theta}}_0) = 4.178$ (corresponding to the normalization by $\|\hat{\boldsymbol{\theta}}_0 - \boldsymbol{\theta}^*\|$ in Table 6.2). These normalized *loss* values are on the order of 10^{-2}. This relative behavior is typical of most optimization problems, reflecting the relative flatness of the loss surface in the area surrounding $\boldsymbol{\theta}^*$. ❑

Table 6.2. Comparison of FDSA and random search algorithm B in accuracy of $\boldsymbol{\theta}$. Algorithms use 1000 loss measurements per replication. Values shown are sample means over 50 replications with approximate 95 percent confidence intervals.

	FDSA: semiautomatic gains	FDSA: manually tuned gains	Random search B
Normalized error $\dfrac{\|\hat{\boldsymbol{\theta}}_k - \boldsymbol{\theta}^*\|}{\|\hat{\boldsymbol{\theta}}_0 - \boldsymbol{\theta}^*\|}$	0.427 [0.411, 0.443]	0.531 [0.502, 0.561]	1.285 [1.190, 1.378]

6.8 SOME EXTENSIONS AND ENHANCEMENTS TO THE FINITE-DIFFERENCE ALGORITHM

The sections above discussed the basic FDSA algorithm. Let us now summarize a few extensions of the basic method.

Fabian (1967, 1971) presents several methods for accelerating the convergence of FDSA-type algorithms, analogous to some of the methods in Subsection 4.5.2 for root-finding SA. The methods are based on taking additional measurements to explore the loss function surface in detail. One such method in Fabian (1971) is a stochastic analogue to second-order algorithms of the generic Newton–Raphson form. This algorithm uses $O(p^2)$ measurements $y(\cdot)$ per iteration for the gradient and Hessian estimation. The gradient is estimated in a standard way (e.g., (6.6)) and the Hessian is estimated using an analogous FD-based double-differencing scheme. Although this method is intuitively sensible, it demands many extra loss measurements if p is even moderately large and is likely to be numerically unstable with even a small level of noise. There appear to be no serious published applications or numerical evaluations of this idea. Further, a more efficient scheme for approximating the Hessian matrix is given in Section 7.8.

The iterate averaging approach of Subsection 4.5.3 may also be used with FDSA. Dippon and Renz (1997) theoretically analyze iterate averaging in the FDSA context. Unlike the root-finding setting, Dippon and Renz (1997) find that because of the biases in the underlying gradient approximations, iterate averaging does *not* generally yield asymptotically optimal performance. However, the difference between the iterate averaging solution and the true optimal solution based on the best (intractable) gain sequence a_k is *quantifiably* bounded. The true optimal gain sequence a_k depends on the unknown third derivatives of L evaluated at the unknown $\boldsymbol{\theta}^*$. This bound suggests that iterate averaging yields "good"—but not the best—asymptotic behavior for FDSA. Further, the practical finite-sample issues limiting the performance of iterate averaging, as discussed in Subsection 4.5.3, apply here as well.

Injected Monte Carlo randomness—as in Property B in Subsection 1.1.3—can be used in combination with FDSA-type methods. This is the essence of the simultaneous perturbation SA method of Chapter 7. Ermoliev (1969) was apparently the first to introduce such an idea via random directions SA (RDSA). The essential idea with RDSA is to use the basic SA recursion in (6.5), but to replace the FD gradient approximation in (6.6) with a more efficient gradient approximation generated with the help of a Monte Carlo-generated perturbation vector. The basic form of this approach requires only two loss measurements to approximate the gradient vector (for any dimension p) and replaces the deterministic perturbations of FDSA (i.e., the $\pm c_k \boldsymbol{\xi}_i$ at each k for all $i = 1, 2, \ldots, p$) with *random* perturbations. Relative to classical two-sided FDSA ((6.5) and (6.6)), RDSA provides the potential for increased efficiency because of the p-fold reduction in loss measurements per iteration ($2p$ for FDSA versus 2 for

RDSA). (Koronacki, 1975, also considers the idea of random perturbations, including some convergence theory. This paper suggests the use of $2m$ loss measurements, with m usually less than p. There is, however, no formal evidence of improved efficiency over basic FDSA.)

The basic form for the ith component of the RDSA gradient approximation at iteration k is

$$\hat{g}_{ki}(\hat{\boldsymbol{\theta}}_k) = \pi_{ki} \frac{y(\hat{\boldsymbol{\theta}}_k + c_k \boldsymbol{\pi}_k) - y(\hat{\boldsymbol{\theta}}_k - c_k \boldsymbol{\pi}_k)}{2c_k},$$

where $\boldsymbol{\pi}_k = [\pi_{k1}, \pi_{k2}, \ldots, \pi_{kp}]^T$ is a vector of Monte Carlo-generated random variables satisfying certain regularity conditions. Although we wait until Chapter 7 to analyze the properties and implications of this type of approach in detail, note that the two $y(\cdot)$ values are reused for all elements of the gradient approximation. Ermoliev (1969) includes analysis of the bias in the gradient approximation and considers one specific distribution (uniform) for the π_{ki}.

Polyak and Tsypkin (1973) and Kushner and Clark (1978, Sect. 2.3.5) perform a more complete theoretical analysis of RDSA-type algorithms, including a demonstration of a.s. convergence. These works consider the RDSA form with perturbations that are uniformly distributed on a p-dimensional sphere. Convergence alone, however, is not a compelling reason to use the RDSA methods, since the FDSA algorithm also converges using the same type of input information (noisy loss measurements). Nevertheless, the p-fold savings in loss measurements *per iteration* provides a tantalizing suggestion that RDSA may provide overall savings *across iterations* in the optimization process if p is large.

Using asymptotic theory, Kushner and Clark (1978, Sects. 2.3.5 and 7.4) provide the first rate-of-convergence analysis for RDSA with the aim of rigorously addressing the efficiency question. They draw a mainly negative conclusion about the efficiency gains of RDSA over FDSA: "... it is conceivable that with such a method [RDSA] the rate of convergence, as a function of the total number of observations used, might be increased [relative to FDSA] for large [dimension]. This hope is somewhat in vain..." (Kushner and Clark, 1978, p. 58). This conclusion, however, may be misleading, appearing to result from an analysis too restrictive in the choice of the gains (using only asymptotically optimal decay rates for the gains) and (generally) too conservative in the bound applied to the bias in the gradient estimate. In fact, significant improvements are possible in practice with this or similar (simultaneous perturbation) algorithms.

Chapter 7 is devoted to an extensive study of SA algorithms with Monte Carlo randomness. The simultaneous perturbation SA method considered there has a gradient approximation with an algebraic form similar to the RDSA form above. More importantly, the chapter considers general perturbation distributions and the associated regularity conditions. The choice of probability distribution for the Monte Carlo perturbations is important in realizing the desired improvements in efficiency.

6.9 CONCLUDING REMARKS

The stochastic approximation framework introduced in Chapters 4 and 5 is a powerful approach for dealing with nonlinear root-finding and optimization problems, especially problems involving noisy measurements of the functions involved. This chapter has continued with the SA-based *optimization* setting emphasized in Chapter 5, but with the significant difference that direct measurements of the gradient are not required. Rather, this chapter has considered methods that build gradient approximations from measurements of the loss function. This *mathematical* distinction has profound *practical* implications.

In particular, it is often difficult or impossible to calculate the stochastic gradient for the methods of Chapter 5. We discussed three examples—parameter estimation with complicated loss functions, model-free control, and simulation-based optimization—illustrating the difficulties. It is in such cases that the FD-based method of this chapter is most useful. While the direct search methods of Chapter 2 are also based on only loss function values (no gradients), their implementation with *noisy* loss measurements is largely ad hoc. This lack of formal justification with noisy measurements also applies to standard implementations of other popular methods, including simulated annealing and genetic algorithms (Chapters 8–10). In contrast, the FDSA method has a deep theoretical justification with noisy measurements.

One of the main shortcomings of FDSA is its inefficiency in high-dimensional problems. That is, the number of loss measurements in each gradient approximation grows directly with the dimension. Because a typical implementation for optimization requires many gradient approximations (one at each iteration), the overall number of loss measurements in the optimization process may become prohibitive. To tackle this problem, some methods have been introduced that create gradient approximations via Monte Carlo schemes. By reducing the number of loss measurements for each gradient approximation, these methods may offer significant improvements in efficiency, as discussed in detail in Chapter 7.

EXERCISES

6.1 Derive the multivariate analogue of (6.4) (i.e., the chain rule form for $\partial Q_k/\partial\boldsymbol{\theta}$) with the relevant gradient expressions expressed in matrix-vector notation (as in Appendix A) when $\boldsymbol{x}_k \in \mathbb{R}^q$ and $\boldsymbol{u}_k \in \mathbb{R}^r$ for some $q > 1$, $r > 1$. Use Q_k as given in (6.2).

6.2 Show that the one-sided FD gradient estimator has an $O(c_k)$ bias under the same conditions used in deriving the $O(c_k^2)$ bias of the two-sided version (eqn. (6.8)). Give at least two reasons why this does not necessarily mean that the one-sided estimator is inferior in practice.

6.3 Consider the function $L(\boldsymbol{\theta}) = t_1^4 + t_1^2 + t_1 t_2 + t_2^2$, $\boldsymbol{\theta} = [t_1, t_2]^T$. Suppose that $\hat{\boldsymbol{\theta}}_k = [1, 1]^T$ for some k. Based on the FD approximation in (6.6), determine upper bounds to the magnitudes of the components of the bias $\boldsymbol{b}_k$ when c_k = 0.2 and c_k = 0.5. Comment on these magnitudes relative to the magnitudes of the corresponding components of $\boldsymbol{g}(\hat{\boldsymbol{\theta}}_k)$; are the biases potentially significant?

6.4 It was noted in Section 6.5 that $\lim_{k\to\infty} E[k^{\beta/2}(\hat{\boldsymbol{\theta}}_k - \boldsymbol{\theta}^*)]$ is not necessarily equal to $\boldsymbol{\mu}_{\text{FD}}$ when $k^{\beta/2}(\hat{\boldsymbol{\theta}}_k - \boldsymbol{\theta}^*) \xrightarrow{\text{dist.}} N(\boldsymbol{\mu}_{\text{FD}}, \boldsymbol{\Sigma}_{\text{FD}})$. Explain this in light of the results in Appendix C.

6.5 Show that gain conditions B.1′, $\beta > 0$, and $3\gamma - \alpha/2 \geq 0$ imply the relationships in (6.14).

6.6 Condition A.1′ (Section 6.4) includes the gain condition $\sum_{k=0}^{\infty} a_k c_k < \infty$. How does A.1′ further restrict (6.14)?

6.7 Let $\boldsymbol{W}_k = k^{-\gamma}\{\hat{\boldsymbol{g}}_k(\hat{\boldsymbol{\theta}}_k) - E[\hat{\boldsymbol{g}}_k(\hat{\boldsymbol{\theta}}_k) | \Im_k]\}$. Assume that the noises ε in the measurements $y(\boldsymbol{\theta}) = L(\boldsymbol{\theta}) + \varepsilon$ are i.i.d. with $\text{var}(\varepsilon) = \sigma^2/2$. A key step in the proof of the asymptotic normality in (6.13) is to show that $E(\boldsymbol{W}_k \boldsymbol{W}_k^T | \Im_k) \to c^{-2}\sigma^2 \boldsymbol{I}_p/4$ (a.s.) as $k \to \infty$, where c appears in the numerator of c_k in (6.11). Sketch the arguments showing that this is true; state any additional conditions needed. (Note: This result also holds under more general conditions on the noise ε.)

6.8 Repeat the study described in Example 6.5 and statistically compare the four sample mean results you obtain with the four sample means in Table 6.1. (Hint: Use the appropriate two-sample *t*-test from Appendix B after "backing out" the approximate sample standard deviations from the confidence intervals shown in Table 6.1.)

6.9 Provide some intuitive rationale for the practical guideline that the numerator c in $c_k = c/(k+1)^{\gamma}$ should be set at approximately the magnitude of the initial standard deviation of the measurement noise in $y(\boldsymbol{\theta})$.

6.10 Given a desirable step size of 0.1 for the elements of $\boldsymbol{\theta}$, use the semiautomatic method of Section 6.6. Show that the values of c, A, and a in Examples 6.6 and 6.7 are $c = 1$, $A = 5$ (using 10 percent of iterations), and, approximately, $0.20 \leq a \leq 0.25$ ($a = 0.25$ is used to generate the relevant curve in Figure 6.2 of Example 6.6 and the relevant column in Table 6.2 in Example 6.7).

6.11 The following coefficient values are used in Examples 6.6 and 6.7:

- FDSA with semiautomatic gains: see Exercise 6.10.
- FDSA with manual tuning: same as Exercise 6.10 except that $a = 0.5$.
- Random search method B: 20 loss measurements are averaged at each iteration, and normally distributed perturbations are used with distribution $N(\boldsymbol{0}, 0.5^2\boldsymbol{I}_{10})$.

Produce a plot analogous to Figure 6.2 (i.e., three curves, each an average of 50 independent runs), but rather than show loss values, show the sample mean of the normalized errors $\|\hat{\boldsymbol{\theta}}_k - \boldsymbol{\theta}^*\| / \|\hat{\boldsymbol{\theta}}_0 - \boldsymbol{\theta}^*\|$ as a function of the number of loss measurements. Comment on the results relative to those in Figure 6.2.

CHAPTER 7

SIMULTANEOUS PERTURBATION STOCHASTIC APPROXIMATION

Chapters 5 and 6 discussed the use of stochastic approximation (SA) for problems of minimizing a loss function $L(\boldsymbol{\theta})$. Chapter 5 considered the case where direct unbiased measurements of the gradient $\boldsymbol{g}(\boldsymbol{\theta})$ are available. Chapter 6, on the other hand, introduced the notion of gradient-free SA, where optimization is carried out with only noisy measurements of the loss function. This is motivated by the many problems where direct measurements of the gradient are not available. This chapter explores a method—simultaneous perturbation stochastic approximation (SPSA)—applicable in both the stochastic gradient and gradient-free settings. SPSA typically offers significant efficiency gains in problems with a large number of variables to be optimized.

Section 7.1 is a brief introduction. Section 7.2 describes the basic SPSA algorithm and contrasts it with the finite-difference SA (FDSA) algorithm of Chapter 6. Sections 7.3 and 7.4 discuss some of the theory associated with the convergence and asymptotic normality of SPSA, much like the theory for the root-finding SA and FDSA algorithms considered earlier. The asymptotic normality of the iterate is used to draw some powerful conclusions about the relative efficiency of SPSA and FDSA. These efficiency results provide the main rationale for using SPSA instead of FDSA in practical applications. Section 7.5 presents a step-by-step guide to implementation that is aimed at helping the reader code the algorithm for his or her specific application. This section also summarizes some additional implementation aspects regarding the choice of algorithm gain sequences. Section 7.6 presents some numerical results, including results that illustrate the theoretical efficiency conclusions in Section 7.4. Section 7.7 briefly discusses some extensions to the basic SPSA algorithm, including modifications to perform discrete optimization (discrete $\boldsymbol{\theta}$), global optimization, and constrained optimization. Section 7.8 summarizes some relatively recent results on a second-order (adaptive) version of SPSA that emulates for stochastic problems the Newton–Raphson algorithm of deterministic optimization. This adaptive SPSA approach applies in either the "standard" gradient-free setting of Chapter 6, where only (noisy) loss function measurements are available, or in the stochastic gradient (Robbins–Monro) setting of Chapter 5, where direct unbiased gradient measurements are available.

Section 7.9 offers some concluding remarks and Section 7.10 is an appendix containing conditions for asymptotic normality of the iterate.

7.1 BACKGROUND

Chapters 5 and 6 demonstrated that stochastic approximation applies to a large number of optimization problems where only noisy measurements of a criterion are available. In the stochastic gradient framework (Chapter 5), a direct unbiased measurement of the gradient $\boldsymbol{g}(\boldsymbol{\theta}) = \partial L/\partial\boldsymbol{\theta}$ is used in the SA algorithm. Chapter 6, on the other hand, works with gradient *approximations* built from (noisy) measurements of L. Such methods are useful when it is very costly or impossible to directly measure the gradient $\boldsymbol{g}(\boldsymbol{\theta})$ at different values of $\boldsymbol{\theta}$. Continuing in the spirit of Chapter 6, where only noisy loss measurements are available, this chapter discusses the simultaneous perturbation stochastic approximation (SPSA) algorithm for stochastic optimization of multivariate systems. Relative to the finite-difference-based methods of Chapter 6, the principal benefit of SPSA is a reduction in the number of loss measurements required to achieve a given level of accuracy in the optimization process.

The central focus with SPSA is the stochastic setting where only measurements of the loss function are available (i.e., no gradient information). That is, the algorithm is based on loss measurements $y(\boldsymbol{\theta}) = L(\boldsymbol{\theta}) + \varepsilon$ at various values of $\boldsymbol{\theta}$; equivalently, as in Chapter 5, one can write $y(\boldsymbol{\theta}) = Q(\boldsymbol{\theta}, V)$, where Q represents some observed cost as a function of the chosen $\boldsymbol{\theta}$ and random effects V (so $\varepsilon = \varepsilon(\boldsymbol{\theta}) = Q(\boldsymbol{\theta}, V) - L(\boldsymbol{\theta})$). More recent results, however, show that the SPSA idea can be extended in a relatively simple way to the stochastic gradient setting, providing an efficient means for building an asymptotically optimal stochastic analogue of the Newton–Raphson algorithm (Section 7.8). Hence, the SPSA principles can be used in either the stochastic gradient (Chapter 5) or gradient-free (Chapter 6) settings. As we have seen, the interest in both the stochastic gradient and gradient-free SA algorithms has been motivated by problems such as adaptive control, model parameter estimation, and simulation-based optimization.

We saw in Chapter 6 that finite-difference SA (FDSA) exhibits convergence properties similar to the stochastic gradient-based stochastic algorithms while requiring only loss function measurements. SPSA shares this property with FDSA. The asymptotic normality of SPSA and FDSA can be used to draw fundamental conclusions on the relative large-sample efficiency of the approaches.

The SPSA Web site (*www.jhuapl.edu/SPSA*; also accessible through this book's Web site) includes references describing applications in areas such as queuing systems, industrial quality improvement, aircraft design, pattern recognition, simulation-based optimization (with applications, e.g., to air traffic management and military planning), bioprocess control, neural network training, chemical process control, fault detection, human-machine interaction, sensor placement and configuration, and vehicle traffic management.

7.2 FORM AND MOTIVATION FOR STANDARD SPSA ALGORITHM

7.2.1 Basic Algorithm

This section is devoted to the "basic" SPSA algorithm, which applies in the gradient-free setting of FDSA. (Section 7.8 considers the adaptive version of SPSA that applies in either the gradient-free or the stochastic gradient-based setting.) As motivated above, we now assume that no direct measurements of $\boldsymbol{g}(\boldsymbol{\theta})$ are available. Following the mold of Chapter 6, the basic unconstrained SPSA algorithm is in the general recursive SA form:

$$\hat{\boldsymbol{\theta}}_{k+1} = \hat{\boldsymbol{\theta}}_k - a_k \hat{\boldsymbol{g}}_k(\hat{\boldsymbol{\theta}}_k), \tag{7.1}$$

where $\hat{\boldsymbol{g}}_k(\hat{\boldsymbol{\theta}}_k)$ is the simultaneous perturbation estimate of the gradient $\boldsymbol{g}(\boldsymbol{\theta}) = \partial L/\partial\boldsymbol{\theta}$ at the iterate $\hat{\boldsymbol{\theta}}_k$ based on the measurements of the loss function and a_k is a nonnegative scalar gain coefficient. (The constrained case is briefly discussed in Section 7.7.)

The essential part of (7.1) is the gradient approximation $\hat{\boldsymbol{g}}_k(\hat{\boldsymbol{\theta}}_k)$. Recall that with FDSA, this gradient approximation is formed by perturbing the components of $\hat{\boldsymbol{\theta}}_k$ one at a time and collecting a loss measurement $y(\cdot)$ at each of the perturbations (in practice, the loss measurements are sometimes noise-free, à la $y(\cdot) = L(\cdot)$). This requires $2p$ loss measurements for a two-sided FD approximation. In contrast, with simultaneous perturbation, all elements of $\hat{\boldsymbol{\theta}}_k$ are randomly perturbed together to obtain two loss measurements $y(\cdot)$. For the two-sided SP gradient approximation, this leads to

$$\hat{\boldsymbol{g}}_k(\hat{\boldsymbol{\theta}}_k) = \begin{bmatrix} \dfrac{y(\hat{\boldsymbol{\theta}}_k + c_k\boldsymbol{\Delta}_k) - y(\hat{\boldsymbol{\theta}}_k - c_k\boldsymbol{\Delta}_k)}{2c_k\Delta_{k1}} \\ \vdots \\ \dfrac{y(\hat{\boldsymbol{\theta}}_k + c_k\boldsymbol{\Delta}_k) - y(\hat{\boldsymbol{\theta}}_k - c_k\boldsymbol{\Delta}_k)}{2c_k\Delta_{kp}} \end{bmatrix}$$

$$= \frac{y(\hat{\boldsymbol{\theta}}_k + c_k\boldsymbol{\Delta}_k) - y(\hat{\boldsymbol{\theta}}_k - c_k\boldsymbol{\Delta}_k)}{2c_k}\left[\Delta_{k1}^{-1}, \Delta_{k2}^{-1}, \ldots, \Delta_{kp}^{-1}\right]^T, \tag{7.2}$$

where the mean-zero p-dimensional random perturbation vector, $\Delta_k = [\Delta_{k1}, \Delta_{k2}, \ldots, \Delta_{kp}]^T$, has a user-specified distribution satisfying conditions discussed in Sections 7.3 and 7.4 and c_k is a positive scalar. Because the numerator is the same in all p components of $\hat{g}_k(\hat{\theta}_k)$, the number of loss measurements needed to estimate the gradient in SPSA is *two*, regardless of the dimension p. Recall that Section 6.8 discussed a similar random directions gradient approximation.

While the number of loss function measurements $y(\cdot)$ needed in each iteration of FDSA grows with p, the number in SPSA is fixed. This measurement savings per iteration, of course, provides only the *potential* for SPSA to achieve large savings (over FDSA) in the total number of measurements required to estimate θ when p is large. This potential is realized if the number of iterations required for effective convergence to an optimum θ^* does not increase in a way to cancel the measurement savings per gradient approximation at each iteration. We would expect this potential to be realized if, roughly speaking, the FD and SP gradient approximations acted the same in some statistical sense *relative to their use in the basic optimization recursion* (which is the same basic form in both FDSA and SPSA).

It is clear that the SP approximation above will *not* act the same as the FD approximation as an estimate of the gradient per se. The FD approximation will generally be superior in that sense. However, the interest is not in the gradient per se. Rather, the interest is in how the approximations operate *when considered in optimization* over multiple iterations with a changing point of evaluation θ. Section 7.4 discusses the efficiency issue further, establishing the fundamental result:

> Under reasonably general conditions (see Section 7.4), the SPSA and FDSA recursions achieve the same level of statistical accuracy for a given number of iterations even though SPSA uses only $1/p$ times the number of function evaluations of FDSA (since each gradient approximation uses only $1/p$ the number of function evaluations).

7.2.2 Relationship of Gradient Estimate to True Gradient

The informal rationale for the strange-looking gradient approximation in (7.2) is quite simple. Consider the mth element of the approximation. Let us sketch how this element is an "almost unbiased" estimator of the mth element of the true gradient. The formal results on the bias to follow and the convergence and asymptotic normality in Sections 7.3 and 7.4 make this more rigorous. Suppose that the measurement noise $\varepsilon(\theta)$ has mean zero and that L is several times differentiable at $\theta = \hat{\theta}_k$. Then, using a simple first-order Taylor expansion,

$$E\left[\hat{g}_{km}(\hat{\boldsymbol{\theta}}_k)\,|\,\hat{\boldsymbol{\theta}}_k\right] = E\left[\frac{y(\hat{\boldsymbol{\theta}}_k + c_k\boldsymbol{\Delta}_k) - y(\hat{\boldsymbol{\theta}}_k - c_k\boldsymbol{\Delta}_k)}{2c_k\Delta_{km}}\middle|\hat{\boldsymbol{\theta}}_k\right] \quad \text{(from (7.2))}$$

$$= E\left[\frac{L(\hat{\boldsymbol{\theta}}_k + c_k\boldsymbol{\Delta}_k) - L(\hat{\boldsymbol{\theta}}_k - c_k\boldsymbol{\Delta}_k)}{2c_k\Delta_{km}}\middle|\hat{\boldsymbol{\theta}}_k\right] \quad \text{(noise terms disappear)}$$

$$\approx E\left[\frac{L(\hat{\boldsymbol{\theta}}_k) + c_k\boldsymbol{g}(\hat{\boldsymbol{\theta}}_k)^T\boldsymbol{\Delta}_k - [L(\hat{\boldsymbol{\theta}}_k) - c_k\boldsymbol{g}(\hat{\boldsymbol{\theta}}_k)^T\boldsymbol{\Delta}_k]}{2c_k\Delta_{km}}\middle|\hat{\boldsymbol{\theta}}_k\right] \quad \text{(first-order expansions of } L(\hat{\boldsymbol{\theta}}_k \pm c_k\boldsymbol{\Delta}_k))$$

$$= E\left[\frac{2c_k\sum_{i=1}^{p} L_i'(\hat{\boldsymbol{\theta}}_k)\Delta_{ki}}{2c_k\Delta_{km}}\middle|\hat{\boldsymbol{\theta}}_k\right] \quad \text{(cancel } L \text{ terms)}$$

$$= L_m'(\hat{\boldsymbol{\theta}}_k) + \sum_{i\neq m} L_i'(\hat{\boldsymbol{\theta}}_k)E\left(\frac{\Delta_{ki}}{\Delta_{km}}\right) \quad \text{(rewrite of above)}, \tag{7.3}$$

where $L_i'(\cdot)$ denotes the ith component of $\boldsymbol{g}(\cdot)$. Assume that Δ_{ki} has mean zero and is independent of Δ_{km} for all $i \neq m$. Then, to ensure that $E(\Delta_{ki}/\Delta_{km})$ in the last line of (7.3) represents a valid expectation, it is necessary that the inverse moment $E(|1/\Delta_{km}|)$ be finite. (See Exercise 7.2, recalling the formal definition of expectation in Section C.1 of Appendix C.) The algorithm convergence results in Sections 7.3 and 7.4 use a more stringent version of this inverse moment condition, but the basic idea is the same; see the comments after conditions B.1″–B.6″ in Section 7.3 regarding the implication of this inverse moment condition. Given the above, $E(\Delta_{ki}/\Delta_{km}) = 0$ for all $i \neq m$.

Hence, the term after the "+" sign in the bottom line of (7.3) disappears, indicating that

$$E[\hat{g}_{km}(\hat{\boldsymbol{\theta}}_k)\,|\,\hat{\boldsymbol{\theta}}_k] \approx L_m'(\hat{\boldsymbol{\theta}}_k), \tag{7.4}$$

as desired. The precise meaning of the "≈" in (7.4) is given below, but one can see from (7.3) that the bias in the gradient approximation (i.e., the difference between an "=" and the indicated "≈") is due to the higher-order terms in the expansion of L appearing on the third line. As with FDSA, the resulting bias is $O(c_k^2)$, although the exact form of the bias differs from FDSA.

Figure 7.1 provides a pictorial representation of SPSA for the case where, for all k and i, the Δ_{ki} are generated by a symmetric Bernoulli ±1 distribution (i.e., with probability 1/2, there is one of two possible outcomes: $\Delta_{ki} = 1$ or $\Delta_{ki} = -1$). As discussed in Sections 7.3 and 7.7, the symmetric Bernoulli distribution is an important special case among the distributions that satisfy the conditions for $\boldsymbol{\Delta}_k$ (symmetric, mean zero, finite variance, and finite inverse moments).

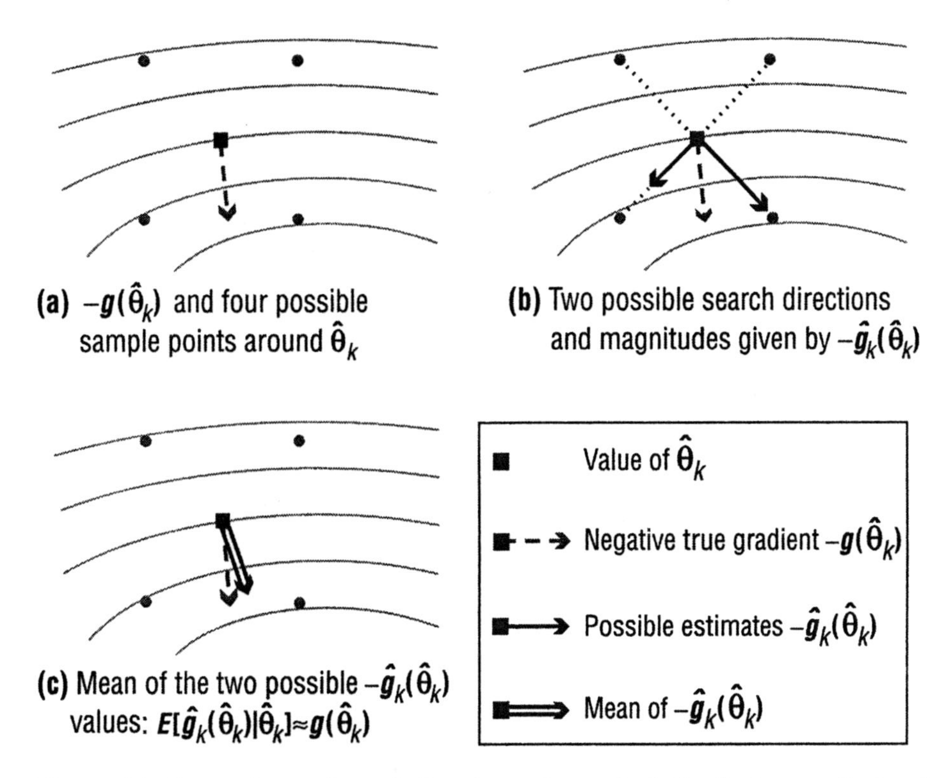

Figure 7.1. Comparisons of search direction and magnitude for the true gradient and SPSA gradient estimate in a low-noise setting with $p = 2$. Part (a) shows the four possible values of $\hat{\theta}_k \pm c_k\Delta_k$ surrounding $\hat{\theta}_k$ when using Bernoulli-distributed perturbations; true gradient is perpendicular to level curve at $\hat{\theta}_k$. Part (b) shows the two possible search directions and magnitudes for $-\hat{g}_k(\hat{\theta}_k)$. Each possibility has probability 1/2. The arrow pointing southeast is longer than the arrow pointing southwest because of the greater change in the function in sampling northwest-southeast points than in sampling southwest–northeast points. Part (c) shows the *mean* search direction and magnitude from the two possibilities in part (b). The slight deviation between the mean and true gradient is due to the bias.

Part (a) of Figure 7.1 shows the four points surrounding $\hat{\theta}_k$ (there are four *unique* possibilities for $\hat{\theta}_k \pm c_k\Delta_k$ from the four possible values for Δ_k). Note that the negative of the true gradient $-g(\hat{\theta}_k)$ points in a direction perpendicular to the level curve at $\hat{\theta}_k$ and in a direction of decreasing L (recall Section 1.4). Part (b) shows the negative of the two possible SP gradient estimates $-\hat{g}_k(\hat{\theta}_k)$ under the assumption that the noise (ε) is small. Note that one vector is longer than the other. This results from the difference in the loss values along the two lines forming the "X" in the figure. Along the northwest/southeast line, there is a greater change in the loss values than along the northeast/southwest line. From (7.2), it is apparent that the greater difference makes for a larger magnitude $-\hat{g}_k(\hat{\theta}_k)$. The difference in loss magnitudes may be masked if the noise effects are relatively large (but it does not alter the *orientation* of the gradient estimates along the two lines in the "X" since the orientations depend solely on the evaluation points $\hat{\theta}_k \pm c_k\Delta_k$).

Finally, part (c) of Figure 7.1 shows the mean of the two possible negative gradient estimates for part (b) (which is one half of the vector sum of the two estimates in part (b)). This mean depends on *both* the orientation and the magnitude of the candidate vectors $-\hat{g}_k(\hat{\theta}_k)$ in part (b). We see that the mean does not quite correspond to the true gradient. The indicated difference in $g(\hat{\theta}_k)$ and $E[\hat{g}_k(\hat{\theta}_k)\,|\,\hat{\theta}_k]$ in part (c) represents the above-mentioned $O(c_k^2)$ bias.

In the manner of the FDSA bias discussion in Subsection 6.4.1, let us more formally analyze the bias $\boldsymbol{b}_k \equiv E\{[\hat{g}_k(\hat{\theta}_k)-g(\hat{\theta}_k)]\,|\,\Im_k\}$ where $\Im_k$ = $\{\hat{\theta}_0, \hat{\theta}_1, \ldots, \hat{\theta}_k;\ \Delta_0, \Delta_1, \ldots, \Delta_{k-1}\}$ for $k \geq 1$. Relative to the definition of $\Im_k$ in Sections 4.3 and 6.4, this definition of $\Im_k$ is expanded to account for the additional randomness introduced through the Δ_k process (which affects both the $\hat{\theta}_k$ and noise ε processes). Let $L'''(\theta) = \partial^3 L / \partial\theta^T \partial\theta^T \partial\theta^T$ denote the $1 \times p^3$ row vector of all possible third derivatives of L. Assume that $L'''(\theta)$ exists and is continuous. By a Taylor expansion analogous to that in Section 6.4, the mth component of $\boldsymbol{b}_k$ is

$$b_{km} = \frac{1}{12} c_k^2 E\left\{\Delta_{km}^{-1}[L'''(\bar{\theta}_k^{(+)}) + L'''(\bar{\theta}_k^{(-)})][\Delta_k \otimes \Delta_k \otimes \Delta_k]\,\Big|\,\Im_k\right\}, \tag{7.5}$$

where $\bar{\theta}_k^{(\pm)}$ denotes points on the (two) line segments between $\hat{\theta}_k$ and $\hat{\theta}_k \pm c_k\Delta_k$ and $\otimes$ denotes the Kronecker product (Appendix A). Suppose that $|\Delta_{ki}| \leq \eta_0$, $E(|1/\Delta_{ki}|) \leq \eta_1$, and $|L'''_{i_1 i_2 i_3}(\theta)| \leq \eta_2$, where the η_i are positive constants and $L'''_{i_1 i_2 i_3}(\theta)$ denotes the element of $L'''(\theta)$ representing the third derivative with respect to the i_1, i_2, and i_3 elements of θ ($i_j = 1, 2, \ldots, p$). Then, from the right-hand side of (7.5),

$$\begin{aligned} |b_{km}| &\leq \frac{\eta_2 c_k^2}{6} \sum_{i_1}\sum_{i_2}\sum_{i_3} E\left(\left|\frac{\Delta_{ki_1}\Delta_{ki_2}\Delta_{ki_3}}{\Delta_{km}}\right|\right) \\ &\leq \frac{\eta_2 c_k^2}{6}\left\{[p^3 - (p-1)^3]\eta_0^2 + (p-1)^3 \eta_1 \eta_0^3\right\} \end{aligned} \tag{7.6}$$

(see Exercise 7.5). The bound in (7.6) provides an explicit form for the $O(c_k^2)$ bias.

7.3 BASIC ASSUMPTIONS AND SUPPORTING THEORY FOR CONVERGENCE

This section presents conditions for convergence of the SPSA iterate ($\hat{\theta}_k \to \theta^*$ a.s. as $k \to \infty$). The proof uses the ordinary differential equation (ODE) approach discussed in Section 4.3 for the root-finding SA algorithm and Section 6.4 for the FDSA algorithm. The conditions here are close to the "engineering" conditions B.1′–B.5′ of Section 6.4 for FDSA. As with FDSA, but unlike stochastic gradient SA in Chapter 5, there are conditions on *two* gain sequences (a_k and c_k). In addition to the conditions for FDSA, however, we must impose conditions on the distribution of $\boldsymbol{\Delta}_k$, and the statistical relationship of $\boldsymbol{\Delta}_k$ to the measurements $y(\cdot)$.

The conditions here ensuring convergence of $\hat{\boldsymbol{\theta}}_k$ to a minimizing point $\boldsymbol{\theta}^*$ are based on the arguments in Spall (1988b, 1992).[1] Following the pattern established in Section 6.4, condition B.*i*″ below is identical to, or closely related to, condition B.*i* in Section 4.3 and condition B.*i*′ in Section 6.4 for $i = 1, 2,\ldots, 5$ ($i = 6$ corresponds to a unique condition for SPSA). Let $\varepsilon_k^{(\pm)} = \varepsilon(\hat{\boldsymbol{\theta}}_k \pm c_k\boldsymbol{\Delta}_k)$. When a condition is identical to one of the previous conditions in Sections 4.3 and/or 6.4 and relatively long to state, we simply refer back to the earlier condition rather than restate the condition here.

B.1″ **(Gain sequences)** Same as condition B.1′ in Section 6.4 (i.e., a_k and $c_k > 0$; a_k and $c_k \to 0$; $\sum_{k=0}^{\infty} a_k = \infty$; and $\sum_{k=0}^{\infty} a_k^2/c_k^2 < \infty$).

B.2″ **(Relationship to ODE)** Same as condition B.2 in Section 4.3 (and condition B.2′ in Section 6.4).

B.3″ **(Iterate boundedness)** Same as condition B.3 in Section 4.3 (and condition B.3′ in Section 6.4). Main requirement: $\sup_{k\geq 0}\|\hat{\boldsymbol{\theta}}_k\| < \infty$ a.s.

B.4″ **(Measurement noise; relationship between the measurement noise and $\boldsymbol{\Delta}_k$)** For all k, $E[(\varepsilon_k^{(+)} - \varepsilon_k^{(-)})|\Im_k, \boldsymbol{\Delta}_k] = 0$ and the ratio of measurement to perturbation is such that $E\left[\left(y(\hat{\boldsymbol{\theta}}_k \pm c_k\boldsymbol{\Delta}_k)/\Delta_{ki}\right)^2\right]$ is uniformly bounded (over k and i).

B.5″ **(Smoothness of L)** L is three-times continuously differentiable and bounded on $\mathbb{R}^p$. (This is slightly stronger than condition B.5′ in Section 6.4, which pertains only to the "unmixed" partials $L_{iii}'''(\boldsymbol{\theta})$ for all i.)

B.6″ **(Statistical properties of the perturbations)** The $\{\Delta_{ki}\}$ are independent for all k, i, identically distributed for all i at each k, symmetrically distributed about zero and uniformly bounded in magnitude for all k, i.

Let us comment on the above conditions. From the point of view of the user's input, conditions B.1″, B.4″, and B.6″ are the most relevant since they govern the gains a_k, c_k and the random perturbations $\boldsymbol{\Delta}_k$. The role of a_k in B.1″ is similar to its role in the root-finding SA algorithm, as discussed in Section 4.3. The square summability in condition B.1″ ($\sum_{k=0}^{\infty} a_k^2/c_k^2 < \infty$) balances the decay

---

[1]The conditions given here differ slightly from those in Spall (1988b, 1992). The conditions here are streamlined to be closer to the minimal required using the proof techniques in Spall (1988b, 1992).

of a_k against the decay of c_k to ensure that the update in moving $\hat{\boldsymbol{\theta}}_k$ to $\hat{\boldsymbol{\theta}}_{k+1}$ is well behaved. In particular, the condition prevents c_k from going to zero too quickly, which prevents the gradient estimate from becoming too wild and overpowering the decay associated with a_k. The motivation behind B.2″ and B.3″ is identical to the motivation for B.2 and B.3 in Section 4.3. These conditions impose the requirement that $\hat{\boldsymbol{\theta}}_k$ (including the initial condition) is close enough to $\boldsymbol{\theta}^*$ so that there is a natural tendency for an analogous deterministic algorithm (manifested in continuous time as an ODE) to converge to $\boldsymbol{\theta}^*$.

Conditions B.4″ to B.6″ on the perturbation distribution and smoothness of L guarantee that the gradient estimate $\hat{g}_k(\hat{\boldsymbol{\theta}}_k)$ is an unbiased estimate of $g(\hat{\boldsymbol{\theta}}_k)$ to within an $O(c_k^2)$ error. This $O(c_k^2)$ bias is small enough so that (as in the stochastic gradient case, where the gradient estimate typically has no bias) the $\boldsymbol{\theta}$ iterate is able to converge to $\boldsymbol{\theta}^*$. Further, the boundedness of $E\left[\left(y(\hat{\boldsymbol{\theta}}_k \pm c_k\Delta_k)/\Delta_{ki}\right)^2\right]$ in B.4″ is used together with $\sum_{k=0}^{\infty} a_k^2/c_k^2 < \infty$ in B.1″ to ensure that a sum of variances is finite, analogous to the proof of convergence of FDSA.

Let us comment on the important relationship of finite inverse moments for the elements of $\boldsymbol{\Delta}_k$ to the condition in B.4″ that $E\left[\left(y(\hat{\boldsymbol{\theta}}_k \pm c_k\Delta_k)/\Delta_{ki}\right)^2\right]$ be bounded. Using Hölder's inequality (see Exercise C.4, Appendix C),

$$E\left[\left(\frac{y(\hat{\boldsymbol{\theta}}_k \pm c_k\Delta_k)}{\Delta_{ki}}\right)^2\right] \le \left[E\left(\left|y(\hat{\boldsymbol{\theta}}_k \pm c_k\Delta_k)\right|^{2+2\eta}\right)\right]^{1/(1+\eta)} \left[E\left(\left|\frac{1}{\Delta_{ki}}\right|^{2+2\tau}\right)\right]^{1/(1+\tau)}, \tag{7.7}$$

where η and τ are any strictly positive values that satisfy $(1+\eta)^{-1} + (1+\tau)^{-1} = 1$ (see Exercise 7.6). The analyst may choose η and τ arbitrarily subject to these conditions. If the measurements $y(\hat{\boldsymbol{\theta}}_k \pm c_k\Delta_k)$ have bounded moments of order $2+2\eta$, then the first $[\cdot]$ term on the right-hand side of (7.7) is bounded. Hence, (7.7) implies that $E\left[\left(y(\hat{\boldsymbol{\theta}}_k \pm c_k\Delta_k)/\Delta_{ki}\right)^2\right]$ is uniformly bounded if there exists a $\tau > 0$ such that $(1+\eta)^{-1} + (1+\tau)^{-1} = 1$ and

$$E\left(\left|\frac{1}{\Delta_{ki}}\right|^{2+2\tau}\right) \le C \tag{7.8}$$

for some $C > 0$. This bounded inverse moments condition for the Δ_{ki} is an important part of SPSA. We saw in the informal arguments of (7.3) and (7.4) an application of a similar (weaker) condition in showing that $\hat{g}_k(\hat{\boldsymbol{\theta}}_k)$ is a nearly unbiased estimate of $g(\hat{\boldsymbol{\theta}}_k)$.

One important—and very simple—distribution that satisfies the inverse moments condition is the symmetric Bernoulli ± 1 distribution. Two common mean-zero distributions that do *not* satisfy the inverse moments condition are symmetric uniform and normal with mean zero. The failure of both of these distributions is a consequence of the amount of probability mass near zero. In particular, with too much probability near zero, the expectation integral of the generic form

$$E\left(\left|\frac{1}{\Delta_{ki}}\right|^{2+2\tau}\right) = \int_{-\infty}^{\infty}\left|\frac{1}{\delta}\right|^{2+2\tau} p_{\Delta}(\delta)\, d\delta \tag{7.9}$$

is undefined (infinite), where $p_{\Delta}(\delta)$ represents the density function for Δ_{ki} (or mass function if the integral is replaced by a sum).

Figure 7.2 shows several probability density or probability mass functions for mean-zero random variables. The top row shows density or mass functions for which the inverse moments $E\left(|1/\Delta_{ki}|^{2+2\tau}\right)$ are finite; the bottom row shows functions for which the inverse moments are not defined. Although the U-shaped density function is shown in the top row, an arbitrary U-shaped density may not have a finite inverse $2 + 2\tau$ moment, depending on the curvature of the "U" and the value of τ. One must evaluate (7.9) to determine if it is finite (Exercise 7.8).

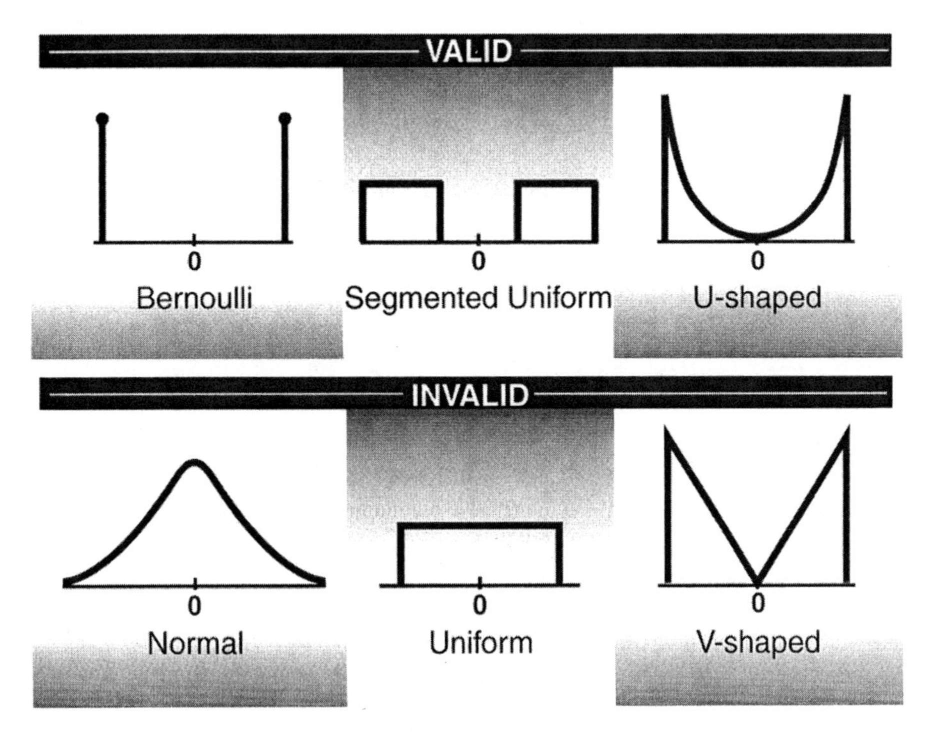

Figure 7.2. Probability density or mass functions that are valid or invalid relative to the inverse moments condition: $E\left(|1/\Delta_{ki}|^{2+2\tau}\right) < \infty$ for any k, i.

Aside from Spall (1992), conditions for convergence of SPSA have been presented in Dippon and Renz (1997), Wang and Chong (1998), Chen et al. (1999), and Gerencsér (1999). While the conditions of these authors tend to be slightly different from one another, in all cases there are conditions similar to the finite inverse moments conditions discussed above for the elements of $\mathbf{\Delta}_k$. The theorem below is proved in Spall (1992).

Theorem 7.1. Suppose that conditions B.1″–B.6″ hold. Further, suppose that $\boldsymbol{\theta}^*$ is a unique minimum (i.e., Θ^* is the singleton $\boldsymbol{\theta}^*$). Then, for the SPSA algorithm in (7.1), $\hat{\boldsymbol{\theta}}_k \to \boldsymbol{\theta}^*$ a.s. as $k \to \infty$.

7.4 ASYMPTOTIC NORMALITY AND EFFICIENCY ANALYSIS

Although the convergence result for SPSA is of some independent interest, the most interesting theoretical results in Spall (1992), and those that provide most of the rationale for using SPSA, are the asymptotic efficiency conclusions that follow from an asymptotic normality result. In particular, in addition to conditions B.1″–B.6″ above, let us suppose that B.7″–B.9″ in the chapter appendix (Section 7.10) hold. The theorem below is proved in Spall (1992).

Theorem 7.2. Suppose that the gains have the standard form $a_k = a/(k+1+A)^{\alpha}$ and $c_k = c/(k+1)^{\gamma}$, $k = 0, 1, 2, \ldots$, with a, c, α, and γ strictly positive, $A \geq 0$, $\beta = \alpha - 2\gamma > 0$ (as in Section 6.5), and $3\gamma - \alpha/2 \geq 0$. Further, suppose that the conditions of Theorem 7.1 plus conditions B.7″ – B.9″ hold. Then, for the SPSA algorithm in (7.1),

$$k^{\beta/2}(\hat{\boldsymbol{\theta}}_k - \boldsymbol{\theta}^*) \xrightarrow{\text{dist.}} N(\boldsymbol{\mu}_{\text{SP}}, \boldsymbol{\Sigma}_{\text{SP}}) \text{ as } k \to \infty, \tag{7.10}$$

where $\boldsymbol{\mu}_{\text{SP}}$ and $\boldsymbol{\Sigma}_{\text{SP}}$ are a mean vector and covariance matrix.

In Theorem 7.2, $\boldsymbol{\mu}_{\text{SP}}$ depends on both the Hessian matrix and the third derivatives of $L(\boldsymbol{\theta})$ at $\boldsymbol{\theta}^*$ and $\boldsymbol{\Sigma}_{\text{SP}}$ depends on the Hessian matrix at $\boldsymbol{\theta}^*$ (notation consistent with $\boldsymbol{\mu}_{\text{FD}}$ and $\boldsymbol{\Sigma}_{\text{FD}}$ from Section 6.5). The specific forms are given in Spall (1992) and Dippon and Renz (1997). In general, $\boldsymbol{\mu}_{\text{SP}} \neq \boldsymbol{\mu}_{\text{FD}}$ and $\boldsymbol{\Sigma}_{\text{SP}} \neq \boldsymbol{\Sigma}_{\text{FD}}$. As in the FDSA result in Section 6.5, $\boldsymbol{\mu}_{\text{SP}} \neq \mathbf{0}$ in general, which contrasts with many well-known asymptotic normality results in estimation, including that in Section 4.4 for the root-finding SA algorithm. As discussed in Section 6.5 for FDSA, $\boldsymbol{\mu}_{\text{SP}} \neq \mathbf{0}$ does *not* imply that $\hat{\boldsymbol{\theta}}_k$ is an asymptotically biased estimator.

Similar to the root-finding SA and FDSA cases, the asymptotic distribution result (7.10) allows one to determine asymptotically optimal gain decay rates (i.e., rates that provide the maximum value of $\beta/2$). Given the restrictions in the theorem statement (including B.1″), these are $\alpha = 1$ and $\gamma = 1/6$, yielding $\beta/2 = 1/3$ (the same as FDSA—see Section 6.5). Hence, the fastest possible stochastic rate at which the error $\hat{\boldsymbol{\theta}}_k - \boldsymbol{\theta}^*$ goes to zero is proportional to $k^{-\beta/2} = 1/k^{1/3}$ for large k. This contrasts with the fastest allowable rate (proportional to $1/\sqrt{k}$) for the root-finding (stochastic gradient) SA algorithm.

Hence, one measure of the value of the gradient information in stochastic gradient SA is the increase in rate of convergence. (Section 14.4 discusses a special case where it also possible to get a $1/\sqrt{k}$ rate of convergence in SPSA through the use of common random numbers in a simulation-based optimization context.) However, as discussed in Section 6.6, it is generally superior in finite-sample practice to have gain sequences a_k and c_k that decay more slowly than the asymptotically optimal gains using $\alpha = 1$ and $\gamma = 1/6$. As with FDSA, $\alpha = 0.602$ and $\gamma = 0.101$ are approximately the lowest possible valid values and are recommended in many practical applications (see the implementation guidelines in Section 7.5).

Spall (1992, Sect. 4) uses the asymptotic normality result in (7.10) (together with the parallel result for FDSA in Section 6.5) to evaluate the relative efficiency of SPSA. This efficiency depends on the shape of L, the values for a_k and c_k, and the distributions of the Δ_{ki} and measurement noise terms $\varepsilon_k^{(\pm)}$. There is no single expression that can be used to characterize the relative efficiency.

As discussed in Spall (1992, Sect. 4) and Chin (1997), however, in most practical problems, SPSA will be asymptotically more efficient than FDSA. For example, if $3\gamma - \alpha/2 > 0$ (as in the guidelines in Section 7.5, with $\alpha = 0.602$ and $\gamma = 0.101$), then by equating the asymptotic mean-squared errors of the parameter estimates, as given by the asymptotic distributions in FDSA and SPSA, we find that

$$\boxed{\frac{\text{number of } y(\boldsymbol{\theta}) \text{ values in SPSA}}{\text{number of } y(\boldsymbol{\theta}) \text{ values in FDSA}} \to \frac{1}{p}} \tag{7.11}$$

as the number of loss measurements in both procedures gets large. (The above result also sometimes applies with the asymptotically optimal condition $3\gamma - \alpha/2 = 0$ holding; see Spall, 1992.) Expression (7.11) implies that the p-fold savings per iteration (per gradient approximation) translates directly into a p-fold savings in the overall optimization process. Note that (7.11) is derived under the assumption that FDSA and SPSA use the same gain sequences a_k and c_k (including, e.g., sequences tuned for optimal FDSA performance).

The above efficiency result is derived under the assumption of noisy loss measurements. Perhaps counterintuitively, the mathematics for efficiency analysis in the noise-free case is more difficult than for the noisy case (it is *not* valid to simply take the limit of the relevant expressions in the noisy case as the noise variance goes to zero). Gerencsér and Vágó (2001) analyze the efficiency in the noise-free case and find a result analogous to the convergence rate for deterministic steepest descent algorithm, as in Section 1.4 (i.e., $\|\hat{\boldsymbol{\theta}}_{k+1} - \boldsymbol{\theta}^*\| =$

$O(\|\hat{\theta}_k - \theta^*\|)$ a.s., or, equivalently, $\|\hat{\theta}_{k+1} - \theta^*\| = O(\lambda^k)$ a.s. for some $0 < \lambda < 1$). This rate is superior to the above-mentioned $1/k^{1/3}$ rate in the noisy case.

By providing theoretical evidence of the superiority of one algorithm over another, it might seem that (7.11) contradicts the no free lunch (NFL) theorems of Subsection 1.2.2. It does not. Recall that the NFL theorems state that the performance of any two algorithms is the same when averaged across *all possible problems*. The asymptotic superiority of SPSA over FDSA as expressed in (7.11) applies to only a sliver of all possible problems—those problems satisfying the conditions of the asymptotic normality. Of course, this sliver is an important practical subset of all possible problems, but, nonetheless, it is only a subset.

7.5 PRACTICAL IMPLEMENTATION

7.5.1 Step-by-Step Implementation

The step-by-step summary below shows how SPSA iteratively produces a sequence of estimates. MATLAB code for implementing the steps below is available at the book's Web site.

Basic SPSA Algorithm

Step 0 **(Initialization and coefficient selection)** Set counter index $k = 0$. Pick initial guess $\hat{\theta}_0$ and nonnegative coefficients a, c, A, α, and γ in the SPSA gain sequences $a_k = a/(k+1+A)^{\alpha}$ and $c_k = c/(k+1)^{\gamma}$. Practically effective (and theoretically valid) values for α and γ are 0.602 and 0.101, respectively; a, A, and c may be determined based on the practical guidelines given in Subsection 7.5.2.

Step 1 **(Generation of the simultaneous perturbation vector)** Generate by Monte Carlo a p-dimensional random perturbation vector $\mathbf{\Delta}_k$, where each of the p components of $\mathbf{\Delta}_k$ are independently generated from a zero-mean probability distribution satisfying the conditions above. An effective (and theoretically valid) choice for each component of $\mathbf{\Delta}_k$ is to use a Bernoulli ± 1 distribution with probability of 1/2 for each ± 1 outcome, although other choices are valid and may be desirable in some applications.

Step 2 **(Loss function evaluations)** Obtain two measurements of the loss function based on the simultaneous perturbation around the current $\hat{\theta}_k$: $y(\hat{\theta}_k + c_k\mathbf{\Delta}_k)$ and $y(\hat{\theta}_k - c_k\mathbf{\Delta}_k)$ with the c_k and $\mathbf{\Delta}_k$ from steps 0 and 1.

Step 3 **(Gradient approximation)** Generate the simultaneous perturbation approximation to the unknown gradient $\boldsymbol{g}(\hat{\theta}_k)$ according to eqn. (7.2). It is sometimes useful to average several gradient approximations at $\hat{\theta}_k$, each formed from an independent generation of $\mathbf{\Delta}_k$. The benefits are especially apparent if the noise effects ε_k are relatively large.

Step 4 **(Update θ estimate)** Use the standard SA form in (7.1) to update $\hat{\boldsymbol{\theta}}_k$ to a new value $\hat{\boldsymbol{\theta}}_{k+1}$. Check for constraint violation (if relevant) and modify the updated $\boldsymbol{\theta}$. (A common way to handle "easy" constraints is to simply map violating elements of $\boldsymbol{\theta}$ to the nearest valid point.)

Step 5 **(Iteration or termination)** Return to step 1 with $k+1$ replacing k. Terminate the algorithm if there is little change in several successive iterates or if the maximum allowable number of iterations has been reached.

In addition to the practical guidelines above, the blocking steps discussed in Subsection 7.8.2 for the adaptive SPSA approach can also be applied. In these steps, the iteration update is blocked (i.e., $\hat{\boldsymbol{\theta}}_{k+1}$ is set to $\hat{\boldsymbol{\theta}}_k$) if there would otherwise be a suspiciously large change in $\boldsymbol{\theta}$ or if the measured loss value at the intended value $\hat{\boldsymbol{\theta}}_{k+1}$ does not show enough improvement relative to the value at $\hat{\boldsymbol{\theta}}_k$. Blocking based on an excessive change in $\boldsymbol{\theta}$ requires no additional loss measurements; blocking based on a check of the loss value requires at least one additional loss measurement per iteration (i.e., measurement(s) at the nonperturbed values $\hat{\boldsymbol{\theta}}_k$ versus measurements at only the perturbed values $\hat{\boldsymbol{\theta}}_k \pm c_k\boldsymbol{\Delta}_k$).

A further practical concern is to attempt to define $\boldsymbol{\theta}$ so that the magnitudes of the $\boldsymbol{\theta}$ elements are similar to one another. This desire is apparent by noting that the magnitudes of all components in the perturbations $c_k\boldsymbol{\Delta}_k$ are identical in the case where Bernoulli perturbations are used. By defining the elements in $\boldsymbol{\theta}$ such that they are of similar magnitude, it is possible to ensure that $c_k\boldsymbol{\Delta}_k$ is being added and subtracted to a vector where all components have a similar magnitude in the course of the search process. Although not always possible, an analyst often has the flexibility to choose the units for $\boldsymbol{\theta}$ to ensure similar magnitudes. Consider the following example.

Example 7.1—Definition of θ for an electrical circuit. Consider the optimization of an electrical circuit with variable resistors, inductors, and capacitors. The elements of $\boldsymbol{\theta}$ are the resistance, inductance, and capacitance of the components in the circuit. Suppose that the "typical" magnitudes of the resistance, inductance, and capacitance of the components are 5 to 20 ohms, 2 to 10 millihenries, and 3×10^{-6} to 15×10^{-6} farads. It is undesirable to have a $\boldsymbol{\theta}$ vector with some components having magnitudes of order 10^0 to 10^1 and other components having magnitude of order 10^{-6}. If we define the units of $\boldsymbol{\theta}$ to be ohms, millihenries, and microfarads, then the magnitude of all components of $\boldsymbol{\theta}$ will be commensurate at between 2 and 20. ❑

7.5.2 Choice of Gain Sequences

Let us summarize some additional implementation aspects regarding the choice of algorithm gain sequences a_k, c_k. The reader should be warned that the guidelines provided here are just that—guidelines—and may not be the best for every application. These guidelines were developed based on many test cases conducted by the author and others and form a reasonable basis for starting if one has no specific reason to follow other guidelines. (Theoretical guidelines, such as discussed in Fabian, 1971, and Chin, 1997, are not generally useful in practical applications since they require the very information on the loss function and its gradients that is assumed unavailable!) The guidelines here are a natural extension of those in Section 6.6 for FDSA.

The choice of the gain sequences is critical to the performance. With α and γ as specified in step 0 of the algorithm above, one typically finds that in a high-noise setting (i.e., poor quality measurements of L) it is necessary to pick a smaller a and larger c than in a low-noise setting. As noted below Theorem 7.2, the asymptotically optimal values of α and γ with noisy loss measurements are 1 and 1/6, respectively. In practice, however, it is usually the case that $\alpha < 1$ yields better finite-sample performance through maintaining a larger step size. Hence the recommendation in step 0 to use values (0.602 and 0.101) that are effectively the lowest allowable subject to satisfying the theoretical conditions mentioned in Sections 7.3 and 7.4. When the algorithm is being run with a large number of iterations, it may be beneficial to convert to $\alpha = 1$ and $\gamma = 1/6$ at some point in the iteration process to take advantage of the asymptotic optimality.

With the Bernoulli ± 1 distribution for the elements of Δ_k and the α and γ specified in step 0, a rule of thumb is to set c at a level approximately equal to the standard deviation of the measurement noise in $y(\boldsymbol{\theta})$. This helps keep the p elements of $\hat{g}_k(\hat{\boldsymbol{\theta}}_k)$ from getting excessively large in magnitude. The standard deviation can be estimated by collecting several $y(\boldsymbol{\theta})$ values at the initial guess $\hat{\boldsymbol{\theta}}_0$. When perfect (noise-free) measurements of $L(\boldsymbol{\theta})$ are available, then c should be chosen as some small positive number.

The values of a, A can be chosen together to ensure effective practical performance of the algorithm. A useful rule of thumb is to choose $A > 0$ such that it is 10 percent or less of the maximum number of expected/allowed iterations. After choosing A, one can choose a such that $a_0 = a/(1+A)^{0.602}$ times the magnitude of elements in $\hat{g}_0(\hat{\boldsymbol{\theta}}_0)$ is approximately equal to the smallest of the desired change magnitudes among the elements of $\boldsymbol{\theta}$ in the early iterations. To do this reliably may require several replications of $\hat{g}_0(\hat{\boldsymbol{\theta}}_0)$. These guidelines for choosing a are similar to those mentioned in Brennan and Rogers (1995, Sect. 2). An example of gain selection is given below.

Example 7.2—Choice of gain coefficients. Suppose that the standard deviation of the noise $\varepsilon_k(\boldsymbol{\theta})$ at $\boldsymbol{\theta}$ near $\hat{\boldsymbol{\theta}}_0$ is approximately 0.5 and that there is a budget of 2000 loss measurements for the search process. This suggests a choice of $c = 0.5$ and $A = 0.10 \times 2000/2 = 100$ (10 percent of $2000/2 = 1000$ iterations). Based on prior information, suppose that the analyst felt that the elements of $\boldsymbol{\theta}$ should typically move by a magnitude 0.1 in the early iterations. After computing several values of $\hat{g}_0(\hat{\theta}_0)$, it is determined that the mean magnitude of the elements in $\hat{g}_0(\hat{\theta}_0)$ (after choosing c as above) is approximately 10. With A = 100, it is found that $a = 0.16$ according to $[0.16/(100+1)^{0.602}]\times 10 = 0.1$. ❑

7.6 NUMERICAL EXAMPLES

This section presents two examples illustrating the relative efficiency of FDSA and SPSA, including the implications of the theoretical comparison in Section 7.4 (especially (7.11)).

Example 7.3—FDSA versus SPSA on Rosenbrock function. Consider a p = 10 version of the Rosenbrock test function introduced in Section 2.2. The function has the fourth-order polynomial form

$$L(\boldsymbol{\theta}) = \sum_{i=1}^{5}\left[100(t_{2i} - t_{2i-1}^2)^2 + (1 - t_{2i-1})^2\right],$$

where $\boldsymbol{\theta} = [t_1, t_2, \ldots, t_{10}]^T$. Note that $L(\boldsymbol{\theta}^*) = 0$ at $\boldsymbol{\theta}^* = [1, 1, \ldots, 1]^T$. Assume that the function measurements are taken in the presence of independent, identically distributed (i.i.d.) noise having distribution $N(0, 0.2^2)$.

Given the desire to test the numerical implications of the *asymptotic* theory in Section 7.4, we consider an initial condition that is close to $\boldsymbol{\theta}^*$ (appropriate since it is assumed that the algorithms have been running a long time). Let $\hat{\boldsymbol{\theta}}_0 = [0.99, 1, 0.99, 1, \ldots, 1]^T$, so $L(\hat{\boldsymbol{\theta}}_0) = 0.1985$. Note that the "one-sigma" value of the noise is slightly greater than the initial loss value and will dominate the $L(\boldsymbol{\theta})$ information contained in $y(\boldsymbol{\theta})$ if (as one hopes!) the iterate moves closer to $\boldsymbol{\theta}^*$. Letting A = 10 and c = 0.05 in the standard gains $a_k = a/(k+1+A)^{0.602}$ and $c_k = c/(k+1)^{0.101}$, we find numerically that $a = 0.002$ is an approximately optimal value for FDSA when using 10,000 to 50,000 loss measurements (the A and c values are smaller than the values suggested in Subsection 7.5.2 to better emulate asymptotic effects). Consistent with the efficiency result in Section 7.4, both FDSA and SPSA are run with the same gain values.

Table 7.1 shows the mean values of the normalized loss function

$$L_{\text{norm}}(\boldsymbol{\theta}) \equiv \frac{L(\boldsymbol{\theta}) - L(\boldsymbol{\theta}^*)}{L(\hat{\boldsymbol{\theta}}_0) - L(\boldsymbol{\theta}^*)}$$

over 50 independent pairs of FDSA and SPSA runs for several run lengths. Note that with the relatively high noise level and closeness of $\hat{\boldsymbol{\theta}}_0$ and $\boldsymbol{\theta}^*$, there is a significant challenge to the algorithms to produce notably improved values of $\boldsymbol{\theta}$

Table 7.1. Sample means of normalized loss $L_{\text{norm}} = L_{\text{norm}}(\hat{\boldsymbol{\theta}}_k)$ at terminal $\hat{\boldsymbol{\theta}}_k$ for FDSA and SPSA over 50 independent replications. Number of loss measurements $y(\boldsymbol{\theta})$ is such that FDSA and SPSA take the same number of iterations in each comparison.

Number of $y(\boldsymbol{\theta})$ values [*number of iterations*]	Mean L_{norm} for FDSA	Mean L_{norm} for SPSA
1000-FDSA; 100-SPSA [*50 iterations*]	0.100	0.111
25,000-FDSA; 2500-SPSA [*1250 iterations*]	0.0014	0.0017
50,000-FDSA; 5000-SPSA [*2500 iterations*]	0.0012	0.0011

that manifest themselves in low values of $L_{\text{norm}}(\boldsymbol{\theta})$. The value for $\boldsymbol{\theta}$ used in the 50 evaluations of $L_{\text{norm}}(\boldsymbol{\theta})$ for each of FDSA and SPSA is the value at the final iteration after the indicated number of loss measurements. Table 7.1 shows three combinations of number of loss evaluations. The ratio of number of loss evaluations in FDSA to SPSA is equal to p, so that the algorithms use the same number of iterations (50, 1250, or 2500).

In the three cases of Table 7.1, the sample means for the terminal $L_{\text{norm}}(\hat{\boldsymbol{\theta}}_k)$ values are relatively close for FDSA and SPSA even though the common gain sequences are tuned to approximately optimize performance for FDSA. This provides numerical support for the efficiency result in Section 7.4. That is, even with the ten-fold savings in number of loss measurements in SPSA per iteration, FDSA and SPSA perform comparably with the same number of iterations. ❑

Example 7.4—Neural network control. Figure 7.3 shows results for a $p = 412$ problem in the control of a wastewater treatment system (see Spall and Cristion, 1997 and 1998, for a complete description of the problem and the results of similar studies). The overall approach is the model-free controller summarized in Section 6.2. A feedforward neural network (Section 5.2) with $p = 412$ weights is used for the control function. FDSA and SPSA are used to estimate the weights. The curves represent an average of 50 independent realizations (an individual realization, of course, is much more jagged than the smooth curves shown in the figure.) The loss function L_k is time-varying, reflecting varying target values for water cleanliness and methane gas. The gain coefficients in FDSA and SPSA are separately tuned to approximately optimize the performance of each algorithm.

Note that FDSA and SPSA perform comparably on an iteration-by-iteration basis. The compelling part of the story, however, is that SPSA uses only two measurements per iteration while FDSA uses 824. This 412-fold savings in noisy loss measurements per iteration leads to the large savings in total

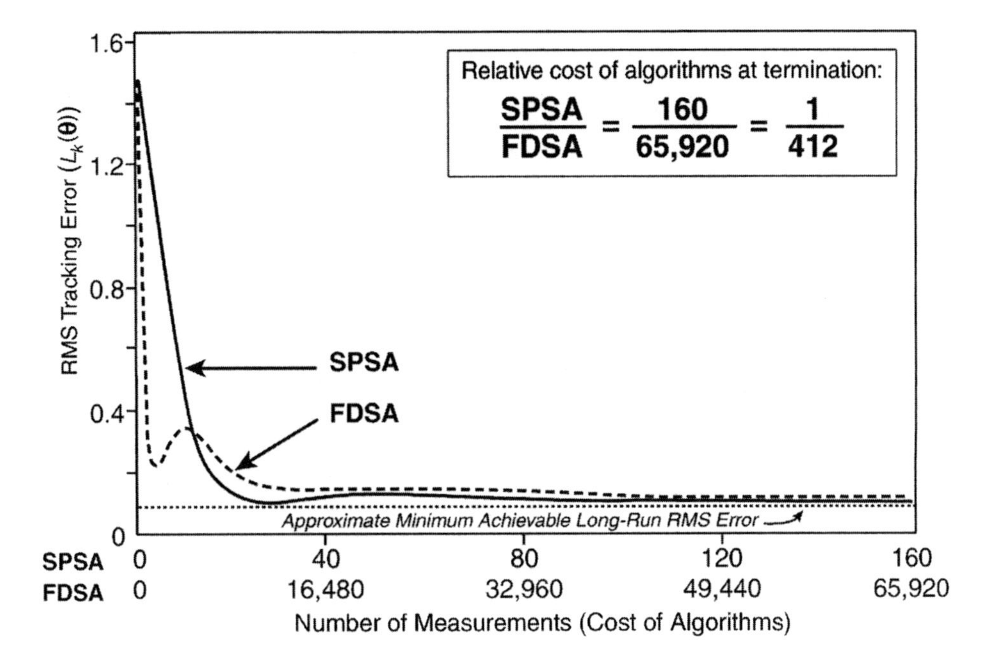

Figure 7.3. Comparison of efficiency for FDSA and SPSA in wastewater treatment problem. Points along the horizontal axis correspond to a common number of iterations for both FDSA and SPSA; 80 iterations are shown. Curves represent sample mean of 50 independent replications.

measurements for the full number of iterations, as shown in the box in the upper right corner. Note that this example is not a direct illustration of efficiency result (7.11) in Section 7.4 because of the time-varying loss functions and noncommon gains. Nevertheless, the relative performance here is essentially the same as predicted in (7.11). ❑

7.7 SOME EXTENSIONS: OPTIMAL PERTURBATION DISTRIBUTION; ONE-MEASUREMENT FORM; GLOBAL, DISCRETE, AND CONSTRAINED OPTIMIZATION

Sadegh and Spall (1998) consider the problem of choosing the best distribution for the Δ_k vector. Based on the asymptotic distribution result in Section 7.4, it is shown that the optimal distribution for the components of Δ_k is symmetric Bernoulli. This simple distribution has also proven effective in many finite-sample practical and simulation examples. The recommendation in step 1 of the algorithm description in Section 7.5 follows from these findings. It should be noted, however, that other distributions are sometimes desirable. Because the user has full control over this choice and since the generation of Δ_k represents a trivial cost toward the optimization, it may be worth evaluating other possibilities in some applications. For example, Maeda and De Figueiredo (1997) use a symmetric segmented uniform distribution (i.e., a uniform distribution with a section removed near zero to preserve the finiteness of inverse moments), in an application for robot control (this density function is depicted in Figure 7.2).

A *one*-measurement form of the SP gradient approximation is considered in Spall (1997). The gradient approximation has the form

$$\hat{g}_k(\hat{\theta}_k) = \begin{bmatrix} \dfrac{y(\hat{\theta}_k + c_k\Delta_k)}{c_k\Delta_{k1}} \\ \vdots \\ \dfrac{y(\hat{\theta}_k + c_k\Delta_k)}{c_k\Delta_{kp}} \end{bmatrix}. \tag{7.12}$$

Although the form above may seem strange in that it does not include explicit information related to the difference of function values, the form shares the nearly unbiased property of the standard two-measurement form in (7.2). In particular, via a Taylor expansion of $L(\hat{\theta}_k + c_k\Delta_k)$, it is found that $E[\hat{g}_k(\hat{\theta}_k)\,|\,\hat{\theta}_k] = g(\hat{\theta}_k) + O(c_k^2)$ (Exercise 7.12). Although it is shown in Spall (1997) that the standard two-measurement form is usually more efficient (in terms of the total number of loss function measurements to obtain a given level of accuracy in the θ iterate), there may be advantages to the one-measurement form in real-time operations. Such real-time applications include target tracking and feedback control, where the underlying system dynamics may change too rapidly to get a credible gradient estimate with two successive measurements.

There are several types of averaging that have been used in the context of SPSA. Dippon and Renz (1997) explore *iterate* averaging (analogous to the idea discussed in Subsection 4.5.3 for root-finding SA), showing that the approach can achieve near-optimal asymptotic mean-squared errors for the iterate average $\overline{\theta}_k$. However, as discussed in Spall (2000), this approach may perform relatively poorly in practical finite-sample problems. This is also discussed in Subsection 4.5.3 in the context of root-finding SA.

Aside from iterate averaging, two methods for averaging the gradient estimate $\hat{g}_k(\hat{\theta}_k)$ have been considered with the aim of reducing the variability of the input at each iteration. The first is simply to average several gradient approximations at *each iteration* (at the cost of additional function measurements). This is mentioned in step 3 of Subsection 7.5.1 and discussed in Spall (1992). The other method (discussed in Spall and Cristion, 1994) is gradient smoothing in a manner analogous to momentum in the neural network literature (momentum is discussed in Section 5.2 in the context of the backpropagation algorithm). Both the simple averaging and the smoothing ideas are aimed at coping with inaccuracies in the gradient estimate resulting from the simultaneous perturbation aspect and the noise in the loss measurements.

The use of SPSA for *global* minimization among multiple local minima is discussed in Maryak and Chin (2001). One of their approaches relies on injecting Monte Carlo noise in the right-hand side of the basic SPSA updating step in (7.1). This approach is a common way of converting SA algorithms to global optimizers (Yin, 1999). Maryak and Chin (2001) also show that basic SPSA *without* injected noise (i.e., eqn. (7.1) without modification) may, under certain conditions, be a global optimizer. Formal justification for this important result follows because the random error in the SP gradient approximation acts in a way that is statistically equivalent to the injected noise mentioned above. Although the injected noise approach is relatively well known, basic SPSA as a global optimizer has a faster asymptotic rate of convergence and reduces the number of user-specified coefficients. Section 8.4 considers in greater detail the subject of global optimization via SA, including the use of SPSA without injected noise.

Discrete optimization problems (where $\boldsymbol{\theta}$ may take on discrete or combined discrete/continuous values) are discussed in Gerencsér et al. (1999). Discrete SPSA relies on a fixed-gain (constant a_k and c_k) version of the standard SPSA method. The loss function is assumed to be convex and, in the process of optimization, is temporarily extended to a unique, continuous convex function. The continuous extension is then used to form a gradient approximation, which is used in a fixed-gain SA algorithm. The parameter estimates produced are constrained to lie on the discrete-valued grid.

The problem of constrained (equality and inequality) optimization with SPSA is considered in Sadegh (1997) and Fu and Hill (1997) using a projection approach. The projection algorithm is a direct analogue of the constrained root-finding SA algorithm in Section 4.1:

$$\hat{\boldsymbol{\theta}}_{k+1} = \Psi_{\Theta}[\hat{\boldsymbol{\theta}}_k - a_k \hat{\boldsymbol{g}}_k(\hat{\boldsymbol{\theta}}_k)],$$

where $\Psi_{\Theta}[\cdot]$ is the mapping that projects any point not in the constraint domain Θ to a new point inside Θ. While the projection approach has an elegant mathematical form, it is quite restricted in the types of constraints that can be handled in practical problems. Essentially, the constraints must be represented explicitly in a "nice" way so as to facilitate the mapping of a constraint violation in $\boldsymbol{\theta}$ to the nearest valid point. A common implementation of projections is to problems with hypercube constraints, where the individual components of $\boldsymbol{\theta}$ are bounded above and below by user-specified constants.

An alternative approach to constrained optimization is given in Wang and Spall (1999). This approach is based on altering the loss function to include a penalty term. In particular, at iteration k, $L(\boldsymbol{\theta})$ is replaced by a modified loss function

$$L(\boldsymbol{\theta}) + r_k P(\boldsymbol{\theta}),$$

where r_k is an increasing sequence of positive scalars ($r_k \to \infty$) and $P(\boldsymbol{\theta})$ is a penalty function that takes on large positive values when the constraints are violated. In many practical problems constraints are only implicit in $\boldsymbol{\theta}$, and the penalty function approach is well suited to handle such cases. For example, if it is required that $0 \leq f(\boldsymbol{\theta}) \leq 1$ for some function $f(\cdot)$, then $P(\boldsymbol{\theta})$ can be designed to take on very large values when $\boldsymbol{\theta}$ is such that $f(\boldsymbol{\theta})$ is outside of [0, 1]. This does not require explicit constraints on the components of $\boldsymbol{\theta}$. Although the penalty function method has broad applicability, the implementation is sometimes a challenge. In general, the specific choice of r_k and $P(\boldsymbol{\theta})$ dramatically affect the performance of the method.

7.8 ADAPTIVE SPSA

We now discuss an adaptive approach based on estimating the Hessian matrix of $L(\boldsymbol{\theta})$ (or, equivalently, the Jacobian matrix of $\boldsymbol{g}(\boldsymbol{\theta})$). The three subsections in this section introduce the adaptive algorithm, discuss some practical implementation issues, and summarize the theory on efficiency.

7.8.1 Introduction and Basic Algorithm

Using the simultaneous perturbation idea, this section presents a general adaptive SPSA (ASP) approach that is based on a simple method for estimating the Hessian (or Jacobian) matrix while, concurrently, estimating the primary parameters of interest ($\boldsymbol{\theta}$). The ASP approach produces a stochastic analogue to the deterministic Newton–Raphson algorithm considered in Section 1.4, leading to a recursion that is optimal or near-optimal in its rate of convergence and asymptotic error. The approach applies in both the gradient-free setting emphasized in this chapter and in the root-finding/stochastic gradient-based (Robbins–Monro) setting considered in Chapters 4 and 5. Like the standard (first-order) SPSA algorithm, the ASP algorithm requires only a small number of loss function (or gradient, if relevant) measurements per iteration—independent of the problem dimension—to adaptively estimate the Hessian and parameters of primary interest.

There are other second-order SA approaches. A more complete discussion on related work is given in Spall (2000). In the gradient-free setting, Fabian (1971) forms estimates of the gradient and Hessian for a Newton–Raphson-type SA algorithm by using, respectively, a finite-difference approximation and a set of differences of finite-difference approximations. This leads to $O(p^2)$ measurements of L per update of the $\boldsymbol{\theta}$ estimate, which is extremely costly when p is large. Ruppert (1985) assumes that direct measurements of the gradient $\boldsymbol{g}(\cdot)$ are available, as in the stochastic gradient setting of Chapter 5. He then forms a Hessian estimate by taking a finite difference of gradient measurements. Hence, $O(p)$ measurements of $\boldsymbol{g}(\cdot)$ are required for each update step in estimating $\boldsymbol{\theta}$. A type of second-order optimal convergence for SA is reported in Ruppert (1991), Polyak and Juditsky (1992), and Dippon and Renz (1997) based on the idea of iterate averaging. This was briefly discussed in Section 7.7.

The algorithm here is in the spirit of adaptive (matrix) gain SA algorithms such as those considered in Benveniste et al. (1990, Chaps. 3 and 4) in that a matrix gain is estimated concurrently with an estimate of the parameters of interest. It differs, however, in the relative lack of prior information required (especially in the gradient-free case) and in the small number of loss and/or gradient measurements needed per iteration. Because the algorithm is a stochastic analogue of the Newton–Raphson method, the algorithm provides a measure of transform invariance, as discussed in Section 1.4. That is, the method automatically scales the $\boldsymbol{\theta}$ updates when there are significant differences in the magnitudes of the elements in $\boldsymbol{\theta}$.

The ASP approach is composed of two parallel recursions: one for $\boldsymbol{\theta}$ and one for the Hessian of $L(\boldsymbol{\theta})$. The two core recursions are, respectively,

$$\hat{\boldsymbol{\theta}}_{k+1} = \hat{\boldsymbol{\theta}}_k - a_k \bar{\bar{\boldsymbol{H}}}_k^{-1} \boldsymbol{G}_k(\hat{\boldsymbol{\theta}}_k), \quad \bar{\bar{\boldsymbol{H}}}_k = \boldsymbol{f}_k(\bar{\boldsymbol{H}}_k), \tag{7.13a}$$

$$\bar{\boldsymbol{H}}_k = \frac{k}{k+1}\bar{\boldsymbol{H}}_{k-1} + \frac{1}{k+1}\hat{\boldsymbol{H}}_k, \quad k = 0, 1, 2, \ldots, \tag{7.13b}$$

where a_k is a nonnegative scalar gain coefficient, $\boldsymbol{G}_k(\hat{\boldsymbol{\theta}}_k)$ is the input information related to $\boldsymbol{g}(\hat{\boldsymbol{\theta}}_k)$ (i.e., the gradient approximation $\hat{\boldsymbol{g}}_k(\hat{\boldsymbol{\theta}}_k)$ from (7.2) in the gradient-free case or the direct observation $\partial Q/\partial\boldsymbol{\theta}$ in the stochastic gradient case of Chapter 5), $\boldsymbol{f}_k$: $\mathbb{R}^{p\times p} \to$ {positive definite $p\times p$ matrices} is a mapping designed to cope with possible nonpositive-definiteness of $\bar{\boldsymbol{H}}_k$, and $\hat{\boldsymbol{H}}_k$ is a per-iteration estimate of the Hessian, $\boldsymbol{H} = \boldsymbol{H}(\boldsymbol{\theta})$, discussed below. (Note that at $k = 0$ in (7.13b), $\bar{\boldsymbol{H}}_{k-1} = \bar{\boldsymbol{H}}_{-1}$ is unspecified—and irrelevant—since the multiplier $k/(k+1) = 0$.)

Eqn. (7.13a) is a stochastic analogue of the Newton–Raphson algorithm. Eqn. (7.13b) is simply a recursive calculation of the sample mean of the per-iteration Hessian estimates.[2] Initialization of the two recursions is discussed in Subsection 7.8.2. Because $\boldsymbol{G}_k(\hat{\boldsymbol{\theta}}_k)$ has a known form, the parallel recursions in (7.13a, b) can be implemented once $\hat{\boldsymbol{H}}_k$ is specified. The remainder of this section focuses on two specific implementations of the ASP approach above: 2SPSA (second-order SPSA) for applications in the gradient-free case and 2SG (second-order stochastic gradient) for applications in the gradient-based case (as in Chapter 5).

---

[2]It is also possible to use a weighted average or sliding window method (where only the most recent $\hat{\boldsymbol{H}}_k$ values are used in the recursion) to determine $\bar{\boldsymbol{H}}_k$. Formal convergence of $\bar{\boldsymbol{H}}_k$ may still hold under such weighting provided that the analogue to (A.10) and (A.13) in the proof of Theorem 2a in Spall (2000) holds.

We now present the per-iteration Hessian estimate $\hat{\boldsymbol{H}}_k$. As with the basic first-order SPSA algorithm, let c_k be a positive scalar (decaying to zero for formal convergence) and $\boldsymbol{\Delta}_k \in \mathbb{R}^p$ be a user-generated mean-zero random vector; conditions on c_k, $\boldsymbol{\Delta}_k$, and other relevant quantities are given in Spall (2000). These conditions are close to those of basic SPSA (e.g., $\boldsymbol{\Delta}_k$ being a vector of independent Bernoulli ± 1 random variables satisfies these conditions, but a vector of uniformly or normally distributed random variables does not). Examples of valid gain sequences are given below. The formula for estimating the Hessian at each iteration is

$$\hat{\boldsymbol{H}}_k = \tfrac{1}{2}\left\{\frac{\delta \boldsymbol{G}_k}{2c_k}\left[\Delta_{k1}^{-1}, \Delta_{k2}^{-1}, \ldots, \Delta_{kp}^{-1}\right] + \left(\frac{\delta \boldsymbol{G}_k}{2c_k}\left[\Delta_{k1}^{-1}, \Delta_{k2}^{-1}, \ldots, \Delta_{kp}^{-1}\right]\right)^T\right\}, \quad (7.14)$$

where

$$\delta \boldsymbol{G}_k = \boldsymbol{G}_k^{(1)}(\hat{\boldsymbol{\theta}}_k + c_k \boldsymbol{\Delta}_k) - \boldsymbol{G}_k^{(1)}(\hat{\boldsymbol{\theta}}_k - c_k \boldsymbol{\Delta}_k),$$

and $\boldsymbol{G}_k^{(1)}$ may or may not equal $\boldsymbol{G}_k$, depending on the setting. (The form in (7.14) is equivalent to the "vector divide" form in Spall, 2000.) In particular, for 2SPSA, there are advantages to using a *one-sided* gradient approximation in order to reduce the total number of function evaluations (vs. the standard two-sided form in (7.2)), while for 2SG, usually $\boldsymbol{G}_k^{(1)} = \boldsymbol{G}_k = \partial Q / \partial \boldsymbol{\theta}$, as in Chapter 5.

Note that all elements of $\hat{\boldsymbol{\theta}}_k$ are varied simultaneously (and randomly) in forming $\hat{\boldsymbol{H}}_k$, as opposed to the finite-difference forms in, for example, Fabian (1971) and Ruppert (1985), where the elements of $\boldsymbol{\theta}$ are changed deterministically one at a time. The symmetrizing operation in (7.14) (the multiple 1/2 and the indicated sum) is convenient in the optimization case being emphasized here to maintain a symmetric Hessian estimate in finite samples. In the general root-finding case, where $\boldsymbol{H}(\boldsymbol{\theta})$ represents a Jacobian matrix, the symmetrizing operation should not be used when the Jacobian is not necessarily symmetric.

For 2SPSA, the core gradient approximation $\boldsymbol{G}_k(\hat{\boldsymbol{\theta}}_k)$ is taken as $\hat{\boldsymbol{g}}_k(\hat{\boldsymbol{\theta}}_k)$ in eqn. (7.2), requiring two measurements of L, $y(\hat{\boldsymbol{\theta}}_k + c_k\boldsymbol{\Delta}_k)$ and $y(\hat{\boldsymbol{\theta}}_k - c_k\boldsymbol{\Delta}_k)$. In addition to this gradient approximation, these two measurements are employed toward generating the one-sided gradient approximations $\boldsymbol{G}_k^{(1)}(\hat{\boldsymbol{\theta}}_k \pm c_k\boldsymbol{\Delta}_k)$ used in forming $\hat{\boldsymbol{H}}_k$. Two additional measurements $y(\hat{\boldsymbol{\theta}}_k \pm c_k\boldsymbol{\Delta}_k + \tilde{c}_k\tilde{\boldsymbol{\Delta}}_k)$ are used in generating the one-sided approximations as follows:

$$G_k^{(1)}(\hat{\boldsymbol{\theta}}_k \pm c_k\boldsymbol{\Delta}_k) = \frac{y(\hat{\boldsymbol{\theta}}_k \pm c_k\boldsymbol{\Delta}_k + \tilde{c}_k\tilde{\boldsymbol{\Delta}}_k) - y(\hat{\boldsymbol{\theta}}_k \pm c_k\boldsymbol{\Delta}_k)}{\tilde{c}_k}\begin{bmatrix} \tilde{\Delta}_{k1}^{-1} \\ \tilde{\Delta}_{k2}^{-1} \\ \vdots \\ \tilde{\Delta}_{kp}^{-1} \end{bmatrix}, \quad (7.15)$$

with $\tilde{\boldsymbol{\Delta}}_k = [\tilde{\Delta}_{k1}, \tilde{\Delta}_{k2}, \ldots, \tilde{\Delta}_{kp}]^T$ generated in the same statistical manner as $\boldsymbol{\Delta}_k$, but independently of $\boldsymbol{\Delta}_k$ (in particular, choosing $\tilde{\Delta}_{ki}$ as independent Bernoulli ± 1 random variables is a valid—but not necessary—choice), and with $\tilde{c}_k$ satisfying conditions similar to c_k.

Let us summarize some examples of gains that satisfy the conditions in Spall (2000) for convergence and asymptotic normality of 2SPSA and 2SG. For both implementations, we can take a_k and c_k in the form given in Theorem 7.2 of Section 7.4. For 2SPSA, we also have $\tilde{c}_k = \tilde{c}/(k+1)^\gamma$, $\tilde{c} > 0$. With these gain forms, examples of specific coefficient values for 2SPSA are: $\alpha = 0.602$, $\gamma = 0.101$ or $\alpha = 1$, $\gamma = 1/6$. For 2SG, $1/2 < \alpha \leq 1$ is valid together with $0 < \gamma < 1/2$.

To illuminate the underlying simplicity of ASP, let us now provide some informal motivation for the $\hat{\boldsymbol{H}}_k$ form in eqn. (7.14). The arguments below are formalized in the theorems of Spall (2000). Suppose that L is four-times continuously differentiable in a neighborhood of $\hat{\boldsymbol{\theta}}_k$. Then, simple Taylor series arguments show that

$$E(\delta\boldsymbol{G}_k \mid \hat{\boldsymbol{\theta}}_k, \boldsymbol{\Delta}_k) = \boldsymbol{g}(\hat{\boldsymbol{\theta}}_k + c_k\boldsymbol{\Delta}_k) - \boldsymbol{g}(\hat{\boldsymbol{\theta}}_k - c_k\boldsymbol{\Delta}_k) + O(c_k^3)$$

$$\equiv \delta\boldsymbol{g}_k + O(c_k^3) \quad (O(c_k^3) = \boldsymbol{0} \text{ in the 2SG case}), \quad (7.16)$$

where this result is immediate in the 2SG case and follows as in Spall (1992, Lemma 1) by a Taylor series argument in the 2SPSA case (where the $O(c_k^3)$ term is the difference of the two $O(c_k^2)$ bias terms in the one-sided SP gradient approximations; see Exercise 7.13). Let δG_{ki} and δg_{ki} be the ith components of $\delta\boldsymbol{G}_k$ and $\delta\boldsymbol{g}_k$. By an expansion of each of $\boldsymbol{g}(\hat{\boldsymbol{\theta}}_k \pm c_k\boldsymbol{\Delta}_k)$ for any i, j,

$$E\left(\frac{\delta G_{ki}}{2c_k\Delta_{kj}}\middle|\hat{\boldsymbol{\theta}}_k, \boldsymbol{\Delta}_k\right) = E\left(\frac{\delta g_{ki}}{2c_k\Delta_{kj}}\middle|\hat{\boldsymbol{\theta}}_k, \boldsymbol{\Delta}_k\right) + O(c_k^2)$$

$$= H_{ij}(\hat{\boldsymbol{\theta}}_k) + \sum_{\ell \neq j} H_{i\ell}(\hat{\boldsymbol{\theta}}_k)\frac{\Delta_{k\ell}}{\Delta_{kj}} + O(c_k^2),$$

where H_{ij} denotes the ijth component of $\boldsymbol{H}$ and the $O(c_k^2)$ term in the second line absorbs higher-order terms in the expansion of δg_{ki}. Then, since $E(\Delta_{k\ell}/\Delta_{kj}) = 0$ for all $j \neq \ell$ by the assumptions for $\boldsymbol{\Delta}_k$,

$$E\left(\frac{\delta G_{ki}}{2c_k\Delta_{kj}}\middle|\hat{\boldsymbol{\theta}}_k\right)=H_{ij}(\hat{\boldsymbol{\theta}}_k)+O(c_k^2),$$

implying that the Hessian estimate $\hat{\boldsymbol{H}}_k$ is nearly unbiased with the bias disappearing at rate $O(c_k^2)$. The addition operation in (7.14) simply forces the per-iteration estimate to be symmetric.

7.8.2 Implementation Aspects of Adaptive SPSA

The two recursions in (7.13a, b) are the foundation for the ASP approach. However, as is typical in all stochastic algorithms, the specific implementation details are important. Eqns. (7.13a, b) do not fully define these details. The five points below have been found important in making ASP perform well in practice. More complete guidelines are given in Spall (2000).

A. **$\boldsymbol{\theta}$ and $\boldsymbol{H}$ initialization.** Typically, eqn. (7.13a) is initialized at some $\hat{\boldsymbol{\theta}}_0$ believed to be near $\boldsymbol{\theta}^*$. One may wish to run standard first-order SA (i.e., (7.13a) without $\bar{\bar{\boldsymbol{H}}}_k^{-1}$) or some other "rough" optimization approach (e.g., the random search methods of Chapter 2) for some period to move the initial $\boldsymbol{\theta}$ for ASP closer to $\boldsymbol{\theta}^*$. The user has the option of initializing or not initializing the $\boldsymbol{H}$ recursion with prior information. If prior information is available, the recursion may be initialized at some value, say, $\bar{\boldsymbol{H}}_0 = \rho \boldsymbol{I}_{p\times p}$, $\rho \geq 0$, or some other positive semidefinite matrix reflecting available prior information (e.g., if one knows that the $\boldsymbol{\theta}$ elements will have very different magnitudes, then $\bar{\boldsymbol{H}}_0$ may be chosen to approximately scale for the differences). Without initializing based on prior information, $\bar{\boldsymbol{H}}_0$ may be computed directly as $\hat{\boldsymbol{H}}_0$.

B. **Numerical issues and choice of $\bar{\bar{\boldsymbol{H}}}_k$.** Since $\bar{\boldsymbol{H}}_k$ may not be positive definite, especially for small k (even if $\bar{\boldsymbol{H}}_0$ is positive definite), it is generally recommended that $\bar{\boldsymbol{H}}_k$ in (7.13b) not be used directly in (7.13a). Hence, as shown in (7.13a), it is recommended that $\bar{\boldsymbol{H}}_k$ be replaced by another matrix $\bar{\bar{\boldsymbol{H}}}_k$ that is closely related to $\bar{\boldsymbol{H}}_k$. Spall (2000) discusses ways in which $\bar{\boldsymbol{H}}_k$ can be transformed to obtain $\bar{\bar{\boldsymbol{H}}}_k$.

C. **Gradient/Hessian averaging.** At each iteration, it may be desirable to average several $\hat{\boldsymbol{H}}_k$ and $\boldsymbol{G}_k(\hat{\boldsymbol{\theta}}_k)$ values despite the additional cost. This may be especially true in a high-noise environment.

D. **Gain selection.** The principles outlined in Section 7.5 are useful here as well for practical selection of the gains a_k, c_k, and, in the 2SPSA case, $\tilde{c}_k$. For 2SPSA and 2SG, the critical gain a_k can simply be picked as $1/(k+1)$ to achieve asymptotic near-optimality or optimality, respectively, although this may not be ideal in practical finite-sample problems (a slower decay is usually preferred in practice). In the 2SPSA case, it may be desirable to choose $\tilde{c}_k$ larger than c_k to enhance numerical stability.

E. **Blocking.** At each iteration, block "bad" steps if the new estimate for $\boldsymbol{\theta}$ fails a certain criterion (i.e., set $\hat{\boldsymbol{\theta}}_{k+1} = \hat{\boldsymbol{\theta}}_k$ in going from k to $k+1$). $\bar{\boldsymbol{H}}_k$ should typically continue to be updated even if $\hat{\boldsymbol{\theta}}_{k+1}$ is blocked. The most obvious blocking applies when $\boldsymbol{\theta}$ must satisfy constraints; an updated value may be blocked or modified if a constraint is violated. There are two methods (say, E.1 and E.2) that one might use to implement blocking when constraints are not the limiting factor, with E.1 based on $\hat{\boldsymbol{\theta}}_k$ and $\hat{\boldsymbol{\theta}}_{k+1}$ directly and E.2 based on loss measurements. Both of E.1 and E.2 may be implemented in a given application. In E.1, the step from $\hat{\boldsymbol{\theta}}_k$ to $\hat{\boldsymbol{\theta}}_{k+1}$ is blocked if $\left\|\hat{\boldsymbol{\theta}}_{k+1} - \hat{\boldsymbol{\theta}}_k\right\| > \text{tol}_{\boldsymbol{\theta}}$, where the norm is any convenient distance measure and $\text{tol}_{\boldsymbol{\theta}} > 0$ is some reasonable maximum distance (tolerance) to cover in one step. The rationale behind E.1 is that a well-behaving algorithm should be moving toward the solution in a smooth manner and very large steps are indicative of potential divergence. The second method, E.2, is based on blocking the step if $y(\hat{\boldsymbol{\theta}}_{k+1})$ (or an average of y values) is not sufficiently near (or lower) than $y(\hat{\boldsymbol{\theta}}_k)$. In a setting where the noise in the loss measurements tends to be large (say, larger than the allowable difference between $L(\boldsymbol{\theta}^*)$ and $L(\hat{\boldsymbol{\theta}}_{\text{final}})$), it is generally undesirable to use E.2 due to the large number of loss measurements that must be averaged to obtain meaningful information about the relative old and new loss values.

Let us close this subsection with a few summary comments about the implementation aspects above. Without the second blocking procedure (E.2) in use, 2SPSA requires *four* measurements y per iteration, *regardless* of the dimension p: two for the standard $\boldsymbol{G}_k(\cdot) = \hat{\boldsymbol{g}}_k(\cdot)$ estimate and two new values for the one-sided SP gradients $\boldsymbol{G}_k^{(1)}(\cdot)$. For 2SG, *three* gradient measurements $\boldsymbol{G}_k(\cdot) = \partial Q/\partial\boldsymbol{\theta}$ are needed, again independent of p. If the second blocking procedure (E.2) is used, one or more additional y measurements are needed for both 2SPSA and 2SG. The use of gradient/Hessian averaging (C) also increases the number of loss or gradient evaluations. E.1 may be used anytime while E.2 is more appropriate in a low- or no-noise setting. While E.1 helps to prevent divergence, it lacks direct insight into whether the loss function is improving. E.2 does provide that insight but requires additional y measurements, the number of which might grow prohibitive in a high-noise setting.

7.8.3 Theory on Convergence and Efficiency of Adaptive SPSA

Spall (2000) presents asymptotic theory showing the a.s. convergence of $\hat{\boldsymbol{\theta}}_k$ and $\bar{\boldsymbol{H}}_k$ to $\boldsymbol{\theta}^*$ and $\boldsymbol{H}(\boldsymbol{\theta}^*)$, respectively, in both the 2SPSA and 2SG settings. Further, conditions are shown for the asymptotic normality of the standardized quantity $k^{\beta/2}(\hat{\boldsymbol{\theta}}_k - \boldsymbol{\theta}^*)$, $\beta > 0$. This normality is then used to analyze the limiting efficiency of the general ASP approach. Let $\boldsymbol{\mu}$ and $\boldsymbol{\Sigma}$ be the mean vector and covariance matrix in the asymptotic distribution for $k^{\beta/2}(\hat{\boldsymbol{\theta}}_k - \boldsymbol{\theta}^*)$. Then the large-sample root-mean-squared (RMS) error of $\hat{\boldsymbol{\theta}}_k$ based on the asymptotic distribution is $\sqrt{[\boldsymbol{\mu}^T\boldsymbol{\mu} + \text{trace}(\boldsymbol{\Sigma})]/k^{\beta}}$. As in the standard SPSA and stochastic gradient algorithms, the best (greatest) β is $\beta = 2/3$ for the 2SPSA case and $\beta = 1$ for the 2SG case (both follow by setting $\alpha = 1$ in the standard gain form $a_k = a/(k+1+A)^{\alpha}$).

To characterize the asymptotic efficiency results based on the asymptotic distributions, let RMS^*_{SPSA} and RMS^*_{SG} represent the *best possible* RMS errors of the normalized $\hat{\boldsymbol{\theta}}_k$ when using the basic SPSA (gradient-free) and stochastic gradient approaches. The best possible RMS errors require gain sequences based on exact information on the third derivative of L (basic SPSA) and the second derivatives of L (stochastic gradient) (Dippon and Renz, 1997). This information, of course, is generally unavailable. Hence, RMS^*_{SPSA} and RMS^*_{SG} represent ideal values that are usually unavailable in practice. (As mentioned in Section 4.4, RMS^*_{SG} is derived from the inverse Fisher information matrix; this matrix is discussed in detail in Section 13.3.) Letting RMS_{2SPSA} and RMS_{2SG} denote the corresponding large-sample RMS errors for the normalized 2SPSA and 2SG estimates when $a_k = 1/(k+1)$ (i.e., $\sqrt{[\boldsymbol{\mu}^T\boldsymbol{\mu} + \text{trace}(\boldsymbol{\Sigma})]/k^{\beta}}$ from above), we find

$$\frac{RMS_{\text{2SPSA}}}{RMS^*_{\text{SPSA}}} < 2 \quad \text{and} \quad \frac{RMS_{\text{2SG}}}{RMS^*_{\text{SG}}} = 1. \tag{7.17}$$

The interpretation of (7.17) is that for the SPSA setting, the 2SPSA algorithm produces an estimate with an asymptotic RMS error *less than* twice the error from the best possible (infeasible) algorithm (requiring the above-mentioned third derivative knowledge). For the stochastic gradient setting, the 2SG algorithm produces an error that is asymptotically *equal* to the best possible. Numerical studies in Spall (2000) show the power of the 2SPSA and 2SG approaches. Luman (2000) applies 2SPSA in a simulation-based optimization problem and demonstrates the improvement possible over basic SPSA when there are very different magnitudes for the elements in $\boldsymbol{\theta}$ (i.e., an illustration of the above-mentioned transform invariance property).

While the ASP approach (via the 2SPSA and 2SG implementations) can be very powerful, there are no guarantees that the small-sample performance will be superior to the basic SPSA or stochastic gradient algorithms. It may take many iterations to accumulate the information needed to produce a credible Hessian estimate, especially in the 2SPSA case where only noisy loss measurements are available. (Zhu and Spall, 2002, present a modification to cope with finite-sample concerns.) Further, as with the deterministic Newton–Raphson algorithm, ASP may be more "brittle" in the sense of being less robust to violations of the formal conditions for convergence. Nevertheless, one of the most important practical implications of (7.17) is that for $\hat{\boldsymbol{\theta}}_0$ sufficiently close to $\boldsymbol{\theta}^*$, 2SPSA or 2SG (as appropriate) will yield a good (or best for 2SG) large-sample solution based on the simple choice of $a_k = 1/(k+1)$. This helps alleviate the potentially nettlesome issue of gain selection subject to some of the practical tips in Subsection 7.8.2.

7.9 CONCLUDING REMARKS

This chapter has described the simultaneous perturbation stochastic approximation approach for stochastic search and optimization. The SPSA method rests on the idea of changing all the parameters in the problem simultaneously to construct gradient estimates. This contrasts with conventional one-at-a-time changes (such as the finite-difference method of Chapter 6). In high-dimensional problems of practical interest, such simultaneous changes admit an efficient implementation by greatly reducing the number of loss function evaluations required to carry out the optimization process.

As with other SA approaches, SPSA is explicitly designed to cope with noisy measurements of the loss function, although it is frequently applied in problems with noise-free measurements. Aside from differences in the mechanics of the algorithms, the formal justification for handling noisy measurements distinguishes SPSA from the random search methods of Chapter 2 and the simulated annealing and evolutionary computation methods of Chapters 8−10.

While the basic method applies in "smooth," unconstrained problems with no extraneous local minima, there are numerous extensions to the algorithm and theory to accommodate other cases. In particular, extensions exist for discrete optimization, constrained problems, and global optimization with multiple local minima. Interestingly, under some conditions, the *basic* SPSA algorithm is guaranteed to converge to a global solution. This appealing property implies that the algorithm automatically combines the broad search needed for global optimization with the localized search that is tied to gradient information (because the algorithm produces gradient information from the loss measurements). There is no need to explicitly transition from a "rough" global optimizer to a refined local optimizer as is common in practice.

Aside from the first-order SPSA algorithm, this chapter summarized a second-order SA approach that is a stochastic analogue of the Newton–Raphson method. This approach applies in either the standard SPSA setting, where only noisy loss function evaluations are available, or in the stochastic gradient-based/root-finding (Robbins–Monro) setting, where noisy gradient evaluations are available. The algorithm produces an asymptotically near-optimal or optimal root-mean-squared error for the iterate. This adaptive simultaneous perturbation algorithm is based on a relatively simple method for estimating the Hessian matrix of the loss function L. The Hessian estimation capability can also be used for computing the Fisher information matrix (see Section 13.3), a quantity important in parameter estimation, model selection, and experimental design.

Further general information and references on theory and applications of SPSA are available at *www.jhuapl.edu/SPSA* (also available through the book's Web site).

7.10 APPENDIX: CONDITIONS FOR ASYMPTOTIC NORMALITY

The conditions below are used in the asymptotic normality result of Section 7.4 (together with conditions B.1″–B.6″ stated in Section 7.3).

B.7″ The continuity and equicontinuity assumptions about $E[(\varepsilon_k^{(+)} - \varepsilon_k^{(-)})^2 | \Im_k]$ in Spall (1992, Prop. 2) hold. (These assumptions are automatically satisfied if the $\varepsilon_k^{(\pm)}$ are independent of $\Im_k$; equicontinuity is a stricter form of continuity.)

B.8″ $\boldsymbol{H}(\boldsymbol{\theta}^*)$ is positive definite where $\boldsymbol{H}(\boldsymbol{\theta})$ is the Hessian matrix of $L(\boldsymbol{\theta})$. Further, let λ_i denote the ith eigenvalue of $a\boldsymbol{H}(\boldsymbol{\theta}^*)$ (the a here is the a in a_k). If $\alpha = 1$, then $\beta < 2\min_i(\lambda_i)$.

B.9″ $E(\Delta_{ki}^2) \to \rho$, $E(\Delta_{ki}^{-2}) \to \rho'$, and $E[(\varepsilon_k^{(+)} - \varepsilon_k^{(-)})^2 | \Im_k] \to \rho''$ for strictly positive constants ρ, ρ', and ρ'' (a.s. in the latter case) as $k \to \infty$ (often, $E(\Delta_{ki}^2)$ and $E(\Delta_{ki}^{-2})$ will be *equal* to ρ and ρ', respectively).

EXERCISES

7.1 Consider the quadratic function $L(\boldsymbol{\theta}) = 2t_1^2 + t_2^2$, $\boldsymbol{\theta} = [t_1, t_2]^T$, and suppose that the measurements of $L(\boldsymbol{\theta})$ have no noise (i.e., $y(\boldsymbol{\theta}) = L(\boldsymbol{\theta})$ for all $\boldsymbol{\theta}$). Perform the following tasks (pencil-and-paper calculations will work fine):

(a) From an initial condition $[1, 1]^T$, with $a_0 = 0.1$, $c_0 = 1.0$, calculate what $\hat{\boldsymbol{\theta}}_1$ will be for all possible combinations of $\boldsymbol{\Delta}_0$ with each component of $\boldsymbol{\Delta}_0$ distributed as Bernoulli ± 1.

(b) Using the $L(\boldsymbol{\theta})$ values as the measure of performance, show that $\hat{\boldsymbol{\theta}}_1$ is guaranteed to be an improvement over $\hat{\boldsymbol{\theta}}_0$.

(c) Show by direct calculation that $\hat{\boldsymbol{g}}_0(\hat{\boldsymbol{\theta}}_0)$ is an unbiased estimator of $\boldsymbol{g}(\hat{\boldsymbol{\theta}}_0)$. (This lack of bias is a consequence of the quadratic $L(\boldsymbol{\theta})$.)

7.2 For $E(\Delta_{ki}/\Delta_{km})$ to exist in (7.3), show that the inverse moment $E(|1/\Delta_{km}|)$ must be finite when $E(\Delta_{ki}) = 0$.

7.3 For the skewed-quartic loss function in Example 6.6 (Section 6.7), compare SPSA with a valid perturbation distribution and two invalid perturbation distributions. In the valid case, let the perturbations (Δ_{ki}) be i.i.d. Bernoulli ± 1 and in the invalid cases, let the perturbations be i.i.d. uniform over

$[-\sqrt{3}, \sqrt{3}]$ and i.i.d. $N(0, 1)$. (Note that the valid and invalid perturbation distributions all have mean 0 and variance 1.) Let $p = 10$ and the gains be in the standard form $a_k = a/(k+1+A)^{\alpha}$ and $c_k = c/(k+1)^{\gamma}$. Suppose that the noise ε in the measurements $y(\boldsymbol{\theta}) = L(\boldsymbol{\theta}) + \varepsilon$ is i.i.d. $N(0, 0.1^2)$, $\hat{\boldsymbol{\theta}}_0 = [1, 1, ..., 1]^T$ (so $L(\hat{\boldsymbol{\theta}}_0) = 4.178$), and the gains satisfy $a = 0.5$, $A = 50$, $c = 0.1$, $\alpha = 0.602$, and $\gamma = 0.101$ (chosen via the gain-selection guidelines in Section 7.5). With no constraints imposed on $\boldsymbol{\theta}$, compare single replications of 2000 measurements for the valid and invalid implementations and comment on the relative performance of the implementations.

7.4 Consider the setting of Exercise 7.3 for the Bernoulli (valid) perturbations *with the exception* of imposing the hypercube constraint $\hat{\boldsymbol{\theta}}_k \in \Theta = [-1, 1]^{10}$ for all k. (The perturbed values $\hat{\boldsymbol{\theta}}_k \pm c_k \boldsymbol{\Delta}_k$ should not be constrained.) Use the appropriate two-sample t-test (Appendix B) to compare the terminal loss values from 20 independent replications with 2000 measurements in the unconstrained and constrained cases. (This exercise demonstrates the advantage of imposing constraints if possible.)

7.5 Prove (7.6) under the stated conditions in Section 7.2. (Hint: Recall from integration theory: $\int_D |f_1(x) f_2(x)| dx \le \sup_{x \in D} |f_1(x)| \int_D |f_2(x)| dx$ for two functions f_1 and f_2 and some domain of integration D.)

7.6 Prove inequality (7.7) and give two specific examples of exponent terms η and τ.

7.7 Consider the problem of minimizing $L(\boldsymbol{\theta}) = t_1^4 + t_1^2 + t_1 t_2 + t_2^2$, $\boldsymbol{\theta} = [t_1, t_2]^T$, where ε is i.i.d. $N(0, 1)$ noise. Define the segmented uniform distribution as in Figure 7.2, where the probability is evenly split that the random variable lies in either the upper interval (u, v) or the lower interval $(-v, -u)$, $v > u > 0$. Let $\hat{\boldsymbol{\theta}}_0 = [1, 1]^T$ and let each experiment below entail 100 replications. Use the gains $a_k = a/(k+1+A)^{0.602}$ and $c_k = c/(k+1)^{0.101}$ as in step 0 in Subsection 7.5.1.

(a) Compute the sample mean of the terminal loss values after 1000 iterations of SPSA when the Δ_{ki} are distributed Bernoulli ± 1 for all k, i. Use the coefficient values $a = 0.05$, $A = 100$, and $c = 1.0$. Next compute the same when the Δ_{ki} have a segmented uniform distribution with $u = 0.4091$ and $v = 1.4909$ (the strange values for u and v are so that the variance is unity, the same as the Bernoulli distribution). Use the coefficient values $a = 0.13$, $A = 100$, and $c = 0.55$. (Each of the two sets of gain coefficients is approximately optimized for the respective SPSA implementation.) Determine a P-value for comparing the two sample means.

(b) Perform the same analysis as in part (a) but with only 10 iterations of SPSA. Use the same gain coefficient values as in part (a), except that $A = 1$. Determine a P-value for comparing the two sample means.

(c) Offer an explanation for the conclusions you reach in parts (a) and (b).

7.8 Create a specific U-shaped density that satisfies (7.8) for an arbitrarily small $\tau > 0$.

Note: Use Bernoulli ±1 perturbations for the components of Δ_k in the numerical studies among the exercises below.

7.9 Given a desirable step size of 0.1 for all elements of $\boldsymbol{\theta}$, use the semiautomatic method of Subsection 7.5.2 to find reasonable gains for SPSA as applied to the loss function $L(\boldsymbol{\theta}) = (\boldsymbol{\theta}^T\boldsymbol{\theta}/5 - 25)^2$, with $p = 10$. Suppose that 1000 noisy loss measurements are used in the search process and $\hat{\boldsymbol{\theta}}_0 = [6, 6,\ldots, 6]^T$.

(a) Find the values of c, A, and a when ε is i.i.d. $N(0, 2^2)$.

(b) Find the values of c, A, and a when $\varepsilon = \varepsilon(\boldsymbol{\theta}) = [\boldsymbol{\theta}^T, 1]\boldsymbol{V}$ and $\boldsymbol{V}$ is an i.i.d. vector with distribution $N(\mathbf{0}, 0.1^2\boldsymbol{I}_{11})$.

(c) Run one replication of SPSA for the noise and gain combination in part (a) and one replication for part (b). How do the loss values at each of the two terminal $\boldsymbol{\theta}$ values compare with $L(\hat{\boldsymbol{\theta}}_0)$ and $L(\boldsymbol{\theta}^*)$ for any $\boldsymbol{\theta}^* \in \Theta^*$? ($\boldsymbol{\theta}^*$ is not unique.)

7.10 Consider the loss function and noise of Exercise 7.9(a):

(a) Use basic SPSA with semiautomatic gains as found in Exercise 7.9(a). Determine the mean terminal loss value and approximate 95 percent confidence interval (using the standard t-distribution approach of Appendix B) based on 40 independent replications of 1000 noisy loss measurements.

(b) Repeat part (a) except for using gains that are manually "tuned" to the problem. Compare the results with the results in part (a).

7.11 There is no unique minimum for the loss function in Exercise 7.9.

(a) Identify Θ^*.

(b) Identify two distinct initial conditions having the same distance to the nearest point in Θ^* as $\hat{\boldsymbol{\theta}}_0 = [6, 6,\ldots, 6]^T$ does to its nearest point.

(c) Run one replication of SPSA for each of the three initial conditions in part (b) (the two new ones plus the original one used in Exercise 7.9). Use 1000 loss measurements with noise as in Exercise 7.9(a) and use the same gains a_k and c_k in the three runs (you may use gains as in Exercise 7.9(a)). How do the three terminal loss values compare and how do the corresponding terminal $\boldsymbol{\theta}$ estimates relate to the initial conditions and to Θ^*?

7.12 Based on the conditions given for (7.5) and (7.6) for the two-measurement gradient approximation, show that the one-measurement gradient approximation in (7.12) has an $O(c_k^2)$ bias (same order as the standard two-measurement form).

7.13 Establish the $O(c_k^3)$ bias in (7.16) when $\boldsymbol{G}_k^{(1)}$ is the one-sided SP gradient approximation in (7.15). (Hint: First, write down an expression for the $O(c_k^2)$ bias in $\boldsymbol{G}_k^{(1)}(\boldsymbol{\theta})$ using arguments such as in Section 7.2. Then, analyze the difference of the two biases for each of $\boldsymbol{G}_k^{(1)}(\hat{\boldsymbol{\theta}}_k \pm c_k\Delta_k)$.)

7.14 Consider the loss function

$$L(\boldsymbol{\theta}) = \sum_{i=1}^{10} i t_i + \prod_{i=1}^{10} t_i^{-1}, \; t_i > 0 \text{ for all } i,$$

where $\boldsymbol{\theta} = [t_1, t_2, \ldots, t_{10}]^T$. Suppose that $\varepsilon = \varepsilon(\boldsymbol{\theta}) = [\boldsymbol{\theta}^T, 1]V$, V is i.i.d. $N(\mathbf{0}, 0.001^2 \boldsymbol{I}_{11})$, and that $\hat{\boldsymbol{\theta}}_0 = 1.1 \times \boldsymbol{\theta}^*$. Carry out the following tasks:

(a) Determine $\boldsymbol{\theta}^*$ by whatever means desired (a deterministic problem since $L(\boldsymbol{\theta})$ is known).

(b) Run basic SPSA with the noise included in the loss measurements. Use the standard gain form and $a = 0.01$, $A = 1000$, $c = 0.015$, $\alpha = 0.602$, and $\gamma = 0.101$ (these gains reflect tuning to achieve good performance). Determine the mean terminal loss value and approximate 90 percent confidence interval (using the standard t-distribution approach of Appendix B) based on 10 replications of 20,000 noisy loss measurements and $\Theta = (0, \infty)^{10}$. Comment on the performance relative to $\hat{\boldsymbol{\theta}}_0$ and $\boldsymbol{\theta}^*$ (e.g., use a measure such as $L_{\text{norm}}(\boldsymbol{\theta})$ in Example 7.3).

(c) For the noisy case of (b), run the 2SPSA approach with a naïve gain $a_k = 1/(k+1)$. Use c_k as in part (b), $\tilde{c} = 2c$, and $\text{tol}_{\theta} = 0.2$. Initialize the Hessian estimate as $\overline{\boldsymbol{H}}_0 = 500\boldsymbol{I}_{10}$. Report results as in part (b) and compare with the performance of the tuned SPSA algorithm in part (b). Comment on the performance relative to $\hat{\boldsymbol{\theta}}_0$ and $\boldsymbol{\theta}^*$.

CHAPTER 8

ANNEALING-TYPE ALGORITHMS

This chapter is devoted to a class of algorithms aimed at the global optimization problem. The common theme in these algorithms is the principle of *annealing*, where the magnitudes of random perturbations are reduced—annealed—in a controlled manner. These random perturbations are injected in a Monte Carlo fashion. The annealing is designed to enhance the likelihood of avoiding local minima en route to a global minimum of $L(\boldsymbol{\theta})$. The injected randomness helps prevent premature convergence to a local minimum by providing a greater "jumpiness" to the algorithm. The term *annealing* comes from analogies to the controlled cooling of physical substances to achieve a type of optimal state for the substance.

This chapter considers two classes of annealing algorithms. Sections 8.1–8.3 discuss the popular simulated annealing algorithm, which is derived from a probability expression (the Boltzmann–Gibbs distribution) governing the energy state of a system at a fixed temperature. The second class of annealing algorithms, considered in Section 8.4, is based on the principles of stochastic approximation. Section 8.5 contains a summary and conclusions and Section 8.6 is an appendix that provides some theoretical justification for the simulated annealing algorithm via connections to stochastic approximation.

8.1 INTRODUCTION TO SIMULATED ANNEALING AND MOTIVATION FROM THE PHYSICS OF COOLING

The simulated annealing (SAN) algorithm continues in the spirit of the stochastic approximation (SA) algorithms of Chapters 6 and 7 in working with only loss function measurements (versus requiring gradient information). Even more generally, it applies in *both* problems of continuous $\boldsymbol{\theta}$ and discrete $\boldsymbol{\theta}$, as introduced in Chapter 1 (although the gradient was not used in the methods of Chapters 6 and 7, it was still assumed to exist). SAN was originally developed for discrete optimization problems, but more recently has found application in continuous optimization problems of the type emphasized in most of the previous chapters. The algorithm is designed to traverse local minima en route to a global minimum of $L = L(\boldsymbol{\theta})$.

The term *annealing* comes from analogies to the cooling of a liquid or solid. A central issue in statistical mechanics is analyzing the behavior of substances as they cool. At high temperatures, molecules have much mobility, but as the temperature decreases, this mobility is lost and the molecules *may* tend to align themselves in a crystalline structure. This aligned structure is the minimum energy state for the system. Note the qualifier "may": Temperature alone does not govern whether the substance has reached a minimum energy state. To achieve this state, the cooling must occur at a sufficiently slow rate. If the substance is cooled at too rapid a rate, an amorphous (or "polycrystalline") state may be reached that is not a minimum energy state of the substance. The principle behind annealing in physical systems is the *slow* cooling of substances to reach the minimum energy state.

In optimization, the analogy to a minimum energy state for a system is a minimizing value of the loss function. The technique of SAN attempts to mathematically capture the process of controlled cooling associated with physical processes, the aim being to reach the lowest value of the loss function in the face of possible local minima. As with the physical cooling process, whereby temporary higher-energy states may be reached as the molecules go through their alignment process, SAN also allows temporary increases in the loss function as the learning process captures the information necessary to reach the global minimum. A more thorough explanation of the analogy between SAN and physical cooling is given, for example, in Kirkpatrick et al. (1983).

It is clear that the critical component of a SAN algorithm is the mathematical analogue of the rate of cooling in physical processes. As with other stochastic search and optimization algorithms, the choice of this implementation- and problem-specific cooling schedule (analogous to the gain coefficients in stochastic approximation) has a strong effect on the success or failure of SAN.

A primary distinction between SAN and the majority of other optimization approaches (including those discussed in earlier chapters) is the willingness to give up the quick gain of a rapid decrease in the loss function by allowing the possibility of temporarily increasing the value of the loss function. SAN derives this property from the Boltzmann–Gibbs probability distribution of statistical mechanics, describing the probability of a system having a particular discrete energy state:

$$P(\text{energy state} = x) = c_T \exp\left(-\frac{x}{c_b T}\right), \tag{8.1}$$

where $c_T > 0$ is a normalizing constant, $c_b > 0$ is known as the Boltzmann constant, and T is the temperature of the system. Note that at high temperatures, the system is more likely to be in a high-energy state than at low temperatures.

The optimization analogy derives from the fact that even at a low temperature (equivalent to the optimization algorithm having run for some time with a decreasing temperature), there is some nonzero probability of reaching a higher energy state (i.e., higher level of the loss function). So, the SAN process sometimes goes uphill, but the probability of an uphill step decreases as the temperature is lowered. Hence, there is the possibility of getting out of a local minimum in favor of finding a global minimum. This possibility is especially prominent in the early iterations when the temperature is high.

It was Metropolis et al. (1953) who first introduced the Boltzmann–Gibbs distribution-based idea into numerical analysis through constructing a means for simulation of a system at some fixed temperature. In particular, if a system is in some current energy state $\mathcal{E}_{\text{curr}}$, and some system aspects are changed to make the system potentially achieve a new energy state $\mathcal{E}_{\text{new}}$, then the Metropolis simulation always has the system go to the new state if $\mathcal{E}_{\text{new}} < \mathcal{E}_{\text{curr}}$. On the other hand, if $\mathcal{E}_{\text{new}} \geq \mathcal{E}_{\text{curr}}$, then the probability of the system going to the new state is

$$\exp\left(-\frac{\mathcal{E}_{\text{new}} - \mathcal{E}_{\text{curr}}}{c_b T}\right). \tag{8.2}$$

Expression (8.2) is known as the *Metropolis criterion.* After a large number of such decisions and outcomes, the system eventually reaches an equilibrium where the system state is governed by the Boltzmann–Gibbs distribution in (8.1). This is predicated on the system being at the fixed temperature T.

Let us outline the arguments showing how repeated application of (8.2) leads to (8.1). This result is central to SAN and to the Metropolis–Hastings version of Monte Carlo sampling in Chapter 16. Suppose that $\mathcal{E}_{\text{new}}$ and $\mathcal{E}_{\text{curr}}$ can each take on one of N values $x_1, x_2, \ldots, x_N$. The aim, therefore, is to show that many applications of (8.2) yield $P(\text{energy state} = x_i) = c_T \exp[-x_i/(c_b T)]$ for each $i = 1, 2, \ldots, N$. The probability of going from the current state i to any other state j is the product of two probabilities: (i) the probability of *choosing* j as the next state given that the process is in state i, and (ii) given that state j is chosen, the probability that the process *actually goes* to state j from i. A common simplification—adopted here—is that the probability in (i) of choosing any one state from any other state is equal $(= 1/N)$. Expression (8.2) provides the acceptance probability in (ii).

Given the above, the aim is to find the state transition probabilities, say p_{ij}, of going from state i to state j. These transition probabilities are the components of a Markov transition matrix (see Appendix E). Let a_{ij} denote the probability of accepting the transition from state x_i to state x_j. In particular, the acceptance probability for $i \neq j$ from (8.2) is

$$a_{ij} = \begin{cases} 1, & x_j < x_i, \\ \exp\left(-\dfrac{x_j - x_i}{c_b T}\right), & x_j \geq x_i. \end{cases} \tag{8.3}$$

The matrix $\boldsymbol{P} = [p_{ij}]$ represents the transition matrix. For $i \neq j$, $p_{ij} = a_{ij}/N$. Note that there is a positive probability that the process stays in state i even when the energy is lower in state j. Eqn. (8.3) does not apply to the diagonal elements p_{ii}. However, the transition probabilities must satisfy the following balance equation since each row in the transition matrix must sum to unity (because there must be a transition from state i to *some* state):

$$\begin{aligned} p_{ii} &= 1 - \sum_{j \neq i} p_{ij} \\ &= 1 - \frac{1}{N} \sum_{j \neq i} a_{ij} . \end{aligned}$$

The transition matrix providing the probabilities of going from state i to state j is therefore

$$\boldsymbol{P} = [p_{ij}] = \begin{bmatrix} 1 - \frac{1}{N} \sum_{k \neq 1} a_{1k} & \frac{1}{N} a_{12} & \cdots & \frac{1}{N} a_{1N} \\ \frac{1}{N} a_{21} & 1 - \frac{1}{N} \sum_{k \neq 2} a_{2k} & \cdots & \frac{1}{N} a_{2N} \\ \vdots & \vdots & \ddots & \vdots \\ \frac{1}{N} a_{N1} & \frac{1}{N} a_{N2} & \cdots & 1 - \frac{1}{N} \sum_{k \neq N} a_{Nk} \end{bmatrix} . \tag{8.4}$$

The transition matrix in (8.4) can be used to find the stationary distribution representing the probabilities of the system energy being in any of the N states. Let this stationary distribution be the N-dimensional vector $\bar{\boldsymbol{p}}$, where the ith element corresponds to the probability of the energy being equal to x_i. As in Appendix E, this vector satisfies $\bar{\boldsymbol{p}}^T = \bar{\boldsymbol{p}}^T \boldsymbol{P}$. This expression yields N equations and N unknowns (the elements $\bar{p}_i$). If it is assumed that the total energy in the system is fixed, that is, $\sum_{i=1}^{N} x_i$ = constant, then the solution to each of these N equations corresponds to the Boltzmann–Gibbs distribution (8.1) (see Exercise 8.1). That is, $\bar{p}_i = P(\text{energy state} = x_i) = c_T \exp[-x_i/(c_b T)]$ for all i, as we set out to show.

8.2 SIMULATED ANNEALING ALGORITHM

8.2.1 Basic Algorithm

Using the principles discussed in Section 8.1, we now present the general form of the SAN algorithm. Kirkpatrick et al. (1983) introduced the modern SAN algorithm through use of the Metropolis criterion in (8.2) for the purpose of optimization. One of the key innovations is to use (8.2) together with the idea of a *changing* temperature that decays according to an annealing schedule (exponentially in the case of Kirkpatrick et al., 1983).

There is no single SAN algorithm; rather there are variations depending on the implementation of the annealing schedule and choice of sampling required for generating a new candidate point. As before, we consider the minimization of some loss function $L(\boldsymbol{\theta})$, $\boldsymbol{\theta} \in \Theta \subseteq \mathbb{R}^p$. Below are the general steps in SAN when there are perfect (noise-free) measurements of L.

SAN Algorithm with Noise-Free Loss Measurements

Step 0 **(Initialization)** Set an initial temperature T and initial parameter vector $\hat{\boldsymbol{\theta}}_0 = \boldsymbol{\theta}_{\text{curr}} \in \Theta$; determine $L(\boldsymbol{\theta}_{\text{curr}})$.

Step 1 Relative to the current value $\boldsymbol{\theta}_{\text{curr}}$, randomly determine a new value of $\boldsymbol{\theta}$, $\boldsymbol{\theta}_{\text{new}} \in \Theta$, and determine $L(\boldsymbol{\theta}_{\text{new}})$.

Step 2 Compare the two L values above via the Metropolis criterion (8.2). Let $\delta = L(\boldsymbol{\theta}_{\text{new}}) - L(\boldsymbol{\theta}_{\text{curr}})$. If $\delta < 0$, accept $\boldsymbol{\theta}_{\text{new}}$. Alternatively, if $\delta \geq 0$, accept $\boldsymbol{\theta}_{\text{new}}$ only if a uniform (0, 1) random variable U (generated by Monte Carlo) satisfies $U \leq \exp[-\delta/(c_b T)]$. (Without loss of generality, one can set $c_b = 1$ since the user has full control over T.) If $\boldsymbol{\theta}_{\text{new}}$ is accepted, then $\boldsymbol{\theta}_{\text{curr}}$ is replaced by $\boldsymbol{\theta}_{\text{new}}$; else, $\boldsymbol{\theta}_{\text{curr}}$ remains as is.

Step 3 Repeat steps 1 and 2 for some period until either the budget of function evaluations allocated for that T has been used or the system reaches some state of equilibrium.

Step 4 Lower T according to the annealing schedule and return to step 1. Continue the process until the total budget for function evaluations has been used or some indication of convergence is satisfied (analogous to the system being frozen in its minimum energy state). The final estimate is $\hat{\boldsymbol{\theta}}_n$ (taken as the most recent $\boldsymbol{\theta}_{\text{curr}}$), representing the $\boldsymbol{\theta}$ value after n iterations ($= n + 1$ loss evaluations).

The specifics of implementation for the steps above can vary greatly. In the annealing schedule, Kirkpatrick et al. (1983), Press et al. (1992, p. 452), and Brooks and Morgan (1995) discuss the case where T decays geometrically in the number of cooling phases (number of times T is lowered according to step 4). Specifically, for some $0 < \lambda < 1$, the new temperature is related to the old temperature according to $T_{\text{new}} = \lambda T_{\text{old}}$. Others recommend temperatures that decay at every iteration (analogous to the decaying gain sequences of SA in Chapters 4–7). One of the most common such forms has the temperature decaying at a rate proportional to $1/\log k$, where k (as usual) is the iteration index (Geman and Geman, 1984; Hajek, 1988). Szu and Hartley (1987) present arguments justifying a faster decay rate proportional to $1/k$. In practice, the analyst will rarely know a priori what temperature decay rate is best for the problem at hand. So the analyst will typically pick a rate relatively arbitrarily and then tune the unknown coefficients associated with that rate (e.g., λ and the length of time a temperature is valid in the geometric case; the constant of proportionality in the continuous-decay rates such as $1/\log k$ or $1/k$).

Another area for different implementations is in step 1, where a new $\boldsymbol{\theta}$ value is generated randomly. Suppose that $\boldsymbol{\theta}_{\text{new}}$ is generated by adding a random perturbation to the current value $\boldsymbol{\theta}_{\text{curr}}$. Probably the most common form of perturbation for continuous optimization is to add a p-dimensional Gaussian random variable to $\boldsymbol{\theta}_{\text{curr}}$ (e.g., Jang et al., 1997, p. 183). Alternative forms include changing only one component of $\boldsymbol{\theta}$ at a time (Brooks and Morgan, 1995) (the component may be chosen either deterministically or randomly), adding spherically uniform distributed perturbations (Bohachevsky et al., 1986), and adding multivariate Cauchy distributed perturbations (Szu and Hartley, 1987; Styblinski and Tang, 1990; the scalar Cauchy distribution is discussed in Exercise C.3 of Appendix C).

Aside from the general variations in implementation above, SAN is critically dependent on the specific values for various algorithm parameters and decision criteria. In particular, these include the initial temperature T, the specific distribution parameters chosen for the perturbation distribution (e.g., the covariance matrix for a Gaussian perturbation), the specific parameter(s) associated with the decay of T (e.g., the λ in $T_{\text{new}} = \lambda T_{\text{old}}$ if one is using a geometric decay), and the criterion for determining when to drop the temperature (e.g., the maximum allowable number of function evaluations before a lowering of T).

There is some formal convergence theory governing the behavior of the SAN algorithm. For problems with discrete $\boldsymbol{\theta}$, Hajek (1988, Theorem 1) gives conditions such that SAN converges in probability to the set of global minima Θ^* for a temperature decaying at a rate proportional to $1/\log k$. For the Hajek result to apply, the constant of proportionality in the temperature decay is chosen in a manner related to the relative closeness of the loss values for the local and global minima. For problems where $\boldsymbol{\theta}$ is continuous, convergence theory for SAN may be built from the theory for stochastic approximation. This theory is similar to that for the SA algorithms in Chapters 4–7, except that the theory pertains to the *global* optimization problem. The chapter appendix (Section 8.6) summarizes this theory. Some useful analytical background for interpreting this theory is given in Section 8.4 on annealing algorithms based on SA algorithms with injected (Monte Carlo) randomness. The essential idea is that SAN is expressed as an SA algorithm where the effective noise in the SA process is given a special interpretation that is tied to the Metropolis accept/reject step (step 2 above). This effective SA noise is not to be confused with the usual SA noise involving corrupted loss or gradient measurements (the SAN theory is for noise-free loss measurements).

8.2.2 Modifications for Noisy Loss Function Measurements

There appears to be no single widely accepted way to accommodate noisy loss function measurements in SAN (similar to the problems in the random search techniques and the Nelder–Mead simplex algorithm of Chapter 2).[1] A limited amount of work has been published on this problem. For example, in the *discrete* $\boldsymbol{\theta}$ setting, Gelfand and Mitter (1989), Fox and Heine (1995), and Alrefaei and Andradóttir (1999) establish some convergence properties with noisy function measurements. Ahmed and Alkhamis (2002) use a ranking and selection method (as in Sections 12.5 and 14.5) embedded within SAN to cope with the randomness.

The primary difficulty with noise arises in the critical decision step 2, where $\boldsymbol{\theta}_{\text{curr}}$ is changed or not changed based on the value of $\delta = L(\boldsymbol{\theta}_{\text{new}}) - L(\boldsymbol{\theta}_{\text{curr}})$ together with the random sampling associated with the Metropolis criterion. With noisy function measurements, the value of δ equals $y(\boldsymbol{\theta}_{\text{new}}) - y(\boldsymbol{\theta}_{\text{curr}})$ rather than $L(\boldsymbol{\theta}_{\text{new}}) - L(\boldsymbol{\theta}_{\text{curr}})$, and, in general, these two differences will not be equal. In fact, even a small level of noise will frequently alter the sign of δ, which is likely to change the decision regarding the acceptance/rejection of the new point $\boldsymbol{\theta}_{\text{new}}$.

The most obvious method for coping with noise is to average several function measurements $y(\cdot)$ at each of the values $\boldsymbol{\theta}_{\text{curr}}$ and $\boldsymbol{\theta}_{\text{new}}$ when performing the comparison in step 2. However, this may dramatically increase the cost of optimization (specifically, the total number of function evaluations required) since a high amount of averaging is generally required to effectively remove the noise, especially in the region around a local or global minimum where the function may be relatively flat and the difference $L(\boldsymbol{\theta}_{\text{new}}) - L(\boldsymbol{\theta}_{\text{curr}})$ is likely to be small. Note also that this type of per-iteration averaging is contrary to the spirit of stochastic approximation, as mentioned in Section 4.1, where efficiencies result from averaging across iterations.

An alternative way to cope with noise is to alter the acceptance criterion to accept the inherent errors that will be made with the noisy measurements. Hence, we replace the criterion $\delta < 0$ or $\delta \geq 0$ with $\delta < \tau$ or $\delta \geq \tau$, where τ is some threshold value that may be positive or negative depending on the circumstances. Likewise, the Metropolis criterion is modified so that $\exp[-\delta/(c_b T)]$ in step 2 is replaced by $\exp[-(\delta - \tau)/(c_b T)]$. (It will sometimes be useful to express τ as a multiple of the measurement noise standard deviation as described in Proposition 8.1 below. This facilitates making approximate probabilistic arguments about the likelihood of a good or bad decision in step 2

[1] One of the particularly appealing features of the stochastic approximation methods in Chapters 4–7 is that they handle noisy function measurements in an essentially seamless manner. Namely, the forms of the algorithms do not have to be altered, the convergence theory applies in the noisy case, and there is a *gradual* degradation in performance as the noise level increases from zero (versus the potentially large degradation in approaches such as SAN that use explicit decision criteria based on loss function measurements).

of the algorithm.) In particular, if it is believed that there are many local minima, then τ should be taken positive. The rationale for such a change is that while we are willing to accept a $\boldsymbol{\theta}_{\text{new}}$ that temporarily increases the loss function (a basic aspect of SAN!), we are less willing to forgo a $\boldsymbol{\theta}_{\text{new}}$ that decreases the loss function. Changing the unconditional acceptance criterion from $\delta < 0$ to $\delta < \tau$ with $\tau > 0$ will allow for a greater number of cases where $L(\boldsymbol{\theta}_{\text{new}}) < L(\boldsymbol{\theta}_{\text{curr}})$ even though $y(\boldsymbol{\theta}_{\text{new}}) \geq y(\boldsymbol{\theta}_{\text{curr}})$ due to the noise. If τ is large enough (say, several times the standard deviation of the measurement noise), one can be sure that most such combinations will be caught. This will be at the inevitable expense of letting some additional $\boldsymbol{\theta}_{\text{new}}$ values that increase the loss function be accepted. Since there are many local minima, this latter fact is tolerable.

On the other hand, if one believes that there are relatively few local minima, then there is less interest in allowing a $\boldsymbol{\theta}_{\text{new}}$ that increases the loss function. Hence, $\tau < 0$, leading to a relatively strict requirement for changing $\boldsymbol{\theta}$ from $\boldsymbol{\theta}_{\text{curr}}$. In particular, in the initial (main) decision in step 2, we now only change $\boldsymbol{\theta}_{\text{curr}}$ if $y(\boldsymbol{\theta}_{\text{new}}) \leq y(\boldsymbol{\theta}_{\text{curr}}) + \tau$, where $\tau < 0$. Picking $\tau << 0$ provides inertia to the algorithm by preventing $\boldsymbol{\theta}$ updates unless there is a strong probability that the new $\boldsymbol{\theta}$ improves the loss value. Of course, even with such a conservative approach, the second part of step 2, which is fundamental to SAN, will always provide some chances of accepting a temporarily poorer $\boldsymbol{\theta}$ in the hopes of escaping local minima.

Proposition 8.1 below provides a probabilistic bound on the likelihood of the events $L(\boldsymbol{\theta}_{\text{new}}) > L(\boldsymbol{\theta}_{\text{curr}})$ and $y(\boldsymbol{\theta}_{\text{new}}) \leq y(\boldsymbol{\theta}_{\text{curr}}) + \tau$ occurring simultaneously, where τ is in terms of some negative multiple of the standard deviations of the overall measurement noise due to both $\varepsilon(\boldsymbol{\theta}_{\text{new}})$ and $\varepsilon(\boldsymbol{\theta}_{\text{curr}})$. Hence, the bound concerns the probability of making a "mistake" via having the data indicate an improved (lowered) loss function even though the actual loss increases. When this probability bound is small, there are strong indications that the acceptance of a new value of $\boldsymbol{\theta}$ (i.e., $y(\boldsymbol{\theta}_{\text{new}}) \leq y(\boldsymbol{\theta}_{\text{curr}}) + \tau$) is associated with the desirable outcome of $L(\boldsymbol{\theta}_{\text{new}}) \leq L(\boldsymbol{\theta}_{\text{curr}})$. This bound is distribution-free in the sense that it relies on a one-sided version of the Chebyshev inequality (sometimes called the Cantelli inequality). The inequality states that $P(X \geq c) \leq \text{var}(X)/[\text{var}(X)+c^2]$, where X is a mean-zero random variable and $c > 0$ (Tong, 1980, p. 155). (This contrasts with the standard two-sided Chebyshev inequality: $P(|X| \geq c) \leq \text{var}(X)/c^2$.)

Proposition 8.1. Suppose that the measurement noise terms in $y(\boldsymbol{\theta}_{\text{new}})$ and $y(\boldsymbol{\theta}_{\text{curr}})$, $\varepsilon(\boldsymbol{\theta}_{\text{new}})$ and $\varepsilon(\boldsymbol{\theta}_{\text{curr}})$, have the same mean and are uncorrelated with finite variances σ_{new}^2 and σ_{curr}^2. If $\tau = -c\sqrt{\sigma_{\text{new}}^2 + \sigma_{\text{curr}}^2}$, $c > 0$, then

$$P\left(\{L(\boldsymbol{\theta}_{\text{new}}) \geq L(\boldsymbol{\theta}_{\text{curr}})\} \cap \{y(\boldsymbol{\theta}_{\text{new}}) \leq y(\boldsymbol{\theta}_{\text{curr}}) + \tau\}\right) \leq \frac{1}{1+c^2}. \tag{8.5}$$

Note. If several, say N_{avg}, values of $y(\cdot)$ are collected at each $\boldsymbol{\theta}$ and averaged for use in the algorithm, then result (8.5) continues to hold provided that τ above is replaced by $\tau = -c\sqrt{(\sigma_{\text{new}}^2 + \sigma_{\text{curr}}^2)/N_{\text{avg}}}$.

Proof. For convenience, let $\delta L = L(\boldsymbol{\theta}_{\text{new}}) - L(\boldsymbol{\theta}_{\text{curr}})$ and $\delta\varepsilon = \varepsilon(\boldsymbol{\theta}_{\text{new}}) - \varepsilon(\boldsymbol{\theta}_{\text{curr}})$. Then the probability of interest is

$$\begin{aligned} P\big(\{L(\boldsymbol{\theta}_{\text{new}}) \geq L(\boldsymbol{\theta}_{\text{curr}})\} \cap \{y(\boldsymbol{\theta}_{\text{new}}) \leq y(\boldsymbol{\theta}_{\text{curr}}) + \tau\}\big) &= P(\delta L \geq 0, \delta L + \delta\varepsilon \leq \tau) \\ &= P(0 \leq \delta L \leq \tau - \delta\varepsilon) \\ &\leq P(0 \leq \tau - \delta\varepsilon). \end{aligned}$$

Note that $\text{var}(\delta\varepsilon) = \sigma_{\text{new}}^2 + \sigma_{\text{curr}}^2$ by the uncorrelatedness assumption. Then, by the above-mentioned one-sided Chebyshev inequality, the last line above satisfies

$$P(\tau - \delta\varepsilon \geq 0) = P(-\delta\varepsilon \geq -\tau) \leq \frac{1}{1+c^2} \; ,$$

which proves (8.5). ❑

An example implication of Proposition 8.1 is that for sufficiently large c, (8.5) implies that $y(\boldsymbol{\theta}_{\text{new}}) \leq y(\boldsymbol{\theta}_{\text{curr}}) + \tau$ is strongly associated with the desirable outcome of $L(\boldsymbol{\theta}_{\text{new}}) < L(\boldsymbol{\theta}_{\text{curr}})$. For instance, if $c = 3$, then the probability is no greater than 0.1 of an incorrect decision in the sense that $L(\boldsymbol{\theta}_{\text{new}}) \geq L(\boldsymbol{\theta}_{\text{curr}})$ happens while $y(\boldsymbol{\theta}_{\text{new}}) \leq y(\boldsymbol{\theta}_{\text{curr}}) + \tau$. Given that the bound in (8.5) is built from the Chebyshev inequality (albeit the stronger one-sided version), it is still quite conservative. If, for example, the noises are known to be Gaussian distributed, then the above-mentioned example with $c = 3$ yields a bound to the probability on the left-hand side of (8.5) of only 0.0013, an 80-fold reduction from the distribution-free Chebyshev bound. Exercise 8.2 is aimed at providing additional insight into the distinction between the one- and two-sided (standard) Chebyshev inequality and the Gaussian assumption.

Unfortunately, while the bound in Proposition 8.1 is of some interest in predicting *per-iteration* behavior of SAN, it is unclear what the implication of the bound is for the full range of a search process. Suppose, for example, that the bound tells us that there is likely to be an error no more than 10 percent of the time in the sense of observing $y(\boldsymbol{\theta}_{\text{new}}) \leq y(\boldsymbol{\theta}_{\text{curr}}) + \tau$ when $L(\boldsymbol{\theta}_{\text{new}}) \geq L(\boldsymbol{\theta}_{\text{curr}})$ is true. It is unclear what this says about the potential convergence to a global solution $\boldsymbol{\theta}^*$.

8.3 SOME EXAMPLES

This section considers three example problems that illustrate the application of SAN. The first is the famous traveling salesperson problem and the next two are small problems involving a comparison of SAN with two random search methods of Chapter 2 and with a deterministic search method.

Before presenting the first example, let us provide some general background on the traveling salesperson problem. This problem is simple to state, but not necessarily simple to solve. A traveling salesperson must visit every city in some territory once and only once. The problem is to find a minimum cost path meeting this goal when the cost between any pair of cities is known. Because there is a finite number of possible paths, this is a discrete optimization problem, although one with a potentially very large number of elements in Θ. A tour with $n \geq 3$ cities has $(n-1)!/2$ possible unique tours, corresponding to the number of elements in Θ. (*Unique* here refers to fundamentally different ordering. For example, if $n = 3$, the tour 1–2–3–1 is considered equivalent to 1–3–2–1 by the symmetry of the cost between cities, where the indicated numbers 1, 2, or 3 represent the labels for the three cities.) For example, finding an optimal tour of the largest cities in all 50 states of the United States entails 3×10^{62} possible routes, far more than the estimated number of atoms in the Earth ($\sim 10^{50}$). In the language of combinatorial optimization, the problem is NP hard ("NP" variously stands for *nonpolynomial* or *nondeterministic polynomial*; general NP problems are believed to have no solution in computing time proportional to a polynomial of the problem size n—see, e.g., Culberson, 1998).

The perturbations in the SAN algorithm are tied to three fundamental operations on this combinatorial problem: inversion, translation, and switching. Inversion selects two cut points from the current tour and reverses (inverts) the order of the cities between these two cut points. Translation selects a subsection of the current tour and inserts it between two randomly selected other cities. Switching randomly selects two cities and switches their positions on the tour. Figure 8.1 illustrates these operations on an eight-city tour.

Example 8.1—Traveling salesperson problem. One of the most successful applications of SAN has been on this well-known problem. Let us consider a relatively modest-sized problem involving 10 cities. Hence, $p = 10$, with each component of $\boldsymbol{\theta}$ representing one of the city labels 1, 2,..., 10. (SAN has been used successfully in much larger problems—in the sense of providing a much-improved tour—but then it is rarely possible to know "truth" for comparison purposes. For example, the seminal 1983 paper of Kirkpatrick et al. considers a 400-city tour.) The constraint set Θ is such that $\boldsymbol{\theta}$ must represent a valid tour; that is, each of the 10 integers must appear once and only once in $\boldsymbol{\theta}$. (The salesperson can start the trip at the city corresponding to any element of $\boldsymbol{\theta}$; hence, there is no need to constrain the first element to be the origin city.)

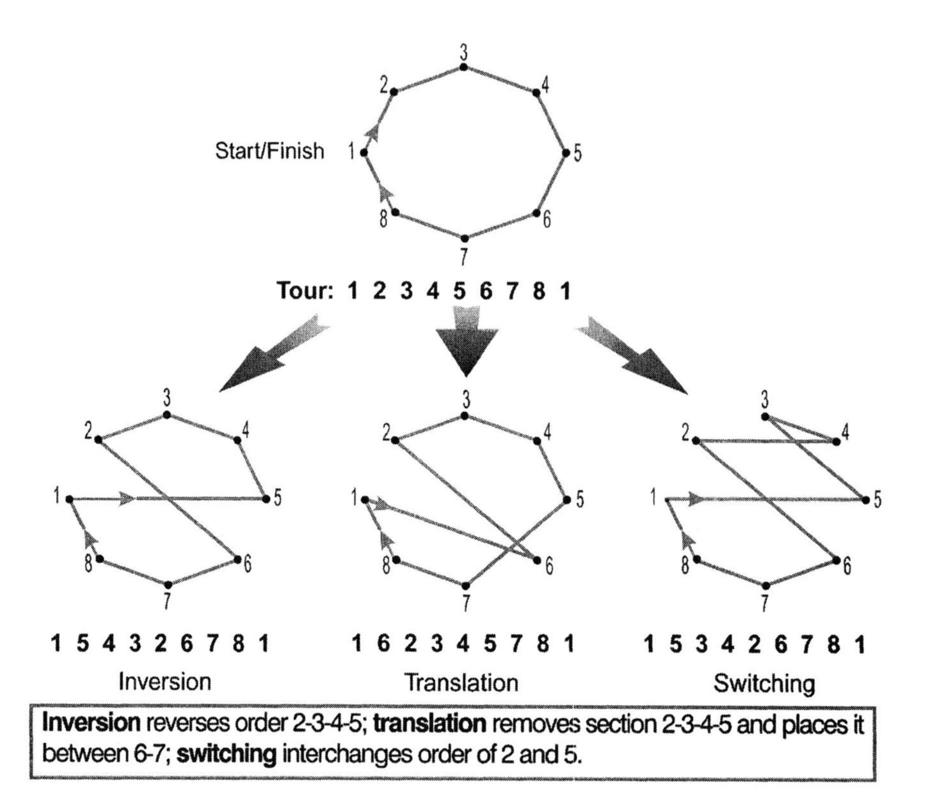

Figure 8.1. Simple network of eight cities to be visited on the tour with the city order listed below each tour. The operations of inversion, translation, and switching are shown relative to the original tour on top.

The city-to-city costs are generated randomly according to a uniform distribution between 10 and 125. Based on the set of randomly generated costs (which are fixed throughout the study), we did an exhaustive search through the $(10-1)!/2 = 181{,}440$ possible tours to determine the optimal tour. The optimal tour has a cost of 440.

Results are determined for 10 independent SAN runs, each initialized at a randomly generated tour and using $N = 8000$ loss evaluations (noise-free in this case). One of the three operations in Figure 8.1 is used to produce each new value $\boldsymbol{\theta}_{\text{new}} \in \Theta$ in step 1 of the SAN algorithm. The inversion, translation, and switching operations are selected randomly with probabilities 0.75, 0.125, and 0.125, respectively. A stepwise temperature decay is used, with 40 iterations occurring at each temperature. The initial temperature T is 70 and the decay factor $\lambda = 0.95$.

Table 8.1 shows representative results for five of the 10 replications. The sample mean loss over the 10 replications is 444.1, which is close to the optimum 440 and a significant improvement over the sample mean of the initial loss values of approximately 700. Eight of the 10 runs found the optimal tour (the optimal runs are equivalent versions of the same tour, as shown in the four

Table 8.1. Representative results for five (of 10) SAN runs on the 10-city traveling salesperson problem. Each run uses 8000 loss measurements (7999 iterations).

$\hat{\boldsymbol{\theta}}_{7999}^T$ (final 10-city tour)	$L(\hat{\boldsymbol{\theta}}_{7999})$
[1, 5, 2, 4, 7, 6, 8, 9, 3, 10]	440
[5, 1, 10, 7, 4, 2, 6, 8, 9, 3]	459
[1, 10, 3, 9, 8, 6, 7, 4, 2, 5]	440
[2, 5, 1, 10, 3, 9, 8, 6, 7, 4]	440
[9, 3, 10, 1, 5, 2, 4, 7, 6, 8]	440

optimal runs of Table 8.1). An adequate frequency of use for the inversion operator is central to the results here. If the probability of inversion is reduced to 0.50 (versus 0.75), none of the 10 runs yielded an optimal tour. (The relatively greater effectiveness of inversion for traveling salesperson problems with symmetric city-to-city costs is explained in Reidys and Stadler, 2002.) ❑

Let us now evaluate SAN in two small-scale problems with continuous $\boldsymbol{\theta}$. Of course, as emphasized in Section 1.2, one should avoid drawing *general* conclusions from the results of such numerical comparisons. Nevertheless, the comparison here sheds at least limited light on the performance of SAN. Section 8.4 also presents a numerical evaluation of SAN relative to a different annealing-type algorithm introduced in that section; the loss function in the evaluation of Section 8.4 is more complex than the losses here.

Example 8.2—Comparison of random search algorithms and SAN. This example compares the outcomes of random search algorithms B and C (Subsection 2.2.2) with SAN for the case of the simple $p = 2$ quartic polynomial loss function first seen as an example in Section 1.4: $L(\boldsymbol{\theta}) = t_1^4 + t_1^2 + t_1 t_2 + t_2^2$, where $\boldsymbol{\theta} = [t_1, t_2]^T$. Each of the random search and SAN results is based on a $N(0, \rho^2 I_2)$ perturbation distribution (to generate the candidate value of $\boldsymbol{\theta}$); ρ is the same for all SAN implementations (i.e., all initial T), but differs in algorithms B, C, and SAN. Although local minima are not a problem with this simple loss function, this evaluation serves as a test of SAN on a simple function with known solution. Only noise-free loss measurements are considered in the algorithms. Table 8.2 presents the results based on an average of 40 independent replications, with each replication initialized at $\hat{\boldsymbol{\theta}}_0 = [1, 1]^T$. Note that SAN reduces to algorithm B when T is very small (i.e., the Metropolis criterion is then near zero, indicating that $\boldsymbol{\theta}_{\text{new}}$ is almost always rejected unless it produces a

Table 8.2. Comparison of random search and SAN algorithms. Sample mean of terminal $L(\boldsymbol{\theta})$ values after N function evaluations for varying initial temperature T (initial loss value = 4.0).

N	Random Search		SAN		
	Alg. B	Alg. C	Initial T = 0.01	Initial T = 0.10	Initial T = 1.0
100	0.00053	0.328	1.86	0.091	0.763
1000	2.8×10^{-5}	1.1×10^{-5}	0.0092	0.067	0.506
10,000	2.7×10^{-6}	2.5×10^{-7}	0.00038	0.0024	0.018

decrease in L). The results shown for SAN here have T large enough to retain the essential character of the Metropolis decision step.

Reasonable efforts were made to tune the SAN parameters to enhance the performance of the algorithm. A standard stepwise temperature decay is used, with 50 iterations occurring at each temperature and a decay factor λ = 0.98. Table 8.2 shows the results for several initial temperatures, T.

While SAN produces a marked improvement in the loss function as N increases, the random search techniques tend to produce even better results. So, this example demonstrates that SAN does not always yield performance superior to the simpler random search algorithms. (Exercise 8.6 considers this example further, including a lower initial T. Results for SAN do not significantly improve with a lower initial T.) ❑

Example 8.3—Evaluation of SAN in problem with multiple local minima. There are many numerical studies in the literature showing favorable results for SAN. Let us summarize one on a constrained problem with continuous $\boldsymbol{\theta}$, as described in Brooks and Morgan (1995). The aim of this study is to compare SAN with a popular method of deterministic optimization in a problem with many local minima. Consider $L(\boldsymbol{\theta}) = t_1^2 + 2t_2^2 - 0.3\cos(3\pi t_1) - 0.4\cos(4\pi t_2)$ with $\boldsymbol{\theta} = [t_1, t_2]^T$ and $\Theta = [-1, 1]^2$. This function has many local minima with a unique global minimum at $\boldsymbol{\theta}^* = \mathbf{0}$. This study compares the popular quasi-Newton algorithm from deterministic optimization (mentioned in Section 1.4) with SAN. (The quasi-Newton algorithm has some of the fast convergence properties of the Newton–Raphson algorithm, but is generally more stable in practice; it is one of the most popular nonlinear programming methods.) Note that, in some sense, this comparison is "apples versus oranges" because the quasi-Newton method uses gradient information while SAN uses only loss information.

The quasi-Newton method is run 100,000 times, each with a different initial condition $\hat{\boldsymbol{\theta}}_0 \in \Theta$. Each initial condition is generated randomly (uniformly) in Θ. It is found that fewer than 20 percent of the quasi-Newton runs yield convergence to points near $\boldsymbol{\theta}^*$. In the other runs, the algorithm settles near one of the local minima. In contrast, all of 1000 runs of SAN with stepwise decaying temperatures settle near $\boldsymbol{\theta}^*$; as with the quasi-Newton evaluation, the initial conditions are generated uniformly for each run. This study shows the relative effectiveness of SAN in finding a global minimum. Additional study details may be found in Brooks and Morgan (1995). ❑

8.4 GLOBAL OPTIMIZATION VIA ANNEALING ALGORITHMS BASED ON STOCHASTIC APPROXIMATION

The general SAN algorithm form discussed in Sections 8.2 and 8.3 is probably what most people involved with stochastic optimization think of when considering annealing algorithms. The basic idea of slow cooling for global optimization, however, can be implemented in other ways as well. One popular way is via connections to the stochastic approximation framework discussed in Chapters 4–7. In particular, let us begin with the basic SA form

$$\hat{\boldsymbol{\theta}}_{k+1} = \hat{\boldsymbol{\theta}}_k - a_k \boldsymbol{G}_k(\hat{\boldsymbol{\theta}}_k), \tag{8.6}$$

where $\boldsymbol{G}_k(\cdot)$ typically represents either a direct (noisy) gradient measurement as in $\partial Q/\partial \boldsymbol{\theta}$ in Chapter 5 or a gradient approximation built from noisy function measurements as in Chapters 6 and 7.

As discussed in Chapters 4–7, the basic algorithm (8.6) is generally for *local* optimization, although one important exception is discussed at the end of this section. To achieve global convergence, the algorithm is modified to include a user (Monte Carlo)-injected random input term $\boldsymbol{w}_k$ scaled by an SA gain coefficient b_k that decays to zero:

$$\hat{\boldsymbol{\theta}}_{k+1} = \hat{\boldsymbol{\theta}}_k - a_k \boldsymbol{G}_k(\hat{\boldsymbol{\theta}}_k) + b_k \boldsymbol{w}_k. \tag{8.7}$$

Typically, $\{\boldsymbol{w}_k\}$ is an independent, identically distributed (i.i.d.) standard Gaussian (i.e., $N(\boldsymbol{0}, \boldsymbol{I}_p)$) sequence, but this distribution is not required. Note that because (8.7) is an SA algorithm, the algorithm is designed to cope with possibly noisy loss or gradient measurements.

Intuitively, one can see how (8.7) offers the possibility of global convergence. Namely, the injected random input $b_k \boldsymbol{w}_k$ provides additional "bounce" to the iteration process and helps avoid premature entrapment in a local minimum. This is analogous to the decision step 2 in the SAN algorithm of Section 8.2, where the algorithm sometimes accepts a poorer value of the loss function in the hopes of leaving a local minimum en route to a global minimum. Further, analogous to SAN, the chances of leaving a minimum as the iteration process proceeds is reduced to allow for reaching a solution that is a global minimum. This explains the requirement that $b_k \to 0$. To establish some guarantees of global convergence, it is important that b_k decay at a sufficiently slow rate. The intuition behind this should be clear: If $b_k \to 0$ too quickly, then for moderately large k, (8.7) would too closely resemble the local algorithm (8.6) and may not have enough "bounce" to ensure escaping a local minimum.

A number of papers have considered the basic algorithm form (8.7)—or its continuous-time analogue—and have established formal conditions for convergence to a global minimum. Among them are Geman and Hwang (1986), Kushner (1987), Gelfand and Mitter (1991, 1993), Fang et al. (1997), and Yin (1999). There is no one standard set of conditions on the gains a_k and b_k in contrast to the reasonably standard conditions on a_k (and, if relevant, c_k) for the local SA algorithms discussed in Chapters 4–7. We list below four different conditions on the gains and the references from which they are drawn. It is assumed that $a > 0$, $b > 0$, $A \geq 0$, and $B > 0$.

Four Possible Gain Conditions for Global Convergence of (8.7)

A. $a_k = a/\log(k+1)$, $b_k = a_k$ (Kushner, 1987).

B. $a_k = a/(k+1+A)$, $b_k = b/\sqrt{(k+1)\log\log(k+1)}$ (Gelfand and Mitter, 1991, 1993).

C. $a_k = a/(k+1+A)^{\alpha}$, $b_k = b/[(k+1)^{\alpha/2}\log(k+1)]$, $0 < \alpha < 1$ (Fang et al., 1997).

D. $a_k = a/(k+1+A)^{\alpha}$, $b_k = b/\sqrt{(k+1)^{\alpha}\log[(k+1)^{1-\alpha}+B]}$, $0 < \alpha < 1$ (Yin, 1999).

While conditions A and B are used exclusively in their respective references, condition C is only one example from a more general set of conditions on a_k and b_k in Fang et al. (1997). Condition D is one of two possible cases (the other pertains to the classical $\alpha = 1$ setting). Using these gain conditions together with other conditions on $L(\boldsymbol{\theta})$, the random noise $\varepsilon(\boldsymbol{\theta})$ or $e(\boldsymbol{\theta})$ (the difference between $y(\boldsymbol{\theta})$ and $L(\boldsymbol{\theta})$ in the gradient-free case; the difference between $\partial Q/\partial\boldsymbol{\theta}$ and $\boldsymbol{g}(\boldsymbol{\theta})$ in the gradient-based case), and the random input $\boldsymbol{w}_k$, the authors are able to establish various forms of convergence of $\hat{\boldsymbol{\theta}}_k$ to the global minimum of $L(\boldsymbol{\theta})$ (or, more generally, to an element in the *set* of global solutions if more than one global minimum exists). Gelfand and Mitter (1991, 1993), for example, establish convergence *in probability* (versus the stronger almost sure convergence for the local SA algorithms).

As with the local SA algorithms, it is of interest to analyze the stochastic *rate* at which $\hat{\boldsymbol{\theta}}_k - \boldsymbol{\theta}^* \to \mathbf{0}$ for (8.7). Yin (1999) finds that this rate of convergence is $O(1/\sqrt{\log k})$ when using the gain combination in D above and

$O(1/\sqrt{\log\log k})$ when using $a_k = a/(k+1)$ and a companion b_k that is analogous to the b_k in condition D. Both of these rates are painfully slow. The convergence rates found by Yin (1999) are typical of the rates for the global SA algorithm in (8.7); similar rates are available, for example, in Gelfand and Mitter (1993). The slow rates of convergence suggest that practical global convergence—at least via algorithms of the form (8.7)—may sometimes be enhanced through the use of a local algorithm (i.e., no injected randomness) once there is evidence that the iterate is near the solution. (Recall that the local algorithm with noisy gradient or loss measurements has the faster rate $O(1/k^{\eta})$, $0 < \eta \le 1/2$ or $0 < \eta \le 1/3$, as discussed in Chapters 4 and 7, respectively.) Note, however, that the slow global rates are asymptotically based; in practical problems, acceptable finite-sample accuracy may often be available with (8.7) directly (see Example 8.4).

The example below evaluates the performance of the SA-based annealing algorithm in (8.7) relative to the SAN algorithm of Section 8.2. This problem is one with a challenging loss function having a large number of local minima (and one global minimum).

Example 8.4—Comparison of global SA in (8.7) and SAN on problem with multiple minima. Given that only noisy loss measurements are available to the algorithms, this example compares global SA and SAN on a challenging problem with many local minima. The global SA algorithm is implemented here using only function evaluations in forming $\boldsymbol{G}_k(\cdot)$ so that we can fairly compare algorithms for a common number of loss functions (versus, say, using the stochastic gradient-based algorithm). In particular, for $\boldsymbol{G}_k(\cdot)$ we use the simultaneous perturbation SA (SPSA) gradient approximation with Bernoulli random perturbations based on the gain sequence combination in item C above (Fang et al., 1997). The SAN algorithm has the standard stepwise decaying temperature form. Both algorithms are run with 20,000 loss measurements per replication. This study considers the loss function:

$$L(\boldsymbol{\theta}) = 0.05\sum_{i=1}^{p} t_i^2 - 40\prod_{i=1}^{p}\cos(t_i), \quad \boldsymbol{\theta} = [t_1, t_2, \ldots, t_p]^T, \tag{8.8}$$

where $p = 10$. This loss has one global minimum at $\boldsymbol{\theta}^* = \boldsymbol{0}$ and a large number of local minima (Styblinski and Tang, 1990, Example 6). The initial condition $\hat{\boldsymbol{\theta}}_0$ is taken at a local minimum $6.2676[1, 1,\ldots, 1]^T$ (not *quite* a multiple of π = 3.14159...). It is assumed that only noisy loss measurements are available, with the noise being i.i.d. $N(0, 1)$.

As with Tables 8.1 and 8.2, efforts are made to tune the algorithm coefficients to achieve approximately optimal performance. The specific values for the gain sequence in global SA (form C above) are $a = 20$, $A = 100$, $b = 0.5$, $c = 10$, and, as recommended in Chapters 6 and 7, $\alpha = 0.602$ and $\gamma = 0.101$. The SAN runs use an initial $T = 20$ with 100 iterations at each temperature; the temperature decay factor is $\lambda = 0.95$. To accommodate the noise, the implementation uses a combination of averaging of loss functions ($N_{\text{avg}} = 2$) and modification to the decision criterion ($\tau = 1.0$) (see Subsection 8.2.2).

Table 8.3 reports on the results of five independent replications for each of global SA and SAN (the algorithms in each of the five pairs are initialized at the same random number seeds). Some interesting patterns emerged from the study. First, the table indicates superior performance for global SA. In the initial tuning of the algorithm coefficients, the use of short (efficient) realizations in global SA provided a good indication of gains that would work well for the longer ($N = 20{,}000$) realization of actual interest. For SAN, this was not the case, as results for short realizations gave little clue about good coefficient values in the realizations based on $N = 20{,}000$. Although poorer than the global SA results, these SAN results are an improvement over those given in Styblinski and Tang (1990), which relied on $N = 50{,}000$ *noise-free* measurements and used a different implementation of the basic SAN steps in Section 8.2. In particular, Styblinski and Tang (1990) use the so-called fast simulated annealing algorithm of Szu and Hartley (1987), which employs an iteration-by-iteration decay of temperature and Cauchy-distributed perturbations to generate the new (candidate) $\boldsymbol{\theta}$ values. ❑

SA with injected randomness (à la (8.7)) is not the only way in which SA and annealing are combined to achieve global optimization. Section 7.7 mentioned that basic SPSA can often be used for global optimization as a result of the effective noise introduced through the "sloppy" gradient approximation. That is, the SPSA recursion can be expressed as

$$\hat{\boldsymbol{\theta}}_{k+1} = \hat{\boldsymbol{\theta}}_k - a_k \mathbf{g}(\hat{\boldsymbol{\theta}}_k) + a_k \text{error}_{\text{noise}} + a_k \text{error}_{\text{perturbation}}, \tag{8.9}$$

Table 8.3. Terminal L value for global SA and SAN with noisy loss measurements and $N = 20{,}000$ function evaluations per replication ($L(\hat{\boldsymbol{\theta}}_0) = -20.3$; $L(\boldsymbol{\theta}^*) = -40.0$)).

Replication	Global SA	SAN
1	−39.2	−29.1
2	−39.2	−32.8
3	−39.5	−28.0
4	−39.2	−29.1
5	−38.6	−31.6
Sample Mean	−39.1	−30.1

where $\text{error}_{\text{noise}}$ is the difference from the true gradient $\boldsymbol{g}(\cdot)$ due to the noise in the loss measurements and $\text{error}_{\text{perturbation}}$ is the difference due to the simultaneous perturbation aspect (which exists even if there are noise-free loss measurements). Maryak and Chin (2001) make the important discovery that the term $a_k\text{error}_{\text{perturbation}}$ on the right-hand side of (8.9) acts in the same statistical way as the Monte Carlo injected randomness $b_k\boldsymbol{w}_k$ on the right-hand side of (8.7). Hence, the basic SPSA provides the needed injected randomness "for free," resulting in global convergence. (While the gain selection guidelines of Subsection 7.5.2 may still have some value, it is generally the case that c should not be near zero to avoid premature convergence. This applies even with noise-free loss measurements.)

There are two important advantages of the global convergence result for basic SPSA: (i) The asymptotic rate of convergence has the faster $O\left(1/k^{\eta}\right)$ form, $0 < \eta \leq 1/3$ (versus the much slower forms mentioned above for (8.7)) and (ii) there are fewer algorithm coefficients to determine since there is no b_k sequence. Nonetheless, as with any theory, there are restrictions to this result. In general, the injected randomness implementation (8.7) has broader applicability for global optimization. Nevertheless, the basic SPSA result may be very effective in some challenging global optimization problems. For example, in 10 of 10 replications with 2500 noise-free loss evaluations per replication, basic SPSA achieves convergence to $\boldsymbol{\theta}$ values having identical optimal losses of -40.0 with the challenging multimodal loss function in (8.8) (Maryak and Chin, 2001).

8.5 CONCLUDING REMARKS

This has been the first of three chapters that explore algorithms with a connection to physical processes. In the case here, the physical process is the cooling of materials via annealing (the next two chapters consider analogies to evolutionary biology).

This chapter considered two broad approaches under the annealing rubric. Most of the discussion focused on the popular simulated annealing algorithm. SAN has become a popular global optimizer for both discrete and continuous optimization problems. It has long been known in metallurgy and other fields that annealing—that is, slowly cooling a substance—provides a crystalline structure with maximum strength. SAN is built on mathematical analogies to this cooling process for physical systems. The Metropolis accept/reject criterion provides a critical part in the SAN algorithm. We also see in Chapter 16 an important application of this criterion in the Markov chain Monte Carlo methods for random number generation and estimation.

The second broad annealing approach is tied to stochastic approximation. SA can be used as a global optimizer with the addition of a Monte Carlo injected random term on the right-hand side of the standard recursion. This injected randomness is annealed (damped) in a manner reminiscent of the temperature cooling in SAN. The injected randomness provides enough "jump" in the algorithm to ensure global convergence (under the appropriate conditions, of course). We also briefly discussed the use of basic SPSA (Chapter 7) as a global optimizer. This algorithm has a built-in random error term that automatically provides a type of injected randomness.

Markov chain theory can be used to establish formal convergence results for the SAN algorithm in discrete (combinatorial) optimization. Analogously, SA theory can be used in the continuous case (see the chapter appendix, Section 8.6). In particular, one can show that special cases of the SA algorithm with injected noise are effectively equivalent to particular SAN algorithms.

As with other approaches, numerical experiments show a variety of results, some very successful and some less so. Broadly speaking, it appears that SAN may be worth trying in discrete multimodal problems with noise-free loss measurements. Of course, SAN may also be successful in other problems as well. In problems with noisy loss measurements, SA algorithms with or without injected randomness may be preferable. Relative to SAN, SA has a stronger theory for convergence with noisy measurements.

8.6 APPENDIX: CONVERGENCE THEORY FOR SIMULATED ANNEALING BASED ON STOCHASTIC APPROXIMATION[2]

Let us begin by summarizing two relatively early works related to convergence of SAN. These are Geman and Geman (1984) and Geman and Hwang (1986). Geman and Geman (1984) go beyond the informal annealing guidelines of Kirkpatrick et al. (1983) and establish conditions on the annealing schedule for formal convergence of an algorithm similar to the standard SAN discussed here; the algorithm is directly motivated by applications to image restoration (see Section 5.4). The temperature variable in Geman and Geman (1984) controls the shape of the distribution of the pixel intensities (rather than the threshold size and perturbation magnitude of the standard SAN), making it more peaked as time proceeds.

More generally, Geman and Hwang (1986) present a framework for continuous-valued $\boldsymbol{\theta}$ optimization. Their approach to optimizing continuous $\boldsymbol{\theta}$ is based on a *continuous-time* stochastic process with random input governed by a continuous-time probability distribution. This continuous-time process is represented as an approximation to SAN via a stochastic differential equation. The random input is a *Brownian motion* process, which is a type of continuous-time representation of a Gaussian process. Geman and Hwang (1986) show that the probability distribution describing the solution to this differential equation will, in the limit, be a uniform distribution (continuous or discrete) on the set Θ^* of global minima to $L(\boldsymbol{\theta})$. In the case of a unique $\boldsymbol{\theta}^*$, this corresponds to

---

[2] This appendix may be skipped or deferred without causing great harm to the overall understanding of implementation aspects of SAN.

convergence in probability (pr.) to $\boldsymbol{\theta}^*$. While there is a strong connection of this analysis to SAN, there is a slight gap between this approach based on continuous-time processes and the standard SAN algorithm, which is based on *discrete* update steps (see Section 8.2). The convergence theory summarized below is more directly tied to the basic SAN form of Section 8.2.

Let us now present a formal convergence analysis for the Metropolis-based SAN algorithm described in Section 8.2, where we let the iteration-dependent decaying temperature sequence $T = T_k$ depend on the current estimate of $\boldsymbol{\theta}$. As we will see, however, this dependence disappears for large k, making the convergence result effectively applicable to a standard state-independent temperature decaying at a (slow) rate T_k that is proportional to $1/\log\log k$. This result relies on a direct correspondence of the SA algorithm (8.7) to the SAN algorithm when $\boldsymbol{\theta}$ is continuous-valued and L is differentiable. The discussion below is based on Gelfand and Mitter (1993).

In showing the correspondence between SA with injected randomness and SAN, assume that the gradient $\boldsymbol{g}(\boldsymbol{\theta})$ exists on $\mathbb{R}^p$ (it does not actually have to be computed). As usual, suppose that $\hat{\boldsymbol{\theta}}_k$ is the Metropolis-type SAN sequence for optimizing L, with k the iteration counter. To define this sequence in terms of the SA recursion (8.7), let $q_k(\boldsymbol{x}, \boldsymbol{y})$ be a normal density function with mean $\boldsymbol{x}$ and covariance matrix $b_k^2\sigma_k^2(\boldsymbol{x})\boldsymbol{I}_p$, where $\sigma_k(\boldsymbol{x}) = \max\{1, a_k^{\tau}\|\boldsymbol{x}\|\}$ and $0 < \tau < 1/4$ ($\boldsymbol{y}$ is the dummy variable for the density). Further, let $s_k(\boldsymbol{x}, \boldsymbol{y}) = \exp\{-[L(\boldsymbol{y}) - L(\boldsymbol{x})]/T_k\}$ if $L(\boldsymbol{y}) > L(\boldsymbol{x})$, and $s_k(\boldsymbol{x}, \boldsymbol{y}) = 1$ otherwise, where $T_k(\boldsymbol{x}) = b_k^2\sigma_k^2(\boldsymbol{x})/(2a_k)$. (As we see below, $\boldsymbol{x}$ and $\boldsymbol{y}$ represent various values of $\boldsymbol{\theta}$.) The function $s_k(\boldsymbol{x}, \boldsymbol{y})$ acts as the *acceptance probability* (analogous to step 2 of the basic SAN algorithm in Section 8.2).

With the appropriate interpretation, the above formulation fits directly into the SAN steps of Section 8.2. The SAN sequence can be obtained through simulation, in a manner similar to the discrete $\boldsymbol{\theta}$ case. Given the current state $\hat{\boldsymbol{\theta}}_k$, generate a candidate solution $\boldsymbol{\theta}_{\text{new}}$ according to (the one-step Markov transition) probability density $q_k(\hat{\boldsymbol{\theta}}_k, \cdot)$. Set the next state $\hat{\boldsymbol{\theta}}_{k+1} = \boldsymbol{\theta}_{\text{new}}$ if $s_k(\hat{\boldsymbol{\theta}}_k, \boldsymbol{\theta}_{\text{new}}) > U_k$, where U_k is an independent $U(0, 1)$ random variable; otherwise, $\hat{\boldsymbol{\theta}}_{k+1} = \hat{\boldsymbol{\theta}}_k$. For any set $S \subseteq \mathbb{R}^p$, the sequence $\hat{\boldsymbol{\theta}}_k$ has Markov transition probabilities

$$P(\hat{\boldsymbol{\theta}}_{k+1} \in S \mid \hat{\boldsymbol{\theta}}_k = \boldsymbol{x}) = \int_S q_k(\boldsymbol{x}, \boldsymbol{y}) s_k(\boldsymbol{x}, \boldsymbol{y})\, d\boldsymbol{y} + r_k(\boldsymbol{x}) I_{\{\boldsymbol{x} \in S\}}, \tag{8.10}$$

where $I_{\{\cdot\}}$ is the indicator function for the set $\{\cdot\}$ (1 if $\{\cdot\}$ is true; 0 otherwise) and $r_k(\boldsymbol{x})$ is the normalization $1 - \int_{\mathbb{R}^p} q_k(\boldsymbol{x}, \boldsymbol{y}) s_k(\boldsymbol{x}, \boldsymbol{y})\, d\boldsymbol{y}$ (see Exercise 8.13 for an equivalent form). Let $\{\boldsymbol{w}_k\}$ be an i.i.d. sequence of $N(\boldsymbol{0}, \boldsymbol{I}_p)$-distributed random vectors and a_k and b_k be suitably chosen sequences.

Following the definition of a noisy root-finding function in Section 1.1 (with noise vector $\boldsymbol{e}_k$), the effective noises $\boldsymbol{e}_k$ here capture the outcome of the Metropolis sampling and decision process. Let $\boldsymbol{e}_k$ be defined by setting

$$\boldsymbol{e}_k = -\boldsymbol{g}(\hat{\boldsymbol{\theta}}_k) + \frac{b_k}{a_k}[1 - \sigma_k(\hat{\boldsymbol{\theta}}_k) d_k] \boldsymbol{w}_k,$$

where the (random) binary decision variable d_k = 0 or 1 according to $P(d_k = 1 | \hat{\boldsymbol{\theta}}_k) = s_k(\hat{\boldsymbol{\theta}}_k, \hat{\boldsymbol{\theta}}_k + b_k \sigma_k(\hat{\boldsymbol{\theta}}_k) \boldsymbol{w}_k)$ and it is assumed that $a_k > 0$ for all k (a standard SA assumption, of course). Given the above definitions, the basic SAN steps can be expressed in the SA-type recursion

$$\hat{\boldsymbol{\theta}}_{k+1} = \hat{\boldsymbol{\theta}}_k - a_k[\boldsymbol{g}(\hat{\boldsymbol{\theta}}_k) + \boldsymbol{e}_k] + b_k \boldsymbol{w}_k. \quad (8.11)$$

(Unlike some SA implementations, note that $\boldsymbol{w}_k$ and $\boldsymbol{e}_k$ are *dependent* for all k in this implementation.)

Gelfand and Mitter (1993) use the SA recursion in (8.11) to show that $\hat{\boldsymbol{\theta}}_k$ converges in probability to the set of global minima Θ^* provided that the initial condition is drawn from an appropriate set near the points in Θ^*. A special case of this result applies when L has a unique minimum at $\boldsymbol{\theta}^*$. Then, $P(\|\hat{\boldsymbol{\theta}}_k - \boldsymbol{\theta}^*\| \geq \eta) \to 0$ as $k \to \infty$ for any $\eta > 0$. That is, the SAN algorithm converges in pr. to $\boldsymbol{\theta}^*$ provided that the a_k and b_k in the temperature sequence $T_k(\hat{\boldsymbol{\theta}}_k) = b_k^2 \sigma_k^2(\hat{\boldsymbol{\theta}}_k)/(2a_k)$ are chosen appropriately. Example values of a_k and b_k that are sufficient for convergence are $a_k = a/(k+1)$ and $b_k = b/\sqrt{(k+1)\log\log(k+1)}$, as discussed in Section 8.4.

Although the above convergence result applies for the indicated *state-dependent* temperature sequence, observe that $\sigma_k^2(\boldsymbol{\theta}) = 1$ for all k sufficiently large when $\boldsymbol{\theta}$ lies in a bounded set Θ. So, the dependence of $\sigma_k^2(\boldsymbol{\theta})$ on $\boldsymbol{\theta}$ disappears for large k. In this sense, (8.11) is equivalent to classical SAN for large enough k. That is, the above convergence result is for an algorithm closely related to classical SAN where the temperatures have the *state-independent* values, $T_k = b^2/[2a \log\log(k+1)]$.

EXERCISES

8.1 Using the transition matrix in (8.4), complete the argument showing that repeated application of (8.2) will lead to a system whose state probability converges to the Boltzmann–Gibbs distribution in (8.1) at a fixed temperature T. That is, show $\overline{p}_i = c_T \exp[-x_i/(c_b T)]$ for all i.

8.2 Consider the setting of Proposition 8.1. For c = 1, 2, and 5, compare the bound in (8.5) with an analogous result based on the standard (two-sided)

Chebyshev inequality (see Exercise C.4 of Appendix C) and an assumption that the noises are normally distributed. In particular, produce a table contrasting the three bounds for the three values of c. Comment on whether there is any valid reason to use the two-sided Chebyshev inequality in this setting and discuss the benefits of making the Gaussian assumption (if valid!).

8.3 For a 10-city tour in the traveling salesperson problem, illustrate (analogous to Figure 8.1) the inversion of cities 4–8, the translation of cities 4–8 between cities 9 and 10, and the switching of cities 3 and 7.

8.4 With the same traveling salesperson setup and SAN coefficients as in Example 8.1, generate a random city-to-city cost matrix (each element being between 10 and 125). For this cost matrix, determine the optimal tour based on exhaustive search. Run SAN for five replications, showing your results as in Table 8.1.

8.5 Consider a traveling salesperson problem where the cost between cities is equal to the Euclidean norm of the distance between the cities. Explain why the optimal solution path cannot include any links that cross each other.

8.6 With the setting of Example 8.2 (indicated loss function, coefficient settings, number of measurements [N = 100, 1000, and 10,000], and number of replications [40]), implement SAN for three initial temperatures: $T = 0.001$, 0.10, and 10.0 (tune the perturbation standard deviation ρ so that your results for $T = 0.10$ are similar to those in Table 8.2). Report results for the three values of N and three initial temperatures. Comment on the results relative to the values in Table 8.2.

8.7 Implement SAN and SPSA for the skewed-quartic loss function in Example 6.6 (Section 6.7) with $p = 20$. Assume that $\hat{\boldsymbol{\theta}}_0 = [1, 1, \ldots, 1]^T$ and a total loss function budget of $N = 2000$ measurements for each replication of SAN and SPSA. The measurements for the algorithms have the form $y(\boldsymbol{\theta}) = L(\boldsymbol{\theta}) + \varepsilon$, where ε is i.i.d. $N(0, \sigma^2)$. Use the stepwise decaying temperature for SAN with 50 iterations at each temperature and tune the coefficients for both SAN and SPSA (use the standard practical gains for SPSA, as described in Section 7.5). Based on the sample mean of the losses at the final iterates over 40 independent replications, compare SAN and SPSA for $\sigma = 0$, 1, and 10.

8.8 Consider the loss function $L(\boldsymbol{\theta}) = \frac{1}{2}[(t_1^4 - 16t_1^2 + 5t_1) + (t_2^4 - 16t_2^2 + 5t_2)]$, $\boldsymbol{\theta} = [t_1, t_2]^T$, as given in Styblinski and Tang (1990, Example 1) (also Example 2.3 here).

(a) Identify the four local minima via any convenient algebraic or numerical means.

(b) With a starting $\boldsymbol{\theta}$ of $[4.0, 6.4]^T$, implement the SAN algorithm with stepwise decaying temperatures to find the global minimum (tune the SAN coefficients as desired). Use a maximum of 5000 function evaluations per replication and perform the experiment for five independent replications. Report the five solutions obtained (terminal $\boldsymbol{\theta}$ and $L(\boldsymbol{\theta})$) and the values of the various SAN coefficients.

8.9 Perform Exercise 8.8(b) with independent $N(0, 4^2)$ noise added to the loss function measurements. Use the techniques of Subsection 8.2.2 as appropriate. If you also did Exercise 8.8, how much degradation is seen in the five solutions?

8.10 Consider a $p = 5$ version of the loss function in eqn. (8.8). Identify one local minimum different from $6.2676\,[1, 1, 1, 1, 1]^T$ and not equal to the global minimum. Run SAN five different times using a maximum of 40,000 function evaluations per SAN replication with the identified local minimum as the initial condition. Report the five solutions.

8.11 Consider the four pairs of gains a_k, b_k in Section 8.4 that are valid for global convergence of SA with injected randomness. For $a = b = A = B = 1$ and $\alpha = 0.8$, contrast the four pairs of gains at $k = 1000$ and $k = 1{,}000{,}000$. Comment on reasons for the observed differences in magnitudes across the four combinations.

8.12 Consider the loss function $L(\boldsymbol{\theta}) = -\cos(t_1 - 100)\cos\left[(t_2 - 100)/\sqrt{2}\right] + \left[(t_1 - 100)^2 + (t_2 - 100)^2\right]/4000$, $\boldsymbol{\theta} = [t_1, t_2]^T$ (sometimes called the *Griewank function*). Let $\Theta = [-200, 400]^2$. The unique global minimum is $\boldsymbol{\theta}^* = [100, 100]^T$.

(a) Analytically or by plotting, provide some sense of roughly how many local minima there are in Θ and specifically identify three of these local minima (other than $\boldsymbol{\theta}^*$).

(b) Based on 50 random initial conditions generated uniformly in Θ, perform 50 replications of SPSA (without the injected noise shown in (8.7)) using 100,000 loss evaluations per replication and the algorithm steps of Subsection 7.5.1 (this implementation is consistent with the discussion at the end of Section 8.4). Assume noise-free loss measurements and use the same gain sequences (a_k and c_k) for all 50 runs. How many of the 50 terminal iterates $\hat{\boldsymbol{\theta}}_{50,000}$ effectively converge to the global minimum in the sense that $\left|L(\hat{\boldsymbol{\theta}}_{50,000}) - L(\boldsymbol{\theta}^*)\right| \le 0.01$ *and* $\left\|\hat{\boldsymbol{\theta}}_{50,000} - \boldsymbol{\theta}^*\right\| \le 0.2$?

8.13 Show that the right-hand side of (8.10) can be written as $\left\{1 - \int_{S^c} q_k(\boldsymbol{x}, \boldsymbol{y}) s_k(\boldsymbol{x}, \boldsymbol{y})\, d\boldsymbol{y} \text{ if } \boldsymbol{x} \in S;\ \int_S q_k(\boldsymbol{x}, \boldsymbol{y}) s_k(\boldsymbol{x}, \boldsymbol{y})\, d\boldsymbol{y} \text{ if } \boldsymbol{x} \notin S\right\}$.

CHAPTER 9

EVOLUTIONARY COMPUTATION I: GENETIC ALGORITHMS

This and the next chapter (Part II) focus on methods of evolutionary computation (EC). The role of evolution in the study of biology and, more generally, the life sciences and demographics is paramount. Essentially, evolution acts as a type of natural optimization process based on the conceptually simple operations of competition, reproduction, and mutation. The term EC (sometimes called *evolutionary algorithms*) refers to a class of stochastic search and optimization methods built on the mathematical emulation of natural evolution.

This chapter focuses on the most popular method of EC—genetic algorithms (GAs). The next chapter includes a summary of other evolutionary approaches, including some commentary on the distinctions between GAs and these other approaches. A warning is in order, however. The area of EC is undergoing rapid development with a concurrent lack of specific "industry standard" approaches. This makes the attempt to summarize the essence of this large and changing field in only two book chapters especially audacious, but let us push on regardless!

After some general remarks in Section 9.1, we summarize some of the history and motivating applications of GAs and other EC methods (Section 9.2). Then, in Section 9.3, we describe the coding process for the parameters being optimized, leading to the main evolution-like operations of the GA (Section 9.4) and the standard form of the algorithm in Section 9.5. Some modern extensions of the algorithm and practical suggestions (e.g., constraints) to enhance performance are presented in Section 9.6 and examples are given in Section 9.7. Concluding remarks are given in Section 9.8.

9.1 INTRODUCTION

Genetic algorithms (GAs) are the most popular of the evolutionary computation (EC) algorithms. Although GAs are general search algorithms, a leading (perhaps *the* leading) application is search as applied to function optimization.

The treatment here will reflect this application. Hence, we consider the familiar problem of minimizing $L(\boldsymbol{\theta})$ subject to $\boldsymbol{\theta} \in \Theta$. We continue in the spirit of Chapter 8 in focusing on the *global* optimization problem. Further, like several other methods considered thus far, the GA is not restricted to only continuous-variable problems. In particular, the elements of $\boldsymbol{\theta}$ can be real-, discrete-, or complex-valued. Whereas the annealing algorithms of Chapter 8 are based on analogies to the physical cooling of substances, the GA is based loosely on principles of natural evolution and survival of the fittest. In fact, in GA terminology, an equivalent *maximization* criterion, such as $-L(\boldsymbol{\theta})$ (or its analogue based on an encoded form of $\boldsymbol{\theta}$), is often referred to as the *fitness function* to emphasize the evolutionary concept of the fittest of a species having a greater likelihood of surviving and passing on its genetic material.

While the dominant use of GAs has been in optimization, it is worth mentioning at least some of the related applications. One is automatic programming (genetic programming), where the algorithm automatically adapts software to perform certain tasks (see Koza, 1992). Another application involves the use of GAs to study human social systems, where one might be interested in investigating the evolution of societies, including the impact of government policies, resource shortages, and human interaction with the environment. A GA might provide a simulation-based means for making policy recommendations. As stated in the preface of the 1992 update to the seminal Holland (1975), "GAs are a tool for investigating the phenomena generated by *complex adaptive systems*—a collective designation for nonlinear systems defined by interactions of large numbers of adaptive agents (economies, political systems, ecologies, immune systems, developing embryos, brains, and the like)." The goal in this chapter is more modest (but *not that* modest!) in focusing on the familiar mathematical problem of minimizing $L(\boldsymbol{\theta})$ (maximizing the fitness function).

A fundamental difference between GAs and all other algorithms considered up to now is that GAs work with a *population* of potential solutions to the problem. The previous algorithms worked with one candidate solution (say, $\hat{\boldsymbol{\theta}}_k$) and moved toward the optimum by updating this one estimate. GAs simultaneously consider multiple candidate solutions to the problem of minimizing L and iterate by moving this population of candidate solutions toward (one hopes) a global optimum. The terms *iteration* and *generation* are used interchangeably to describe the process of transforming one population of solutions to another. If the GA is successful, the population of solutions will cluster at the global optimum after some number of iterations, as illustrated in Figure 9.1 for a population of size 12 with problem dimension $p = 2$.

One iteration of a GA involves obtaining (possibly noisy) loss function measurements (or fitness function values) at each of the candidate solutions in the population and then using this collection of loss measurements to move the population to a new set of candidate solutions. The population-based approach lends itself to parallel processing since any loss measurement for a candidate

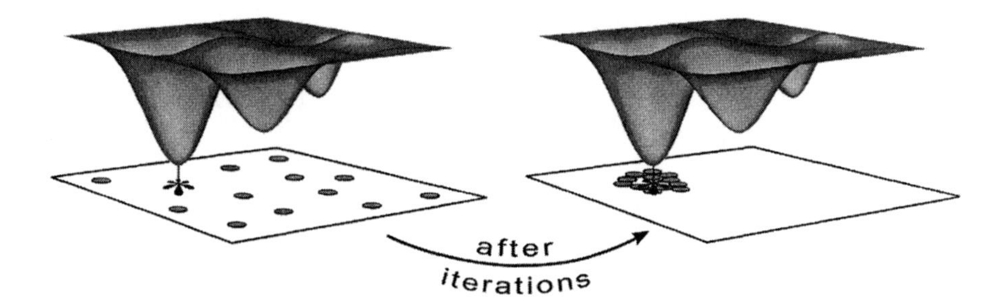

Figure 9.1. Minimization of multimodal loss function. Successful operations of a GA with a population of 12 candidate solutions clustering around the global minimum after some number of iterations (generations).

solution can be determined independently of the loss measurements for the other solutions. With this population-based structure in mind, the fundamental steps of a GA or other EC algorithm may be reduced to:

0. Initialization. Select an initial population size and values for the elements of the population; encode the elements of $\boldsymbol{\theta}$ in a manner convenient for the algorithm operations (e.g., encode in bit form).

1. Mixing. Mix the current population elements according to algorithmic analogues of principles of evolution and produce a new set of population values.

2. Evaluation. Measure the performance of the new population values and determine if the algorithm should be terminated. Return to step 1 if further improvement is required and/or the budget of function evaluations has not been expended.

More detailed steps for the GA are given in Section 9.5, after the basic encoding and mixing operations are defined.

The use of a population versus a single solution fundamentally affects the range of practical problems that can be considered. In particular, the GA tends to be best suited to problems where the loss function evaluations are computer-based calculations such as complex function evaluations or simulations. This contrasts with the single-solution approaches discussed in the prior chapters, where the loss function (or possibly gradient) evaluations may represent computer-based calculations *or* physical experiments. Population-based approaches are not generally suited to working with real-time physical experiments because of the parallel structure of the algorithm and the fact that all of the function evaluations are assumed to come from the same generating mechanism. Implementing a GA with physical experiments requires that either there be multiple identical experimental setups (parallel processing) or that the single experimental apparatus be set to the same state prior to each population member's loss evaluation (serial processing). These situations occur in relatively few practical settings.

As an example, consider the estimation of parameters associated with a feedback control mechanism. If there are many identical control systems available for training, then a population-based approach is feasible. On the other hand, if there is only one system available for training, then to obtain meaningful loss (or fitness) values it is expected that the system can be returned to the same state prior to running the experiment for each element of the population. This is not the usual situation given the inherent dynamic nature of the process. Note that some uses of GAs in control have been reported (e.g., Krishnakumar and Goldberg, 1992; Lennon and Passino, 1999). These approaches typically use the GA as an off-line mechanism for system identification (model estimation) or they replace physical experiments with computer-based simulations, hence putting them in the class of problems with computer-based loss evaluations for which population-based methods are well suited. Note also that nonpopulation methods (e.g., stochastic approximation) must also cope with the inability to return to the same state, but their nonparallel nature makes this somewhat easier (see, e.g., the control applications in Sections 3.2, 6.2, and 7.6).

Specific values of $\boldsymbol{\theta}$—often in some encoded form—are referred to as *chromosomes*. A chromosome is composed of genes, which play a role in the algorithm's iteration process in a way analogous to the biological role of genes in natural evolution. For our purposes , we consider a gene as being those encoded parts of the chromosome (typically, bits or groups of bits) that are associated with each element in $\boldsymbol{\theta}$. Hence a p-dimensional $\boldsymbol{\theta}$ has p genes in the chromosome representation. (Some GA references also employ other biologically motivated terms such as allele, phenotype, genotype, and locus, but we will not need those terms here.) The central idea in a GA is to move a set (population) of chromosomes from an initial collection of values to a point where the fitness function is optimized. The population size (number of chromosomes in the population) will be taken as N.

Initial attention toward the GA came largely with those in the fields of computer science and artificial intelligence, but more recently interest has extended to essentially all branches of business, engineering, and science where search and optimization are of interest. To a greater extent than other methods in stochastic search and optimization, GAs are shrouded in a certain mystique. The widespread interest in GAs appears to be due to this mystique, to the success in solving some difficult optimization problems, and to the intriguing and sometimes quite surprising results that arise from the approximate computer-based emulation of natural evolution. Unfortunately, some interest also appears to be due to a regrettable amount of "hype" and exaggerated claims. While GAs (and more generally, EC) are important tools within stochastic optimization, there appears to be no formal evidence of consistently superior performance—relative to other appropriate types of stochastic algorithms—in any broad, identifiable class of problems.

9.2 SOME HISTORICAL PERSPECTIVE AND MOTIVATING APPLICATIONS

9.2.1 Brief History

There has been interest in the application of biological evolutionary principles to mathematical search and optimization since at least the 1950s, with several early papers bearing a loose resemblance to more recent methods of EC. The cornerstones of modern EC—evolution strategies, evolutionary programming, and GAs—were developed independently of each other in the 1960s and 1970s. Rechenberg (1965) introduced evolution strategies for optimization of continuous variables, with subsequent development by Schwefel (1977) and others. This approach has gained a considerable following. Evolutionary programming, as described in Fogel et al. (1966), treats the candidate solutions in the population as symbols via finite-state machines. In its original configuration, this approach was largely limited to small problems. Subsequent development by the Fogel father–son team (Lawrence and David Fogel) and others has strengthened the approach.

Chapter 10 provides a more detailed discussion of the contrasts between GAs and other EC approaches. We now mention a few of the historical highlights associated with GAs, directing the reader to De Jong (1988), Goldberg (1989, Chap. 4), Mitchell (1996, Chap. 1), Michalewicz (1996, pp. 1–10), and Fogel (2000, Chap. 3) for more complete historical discussions.

In the 1960s and 1970s, John Holland and his colleagues at the University of Michigan introduced GAs in the notation and formulation of today, leading to publication of the seminal monograph *Adaptation in Natural and Artificial Systems* (Holland, 1975). Although it was recognized from the outset that GAs were suited to optimization problems, Holland and his associates considered even broader issues. They developed GAs as algorithmic representations of complex adaptive processes, ranging from biological systems to economies to political systems. In fact, in the preface to the 1992 reprinting/update of the 1975 monograph, Holland states in reference to the 1975 monograph "About the only change I would make would be to put more emphasis on improvement and less on optimization."

Nevertheless, current applications for GAs appear to be overwhelmingly aimed at optimization. This and the next chapter reflect that emphasis. (And, of course, there may be little practical difference between "improvement" and optimization since optimization ultimately underlies the analysis and modeling associated with adaptation and learning.) The main difference is one of emphasis, where optimization often focuses on the question of convergence of an algorithm, while Holland's adaptive system emphasizes the intermediate behavior as learning is taking place.

On the heels of the original Holland monograph was the doctoral dissertation of one of Holland's students, De Jong (1975). Among the results in this dissertation were a large number of numerical test cases. De Jong systematically studied the tuning process for GAs, establishing some reasonable ranges for some of the critical algorithm coefficients. These ranges continue to be cited today. One of the major innovations from the Holland team was the attempt to put GAs on a theoretically strong footing via the notion of schemas. This sometimes-controversial notion is discussed in Chapter 10. After the 1975 Holland monograph and De Jong dissertation, there was a relatively quiescent period of about a decade, with only a trickle of publications on GAs, most apparently by first- or second-generation students of Holland.

In the mid-1980s, activity picked up on GAs in parallel with activity in other dormant artificial intelligence concepts such as neural networks and fuzzy logic. The first large-scale conference devoted to GAs, the International Conference on Genetic Algorithms, was held in 1985 in Pittsburgh, Pennsylvania. The first full-fledged textbook on GAs, Goldberg (1989), was by one of Holland's most prolific students. This book is a lucid and accessible treatment of the subject based on information through the late 1980s. Much of the insight in this book continues to be relevant to the practical implementation of GAs (an update to parts of the 1989 book is Goldberg, 2002).

9.2.2 Early Motivation

The early interest in general adaptation and learning typically resulted in an optimization problem of some form. Holland (1975) and De Jong (1975) considered adaptation and learning problems in game playing, pattern recognition, and automatic programming, all areas remaining of great interest today despite the huge gains in algorithmic understanding and computational power. As an illustration of some of the issues that early GA researchers encountered, let us elaborate a bit on the problem of game playing.

The seemingly lighthearted task of building algorithms for game playing (checkers, chess, etc.) has led to many advances in artificial intelligence and adaptive control. Such problems have provided a good testbed for GAs and other EC algorithms as well as for machine learning methods such as the temporal difference algorithm (Chapter 11). Serious applications of game playing arise in areas such as law enforcement, strategic planning for the military, and economics. GAs were recognized in Holland (1975) and De Jong (1975) as appropriate algorithms for constructing game-playing strategies. One of the major difficulties in such tasks is the need to adapt to the strategy of an opponent. One typically will not know the opponent's strategy in advance, and hence the *adaptive* aspect of GAs for adjusting to an opponent's strategy—as revealed in the course of the game—is especially relevant.

The solution to game-playing problems typically runs into the same combinatorial problems seen with the traveling salesperson problem in Section 8.3. Even in the idealized case where one knows the opponent's strategy in advance, the evaluation of all possible solutions quickly gets out of hand. For example, there are $\boldsymbol{o}^m$ possible strategies that need to be tested in playing a game involving m moves and $\boldsymbol{o}$ options at each move. Even a small game, say $m = 20$

and $\boldsymbol{o} = 10$, would require over 3000 years of computation if each strategy can be evaluated in only 10^{-9} second. A typical chess game, of course, is much more complex than this (greater m and $\boldsymbol{o}$) and has the great complicating aspect of imperfect knowledge of an opponent's strategy.

The population-based aspect of GAs is consistent with a parallel evaluation of different strategies. The GA can consider a range of different strategies and combine aspects of successful strategies via its mixing operations (the mixing is referred to more formally as crossover). An intelligent evolution-based search can yield a good—and possibly optimal—solution with the explicit evaluations of only a fraction of the number of possible strategies.

9.3 CODING OF ELEMENTS FOR SEARCHING

9.3.1 Introduction

As mentioned in Section 9.1, an essential aspect of GAs is the encoding of $\boldsymbol{\theta}$ for performing the GA operations and the associated decoding to return to the natural problem space in $\boldsymbol{\theta}$. Usually, one works with a string representation to facilitate the operations of the GA (a string being a list of numbers). There are many ways that $\boldsymbol{\theta}$ can be encoded in strings. Standard binary $(0, 1)$ bit strings have historically been the most common (bit = ***b**inary dig**it***). Other encoding methods include gray coding (which also uses $(0, 1)$ strings, but differs in the way the bits are arranged) and multiple character encodings (> 2 elements in the string alphabet), including the full 10-character representation associated with the computer-based floating-point representation of the real numbers in $\boldsymbol{\theta}$. The 10-character coding is often referred to as *real-number coding* since it operates as if working with $\boldsymbol{\theta}$ directly. Based largely on successful numerical implementations, this natural representation of $\boldsymbol{\theta}$ has grown more popular over time.

The issue of coding in GAs is an area of active research and there is some controversy regarding the prominent role assigned to bit-string encoding. Some references discussing alternative coding schemes include Davis (1996, Chap. 4), Michalewicz (1996, Chap. 5), Tang et al. (1996), Mitchell (1996, Sects. 5.2 and 5.3), Koehler et al (1998), and Fogel (2000, Sects. 3.5 and 4.3). In this section we first consider the standard bit coding and then consider the other two approaches mentioned above, gray coding and real-number coding.

9.3.2 Standard Bit Coding

The binary $(0, 1)$ (i.e., bit) representation has traditionally been the most popular coding (Mitchell, 1996, p. 156; Davis, 1996, pp. 62–64). There appear to be several technical and other reasons for this popularity: (i) historical precedent, beginning with the seminal work of Holland (1975); (ii) relative ease of implementation of the genetic operations described in Section 9.4; (iii) familiarity with binary $(0, 1)$ manipulations, especially in light of their prominence in computer science where much of the work on GAs has taken place; (iv) rules of thumb that have developed for implementation under bit coding, including picking some of the important algorithm coefficients; and (v) theory associated with schemas suggesting that binary coding maximizes the amount of useful information being processed at each generation (versus other coding schemes). Items (i) through (iv) are largely self-explanatory. The schema theory mentioned in (v) is taken up in Section 10.3.

Because few definitions of $\boldsymbol{\theta}$ in practice are naturally in terms of bits, we now discuss the bit encoding ($\boldsymbol{\theta} \rightarrow$ bits) and decoding (bits $\rightarrow \boldsymbol{\theta}$) operations based on the elements of $\boldsymbol{\theta}$ being defined in terms more natural for the problem at hand. Of course, since one is always working with finite accuracy in solving an actual problem, a real-number definition for the elements of $\boldsymbol{\theta}$ is equivalent in practice to a discrete-valued definition.

Strictly speaking, the *encoding* process may not be needed in some applications since an implementation can be randomly initialized in bit space with all subsequent operations carried out in the bit space until the final decoding into $\boldsymbol{\theta}$. Encoding is, however, useful to understand as a guide to the relationship between $\boldsymbol{\theta}$ and the chromosomes in a GA. Further, encoding will be required for GA implementations when it is desirable to initialize certain elements in the population based on prior knowledge of good values of $\boldsymbol{\theta}$. Encoding can also be used to enforce constraints specified in the natural $\boldsymbol{\theta}$ space. The decoding operation is, of course, always required in converting a final GA solution to a physically meaningful $\boldsymbol{\theta}$.

We discuss one approach for both encoding and decoding below (many other approaches are possible). As is customary in GAs, let us assume that the individual elements of $\boldsymbol{\theta}$ are bounded above and below, and that the analyst knows the values of such bounds (hence, the constraint domain Θ is known to be a p-dimensional hypercube). Note that most of the other algorithms presented in this book have not had such a requirement (which is particularly relevant in conducting a fair comparison of algorithms). For ease of notation, we present the encoding/decoding procedures for a scalar θ. Obviously, the same procedures apply to an individual element of a vector $\boldsymbol{\theta}$, in which case the procedures are associated with one gene in the chromosome. Following the procedure described below generally leads to a different number of bits representing each of the genes (corresponding to each of the elements in $\boldsymbol{\theta}$). This is standard in GA implementation.

The approach described below maps the minimum value of θ (scalar element) to $[0\ 0\ 0 \ldots 0]$ and the maximum value to $[1\ 1\ 1 \ldots 1]$. A direct result of this desirable framework is that, in general, the mapping between a floating-point number and the bit representation is not one-to-one. In particular, a given floating-point number (to within the specified accuracy) may be represented by more than one bit representation (Exercise 9.4). In the description below, let b be the number of bits representing a scalar floating-point number (such as one element in a vector $\boldsymbol{\theta}$). Later we use B to represent the number of bits in a chromosome, which encompasses the p genes corresponding to all of the elements of $\boldsymbol{\theta}$; so $B \geq b$.

Encoding (scalar θ)

1. Let θ_{min} and θ_{max} be such that $\theta_{min} \leq \theta \leq \theta_{max}$ and let m represent the maximum number of positions after the decimal point that is relevant to the problem at hand ($m < 0$ denotes positions before the decimal). Choose a number of bits $b \geq 1$ such that b is the smallest number satisfying $10^m(\theta_{max} - \theta_{min}) \leq 2^0 + 2^1 + 2^2 + \ldots + 2^{b-1} = 2^b - 1$, the maximum value in a standard nonnegative integer representation for a string of length b bits.
2. Let $d = (\theta_{max} - \theta_{min})/(2^b - 1)$. The bit representation varies from [0 0 0...0] for θ_{min} to [1 1 1...1] for θ_{max}. Each increase in θ by an amount d increases the bit representation by one unit.
3. For the given θ, calculate $round[(\theta - \theta_{min})/d]$, where the operator $round[\cdot]$ rounds off the argument to the nearest integer. Represent the integer $round[(\theta - \theta_{min})/d]$ using the standard binary representation $[a_1\ a_2 \ldots a_b]$, where the elements a_i are either 0 or 1.

Decoding (scalar θ)

1. Assume a b-bit representation $[a_1\ a_2 \ldots a_b]$ derived as in the steps above.
2. To within the specified accuracy (m), the value for θ is given by

$$\theta = \theta_{min} + \frac{\theta_{max} - \theta_{min}}{2^b - 1} \sum_{i=1}^{b} a_i 2^{b-i} . \tag{9.1}$$

Let us now give an example of an encoding and a decoding for the elements of $\boldsymbol{\theta}$ when $p = 2$.

Example 9.1—Encoding and decoding. Let $\boldsymbol{\theta} = [t_1, t_2]^T \in [-4.00, 10.00] \times [1000, 4500]$ (so t_1 lies between -4.00 and 10.00 while t_2 lies between 1000 and 4500); $m = 2$ for the first element of $\boldsymbol{\theta}$ and $m = -2$ for the second element. For the first element, $b = 11$ since $2^{10} - 1 = 1023 \leq 10^m(\theta_{max} - \theta_{min}) = 10^2(10.00 - (-4.00)) \leq 2^{11} - 1 = 2047$. For the second element, we pick $b = 6$ since $2^5 - 1 = 31 \leq 10^{-2}(4500 - 1000) \leq 2^6 - 1 = 63$. Now if $\boldsymbol{\theta} = [-2.31, 4300]^T$, an encoding would be [0 0 0 1 1 1 1 0 1 1 1; 1 1 1 0 1 1], where the semicolon separates the genes for the two elements of $\boldsymbol{\theta}$. So, for example, the gene corresponding to the first element of $\boldsymbol{\theta}$ has $d = 0.00684$ and an integer value $round[(-2.31 - (-4.00))/d] = 247 = 2^0 + 2^1 + 2^2 + 2^4 + 2^5 + 2^6 + 2^7$. Applying the decoding formula (9.1), we recover the values of $\boldsymbol{\theta}$: -4.00 +

$(10.00-(-4.00))(2^0+2^1+2^2+2^4+2^5+2^6+2^7)/(2^{11}-1) = -2.311$ and $1000 + (4500 - 1000)(2^0+2^1+2^3+2^4+2^5)/(2^6-1) = 4277.8$. To the specified levels of accuracy (expressed via m), these recovered values are identical to the values of interest, −2.31 and 4300. ❑

9.3.3 Gray Coding

Gray coding also uses the (0, 1) alphabet in its string representation. It differs from the standard bit coding above in that floating-point values that are adjacent to each other, where adjacent is relative to the decimal accuracy chosen, vary by only one bit in the (0, 1) string. The primary motivation for gray coding is that adjacent floating-point values (i.e., the natural values in $\boldsymbol{\theta}$) may have very different representations when using the standard bit form above. This may cause the GA to need an undue number of iterations to move only a small distance in the natural problem space. For example, if a scalar θ is integer-valued, a move of one unit from $\theta = 7$ to $\theta = 8$ requires that all four bits in the representation [0 1 1 1] be changed. Given that the GA operates by flipping individual bits within the chromosome string (see Sections 9.4 and 9.5), the probability of simultaneously changing several bits to produce a small desired change is typically small.

A dual consequence of the requirement to change many bits to move a small amount in the space of floating-point numbers is the Hamming cliffs problem. Here, one finds that small changes to a binary code can cause very large changes in the elements of $\boldsymbol{\theta}$. For instance, in the simple four-bit example of the preceding paragraph, a change of only one bit from [0 0 0 0] to [1 0 0 0] causes the scalar θ to change over half its available range of [0, 15]. Although this may occasionally work to the algorithm's advantage (by forcing the search to cover a wider area and possibly jump away from local minima), it is generally accepted that Hamming cliffs introduce undesirable instability in the search process.

Gray coding is an alternative scheme that tends to make the bit representation more closely match the characteristics of the natural problem space. In gray coding, adjacent floating-point values differ by only one bit in the chromosome representation. Hence, it is expected that small changes in $\boldsymbol{\theta}$ can be achieved more easily. Further, a gray code tends to moderate (but not completely eliminate) the Hamming cliffs problem.

There is no unique gray code. In fact, Rana and Whitley (1997) conjecture that there are $O(b!2^b)$ different gray codes, where b is the number of bits in the representation. One convenient procedure for translating standard binary coding to gray coding is described in Michalewicz (1996, p. 98). Table 9.1 shows the binary and gray codes for integers 0 through 10 using this translation procedure. Relative to the standard code, it is apparent that there is a more gradual change in the gray-coded representation as the integer representation changes. Unlike some of the changes in the standard code, each change in integer by one unit causes a change in the gray code of only one bit.

Table 9.1. Comparison of standard binary code and gray code.

Integer	Standard Binary	Gray
0	0 0 0 0	0 0 0 0
1	0 0 0 1	0 0 0 1
2	0 0 1 0	0 0 1 1
3	0 0 1 1	0 0 1 0
4	0 1 0 0	0 1 1 0
5	0 1 0 1	0 1 1 1
6	0 1 1 0	0 1 0 1
7	0 1 1 1	0 1 0 0
8	1 0 0 0	1 1 0 0
9	1 0 0 1	1 1 0 1
10	1 0 1 0	1 1 1 1

9.3.4 Real-Number Coding

We conclude this section with a discussion of direct real-number (floating-point) coding. Here, of course, there is no coding per se because we work directly with the floating-point-valued elements of $\boldsymbol{\theta}$ as implemented on a computer. So a convenient real-number coding scheme is to simply have each of the N chromosomes in the population correspond directly to one $\boldsymbol{\theta}$ vector (hence each of the p genes in the chromosome correspond to the floating-point number for one element of $\boldsymbol{\theta}$). This is the form of coding used in other population-based methods to be covered in Chapter 10 (evolution strategies and evolutionary programming). There is accumulating evidence that this natural form often offers performance equal or superior to that available with the binary-based coding methods described above.

Several of the advantages of the real-number implementation are: (i) the natural physical interpretation of the solution; (ii) the relative ease of handling constraints more general than the component-wise constraints described above for binary coding (see Section 9.6); (iii) the simplification of several implementation issues; and (iv) the growing number of numerical studies pointing to superior performance (e.g., Davis, 1996, Chap. 5; Michalewicz, 1996, Chap. 5; Salomon, 1996; and Fogel, 2000, Sect. 3.5). In addition to the above, some partial theoretical justification for real-number coding is given in Salomon (1996). This reference considers multimodal continuous $\boldsymbol{\theta}$ optimization problems with loss functions that are separable:

$$L(\boldsymbol{\theta}) = L_1(t_1) + L_2(t_2) + \ldots + L_p(t_p),$$

where $\boldsymbol{\theta} = [t_1, t_2, \ldots, t_p]^T$ and the $L_i(\cdot)$ are some nonlinear functions. In the parlance of GAs (and biology), such separable loss functions are said to have no *epistasis* (epistasis refers to interaction among genes in a chromosome, tantamount here to interaction among the elements of $\boldsymbol{\theta}$). For such L, real-number coding can reduce run times by a factor of B^{b-1} (relative to standard bit coding), where b is the same for all parameter elements t_i and $B = pb$ is the length of the chromosome (number of bits). Salomon (1996) also suggests qualitatively similar behavior for nonseparable loss functions.

9.4 STANDARD GENETIC ALGORITHM OPERATIONS

Before presenting the steps of the algorithm, we must define the basic operations that go into the iterative process. For consistency with the GA literature, assume that $L(\boldsymbol{\theta})$ has been transformed to a fitness function with higher values being better. A common transformation is to simply set the fitness function to $-L(\boldsymbol{\theta}) + C$, where $C \geq 0$ is a constant that ensures that the fitness function is nonnegative on Θ (nonnegativity is only required in some GA implementations). Hence, the operations are described for a *maximization* problem. (Section 9.6 includes some discussion about converting from loss functions to fitness functions in nonstandard cases where the simple transformation, fitness = $-L(\boldsymbol{\theta}) + C$, is not sufficient.) Further, it is assumed that the fitness evaluations are noise-free. Unless otherwise noted, the operations below apply with any coding scheme for the chromosomes.

9.4.1 Selection and Elitism

After evaluating the fitness function for the current population of chromosomes, a subset of chromosomes is selected to use as parents for the succeeding generation. This operation is where the survival of the fittest principle arises, as the parents are chosen according to their fitness value. While the aim is to emphasize the fitter chromosomes in the selection process—so that the offspring tend to have even higher fitness—one must be careful about overdoing this emphasis. Too much priority given to the fittest chromosomes early in the optimization process tends to reduce the diversity needed for an adequate search of the domain of interest, possibly causing premature convergence and preventing the algorithm from uncovering chromosomes that are globally optimal.

Associated with the selection step is the optional "elitism" strategy (first described in De Jong, 1975, pp. 101–106), where the $N_e < N$ best chromosomes (as determined from their fitness evaluations) are placed directly into the next generation. This guarantees the preservation of the current N_e best chromosomes. Note that the elitist chromosomes in the original population are also eligible for selection and subsequent recombination. De Jong's method appended the elitist chromosomes to the current population, creating a temporary population of size larger than N. We use the more common method where the elitist elements are included *within* the N chromosomes.

As with the coding operation for $\boldsymbol{\theta}$, many schemes have been proposed for the selection process. We now describe one of the most popular methods—*roulette wheel selection* (also called *fitness proportionate selection*). In this selection method, the fitness functions must be nonnegative on Θ. Other approaches are summarized in Section 9.6. In the roulette wheel approach, an individual's slice of a Monte Carlo-based roulette wheel is an area proportional to its fitness. The wheel is spun $N - N_e$ times and the parents are chosen based on where the pointer stops. For a given generation (iteration), this can be implemented in software as follows:

1. Sum the total of the N fitness values; call this sum S_f.
2. Generate a uniformly distributed random variable on the interval $[0, S_f]$ (recall that the fitness values are nonnegative).
3. Go through the population, returning the chromosome whose fitness added to the sum of the previous fitnesses is greater than or equal to the random number. Perform this roulette sampling $N - N_e$ times.

The following example illustrates roulette selection.

Example 9.2—Roulette selection. Table 9.2 depicts an example of the roulette selection for ten chromosomes with two chromosomes set aside as elite chromosomes ($N = 10$, $N_e = 2$). Part (a) of the table shows the fitness value and running total, leading to $S_f = 3.50$. Part (b) shows the $N - N_e = 8$ draws from the uniform [0, 3.50] distribution and the corresponding selected chromosomes.

Table 9.2. Roulette parent selection process.

(a) Example chromosomes and their fitness values when $N = 10$.

Chromosome	1	2	3	4	5	6	7	8	9	10
Fitness	0.10	0.20	0.05	0.45	0.25	1.00	0.10	0.80	0.05	0.50
Cumulative sum	0.10	0.30	0.35	0.80	1.05	2.05	2.15	2.95	3.00	3.50

(b) Roulette choice for $N - N_e = 8$ parent chromosomes to form next generation. Elite chromosomes 6 and 8 are placed directly into the next population.

Random no.	0.34	2.96	0.86	3.38	2.27	1.33	1.72	0.36
Selected chromosome	3	9	5	10	8	6	6	4

Note that the chromosome with the greatest fitness value (chromosome 6) received multiple selection, reflecting its fitter status, whereas there are only two selections (3 and 9) from among the five lowest-rated chromosomes (1, 2, 3, 7, and 9). The individual parent pairs for forming the next generation of eight chromosomes to be added to the two elite chromosomes would be chromosomes 3 and 9, 5 and 10, 8 and 6, and 6 and 4 (note that it is possible in general for both parents to be the same chromosome). ❑

9.4.2 Crossover

The crossover operation creates offspring of the pairs of parents from the selection step. A crossover probability, say P_c, is used to determine if the offspring will represent a blend of the chromosomes of the parents. If no crossover takes place, then the two offspring are clones of the two parents. If crossover does take place, then the two offspring are produced according to an interchange of parts of the chromosome structure of the two parents. Figure 9.2 illustrates this for the case of a nine-bit representation of the chromosomes. Case A shows one-point crossover, where the bits appearing after one randomly chosen dividing point in the chromosome are interchanged. Case B shows two-point crossover, where only the middle section is interchanged. In general, one can have a number of splice points up to the number of bits minus one, $B - 1$.

Note that the crossover operator also applies directly with real-number coding since there is nothing intimately connected to binary coding in single- or multi-point crossover. All that is required are two lists of compatible symbols. For example, one-point crossover applied to the chromosomes ($\boldsymbol{\theta}$ values) [2.1, −7.4, 4.0, 3.9 | 6.2, −1.5] and [−3.8, 3.3, 9.2, −0.6 | 8.4, −5.1] yields the two children: [2.1, −7.4, 4.0, 3.9, 8.4, −5.1] and [−3.8, 3.3, 9.2, −0.6, 6.2, −1.5]. Another possible crossover operator under real-number coding is to average the two parents, yielding one child (Davis, 1996, pp. 66–69, uses this with success).

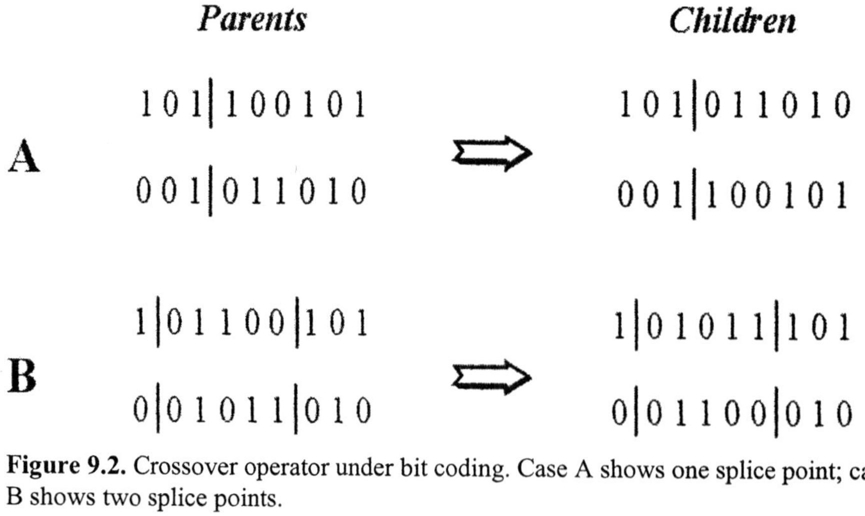

	Parents		*Children*
A	1 0 1 | 1 0 0 1 0 1	⇨	1 0 1 | 0 1 1 0 1 0
	0 0 1 | 0 1 1 0 1 0		0 0 1 | 1 0 0 1 0 1
B	1 | 0 1 1 0 0 | 1 0 1	⇨	1 | 0 1 0 1 1 | 1 0 1
	0 | 0 1 0 1 1 | 0 1 0		0 | 0 1 1 0 0 | 0 1 0

Figure 9.2. Crossover operator under bit coding. Case A shows one splice point; case B shows two splice points.

9.4.3 Mutation and Termination

Since the initial population may not contain encoded information rich enough to find the solution via crossover operations alone, the GA also uses a mutation operator where the chromosomes are randomly changed. For the binary coding, the mutation is usually done on a bit-by-bit basis where a chosen bit is flipped from 0 to 1, or vice versa. Mutation of a given bit only occurs with small probability P_m to preserve the good chromosomes created through crossover. Figure 9.3 depicts a mutation in the third bit for the first child chromosome appearing in Figure 9.2.

Real-number coding requires a fundamental change in the mutation operator above. This follows from the loss of a uniqueness property associated with binary coding. With a (0, 1)-based coding, an opposite is uniquely defined, but with a real number, there is no clearly defined opposite (e.g., it does not make sense to "flip" the 7.63 element). One type of mutation operator is simply to add small independent normal (or other) random vectors to each of the chromosomes (the $\boldsymbol{\theta}$ values) in the population (see, e.g., Salomon, 1996). In a manner more closely corresponding to the bit-based mutation above, some real-number mutation operators work on the individual elements in each of the $\boldsymbol{\theta}$, changing a given element if a particular probability threshold is crossed (e.g., Michalewicz, 1996, p. 103; Davis, 1996, pp. 66–69). There appears to be little evidence that these per-element methods are superior to the simpler approach of adding a random vector to the $\boldsymbol{\theta}$.

As discussed in Section 1.2, there is no easy method for automatically stopping most (all?) stochastic search and optimization algorithms with a guarantee of being close to an optimum. GAs are no exception. The one obvious means of stopping a GA is to end the search when a budget of fitness (equivalently, loss) function evaluations has been spent. Alternatively, termination may be performed heuristically based on subjective and objective impressions about convergence. In the case where noise-free fitness measurements are available, criteria based on fitness evaluations may be most useful.

For example, suppose that F^* and F_* represent the maximum and minimum fitness values over the N population values within a generation. Then a criterion suggested in Schwefel (1995, p. 145), which is based on the population elements yielding nearly the same fitness value, is to stop the search when $|F^* - F_*| \le \eta$ for some $\eta > 0$. Alternative criteria based on the decoded population elements themselves (i.e., the $\boldsymbol{\theta}$ values) may be useful, especially if the fitness measurements are only available with noise (where it would be too

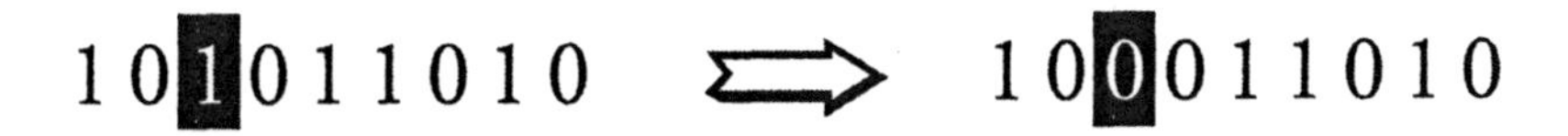

Figure 9.3. Mutation operator affecting one bit in a binary coding.

costly to use averaging to approximate the values F^* and F_* at each iteration). For example, terminate if the normed difference between all decoded population elements within one generation is less than some $\eta > 0$, or terminate if the best $\boldsymbol{\theta}$ value in the population changes negligibly over several generations.

9.5 OVERVIEW OF BASIC GA SEARCH APPROACH

There are *many* variations of the GA, making it difficult to present one "standard" form. Nevertheless, this section presents the steps of a basic form of the algorithm. These are essentially the steps described in Holland (1975, Chap. 6) with the following two exceptions. First, we do not include the inversion operator due to its relative lack of effectiveness and resulting lack of application in modern GAs (see, e.g., Davis, 1996, p. 21; Mitchell, 1996, p. 161). Second, we include the elitist strategy described in Subsection 9.4.1 as part of the basic GA (elitism did not appear in Holland, 1975). Section 9.6 discusses some popular enhancements to the basic algorithm.

Core GA Steps for Noise-Free Fitness Evaluations

Step 0 **(Initialization)** Randomly generate an initial population of N chromosomes and evaluate the fitness function (the conversion of $L(\boldsymbol{\theta})$ to a function to be maximized for the encoded version of $\boldsymbol{\theta}$) for each of the chromosomes.

Step 1 **(Parent selection)** Set $N_e = 0$ if elitism strategy is not used; $0 < N_e < N$ otherwise. Select with replacement $N - N_e$ parents from the full population (including the N_e elitist elements). The parents are selected according to their fitness, with those chromosomes having a higher fitness value being selected more often.

Step 2 **(Crossover)** For each pair of parents identified in step 1, perform crossover on the parents at a randomly (perhaps uniformly) chosen splice point (or *points* if using multi-point crossover) with probability P_c. If no crossover takes place (probability $1 - P_c$), then form two offspring that are exact copies (clones) of the two parents.

Step 3 **(Replacement and mutation)** While retaining the N_e best chromosomes from the previous generation, replace the remaining $N - N_e$ chromosomes with the current population of offspring from step 2. For the bit-based implementations, mutate the individual bits with probability P_m; for real coded implementations, use an alternative form of "small" modification (in either case, one has the option of choosing whether to make the N_e elitist chromosomes candidates for mutation).

Step 4 **(Fitness and end test)** Compute the fitness values for the new population of N chromosomes. Terminate the algorithm if the stopping criterion is met or if the budget of fitness function evaluations is exhausted; else return to step 1.

The above steps are general enough to govern many (perhaps most) implementations of GAs. Many details remain to be specified before one can actually implement a GA in software. Several versions of the GA are available at the book's Web site. The performance of a GA typically depends greatly on the implementation details, just as we have seen with other stochastic optimization algorithms. Some of these practical implementation issues are taken up in the next section.

9.6 PRACTICAL GUIDANCE AND EXTENSIONS: COEFFICIENT VALUES, CONSTRAINTS, NOISY FITNESS EVALUATIONS, LOCAL SEARCH, AND PARENT SELECTION

As suggested above, the core steps in Section 9.5 leave much to be specified before implementation. Further, a straightforward implementation of the core algorithm is often insufficiently powerful to address serious, "industrial strength" problems. This section summarizes some of the modifications, enhancements, and tricks that have evolved to address implementation needs. It is obviously impossible to cover all but a sliver of these variations in this section, but we hope to convey the flavor of the kinds of variations possible. More detailed discussions are given in Mitchell (1996, Chap. 5), Michalewicz (1996, Chaps. 4–6), Davis (1996, Part I), Fogel (2000, Chaps. 3 and 4), Goldberg (2002, Chap. 12), and other references mentioned below.

As with other stochastic optimization methods, the choice of algorithm-specific coefficients has a significant impact on performance. In fact, with the GA, there are more coefficients to be set and decisions to be made than with any of the algorithms seen thus far. Included in the decisions and coefficients to be set to run a GA are: the choice of chromosome encoding, the population size (N), the probability distribution generating the initial population, the strategy for parent selection (roulette wheel or otherwise), the number of splice points in the crossover, the crossover probability (P_c), the mutation probability (P_m), the number of retained chromosomes in elitism (N_e), and some termination criterion. De Jong (1975, especially Chaps. 3 and 4) devoted a considerable fraction of his dissertation to exploring these issues, and developed broad guidelines for picking some of these quantities. These and other guidelines are reviewed in Mitchell (1996, pp. 175–177) and Weile and Michielssen (1997).

Obviously, there is not complete unanimity on the selection of the GA coefficients, but there is some consensus that the population size, crossover probability, and mutation rate for a standard bit implementation satisfy $20 \leq N \leq 100$, $0.60 \leq P_c \leq 0.95$, and $0.001 \leq P_m \leq 0.01$, respectively. Absent prior information suggesting other values, these ranges may provide reasonable starting points for an implementation. Ideally the values for these (and possibly other) GA coefficients should be allowed to evolve over the iterations. Davis (1996, Chap. 7) presents an adaptive method that changes some of the coefficient settings over the course of a run. Of course, such adaptation complicates the algorithm and may not yield a benefit worth the complication.

Constraints on $L(\boldsymbol{\theta})$ (or the equivalent fitness function) and/or $\boldsymbol{\theta}$ are obviously of major importance in practical problems, as discussed many times earlier in this book. The bit-based implementation of GAs provides one natural (but limited) way of imposing component-wise lower and upper bounds on the elements of $\boldsymbol{\theta}$ (i.e., a hypercube constraint). More general approaches to handling constraints are discussed in Michalewicz (1996, Chap. 7 and Sects. 4.5 and 15.3). This reference shows how certain techniques in deterministic nonlinear programming can be extended to the GA context. Included in the suite of approaches are many variations of the penalty function method that was mentioned in Section 7.7 here. Real-number-based implementations of GAs are typically more adaptable to general constraints using techniques similar to those of nonlinear programming. A summary of available methods for handling constraints in GAs is given in Petridis et al. (1998) and Michalewicz and Fogel (2000, Chap. 9).

One of the major themes in this book has been optimization with noise in the loss function measurements. Unfortunately, there appears to be relatively little formal analysis of GAs in the presence of noise, although the application and testing of GAs in such cases was carried out as early as De Jong (1975, p. 203). Several of the references considering the problem are Rana et al. (1996), Nissen and Propach (1998), and Stroud (2001), all of which present numerical results from test cases. As with the search algorithms considered in earlier chapters, there is a fundamental tradeoff of (i) more accurate information for each function input (typically, via an averaging of the inputs) and fewer function inputs versus (ii) less accurate ("raw") information to the algorithm together with a greater number of inputs to the algorithm.

The above-mentioned references reveal performance characteristics that we have seen in other contexts with noisy loss measurements: namely, that short- and long-run results can be significantly degraded and that results can be surprising. GAs are sometimes claimed as being especially appropriate for noisy environments due to implicit averaging through the population-based structure. However, some such claims seem to ignore the additional cost per iteration (in loss evaluations) of a population-based method relative to the cost of a standard nonpopulation method (such as a random search algorithm [Subsection 2.2.2]).

There appears to be no rigorous comparison of GAs with other algorithms regarding relative robustness to noise. In fact, because of the discrete nature of certain decision steps in the algorithm, GAs may suffer relative to stochastic approximation-type algorithms that are designed for noise. That is, as with random search in Chapter 2 and simulated annealing in Chapter 8, noise fundamentally changes decisions made within the algorithm when the decisions are based on testing whether one fitness value is better than another. For example, the presence of nontrivial noise may alter parent selection and choices for elitism (see Exercise 9.3). Although there have been successful implementations of GAs with noisy measurements, these successes are problem specific and/or involve noise with a very special structure (e.g., Baum et al., 2001). Regarding noise, Michalewicz and Fogel (2000, p. 325) state: "There really are no effective heuristics to guide the choices to be made that will work in general." Some numerical comparisons are given in Section 9.7.

One of the most powerful extensions of GAs is to strengthen their local search capabilities. In this way, one aims to cope with the common problem of a GA bogging down after doing a credible job of isolating the approximate location of a global optimum. Michalewicz (1996, Chap. 6) suggests a means by which the GA mutation operator can be modified to strengthen local search capabilities and provides the successful results of three numerical experiments on optimal control problems. As described in the following paragraph, an even easier and more powerful approach for continuous $\boldsymbol{\theta}$ problems may simply be to append a local optimization technique such as one of the classical deterministic methods mentioned in Chapter 1. Alternatively, one of the stochastic approximation methods of Chapters 4–7 can be used, especially if the loss (or gradient) measurements are only available in the presence of noise.

In particular, the GA can be turned off when it has stopped making significant progress toward a solution. Then, one of the local algorithms may be invoked using the GA solution (perhaps the mean of the population values) as the initial condition. These local algorithms are likely to be faster since they work directly (e.g., conjugate gradient or Robbins–Monro/stochastic gradient) or indirectly (e.g., simultaneous perturbation stochastic approximation [SPSA]) with the important gradient information that can dramatically speed convergence. An interesting hybrid form of GA/gradient descent is given in Blackmore et al. (1997). Their *congregational descent* algorithm is based on running a *population* of gradient descent algorithms. A significant benefit is that the differential equation theory discussed in Section 4.3 can be used to establish convergence properties of the algorithm. This includes theory for the *rate* of convergence.

Considerable effort has gone into alternative schemes for the selection of the chromosomes in the population that will produce offspring for the next generation. The essential aim is to favor those chromosomes producing better fitness values while preserving enough diversity to allow for an adequate search of the domain of interest $\Theta \subseteq \mathbb{R}^p$ to uncover possible hidden optima. The roulette wheel approach was discussed in Subsection 9.4.1. Part of the motivation for alternatives to the roulette wheel is to cope with problems resulting from large differences in the relative magnitudes of the fitness function, including possible premature convergence. This arises from the selection scheme putting too much emphasis on the better chromosomes encountered early in the iteration process.

One way of dealing with this is to rescale the fitness values so that they do not vary many orders of magnitude over the domain of interest (e.g., if $-L(\boldsymbol{\theta}) > 0$ for all $\boldsymbol{\theta} \in \Theta$ is the fitness value, then using $\log(-L(\boldsymbol{\theta}))$ instead of $-L(\boldsymbol{\theta})$ may be effective). Methods such as *rank selection* (Baker, 1985) and *tournament selection* (Goldberg and Deb, 1991) are also designed to help cope with possible premature convergence. In rank selection, only the ranks 1 to N of the chromosomes (instead of their fitness values) are used in a proportional selection scheme. Analogous to the median (versus the mean) in statistics, rank selection is less sensitive to extreme values and thereby less likely to be dominated by such values early in the iteration process.

Tournament selection is similar to rank selection in that fitness values are only used to determine whether one chromosome is better than another. It differs in that chromosomes are compared in a "tournament," with the better chromosome being more likely to win. The tournament process is continued by sampling (with replacement) from the original population until a full complement of parents has been chosen. The most common tournament method is the binary approach, where one selects two pairs of chromosomes and chooses the two parents according to which chromosome has the higher fitness in each of the two pairs. Example 9.3 illustrates the process and Examples 9.4 and 9.5 in Section 9.7 use this method with success in some numerical studies. Mitchell (1996, Sect. 5.4) provides a good survey of several other selection methods.

Example 9.3—Binary tournament selection. Let us use the example in Table 9.2. To pick the first parent, we uniformly generate two integers in $\{1, 2, \ldots, 10\}$, say 2 and 7. For the second parent, the same process yields (say) 6 and 7. Then using the fitness values in Table 9.2(a), we find parent 1 = chromosome 2 and parent 2 = chromosome 6. ❑

9.7 EXAMPLES

This section reports on three example implementations with GAs. The first two of these are numerical studies and the third is a conceptual problem. As is certainly clear to the reader by now, any set of numerical studies provides only limited insight into the general performance characteristics of an algorithm. Nevertheless, with the dearth of theory providing usable performance insight for GAs, simulation experiments seem to be the primary tool available for gaining insight into expected performance. This section attempts to use representative implementations of GAs on problems that span a range of the kind of problems encountered in practice. Obviously, countless other references contain numerical studies of GAs, including many of the citations in this chapter. There is no pretense that the numerical studies here represent a comprehensive evaluation of GAs. There are three versions of the GA used: two versions are bit-coded, one with roulette selection (Section 9.4) and one with tournament selection (Section 9.6), and the other version is real-coded with tournament selection. The GA studies here are conducted with MATLAB programs available at the book's Web site (**`GAbit_roulette`**, **`GAbit_tourney`**, and **`GAreal_tourney`**).

To be consistent with the remainder of this book, we describe the problems in terms of minimization for loss functions, even though the internal mechanics of the GAs are in terms of maximization of fitness functions. The examples here are representative of three important classes of optimization problems: smooth (continuous) loss functions with one minimum, smooth loss functions with many minima, and nonsmooth (discrete) loss functions.

Example 9.4—Skewed-quartic loss function. Consider the skewed-quartic loss function in Example 6.6 (Section 6.7) with $p = 10$: $L(\boldsymbol{\theta}) = \boldsymbol{\theta}^T\boldsymbol{B}^T\boldsymbol{B}\boldsymbol{\theta} + 0.1\sum_{i=1}^{10}(\boldsymbol{B}\boldsymbol{\theta})_i^3 + 0.01\sum_{i=1}^{10}(\boldsymbol{B}\boldsymbol{\theta})_i^4$, where $(\cdot)_i$ represents the ith component of the argument vector $\boldsymbol{B}\boldsymbol{\theta}$, and $\boldsymbol{B}$ is such that $p\boldsymbol{B} = 10\boldsymbol{B}$ is an upper triangular matrix of 1's. This unimodal function has a high degree of parameter interaction (epistasis) and a high degree of skewness (the ratio of maximum to minimum eigenvalue of the Hessian at $\boldsymbol{\theta}^*$ is approximately 65). A single minimum occurs at $\boldsymbol{\theta}^* = \boldsymbol{0}$. The problem of determining a global minimum from among multiple local minima is not relevant here.

This study compares several implementations of GAs against SPSA (Chapter 7) in cases with both noise-free and noisy loss measurements. As discussed in Schwefel (1995, pp. 158–159), a difficulty in performing a fair comparison of GAs to many other algorithms (including basic SPSA) is that the GA may require more prior information to implement. In particular, with GAs, one specifies a hypercube containing the solution, information that may not be available to many other algorithms. To compensate for this advantage, we use a constrained version of SPSA.

The GAs are initialized with N values of $\boldsymbol{\theta}$ uniformly distributed in the hypercube $\Theta = [-1.6383, 1.6384]^{10}$, with the endpoints chosen so that $m = 4$ and $b = 15$ (postdecimal accuracy and number of bits, respectively) for all elements of $\boldsymbol{\theta}$. The slightly peculiar endpoints are chosen so that each increment in the bit representation accounts for exactly 0.0001 in the corresponding element in $\boldsymbol{\theta}$. It is not obvious how one initializes a single-path search method (such as SPSA) to provide a fair comparison with a population-based method such as a GA. The approach here is to pick an initial condition for the single-path method that has the same mean distance to $\boldsymbol{\theta}^*$ as the elements in the initial population in a GA. In this vein, the SPSA solution is constrained to lie in Θ and the algorithm is initialized by picking $\hat{\boldsymbol{\theta}}_0$ such that the distance to $\boldsymbol{\theta}^*$ is the same as the mean distance of the GA population to $\boldsymbol{\theta}^*$ ($\hat{\boldsymbol{\theta}}_0 = 0.936\times[1, 1,\ldots, 1]^T$ is used here).

Table 9.3 summarizes the results of the study based on 50 independent replications of optimization runs, each based on 2000 loss function measurements. All algorithm coefficients for the GAs and SPSA (N, P_c, a_k, etc.) are numerically tuned to achieve approximately optimal performance. Mutation for the real-coded GA is implemented by adding a small p-variate normal random vector to the nonelite members of the population (in the same general manner as Example 9.5 below). For the bit-based GAs, we used the formula in Exercise 9.6 to set the number of generations at the smallest number such that the total number of *expected unique* function evaluations is at least 2000. This accounts for the chromosomes retained through elitism and the fact that some offspring are identical to the parents. For the roulette selection implementation, the fitness values are of the form $-L(\boldsymbol{\theta}) + C$, where C is changed at each iteration to be the highest loss value in the population (so all fitness values at each iteration are nonnegative) and $\boldsymbol{\theta}$ is the decoded form of the bit representation. The tournament selection implementation simply uses fitness = $-L(\boldsymbol{\theta})$ since there is no requirement of nonnegativity.

Table 9.3 includes results for both noise-free and noisy loss measurements. In the latter, the noise in the loss function measurements at any value of $\boldsymbol{\theta}$ is given by $[\boldsymbol{\theta}^T, 1]\boldsymbol{z}$, where $\boldsymbol{z} \sim N(\boldsymbol{0}, \sigma^2 \boldsymbol{I}_{11})$ is independently generated at each $\boldsymbol{\theta}$. This relatively simple noise structure represents the usual scenario where the noise values in the loss measurements are dependent on $\boldsymbol{\theta}$; the last (z_{11}) term in $\boldsymbol{z}$ provides some degree of independence at each noise contribution and ensures that the loss measurements (i.e., the $y(\cdot)$) always contain noise of variance at least σ^2 (even if $\boldsymbol{\theta} = \boldsymbol{0}$). Approximate 90 percent confidence regions are shown below the loss values, computed according to the t-distribution method of Appendix B.

All algorithms show significant improvement relative to the loss values in the first iteration. For the noise-free case, the terminal GA solutions (corresponding to the values in the table) are the best loss values in the population at the last generation. For the noisy case, the GA solution is the population element having the lowest noisy measurement in the final population (the imperfect information resulting from the noise is not too damaging when the population has clustered around the solution). As usual, for evaluation purposes, the values in the table are the true (noise-free) loss values.

Table 9.3 indicates that the two bit-based GA forms are inferior to the real-coded form in the no-noise case; all the GAs are similar in the noisy case. SPSA outperforms the GAs in both the noise-free and noisy cases. Other studies

Table 9.3. Sample mean of terminal loss values for GAs and SPSA with skewed-quartic loss function; noise-free and noisy measurement cases; approximate 90 percent confidence intervals shown in [·]. Initial values for the GA populations include loss values both smaller and larger than $L(\hat{\boldsymbol{\theta}}_0) = 3.65$ for SPSA.

Noise σ	GA Bit-coded roulette selection	GA Bit-coded tournament selection	GA Real-coded tournament selection	Constrained SPSA
0	0.0036 [0.0031,0.0041]	0.0031 [0.0026,0.0036]	8.5×10^{-5} $[7.8\times10^{-5}, 9.2\times10^{-5}]$	2.7×10^{-5} $[2.1\times10^{-5}, 3.3\times10^{-5}]$
0.1	0.065 [0.056,0.074]	0.052 [0.047,0.057]	0.057 [0.051,0.063]	0.017 [0.006,0.027]

of GA relative to SPSA have been conducted; for example, Nandi et al. (2001) show that the two methods perform comparably in a chemical process control application involving neural network training. ❑

Example 9.5—Many-minima continuous problem. Consider the loss function $L(\boldsymbol{\theta}) = t_1\sin(4t_1)+1.1t_2\sin(2t_2)$, $\boldsymbol{\theta} = [t_1, t_2]^T$. Let $\Theta = [0, 10]^2$. As shown in Figure 9.4, this function has many local minima, with one global minimum at $\boldsymbol{\theta}^* = [9.039, 8.668]^T$, corresponding to $L(\boldsymbol{\theta}^*) = -18.555$. Suppose that noise-free loss measurements are available. Let us compare stochastic approximation with injected randomness (Section 8.4) to a GA with real-number coding and tournament selection.

The SA algorithm with injected $N(\mathbf{0}, \boldsymbol{I}_{10})$ randomness is used in conjunction with the SPSA gradient approximation for the input $\boldsymbol{G}_k(\cdot)$. The algorithm uses tuned gains of form C in Section 8.4 (due to Fang et al., 1997): $a_k = a/(k+20)^{\alpha}$ and $b_k = b/[(k+1)^{\alpha/2}\log(k+1)]$. The specific parameter values in the gains are $a = 0.1$, $b = 5$, and $\alpha = 0.602$; the SPSA gradient estimate is formed with $c_k = c/(k+1)^{\gamma}$, $c = 0.001$, and $\gamma = 0.101$ (the indicated α and γ are the standard values discussed in Section 7.5 and satisfy conditions C in Section 8.4). In contrast to the fixed initial condition in Example 9.4, the SA replications are initialized randomly (uniformly) within Θ. The GA uses binary tournament selection with coefficients determined from manual tuning. The following settings work well: $N = 80$, $N_e = 2$, $P_c = 0.80$, with a mutation given by an independent, additive $N(\mathbf{0}, 0.05^2\boldsymbol{I}_{10})$ perturbation applied to the $N - N_e = 78$ nonelite elements in the population (mutation is not applied to the N_e elite

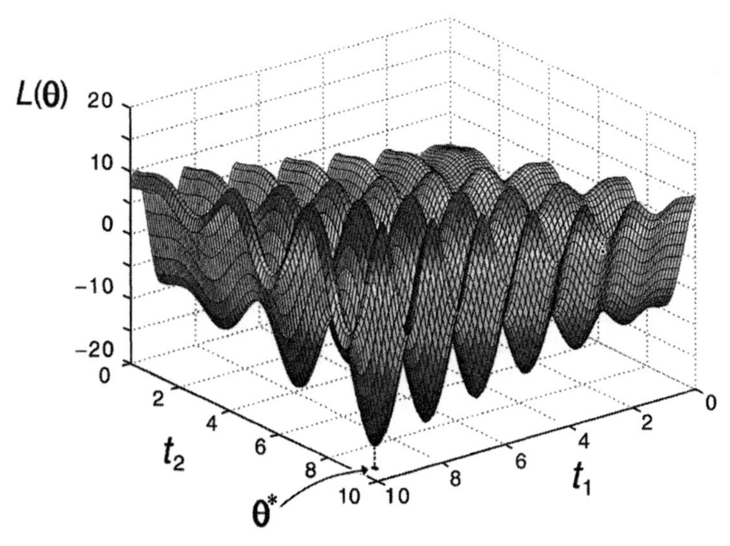

Figure 9.4. Multimodal loss function with unique global minimum at $\boldsymbol{\theta}^*$ near $[9.0, 8.7]^T$ (used in Example 9.5). The function is plotted over the domain Θ.

elements). The initial population is distributed uniformly in Θ. The fitness function is simply $-L(\boldsymbol{\theta})$.

The GA does well on this problem, significantly better than the SA algorithm. In 39 of 40 independent replications of 4000 loss measurements per replication, the GA produced a best final chromosome value having a loss value within 0.0005 of $L(\boldsymbol{\theta}^*)$ (the lone bad replication had a loss value greater than 1.5 units away from $L(\boldsymbol{\theta}^*)$). In contrast, in only two of 40 replications is the SA algorithm within 0.0005 of $L(\boldsymbol{\theta}^*)$ and in only five additional replications does the algorithm come within 0.01 of $L(\boldsymbol{\theta}^*)$. These seven "brushes" with $L(\boldsymbol{\theta}^*)$ (the values within 0.01 of $L(\boldsymbol{\theta}^*)$) generally come *before* the final iteration. That is, because there is no analogue of elitism with basic SA as implemented, six of these seven promising solutions are lost before the final iteration. (Of course, in practice, if one has a record of the iterates, it is possible to recover the best values in the iteration process.)[1] ❑

Example 9.6—Traveling salesperson problem. This famous problem is described in Section 8.3, including Example 8.1, where simulated annealing is discussed as a solution method. A number of authors have also pursued GA solutions to this problem. Let us summarize some of the issues associated with implementing a GA in the traveling salesperson problem. Numerical studies may be found in countless references, some of which are summarized in Michalewicz and Fogel (2000, Chap. 8). (Among the best numerical results cited is one where a GA finds a tour in a 532-city problem that has a cost within 0.06 percent of the known optimal cost.) To implement a GA, one needs to determine the appropriate coded representation for $\boldsymbol{\theta}$ and the appropriate operations (or analogies) for selection, crossover, and mutation. A central issue in GAs is implementing these steps in a way to maintain legal tours (recall that each city must be visited once and only once).

With respect to representing $\boldsymbol{\theta}$, binary codes are undesirable because of the ease with which they produce illegal tours. A number of possible nonbinary representations for $\boldsymbol{\theta}$ are discussed in Michalewicz and Fogel (2000, Sect. 8.1). The most natural is simply the ordered list of cities to be visited. For example, $\boldsymbol{\theta}$ = [10, 2, 28,...] says that city 10 is followed by city 2, which is followed by city 28, and so on. Once the $\boldsymbol{\theta}$ representation has been determined, there is nothing

---

[1] As an illustration of the dangers of drawing broad conclusions from numerical studies, a study with similar general characteristics is carried out in Maryak and Chin (2001). In particular, as in Example 9.5, $L(\boldsymbol{\theta})$ is continuous with many local minima and one global minima, and the loss measurements are noise-free (L is the function used in Example 8.4). Here, SPSA without injected randomness is compared with a real-coded GA of the same general form in Example 9.5 (note that SPSA without injected randomness performed more poorly than SPSA with injected randomness in the problem of Example 9.5). In this study, SPSA achieved effective convergence in 10 of 10 replications using 2500 loss measurements per replication; a tuned GA achieved effective convergence in only one of 10 replications using 50,000 loss measurements per replication.

special about the structure of the traveling salesperson problem that requires altering the traditional selection methods (e.g., roulette or tournament) or elitism.

A standard crossover, such as described in Subsection 9.4.2, can easily lead to illegal solutions involving no—or multiple—visits to one city. As with the $\boldsymbol{\theta}$ representation, many proposals have been made for modified crossover operators that are guaranteed to provide valid tours (see the summary in Michalewicz and Fogel, 2000, Sect. 8.1). However, there does not appear to be any compelling reason to use one of these crossover operations in lieu of the simple operations described in Section 8.3 (inversion, translation, and switching). Rather, through use of a *single-parent* inversion operator, offspring can be created that are guaranteed to represent legal tours. In some numerical studies, this choice is shown to be more effective than several different crossover approaches (Michalewicz, 1996, Chap. 10). In particular, two cut points are randomly chosen in the chromosome and the ordering of the tour between these cut points is reversed. For example,

$$[3, 6, 11, 2 \mid 1, 9, 12, 5, 4 \mid 7, 10, 8]$$

is replaced by

$$[3, 6, 11, 2 \mid 4, 5, 12, 9, 1 \mid 7, 10, 8].$$

Finally, there are many options for mutation operators in traveling salesperson problems. One is to use mutation in conjunction with the switching operation in Section 8.3. In particular, mutation may be applied to each element (city) in $\boldsymbol{\theta}$ with probability P_m; if mutation is invoked, then the given element is interchanged with a randomly selected other element. ❑

9.8 CONCLUDING REMARKS

The genetic algorithms emphasized in this chapter are one important implementation of general evolutionary computation. GAs are based on an intriguing algorithmic analogue of the physical evolution of species in a population. The fundamental distinction between GAs and the methods considered in previous chapters is the use of a *population* of solutions. With a population, each algorithm iteration simultaneously updates a number of $\boldsymbol{\theta}$ values (or their bit-coded equivalent). This contrasts with the updating of a single $\boldsymbol{\theta}$.

Despite its intriguing form, a researcher or practitioner might ask whether a GA is going to help solve a problem better than other approaches. That is, with a comparable amount of effort in algorithm design and coding, will a GA be more efficient or otherwise outperform other methods we have studied? There is no easy answer to this question. Certainly, there have been many successful implementations of GAs (and other evolutionary methods) to challenging practical problems. However, while the connection to natural evolution might suggest that the GA is somehow an optimized algorithm, this line of reasoning appears wrong on at least three counts: (i) There is no proof that natural evolution is an optimized process for species' development (it is difficult to see how a peacock's tail can be the result of an optimized process!); (ii) the no free lunch theorems (Subsection 1.2.2 and Section 10.6) guarantee that "on average" a GA can work no better than other algorithms; and (iii) many numerical studies exist on reasonable problems where other methods soundly outperform GAs.

Certainly, GAs are important tools in the collection of stochastic search and optimization methods. With their population-based structure, they are well suited to parallel processing. However, while there is some theory related to convergence and convergence rates (see Chapter 10), the theory is not as rich as for stochastic approximation and some other stochastic methods. This is unsurprising given the more complex structure of GAs (this relative complexity is evident by comparing the lengths of the GA codes and the codes for some other algorithms at the book's Web site). Further, the justification for the use of GAs with noisy measurements of the loss (fitness) function is thin. As with simulated annealing, the built-in binary decisions (e.g., whether or not to select a particular population element as a parent) suggest a possible lack of robustness to noise, although the redundancy in the population structure might ameliorate this effect at the cost of additional loss measurements.

In summary, GAs and other evolutionary methods provide an effective means for solving some difficult search and optimization problems, including some problems that vex other methods. Nonetheless, in contrast to the implications of some statements in the literature, GAs are not necessarily the only algorithms—or even the best algorithms—for solving certain difficult problems.

EXERCISES

9.1 Prove that the bit coding scheme in Section 9.3 yields the stated level of accuracy (i.e., that an encoding–decoding process on an arbitrary number is guaranteed to yield an accuracy neither greater than nor less than the m digits specified).

9.2 Let $\boldsymbol{\theta} \in \Theta = [-15.000, 15.000] \times [1.2000\times10^6, 2.4000\times10^6]$. Using the procedure of Subsection 9.3.2, determine the bit encoding for $\boldsymbol{\theta} = [-7.222, 2.1000\times10^6]^T$ when $m = 3$ for the first component and $m = -2$ for the second component. Apply the decoding process to recover $\boldsymbol{\theta}$ to within the specified accuracy. (Note: The zeros shown in Θ and $\boldsymbol{\theta}$ are consistent with the values of m.)

9.3 Suppose that fitness values are observed only in the presence of independent $N(0, 1)$ noise and that $N = 15$ and $N_e = 1$. Suppose that one chromosome in the current population has a true (noise-free) fitness value of 11, while the other 14 chromosomes have true fitness values of 10. Based on picking the chromosome with the highest noisy fitness measurement as the elite element, what is the probability of choosing one of the "wrong" chromosomes as the elite element for the next generation's population?

9.4 Suppose that a scalar θ satisfies $-14 \le \theta \le 14$ and that $m = 3$ and $b = 15$. Compare the number of floating-point values to be encoded with the number of possible binary encodings using the procedure of Subsection 9.3.2. How many surplus encodings are there? Give one example of two encodings mapping to the same floating-point number.

9.5 Using the numbers in Table 9.2(a), perform 500 Monte Carlo replications of (binary) tournament selection. What is the allocation of the resulting 1000 parents among the 10 chromosomes? Compare this allocation with the probabilities associated with choosing each of the chromosomes using roulette selection. Comment on any significant differences.

9.6 The bit-based GA implementations in Section 9.7 were terminated when the expected number of unique fitness function evaluations first exceeded a specified value, say n_{fit} (e.g., $n_{\text{fit}} = 2000$ in Example 9.4). Let *ident* represent the cumulative number of times (over all iterations) that a chromosome involved in crossover in the parent generation is identical to a chromosome in the offspring.[2] Assume that the probability of the mutation process returning any chromosome to a value previously obtained is negligible. As in Section 9.3, let B represent the total number of bits in the chromosomes ($B = pb$ if there are b bits per gene). For the bit-based forms of the GA, verify that the algorithm must perform at least

$$\frac{n_{\text{fit}} - N + E(ident)(1-P_m)^B}{(N-N_e)[P_c + (1-(1-P_m)^B)(1-P_c)]}$$

iterations to ensure that the expected number of function evaluations meets or exceeds n_{fit}. (The implementations in **`GAbit_roulette`** and **`GAbit_tourney`** at the book's Web site use the actual observed *ident* instead of $E(ident)$ for determining the number of generations to process to achieve an expected number of fitness evaluations of n_{fit}.)

9.7 Given that a particular nonelite, binary-coded chromosome has been uniquely selected as a parent (i.e., there is one and only one copy of this chromosome among the parents in one generation), what is a (generally) nonzero lower bound to the probability of this chromosome passing intact to the next generation? (Ignore any negligible probability of mutating back to the

---

[2]So, if in one iteration (generation), two parents are identical through the selection process (guaranteed to produce two identical offspring) and in the next iteration another two selected parents are identical (usually different parents from the previous identical parents), then the total contribution toward *ident* for these two generations is four. The counter *ident* accumulates the identical parents over all iterations. Note that when two parents are not identical, the crossover probability provides an *upper bound* to the probability of producing two offspring different from the parents.

original chromosome after crossover.) Express the answer in terms of P_c, P_m, and B. Provide a specific calculation of this lower bound when $P_c = 0.70$, $P_m = 0.005$, and $B = 50$.

9.8 For the $p = 10$ Rosenbrock function first considered in Example 2.5, implement a GA with bit coding and binary tournament selection. Assume that Θ is given by $[-2.048, 2.047]^{10}$ with $m = 3$ in the bit coding (the strange-looking endpoints yield no "excess" bit representations). Report the mean terminal loss value for 50 independent replications of 4000 loss (fitness) measurements at each replication (the terminal loss for one replication is the lowest loss value in the population at the last iteration). Also report the 90 percent confidence interval (based on the t-test of Appendix B) and the values for the GA parameters N, P_c, P_m, and N_e. Make some efforts to tune the GA coefficients to enhance performance.

9.9 Assume the loss function and Θ as given in Exercise 9.8. As in Example 9.4, let the noise in the loss function measurements used as input to the algorithm be given by $[\boldsymbol{\theta}^T, 1]z$, where $z \sim N(\mathbf{0}, \sigma^2 \boldsymbol{I}_{11})$ is independently generated at each $\boldsymbol{\theta}$. Consider the three noise levels $\sigma = 0$, 0.1, and 2.0. At each noise level, perform 50 replications based on 4000 loss (fitness) function measurements for each replication. Use a GA with real coding and binary tournament selection. Make some efforts to tune the GA coefficients to enhance performance at each noise level. At each noise level, report the sample mean of the (noise-free) terminal loss values, the 90 percent confidence interval (based on the t-test of Appendix B), and the values for the GA parameters N, P_c, P_m, and N_e. If you did Exercise 9.8, contrast the results at $\sigma = 0$ here with the results for the GA with bit coding.

9.10 Consider the method in Example 9.4 for initializing a single-path search method (such as stochastic approximation) to provide a fair comparison with a population-based method such as a GA. Consider the domain Θ and optimum $\boldsymbol{\theta}^*$ of Exercises 9.8 and 9.9.

(a) For a uniformly distributed initial GA population on Θ, determine by Monte Carlo or analytical methods the mean (Euclidean) distance of the initial population to $\boldsymbol{\theta}^*$.

(b) Determine an initial condition of the form $[-c, 1, -c, \ldots, 1]^T$ that has the same distance to $\boldsymbol{\theta}^*$, where $c > 0$.

(c) Determine an alternative initial condition of the form $-[c, c, \ldots, c]^T$ that has the same distance to the solution, where $c > 0$.

CHAPTER 10

EVOLUTIONARY COMPUTATION II: GENERAL METHODS AND THEORY

This chapter continues the study of evolutionary computation (EC) begun in Chapter 9. The genetic algorithms (GAs) emphasized in Chapter 9 are the most popular methods of EC. Among the issues not addressed in that chapter are the relationship of GAs to other EC methods and the theoretical justification for EC. This chapter addresses these issues.

Section 10.1 provides some background related to EC. Section 10.2 summarizes some of the EC approaches that differ slightly from the GA emphasized in Chapter 9, including evolution strategies and evolutionary programming. Section 10.3 discusses the schema theory that has developed in support of GAs and Section 10.4 uses this and other theory to comment on what makes a problem inherently difficult. Section 10.5 discusses some of the theory available for convergence of EC methods, with an emphasis on theory based on Markov chains. In light of what has been covered in EC and other methods, Section 10.6 revisits the no free lunch theorems that were introduced in Section 1.2. Section 10.7 offers some concluding remarks.

10.1 INTRODUCTION

Chapter 9 introduced evolutionary computation (EC) methods for search and optimization with an emphasis on the genetic algorithm form of such methods. EC is the umbrella term that is used to refer to most population-based search methods. EC algorithms share the principle of being computer-based approximate representations of natural evolution. These algorithms alter the population solutions over a sequence of generations according to statistical analogues of the processes of evolution.

Although there are currently many varieties of EC, there have historically been three general approaches that fall under the umbrella EC heading: GAs, evolution strategies (ES), and evolutionary programming (EP). The three approaches differ in the types of generation-to-generation alterations and the form of computer representation for the population elements. As noted by Fogel (2000, p. 85), however, "an iterative blending has occurred such that all classes of evolutionary algorithms now appear quite similar." Each area within EC has borrowed and modified ideas from the others.

Two broad approaches to theoretical analysis of EC algorithms are the schema-based analysis and Markov chains. Schema-based theory was the first serious attempt at putting EC methods on a rigorous footing. More recently—and with greater success—Markov chains have been used. In fact, the Markov-based approach has revealed significant shortcomings in the schema-based approach and has led to theory that contradicts the implications of the more-informal schema-based methods.

Associated with the theory for EC are the no free lunch theorems introduced in Subsection 1.2.2. The NFL theorems have had their greatest impact on the EC community, partly because the bulk of the NFL work has been published in the EC literature and partly because some of the most extravagant claims of "universality"—which NFL negates—have been associated with EC algorithms.

Let us now begin our discussion of the non-GA methods for EC, the theory for EC, and the NFL theorems. Consistent with the vast majority of EC literature, this chapter considers only noise-free loss (or fitness) measurements, unless noted otherwise.

10.2 OVERVIEW OF EVOLUTION STRATEGY AND EVOLUTIONARY PROGRAMMING WITH COMPARISONS TO GENETIC ALGORITHMS

As discussed in Section 10.1, there are numerous approaches to EC other than GAs, two of which are ES and EP. The aim of this section is to summarize the main distinctions between these two other population-based approaches and GAs. Recall, however, that over time the distinctions have been getting less important as more practical implementations borrow elements from more than one of the EC methods. Nonetheless, it is worth being exposed to the distinctions in order to understand the literature in the field and in order to recognize features that may be useful in practical implementations. More detailed discussion on these distinctions is given in Michalewicz (1996, pp. 159–168, 283–285) and Schwefel (1995, pp. 151–160). An additional relatively popular approach similar to ES is differential evolution, as described in Storn and Price (1997).

ES was originally designed for constrained continuous variable optimization problems, in contrast to the broader aims of GAs with their use in studying adaptive systems. (Of course, in practice, GAs have also been most often used in optimization in both discrete and continuous variable problems.) Like GAs, the ES moves a population of candidate solutions from generation-to-generation with the aim of converging to a global minimum $\boldsymbol{\theta}^*$. Although the original ES in Rechenberg (1965) worked with only a population size $N = 1$, more modern implementations have emphasized the $N > 1$ setting. The two general forms of ES in most widespread use are referred to by the notation $(N+\lambda)$-ES and (N, λ)-ES; these will be summarized in the basic steps below and discussion to follow. ES works directly with $\boldsymbol{\theta}$; there is no coding of $\boldsymbol{\theta}$ as often occurs with GAs.

Core ES Steps for Noise-Free Fitness Evaluations

Step 0 **(Initialization)** Randomly or deterministically generate an initial population of N values of $\boldsymbol{\theta} \in \Theta$ and evaluate L for each of the values.

Step 1 Generate λ offspring from the current population of N candidate $\boldsymbol{\theta}$ values such that all λ values satisfy direct or indirect constraints on $\boldsymbol{\theta}$.

Step 2 For an $(N+\lambda)$-ES, select the N best values from the combined population of N original values plus λ offspring; for an (N, λ)-ES, select the N best values from the population of $\lambda > N$ offspring only.

Step 3 Terminate the algorithm if a stopping criterion is met or the budget of fitness function evaluations is exhausted; else return to step 1.

The central operation in moving from generation to generation in ES is the creation of the λ offspring (child) $\boldsymbol{\theta}$ values in step 1 above. The offspring are generated from parent $\boldsymbol{\theta}$ values according to the formula

$$\boldsymbol{\theta}_{\text{child}} = \boldsymbol{\theta}_{\text{parent}} + N(\mathbf{0}, \boldsymbol{D}_{\text{parent}}), \tag{10.1}$$

where the "$+ N(\mathbf{0}, \boldsymbol{D}_{\text{parent}})$" part refers to the addition of a p-dimensional normal random vector with a diagonal covariance matrix $\boldsymbol{D}_{\text{parent}}$. If $\boldsymbol{\theta}_{\text{child}}$ does not satisfy the constraints in the problem (on $\boldsymbol{\theta}$ and/or on $L(\boldsymbol{\theta})$), then the solution is discarded and another $\boldsymbol{\theta}_{\text{child}}$ is generated in its place. Optionally, the $i = 1, 2, \ldots, p$ diagonal elements of the matrix $\boldsymbol{D}_{\text{parent}}$ can themselves be updated according to

$$D_{\text{child},i} = D_{\text{parent},i} \exp[N(0, \sigma^2)], \tag{10.2}$$

where $\exp[N(0, \sigma^2)]$ refers to the exponentiation of a $N(0, \sigma^2)$ random variable with user-specified variance σ^2. The matrix $\boldsymbol{D}_{\text{parent}}$ in the subsequent generation is replaced by $\boldsymbol{D}_{\text{child}}$ (which becomes $\boldsymbol{D}_{\text{parent}}$ in the next iteration).

The parent $\boldsymbol{\theta}$ in (10.1) can be chosen in any of several ways from the current population of N elements. One of the most popular methods is via an analogue to the crossover operation of a GA. Essentially, two $\boldsymbol{\theta}$ vectors are chosen (with or without replacement) from the current population of size N according to a uniform distribution. Then each of the p components of $\boldsymbol{\theta}_{\text{parent}}$ is created by randomly choosing from the corresponding component of either the first or second $\boldsymbol{\theta}$ vector. For example, when $p = 3$ and the first and second $\boldsymbol{\theta}$ vectors are $[0.1, 10.4, 2.3]^T$ and $[0.7, 46.3, 1.3]^T$, respectively, we might have $\boldsymbol{\theta}_{\text{parent}} = [0.1, 10.4, 1.3]^T$. The same type of interchange operation used to create $\boldsymbol{\theta}_{\text{parent}}$ can be used to create $\boldsymbol{D}_{\text{parent}}$. Relevant constraints must be satisfied before a $\boldsymbol{\theta}_{\text{child}}$ will be considered for survival into the next generation. The process of drawing from the population and creating offspring is continued until a full population for the next generation is created.

Steps 2 and 3 of the ES are largely self-explanatory. Note that in the (N, λ) strategy, the life of each population element is limited to one generation since the selection of elements for the succeeding generation is from the λ offspring only. This is believed to have advantages in nonstationary systems where $\boldsymbol{\theta}^*$ may be changing in time (Michalewicz, 1996, p. 162). As with other stochastic algorithms, termination occurs using ad hoc stopping principles or when the budget of loss evaluations is expended (see Subsection 9.4.3).

The third cornerstone of EC, evolutionary programming (EP), was described in its early form by L. Fogel et al. (1966). Some more recent incarnations are described in D. Fogel (2000, Sect. 3.3), among other references. EP is strongly motivated by problems in artificial intelligence. The original goal of EP was slightly different from the $\boldsymbol{\theta}$-based search and optimization goals emphasized above.

Namely, the EP was aimed at evolving artificial intelligence by creating *finite-state machines* that are adept at prediction. (Chapter 11 also treats prediction problems via learning and the temporal difference method.) A finite-state machine can be represented as a directed graph (a representation with nodes and edges); this representation can be used to predict the next symbol in a sequence of symbols. For example, if some real system has generated output $\boldsymbol{s}_1$, $\boldsymbol{s}_2, \ldots, \boldsymbol{s}_n$, then the machine can be used to predict $\boldsymbol{s}_{n+1}$. EP works by evolving a population of finite-state machines, much as the other evolutionary methods evolve some representation of the $\boldsymbol{\theta}$ vector. Each finite-state machine in the population is typically represented in matrix form. In evaluating the fitness of a machine in the population, the predictions from the machine are compared with real outcomes according to some fitness function. As in ES above, EP first creates offspring through a mutation operator and then selects individuals for the next generation. An extensive list of references on EP, including references dealing with our main interest in minimizing $L(\boldsymbol{\theta})$ with respect to a vector $\boldsymbol{\theta}$, is given in Fogel (2000, Sect. 3.3).

Let us now contrast the three EC approaches. The main commonality is that they are population-based methods. GAs have traditionally relied on bit coding, whereas ES and EP have operated with the more natural floating-point representations (although, more recently, there has also been a drift in GA applications toward floating-point implementations). Another difference between GAs and both ES and EP is the ordering of the main operations. In GAs, the selection step precedes the crossover and mutation operations, while with ES and EP the opposite is true. For example, in GAs, candidate parents are selected and then crossover and mutation are applied. In ES, first, a $\boldsymbol{\theta}_{\text{parent}}$ is formed by a crossover operation; then the mutation step is applied, leading to $N+\lambda$ or $\lambda > N$ offspring; finally, selection is applied to bring the population size back to N.

A final contrast is the emphasis on general constrained problems in the ES and EP. These algorithms allow for a direct check on constraint violation and the exclusion of an offspring that violates the constraints. The coefficients of the algorithms may automatically be adjusted if the constraints are violated too frequently (e.g., by lowering $\boldsymbol{D}_{\text{parent}}$ in an ES). In contrast, the GA is largely used with simple hypercube constraints, although it is possible to modify the fitness function to include a penalty function as a way of handling more general constraints. Other distinctions are discussed by Michalewicz (1996, Sects. 8.2 and 13.1) and Fogel (2000, Chap. 3).

As mentioned above, however, the distinctions among these various EC methods are washing away over time as various desirable elements of the evolutionary methods are combined into hybrid forms (e.g., real-coded GAs with adaptable algorithm parameters). All EC methods involve random variation applied to a population of candidate solutions and some method for selection that tends to favor the best solutions.

10.3 SCHEMA THEORY

This and the next two sections are devoted to some of the theory behind EC, focusing on GAs in particular. The schema theory of this section is both illuminating and controversial, illuminating because it provides some of the intuitive basis for GAs and controversial because it has led to some "leaps of faith" that later proved misguided. Although EC ideas have been around since at least the 1950s, schema theory was later developed as the first serious attempt to put EC—and GAs in particular—on a rigorous footing.

Holland (1975) pioneered this concept as applied to bit-based GAs, and a large fraction of his book is devoted to the discussion of schemas. A more recent description of the theory is given in Michalewicz (1996, Chap. 3). Essentially, a schema is a template for the chromosome that constrains certain elements to take on fixed bit values while allowing the other elements to be free. An example would be the eight-bit chromosome with template [* 1 0 * * * * 1], where the * symbol represents a *don't care* (or free) element. The chromosomes [0 1 0 1 1 0 1 1] and [1 1 0 0 1 1 0 1] are two specific *instances* of this schema. On the other hand, [1 1 1 0 1 1 0 1] is not an instance. The *order* of a schema is the number of defined elements (0 or 1). Schemas are sometimes referred to as the *building blocks* of GAs because it is believed by some that a GA constructs its solution by the manipulation of schemas. There are two main theoretical results associated with schemas. The first is sometimes referred to as the *schema theorem* and the second goes by the name of *implicit parallelism*. We summarize each of these results below. Unless noted otherwise, we restrict ourselves to the standard bit coding introduced in Subsection 9.3.2.

The essence of the schema theorem is that templates with better-than-average fitness values (where the template average is taken over all $2^{\text{no. of free elements}}$ possible chromosomes) will dominate the population as generations proceed (and conversely, the influence of templates with below-average fitness values will dwindle over generations). For consistency with prior work in this area, we suppose that the fundamental problem of minimizing $L(\boldsymbol{\theta})$ has been mapped into comparing *strictly positive* fitness values with the higher fitness value being better. Holland (1975, Corollary 6.4.1) and Michalewicz (1996, expression (3.3)) give a lower bound to the rate at which the domination will occur across generations under the basic GA operations of roulette selection, crossover, and mutation that were discussed in Section 9.4. Let $\varphi(S, k) \leq N$

denote the number of chromosomes in the population of size N that are instances of the schema template S at iteration k and *fitness*(S) represent the mean fitness of all the chromosomes that are instances of S.

The bound from the schema theorem under the basic GA of Section 9.5 with $N_e = 0$ (i.e., no elitism), crossover rate $0 < P_c \leq 1$, and mutation rate $P_m \approx 0$ is

$$\min\left\{\varphi(S,k)\frac{fitness(S)}{(mean\ fitness)_k}(1 - c_0 P_c - c_1 P_m),\ N\right\} \leq \varphi(S,k+1) \leq N, \quad (10.3)$$

where c_0 and c_1 denote constants that depend on the bit string length and characteristics of the schema configuration (e.g., the order of the schema), and the indicated ratio measures the fitness of the specified schema relative to the mean fitness of all the chromosomes in the population at iteration k, $(mean\ fitness)_k$. (The "min" operator in (10.3) ensures that $\varphi(S, k+1) = N$ if the bound would otherwise be larger than N.) As a consequence of the definitions of c_0 and c_1, the $-c_0 P_c - c_1 P_m$ term will be nearly zero in most circumstances when there are relatively few "don't care" symbols (*) in the schema of interest, S. Hence, (10.3) implies that a schema S that consistently has a fitness value higher (better) than the possibly increasing average of all the chromosomes (i.e., $fitness(S)/(mean\ fitness)_k > 1$) will dominate in the sense that more and more chromosomes will be instances of that schema over generations.

The GA literature often refers to (10.3) as indicating that good schemas will have an exponentially growing influence across generations. However, the use of *exponentially growing* in this context is not fully justified: (i) the time-varying $(mean\ fitness)_k$ term may (at least partly) neutralize the exponential growth that would result if $(mean\ fitness)_k$ were to remain constant (i.e., a product such as $(1+1/2)(1+1/3)(1+1/4)\ldots$ does not represent exponential growth; see Stephens and Waelbroeck, 1999, for a specific GA example) and (ii) there is a Taylor series approximation associated with the role of the mutation probability (see Holland, 1975, p. 111) that makes the bound in (10.3), strictly speaking, only an approximation. Despite these potential shortcomings, the schema theorem as expressed in (10.3) provides some insight into the underlying mechanics of the standard GA (sans elitism) described in Section 9.5.

The second widely cited schema result, implicit parallelism (or *intrinsic* parallelism in Holland, 1975), states that the number of schemas processed by the algorithm in one generation is much larger than N. This suggests that the GA has powerful capabilities to process a greater amount of information at each iteration than would be suggested by the population size alone. Further, this implicit information is available without additional storage and/or processing requirements. Information is obtained about a number of schemas much larger than the population size because a given chromosome can be an instance of many different schemas; each chromosome, therefore, reveals information about the worth of many specific schemas.

It is widely reported in the GA literature that implicit parallelism indicates that $O(N^3)$ schemas are processed at each iteration. However, the conditions for this result are more stringent than much of this literature would suggest, as discussed below. The claimed $O(N^3)$ result is also frequently cited as a basis for the optimality of coding with the standard bit (binary) sequence (via the claim that this coding maximizes the number of schemas that are implicitly processed). However, this claim has been shown to be incorrect, as summarized in Fogel (2000, pp. 75–76); this claim is also wrong in the sense of violating the no free lunch theorems of Subsection 1.2.2 and Section 10.6. Fogel (2000, p. 76) notes that bit coding has been formally shown to offer no particular advantage unless the problem is well mapped to a sequence of binary decisions.

There are also many numerical examples where bit coding performs worse than other coding schemes, primarily real-number coding. Some of these studies are summarized in Section 9.7 and in Davis (1996, Chap. 4), Michalewicz (1996, p. 54), and Mitchell (1996, Chap. 4). Davis, for example, notes that in nine real-world applications he has solved with GAs, the bit string encoding was always outperformed by other coding schemes. The examples in Michalewicz and Mitchell are of a more conceptual (versus empirical) nature, pointing to potential problems with the underlying "building block" basis of schema theory when the chromosome-based equivalent of a loss function evaluation involves a large amount of interaction among the genes in a chromosome (i.e., high epistasis, as discussed in Subsection 9.3.4).

A more careful derivation of the implicit parallelism bound on schemas is given in Bertoni and Dorigo (1993). Their results apply to the standard bit coding and to a GA with crossover but no mutation (it appears that including a small mutation probability would not substantially change the bounds reported). We now summarize these results. Let $\beta > 0$ be such that $N = 2^{\beta t}$, where t is proportional to the string length B according to $t \equiv BP^u/2$, with P^u an upper bound to the probability that one of the schemas gets destroyed by crossover (not to be confused with an upper bound to the crossover probability P_c). Bertoni and Dorigo (1993) show that there is a strong relationship between β and the implicit parallelism bound, which is tantamount to the relative size of the population and the string length having a strong influence on the degree of implicit parallelism.

The central result in Bertoni and Dorigo (1993) states that a lower bound on the expected number of disjoint schemas to be processed at each iteration is of the form $N^{f(\beta)}/\sqrt{\log_2 N}$, where $f(\beta) > 0$ is defined differently for each of three ranges of β.[1] Only when $\beta = 1$ does the resulting bound correspond to the widely cited $O(N^3)$ bound. As an illustration of the large differences from the widely cited case, if $\beta = 0.1$ (i.e., a small population size relative to the string length), then the bound is $N^{21}/\sqrt{\log_2 N}$, while if $\beta = 10.0$ (a large population size

[1] This result relies on the hypothesis of a uniformly random population. Because the GA operations of selection and reproduction will skew the population after the initial random population, the results do not strictly hold in subsequent generations.

relative to the string length) the bound is $N^{0.317}/\sqrt{\log_2 N}$. A further consequence of the results in Bertoni and Dorigo (1993) is that it is misleading to refer to any of these bounds as "big *O*" type bounds. When *N* changes, β changes, and with that, the *form* of dependence on *N* changes (so a dependence, say, on N^3 may change to a dependence on N^2 when the value of *N* increases). Hence, while the GA literature discussing specific order bounds for implicit parallelism is not entirely correct, the literature may be *qualitatively* correct in arguing that the GA tends to implicitly process a larger number of schemas than the population size.

There is considerable controversy about the implications of the above schema results on practical implementations of GAs. It is clear that the notion of a schema is a somewhat removed from the actual issues an analyst faces in solving an optimization problem. The importance of good overall templates on specific solutions must be accepted with a degree of faith. The fact that many good schemas are processed in a particular generation of a GA may or may not be relevant to reaching a good solution with the chromosomes actually processed. For example, Baum et al. (2001) discuss a problem where the schema theorem in (10.3) applies, but where a standard GA is very inefficient (some modifications to the GA can make it much more efficient in this specific problem).

Although the theory discussed in some of the earlier chapters had limitations (generally due to the asymptotic nature), it was more germane to actual algorithm performance since it dealt directly with convergence and rates of convergence of the **θ**-based iterates. In contrast, schema theory describes a notion that is related only indirectly to algorithm performance. Further, in convergence theory, seemingly minor approximations must be carefully handled due to their potential impact over the course of the many iterations of an asymptotic analysis. Despite these reservations, schema theory provides some of the intuitive rationale for why GAs are often effective on challenging problems (Stephens and Waelbroeck, 1999).

10.4 WHAT MAKES A PROBLEM HARD?

One of the most interesting collateral aspects of the field of EC in general and GAs in particular has been the study of what makes a problem inherently difficult. Of course, this has also long been of concern to those involved in the *general* theory and practice of search and optimization (e.g., Hromkovič, 2001).

The motivation for researchers in EC has arisen largely from observed performance. Despite their success in a wide variety of applications, GAs have also failed in a large number of problems. In fact, GAs may perform significantly poorer than even a simple random bit-flipping algorithm (e.g., Davis, 1991;

Mitchell, 1996, pp. 129–130).[2] This has prompted researchers and analysts to try to analyze the essential characteristics of problems in which GAs fail. Not surprisingly, some of the difficulties seen in earlier chapters (high dimensionality, challenging nonlinearities, noise in the loss/fitness evaluations, etc.) also hamper GA performance (see, e.g., Kargupta and Goldberg, 1997; Goldberg, 2002, pp. 76–100). Beyond some of these obvious hindrances, GA researchers have looked for problem aspects that may uniquely hamper a GA. Unfortunately, identifying what makes a problem "GA hard" is hard itself! If this were not so, we would have insight into the dual question of what makes a problem appropriate for a GA, contradicting the statements in Section 9.8 on the difficulty of knowing a priori whether a GA will be a good match to a problem.

One of the characteristics long associated with problem difficulty is *deception*, a term introduced by Goldberg (1987). This term is motivated by the schema-based interpretation of how GAs work. Namely, in the early iterations of the algorithm, lower-order schemas (those with a large proportion of * symbols) should provide useful information about subsequent higher-order schemas that are closer to the full chromosome representations. This process is sometimes referred to as the *building block hypothesis* (see, e.g., Goldberg, 1989, pp. 41–45; Stephens and Waelbroeck, 1999). Deception occurs when the lower-order schemas provide misleading information about what are the best chromosomes.

As an extreme example of deception, suppose that any schema of order less than B whose defined (non-*) bits are all 1's is the best among schemas of the same order but that in the full-order (no * bits) schema, [0 0 0 … 0] offers the best fitness. Then, in principle, it should be difficult for the GA to find the optimal value [0 0 0 … 0] since all lower-order schemas suggest that having many 1's is good and the schema theorem (Section 10.3) will tend to emphasize the schema with many 1's. In fact, in analyzing a test suite of problems where the GA performed poorly, Forrest and Mitchell (1993) and Goldberg et al. (1993) offered the lack of information from lower-order schemas and consequent hampering of the crossover process as a reason for a GA's poor performance. However, despite the intuition behind this schema-based explanation of difficulty through deception, there is ample evidence that deception alone is neither necessary nor sufficient to cause difficulties for a GA. Some discussion of this is given in Mitchell (1996, pp. 125–127).

Epistasis is another problem characteristic that has been cited as creating difficulties in GAs (e.g., Reeves and Wright, 1994). However, as with deception, it is not clear that epistasis will always have negative consequences (other things being equal). Heckendorn and Whitley (1999), for example, point out that the effect of epistasis is subtler than previously believed. Their analysis is based on the concept of *Walsh polynomials*, which can be used as alternative

---

[2] Random bit flipping refers to a class of bit-based analogues to the random search algorithms of Chapter 2. As with random search, there are many variations of random bit flipping. The most basic form is to start with a single chromosome (not a population) and randomly choose a bit (or bits) to change in value, accepting the new chromosome if the loss (fitness) is improved.

representations of fitness functions defined on the domain $\{0, 1\}^B$ (i.e., the fitness function relies on a standard bit-based, Bth-order chromosome as the argument). Naudts and Kallel (2000) show by theory and example that the traditional measures for "GA easy" or "GA hard" are easily contradicted. That is, a function that appears difficult to optimize based on measures such as deception may, in fact, be easy. The opposite is also true. They conclude that any static measure of problem difficulty will always be severely limited.

10.5 CONVERGENCE THEORY

We have yet to analyze the formal convergence of EC algorithms along the lines of the probabilistic convergence analysis of Chapters 2–8. Although the schema theory in Section 10.3 has played a role in understanding why GAs may sometimes be effective, it has not been especially useful in addressing the question of convergence. There have, however, been convergence results established by other means. This section summarizes some of these results.

One of the reasons that schema theory has not been used toward formal convergence analysis is the following negative result due to Rudolph (1994). This result applies to a *canonical GA* having strictly positive (noise-free) fitness values, bit coding, crossover, and mutation, but no elitism. The canonical GA is the basic algorithm of Holland (1975) and one for which the schema theorem and implicit parallelism (Section 10.3) apply. Let $\hat{L}_{\min,k}$ denote the lowest of the N loss values within the population at iteration k. That is, $\hat{L}_{\min,k}$ represents the loss value for the $\boldsymbol{\theta}$ in population k that has the maximum fitness value. The following negative result applies.

Theorem 10.1. A canonical GA with roulette wheel selection (Subsection 9.4.1), $0 \leq P_c \leq 1$, and $0 < P_m < 1$ does not converge in the sense that

$$\lim_{k\to\infty} P\left(\hat{L}_{\min,k} = L(\boldsymbol{\theta}^*)\right) \neq 1 .$$

Comments on proof. The proof is given in Rudolph (1994, Theorems 3 and 4). The proof is based on Markov chains (see Appendix E). The GA can be represented as a Markov chain with 2^{NB} states, each representing a possible population. ❑

One implication of Theorem 10.1 is that in the convergence sense, the canonical GA performs even more poorly than a brute-force enumeration or blind random search (algorithm A in Subsection 2.2.2), both of which are notoriously poor algorithms in all but the simplest practical problems. The essential problem with the canonical GA is that it has no way to guarantee keeping the best solution it finds; even if the algorithm finds an optimum, it is likely to be lost through subsequent crossover and/or mutation. On the other hand, enumeration is guaranteed to converge since the search space containing the solution has a finite number of elements (i.e., the bit representation allows only a finite number of possible outcomes for each chromosome) and the best solution is retained. The search will definitely "hit" (and keep) the best solution at some finite number of iterations, although "finite" in this case must be interpreted judiciously. Achieving the required number of iterations may take millennia on the fastest available computers. Likewise, modest conditions are given in Subsection 2.2.2 for convergence (a.s.) of blind random search.

Although Theorem 10.1 is interesting, and points to one of the reasons that people have not been able to use schema theory (or any other method!) to show convergence of the canonical GA, it has limited practical implications. Almost any serious GA implementation will include elitism and possibly other enhancements (such as in Section 9.6). Of course, such enhancements complicate the theoretical analysis, but some positive results are available, as outlined later. Further, Theorem 10.1 says nothing about *finite-sample* performance of a GA, and, in fact, a simple canonical GA may in some cases produce a satisfactory finite-iteration solution (see Holland, 1975, pp. 161–164; De Jong, 1975, pp. 96–101).

The following small-scale example provides some intuitive sense of how a canonical GA may fail to converge. The same principles apply to general problems (larger N and/or B), as illustrated in Exercise 10.9.

Example 10.1—Failure of convergence for a GA. Consider a very small scale canonical GA with $N = 2$ and $B = 3$. Let $P_c > 0$ and P_m be negligibly small. The total number of possible populations is $2^{NB} = 64$ (i.e., at every iteration, the GA will produce one of these 64 populations). Among these 64 states are $2^B = 8$ states where the two population elements are identical. In these eight states, crossover between the two population elements will not change the elements. For example, if the population has the two chromosomes {[1 0 1], [1 0 1]}, the population after crossover is also guaranteed to be {[1 0 1], [1 0 1]}. By the principles of roulette selection there is a nonzero probability of reaching any one of these eight absorbing states because they all have nonzero fitness values. Because the mutation probability is negligible, the GA becomes "stuck" after reaching one of these states. (Even if the population should eventually change due to a mutation, the selection/crossover process will repeat itself and continue pushing the GA to the absorbing states.) Unless the chromosomes in the absorbing states correspond to $\boldsymbol{\theta}^*$ upon decoding, the GA does not converge to the optimum. ❑

On a more positive note, conditions for the formal convergence of EC algorithms to an optimal $\boldsymbol{\theta}^*$ are presented in a number of references. Qi and Palmeiri (1994), Hart (1997), Rudolph (1997a, 1998), Vose (1999, Chaps. 13 and 14), Fogel (2000, Chap. 4), and Kallel et al. (2001) are several of the references that present conditions for EC convergence.

Let us summarize an approach to convergence and convergence rates based on Markov chains. Consider a bit-coded GA with (as usual) a population size of N and a string length of B bits per population element (chromosome). Hence, there are 2^B possible strings for an *individual* chromosome. Then the total number of possible *unique* populations is "$N + 2^B - 1$ choose N":

$$N_P \equiv \binom{N+2^B-1}{N} = \frac{(N+2^B-1)!}{(2^B-1)!N!} \tag{10.4}$$

(Exercise 10.7). (Eqn. (10.4) differs from the 2^{NB} states mentioned in the context of Theorem 10.1 because of redundancy among the 2^{NB} states. For example, with $N = 2$ and $B = 3$, the two populations {[1 1 0], [0 1 0]} and {[0 1 0], [1 1 0]} are counted only once in (10.4).) One of the ways to analyze the performance of GA is to determine the probability that a particular population contains a chromosome corresponding to an optimum $\boldsymbol{\theta}^*$.

It is possible to construct an $N_P \times N_P$ Markov transition matrix $\boldsymbol{P}$, where the ijth element is the probability of transitioning from the ith population of N chromosomes to the jth population of the same size. Let $\boldsymbol{p}_k$ be an $N_P \times 1$ vector having jth component $p_k(j)$ equal to the probability that the kth generation will result in population j, $j = 1, 2, \ldots, N_P$. From basic Markov chain theory (Appendix E),

$$\boldsymbol{p}_{k+1}^T = \boldsymbol{p}_k^T \boldsymbol{P} = \boldsymbol{p}_0^T \boldsymbol{P}^{k+1},$$

where $\boldsymbol{p}_0$ is an initial probability distribution. If the chain is irreducible and ergodic, the limiting distribution of the GA (i.e., $\bar{\boldsymbol{p}}^T = \lim_{k\to\infty} \boldsymbol{p}_k^T = \lim_{k\to\infty} \boldsymbol{p}_0^T \boldsymbol{P}^k$) exists and satisfies the stationarity equation $\bar{\boldsymbol{p}}^T = \bar{\boldsymbol{p}}^T \boldsymbol{P}$. (Recall from Appendix E that irreducibility indicates that any state is accessible from any other state.) An individual element in $\boldsymbol{P}$ can be computed according to the formulas in Suzuki (1995) and Stark and Spall (2001). These elements depend in a nontrivial way on N, the crossover rate, and the mutation rate; the number of elite chromosomes is assumed to be $N_e = 1$.

Suppose that $\boldsymbol{\theta}^*$ is unique (i.e., Θ^* is the singleton $\boldsymbol{\theta}^*$). Let $J \subseteq \{1, 2, \ldots, N_P\}$ be the set of indices corresponding to the populations that contain at least one chromosome representing $\boldsymbol{\theta}^*$. So, for example, if $J = \{4, N_P - 2\}$, then each of the two populations indexed by 4 and $N_P - 2$ contains at least one chromosome that, when decoded, is equal to $\boldsymbol{\theta}^*$. Under the above-mentioned assumptions of irreducibility and ergodicity, $\sum_{i\in J} \bar{p}_i = 1$, where $\bar{p}_i$ is the ith element of $\bar{\boldsymbol{p}}$. Hence, a GA with $N_e = 1$ and a transition matrix that is irreducible and ergodic converges in probability to $\boldsymbol{\theta}^*$. This result should not be surprising. The assumptions on $\boldsymbol{P}$ imply that the GA will eventually visit every state. Because the algorithm is always saving the best population chromosome encountered (knowable here because of the noise-free loss/fitness measurements), the algorithm will ultimately be in a state containing the best possible chromosome, a chromosome equaling $\boldsymbol{\theta}^*$ upon decoding.

To establish the fact of convergence alone, it may not be necessary to compute the $\boldsymbol{P}$ matrix. Rather, it suffices to know that the chain is irreducible and ergodic. (For example, Rudolph, 1997a, p. 125, shows that the Markov chain approach yields convergence when $0 < P_m < 1$.) However, $\boldsymbol{P}$ must be explicitly computed to get the *rate* of convergence information that is available from $\boldsymbol{p}_k$. Unfortunately, this is rarely possible in practice. The dimension N_P grows very rapidly with increases in the number of bits B and/or the population size N. An estimate of N_P can be obtained by Stirling's approximation to a factorial (see Exercise 10.8). Even a modest-sized GA can result in a very large $\boldsymbol{P}$. For example, if $N = B = 6$, $\boldsymbol{P}$ is larger than a $10^8 \times 10^8$ matrix. Nevertheless, even with the very rapid growth in N_P, it is possible to use the Markov chain analysis to analyze the convergence rate in some problems, as we now illustrate.

Example 10.2—Convergence rate for a GA. Consider the loss function with scalar θ from Schwefel (1995, pp. 328–329): $L(\theta) = -|\theta \sin(\sqrt{\theta})|$, $\theta \in \Theta = [0, 15]$. The function is shown in Figure 10.1. There is a local minimum near 5.2 and a global minimum at the boundary point, $\theta^* = 15$. Using the formulas for the elements of $\boldsymbol{P}$ in Suzuki (1995) and Stark and Spall (2001), Table 10.1 gives the probability of the GA finding one of the unique populations containing an optimal chromosome from within the N_P unique populations. Given the decoding process in Subsection 9.3.2, an optimal chromosome is correct to within an

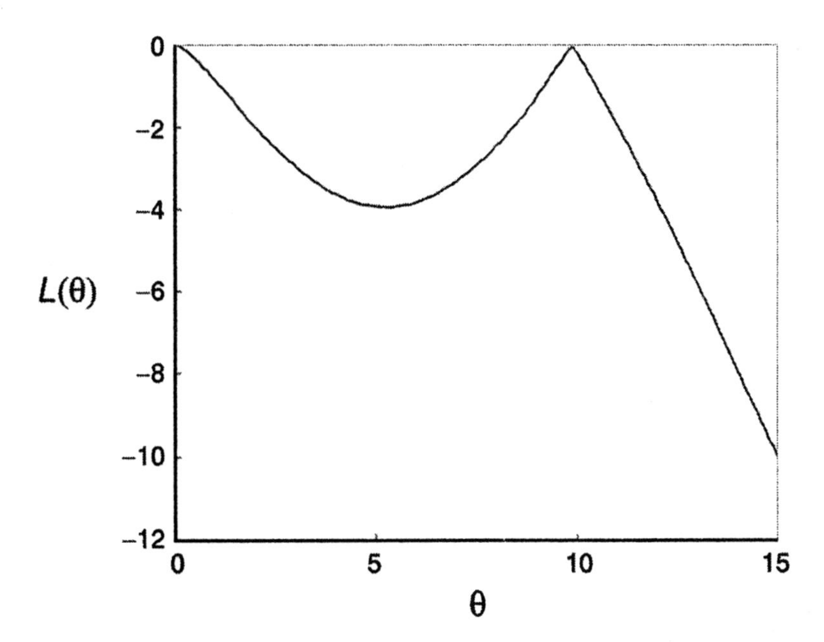

Figure 10.1. Loss function for use in Example 10.2. With $\Theta = [0, 15]$, local minimum is at $\theta \approx 5.2$; global minimum is at $\theta^* = 15$.

Table 10.1. Probability of the GA finding a population that contains an optimal chromosome.

GA coefficients	Iteration number						
	0	5	10	20	30	50	100
$P_c = 1.0$ and $P_m = 0.05$; $N = 2$ and $B = 6$	0.03	0.08	0.15	0.32	0.48	0.74	0.97
$P_c = 1.0$ and $P_m = 0.05$; $N = 4$ and $B = 4$	0.21	0.51	0.69	0.92	1.00	1.00	1.00

accuracy related to N and B (the resulting quantity d in Subsection 9.3.2 governs the accuracy). It is assumed that the initial population is drawn according to a uniform distribution over the unique populations (see Exercise 10.11). The values for N and B are very small by the standards of most practical GA implementations. This was done to keep the computations for the transition matrix $\boldsymbol{P}$ manageable (the calculations for each element of $\boldsymbol{P}$ are significant and there are many elements). In particular, $\boldsymbol{P}$ has dimension 2080×2080 when $N = 2$ and $B = 6$ and dimension 3876×3876 when $N = 4$ and $B = 4$. ❑

Let us close this section with a brief discussion of other approaches to convergence rates. Some of these approaches use methods other than Markov chains. For example, when the algorithm applies directly to $\boldsymbol{\theta}$ without a coding process and with real numbers for at least some elements of $\boldsymbol{\theta}$, there is no longer a finite number of states, making it less amenable to Markov chain analysis. Stochastic approximation methods (Section 4.3) have been used for convergence analysis of some forms of EC (e.g., Yin et al., 1996).

Some of the available results for convergence *rates* are for EC algorithms using mutation and selection only, or using crossover and selection only. Both Beyer (1995) and Rudolph (1997a) examine ES algorithms that include selection, mutation, and crossover. The function analyzed in both cases is the simple spherical loss $L(\boldsymbol{\theta}) = \boldsymbol{\theta}^T\boldsymbol{\theta}$. Convergence rates based on the spherical fitness function are somewhat useful in practice if it is assumed that the sphere approximates a local basin of attraction for the loss function of real interest. A number of other convergence rate results are also available for this simple function—for example, Qi and Palmeiri (1994) for real-valued GA. Rudolph (1997b) considers convergence rates for an ES algorithm applied to a subclass of convex functions (but a subclass broader than only $L(\boldsymbol{\theta}) = \boldsymbol{\theta}^T\boldsymbol{\theta}$). He establishes conditions such that $\hat{L}_{\min,k} - L(\boldsymbol{\theta}^*) = o(c^k)$ a.s. for some $1/2 < c < 1$. This implies that the best loss value in the population at each k converges to $L(\boldsymbol{\theta}^*)$ at a rate *faster* than c^k goes to zero as $k \to \infty$.

10.6 NO FREE LUNCH THEOREMS

Subsection 1.2.2 introduced the basics of the no free lunch theorems, showing that no algorithm can be universally more efficient than other algorithms. We now build on this earlier discussion by presenting some aspects of the NFL theory in a more formal manner. Although the implications of this theory extend to arbitrary search and optimization methods, this discussion is appearing in a chapter devoted to EC algorithms because most of the critical work has come from those in the EC field. A premonition of the NFL results appeared in Goldberg et al. (1993): "...the fiddling with codes, operators, and parameters that characterize so much of the GA literature seriously challenges any claims that the promised land of robustness has been achieved." The NFL theorems indicate that this "promised land" can never be achieved.

In their NFL theory, Wolpert and Macready (1997) present a formal analysis of general search algorithms for optimization, including, of course, the EC class considered here, but also encompassing simulated annealing, random search, and so on. (Some related philosophical discussion and a summary of earlier references are given in Goldberg, 2002, pp. 74–75.) Wolpert and Macready (1997) present several theorems linked to two general approaches to the question of general algorithm efficiency. One of their approaches is to compare the performance of algorithms over the set of possible optimization problems; the other is to compare performance for a particular problem over a specified collection of algorithms. Most of their results (and the ones we discuss here) focus on the former type of comparison. The essence of the NFL theorems is that the performance of any optimization algorithm, when averaged across all possible problems, is identical to the performance of any other algorithm. Of course, these results do not reflect the usual types of prior information that might be available to the algorithms and thus may not adequately reflect the performance of algorithms as they are actually applied. Nevertheless, the NFL results are an antidote to inflated claims of efficiency that have appeared in some of the EC and other literature.

Recall from Subsection 1.2.2 that the NFL results apply to discrete problems where there are $N_{\boldsymbol{\theta}} < \infty$ possible values for $\boldsymbol{\theta}$ (i.e., $N_{\boldsymbol{\theta}}$ points in Θ) and $N_L < \infty$ possible values for the loss (fitness) function. (We discuss NFL in the context of noise-free loss measurements; the same ideas apply when there are N_L possible noisy values y.) Note that the finiteness assumptions for the search space Θ and associated space of possible loss values are met in practice for algorithms implemented on digital computers (i.e., 32- or 64-bit representations of real numbers). Let Λ represent the set of $(N_L)^{N_{\boldsymbol{\theta}}}$ possible mappings $L(\boldsymbol{\theta})$. Each element in Λ represents *one* set of rules that takes the $N_{\boldsymbol{\theta}}$ possible values for $\boldsymbol{\theta}$ and maps them into the N_L possible values for the loss. Example 1.7 (Subsection 1.2.2) illustrates a simple case where Λ has eight possible mappings. The number of possible mappings is often huge ($>10^{1000}$ is common when considering digital computer implementations of continuous problems).

Suppose that an algorithm $\mathcal{A}$ is applied to a loss function L. Let $\hat{L}^{(n)}$ represent the loss value reported out of the algorithm after n unique loss evaluations (multiple evaluations that result when an algorithm revisits a point in Θ are only counted once). For example, if $\mathcal{A}$ represents a particular EC algorithm, $\hat{L}^{(n)}$ may represent the best (lowest) loss value from the most recent N population elements after the algorithm has performed $n \geq N$ unique loss evaluations. If elitism is included, then $\hat{L}^{(n)}$ is guaranteed to be the lowest loss value encountered over *all* of the n loss evaluations.

For any given problem, it is likely that the performance of one algorithm will be superior to others. A priori, of course, it is rarely possible to know which algorithm is superior. NFL theorems do not address the performance of a specific algorithm applied to a specific loss function. Rather, they compare the performance of algorithms over *all* problems, where each problem (each mapping L) is considered equally likely. In particular, assume that the prior probability of encountering any given problem is the same as the probability for any other problem (i.e., the prior distribution on the set of all mappings [loss functions] in Λ is the uniform distribution). An NFL theorem based on summing over Λ is:

Theorem 10.2 (NFL). Consider any pair of algorithms $\mathcal{A}_1, \mathcal{A}_2$ relying on a fixed number of unique loss (fitness) evaluations n. For the discrete optimization structure outlined above and any point λ in the set of N_L allowable values for the loss:

$$\sum_{L\in\Lambda} P\left(\hat{L}^{(n)}=\lambda \mid L,\mathcal{A}_1\right) = \sum_{L\in\Lambda} P\left(\hat{L}^{(n)}=\lambda \mid L,\mathcal{A}_2\right),$$

where the probability $P\left(\hat{L}^{(n)}=\lambda \mid L,\mathcal{A}_i\right)$ is conditional on one of the $(N_L)^{N_\Theta}$ mappings and on the specific algorithm.

An important special case is where λ is the minimum of the possible values for the loss (this minimum is achievable in only those mappings where it lies in the range space). Then, we are considering the probabilities of $\hat{L}^{(n)}$ achieving the optimum for those mappings where λ is in the range space; in other mappings, the probability $P\left(\hat{L}^{(n)}=\lambda \mid L,\mathcal{A}_i\right)=0$.

According to the above NFL theorem, the average efficiency of all algorithms is the same in the sense that $\sum_{L\in\Lambda} P\left(\hat{L}^{(n)}=\lambda \mid L,\mathcal{A}_i\right)$ is independent of $\mathcal{A}_i$. A self-evident implication of the above theorem is: If $\mathcal{A}_1$ has a faster rate of convergence than $\mathcal{A}_2$ for one set of problems, then there is a set of problems for which $\mathcal{A}_2$ has a faster rate of convergence than $\mathcal{A}_1$. A related implication is that no algorithm can have better overall efficiency than blind random search (algorithm A in Subsection 2.2.2). Simply knowing that a loss function has a particular structure does not generally make it a priori preferable to use one algorithm over another. However, if the problem structure is used in the algorithm design or if it is known that the algorithm and problem are well matched, then it *is* possible to overcome the limits of NFL in the sense that the algorithm may outperform blind random search and other methods.

10.7 CONCLUDING REMARKS

This completes the second of two chapters devoted to evolutionary computation. EC algorithms represent an abstraction of natural evolutionary processes. While various EC approaches have historically been placed in "bins" (GA, ES, EP, etc.), the distinctions have become less important. Effective implementations of EC often borrow aspects from several of the named approaches to achieve effective practical performance.

We summarized some of the theory available for EC, focusing on the traditional schema theory and on the Markov chain-based approach. Over the last several years, schema theory has fallen somewhat into disfavor. This seems to have happened because some used the theory to make performance claims that, under more careful scrutiny, were not actually supported by the theory. That is, while the theory *itself* is largely correct (subject to some "adjustments" and caveats), many of the stated *implications* have not been correct. Nevertheless, schema theory has historical significance in EC development and, in an appropriately restricted sense, provides some intuitive justification for the good performance that is frequently observed. Theory based on Markov chains and other forms of probabilistic analysis is more directly connected to the performance of EC methods. Using such theory, it can be shown that typical EC algorithms *with elitism* converge to an optimal point.

Because EC is based on analogies to natural evolution, it was long believed by some that EC algorithms are "best" in some sense. This is now known to be false. Aside from the fact that evolution itself does not appear to be an optimized process, the no free lunch theorems formalize the intuitively sensible notion that no algorithm can be universally preferred.

What is one to conclude about the efficacy of EC algorithms for challenging optimization problems? Essentially, there is no reason to believe that GAs or other EC algorithms will be generally superior to other methods. In particular, simpler methods such as random search in Chapter 2 (or the FDSA or SPSA methods of Chapters 6 and 7) may be as (or more) numerically efficient on a particular problem. We saw evidence of this in the numerical studies of Chapter 9. In fact, the appeal of the simpler approaches goes beyond numerical efficiency as analyzed via the NFL theorems; one should also consider the human cost associated with the relative complexity of many EC implementations. In spite of NFL, Culberson (1998) and Baum et al. (2001) discuss some cases where the EC framework may be beneficial. One area is in combining the results from various local searches. However, much further research is required to identify a broad class (or classes) of problems for which

EC algorithms are especially powerful. It does not appear that such results will be available in the near future.

Nevertheless, despite the issues mentioned above, EC methods have proven very effective in many challenging practical problems. Their popularity is expected to grow in the years to come.

EXERCISES

10.1 Suppose that $B = 4$ and that a standard bit representation (Subsection 9.3.2) is used. Suppose the fitness function is the integer represented by the binary argument (so, e.g., the fitness of [0 1 0 1] is 5). Contrast the mean fitness of the schemas [1 * * *] and [0 * * *].

10.2 Suppose that $B = 7$ and consider two schemas [* 1 * * * * 0] and [* * * 1 0 * *]. Identify one chromosome that is an instance of both of the schemas. Illustrate with this chromosome why the schema with the greater number of successive * positions (the first schema here) is more likely to be destroyed by single-point crossover. Now suppose that two matings will take place and that for the first mating, one of the chromosome parents is an instance of the first schema and the other is not. Likewise for the second mating and schema. What is the probability (conditional on the supposition) of each of the two schemas surviving a single-point crossover? (This problem illustrates that schemas where the non-* positions are clustered together are more likely to survive into future generations than other schemas.)

10.3 Prove that a specific chromosome of string length B is an instance of 2^B schemas (the null representation [* * * ... *] counts as a schema; likewise, a specific chromosome is a schema with zero "don't care" symbols).

10.4 An intermediate calculation for the implicit parallelism bound of Section 10.3 is the maximization of $\binom{2t}{x}$ with respect to x (t and x are positive integers). Show that $x = t$ is the solution to this maximization problem.

10.5 With noisy loss (fitness) measurements or other randomness, it can be shown that the proportion of a particular schema in a future population is, under the appropriate conditions, $S/(S+S')$, where S and S' are independent random variables describing, respectively, the fitness of the schema and the fitness of its complement (i.e., the fitness of all schemas disjoint from the schema under consideration). (Fogel, 2000, pp. 119–120, considers this formulation.) For $S \sim U(0, \mu)$ and $S' \sim U(0, \mu')$, with $\mu > 0$ and $\mu' > 0$, compute the *expected* proportion for the particular schema.

10.6 Consider the function $L(\boldsymbol{\theta}) = \sum_{i=1}^{15} |t_i| + \prod_{i=1}^{15} |t_i|$, where $\boldsymbol{\theta} = [t_1, t_2, \ldots, t_{15}]^T$ and $\Theta = [-1.5, 1.5]^{15}$. Implement two versions of a (10, 100)-ES algorithm: one with a constant covariance matrix $\boldsymbol{D}_{\text{parent}}$, and one with a dynamic matrix changed according to (10.2). Tune the two algorithms as appropriate and enforce the constraints by regenerating $\boldsymbol{\theta}_{\text{child}}$ values as needed. Generate the initial populations randomly (uniformly) in Θ and generate the $\boldsymbol{\theta}_{\text{parent}}$

according to the idea in the discussion below (10.2). Run both algorithms for 40 independent replications of 10 iterations per replication. Determine whether there is statistical evidence for the superiority of the form using dynamic matrix updating; use the appropriate two-sample test (Appendix B) applied to the lowest loss value in the final population.

10.7 (More difficult.) Derive the formula for N_P shown in (10.4) (i.e., why "$N+2^B-1$ choose N"?).

10.8 Stirling's approximation to a factorial is $K! \approx K^K e^{-K}\sqrt{2\pi K}$. Use this formula to derive an approximation to N_P in (10.4). Compare the exact formula for N_P and the approximation at $N = 8$ and $B = 6$.

10.9 A generic form of a Markov transition matrix for a GA with selection, crossover, and possibly elitism is $\begin{bmatrix} \boldsymbol{I} & \boldsymbol{0} \\ \boldsymbol{R} & \boldsymbol{Q} \end{bmatrix}$, where $\boldsymbol{I}$ is an identity matrix for the absorbing states (populations), $\boldsymbol{R}$ is a matrix of probabilities for the transition from transient states to the absorbing states, and $\boldsymbol{Q}$ is a matrix of probabilities to the transient states. Let $\max_i|\lambda_i| < 1$, where λ_i is the ith eigenvalue of $\boldsymbol{Q}$. Show that the GA converges to the absorbing states as the number of iterations increase. (This can be used to show convergence or nonconvergence to an optimum, depending on whether the absorbing states include a population with an optimal chromosome.) (Hint: The matrix relationships in Appendix A [Section A.2] may be useful.)

10.10 Consider a very small problem where $N = 2$ and $B = 2$ for a bit-coded GA.
(a) Verify by enumeration that there are 2^{NB} possible states (populations).
(b) List the N_P unique states (eqn. (10.4)).
(c) Identify the absorbing states when there is no mutation.

10.11 For the two settings represented in Table 10.1 ($N = 2$, $B = 6$ and $N = 4$, $B = 4$), compute the probability of the initial population containing the optimal chromosome *without* restricting to only the N_P unique populations. That is, assume that the initial population is drawn according to a uniform distribution over all 2^{NB} possible populations. Comment on why the probabilities here may differ slightly from the corresponding probabilities in Table 10.1.

10.12 Suppose that $\Theta = \{\boldsymbol{\theta}_1, \boldsymbol{\theta}_2\}$ and that there are four possible outcomes for the noise-free loss measurements, $\{L_1, L_2, L_3, L_4\}$. Demonstrate the implications of the NFL theorems by showing that the mean performance of any two algorithms is the same across all possible mappings.

CHAPTER 11

REINFORCEMENT LEARNING VIA TEMPORAL DIFFERENCES

The fields of artificial intelligence, computer science, and control have been a source of numerous methods in stochastic search and optimization. The evolutionary algorithms of Chapters 9 and 10 are one such class of methods. We continue in this vein here by describing an approach in the general area of "learning." At some level, learning is devoted to the broadest of issues in mathematical modeling and understanding from data: How can one best learn about a system from data? Chapter 13 also deals with some aspects of this question by addressing the tradeoff in bias and variance for a model, the choice of model form, and the quantification of the information available for estimation through the (Fisher) information matrix.

This chapter takes the model form as a given and describes methods in reinforcement (trial and error) learning. The focus in this chapter is one important instance of reinforcement learning—the temporal difference method. This algorithm is useful for prediction problems arising in, say, machine learning, artificial intelligence, forecasting, and optimal control. As the name would suggest, this method is based on information available from certain differences over time (with these differences related to model predictions or actual system performance). The reader should be warned that the results here represent only a sliver of the available results in general learning.

Section 11.1 gives some general background on reinforcement learning. Section 11.2 connects reinforcement learning to the idea of temporal differences and Section 11.3 presents the basic temporal difference algorithm for parameter estimation. Section 11.4 discusses extensions of the temporal difference algorithm to batch and online implementations. Section 11.5 presents some examples of the temporal difference algorithm and Section 11.6 makes some connections to the root-finding stochastic approximation framework of Chapters 4 and 5. Section 11.7 offers some concluding remarks.

11.1 INTRODUCTION

The term *learning* has been used in a number of fields to refer to computational approaches to understanding input–output relationships in a system of interest. Learning takes place when input–output data are collected on the system and processed in a meaningful way with an appropriate algorithm. This, of course, does not differ fundamentally from the aim of other algorithms studied in this book. Nevertheless, the term *learning* has often been associated with algorithms popularized in the fields of computer science and artificial intelligence. In fact, the term *machine learning* is often used as a synonym for learning, motivating the computer science connection even more strongly. Certainly, statisticians have been attacking learning problems for as long as the field of statistics has existed, although the terminology and focus differ somewhat from those in computer science and related areas. For a variety of historical and technical reasons, areas such as data mining, system identification, supervised and unsupervised training (as for a neural network), and pattern recognition—all of which heavily involve statistical notions—have not been the traditional province of statisticians.

Reinforcement learning is a fundamental set of methods within the broader class of learning approaches. Reinforcement learning is a formal means of doing trial-and-error learning. Of course, most members of the animal kingdom (including humans) implement trial and error without the formalized structure of a mathematical learning algorithm. In developing a formal algorithmic implementation of this principle, one aims for an approach satisfying the following:

> Reinforcement learning is based on the common-sense idea that if an action is followed by a satisfactory state of affairs, or by an improvement in the state of affairs (as determined in some clearly defined way), then the tendency to produce that action is strengthened (i.e., reinforced) (Sutton et al., 1992).

Conversely, an undesirable—or worsening—result leads to a weakened tendency to follow the action.

Within the field of machine learning, the reinforcement learning principle is used in general artificial intelligence applications through tools such as computational agents, reinforcement signals, and learning agents. We do not need the full generality of these tools here as our goals are focused on the more specific (familiar) problem of minimizing a loss function $L(\boldsymbol{\theta})$. In particular, we will not consider rule-based or symbolic estimation and prediction; rather, we focus on the familiar problem of *numerical* estimation. Some specific reinforcement learning methods include adaptive critic methods (Barto, 1992), dynamic programming (Bertsekas, 1995a), *Q*-learning (Watkins and Dayan, 1992), and temporal difference learning. The latter of these is considered in the remainder of this chapter.

The reader interested in more general principles of learning may wish to consult Russell and Norvig (1995, Chaps. 18–21), Jang et al. (1997), Mitchell (1997), Cherkassky and Mulier (1998), Michalski et al. (1998), and Sutton and Barto (1998).

11.2 DELAYED REINFORCEMENT AND FORMULATION FOR TEMPORAL DIFFERENCE LEARNING

One of the major settings for machine learning is the delayed reinforcement problem. The prototype form for this problem involves a sequence of actions, with no feedback until after the last action as to whether the sequence was good or not.

As an illustration of this problem, suppose that a sequence of estimates of one quantity is formed over time, and only after the full sequence is formed is the true value of the quantity being estimated revealed. Hence, for all but the last estimate in the sequence, there is a delay in determining if a particular estimate is "good" in the sense that at least one intervening estimate must be formed. In particular, suppose that a scalar random outcome Z occurs at some time in the future, and that prior to observing Z, there exists a sequence of $n + 1$ predictions $\hat{z}_0, \hat{z}_1, \ldots, \hat{z}_n$, where $\hat{z}_0$ occurs before $\hat{z}_1$, $\hat{z}_1$ occurs before $\hat{z}_2$, and so on, and $\hat{z}_n$ is the final prediction before the event yielding Z. As we will see below, it is also convenient to define a "null" prediction $\hat{z}_{n+1} = Z$. Figure 11.1 depicts the process.

This delayed reinforcement problem is pervasive. Sutton (1988) discusses weather forecasting as one example. Suppose that for each week, one is interested in issuing a forecast on each day of the week for the weather on the following Saturday. The input information for each day's forecast would be current and recent past weather and other meteorological indicators. Conventional training algorithms (e.g., "supervised learning" as in the backpropagation algorithm of Section 5.2) perform the parameter estimation based on comparisons of predictions and actual outcomes. That is, in developing the function for predicting Saturday's weather on Wednesday, one would compare a sequence of Wednesday predictions and Saturday outcomes. The temporal difference (TD) learning procedure, on the other hand, builds up its parameter estimates by looking at successive predictions. For example, if Wednesday's prediction is for a 50 percent chance of rain on Saturday and Thursday's prediction is for a 75 percent chance, then subsequent days similar to Wednesday will have a prediction greater than 50 percent. The intuitive rationale for this adjustment is that predictions closer to the actual event will be more reliable.

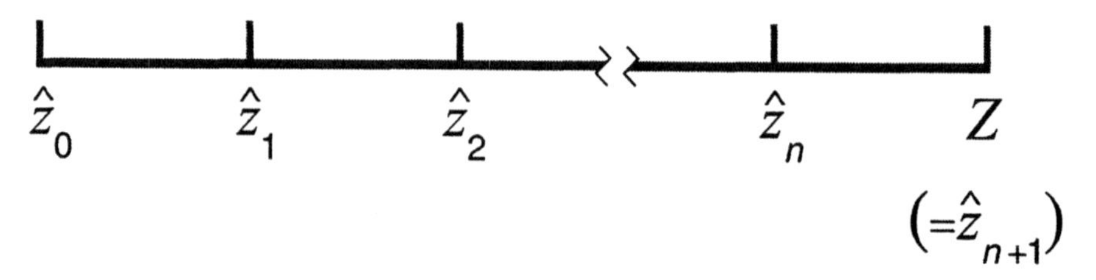

Figure 11.1. Relationship between the predictions $\hat{z}_0, \hat{z}_1, \ldots, \hat{z}_n$ and event Z in the delayed reinforcement problem. Time is depicted as moving left to right.

Many other examples of delayed reinforcement exist. For instance, when humans see or hear something, they receive a stream of input, constantly updating their perception of the object(s) being seen or heard. Humans typically do not wait until they have a final resolution of the nature of the object to update their perception of the object. Rather, a person takes the stream of input together with the previous predictions to continuously (or periodically) recalibrate the prediction of the nature of the object. Similarly, a business selling a new product does not wait until the end of the fiscal year to determine if the product is a success; the assessment of the product is made continuously from the product's introduction based on current input and past predictions. The expected outcome of a chess game or golf match is similarly updated throughout the competition.

The aim here is to construct a function for producing the predictions $\hat{z}_0$, $\hat{z}_1, \ldots, \hat{z}_n$. As with any such function, there are adjustable parameters ($\boldsymbol{\theta}$) to be estimated. Following the general regression framework of Section 5.1, let $\hat{z}_\tau = h_\tau(\boldsymbol{\theta}, \boldsymbol{x}_\tau)$ where $h_\tau(\cdot)$ is a (possibly nonlinear) regression function and $\boldsymbol{x}_\tau$ is a vector of input observations available at time τ to be used in forming the prediction. The subscript τ is being used to emphasize time (rather than the more generic k in previous chapters). Section 5.2 focused on the common case where the regression function is a time-invariant neural network (i.e., $h(\cdot) = h_\tau(\cdot)$). We will, however, consider the time-*varying* form here (see, e.g., Examples 11.1 and 11.2 in Section 11.5).

There is a countless variety of general forms for the regression functions. Chapter 3 considered linear regression and Section 5.2 considered a notable nonlinear case—neural networks—where $\boldsymbol{\theta}$ represents the connection weights. Suppose that the functional forms of the prediction functions have been fixed (e.g., feedforward NNs with a specified number of hidden layers and nodes per layer). Hence, we must cope with the problem of the delayed response in estimating the unknown parameters $\boldsymbol{\theta}$. Note the contrast with conventional supervised learning, such as backpropagation as seen in Section 5.2. In supervised learning, the algorithm compares a predicted output $\hat{z}_\tau$ directly with an actual output z_τ, using the difference $z_\tau - \hat{z}_\tau$ in the training process of updating the parameters in the prediction function. For each $\hat{z}_\tau$, there is no significant delay in the sense that there are no subsequent predictions of z_τ that arrive before forming the comparison of $\hat{z}_\tau$ and z_τ for use in training $\boldsymbol{\theta}$. This contrasts with the delayed reinforcement setting for TD learning.

The essential characteristic of the TD algorithm is that, rather than use the difference between predictions ($\hat{z}_\tau$) and the actual outcome (Z), it uses the temporally successive collection of predictions $\hat{z}_{\tau+1} - \hat{z}_\tau$. Samuel (1959) used the idea in the problem of automated checkers playing; Sutton (1988) provided a more general setting and coined the term *temporal difference learning*. The TD method applies to multistep prediction problems, and is a prominent example of a delayed reinforcement algorithm. TD learning is closely connected to another form of reinforcement learning called *Q-learning*, which itself is a type of stochastic dynamic programming. The connections between TD learning, *Q*-learning, and dynamic programming are explored, for example, in Jaakkola et al. (1994) and Sutton and Barto (1998, Chaps. 4–7).

Further, there are strong connections of these learning methods to problems in control. In control, one attempts to find an optimal policy based on predictions of future system behavior. Sutton et al. (1992), Jaakkola et al. (1994), and Sutton and Barto (1998, Chaps. 4–7) are among the many references exploring the connection of reinforcement learning methods to control. In particular, TD has a strong connection to control-oriented methods such as dynamic programming where one is trying to pick a sequence of optimal policies to cause a delayed reinforcement system to behave acceptably until some future termination event occurs (an excellent overview of dynamic programming is Bertsekas, 1995a). This chapter, however, does not dwell on the control aspects of TD. The control-oriented implementations rely extensively on the methods for the straight prediction problems considered here.

Why should one use TD? First, of course, TD is especially appropriate in the delayed reinforcement problem, a setting not explicitly considered in the other approaches of this book. Further, as shown in the sections to follow, TD methods are suitable for exploiting prior information about the behavior of a dynamic system. The prior information is embedded in past predictions (and the knowledge used to create the past predictions). TD uses this prior information in forming current and future predictions.

In contrast, traditional training methods for prediction modeling (e.g., the supervised learning of backpropagation—see Section 5.2) use only the input variables and the outcomes (Z), ignoring the predictions that contain useful prior information about the process. Further, the supervised learning methods must wait until an outcome is observed before an update is possible to $\boldsymbol{\theta}$, which can delay learning considerably. In contrast, some versions of TD (e.g., the online version in Section 11.4) allow for updates at every time step. Finally, although there is apparently no theoretical comparison of efficiency of TD to other learning methods (even in the linear problems where TD is known to converge), numerical evidence suggests its superiority in a range of prediction problems (e.g., Sutton and Barto, 1998, pp. 138–139).

As with any other method, of course, TD has its drawbacks as well. Among the relative shortcomings is the weaker convergence theory for TD in comparison to the stochastic gradient methods for supervised learning (Chapter 5). While root-finding (Robbins–Monro) SA can be used to characterize the convergence of both TD and the stochastic gradient methods, the conditions for TD are more stringent than the conditions for the stochastic gradient methods. Most important, as discussed in Section 11.6, the current conditions for convergence of TD are directed toward models that are *linear* in $\boldsymbol{\theta}$; no such restriction exists for stochastic gradient methods such as backpropagation for neural networks, which are fully justified for nonlinear models. (Despite the lack of formal theory, TD methods are sometimes *used*, quite successfully, in nonlinear systems.) There are also published counterexamples to TD convergence in linear and nonlinear models (e.g., Bertsekas, 1995b; Tsitsiklis and Van Roy, 1997). TD is not likely to perform as well as supervised learning when a given set of inputs (the $\boldsymbol{x}_\tau$) will repeatedly produce the *same* outcome Z (i.e., there is no noise in the process). In this relatively rare noise-free setting, there is no better source of information about the process than the observed Z.

11.3 BASIC TEMPORAL DIFFERENCE ALGORITHM

The fundamental problem is to train a predictor of future outcomes from one or more "trial" sequences of inputs and outputs $\{\boldsymbol{x}_0, \boldsymbol{x}_1, ..., \boldsymbol{x}_n; Z\}$. This training process estimates the parameters $\boldsymbol{\theta}$ appearing in the prediction functions $h_\tau(\cdot)$, $\tau = 0, 1, \ldots, n$. The final $\boldsymbol{\theta}$ estimate will be used with the predictors $h_\tau(\cdot)$ as they apply in future problems having their own inputs ($\boldsymbol{x}_\tau$) and outputs (Z). There are several ways in which TD learning for updating $\boldsymbol{\theta}$ can be implemented. Let us first assume that $\boldsymbol{\theta}$ is updated only once after one full sequence of measurements $\{\boldsymbol{x}_0, \boldsymbol{x}_1, ..., \boldsymbol{x}_n; Z\}$. We call this the *basic TD algorithm* (this form is the focus of Sutton, 1988). Alternative batch and online methods for TD are discussed in the next section. The basic TD method is a special case of batch TD, with the number of batches being one.

As mentioned above, all predictions during the training in basic TD are based on a fixed $\boldsymbol{\theta}$, taken as an initial estimate $\hat{\boldsymbol{\theta}}_0$. So, $\hat{z}_\tau = h_\tau(\hat{\boldsymbol{\theta}}_0, \boldsymbol{x}_\tau)$ for all $\tau \leq n$ here. After passing through the sequence $\{\boldsymbol{x}_0, \boldsymbol{x}_1, ..., \boldsymbol{x}_n; Z\}$, basic TD produces a new value of $\boldsymbol{\theta}$, say $\hat{\boldsymbol{\theta}}_{\text{new}}$. The new value of $\boldsymbol{\theta}$ is related to the initial value of $\boldsymbol{\theta}$ by the following equation:

$$\hat{\boldsymbol{\theta}}_{\text{new}} = \hat{\boldsymbol{\theta}}_0 + \sum_{\tau=0}^{n} (\delta\boldsymbol{\theta})_\tau\,, \tag{11.1}$$

where $(\delta\boldsymbol{\theta})_\tau$ is an increment in $\boldsymbol{\theta}$ computed based on the difference in predictions $\hat{z}_{\tau+1} - \hat{z}_\tau$ as discussed below. So the aim is to calculate the sequence of increments, $(\delta\boldsymbol{\theta})_\tau$, by which (11.1) produces a new parameter value.

The stochastic gradient algorithm in Section 5.1 forms the starting point for deriving the increments $(\delta\boldsymbol{\theta})_\tau$. Recall that the stochastic gradient algorithm was applied to stochastic nonlinear regression problems. Let us consider the standard mean-squared-error criterion, $\frac{1}{2}E\{[Z - h_\tau(\boldsymbol{\theta}, \boldsymbol{x}_\tau)]^2\}$. From the basic stochastic gradient algorithm in Section 5.1, an updated value of $\boldsymbol{\theta}$ is produced by adding the quantity below to the original value $\hat{\boldsymbol{\theta}}_0$:

$$a(Z - \hat{z}_\tau)\left[\frac{\partial h_\tau(\boldsymbol{\theta}, \boldsymbol{x}_\tau)}{\partial \boldsymbol{\theta}}\right]_{\boldsymbol{\theta}=\hat{\boldsymbol{\theta}}_0}, \tag{11.2}$$

where a is the familiar nonnegative gain coefficient, usually taken as a constant across τ in applications here. This contrasts with the indexed form of gain coefficient considered in most of Chapters 4–7. We will, however, discuss an indexed form in Section 11.6 in making some connections to stochastic approximation. Expression (11.2) arises in the traditional supervised learning setting where $\boldsymbol{\theta}$ is updated based on the outcome of a process. Note that (11.2) cannot be implemented prior to observing the final outcome because it requires that Z be available. This is where TD comes to the fore. The TD procedure *can* be implemented in transitioning between time points prior to the final outcome.

Let us now derive the form for the TD increments $(\delta\boldsymbol{\theta})_\tau$. Following the framework of (11.1), an update of $\boldsymbol{\theta}$ is available by summing the stochastic gradient-based increments in (11.2) over $0 \leq \tau \leq n$, yielding

$$\hat{\boldsymbol{\theta}}_{\text{new}} = \hat{\boldsymbol{\theta}}_0 + a\sum_{\tau=0}^{n}\left[(Z - \hat{z}_\tau)\frac{\partial h_\tau(\boldsymbol{\theta}, \boldsymbol{x}_\tau)}{\partial \boldsymbol{\theta}}\right]_{\boldsymbol{\theta}=\hat{\boldsymbol{\theta}}_0}. \tag{11.3}$$

It is possible to rewrite $Z - \hat{z}_\tau$ in (11.3) as a telescoping sum involving the full set of predictions beginning at time τ:

$$Z - \hat{z}_\tau = \sum_{i=\tau}^{n}(\hat{z}_{i+1} - \hat{z}_i) \tag{11.4}$$

(recall that $\hat{z}_{n+1} = Z$). Substituting (11.4) into (11.3) and rearranging the double sums (Exercise 11.1) yields

$$\hat{\boldsymbol{\theta}}_{\text{new}} = \hat{\boldsymbol{\theta}}_0 + a\sum_{\tau=0}^{n}(\hat{z}_{\tau+1} - \hat{z}_\tau)\sum_{i=0}^{\tau}\left[\frac{\partial h_i(\boldsymbol{\theta}, \boldsymbol{x}_i)}{\partial \boldsymbol{\theta}}\right]_{\boldsymbol{\theta}=\hat{\boldsymbol{\theta}}_0}. \tag{11.5}$$

From (11.5), one can peel off the incremental change $(\delta\boldsymbol{\theta})_\tau$ to be used in (11.1). In particular,

$$(\delta\boldsymbol{\theta})_\tau = a(\hat{z}_{\tau+1} - \hat{z}_\tau)\sum_{i=0}^{\tau}\left[\frac{\partial h_i(\boldsymbol{\theta}, \boldsymbol{x}_i)}{\partial \boldsymbol{\theta}}\right]_{\boldsymbol{\theta}=\hat{\boldsymbol{\theta}}_0} \tag{11.6}$$

for $\tau = 0, 1, \ldots, n$. Eqn. (11.6) represents a special case of the TD learning algorithm (the TD(1) algorithm according to the notation below). Unlike (11.2), which requires Z at all values of τ, the incremental update in (11.6) depends only on the difference of predictions (at the parameter value $\hat{\boldsymbol{\theta}}_0$) and the sum of the gradients $\partial h_i/\partial\boldsymbol{\theta}$ at the various $\boldsymbol{x}_i$ inputs. Hence, (11.6) can be computed without Z until the final increment at $\tau = n$.

There is an obvious generalization of expression (11.6) for estimating the increment $(\delta\boldsymbol{\theta})_\tau$ to allow for more weight to be assigned to more recent system measurements. This generalization is called the TD(λ) family of learning algorithms. In particular, we might expect that past information (past $\boldsymbol{x}_i$) should be given less weight than more recent information in computing a current increment to $\boldsymbol{\theta}$. Suppose that we damp past information according to geometric decay, with $0 \leq \lambda \leq 1$ being the decay factor. Then, the sum on the right-hand side of (11.6) can be replaced with a weighted sum according to

$$(\delta\boldsymbol{\theta})_\tau = a(\hat{z}_{\tau+1} - \hat{z}_\tau)\sum_{i=0}^{\tau}\lambda^{\tau-i}\left[\frac{\partial h_i(\boldsymbol{\theta}, \boldsymbol{x}_i)}{\partial \boldsymbol{\theta}}\right]_{\boldsymbol{\theta}=\hat{\boldsymbol{\theta}}_0}. \tag{11.7}$$

Eqn. (11.7) is called the *TD(λ) learning rule*. Together with (11.1), this rule can be used to generate an updated $\boldsymbol{\theta}$ value, $\hat{\boldsymbol{\theta}}_{\text{new}}$. For computational purposes, it may be convenient to accumulate the sum on the right-hand side of (11.7). In particular, for $\tau \geq 0$, let

$$s_{\tau+1} = \lambda s_\tau + \left[\frac{\partial h_{\tau+1}(\boldsymbol{\theta}, \boldsymbol{x}_{\tau+1})}{\partial \boldsymbol{\theta}}\right]_{\boldsymbol{\theta}=\hat{\boldsymbol{\theta}}_0},$$

where $s_0 = \left[\partial h_0(\boldsymbol{\theta}, \boldsymbol{x}_0)/\partial \boldsymbol{\theta}\right]_{\boldsymbol{\theta}=\hat{\boldsymbol{\theta}}_0}$. Then, (11.7) can be written as

$$(\delta\boldsymbol{\theta})_\tau = a(\hat{z}_{\tau+1} - \hat{z}_\tau)s_\tau. \tag{11.8}$$

The extreme versions, TD(0) and TD(1), are important special cases, with TD(0) being defined using the convention $0^0 = 1$. In the TD(0) algorithm, only the most recent measurement is used:

$$\text{TD(0):}\ (\delta\boldsymbol{\theta})_\tau = a(\hat{z}_{\tau+1} - \hat{z}_\tau)\left[\frac{\partial h_\tau(\boldsymbol{\theta}, \boldsymbol{x}_\tau)}{\partial \boldsymbol{\theta}}\right]_{\boldsymbol{\theta}=\hat{\boldsymbol{\theta}}_0}.$$

In the TD(1) algorithm, all the past prediction gradients are weighted equally:

TD(1): See eqn. (11.6).

The TD(0) algorithm, where only the most recent prediction affects the update to $\boldsymbol{\theta}$, corresponds to a conventional *dynamic programming* formulation for stagewise optimization. Dynamic programming is important in stochastic control and learning theory, with entire books devoted to the subject (e.g., Bertsekas, 1995a). The TD(1) algorithm, at the other extreme, assigns equal weight to all previous gradient evaluations. This implies the equivalence of TD(1) to supervised learning in the following sense.

From (11.3)—which is equivalent to the TD(1) substitution of $(\delta\boldsymbol{\theta})_\tau$ appearing in (11.6) into (11.1)—the update from $\hat{\boldsymbol{\theta}}_0$ to $\hat{\boldsymbol{\theta}}_{\text{new}}$ is the same as the sum of stochastic gradient increments in (11.2). Further, the sum of the stochastic gradient increments is the solution corresponding to *one step* of a supervised learning algorithm where the loss function is the sum of mean-squared errors:

$$L(\boldsymbol{\theta}) = \frac{1}{2}\sum_{\tau=0}^{n} E\left\{[Z - h_\tau(\boldsymbol{\theta}, \boldsymbol{x}_\tau)]^2\right\}. \tag{11.9}$$

(This supervised learning approach is directly comparable to the solution associated with the batch stochastic gradient form in Subsection 5.1.3 with the exception that there is a *common* output Z here—the delayed response—as opposed to the separate responses for each input in conventional stochastic gradient implementations, as in Chapter 5.) In particular, the TD(1) algorithm ((11.6) into (11.1)) and one step of the supervised learning algorithm applied to $L(\boldsymbol{\theta})$ above yield the same update to $\boldsymbol{\theta}$ (from $\hat{\boldsymbol{\theta}}_0$ to $\hat{\boldsymbol{\theta}}_{\text{new}}$). On the other hand, the individual TD(1) increments $(\delta\boldsymbol{\theta})_\tau$ from (11.6) are *not* the same as the stochastic gradient increments in (11.2) (why?)[1], although the *sums* of the increments from $\tau = 0$ to $\tau = n$ ((11.1) for TD(1); (11.3) for stochastic gradient) are by construction identical. The fundamental difference in implementation is that the supervised learning formulation requires the final outcome (Z) for every increment, while TD(1) needs Z only at the final increment. The latter is useful in real-time applications, as it allows the $\boldsymbol{\theta}$ increments to be computed in concert with the inputs $\boldsymbol{x}_\tau$ and associated predictions $\hat{z}_\tau$ without having to wait until the final outcome is available.

Note that in the special case where $\hat{z}_\tau$ is a linear predictor with time-invariant regression function (i.e., $\hat{z}_\tau = h(\boldsymbol{\theta}, \boldsymbol{x}_\tau) = \boldsymbol{\theta}^T\boldsymbol{x}_\tau$), TD(1) is equivalent to the least-mean-squares (LMS) representation of the stochastic gradient algorithm of Section 5.1 in the following sense: The sum of updates $(\delta\boldsymbol{\theta})_\tau$ under TD(1) corresponds to the stochastic gradient as applied to the loss function (11.9). In particular, each of the increments according to the stochastic gradient form (11.2) is

$$a(Z - \boldsymbol{\theta}^T\boldsymbol{x}_\tau)\boldsymbol{x}_\tau\,,$$

which is the same as the increment in Subsection 5.1.4 for LMS after making the obvious notational changes and accounting for the "–" sign preceding the increment in Subsection 5.1.4 versus the "+" sign here. One difference from conventional LMS is that each input here ($\boldsymbol{x}_\tau$) is associated with the *same* output Z (the delayed response of delayed reinforcement learning). In contrast, in

[1] The nonequality of the increments is apparent by noting that only the final TD increment requires Z, while *all* of the stochastic gradient increments require Z.

conventional LMS, each input produces a unique output. From the equivalence of the sum in (11.3) to the sum represented by the substitution of $(\delta\boldsymbol{\theta})_\tau$ into (11.1), there is equality of the sums of the increments

$$\underbrace{a\sum_{\tau=0}^{n}(Z-\boldsymbol{\theta}^T\boldsymbol{x}_\tau)\boldsymbol{x}_\tau}_{\text{LMS}} = \underbrace{\sum_{\tau=0}^{n}(\delta\boldsymbol{\theta})_\tau = a\sum_{\tau=0}^{n}(\hat{z}_{\tau+1}-\boldsymbol{\theta}^T\boldsymbol{x}_\tau)\sum_{i=0}^{\tau}\boldsymbol{x}_i}_{\text{TD(1)}}, \tag{11.10}$$

where $\hat{z}_{\tau+1} = \boldsymbol{\theta}^T\boldsymbol{x}_{\tau+1}$, except that $\hat{z}_{n+1} = Z$. So, TD(1) is the same as LMS in the linear case when applied to a delayed reinforcement system.

What is the best value of λ to use in practice? As with the selection of coefficients for other algorithms we have seen, there is no unique "best" value. Let us summarize why. By making a connection to maximum likelihood estimation, Sutton and Barto (1998, Sect. 6.3) establish the superiority of $\lambda = 0$ for the batch class of TD implementations, which includes the basic TD form above together with the batch generalization in the next section. On the other hand, Bertsekas (1995b) considers a class of nonlinear problems for which $\lambda = 1$ is optimal, with behavior getting progressively worse as λ decreases to zero. Numerical studies on one example in Sutton (1988) reveal that the best performance is found with $\lambda \approx 0.3$. These results are consistent with the types of behavior we have seen with other algorithms, where the choice of optimal algorithm coefficients is typically problem-specific. In practice, experimentation is likely to be needed to pick the best value of λ (and step-size coefficient a) for individual problems.

11.4 BATCH AND ONLINE IMPLEMENTATIONS OF TD LEARNING

The discussion above focused on *one* update of $\boldsymbol{\theta}$ (from $\hat{\boldsymbol{\theta}}_0$ to $\hat{\boldsymbol{\theta}}_{\text{new}}$). In practice, it is sometimes possible to repeat the process several times with new values of the input variables. For example, in the weather forecasting problem above, we may be able to compare the weekday forecasts and Saturday outcomes for several successive weeks, with each week having its own input and outcome conditions. In this case, each realization of the process represents the update from one $\boldsymbol{\theta}$ value to another, which occurs after each Saturday weather outcome. So $\hat{\boldsymbol{\theta}}_{\text{new}}$ at the conclusion of one repetition becomes $\hat{\boldsymbol{\theta}}_0$ at the next.

The combination of (11.1) and (11.7) in this iterative manner is sometimes referred to as the *batch TD* method. This label springs from the batch of realizations used in forming the *total* update of $\boldsymbol{\theta}$. A sequence of predictions is made at a fixed value of $\boldsymbol{\theta}$ during each realization, leading to a *partial* update of $\boldsymbol{\theta}$ at the conclusion of the realization. For each realization the predictions are made for the inputs at $\tau = 0, 1,\ldots, n_k$, where n_k denotes the time of the last prediction during the kth realization (so n_k corresponds to n in the basic TD method above). One iteration of batch TD corresponds to one complete pass through the sequence of predictions, resulting in the partial update of $\boldsymbol{\theta}$. For indexing purposes, τ is assumed to restart at 0 for each realization.

In place of the one update in basic TD from the prior value $\hat{\boldsymbol{\theta}}_0$ to the updated value $\hat{\boldsymbol{\theta}}_{\text{new}}$, the batch form is based on updates from $\hat{\boldsymbol{\theta}}_{k-1}$ to $\hat{\boldsymbol{\theta}}_k$ as we move from the $(k-1)$st process realization to the kth realization. In the update, $\hat{\boldsymbol{\theta}}_{k-1}$ and $\hat{\boldsymbol{\theta}}_k$ are, respectively, analogous to $\hat{\boldsymbol{\theta}}_0$ and $\hat{\boldsymbol{\theta}}_{\text{new}}$. For the batch generalization of the predictions used in the basic TD algorithm of (11.1) with (11.7), we define $\hat{z}_\tau^{(k)} = h_\tau(\hat{\boldsymbol{\theta}}_{k-1}, \boldsymbol{x}_\tau^{(k)})$, where $\boldsymbol{x}_\tau^{(k)}$ represents the input at time $\tau = 0, 1,\ldots, n_k$ during the kth batch. This notation is an extension of the notation for basic TD. Analogous to $\hat{z}_{n+1}$ in basic TD, $\hat{z}_{n_k+1}^{(k)}$ represents the outcome $Z^{(k)}$ for the kth realization (so the arguments in the function $h_{n_k+1}(\hat{\boldsymbol{\theta}}_{k-1}, \boldsymbol{x}_{n_k+1}^{(k)})$ are superfluous). Figure 11.2 depicts the batch TD process as an extension of the updating process in basic TD, showing a total of N realizations of the process.

An alternative to batch TD—sometimes called *online TD* or *intra-sequence updating*—is for $\boldsymbol{\theta}$ to be updated after each input. In this algorithm, an updated value of $\boldsymbol{\theta}$ is used for each prediction. In particular, the increment in $\boldsymbol{\theta}$ generated at each measurement is used immediately afterward to update $\boldsymbol{\theta}$. This contrasts with the batch method where the increments are accumulated over all of the measurements at $\tau = 0, 1,\ldots, n_k$, with $\boldsymbol{\theta}$ being updated only after the entire sequence of measurements in each realization has been processed. In the weather problem, for example, the online method updates $\boldsymbol{\theta}$ after each day's forecast and after the Saturday outcome for as many weeks as data are collected (recall, as in Figure 11.1, that the Saturday outcome is also considered a forecast).

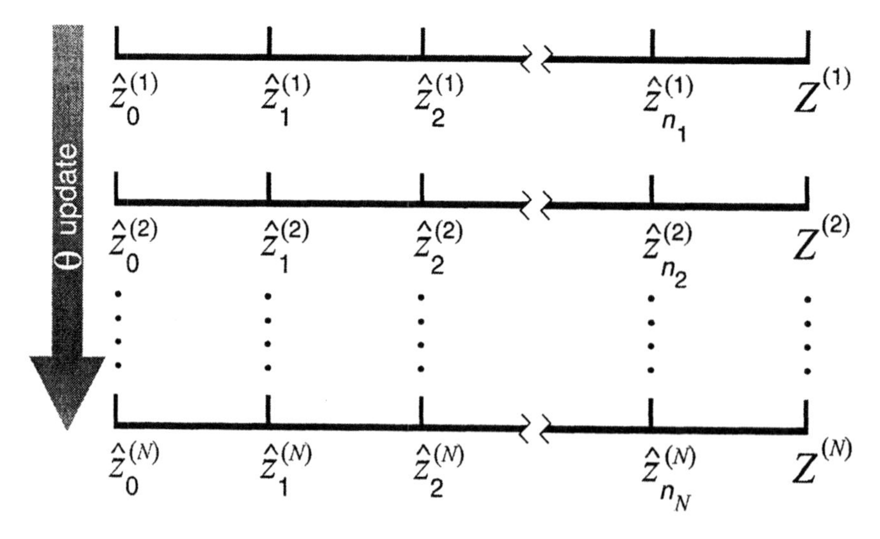

Figure 11.2. Relationship between the predictions, outcomes, and $\boldsymbol{\theta}$ updating in a batch TD process with N realizations.

Unlike the batch method, the online process is not restarted with each new week. Rather, the days of the weeks under evaluation are strung together with $\boldsymbol{\theta}$ being updated continuously. Intuitively, the online method seems more natural, as it promptly uses the information available to (one hopes) improve the subsequent predictions. In fact, as noted in Sutton (1988), "all previously studied TD methods have operated in this more fully incremental way." A further advantage of online learning is that $\boldsymbol{\theta}$ can be updated without observing an actual outcome; rather, an update occurs after each successive prediction. A downside of the online approach is the more difficult theory. Now, each temporal difference in the predictions depends on changes in the inputs *and* the value of $\boldsymbol{\theta}$.

Aside from the more frequent updating of $\boldsymbol{\theta}$, the overall structure of online TD is closely related to the basic TD form of (11.1) and (11.7). In particular, online TD is based on modifying (11.7) to reflect the current parameter value at every iteration as follows:

$$\hat{\boldsymbol{\theta}}_{\tau+1} = \hat{\boldsymbol{\theta}}_{\tau} + a_{\tau}\left[h_{\tau+1}(\hat{\boldsymbol{\theta}}_{\tau}, \boldsymbol{x}_{\tau+1}) - h_{\tau}(\hat{\boldsymbol{\theta}}_{\tau}, \boldsymbol{x}_{\tau})\right]\sum_{i=0}^{\tau}\lambda^{\tau-i}\left[\frac{\partial h_i(\boldsymbol{\theta}, \boldsymbol{x}_i)}{\partial \boldsymbol{\theta}}\right]_{\boldsymbol{\theta}=\hat{\boldsymbol{\theta}}_{\tau}}, \quad (11.11)$$

where a_{τ} is a potentially time-varying gain (step size) coefficient. This contrasts with the use of $\hat{z}_{\tau+1} = h_{\tau+1}(\hat{\boldsymbol{\theta}}_0, \boldsymbol{x}_{\tau+1})$, $\hat{z}_{\tau} = h_{\tau}(\hat{\boldsymbol{\theta}}_0, \boldsymbol{x}_{\tau})$, and $[\partial h_i(\boldsymbol{\theta}, \boldsymbol{x}_i)/\partial\boldsymbol{\theta}]_{\boldsymbol{\theta}=\hat{\boldsymbol{\theta}}_0}$ for all τ and i in the generation of the increments to $\boldsymbol{\theta}$ in the basic algorithm of (11.7) (equivalently (11.8)) together with (11.1) (Exercise 11.3 discusses another version of online TD). The difference $\hat{\boldsymbol{\theta}}_{\tau+1} - \hat{\boldsymbol{\theta}}_{\tau}$ in (11.11) (i.e., the term to the right of the "+" sign) is analogous to the increment $(\delta\boldsymbol{\theta})_{\tau}$ in (11.7) and (11.8). Unlike the basic form in (11.8) with its recursion on $\boldsymbol{s}_{\tau}$, there is no nice recursion to replace the sum over i appearing on the right-hand side of (11.11) because all summands depend on the most recent parameter estimate. (An exception to this is when the regression functions are linear in $\boldsymbol{\theta}$ since the gradients then do not depend on $\boldsymbol{\theta}$; see, e.g., Tsitlikis and Van Roy, 1997, for such a recursion.)

11.5 SOME EXAMPLES

It may seem counterintuitive that TD methods can work better than supervised learning. After all, supervised learning is based on comparing the predictions with the real outcome, and what could be better than using the real outcome in training? The examples below are meant to provide an intuitive understanding of how the use of previous predictions for learning in TD can actually be better than the use of the real outcome. The essence of these examples is that the predictions used in TD learning better reflect historical information than the usually limited information available via the outcome Z. In this sense, TD provides a ready way to incorporate prior information, whereas supervised learning in its standard form does not.

On the other hand, if Z is formed from an average of many independent experiments—or is otherwise very reliable in indicating typical performance for the inputs ($\boldsymbol{x}_\tau$) of interest—TD learning may offer no advantages over supervised learning. In fact, TD learning may be distinctly inferior in such a setting because it does not give full weight to the Z values. The algorithm considered in the first two examples (Examples 11.1 and 11.2) is basic TD (eqns. (11.1) and (11.7)), which is equivalent to one iteration through the batch method of TD in Figure 11.2. The third illustration (Example 11.3) uses batch TD. One can illustrate the power of TD methods using more sophisticated delayed reinforcement examples—as in the backgammon problem of Tesauro (1995)—but the basic message on the value of using successive predictions in the training process remains the same.

Example 11.1—Conceptual example in traffic modeling. Consider a major artery—Easy Street—in a traffic network. We are interested in establishing a model for predicting whether traffic conditions befit the artery's moniker. In particular, the model can be used to predict the likelihood of benign traffic conditions consistent with an easy traverse during the evening rush period given traffic patterns earlier in the day at test locations in the traffic network. The model produces predictions based on traffic patterns up to two hours before the beginning of the rush period. Let Z be a binary random variable, where $Z = 1$ denotes severe congestion by some specific criterion and $Z = 0$ denotes the desired low-congestion state. Suppose that there are two prediction functions,

$$\hat{z}_0 = h_0(\boldsymbol{\theta}, \boldsymbol{x}_0),$$

$$\hat{z}_1 = h_1(\boldsymbol{\theta}, \boldsymbol{x}_1),$$

where $\hat{z}_\tau$, $\tau = 0$ or 1, represents the probability of severe congestion given the input $\boldsymbol{x}_\tau$ at time τ, with $\tau = 0$ representing a prediction two hours before evening rush period and $\tau = 1$ representing a prediction one hour before. The input represents some indicator of the level of traffic in the network at one or two hours before the rush period. In all instances below, $\hat{z}_\tau$ depends on $\boldsymbol{\theta} = \hat{\boldsymbol{\theta}}_0$.

Consider the problem of updating $\boldsymbol{\theta}$ from one day's worth of data (i.e., one iteration of the batch method for TD; the online version of TD in (11.11) is considered in Exercise 11.7). We compare the TD method with a standard supervised learning method applied to the instantaneous error loss functions $\frac{1}{2}E\{[Z - h_\tau(\boldsymbol{\theta}, \boldsymbol{x}_\tau)]^2\}$. Note that the supervised learning here is the "standard" recursive stochastic gradient formulation of updating each regression (prediction) function with the experimental outcome. This is not the same as the summation-based loss function in (11.9); hence the supervised learning here is not the same as TD(1). In carrying out the supervised learning, one creates pairs based on the two inputs $\boldsymbol{x}_0$ and $\boldsymbol{x}_1$. In particular, the τth increment to $\boldsymbol{\theta}$ has the form

$$(\delta\boldsymbol{\theta})_\tau^{\text{SL}} \equiv a(Z - \hat{z}_\tau)\left[\frac{\partial h_\tau(\boldsymbol{\theta}, \boldsymbol{x}_\tau)}{\partial\boldsymbol{\theta}}\right]_{\boldsymbol{\theta}=\hat{\boldsymbol{\theta}}_0} \tag{11.12}$$

for $\tau = 0$ or 1, which is the same as the stochastic gradient update of (11.2). Hence, the supervised learning estimate for $\boldsymbol{\theta}$ is $\hat{\boldsymbol{\theta}}_{\text{new}}^{\text{SL}} \equiv \hat{\boldsymbol{\theta}}_0 + (\delta\boldsymbol{\theta})_0^{\text{SL}} + (\delta\boldsymbol{\theta})_1^{\text{SL}}$.

In contrast, a TD method forms pairs based on the differences of predictions. For example, with TD(0),

$$(\delta\boldsymbol{\theta})_0 = a(\hat{z}_1 - \hat{z}_0)\left[\frac{\partial h_0(\boldsymbol{\theta}, \boldsymbol{x}_0)}{\partial\boldsymbol{\theta}}\right]_{\boldsymbol{\theta}=\hat{\boldsymbol{\theta}}_0} \tag{11.13}$$

and

$$(\delta\boldsymbol{\theta})_1 = a(Z - \hat{z}_1)\left[\frac{\partial h_1(\boldsymbol{\theta}, \boldsymbol{x}_1)}{\partial\boldsymbol{\theta}}\right]_{\boldsymbol{\theta}=\hat{\boldsymbol{\theta}}_0}. \tag{11.14}$$

Note that Z in (11.14) is equivalent to the "prediction" $\hat{z}_2$. From (11.1), the new estimate for $\boldsymbol{\theta}$ is $\hat{\boldsymbol{\theta}}_{\text{new}} = \hat{\boldsymbol{\theta}}_0 + (\delta\boldsymbol{\theta})_0 + (\delta\boldsymbol{\theta})_1$. Conditioned on historical data providing a prior value of $\boldsymbol{\theta}$, $(\delta\boldsymbol{\theta})_0$ is purely deterministic, while $(\delta\boldsymbol{\theta})_1$ is random because it depends directly on the outcome Z.

To see how the TD(0) formulation can provide a better estimate, suppose without loss of generality that the components $\partial h_\tau(\boldsymbol{\theta}, \boldsymbol{x}_\tau)/\partial\boldsymbol{\theta}$ are positive (so an increase in the value of the elements of $\boldsymbol{\theta}$ cause an increase in the predicted probability of congestion on Easy Street).[2] Based on historical data, it is known that if the traffic network has a general set of bad conditions one hour prior to the evening rush period (reflected in $\boldsymbol{x}_1$), then there is an 80 percent chance that there will be severe congestion on Easy Street during the evening rush period. Let us call this state BAD, in contrast to two other possible states, GOOD and MODERATE.

Suppose a development company wishes to alter the entrance and exit patterns at a shopping mall in the traffic network, and there is concern about the effect of this change on the congestion on Easy Street. Before making the alterations permanent, the local government demands that a test be performed to determine if the changes in the entrance/exit patterns are likely to increase the likelihood of severe congestion on Easy Street. Because of the risks of serious disruption to the network traffic patterns during the testing phase, the developer is restricted to collecting data for only one day. Suppose during this test, that state BAD occurs, but that there is *not* severe traffic congestion on Easy Street during the evening rush period. Hence, $Z = 0$, while from the historical data, $\hat{z}_1 = 0.80$. Note that $(\delta\boldsymbol{\theta})_1^{\text{SL}} = (\delta\boldsymbol{\theta})_1$, but as a consequence of the difference between Z and $\hat{z}_1$, $(\delta\boldsymbol{\theta})_0^{\text{SL}} \neq (\delta\boldsymbol{\theta})_0$. In particular, the elements of $(\delta\boldsymbol{\theta})_0^{\text{SL}}$ are less than the corresponding elements of $(\delta\boldsymbol{\theta})_0$, indicating that the elements of $\hat{\boldsymbol{\theta}}_{\text{new}}^{\text{SL}}$ are lower than the elements of $\hat{\boldsymbol{\theta}}_{\text{new}}$.

---

[2]If some elements of $\boldsymbol{\theta}$ have a decreasing effect on $h_\tau(\cdot)$, the results to follow are unaffected because the TD learning process automatically alters the estimate of those elements to compensate for the opposite sign in the gradient elements.

Because the supervised learning parameter estimates are artificially lowered by the anomalous outcome of nonsevere congestion following the BAD state for $\boldsymbol{x}_1$, supervised learning produces predictions for severe congestion that are too low (recalling the monotonic relationship between the elements of $\boldsymbol{\theta}$ and h_τ). This assumes, of course, that the shopping center alterations introduce no other effects that may mitigate the BAD state. The TD(0) method, on the other hand, more accurately reflects the historical connection between the BAD state and severe congestion, producing parameter values (and predictions) that are greater than the supervised learning method.

What if the experimental outcome involved the BAD state for $\boldsymbol{x}_1$ (as above), but severe congestion on Easy Street? In this case the TD(0) method still produces a better estimate. The supervised learning method associates the BAD state fully with the outcome of severe congestion, while the TD(0) method more accurately reflects that, historically, 20 percent of the time, severe congestion does *not* follow the BAD state. Hence, in this case the TD(0) parameter estimates (and predictions) are lower than those from supervised learning. ❑

Example 11.2—Numerical results for example in traffic modeling. Continuing with Example 11.1, suppose that the prediction function for the probability of severe congestion on Easy Street is

$$\hat{z}_\tau = h_\tau(\theta, x_\tau) = \frac{1}{1+10\exp[-\theta(\tau+1)x_\tau]}, \tag{11.15}$$

where the domain for the scalar θ satisfies $\Theta = [0, \infty)$. The scalar input x_τ can take on one of three values: 0, 1, or 2, representing network traffic conditions that are GOOD, MODERATE, or BAD, respectively (representing low congestion to high congestion). From Example 11.1, historical data indicate that $h_1(\theta, 2) = 0.80$. From (11.15), this uniquely determines a value for θ, taken as the prior estimate $\hat{\theta}_0 = 0.922$. Note that for a given θ and $x_0 = x_1$, the prediction $\hat{z}_0$ is less than $\hat{z}_1$ as a reflection of: (i) the normal state of no congestion on Easy Street (hence the eponymous name!) and (ii) the reduced ability to use current conditions to predict congestion two hours in advance versus one hour in advance (e.g., with a MODERATE congestion input, $\hat{z}_0 = h_0(\hat{\theta}_0, 1) = 0.201$ versus $\hat{z}_1 = h_1(\hat{\theta}_0, 1) = 0.387$). For use in the learning algorithms, note that

$$\frac{\partial h_\tau(\theta, x_\tau)}{\partial \theta} = \frac{10(\tau+1)x_\tau \exp[-\theta(\tau+1)x_\tau]}{\{1+10\exp[-\theta(\tau+1)x_\tau]\}^2}.$$

From Example 11.1, $x_1 = 2$ and $Z = 0$; also suppose that traffic conditions are MODERATE two hours before rush period (i.e., $x_0 = 1$). Letting $a = 1$, expression (11.12) for the increments in θ implies that the two supervised training increments are

$$(\delta\theta)_0^{\mathrm{SL}} = (0.0 - 0.201)\frac{3.9764}{(1+3.9764)^2} = -0.032,$$

$$(\delta\theta)_1^{\mathrm{SL}} = (0.0 - 0.800)\frac{1.0001}{(1+0.2500)^2} = -0.512,$$

whereas from (11.13) and (11.14), the TD(0) increments are

$$(\delta\theta)_0 = (0.800 - 0.201)\frac{3.9764}{(1+3.9764)^2} = 0.096$$

$$(\delta\theta)_1 = (\delta\theta)_1^{\mathrm{SL}} = -0.512.$$

Using these increments, the new values of θ from supervised learning and TD(0) are

$$\hat{\theta}_{\mathrm{new}}^{\mathrm{SL}} \equiv \hat{\theta}_0 + (\delta\theta)_0^{\mathrm{SL}} + (\delta\theta)_1^{\mathrm{SL}} = 0.922 - 0.032 - 0.512 = 0.378,$$

$$\hat{\theta}_{\mathrm{new}} = \hat{\theta}_0 + (\delta\theta)_0 + (\delta\theta)_1 = 0.922 + 0.096 - 0.512 = 0.506.$$

The TD(0) method produces a more accurate parameter estimate in the sense that it better preserves the prior information associating the BAD state one hour before the rush period with congestion during the rush period. For $\tau = 1$ (i.e., one hour ahead), Table 11.1 compares the predictions based on $\hat{\theta}_{\mathrm{new}}^{\mathrm{SL}}$ and $\hat{\theta}_{\mathrm{new}}$ with those from historical data.

Relative to supervised learning, the TD(0) estimates are closer to the historical estimates for all input state values, reflecting some level of congestion

Table 11.1. Values of the prediction $\hat{z}_1$ for the three levels of input (x_1) under the parameter values $\hat{\theta}_0$, $\hat{\theta}_{\mathrm{new}}^{\mathrm{SL}}$, and $\hat{\theta}_{\mathrm{new}}$.

Input state x_1	Historical estimate (from $\hat{\theta}_0$)	Supervised learning (from $\hat{\theta}_{\mathrm{new}}^{\mathrm{SL}}$)	Basic TD(0) (from $\hat{\theta}_{\mathrm{new}}$)
0 (GOOD)	0.09	0.09	0.09
1 (MODERATE)	0.39	0.18	0.22
2 (BAD)	0.80	0.31	0.43

in the network one hour prior to the rush period ($x_1 = 1$ or 2). As a consequence of the experimental outcome (Z), however, the supervised learning and TD(0) predictions lie below the historical predictions for these nonzero input states. ❑

Example 11.3—Random-walk model. Let us summarize an interesting example from Sutton and Barto (1998, pp. 139–141). Consider a process with seven possible states as shown in Figure 11.3. The random walk is initialized at state S_3 and proceeds until it reaches one of the termination states T_{left} or T_{right}. A possible walk might be S_3–S_4–S_3–S_4–S_5–T_{right}. The aim is to estimate the probability of the walk ending at T_{right} given that the walk is in any one of the five nonterminal states. For example, if the process is in state S_2, we are interested in the probability of the process ultimately reaching T_{right}. The outcome is $Z = 1$ if termination occurs at T_{right}; $Z = 0$ if termination occurs at T_{left}. Hence, $P(Z=1|S_i) = E(Z|S_i)$. Consider the simple linear prediction function $\hat{z}_\tau = h_\tau(\boldsymbol{\theta}, \boldsymbol{x}_\tau) = \boldsymbol{\theta}^T\boldsymbol{x}_\tau$, where $\boldsymbol{x}_\tau$ is a vector of five elements, four of which are zero and one of which is unity as an indicator of the state in which the walk resides. For example, if the walk is in state S_2 at time τ, then $\boldsymbol{x}_\tau = [0, 1, 0, 0, 0]^T$. Let $\boldsymbol{\theta} \in \Theta = [0, 1]^5$ be a five-dimensional vector of the probabilities of interest for the five nonterminal states.

Batch TD (Section 11.4) is applied to estimating these five probabilities based on an initial condition $\hat{\boldsymbol{\theta}}_0 = [0.5, 0.5, 0.5, 0.5, 0.5]^T$. From Sutton and Barto (1998, Fig. 6.6), Table 11.2 shows the estimated probabilities after batch sizes of $N = 1$, 10, and 100 using a constant step size $a = 0.1$. The true probabilities are found in Sutton (1988) as $\boldsymbol{\theta}^* = [1/6, 1/3, 1/2, 2/3, 5/6]^T$ (see Exercise 11.8). That only one parameter (the first element of $\boldsymbol{\theta}$, corresponding to state S_1) changes with $N = 1$ is a direct consequence of having only one outcome Z ($= Z^{(0)}$) in the training process (see Exercise 11.9). As expected, the estimates improve with increasing N. The estimate at $N = 100$, however, is about as close to truth as the algorithm can ever get. This is a consequence of the constant (nondecaying) gain causing the $\boldsymbol{\theta}$ estimates to fluctuate indefinitely with each new realization in the batch process (see Exercises 11.10 and 11.11). ❑

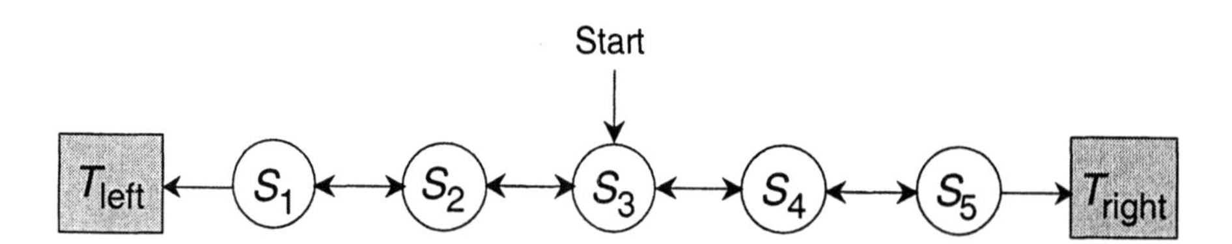

Figure 11.3. Schematic of process for generating random walks. All walks begin in the center state S_3. For each of states S_1 to S_5, the walk has a 50 percent chance of moving either left or right. The walk stops if the process reaches either of the termination states T_{left} or T_{right}.

Table 11.2. True probabilities and estimated probabilities of reaching T_{right} for the five states based on N realizations in batch TD(0).

State	True probability	$N = 1$	$N = 10$	$N = 100$
S_1	0.17	0.45	0.35	0.13
S_2	0.33	0.50	0.50	0.30
S_3	0.50	0.50	0.52	0.48
S_4	0.67	0.50	0.56	0.71
S_5	0.83	0.50	0.71	0.80

11.6 CONNECTIONS TO STOCHASTIC APPROXIMATION

The TD formulation has a ready connection to root finding and the stochastic gradient form of SA that was discussed in Chapter 5. This connection has been exploited to make rigorous statements about the convergence of TD methods. Using methods other than SA, there were prior attempts at showing formal convergence (e.g., Sutton, 1988), but these were largely unsatisfactory as a result of the weak mode of convergence shown[3] and the conditions being so restrictive that few practical settings would qualify. Once the connection to SA was identified, more satisfying convergence results were possible in terms of the strength of convergence (a.s.), the rigor of the analysis, and the general applicability of the results. The TD conditions are directly related to the type of conditions seen in root-finding (Robbins–Monro) SA (Section 4.3). This section will only outline the connections of TD to the SA formulation. The interested reader is directed to the indicated references for the formal convergence conditions and proofs.

Let us first consider the batch form of TD represented in Figure 11.2. We comment on the online version (the form in (11.11)) below. Dayan and Sejnowski (1994), Jaakkola et al. (1994), and Pineda (1997) use SA to show formal (a.s.) convergence of batch TD. Replacing the fixed a with the typical gain coefficient indexed by k, the batch TD(λ) algorithm can then be written in SA form as

$$\hat{\boldsymbol{\theta}}_k = \hat{\boldsymbol{\theta}}_{k-1} - a_{k-1} \boldsymbol{Y}_{k-1}(\hat{\boldsymbol{\theta}}_{k-1}) \tag{11.16}$$

---

[3]For example, Sutton (1988) shows only that the mean of the prediction $h_\tau(\cdot)$ using TD(0) training converges. This convergence *of* mean is a very weak mode of convergence (see Corollary 1 to Theorem C.1 in Appendix C), far weaker than convergence *in* mean (Theorem C.1 with $q = 1$). For example, an i.i.d. sequence of nondegenerate random variables will never converge in any meaningful way, but the mean of the sequence is trivially convergent from the very first element in the sequence.

(relative to the usual SA form in this book, we are showing a form lagged by one in k to avoid messy super/subscript notation below). From (11.1) and (11.7),

$$\boldsymbol{Y}_{k-1}(\hat{\boldsymbol{\theta}}_{k-1}) = -\sum_{\tau=0}^{n_k}\left(\hat{z}_{\tau+1}^{(k)} - \hat{z}_{\tau}^{(k)}\right)\sum_{i=0}^{\tau}\lambda^{\tau-i}\left[\frac{\partial h_i(\boldsymbol{\theta}, \boldsymbol{x}_i^{(k)})}{\partial\boldsymbol{\theta}}\right]_{\boldsymbol{\theta}=\hat{\boldsymbol{\theta}}_{k-1}} . \quad (11.17)$$

Note that at any $\boldsymbol{\theta}$, $\boldsymbol{Y}_k(\boldsymbol{\theta})$ is an unbiased measurement of the *time-varying* function (e.g., Subsection 4.5.4) for which we wish to find a zero:

$$\boldsymbol{g}_{k-1}(\boldsymbol{\theta}) \equiv -E\left\{\sum_{\tau=0}^{n_k}\left[h_{\tau+1}(\boldsymbol{\theta}, \boldsymbol{x}_{\tau+1}^{(k)}) - h_{\tau}(\boldsymbol{\theta}, \boldsymbol{x}_{\tau}^{(k)})\right]\sum_{i=0}^{\tau}\lambda^{\tau-i}\frac{\partial h_i(\boldsymbol{\theta}, \boldsymbol{x}_i^{(k)})}{\partial\boldsymbol{\theta}}\right\}, \quad (11.18)$$

where the expectation is computed with respect to the random outcome (the $Z^{(k)} = \hat{z}_{n_k+1}^{(k)} = h_{n_k+1}(\cdot)$) values at the end of the batches), the potential randomness in the inputs values $\boldsymbol{x}_0^{(k)}, \boldsymbol{x}_1^{(k)}, \ldots, \boldsymbol{x}_{n_k}^{(k)}$, and the potential randomness in the realization length n_k (e.g., random realization lengths as in Example 11.3).

The arguments of Section 5.1 provide the justification for $\boldsymbol{Y}_k(\boldsymbol{\theta})$ being an unbiased measurement of $\boldsymbol{g}_k(\boldsymbol{\theta})$. Recall that these arguments rely on the probability distribution of the fundamental randomness in $\boldsymbol{Y}_k(\boldsymbol{\theta})$ being independent of $\boldsymbol{\theta}$. That is, for the application here, the distribution of the $Z^{(k)}$, $\boldsymbol{x}_{\tau}^{(k)}$, and n_k should not depend on $\boldsymbol{\theta}$ (which is entirely reasonable since the *physical* $Z^{(k)}$, $\boldsymbol{x}_{\tau}^{(k)}$, and n_k may be generated without regard to the *model* parameters $\boldsymbol{\theta}$). Then, from Section 5.1, an unbiased measurement of $\boldsymbol{g}_k(\boldsymbol{\theta})$ is the argument inside the $\{\cdot\}$ braces in (11.18), which is the same as $\boldsymbol{Y}_k(\hat{\boldsymbol{\theta}}_k)$ in (11.17). (These ideas provide the basis for some of the counterexamples to TD learning mentioned at the end of Section 11.2. In control applications, the control actions $\boldsymbol{x}_i^{(k)}$ depend on $\boldsymbol{\theta}$ through feedback. Hence, the probability distribution of the fundamental randomness entering $\boldsymbol{Y}_k(\boldsymbol{\theta})$ depends on $\boldsymbol{\theta}$.)

Relative to batch TD, online TD has a more direct connection to the traditional SA formulation since $\boldsymbol{\theta}$ is naturally updated as every new input value is collected. The formal convergence proofs, however, require a modification to the basic online form in (11.11). In particular, there is a need for frequent examples of "truth" to ensure convergence to a *meaningful* $\boldsymbol{\theta}$ (convergence to *some* value of $\boldsymbol{\theta}$ may be possible without the true outcomes, but this $\boldsymbol{\theta}$ may not be a "good" value relative to the true system). Note that in (11.11), it is possible for the algorithm to run arbitrarily long with only one actual outcome; this is an insufficient instance of truth for meaningful convergence. The available online convergence proofs (e.g., Jaakkola et al., 1994; Tsitsiklis and Van Roy, 1997, 1999) overcome this difficulty by posing the problem in the framework of dynamic programming, with the goal of predicting some type of average or cumulative cost for a process based on a sequence of observed instantaneous costs. The observed costs at each iteration represent measurements of the actual physical system. The cost inputs provide the necessary information to ensure meaningful convergence.

The online version of TD once again uses the basic SA recursion (11.16), but rather than input (11.17), the input is

$$\boldsymbol{Y}_k(\hat{\boldsymbol{\theta}}_k) = -\left[h_{k+1}(\hat{\boldsymbol{\theta}}_k, \boldsymbol{x}_{k+1}) - h_k(\hat{\boldsymbol{\theta}}_k, \boldsymbol{x}_k) + \text{cost}_k\right]\sum_{i=0}^{k}\lambda^{k-i}\left[\frac{\partial h_i(\boldsymbol{\theta}, \boldsymbol{x}_i)}{\partial \boldsymbol{\theta}}\right]_{\boldsymbol{\theta}=\hat{\boldsymbol{\theta}}_k}, \tag{11.19}$$

where the temporal difference of predictions includes the instantaneous cost measurement, cost_k. (Jang et al., 1997, pp. 268–270, includes a brief tutorial on the use of TD methods in such cost estimation.) As with the batch form, $\boldsymbol{Y}_k(\boldsymbol{\theta})$ represents a measurement of some time-varying function $\boldsymbol{g}_k(\boldsymbol{\theta})$.

While the form in (11.16) with input of either (11.17) or (11.19) fits well within the framework of root-finding SA with a time-varying nonlinear functions, existing convergence results are only for the case where the $h_\tau(\boldsymbol{\theta}, \boldsymbol{x})$ are linear functions in $\boldsymbol{\theta}$. The simplest such case is the pure linear form $\boldsymbol{\theta}^T\boldsymbol{x}$ considered, for example, in Dayan and Sejnowski (1994). The more general curvilinear case (linear in $\boldsymbol{\theta}$, nonlinear in $\boldsymbol{x}$ with, as in Section 3.1, $\dim(\boldsymbol{\theta})$ not necessarily equal to $\dim(\boldsymbol{x})$) is considered in Tsitsiklis and Van Roy (1997, 1999). Significant difficulties exist for proving convergence in more general functions nonlinear in $\boldsymbol{\theta}$ (e.g., neural networks with weights $\boldsymbol{\theta}$). One of the obvious difficulties to showing convergence in general nonlinear problems is that the function $\boldsymbol{g}_k(\cdot)$ is time varying. To illustrate the difficulties in convergence for general nonlinear functions, Tsitsiklis and Van Roy (1997, Sect. 10) provides a counterexample to convergence using a fairly simple nonlinear form for the $h_\tau(\boldsymbol{\theta}, \boldsymbol{x})$.

11.7 CONCLUDING REMARKS

We have looked at aspects of the delayed reinforcement problem. A fundamental challenge is that there is no system output that indicates "truth" at each time step of the learning process. The temporal difference algorithm is one powerful method for addressing this challenge by processing the differences of model-based *predictions* of the process. The temporal differences—although not based on the final outcome of the process until the last time step—contain valuable information for improving the prediction process. In the language of machine learning and artificial intelligence, the learning system is to predict the final state of the system (or final reward) without the benefit of a teacher signal indicating the correct value at each time step.

The TD algorithm is able to develop improved final predictions based on the prior information contained in the current model with its associated intermediate predictions. The basic and enhanced TD algorithms are built from connections to the stochastic gradient algorithm (Chapter 5). There are additional connections to stochastic approximation in the batch and online versions of TD learning; these connections are useful in establishing formal convergence.

A number of papers in the literature pertain to the connection of TD learning to Markov chains and dynamic programming, including the use in stochastic control problems (e.g., Jaakkola et al., 1994; Tsitsiklis and Van Roy, 1997). This chapter did not explore these connections. The focus here was pure prediction problems. In control, prediction is used to determine the states to which the system is expected to go with specific control inputs. The control values can be adjusted over time to produce predictions that are close to desired values. The combined control and delayed reinforcement learning problem, of course, introduces complications in theory and implementation not present in the pure prediction problems. Sutton et al (1992) and Tesauro (1992), among others, consider control in TD or more general reinforcement learning.

EXERCISES

11.1 Fill in the missing steps in the double-sum formula (11.5) used in obtaining the basic TD algorithm.

11.2 Show by direct calculation the validity of the equality of LMS and TD(1) in (11.10).

11.3 Relative to (11.11), an alternative online form of TD learning for constant gains $a = a_\tau$ is

$$\hat{\boldsymbol{\theta}}_{\tau+1} = \hat{\boldsymbol{\theta}}_\tau + a\left[h_{\tau+1}(\hat{\boldsymbol{\theta}}_\tau, \boldsymbol{x}_{\tau+1}) - h_\tau(\hat{\boldsymbol{\theta}}_{\tau-1}, \boldsymbol{x}_\tau)\right]\sum_{i=0}^{\tau}\lambda^{\tau-i}\left[\frac{\partial h_i(\boldsymbol{\theta}, \boldsymbol{x}_i)}{\partial \boldsymbol{\theta}}\right]_{\boldsymbol{\theta}=\hat{\boldsymbol{\theta}}_{i-1}},$$

where $\hat{\boldsymbol{\theta}}_{-1} = \hat{\boldsymbol{\theta}}_0$. This form is natural in the sense that each regression function is associated with its most recent parameter estimate.

(a) Derive a recursion analogous to the $\boldsymbol{s}_\tau$ recursion used in (11.8) so that the sum over i does not have to be explicitly computed at each iteration.

(b) Provide at least two reasons why this online form of TD is likely to be inferior to the form in (11.11).

11.4 In the manner of Example 11.1, design a *conceptual* example illustrating the potential power of the TD method to produce more reliable estimates than the supervised learning method. The example should be substantively different from the traffic problem in Example 11.1.

11.5 Provide a conceptual example illustrating how it is possible for the TD method to tend to provide *poorer* estimates than supervised learning.

11.6 Consider the setting of Example 11.2 with TD(0.5) replacing TD(0) while continuing with the basic algorithm of (11.1) and (11.7):

(a) Produce a table (analogous to Table 11.1) that includes the historical, supervised learning, and TD(0.5) predictions two hours (versus one hour in Table 11.1) in advance.

(b) Update Table 11.1 (based on the possible values of x_1) to reflect the change to TD(0.5). Comment on the relative performance of TD(0.5) and TD(0).

11.7 Compare the predictions for basic TD(0) in Table 11.1 with corresponding predictions from online TD(0) (Section 11.4) using the information in Example 11.2.

11.8 Construct the 7×7 transition matrix (see Appendix E) describing the probability of going from one state to another in Example 11.3. Use this transition matrix to obtain the true probabilities shown in Table 11.2.

11.9 As shown in Table 11.2, only the first element of $\boldsymbol{\theta}$ changes when $N = 1$ for the random walk example.

(a) Illustrate how this result occurred.

(b) Discuss the other possible estimate for $\boldsymbol{\theta}$ that could have occurred.

(c) How do the results in parts (a) and (b) differ fundamentally from those that would be obtained if TD(λ), $\lambda > 0$, were used instead of TD(0)?

11.10 Consider Example 11.3. With $a = 0.10$ and $\hat{\boldsymbol{\theta}}_0 = [0.5, 0.5, 0.5, 0.5, 0.5]^T$, as in the example, calculate terminal $\boldsymbol{\theta}$ estimates (a single realization) using $N = 1000$ and $N = 2000$ batches with TD(0). Verify that these estimates based on a constant gain improve little from the $N = 100$ values in Table 11.2.

11.11 Consider the setup in Exercise 11.10 with the exception of using *decaying* gains a_k of the standard harmonic-sequence form $a_k = a/(k+1)$ for some $a > 0$ (Section 4.3). The gain sequence should decay from realization to realization (not within the realization), so N realizations in the batch updating requires $a_0, a_1, \ldots, a_{N-1}$. As an example of the convergence results for batch TD (Section 11.6), show numerical convergence as k ranges from $k = 0$ to $k = N - 1 = 1999$.

CHAPTER 12

STATISTICAL METHODS FOR OPTIMIZATION IN DISCRETE PROBLEMS[1]

This chapter is unique in this book in focusing exclusively on problems where $\boldsymbol{\theta}$ takes on only discrete values. We discuss some statistical methods of comparing the loss values for the possible outcomes for $\boldsymbol{\theta}$, the aim being to find the values of $\boldsymbol{\theta}$ yielding the lowest values of the loss function. The use of statistical testing here differs from most of the rest of the book in that testing is being used to *solve* the optimization problem rather than to *evaluate* other optimization methods. Further, the methods here differ in not generally relying on the notion of updating a $\boldsymbol{\theta}$ estimate in an iterative manner. Rather, the methods sift through the candidate values of $\boldsymbol{\theta}$ to produce an estimate in a batch manner.

The central challenge is that classical statistical testing is best suited to *pairwise* comparisons (is option 1 better than option 2?), yet the primary interest here is a many-way comparison. Numerous approaches have been developed for extending notions of pairwise comparisons to general multiple comparisons. Some of these will be discussed here. Although there may be uniquely "obvious" approaches to pairwise testing—such as the pairwise t-tests in Appendix B—there will rarely be such a unique approach when the test involves a comparison of at least three options (i.e., at least three values in the domain Θ for $\boldsymbol{\theta}$).

This chapter examines several approaches to multiple comparisons tests as they apply in the problem of stochastic optimization. Section 12.1 provides some general background on the testing framework. Section 12.2 describes a popular test for performing an initial statistical assessment of whether there exist some candidate $\boldsymbol{\theta}$ values that are better than other $\boldsymbol{\theta}$ values. Sections 12.3 and 12.4 assume that prior information is available to suggest that a particular $\boldsymbol{\theta}$ is a candidate for being the optimum. This allows for a potentially dramatic reduction in the number of comparisons that need to be tested, thereby improving the ability to make concrete statistical statements. Section 12.3 assumes that the

---

[1]Without loss of continuity, this chapter may be skipped or postponed by those whose interests focus on algorithms for search and optimization where the estimate is updated recursively. Although this chapter contains important material for discrete problems, the material here is not a prerequisite to the material in other chapters, with the possible exception of Section 14.5 on selection methods for optimization via Monte Carlo simulations.

measurement noise variance is known, whereas Section 12.4 does not. Section 12.5 summarizes some other statistical tests for aiding in the discrete optimization task. These tests are sometimes referred to as ranking and selection methods in the statistics and other literature. Section 12.6 offers some concluding remarks, including some comments about sequential analysis methods—such as multi-armed bandit problems—for optimizing over a discrete set Θ.

Roughly speaking, the multiple comparisons tests that dominate this chapter provide a means for combining *pairwise* comparisons of options in a meaningful way to perform general *multiple* comparisons. The ranking and selection methods, in contrast, are not based directly on the combination of pairwise comparisons. Section 14.5 covers multiple comparisons tests in the special case where Monte Carlo simulations are generating the measurements of L. In particular, the property of common random numbers discussed in that chapter is used to advantage in the simulation-based setting.

12.1 INTRODUCTION TO MULTIPLE COMPARISONS OVER A FINITE SET

While much of this book concentrates on search and optimization when the elements of $\boldsymbol{\theta}$ can, in principle, take on any real-numbered value satisfying possible constraints in Θ, there are many important problems where there are a finite number of discrete options. (The "in principle" qualifier pertains to the fact that a digital computer implementation is necessarily discrete, although for all practical purposes may be treated as continuous.) Problems having a finite number of options can arise in many ways, including as a natural part of the problem definition or as a simplification of a large-scale continuous or discrete problem with an unbounded number of possibilities. Suppose that the interest is in finding $\boldsymbol{\theta}^* = \arg\min_{\boldsymbol{\theta}\in\Theta} L(\boldsymbol{\theta})$, where $\Theta = \{\boldsymbol{\theta}_1, \boldsymbol{\theta}_2,\ldots, \boldsymbol{\theta}_K\}$ represents the K possible values for $\boldsymbol{\theta}$. It is possible that more than one value of $\boldsymbol{\theta}_i$ can serve as $\boldsymbol{\theta}^*$ (i.e., the set Θ^* from expression (1.1) contains more than one point), but for convenience in discussion, we generally make singular (versus plural) references to $\boldsymbol{\theta}^*$. If at any value of $\boldsymbol{\theta}$, one can measure L exactly (i.e., without noise), then the obvious means by which one can determine $\boldsymbol{\theta}^*$ is to calculate $L(\boldsymbol{\theta}_i)$ for all $i = 1, 2,\ldots, K$, taking $\boldsymbol{\theta}^*$ as the $\boldsymbol{\theta}_i$ yielding the lowest value of L. (Of course, if K is a very large number, even this simple approach may not be feasible.)

We are interested in the more challenging case where L can be observed only in the presence of noise (i.e., Property A in Section 1.1). Hence, simply collecting measurements of L and comparing values—without further analysis—is generally inadequate for making rigorous statements about the optimality of a point. The job of the statistical comparison procedures here is to sort through potentially confusing or ambiguous measurements in an attempt to provide a formal means of estimating $\boldsymbol{\theta}^*$. Because the number of tests required in the statistical procedures grows rapidly with the number of options to be considered, statistical procedures such as these are appropriate when the number of options, K, is not too large. Goldsman and Nelson (1998), for example, suggest $2 \le K \le 20$ as a reasonable range.

In problems with a larger K, it is possible to combine several methods for screening as a way of narrowing the options (e.g., Nelson et al., 2001). Alternatively, methods such as the direct search techniques of Chapter 2, simulated annealing and related methods (Chapter 8), or evolutionary computation (Chapters 9 and 10) *may* be useful. However, the application of such methods with noisy loss measurements (as we have here) is largely ad hoc. Further, there is a loss of statistical interpretation for the solution.

There also exist sequential methods for optimizing over a discrete set Θ based on principles of decision theory (see the survey paper by Lai, 2001). These sequential methods involve an adaptive choice of the sampling strategy for determining the optimal $\boldsymbol{\theta}$. As with the methods of this chapter, the sequential methods provide statistical guarantees in the form of hypothesis tests. Important examples of such alternative methods are multiarmed bandit problems, so named by statistical analogies to slot machines. While the sequential methods are powerful and possibly more efficient in terms of the sample sizes required to guarantee certain levels of statistical accuracy, they tend also to be more cumbersome to implement than the relatively simple methods of this chapter.

The conceptual example below illustrates one type of application for the statistical comparisons methods of this chapter. There are countless other types as well. Although the example below is based on Monte Carlo simulations, the comparisons methods can also apply when physical experiments are used to produce the measurements of L. Section 14.5 demonstrates how certain properties unique to simulations can be exploited to enhance some types of statistical comparisons procedures.

Example 12.1—Decision making via Monte Carlo simulation. Let us summarize a conceptual example where multiple comparisons of the type here might be used. Suppose that a decision maker for a firm is attempting to choose the best alternative from among K possible options. Each of these options represents a significant investment for which it is not possible to run actual field experiments. For example, an option may involve the acquisition of another firm or the construction of a new manufacturing plant. Suppose that the firm has developed a Monte Carlo simulation that provides a realistic evaluation of alternatives. Each run of the simulation produces an estimate of the firm's annual profit under a particular option. To run the simulation, values for system parameters $\boldsymbol{\theta}$ must be specified. The ith option is represented in the simulation by setting $\boldsymbol{\theta} = \boldsymbol{\theta}_i$. As a reflection of randomness in the real world, multiple runs of the Monte Carlo simulation at a *fixed* scenario (a fixed $\boldsymbol{\theta}$) produce *different* estimates for the profit.

To help decide which option is best—that is, to find $\boldsymbol{\theta}^*$—the decision maker runs the simulation multiple times at each $\boldsymbol{\theta}_i$. By averaging the simulation outputs under each scenario, the decision maker has information to judge whether one or more scenarios are clearly superior or inferior in a statistical sense given the inherent variability at each $\boldsymbol{\theta}_i$. This comparison can be carried out using statistical tests of the type in this chapter (and Section 14.5). ❑

One will find in the methods below that determining a likely $\boldsymbol{\theta}^*$ with the comparison-based statistical methods does not yield a "cut and dried" approach (just like most other methods of stochastic search and optimization!). Although there are some rigorous methods that can point to some values of $L(\boldsymbol{\theta}_i)$ as probably being better (lower) than other values, it is difficult to establish a single rigorous probabilistic statement about the likelihood of a specific $\boldsymbol{\theta}_i$ being the optimal value. Rather, one generally has to be satisfied with statistical *indicators* of an optimum. These indicators—though imperfect—provide more quantification of the likelihood of reaching $\boldsymbol{\theta}^*$ than the information that is available to the analyst in the other discrete search methods mentioned above.

The fundamental difficulty arises from the ambiguity inherent in making more than one comparison: If statistical indications suggest that one specific hypothesis is not likely, then there may be many plausible alternatives that need to be considered, only one of which corresponds to the (unknown) truth when $\boldsymbol{\theta}^*$ is unique. This contrasts with pairwise statistical testing, where if the data collected are strongly inconsistent with a null hypothesis of, say, $L(\boldsymbol{\theta}_1) \geq L(\boldsymbol{\theta}_2)$, then there are indications that the *unique* alternative hypothesis $L(\boldsymbol{\theta}_1) < L(\boldsymbol{\theta}_2)$ is true (i.e., $\boldsymbol{\theta}_1$ is optimal). Furthermore, the error in this conclusion is explicitly quantifiable at the false alarm rate of the test. In some cases, it is possible to introduce additional assumptions—such as knowledge that the optimal point produces a loss function value at least δ units better than all other values—as a way of circumventing the difficulties with a lack of unique alternatives. With such assumptions, there is sometimes an explicit quantifiable error as in the pairwise case. An example is given in Section 12.5. The broad philosophy below is to test a general (null) hypothesis that all points in Θ are equivalent, in the hopes that this hypothesis can be *rejected*. One must then perform detailed analysis to attempt to isolate the element in Θ most likely to be $\boldsymbol{\theta}^*$.

To determine $\boldsymbol{\theta}^*$, the statistical approaches here are based on the principle of multiple comparisons. Multiple comparisons is fundamental when one needs to compare more than two quantities. The methods below are based on two broad steps: (i) a test where we simultaneously determine if there is evidence that not all candidate points $\boldsymbol{\theta}_i$ produce the same loss value and (ii) given that there is such evidence, refined analysis on a subset of points as necessary to resolve ambiguities that may remain after the full simultaneous test in (i). The ultimate aim, of course, is to determine clear winners, that is, particular $\boldsymbol{\theta}_i$ that are statistically better than all other $\boldsymbol{\theta}_j$, $j \neq i$. In practice, the analyst may have to settle for less information than clearly identified winners, perhaps instead a strong indication of a likely optimal point or small set of likely optimal points. The two-sample tests discussed in Appendix B are fundamental in conducting the refined analysis. The treatment of multiple comparisons here is relatively brief. The reader with a serious interest in this branch of search and optimization is directed to more comprehensive references such as Gupta (1965), Miller (1981), Hochberg and Tamhane (1987), Hsu (1996), and Goldsman and Nelson (1998). Although some of these references are of an older vintage, the points raised are still largely relevant.

Statistical hypothesis testing plays a prominent role in executing the broad steps mentioned above. Such testing involves a *null* hypothesis and an *alternative* hypothesis. The null hypothesis in all the tests being considered here is that all candidate $\boldsymbol{\theta}_i$ produce loss values that are effectively equal. The aim in trying to find an optimal $\boldsymbol{\theta}_i$ is, of course, to *reject* this null hypothesis.

One might wonder why the null hypothesis is not taken to be the primary hypothesis of interest (that a particular $\boldsymbol{\theta}_i$ is the best $\boldsymbol{\theta}$ value). It is customary to make the null hypothesis a nominal state of nature—in this case that $\boldsymbol{\theta}_i$ is no better than the other $\boldsymbol{\theta}$ values—because it is almost always easier to characterize the distribution of the test statistic under the null hypothesis. Then, based on this characterization, one can guarantee that a null hypothesis will only be mistakenly rejected with a small probability (the *type I error* or *false alarm rate* in common vernacular). In other words, rejection of the null hypothesis is strong evidence that the alternative hypothesis is, in fact, true since it is unlikely that the null hypothesis will be rejected when it is true. On the other hand, failing to reject the null hypothesis is usually only weak evidence that the null is true because one usually has only limited (or no!) control over the *type II error*, that of failing to reject the null when the alternative is, in fact, true.

With multiple options, hypothesis testing is complicated in a significant way: the null hypothesis may be false in many ways. For example, $\boldsymbol{\theta}_1$ may yield a loss value different from all other $\boldsymbol{\theta}_i$; or only $\boldsymbol{\theta}_1$ and $\boldsymbol{\theta}_2$ may yield different loss values; or only $\boldsymbol{\theta}_1$ and $\boldsymbol{\theta}_3$ may yield different values; and so on. Only one of the many alternative hypotheses corresponds to a specified $\boldsymbol{\theta}_i$ being the best $\boldsymbol{\theta}$ value. Hence the second broad step mentioned above: to attempt to isolate the cause of the rejected null hypothesis. This will often involve an attempt to determine whether the *one* alternative hypothesis demonstrating the superiority of a *specified* element in Θ is, statistically, the correct alternative hypothesis. That will be the focus of Sections 12.3 and 12.4, following the general test of Section 12.2.

As is typical in statistical practice, acceptance regions are used for testing the null hypothesis. These regions are designed to contain the chosen test statistic with high probability *if* (an important "if") the null hypothesis of equality of the loss values for the various elements of Θ is true. If the chosen test statistic does not land in the acceptance region, the null hypothesis is rejected. To obtain concrete solutions, to be consistent with the vast majority of multiple comparisons approaches, and to aid in isolating the effects of individual $\boldsymbol{\theta}_i$ in Θ, the acceptance *region* will be restricted to a set of acceptance *intervals*. Each interval corresponds to a pairwise comparison between two of the elements in Θ.

Hence, the acceptance regions of interest here will be multidimensional rectangles.

A fundamental property in multiple comparisons tests is that such intervals must be wider than their single comparison counterpart (as in the pairwise tests of Appendix B). The wider intervals guard against the greater chance of an "extreme" event following from the greater number of opportunities to find significant occurrences. In particular, when there are many possible events, it is not surprising to find at least one oddball! Interestingly, this *multiplicity effect* is sometimes ignored, leading to false conclusions about whether particular events are statistically significant. For example, Benjamini and Yekutieli (2001) discuss a published medical study where clinical trials suggest that a particular treatment for breast cancer provides 6 of 18 possible medical benefits. This result is based on a claim of statistically significant results in 6 of 18 comparisons (P-values ≤ 0.05), but where the multiplicity is ignored. In fact, as noted by Benjamini and Yekutieli (2001), only 2 to 4 (versus 6) of the 18 possible benefits are supported by the data at an overall P-value of 0.05 (the range of 2 to 4 benefits depends on the details of the testing procedure).[2]

In carrying out the tests below, at each value of $\boldsymbol{\theta} \in \Theta = \{\boldsymbol{\theta}_1, \boldsymbol{\theta}_2, \ldots, \boldsymbol{\theta}_K\}$, let $\overline{L}_i$ represent a sample mean of measured values of L at $\boldsymbol{\theta} = \boldsymbol{\theta}_i$. In particular, let $y_k(\boldsymbol{\theta}_i) = L(\boldsymbol{\theta}_i) + \varepsilon_{ki}$, represent the kth measurement of L at the given $\boldsymbol{\theta}_i$, $k = 1, 2, \ldots, n_i$ (say), and ε_{ki} represent the mean-zero noise term. Then, $\overline{L}_i = n_i^{-1}\sum_{k=1}^{n_i} y_k(\boldsymbol{\theta}_i)$. A fundamental quantity in the tests below is the difference

$$\delta_{ij} \equiv \overline{L}_i - \overline{L}_j \tag{12.1}$$

for any pair i, j. We will make varying assumptions about the distribution of the noise terms, beginning with independent, identically distributed (i.i.d.) normal in Section 12.2 and relaxing this condition in Section 12.3. As discussed in Section 1.1 and illustrated in numerous places throughout this book, many practical problems have noise that is non-i.i.d., often because of a dependence of the noise on the value of $\boldsymbol{\theta}$. In some of the tests below, it is assumed that it is possible to do a table lookup to find critical values of a relevant distribution; these critical values are often based on the noises in the loss measurements (the ε) having desirable properties such as independence and normality. In some applications, such assumptions will clearly not hold, as discussed in Section 1.1. Somerville (1997) provides an automated way of computing the required constants in cases where simple table lookups do not apply. The method is a combination of numerical integration and Monte Carlo sampling.

---

[2]Benjamini and Yekutieli (2001) point out that the *New England Journal of Medicine* (which is not where the study they discussed appeared) requires that researchers account for the multiplicity effect in reporting results. This appears to be a rare requirement in medical journals.

12.2 STATISTICAL COMPARISONS TEST WITHOUT PRIOR INFORMATION

A classical—and popular—method for performing simultaneous comparisons was introduced in 1953 by J. W. Tukey in an unpublished technical report from Princeton University entitled "The Problem of Multiple Comparisons." The method was independently introduced in Kramer (1956). This test is a generalization of the familiar two-sample t-tests in Appendix B. In fact, Goldsman and Nelson (1998) state: "The most widely used method for forming [simultaneous] intervals is Tukey [–Kramer]'s method, which is implemented in many statistical software packages." The interest here is in constructing acceptance intervals for all $K(K-1)/2$ differences $L(\boldsymbol{\theta}_i) - L(\boldsymbol{\theta}_j)$ (Exercise 12.1). If prior information suggests that a particular $\boldsymbol{\theta}_i$ is a strong candidate for being $\boldsymbol{\theta}^*$, then the Tukey–Kramer procedure may be unnecessary and one should proceed to a more focused test such as in Section 12.3 or 12.4. On the other hand, in the absence of prior information, the construction of the $K(K-1)/2$ simultaneous intervals is a useful step in identifying candidate values $\boldsymbol{\theta}_i$ for consideration as the optimal point $\boldsymbol{\theta}^*$.

The Tukey–Kramer technique is derived under the assumptions that the noises ε_{ki} in the measurements $y_k(\boldsymbol{\theta}_i) = L(\boldsymbol{\theta}_i) + \varepsilon_{ki}$ are i.i.d. normal with mean zero across k and i. In many practical problems, however, these noise assumptions do not hold, motivating the alternative inequality-based approaches in Sections 12.3 and 12.4. Nevertheless, the assumptions are sometimes reasonable. Furthermore, the Tukey–Kramer method is sometimes used irrespective of the distribution of the noise terms, as it is "sensible" in much the way that the t-distribution is often used when the parent population is nonnormal as discussed in Appendix B, Mardia et al. (1979, p. 147), and Goldsman and Nelson (1998). If being used in a problem where the conditions are seriously violated, however, one should exercise caution, especially if the inference leads to some "close calls."

For this method, suppose that one is interested in constructing multiple comparison acceptance intervals such that when the null hypothesis, $L(\boldsymbol{\theta}_i) = L(\boldsymbol{\theta}_j)$ for all i, j, is true (i.e., all points in Θ yield the same loss value), then it is known with probability at least $1 - \alpha$ (say) that *all* $K(K-1)/2$ test statistics, each corresponding to one of the differences $L(\boldsymbol{\theta}_i) - L(\boldsymbol{\theta}_j)$, will lie in the joint acceptance region. Hence, under the null hypothesis, the probability is no more than α that at least one of the test statistics will lie outside its acceptance interval. Typical values for α—the false alarm rate—from classical hypothesis testing are 0.01, 0.05, or 0.10.

Given that the noises are assumed i.i.d, it is valid to pool all measurements in forming the estimate of the common variance. Let the pooled sample variance from the $n_1 + n_2 + \ldots + n_K$ sample values of the loss function be

$$s^2 \equiv \frac{\sum_{i=1}^{K}\sum_{k=1}^{n_i}[y_k(\boldsymbol{\theta}_i)-\bar{L}_i]^2}{\sum_{i=1}^{K}(n_i-1)}. \tag{12.2}$$

The above is a natural extension of the classical variance estimate for one sample (see Appendix B). Note that s^2 is an unbiased estimate of the variance (another unbiased—but inferior—variance estimate is given in Exercise 12.12). The simultaneous acceptance intervals for the differences δ_{ij} are then

$$\left[-Q_{K,\nu}^{(\alpha)}\frac{s\sqrt{n_i^{-1}+n_j^{-1}}}{\sqrt{2}},\ Q_{K,\nu}^{(\alpha)}\frac{s\sqrt{n_i^{-1}+n_j^{-1}}}{\sqrt{2}}\right], \tag{12.3}$$

where $Q_{K,\nu}^{(\alpha)}$ is the $1-\alpha$ quantile of the studentized range distribution with parameter K and degrees of freedom $\nu = n_1 + n_2 + \ldots + n_K - K$ (the quantile is the point such that the probability of being less than or equal to this point is $1-\alpha$). A table with the critical values of this distribution is given in Miller (1981, pp. 234–237), Hochberg and Tamhane (1987, App. 3, Table 8), and, in condensed form, Goldsman and Nelson (1998). Note that the interval in (12.3) corresponds exactly to the identical variances pairwise t-test interval in (B.3b) in Appendix B in the special case where $K = 2$ (i.e., $Q_{2,\nu}^{(\alpha)}/\sqrt{2} = t_{\nu}^{(\alpha/2)}$).

Hayter (1984) made the important discovery that the intervals in (12.3) are *at least* as wide as necessary to provide the probability $1-\alpha$ coverage. That is, if the null hypothesis of $L(\boldsymbol{\theta}_1) = L(\boldsymbol{\theta}_2) = \ldots = L(\boldsymbol{\theta}_K)$ is true, then the probability of all δ_{ij} simultaneously lying in their respective intervals from (12.3) is at least $1-\alpha$. So, in conducting an experiment, the probability that at least one δ_{ij} will not lie in its corresponding interval is less than or equal to α. Hence, there is potentially strong evidence that the null hypothesis is false if at least one δ_{ij} does not lie in its corresponding interval. The formal statement of the result is below.

Theorem 12.1 (Hayter, 1984). Suppose that the measurement noises ε_{ki} are i.i.d., normal, with mean zero across k and i. Under the null hypothesis of equal $L(\boldsymbol{\theta}_i)$, the probability of all δ_{ij} simultaneously lying in the random intervals in (12.3) is greater than or equal to $1-\alpha$. The probability is equal to $1-\alpha$ *only* when $K = 2$ or $n_1 = n_2 = \ldots = n_K$ for $K \geq 3$. If any $n_i \neq n_j$ when $K \geq 3$, then the actual probability is strictly above $1-\alpha$.

Section 12.1 mentioned one of the fundamental properties of multiple comparisons tests. That is, the acceptance intervals need to be wider than their pairwise counterparts in order to guard against the greater likelihood of an extreme event when there is more than one event (i.e., $K \geq 3$ or, equivalently, an evaluation of more than one $\delta_{ij} = \bar{L}_i - \bar{L}_j$). Figure 12.1 illustrates this property for the Tukey–Kramer test. It is assumed that $\alpha = 0.05$, $n_i = 10$ for all i, and that $s^2(n_i^{-1} + n_j^{-1}) = 1$ for all K. The latter assumption is simply a convenient normalization that follows from s^2 being an unbiased estimator of the variance at any K. That is, $E[s^2(n_i^{-1} + n_j^{-1})]$ is the same for all K. Figure 12.1 shows intervals for four values of K. As expected, the intervals grow in width with K, although at a slowing rate as K gets larger.

Example 12.2 illustrates the above approach for a simple problem with $K = 4$. This example depicts a realistic experiment where the null hypothesis $L(\boldsymbol{\theta}_1) = L(\boldsymbol{\theta}_2) = L(\boldsymbol{\theta}_3) = L(\boldsymbol{\theta}_4)$ is rejected, but where the optimal point does not unambiguously leap out.

Example 12.2—Simultaneous acceptance intervals. Suppose that the true loss values unknown to the analyst are $L(\boldsymbol{\theta}_i) = 2.5, 3.0, 3.5$, and 2.0 for $i = 1, 2, 3$, and 4 respectively, and that the measurement noise is i.i.d. normal with mean zero and unknown variance. The aim is to construct acceptance intervals for the $K(K-1)/2 = 6$ pairs δ_{ij} when $\alpha = 0.05$. Suppose that each loss measurement represents an expensive evaluation, so that the analyst can afford only 40 measurements for the analysis, allocated as $n_1 = n_2 = n_3 = n_4 = 10$. The following matrix of data is collected, where the four columns, in order, contain the 10 measurements of the loss function at $\boldsymbol{\theta}_1$, $\boldsymbol{\theta}_2$, $\boldsymbol{\theta}_3$, and $\boldsymbol{\theta}_4$:

$$\begin{bmatrix} 2.69 & 1.99 & 3.25 & 3.30 \\ 3.11 & 1.23 & 2.97 & 3.25 \\ 2.07 & 5.20 & 4.93 & 0.80 \\ 1.93 & 2.90 & 4.81 & 2.21 \\ 2.42 & 3.83 & 2.99 & 2.77 \\ 2.71 & 3.09 & 3.51 & 0.42 \\ 1.66 & 1.12 & 2.36 & 1.64 \\ 3.33 & 3.15 & 5.17 & 1.43 \\ 3.44 & 3.69 & 2.53 & 1.98 \\ 2.05 & 4.48 & 3.87 & 1.40 \end{bmatrix}.$$

We find that $\bar{L}_i = 2.54, 3.07, 3.64$, and 1.92 for $i = 1, 2, 3$, and 4, respectively. (These sample means—and certain other data-derived values here—may differ slightly from quantities based on the data shown above due to round-off error; the reported means are calculated from original data with more digits of accuracy.) Further, $s = 1.017$ using (12.2) and $Q_{4,36}^{(0.05)} = 3.82$, the latter from Table 8.1 in Goldsman and Nelson (1998). From (12.3), the (common) acceptance interval for use in evaluating the six differences from (12.1) is

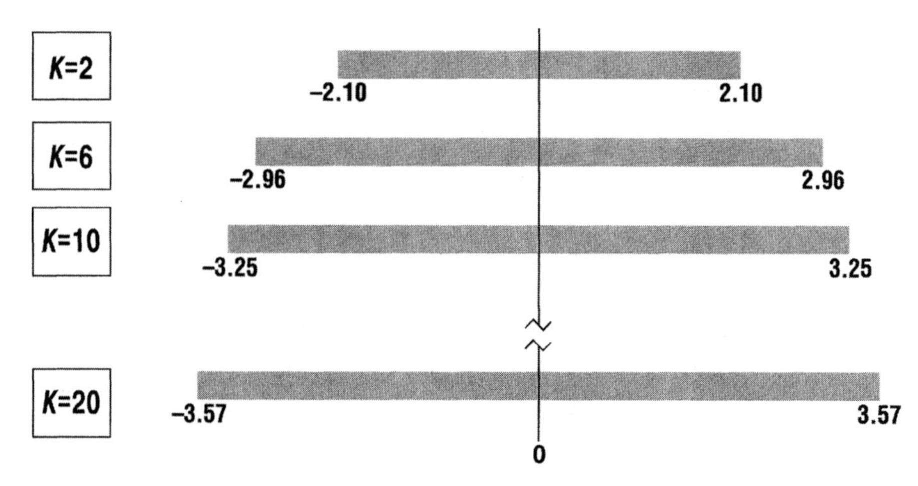

Figure 12.1. Relative widths of acceptance intervals (12.3) for varying K in Tukey–Kramer test ($n_1 = n_2 = \ldots = n_K = 10$). Coverage probability is $1 - \alpha = 0.95$.

[−1.23, 1.23]. (For comparison purposes, with only one pairwise test—as in Appendix B—the interval with the same s and sample size n_i is [−0.96, 0.96], illustrating the greater width required to protect against the greater number of possible events. See also Figure 12.1 above.) Table 12.1 shows the values of the differences δ_{ij} that are to be tested for inclusion in the acceptance interval.

Because δ_{34} lies outside of the acceptance region (corresponding to the null hypothesis $L(\boldsymbol{\theta}_1) = L(\boldsymbol{\theta}_2) = L(\boldsymbol{\theta}_3) = L(\boldsymbol{\theta}_4)$), one can reject the null hypothesis (at level 0.05) of equality in the loss values. Unfortunately, this test does not provide an "automatic" means of identifying the cause of the rejection, although an objective examination of the table below does correctly suggest that $\boldsymbol{\theta}_4$ is a possible cause of the rejection. The point $\boldsymbol{\theta}_4$ is also associated with another near rejection in comparison to $i = 2$. Note, however, that the test statistic in Table 12.1 for $i = 1$ and $j = 3$ provides some evidence that $\boldsymbol{\theta}_1$ should be evaluated

Table 12.1. Difference values for testing hypothesis.

Indices i, j	δ_{ij}
1, 2	−0.53
1, 3	−1.10
1, 4	0.62
2, 3	−0.57
2, 4	1.15
3, 4	1.72

further given that it nearly falls outside of the acceptance interval. Further analysis, possibly of the type discussed in Sections 12.3 and 12.4, would be required to solidify the conclusion that $\boldsymbol{\theta}^* = \boldsymbol{\theta}_4$. ❑

12.3 MULTIPLE COMPARISONS AGAINST ONE CANDIDATE WITH KNOWN NOISE VARIANCE(S)

The Tukey–Kramer multiple comparisons method above is useful for obtaining broad information about possible optimal points and likely nonoptimal points. However, because it does not assume any prior information about a likely optimum, the method suffers in needing to produce acceptance intervals that are wide enough to cover the number of expected anomalies in all $K(K-1)/2$ pairs i, j. A reduction in the number of comparisons can lead to a sharper analysis by producing tighter acceptance intervals. That is one of the goals of the general approach in this section. The other main goal is to consider some more general noise conditions under which multiple comparisons can be carried out. The aims and general approach of this section and the next are identical, the only difference being that this section assumes that the variances of certain differences δ_{ij} are known, whereas the next section does not.

The testing approach described below rests on the premise that one has prior information suggesting that one of the K points, say $\boldsymbol{\theta}_m$, is a candidate for being the optimal point. This may have resulted from preliminary testing based on an initial data set different from the data to be used in the formal testing below[3] or it may be a consequence of a "hunch" or physical understanding of the process. The Tukey–Kramer method in Section 12.2 is one of the methods that may be useful for the initial culling to identify a candidate $\boldsymbol{\theta}_m$. In contrast to having no prior insight (where any one of the K values is a candidate to be tested), testing that revolves around one candidate point dramatically reduces the number of comparisons that need to be tested from all $K(K-1)/2$ pairs i, j to just the $K-1$ pairs of $\boldsymbol{\theta}_m$ against each of the other $\boldsymbol{\theta}_j$. (A further reduction may be possible if some of the original K candidates can be clearly eliminated from consideration; in that case the K that is discussed below will be less than the K in the sections above. For ease of discussion, however, we will not distinguish between these "before and after" values of K, proceeding under the assumption that all of the original K candidates are viable candidates for consideration.)

Other aspects being equal, this reduction to only $K-1$ pairs to be tested sharpens the inference considerably. In addition, further sharpening is possible since we now consider only *one-sided* acceptance regions. That is, since there is

---

[3]It is statistically invalid to reuse the data employed in such preliminary analysis in the formal hypothesis tests to follow. The same data should not be used to both construct a hypothesis for testing and to test the hypothesis. Formally, such reuse invalidates the null distribution forming the basis for the hypothesis test.

an identified candidate optimal point, it is only necessary to consider deviations in one direction for each of the differences δ_{mj}. (In the absence of a candidate point, one must consider wider *two-sided* intervals, as in the Tukey–Kramer method.) The "other aspects being equal" qualifier above is relevant here since we include some more general noise conditions than the i.i.d. normal assumption of the Tukey–Kramer method, in which case one may not always see a tightening of the confidence intervals. Methods involving a comparison against one candidate point are called *many-to-one* in Tong (1980, p. 187) and *multiple comparisons with control* or *multiple comparisons with best* in Hsu (1996) and Goldsman and Nelson (1998). We use the term *many-to-one*.

Let us assume that the noise terms have a common (zero or nonzero) mean for all i, j; the common mean is needed to ensure that the differences δ_{mj} have zero mean. Except as noted below, it is not assumed that the noises are mutually independent (mutual independence is stronger than pairwise independence, which in turn is stronger than uncorrelatedness). Further, the methods here are based on knowing the variances of the differences δ_{mj}. From (12.1), knowing the variance—and, if relevant, the correlations—of the noises provides the variance of δ_{mj} via the relationship

$$\text{var}(\overline{L}_m - \overline{L}_j) = \text{var}\left[\frac{\varepsilon_{m1}+\varepsilon_{m2}+\ldots+\varepsilon_{mn_m}}{n_m} - \frac{\varepsilon_{j1}+\varepsilon_{j2}+\ldots+\varepsilon_{jn_j}}{n_j}\right].$$

The next section extends the methods here to cases where these variances are not known. If, for some m, the δ_{mj} are sufficiently negative for all $j \neq m$, then there is evidence that $\boldsymbol{\theta}_m$ is the optimal point. Unfortunately, it is usually difficult in practice to make a precise statistical statement about the likelihood of $\boldsymbol{\theta}_m$ being the optimal. The fundamental difficulty is that

$$\boldsymbol{\delta}_{m|K} \equiv [\delta_{m1}, \delta_{m2}, \ldots, \delta_{m,m-1}, \delta_{m,m+1}, \ldots, \delta_{mK}]^T$$

represents a vector of $K-1$ *dependent* random variables through the common presence of $\overline{L}_m$ in all of the δ_{mj} (and possibly through dependence introduced if the noise terms ε_{jk} are not mutually independent).

As mentioned in Section 12.1, multiple comparisons generally involve two broad steps. If the Tukey–Kramer approach of Section 12.2 is used for general determination of nonequivalence of the $\boldsymbol{\theta}_i$, the many-to-one approaches of this section may broadly be considered as addressing the isolation problem of the second step. Even within this isolation problem, however, there is a similar pattern of steps: (i) a test simultaneously determining if the suspected $\boldsymbol{\theta}_m$ is significantly different from the other points $\boldsymbol{\theta}_j, j \neq m$, and (ii) refined analysis on a subset of points, as necessary, to resolve ambiguities that may remain after the full simultaneous test in (i). The first step relies on the construction of a multiple comparison test for simultaneously testing the null hypothesis that $L(\boldsymbol{\theta}_m) \geq L(\boldsymbol{\theta}_j)$

for all $j \neq m$. In other words, the null hypothesis is that all $\boldsymbol{\theta}_j$ produce loss values at least as low as $\boldsymbol{\theta}_m$. The aim in showing the superiority of $\boldsymbol{\theta}_m$ is to *reject* this null hypothesis.

There are, however, many ways in which the null hypothesis that $\boldsymbol{\theta}_m$ is no better than $\boldsymbol{\theta}_j$ may not be true. Only one of the ways corresponds to $\boldsymbol{\theta}_m$ being the best $\boldsymbol{\theta}$ value. That is the role of the second part of the test: to attempt to isolate the cause of the rejected null hypothesis as a means of determining whether the one alternative hypothesis demonstrating the superiority of $\boldsymbol{\theta}_m$, $L(\boldsymbol{\theta}_m) < L(\boldsymbol{\theta}_j)$ for all $j \neq m$, is, statistically, the correct alternative hypothesis. For the first part, suppose that the interest is in constructing a multiple comparison region such that when $L(\boldsymbol{\theta}_m) \geq L(\boldsymbol{\theta}_j)$ for all j (i.e., the conjectured $\boldsymbol{\theta}_m$ is no better than any other point), then the probability is at least $1 - \alpha$ (say) that $\boldsymbol{\delta}_{m|K}$ will lie in this $(K - 1)$-dimensional acceptance region.

Let $\overline{\delta}_{mj} < 0$ represent a critical value for the random variable δ_{mj}. This critical value defines the lower limit in the acceptance region for δ_{mj}. That is, $[\overline{\delta}_{mj}, \infty)$ is the acceptance region. Associated with this region is the acceptance event

$$E_{mj} \equiv \left\{\delta_{mj} \geq \overline{\delta}_{mj}\right\}.$$

As in the Tukey–Kramer test, the critical values in a multiple comparison test must be chosen more conservatively (more negative here) than in a conventional one-comparison test to guard against the greater chance of an extreme event. If the events E_{mj} are true for all $j \neq m$, then there is insufficient evidence to reject the null hypothesis that $L(\boldsymbol{\theta}_j) \leq L(\boldsymbol{\theta}_m)$ for all $j \neq m$. That is, there is insufficient evidence that $\boldsymbol{\theta}_m$ is a possible optimal point at the confidence level $1 - \alpha$. The fact that $\overline{\delta}_{mj}$ is *strictly* negative suggests that the evidence for rejecting the null hypothesis is not sufficient until $\overline{L}_m$ is "considerably" less than $\overline{L}_j$.

The test philosophy outlined above is in the spirit of the *union–intersection principle* of multivariate testing (e.g., Mardia et al., 1979, pp. 127–131). In this principle, the null hypothesis is composed of the *intersection* of events, as in

$$E_{m|K} \equiv E_{m1} \cap E_{m2} \cap \ldots \cap E_{m,m-1} \cap E_{m,m+1} \cap \ldots \cap E_{mK}.$$

For the null hypothesis to be true, all of the events E_{mj}, $j = 1, 2, \ldots, m - 1, m + 1, \ldots, K$, must be true. Hence, the null hypothesis is rejected if any one of the events E_{mj} is not true. This leads to a rejection region (corresponding to the alternative hypothesis) that is the *union* of the events $E_{mj}^c = \{\delta_{mj} < \overline{\delta}_{mj}\}$ (i.e., the rejection region is $\bigcup_{j \neq m} E_{mj}^c$, where the superscript c denotes set complement).

An essential aspect of this union–intersection principle is that when the null hypothesis is rejected, one can try to determine the cause of the rejection by determining which of the events E_{mj}^c occurred. This contrasts with some other well-known hypothesis testing approaches (e.g., the likelihood ratio test, as in Mardia et al., 1979, pp. 123–127; not to be confused with the likelihood ratio gradient estimation method of Chapter 15). In these other approaches, the rejection of the null hypothesis provides little or no guidance about the cause of the rejection. In this section, we are especially interested in the cause of the rejection since one particular combination of events, $\bigcap_{j\neq m} E_{mj}^c \subseteq \bigcup_{j\neq m} E_{mj}^c$, corresponds to the outcome most consistent with the hypothesis of principal interest: that $\boldsymbol{\theta}_m$ is the best value of $\boldsymbol{\theta}$. The union–intersection principle is ideal for this because we can simply determine whether each of the events E_{mj} occurred or did not occur. A further advantage of the union–intersection approach is that there is no need to know the full (joint) distribution of the $K-1$ elements in $\boldsymbol{\delta}_{m|K}$ forming the basis of the test. Through the use of probability inequalities discussed below, less than complete distributional information is sufficient for conducting a test.

The following joint probability is related to conducting a hypothesis test of $\boldsymbol{\theta}_m$ possibly being the optimal $\boldsymbol{\theta}$ value:

$$P(E_{m|K}) = P(E_{m1} \cap E_{m2} \cap \ldots \cap E_{m,m-1} \cap E_{m,m+1} \cap \ldots \cap E_{mK}). \quad (12.4)$$

From (12.4) it is possible to determine the above-mentioned critical values $\overline{\delta}_{mj}$. In an ideal case where the *joint* distribution of the elements in $\boldsymbol{\delta}_{m|K}$ is known to be normal and the noises are independent, one can, in principle, compute the probability and determine the critical values using the fact that

$$\text{cov}(\delta_{mi}, \delta_{mj}) = \text{var}(\overline{L}_m) \text{ for all } i \neq j.$$

(See Exercise 12.4; see also Appendix C for a discussion of joint normality. Recall that the elements of a random vector $\boldsymbol{X}$ are jointly normally distributed if, and only if, $\boldsymbol{a}^T\boldsymbol{X}$ has a univariate normal distribution for all conformable nonrandom vectors $\boldsymbol{a} \neq \boldsymbol{0}$. Each element of $\boldsymbol{X}$ having a normal distribution is not alone sufficient for joint normality.)

Because of the dependence among the elements in $\boldsymbol{\delta}_{m|K}$, no table lookup is available for readily determining the critical values. That is, since $P(E_{m|K})$ is not a product of the $P(E_{mj})$, $j \neq m$, it is not possible to simply compute critical values from a normal table for the marginal probabilities $P(E_{mj})$ and multiply the marginal probabilities to obtain the chosen probability $1-\alpha$ (although see the Slepian inequality below). Hence, in practice, a numerical search procedure is typically needed to solve for the $\overline{\delta}_{mj}$. The numerical search would work by setting the expression in (12.4) to the chosen $1-\alpha$ level and solving for the $\overline{\delta}_{mj}$. If $\boldsymbol{\delta}_{m|K}$ is jointly normally distributed, then deterministic search methods may be used to find the $\overline{\delta}_{mj}$.

If the full joint normality assumption above does not apply or if one seeks an analytical solution for analysis purposes, there are two inequalities that provide partial insight into the probability in (12.4), and, in the course of this, allow for a relatively easy computation of critical values $\bar{\delta}_{mj}$. These are the Bonferroni and Slepian inequalities (Tong, 1980, pp. 143–144 for Bonferroni; pp. 8–12 for Slepian). The Bonferroni and Slepian inequalities are both means of converting this difficult-to-compute joint probability into a much-easier-to-compute expression based only on the marginal probabilities, $P(E_{mj})$, or bounds to these probabilities. These inequalities yield only partial insight because, as inequalities, they provide conservative (lower) bounds to the probability of interest $P(E_{m|K})$, yielding critical values that are more negative than strictly required (i.e., wider-than-necessary acceptance intervals for the δ_{mj}). Hence, rejection of the null hypothesis under these critical values is an even stronger indication of the validity of the alternative hypothesis than would be suggested by the false alarm rate α. On the other hand, some rejections may be missed due to the conservatism.

The *Bonferroni inequality* for converting the joint probability $P(E_{m|K})$ into a function of the marginal probabilities is

$$P(E_{m|K}) \geq 1 - \sum_{\substack{j=1 \\ j \neq m}}^{K} [1 - P(E_{mj})]. \tag{12.5}$$

Prior information, if available, should be used to determine $P(E_{mj})$. Ideally, the marginal distribution (typically, normal) of each δ_{mj} will provide the values of $P(E_{mj})$. If a precise calculation is not available, a one-sided Chebyshev inequality can be used to produce a conservative bound for $P(E_{mj})$ using only information about the variance of the δ_{mj}. The Chebyshev bound is $P(X \leq c) \geq 1 - \text{var}(X)/[\text{var}(X) + c^2]$, where X is a mean-zero random variable and $c > 0$ (see Tong, 1980, p. 155 or Subsection 8.2.2 here). This bound yields the following lower bound to the probability of interest:

$$P(E_{mj}) = P(\delta_{mj} \geq \bar{\delta}_{mj}) = P(-\delta_{mj} \leq -\bar{\delta}_{mj}) \geq 1 - \frac{\text{var}(\delta_{mj})}{\text{var}(\delta_{mj}) + \bar{\delta}_{mj}^2}. \tag{12.6}$$

Note that the second equality in (12.6) is employed because the Chebyshev bound pertains to probabilities of the form $P(X \leq c)$, where c, like $-\bar{\delta}_{mj}$, is *positive*. In the absence of an exact value for $P(E_{mj})$, we seek a *lower* bound to $P(E_{mj})$ to retain the validity of the lower bound in (12.5). Hence, the Chebyshev bound of (12.6) is in the correct direction to maintain the validity of the lower bound in (12.5).

The *Slepian inequality* applies when the elements of $\boldsymbol{\delta}_{m|K}$ are *jointly* normally distributed with mean zero:

$$P(E_{m|K}) \geq \prod_{\substack{j=1 \\ j \neq m}}^{K} P(E_{mj}) . \qquad (12.7)$$

(Recall the comment above that it is necessary—but *not sufficient*—for joint normality that the marginal distributions of all components be normal.) The mean-zero condition holds under the above-mentioned assumption that the noise terms ε_{km} and ε_{kj} have a common mean since this common mean will subtract out in forming the δ_{mj}. The joint normality for $\boldsymbol{\delta}_{m|K}$ holds if the noise terms are jointly normally distributed. More generally, it approximately applies in other cases, such as when the noise terms in all K loss measurements are mutually independent (not necessarily normal) and satisfy the modest conditions for one of the central limit theorems (e.g., the variances are uniformly bounded above and uniformly bounded away from zero over $k = 1, 2,\ldots, \infty$; see Exercise 12.5). So, the Slepian inequality applies (at least approximately) more broadly than the narrow case where the noise terms are jointly normally distributed. Note, however, that when α is very small, the Slepian bound offers only a negligible improvement over the Bonferroni bound (see Exercise 12.7).

To determine a bound for the probability $P(E_{m|K})$, one can fix α and then use (12.5) or (12.7) as appropriate to determine the critical values $\overline{\delta}_{mj}$. Note that when the noise terms ε_{kj} are i.i.d., the critical values may be chosen identically (i.e., $\overline{\delta} \equiv \overline{\delta}_{m1} = \overline{\delta}_{m2} = \ldots$) as a way of forcing equality of either the marginal probabilities $P(E_{mj})$ or the Chebyshev bounds to the marginal probabilities according to (12.6). In other cases, it may be necessary to vary the critical values to preserve equality of the marginal probabilities $P(E_{mj})$ (or their bounds). For all $j \neq m$ and a level α test, (12.5) implies that $P(E_{mj}) = 1 - \alpha/(K-1)$ in the Bonferroni case and (12.7) implies that $P(E_{mj}) = (1-\alpha)^{1/(K-1)}$ in the Slepian case. After collecting the sample means $\overline{L}_1, \overline{L}_2,\ldots, \overline{L}_K$, the vector of differences $\boldsymbol{\delta}_{m|K}$ is formed. We then determine if the null hypothesis is rejected by determining if $\boldsymbol{\delta}_{m|K}$ lies in the acceptance region $E_{m|K}$. If the vector does not lie in $E_{m|K}$, the null hypothesis is rejected. If the vector lies in $E_{m|K}$, there is insufficient evidence to consider $\boldsymbol{\theta}_m$ as a candidate optimal point.

Alternatively, if the prior information suggesting the optimality of $\boldsymbol{\theta}_m$ is strong enough, we may increase the sample sizes (n_j, $j = 1, 2,\ldots, K$) and repeat the experiment. One should be aware, however, that repeating the experiment increases the effective false alarm rate from that implied by α since by random chance we are sure to reject the null hypothesis if the experiment is repeated often enough. If the null hypothesis has been rejected and $\boldsymbol{\delta}_{m|K} \in \bigcap_{j \neq m} E_{mj}^{c}$, then we are in the ideal unambiguous situation providing strong evidence that $\boldsymbol{\theta}_m$ is, in fact, the optimal point.

Even more information is available by evaluating the P-values (see Appendix B) associated with each δ_{mj}. Very small values provide strong evidence in support of the hypothesis that $\boldsymbol{\theta}_m$ is the optimal point. More likely than this ideal situation, perhaps, is the ambiguous case where the null hypothesis has been rejected but not all δ_{mj} lie in their respective rejection regions E_{mj}^c. In such cases, one can subject the subset of $\{\boldsymbol{\theta}_1, \boldsymbol{\theta}_2, \ldots, \boldsymbol{\theta}_K\}$ corresponding to the nonrejections to more detailed analysis, perhaps involving an increase in sampling (the n_i) and a repeat of the experimental approach above.

Unfortunately, even in the ideal situation, $\boldsymbol{\delta}_{m|K} \in \bigcap_{j \neq m} E_{mj}^c$, it is difficult to assign an easy probabilistic interpretation to the rejection. The reason again comes back to the fundamental problem of lack of uniqueness: The "strength" of one's conclusions comes largely from the *P*-value of some test statistic under a unique null distribution. However, if $\bigcap_{j \neq m} E_{mj}^c$ is the only alternative hypothesis of interest, there is no unique complementary null hypothesis under which a distribution for the test statistic can be derived. (This is the dual relationship to the nonuniqueness of alternative hypotheses for a fixed null hypothesis, as discussed in Section 12.1.) Hence, no simple *P*-value is possible. Note also that the Bonferroni inequality applied directly to $P\left(\boldsymbol{\delta}_{m|K} \in \bigcap_{j \neq m} E_{mj}^c\right)$ is not helpful in characterizing the *P*-value since it provides a *lower* bound to the probability, which is the wrong direction for an event intended to have a small probability.

One way to cope with the above is to eliminate all competitors to $\boldsymbol{\theta}_m$ except the one that appears to be the strongest challenger (this information should be available based on the marginal *P*-values for each δ_{mj} or, equivalently, the degree to which each δ_{mj} lies outside of its acceptance interval $[\bar{\delta}_{mj}, \infty)$). Then, one can perform a final pairwise test between $\boldsymbol{\theta}_m$ and this competitor. With methods such as in Appendix B, a unique null and alternative hypothesis can be determined, leading to a *P*-value. This approach is illustrated in Example 12.7 for the case where the variances are estimated. The equivalent approach would be used here when the variances are assumed known.

Below we present three related examples. Example 12.3 shows how the Bonferroni and Slepian inequalities compare in the calculation of critical values $\bar{\delta}_{ij}$. Examples 12.4 and 12.5 depict testing outcomes of the "unambiguous" and "ambiguous" types. These examples illustrate the relative ease with which the critical values can be obtained via the inequalities.

Example 12.3—Calculation of critical values. Suppose that $K = 4$, $n = 15$ (identical sample sizes for the four options), and that, based on prior information, we have reason to believe that $\boldsymbol{\theta}_2$ may be the optimal point. Further, suppose that the measurement noise variance has been reliably estimated as $\text{var}(\varepsilon_{kj}) = 2.0$ for all k, j, and that the noise terms are mutually independent. The aim is to determine the critical values $\bar{\delta}_{21}$, $\bar{\delta}_{23}$, and $\bar{\delta}_{24}$ for testing the alternative hypothesis that $\boldsymbol{\theta}_2$ is, in fact, the best point. Note that with $n = 15$, the standard deviation of δ_{21}, δ_{23}, and δ_{24} is $\sqrt{(2.0+2.0)/15} = 0.516$. Let us compare the critical values and results from a hypothesis test under three different options:

1. The Bonferroni test with Chebyshev inequality-based calculation for $P(E_{ij})$.
2. The Bonferroni test with a normal distribution-based calculation for $P(E_{ij})$.
3. The Slepian inequality test.

These options are listed in increasing order of the prior information required. Given the equal variances for all of δ_{21}, δ_{23}, and δ_{24}, we seek a common critical value $\bar{\delta} \equiv \bar{\delta}_{21} = \bar{\delta}_{23} = \bar{\delta}_{24}$. (In some applications—not this one—the noise terms will depend on $\boldsymbol{\theta}$, suggesting that the critical values should vary along the lines of the discussion above.)

Consider a test at level $\alpha = 0.05$. Under option 1, which makes no assumptions about the forms of the distributions for δ_{21}, δ_{23}, and δ_{24},

$$P(E_{2|4}) \geq 1 - \sum_{\substack{j=1 \\ j \neq 2}}^{4} \frac{1}{1+\bar{\delta}_{2j}^2 / \mathrm{var}(\delta_{2j})} = 1 - \frac{3}{1+\bar{\delta}^2/0.516^2} = 1 - 0.05. \quad (12.8)$$

From the last equality in (12.8), $\bar{\delta} = -3.96$.

Option 2 is useful when the marginal probabilities $P(E_{2j})$ can be computed (e.g., the marginal distributions of δ_{21}, δ_{23}, and δ_{24} are normal), but the *joint* distribution of δ_{21}, δ_{23}, and δ_{24} is not known to be normal. One way this might occur is if the noise terms $\{\varepsilon_{k2}\}$ are independent (with conventional central limit theory implying the approximate normality of $\bar{L}_2$) while the noise terms $\{\varepsilon_{k1}, \varepsilon_{k3}, \varepsilon_{k4}\}$ are *not* mutually independent but are independent of $\{\varepsilon_{k2}\}$. Suppose a central limit theorem for dependent random variables (e.g., Laha and Rohatgi, 1979, p. 355) implies that $\bar{L}_j$, $j \neq 2$, is approximately normal. The independence of $\{\varepsilon_{k2}\}$ and $\{\varepsilon_{kj}\}$, $j \neq 2$, implies that each of δ_{21}, δ_{23}, and δ_{24} is approximately normal but the dependence of $\{\varepsilon_{k1}, \varepsilon_{k3}, \varepsilon_{k4}\}$ destroys the *joint* normality needed in the Slepian inequality. By the identical marginal distributions of δ_{21}, δ_{23}, and δ_{24}, (12.5) implies that

$$P(E_{2|4}) \geq 1 - 3P\left(Z < \frac{\bar{\delta}}{0.516}\right) = 1 - 0.05, \quad (12.9)$$

where $Z \sim N(0, 1)$. From the equality part of (12.9), it is found that $\bar{\delta} = -1.10$.

Option 3 has the strongest requirements, assuming all of the noise terms $\{\varepsilon_{k1}, \varepsilon_{k2}, \varepsilon_{k3}, \varepsilon_{k4}\}$ to be mutually independent, with the noises either being all normally distributed or with the sample size $n = 15$ being large enough so that central limit theorem effects have fully taken hold. Then from the Slepian inequality (12.7),

$$P(E_{2|4}) \geq P\left(Z \geq \frac{\overline{\delta}}{0.516}\right)^3 = 1 - 0.05, \tag{12.10}$$

where Z is the standard normal variable as above. Solving for $\overline{\delta}$ in the right-hand equality in (12.10) yields $\overline{\delta} = -1.09$. There is little change in $\overline{\delta}$ from the value under option 2; this small change is consistent with the relatively small change in the regularity conditions (see also Exercise 12.7). (See Exercise 12.8 for a numerical computation of $\overline{\delta}$ in the setting of option 3 where the noises are i.i.d. normally distributed.) ❑

Example 12.4—Results of unambiguous hypothesis test. Based on the setting in Example 12.3, suppose that $n = 15$ values of the loss function are measured at each of the four candidate points. The sample mean values are $\overline{L}_i = 2.3$, 1.0, 3.7, and 3.2 for $i = 1$, 2, 3, and 4, respectively. This yields $\delta_{21} = -1.3$, $\delta_{23} = -2.7$, and $\delta_{24} = -2.2$. If it is not possible to make assumptions about the distributional form of the random variables δ_{21}, δ_{23}, and δ_{24}, then (12.8) applies. Since δ_{21}, δ_{23}, and δ_{24} are all less negative than the value $\overline{\delta} = -3.96$, we cannot reject the null hypothesis that $\boldsymbol{\theta}_2$ is indistinguishable from the other three points.

On the other hand, stronger results follow when the above partial or full independence conditions for the noises hold. This allows the use of (12.9) or (12.10) given the further assumptions that the central limit theorem holds or that the noises are jointly normally distributed. Then, it is possible to reject the null hypothesis in favor of the alternative hypothesis that $\boldsymbol{\theta}_2$ is better than at least one other $\boldsymbol{\theta}_j$ since the values of δ_{21}, δ_{23}, and δ_{24} are all less than $\overline{\delta} = -1.10$. A more precise conclusion than this nebulous alternative hypothesis is available by looking at the marginal P-values. That is, each observed δ_{2j}, $j = 1$, 3, and 4, provides information about the strength of the rejection. For instance, based on the normal distribution and the fact that the standard deviation of δ_{2j} is 0.516 for $j = 1$, we have $P(\delta_{21} \leq -1.3) = P(\delta_{21}/0.516 \leq -1.3/0.516) = P(Z \leq -2.52) = 0.006$. For $j = 3$, 4, the P-values are both less than 10^{-5}. The small marginal P-values provide strong evidence that $\boldsymbol{\theta}_2$ is, in fact, the best point. (If desired, a standard pairwise statistical test can be used to compare $\boldsymbol{\theta}_2$ and $\boldsymbol{\theta}_4$ in a final confirmation of this conclusion.) ❑

Example 12.5—Results of ambiguous hypothesis test. Continuing with the setting of Examples 12.3 and 12.4, now assume that the data yield $\overline{L}_i = 5.4$, 1.0, 0.8, and 3.2 for $i = 1$, 2, 3, and 4, respectively. Based on the formulation of Example 12.3, we would reject the null hypothesis that $\boldsymbol{\theta}_2$ is indistinguishable from the other three points under any of the distributional assumptions behind (12.8), (12.9), or (12.10) because $\delta_{21} = -4.4$ is more negative than any one of the three values of $\overline{\delta}$. This is even stronger than the conclusion of Example 12.4, where it was not possible to reject the null hypothesis under the weak assumptions of (12.8). However, since $\delta_{23} = 1.0 - 0.8 > 0$, there is clearly insufficient evidence to claim that $\boldsymbol{\theta}_2$ is optimal, although under the normality assumption of (12.9) and (12.10), the marginal P-value of $P(\delta_{23} \leq 0.2) = P(Z \leq 0.39) = 0.65$ is not close enough to 1.0 to suggest that $\boldsymbol{\theta}_3$ is clearly better than $\boldsymbol{\theta}_2$. The other P-values, being less than 10^{-5}, provide strong evidence of $\boldsymbol{\theta}_2$ being better than $\boldsymbol{\theta}_1$ or $\boldsymbol{\theta}_4$. This suggests that further analysis is required comparing $\boldsymbol{\theta}_2$ and $\boldsymbol{\theta}_3$, perhaps by increasing the sample size from the current $n = 15$. ❑

12.4 MULTIPLE COMPARISONS AGAINST ONE CANDIDATE WITH UNKNOWN NOISE VARIANCE(S)

The many-to-one approaches of Section 12.3 assume that knowledge of the variances of the δ_{mj} is available for $j \neq m$. Let us now consider the case where the variances are estimated. The general philosophical and solution strategy issues raised in Section 12.3 do not change in this setting. Consistent with standard practice when variances need to be estimated, the well-known t-distribution is used here. As in the Tukey–Kramer method of Section 12.2, the noise distributions must be normal to guarantee the validity of the technique, although, as noted in that section, t-distribution-based techniques have a long history of success in practical nonnormal settings. As in the known variance case, the events $E_{mj} \equiv \{\delta_{mj} \geq \overline{\delta}_{mj}\}$ and joint probability in (12.4) are central to determining the acceptance region of a test for the vector of differences $\boldsymbol{\delta}_{m|K}$. The main distinction here is that the values $\overline{\delta}_{mj}$ are *random* because of their dependence on the appropriate variance estimate (versus the exact variance above).

The Bonferroni inequality of (12.5) applies here as well with the marginal probabilities determined from one of the two-sample t-tests described in Appendix B. In particular, suppose that the measurement noise sequences $\{\varepsilon_{km}\}$ and $\{\varepsilon_{kj}\}$, $j \neq m$, are mutually independent with $\{\varepsilon_{km}\}$ being identically distributed and $\{\varepsilon_{kj}\}$ being identically distributed; note that $\text{var}(\varepsilon_{km})$ is not necessarily equal to $\text{var}(\varepsilon_{kj})$. The two tests described in (B.3b) and (B.3c) are relevant here, (B.3b) when the measurement noises in the two sequences have identical variances and (B.3c) when they do not. The advantage in the former case is that the estimated variance needed in the t-test is an estimate from pooling both data sets together to form a set of size $n_m + n_j$. In the latter case where the variances are not equal, the two required estimates are formed from either the set of size n_m or the set of size n_j. Let the individual variance estimates s_i^2, $i = 1, 2, \ldots, K$ be computed in the standard way:

$$s_i^2 \equiv \frac{1}{n_i - 1} \sum_{k=1}^{n_i} \left[y_k(\boldsymbol{\theta}_i) - \overline{L}_i \right]^2 .$$

When the noise sequences have identical variances, then, as in (B.2), the pooled variance estimate for the samples at $\boldsymbol{\theta}_m$ and $\boldsymbol{\theta}_j$, say s_{mj}^2, is given by

$$s_{mj}^2 = \frac{n_m - 1}{n_m + n_j - 2} s_m^2 + \frac{n_j - 1}{n_m + n_j - 2} s_j^2, \tag{12.11}$$

which represents a weighted sum of the two constituent sample variances.

In forming the acceptance regions for the tests under the identical or nonidentical variance case, the above formulas are used to construct the acceptance regions E_{mj}. In the identical variance case,

$$\overline{\delta}_{mj} = -t_{n_m+n_j-2}^{(\alpha')} s_{mj} \sqrt{n_m^{-1} + n_j^{-1}}, \tag{12.12}$$

where $t_{n_m+n_j-2}^{(\alpha')}$ is the $1-\alpha'$ quantile of the t-distribution with degrees of freedom $n_m + n_j - 2$ and probability level $\alpha' = \alpha/(K-1)$. Note that with this probability level, the sum of the terms, $1 - P(E_{mj})$, in the Bonferroni inequality will equal α as required (i.e., $\sum_{j\neq m}[1 - P(E_{mj})] = (K-1)\alpha' = \alpha$). A P-value can be computed if desired by appealing to the test statistic in (B.4). In the nonidentical variance case, there is no known exact test statistic, as discussed in Appendix B. However, the formula in (B.3c) can be used to construct an event E_{mj} with associated test statistic that has an approximate t-distribution.

Relative to the simultaneous comparison of all $K(K-1)/2$ pairs δ_{ij} in the Tukey–Kramer test, Figure 12.2 shows the tightening in the acceptance intervals

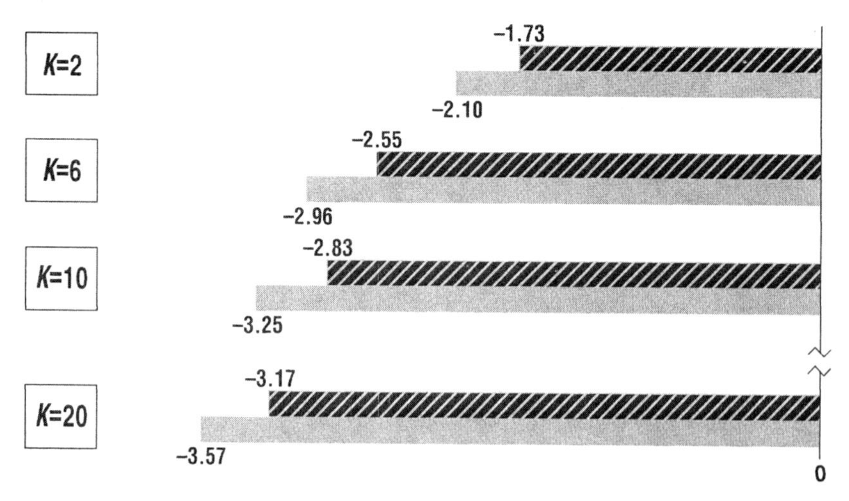

Figure 12.2. Relative widths of acceptance intervals (< 0) for Tukey–Kramer and many-to-one tests ($n_1 = n_2 = \ldots = n_K = 10$ for all K).

possible by considering only the $K-1$ comparisons in the Bonferroni-based many-to-one approach. The figure shows only the portions of the intervals to the left of zero, which are the relevant parts of the intervals for the many-to-one test since that comparison is based on testing whether δ_{mj} is sufficiently negative to reject the hypothesis $L(\boldsymbol{\theta}_m) \geq L(\boldsymbol{\theta}_j)$. We consider the same setting as Figure 12.1 ($\alpha = 0.05$, $n_i = 10$ for all i, and $s\sqrt{n_i^{-1}+n_j^{-1}} = s_{mj}\sqrt{n_m^{-1}+n_j^{-1}} = 1$ for all K and i, j, m, where s^2 is the pooled variance estimate from all samples in the Tukey–Kramer test). The four intervals for the Tukey–Kramer test are the same as the intervals in Figure 12.1. The figure illustrates the smaller intervals at each K for the many-to-one approach and the fact that the intervals grow with K at approximately the same rate as the intervals in the Tukey–Kramer approach. Nontabled values associated with the t-distribution are calculated via the TINV and TDIST functions in MS EXCEL.

Examples 12.6 and 12.7 illustrate the Bonferroni approach outlined above in the context of a problem where the variances must be estimated.

Example 12.6—Multiple comparisons against one candidate with estimated variances. Let us return to the setting of Example 12.2, where it was found that sufficient evidence exists to know at level $\alpha = 0.05$ that all four values in Θ do not provide the same loss value. Further, the data suggested that $\boldsymbol{\theta}_4$ is possibly better than the other values of $\boldsymbol{\theta}$, but ambiguities remained, suggesting a more refined test is needed. Let us again consider a level $\alpha = 0.05$ test, indicating that the critical values from the t-distribution will be chosen such that $P(E_{4j}) = \alpha/(K-1) = 0.05/3 = 0.0167$ for $j = 1, 2$, and 3 since the three probabilities are being combined in the Bonferroni inequality.

Suppose that we collect a *new* data set of size $n_1 = n_2 = n_3 = n_4 = 30$ (recalling the need to have data independent of the data forming the basis for the hypothesis). From these data we find sample means $\bar{L}_i = 2.52, 3.44, 3.83$, and 1.91 and estimated standard deviations $s_i = 0.96, 0.89, 0.92$, and 1.04 for $i = 1, 2$, 3, and 4, respectively. With $m = 4$, from (12.11), we find $s_{4j} = 1.00, 0.97$, and 0.98 and $\delta_{4j} = -0.61, -1.53$, and -1.92 for $j = 1, 2$, and 3, respectively. Given the identical sample sizes for each of the candidate points, the common critical value from the t-distribution at level 0.0167 for use in (12.12) is $t_{58}^{(0.0167)} = 2.18$ (with the degrees of freedom equaling $30 + 30 - 2$). From (12.12), it is found that $\bar{\delta}_{4j} = -0.56, -0.55$, and -0.55 for $j = 1, 2$, and 3, respectively.

Because at least one value of δ_{4j} is more negative than its corresponding value of $\bar{\delta}_{4j}$, we can reject the null hypothesis of no difference between $\boldsymbol{\theta}_4$ and the other $\boldsymbol{\theta}_j$. Moreover, since *all* values of δ_{4j} are more negative than the corresponding values of $\bar{\delta}_{4j}$, we have strong evidence that $\boldsymbol{\theta}_4$ is, in fact, the optimal $\boldsymbol{\theta}$. That is, the vector $\boldsymbol{\delta}_{m|K} = \boldsymbol{\delta}_{4|4} \in \bigcap_{j\neq 4} E_{4j}^c$, which is the ideal unambiguous situation providing strong evidence that $\boldsymbol{\theta}_m = \boldsymbol{\theta}_4$ is the optimal point. ❑

Example 12.7—Pairwise comparison. Recall from Section 12.3 that in the absence of additional information, it is difficult to attach a precise statistical interpretation to the "strong evidence" in Example 12.6 since there is no unique null hypothesis (and associated null distribution) complementary to this alternative hypothesis. One way around this problem is to perform an appropriate final pairwise test. Here, the rejection of a null hypothesis provides a clearer statistical interpretation. From the analysis above, it is apparent that if $\boldsymbol{\theta}_4$ is not the optimum, then $\boldsymbol{\theta}_1$ is the most likely choice. Therefore, as a final test, suppose that we collect 30 new measurements at each of $\boldsymbol{\theta}_1$ and $\boldsymbol{\theta}_4$ and apply the standard two-sample test (B.3b). ((B.3b) can be used because the underlying true variances are assumed identical via Examples 12.6 and 12.2.) For these new measurements, it is found that $\overline{L}_1 = 2.50$, $\overline{L}_4 = 2.01$, $s_1 = 0.98$, and $s_4 = 1.10$, leading to a pooled standard deviation from (12.11) of 1.04. The value of the test statistic using (B.4) is −1.79, which corresponds to a one-sided P-value of 0.039 using a t-distribution with 30 + 30 − 2 = 58 degrees of freedom. This provides evidence that $\boldsymbol{\theta}_4$ is the optimum. ❑

Note that the Slepian bound of Section 12.3 will also sometimes apply in the case where the variances are to be estimated, as discussed in Tong (1980, pp. 37–39) and Miller (1981, pp. 254–255). We do not pursue this t-distribution analogue of the Slepian inequality here because the improvement to the Bonferroni bound is slight for the typically small values of α used in practice (see Exercise 12.7). However, Exercise 12.13 provides some conditions under which the Slepian bound applies.

12.5 EXTENSIONS TO BONFERRONI INEQUALITY; RANKING AND SELECTION METHODS IN OPTIMIZATION OVER A FINITE SET

The approaches described above represent some of the fundamental methods for determining solutions to arg $\min_{\boldsymbol{\theta}\in\Theta} L(\boldsymbol{\theta})$ when there are a finite—and relatively small—number of values in the domain Θ and only noisy measurements of L are available. (Some of the other approaches in this book apply to discrete problems with a larger—poss.bly unbounded—number of options. The price to be paid is that one loses the statistical guarantees associated with the statistical comparisons methods.) As in the discussion above, we focus on the case where $\boldsymbol{\theta}^*$ is unique. This section is a brief discussion of some other methods related to multiple comparisons, including improvements to the Bonferroni inequality, an indifference zone method for ensuring with high probability that a good (if not best) solution is obtained, and a method for culling the original pool of K candidates to narrow the search to a smaller number of options. These latter two topics are often labeled *ranking and selection* methods and are sometimes treated as distinct entities from the multiple comparisons methods that have dominated this chapter (Fu, 1994, e.g., discusses this dichotomy).

The methods for multiple comparisons against one candidate in Sections 12.3 and 12.4 rest principally on the Bonferroni inequality of probability theory with small improvement possible via the Slepian inequality if its conditions hold. Although the Bonferroni inequality is simple and has been known from the beginnings of formal probability theory, it generally provides a surprisingly tight bound to the joint probability of interest in applications to the types of multiple comparisons considered here. In fact, Miller (1981, p. 254) states: "I knew the Bonferroni inequality was very useful, but over the course of the past ten years, I have become even more impressed with the tightness of the bound....Although special techniques and distribution theory can improve on it, the improvement is very often only minor." Note that this sentiment on the power of the Bonferroni inequality is focused on the multiple comparisons framework here. In *other* applications, the Bonferroni inequality may perform poorly, especially when the underlying random variables are strongly dependent (e.g., Naiman and Priebe, 2001, for problems in genetics and medical image analysis; Hill and Spall, 2000, for a problem in multiple component reliability).

In some critical applications, even slight improvements to the Bonferroni inequality are welcome. Naiman and Wynn (1992) pursue enhancements to the Bonferroni inequality via the introduction of *joint* probability information, à la $P(E_{mi} \cap E_{mj})$, $P(E_{mi} \cap E_{mj} \cap E_{mk})$, and so on. The authors pursue both geometric and combinatoric approaches, with the geometric approach providing interesting insight and computer-implementable algorithms. Monte Carlo sampling methods are used to compute the required joint probabilities, which are demonstrated in the determination of critical values (analogous to $\overline{\delta}_{mj}$ here). Nakayama (1997) exploits structure in multiple comparisons problems where Monte Carlo simulations are being used to generate the noisy loss measurements. Using time series methods, the bounds produced in Nakayama (1997) are guaranteed to have a false alarm rate no higher than α as the simulation run length approaches infinity and to be more precise than the Bonferroni bounds. Hill and Spall (2000) provide a summary of special cases of improvements to the Bonferroni inequality that are based on only pairwise (not higher-order) joint probabilities, such as $P(E_{mi} \cap E_{mj})$ here.

Goldsman and Nelson (1998) provide a summary of several procedures based on the notion of an *indifference zone*. Section 14.5 discusses the use of indifference zone methods when Monte Carlo simulations are used to generate the loss measurements $y_k(\boldsymbol{\theta}_i)$. Based on some results in Rinott (1978), indifference zone approaches are guaranteed with probability at least $1 - \alpha$ to provide a solution equal to $\boldsymbol{\theta}^*$ when the true loss value at $\boldsymbol{\theta}_i \neq \boldsymbol{\theta}^*$ is at least δ units greater than $L(\boldsymbol{\theta}^*)$. The term *indifference* arises because the user is willing to tolerate any solution $\boldsymbol{\theta}_i$ such that the loss $L(\boldsymbol{\theta}_i)$ is in the range $[L(\boldsymbol{\theta}^*),\, L(\boldsymbol{\theta}^*)+\delta)$, sometimes called the indifference zone.

The basic steps in indifference zone procedures are as follows: (i) Collect a first stage of data at each element of Θ and calculate sample means and variances; (ii) use the specified value of δ, the sample variances calculated at the first stage, and an appropriate distributional table to determine the number of measurements needed at each $\boldsymbol{\theta}_i$ in the second stage (the smaller the value of δ, the larger the required sample sizes); (iii) combine the data collected in the two stages to form sample means at each element of Θ; and (iv) take as an estimate of $\boldsymbol{\theta}^*$ the $\boldsymbol{\theta}_i$ producing the lowest sample mean in step (iii).

In a university–industry collaboration, Goldsman et al. (1999) illustrate the application of indifference zone methods in three nontrivial applications: inventory control, queuing systems, and project management. The indifference zone techniques may sometimes be used to circumvent the general difficulty discussed in Section 12.1 of giving a precise statistical interpretation (e.g., false alarm rate) for the alternative hypothesis of one, and only one, of the $\boldsymbol{\theta}_i$ being optimal. In particular, *if* it is known a priori that the unique unknown $\boldsymbol{\theta}^*$ yields a loss value at least δ units better than the other $\boldsymbol{\theta}_i$, (i.e., $L(\boldsymbol{\theta}^*) \le L(\boldsymbol{\theta}_i) - \delta$ for $\boldsymbol{\theta}_i \ne \boldsymbol{\theta}^*$), then the indifference zone techniques provide a "clean," quantifiable error rate. In particular, the unknown optimal point will be correctly identified with probability of at least $1 - \alpha$.

Although the indifference zone methods have the advantage of quantifiable error rate and the advantage of not requiring the noise variances to be known (or even to be equal at the different $\boldsymbol{\theta}$ values), they are based on the assumption of normally distributed data and they require a potentially large number of measurements (proportional to δ^{-2}) at *all* of the K values of $\boldsymbol{\theta}$. (As discussed in Section 14.5, the use of common random numbers in simulation can help mitigate this potentially large number of measurements.) Further, in general, the user must select the size of the indifference zone via picking δ.

Related to the above methods for selecting an optimal solution are methods for screening the K candidates to find a subset likely to contain the optimum. These are sometimes called *ranking and selection* methods. Gupta (1965) describes such a procedure when the sample sizes satisfy $n_1 = n_2 = \ldots = n_K = n$ and the measurement noises are i.i.d. normal across all measurements (see also Goldsman and Nelson, 1998; Nelson et al., 2001). This is a stronger set of assumptions than the Tukey–Kramer procedure, where the sample sizes and measurement noise distributions need not be identical. Under these assumptions, this screening procedure is guaranteed with probability at least $1 - \alpha$ to produce a subset that contains $\boldsymbol{\theta}^*$. The subset is chosen to include all points $\boldsymbol{\theta}_i$ satisfying

$$\overline{L}_i \le \min_{1\le j\le K} \overline{L}_j + s\, t^{(\alpha)}_{K-1,K(n-1)}\sqrt{\frac{2}{n}}\,, \tag{12.13}$$

where s is as in (12.2) (simplified here due to the identical n_i) and $t^{(\alpha)}_{K-1,K(n-1)}$ is a critical value from a multivariate t-distribution tabled in Goldsman and Nelson (1998, Table 8.2) and Hochberg and Tamhane (1987, App. 3, Table 4). The following example illustrates (12.13).

Example 12.8—Subset selection. Consider an expensive-to-run system where only $n = 6$ measurements are available at each of $K = 5$ options, and where it is believed that the noises are i.i.d. normal. The following data are obtained: $\bar{L}_i$ = 70.6, 76.1, 84.9, 63.0, and 79.5 for i = 1, 2, 3, 4, and 5, respectively, with s = 11.5 using (12.2). We seek a subset of the five options that is guaranteed to contain $\boldsymbol{\theta}^*$ with probability 0.95. From the above-mentioned table in Goldsman and Nelson (1998), $t_{4,25}^{(0.05)} = 2.27$. Therefore, the procedure selects those systems satisfying

$$\bar{L}_i \leq 63.0 + 11.5 \times 2.27\sqrt{\frac{2}{6}} = 78.1\,.$$

Hence, with a probability of at least 0.95, it is known that $\boldsymbol{\theta}^*$ corresponds to one of options 1, 2, and 4. At this point, the test of multiple comparisons against one candidate in Section 12.4 might be used to isolate the optimum from the three candidates. Alternatively, one could carry out additional subset selection from options 1, 2, and 4 using a *new* set of data. This represents a recursive application of subset selection. ❑

12.6 CONCLUDING REMARKS

This chapter presented methods for statistically characterizing the relative quality of the multiple options in a discrete set $\Theta = \{\boldsymbol{\theta}_1, \boldsymbol{\theta}_2, \ldots, \boldsymbol{\theta}_K\}$. The approaches here differ in character from most of those in the rest of this book. Most of the other methods in this book are based on iteratively updating $\boldsymbol{\theta}$ from an initial guess (or *set* of guesses in the population-based methods of Chapters 9 and 10). In contrast, the statistical approaches here are based on collecting loss measurements at all candidate $\boldsymbol{\theta}$ values and comparing in some formal manner the resulting sample means at each $\boldsymbol{\theta}$. In the other iterative search and optimization methods, all possible values of $\boldsymbol{\theta}$ are typically *not* evaluated en route to finding an optimum. In fact, the very concept is nonsensical when considering continuous search spaces. (In the other search methods, it is not necessary to evaluate all possible $\boldsymbol{\theta}$ values to ensure convergence because of assumptions made about the form of the loss $L = L(\boldsymbol{\theta})$.)

A type of recursive implementation is possible with the statistical methods here. Namely, the comparisons methods may first be used to screen all K options to determine if there is evidence that all options are not of equal value in terms of the loss function (à la the Tukey–Kramer test of Section 12.2). Given evidence that at least some options are better than others, the options can be pared via the selection and screening methods of Sections 12.3–12.5. These paring methods require stronger assumptions, such as prior knowledge of likely optimal $\boldsymbol{\theta}_i$. The above approach of testing and paring can be repeated after reducing the number of options.

Most of the methods here are relatively mature. Commercial software is available that implements most of the techniques (see, e.g., Goldsman et al., 1999). While the methods are presented here in a generic fashion, problem structure can sometimes enhance performance. As discussed in Section 14.5, for instance, common random numbers in simulations can reduce the sample sizes required to attain guaranteed levels of statistical performance when simulation runs provide the measurements of L.

The goals of the statistical methods in this chapter are connected to the goals of certain methods in the field of sequential analysis, particularly to the multiarm bandit problem (so named by analogies to slot machines). In such methods, one adaptively determines how the sampling should be done with the aim of optimizing some criterion. As above, the sampling may be assumed to come from a population of K possible generating mechanisms. So, if (as above) one's aim is to identify the $\boldsymbol{\theta}_i$ that minimizes L, the sequential sampling methods provide rules governing how many samples should be taken at each $\boldsymbol{\theta}_i$. If evidence is acquired early in the sampling process that certain values of $\boldsymbol{\theta}$ are clearly inferior, then sampling can be focused on the remaining $\boldsymbol{\theta}$ that are legitimate candidates to be $\boldsymbol{\theta}^*$. This contrasts with the comparisons methods in this chapter, where the sampling strategy is generally specified *prior* to collecting any information about the competing alternatives. Such sequential sampling may increase the efficiency in the use of resources.

Lai (2001) provides a *tour de force* of sequential methods, including the multiarmed bandit problem. He notes that one of the major areas of application for such methods is stochastic adaptive control based on Markov chains. There are also close connections to stochastic approximation of the type introduced in Chapter 4 and to experimental design as discussed in Chapter 17.

EXERCISES

12.1 Prove that if Θ contains K elements, there are $K(K-1)/2$ differences $L(\boldsymbol{\theta}_i) - L(\boldsymbol{\theta}_j)$ to be tested in the absence of any prior information eliminating some of the elements in Θ.

12.2 Suppose that the noise conditions for the Tukey–Kramer method of multiple comparisons apply and $K = 5$. Suppose that $\overline{L}_i = 71, 86, 72, 63$, and 70 for $i = 1, 2, 3, 4$, and 5, respectively. Let $n_1 = n_2 = n_3 = 10$ and $n_4 = n_5 = 6$. Further, suppose that $s = 9.8$ using (12.2).

(a) Using the level 0.05 quantile value $Q_{5,37}^{(0.05)} = 4.06$, discuss why the null hypothesis of equality of loss values is rejected at some level *less* than $\alpha = 0.05$.

(b) Demonstrate that $\boldsymbol{\theta}_2$ can be effectively ruled out as a candidate for $\boldsymbol{\theta}^*$.

12.3 For $K = 2$ with $n_1 = 12$ and $n_2 = 8$, suppose that $s^2 = 1.0$. For $\alpha = 0.05$, verify that the Tukey–Kramer interval computed via (12.3) is the same as the classical two-sample t-test interval in (B.3b) in Appendix B. (Note: $Q_{2,18}^{(0.05)}$ = 2.971 from Miller, 1981, p. 234.)

12.4 In addition to having a common mean, assume that all noise terms are mutually uncorrelated. Show that $\text{cov}(\delta_{mi}, \delta_{mj}) = \text{var}(\overline{L}_m)$ for $i \neq m$, $j \neq m$, and $i \neq j$.

12.5 In the setting of Section 12.3, suppose that the noises $\{\varepsilon_{11}, \varepsilon_{21}, \ldots, \varepsilon_{n1}, \varepsilon_{12}, \ldots, \varepsilon_{nK}\}$ are mutually independent and that a central limit theorem shows that the normalized sample means, $\sqrt{n}[\overline{L}(\theta_i) - E(\overline{L}(\theta_i))]$, have a limiting normal distribution for each $i = 1, 2, \ldots, K$. Given that the common sample size n is large, establish that $\boldsymbol{\delta}_{m|K} = [\delta_{m1}, \delta_{m2}, \ldots, \delta_{m,m-1}, \delta_{m,m+1}, \ldots, \delta_{mK}]^T$ is approximately *jointly* normally distributed.

12.6 (More difficult.) The basic Slepian inequality states: If $\boldsymbol{X} \sim N(\boldsymbol{0}, \boldsymbol{\Sigma})$ and $\boldsymbol{R} = [r_{ij}]$ and $\boldsymbol{S} = [s_{ij}]$ are two correlation matrices with $r_{ij} \geq s_{ij}$ for all i, j, then $P_{\Sigma=\boldsymbol{R}}\left(\bigcap_j \{X_j \leq a_j\}\right) \geq P_{\Sigma=\boldsymbol{S}}\left(\bigcap_j \{X_j \leq a_j\}\right)$ holds for all $\{a_j\}$, where the subscript on the probability indicates whether the probability is computed under the $N(\boldsymbol{0}, \boldsymbol{R})$ or $N(\boldsymbol{0}, \boldsymbol{S})$ distribution. (Note: A correlation matrix is a covariance matrix scaled such that all diagonal elements are equal to 1 and each off-diagonal element is the correlation between a given pair of random variables.) Show that this inequality implies (12.7) in the nondegenerate case where $\text{cov}(\boldsymbol{\delta}_{m|K}) > \boldsymbol{0}$ (i.e., is positive definite).

12.7 Suppose that conditions for the Slepian probability bound apply and that $P(E_{mj}) = \beta$ for all $j \neq m$. Show that:

(a) The Slepian probability bound is greater than the Bonferroni bound.

(b) The two bounds will be close to each other when $\beta \approx 1$.

12.8 In the setting of Example 12.3, assume that $\{\varepsilon_{k1}, \varepsilon_{k2}, \varepsilon_{k3}, \varepsilon_{k4}\}$ are jointly normally distributed. Calculate the value of $\overline{\delta}$ via one of the direct search methods in Chapter 2 or any other method such as a deterministic search technique. Contrast this value with the three values based on probability inequalities as given in Example 12.3. (Hint: Recall that $\text{cov}(\delta_{mi}, \delta_{mj}) = \text{var}(\overline{L}_m)$ for $i \neq j$. This problem requires the evaluation of a trivariate integral, which can be done numerically via deterministic or Monte Carlo means.)

12.9 Suppose that measurements $y_k(\theta_i) = L(\theta_i) + \varepsilon_{ki}$, $k = 1, 2, \ldots, n$, are available with the ε_{ki} being i.i.d. with mean 0 and variance 1 for all i, k. Let $n = 10$, $K = 6$, and the false alarm rate be $\alpha = 0.10$. Determine a common critical value $\overline{\delta}$ for using the many-to-one comparisons test in Section 12.3 to help determine whether the mth element of Θ is the optimal point at the chosen α in the following three manners:

(a) Use the distribution-free approach based on the Chebyshev and Bonferroni inequalities.

(b) Use the Bonferroni inequality together with exact knowledge of $P(E_{ij})$ when it is assumed that the noises are normally distributed.

(c) Use the approach based on the Slepian inequality under the same normal distribution assumption.

12.10 Consider the setting of Exercise 12.9 with $\boldsymbol{\theta} = [t_1, t_2]^T$, $L(\boldsymbol{\theta}) = |t_1 - t_2|$, and domain $\Theta = \{-1, 0, 1\} \times \{1, 2\}$. Using a normal distribution for the noise, generate values of $\overline{L}(\boldsymbol{\theta}_j)$ by Monte Carlo for each of the six possible values of $\boldsymbol{\theta}_j$. Use $n = 10$. Suppose that one correctly picks $\boldsymbol{\theta}_m = \boldsymbol{\theta}^*$. Based on the three critical values computed in Exercise 12.9, test the null hypothesis that $\boldsymbol{\theta}_m$ is no better than any other $\boldsymbol{\theta}_j$. Report P-values for the δ_{mj} generated under the (correct) assumption of normally distributed noise. Comment on whether these tests allow one to statistically conclude that $\boldsymbol{\theta}_m$ is the optimal point.

12.11 In the setting of Example 12.2, suppose that one had prior knowledge that $\boldsymbol{\theta}_4$ may provide the lowest loss value. Using the data of Example 12.2, carry out an analysis based on multiple comparisons against one candidate when the variances must be estimated. Use $\alpha = 0.10$.

12.12 (More difficult.) A unbiased variance estimator different from (12.2), which sometimes appears in the literature, is

$$\tilde{s}^2 \equiv \frac{1}{K}\sum_{i=1}^{K}\frac{1}{n_i - 1}\sum_{k=1}^{n_i}[y_k(\boldsymbol{\theta}_i) - \overline{L}_i]^2 .$$

This estimate has a natural interpretation as an average variance because each of the summands in $\sum_{i=1}^{K}(\cdot)$ is the sample variance for the sample of size n_i. Show that s^2 in (12.2) is a better variance estimator in the sense that $\text{var}(s^2) \le \text{var}(\tilde{s}^2)$.

12.13 When the variances must be estimated (versus being known), Tong (1980, pp. 37–38) gives conditions such that the t-distribution analogue of the Slepian inequality for a random vector $\boldsymbol{X}$ will hold. Two of the conditions are: (i) $\boldsymbol{X}$ is distributed according to a $N(\boldsymbol{0}, \sigma^2\boldsymbol{R})$ distribution where $\boldsymbol{R}$ is a correlation matrix (see definition in Exercise 12.6), and (ii) there exists an s constructed from the components of $\boldsymbol{X}$ such that Ms^2/σ^2 has a chi-squared distribution with M degrees of freedom for some M. (A special case of the Slepian-like result most useful to us is $P(\bigcap_i\{T_i \ge a_i\}) \ge \prod_i P(T_i \ge a_i)$, where $T_i = X_i/s$, X_i is the ith component of the vector X, and the a_i are constants.) Prove that conditions (i) and (ii) hold in the special case where $\boldsymbol{X} = \boldsymbol{\delta}_{m|K}$ has a multivariate normal distribution with $\{\overline{L}_1, \overline{L}_2, \ldots, \overline{L}_K\}$ being i.i.d. and having distribution $\overline{L}_i \sim N(0, \sigma^2/2)$.

12.14 Consider a system where 10 measurements are available at each of $K = 6$ options and where the noises are i.i.d. normal. The following data are obtained: $\overline{L}_i = 2.3, 1.1, 0.8, 1.9, 2.0$, and 1.7 for $i = 1, 2, 3, 4, 5$, and 6, respectively. Further, $s^2 = 1.6$ using (12.2). Identify the subset of options that is guaranteed to contain $\boldsymbol{\theta}^*$ with probability 0.95. (Note: $t_{5,54}^{(0.05)} = 2.29$ from Table 8.2 in Goldsman and Nelson, 1998.)

CHAPTER 13

MODEL SELECTION AND STATISTICAL INFORMATION

All models are wrong; some are useful.

—George E. P. Box, Professor of Statistics and Professor of Industrial Engineering, University of Wisconsin

Mathematical modeling is a fundamental topic underlying many aspects of stochastic search and optimization. Up to now, however, we have largely suppressed important aspects of modeling. In fact, we will continue to suppress important aspects of modeling, as the subject is huge and this book is not!

The focus in this text has been on the configuration, theory, and mechanics of the procedures for determining the best values for some quantities when the *loss function and parameter definitions have been taken as a given.* The definition of the loss function is fundamentally connected to assumptions about the mathematical structure representing the system. Even when minimal assumptions about the model are made, *some* information must be available to even state the problem and define the parameters $\boldsymbol{\theta}$.

At some level, modeling is devoted to the broadest of issues in mathematical analysis: How can one best use data to learn about a system? Because many of the issues relevant to such a question are far afield of our focus on search algorithms, we address only a sliver of the available results. The limited—but important—results here on model selection and the Fisher information matrix connect to the basic focus of this book in at least three ways. First, choosing the model is a stochastic search and optimization process itself ("find the *best* model"), although it is a process of a different type than that for estimating $\boldsymbol{\theta}$. Second, the definitions of $\boldsymbol{\theta}$ and $L(\boldsymbol{\theta})$ are intimately connected to the choice of model. Third, the Fisher information matrix, as studied here, provides a measure of accuracy for the estimate for $\boldsymbol{\theta}$ and provides a mechanism for choosing inputs to enhance this accuracy.

Section 13.1 treats the fundamental tradeoff between the bias and variance in choosing a model form. This provides a structure for balancing the need to have a relatively simple model that is easy to interpret and the need to have a model sufficiently rich to capture all relevant linear or nonlinear effects. The bias–variance tradeoff in selecting a model is related to determining the value for p in the search and optimization and for determining the analytical form of $L(\boldsymbol{\theta})$ and/or $\boldsymbol{g}(\boldsymbol{\theta})$. Section 13.2 focuses on cross-validation, one of the most popular and flexible means of realizing an optimal tradeoff between bias and variance. Cross-validation provides a mechanism for choosing between candidate model forms.

Section 13.3 discusses applications and the computation of the Fisher information matrix. Among other uses, the information matrix can be applied to construct uncertainty bounds (e.g., confidence intervals) for the estimates of $\boldsymbol{\theta}$. (Chapter 17 discusses an application of this matrix in the problem of optimal experimental design.) Section 13.3 also presents a Monte Carlo method for approximating the information matrix in complex estimation problems where an analytical derivation may be difficult or impossible. Finally, Section 13.4 offers some concluding remarks.

13.1 BIAS–VARIANCE TRADEOFF

Pluralitas non est ponenda sine necessitate. (Plurality should not be assumed without necessity.)

—The "Occam's razor" principle, William of Occam, English philosopher, 1285–1349 (approx.).

13.1.1 Bias and Variance as Contributors to Model Prediction Error

The bias–variance tradeoff is a fundamental principle in comparing the quality of different mathematical models. In most of this book, we have taken the dimension p and the form of loss function $L(\boldsymbol{\theta})$ and/or root-finding function $\boldsymbol{g}(\boldsymbol{\theta})$ as a given (although $L(\boldsymbol{\theta})$ and/or $\boldsymbol{g}(\boldsymbol{\theta})$ may not be directly available, corresponding to cases with noisy function measurements). We then described methods by which $\boldsymbol{\theta}$ can be estimated. This estimation will continue to be the focus of the book. A closely related issue, however, is determining the mathematical *form* (and associated dimension of $\boldsymbol{\theta}$) of the functions $L(\boldsymbol{\theta})$ or $\boldsymbol{g}(\boldsymbol{\theta})$ prior to estimating $\boldsymbol{\theta}$. In the linear or nonlinear regression context (Chapters 3, 5, 11, and 17), this is essentially tantamount to determining the form of an underlying model describing the input–output relationship of the system of interest. Parameter estimation (à la nonlinear regression) is also critical for simulation modeling (Chapters 14 and 15) to determine the internal coefficients of the simulation *prior* to application of the simulation for analyzing the system under study (the latter application is the focus of Chapters 14 and 15). A typical question might be: How many terms should be included in a polynomial approximation to an unknown function of interest?

We encountered in Section 5.2 a taste of the bias–variance tradeoff in the discussion of over- and under-fitting for neural networks. The bias–variance tradeoff also connects to the message in the *Occam's razor* principle, which states, in essence, that one should seek simpler models over more complex models. Of course, to achieve optimal predictive power from a mathematical model, it is also necessary to include sufficient richness in the model to capture the essential characteristics of the process. Hence, one should not use a model that is *too* simple. ("Everything should be made as simple as possible, but not simpler."—Albert Einstein.) The bias–variance tradeoff provides a formal structure for interpreting the Occam's razor principle. The formal structure, however, does not lead directly to implementable algorithms for realizing an optimal tradeoff between the bias and the variance. That will await the next section, which considers into the cross-validation method for model selection.

Let z represent the scalar output for some system based on an input vector $\boldsymbol{x}$. Suppose, as in previous chapters, that we model this input–output process with a regression function $h(\boldsymbol{\theta},\boldsymbol{x})$ and a noise term. In particular, the *model* for the actual output z is the right-hand side of

$$z \overset{\text{model}}{=} h(\boldsymbol{\theta},\boldsymbol{x}) + v,$$

where v is a noise term that may or may not have mean zero. The model on the right-hand side above does not generally correspond to the actual mechanism for generating the *true* output z.

We are emphasizing this distinction by writing $\overset{\text{model}}{=}$ in place of the usual equal sign, indicating that true equality only holds in the idealized case where the model is a precise description of the process. Elsewhere in the literature and in this book, the standard equal sign is used instead of the modified form above, where the equality should be interpreted to mean "the left-hand side is *modeled* to equal the right-hand side" (analogous to an assignment operator in computer programming). While the distinction may seem pedantic, the reader should understand that the equality here does not have the same meaning as an equality such as $10 = 6 + 4$. As we saw in Chapters 3 and 5 (and will see again in the experimental design discussion of Chapter 17), the loss and gradient functions, $L(\boldsymbol{\theta})$ and $\boldsymbol{g}(\boldsymbol{\theta})$, in a regression context are directly related to $h(\boldsymbol{\theta},\boldsymbol{x})$. So the choice of model directly affects the optimization process.

A natural measure of effectiveness of the regression function (with specific $\boldsymbol{\theta}$) as a predictor of z given the current value of $\boldsymbol{x}$ is the conditional mean-squared error (MSE), $E\left[(h(\boldsymbol{\theta},\boldsymbol{x})-z)^2 \,|\, \boldsymbol{x}\right]$, where the expectation is computed with respect to the random variable z.[1] Conditional on $\boldsymbol{x}$ and, for the moment, assuming a fixed $\boldsymbol{\theta}$,

[1]For any function $f(z,\boldsymbol{x})$, the conditional expectation $E[f(z,\boldsymbol{x})|\boldsymbol{x}]$ is the average of $f(z,\boldsymbol{x})$ taken with respect to the conditional probability measure $P(\cdot\,|\boldsymbol{x})$ for the random variable z.

$$
\begin{aligned}
E\left[(h(\boldsymbol{\theta},\boldsymbol{x})-z)^2\big|\boldsymbol{x}\right] &= E\left[(h(\boldsymbol{\theta},\boldsymbol{x})-E(z\,|\,\boldsymbol{x})+E(z\,|\,\boldsymbol{x})-z)^2\big|\boldsymbol{x}\right] \\
&= E\left[(E(z\,|\,\boldsymbol{x})-z)^2\big|\boldsymbol{x}\right]+[h(\boldsymbol{\theta},\boldsymbol{x})-E(z\,|\,\boldsymbol{x})]^2 \\
&\quad +2[h(\boldsymbol{\theta},\boldsymbol{x})-E(z\,|\,\boldsymbol{x})]E[(z-E(z\,|\,\boldsymbol{x}))|\boldsymbol{x}] \\
&= \underbrace{E\left[(z-E(z\,|\,\boldsymbol{x}))^2\big|\boldsymbol{x}\right]}_{\substack{\text{process variance}\\ \text{(nonmodel)}}}+\underbrace{[h(\boldsymbol{\theta},\boldsymbol{x})-E(z\,|\,\boldsymbol{x})]^2}_{\text{model error}}, \qquad (13.1)
\end{aligned}
$$

where the last line follows by $E[(z-E(z\,|\,\boldsymbol{x}))\,|\,\boldsymbol{x}]=0$. Expression (13.1) states that the MSE for a regression prediction can be decomposed into a part due to the inherent variability of the process (i.e., $E[(z-E(z\,|\,\boldsymbol{x}))^2|\boldsymbol{x}]$) and a part due to the error in the model at a specified $\boldsymbol{\theta}$ (i.e., $[h(\boldsymbol{\theta},\boldsymbol{x})-E(z\,|\,\boldsymbol{x})]^2$). The first of these two parts does not depend on the model—it simply reflects the conditional variance of the *true process*. The analyst, presumably, has no control over that. The second part, on the other hand, is directly related to the model. It is for this second part that we are interested in a bias–variance analysis. Moreover, the bias–variance analysis is fundamentally based on *estimated* values of $\boldsymbol{\theta}$, rather than the fixed value above, as we now discuss.

An analysis of the squared error of the model $[h(\boldsymbol{\theta},\boldsymbol{x})-E(z\,|\,\boldsymbol{x})]^2$ appearing as the second term in the last line of (13.1) provides direct insight into the quality of the regression function $h(\boldsymbol{\theta},\boldsymbol{x})$ as a predictor of z given $\boldsymbol{x}$ for a fixed $\boldsymbol{\theta}$. Of course, in practice, $\boldsymbol{\theta}$ is not usually fixed, but is *estimated* from a set of input–output data. That is, although the decomposition in (13.1) is of interest for analyzing the overall prediction error, it does not provide direct insight into the connection of the input–output fitting (training) data and the quality of final regression function. Hence, we must average the squared error over reasonable values of $\boldsymbol{\theta}$, which are the possible values estimated from input–output data.

Let $\{(\boldsymbol{x}_1, z_1), (\boldsymbol{x}_2, z_2),\ldots, (\boldsymbol{x}_n, z_n)\}$ be n input–output data pairs that will be collected to form the estimate of $\boldsymbol{\theta}$, say $\hat{\boldsymbol{\theta}}_n$ (recursive) or $\hat{\boldsymbol{\theta}}^{(n)}$ (batch), based on an appropriate search and optimization algorithm. The data are processed in a sequential or batch manner—as appropriate—to form the estimate of $\boldsymbol{\theta}$. If the data are processed recursively (i.e., one at a time), $\hat{\boldsymbol{\theta}}_n$ is an estimate of $\boldsymbol{\theta}$ after n iterations of an algorithm, notationally corresponding to the usage in previous chapters. If the data are processed en masse, as in the classical batch least-squares solution (Subsection 3.1.2), $\hat{\boldsymbol{\theta}}^{(n)}$ is the estimate. To avoid the cumbersome need to include multiple versions of particular equations and formulas, we also use $\hat{\boldsymbol{\theta}}_n$ when making a *generic* reference to an estimate of $\boldsymbol{\theta}$ (without specifying if it is recursive or batch). It should always be clear from the context if $\hat{\boldsymbol{\theta}}_n$ is being used to denote a recursive estimate or a generic estimate.

Following the guidelines above for a generic $\boldsymbol{\theta}$ estimate, an expectation of the form $E[h(\hat{\boldsymbol{\theta}}_n, \boldsymbol{x}) | \boldsymbol{x}]$ represents an expectation of $h(\cdot)$ conditioned on an input $\boldsymbol{x}$, where the expectation is with respect to the randomness in the input–output data as they manifest themselves in $\hat{\boldsymbol{\theta}}_n$ (since $\hat{\boldsymbol{\theta}}_n$ is a function of the $\boldsymbol{x}_i$, z_i pairs). The input $\boldsymbol{x}$ represents some fixed value of interest, perhaps corresponding to some future value that the analyst expects to encounter ($\boldsymbol{x}$ does not generally correspond to any previously observed input $\boldsymbol{x}_i$). It is possible that a specific $\boldsymbol{x}$ and set of input–output data will yield an $h(\hat{\boldsymbol{\theta}}_n, \boldsymbol{x})$ providing a good prediction, while the same $\boldsymbol{x}$ and another set of data (a different estimate $\hat{\boldsymbol{\theta}}_n$) yields a poor prediction.

A useful approach to analyzing the inherent model quality is to take the mean of the squared model error $[h(\hat{\boldsymbol{\theta}}_n, \boldsymbol{x}) - E(z | \boldsymbol{x})]^2$ based on averaging with respect to the distribution for the fitting data. This is equivalent to averaging with respect to the resulting distribution for $\hat{\boldsymbol{\theta}}_n$. It can be shown (Exercise 13.1) that for any future $\boldsymbol{x}$ this MSE is

$$E\left\{\left[h(\hat{\boldsymbol{\theta}}_n, \boldsymbol{x}) - E(z | \boldsymbol{x})\right]^2 \middle| \boldsymbol{x}\right\}$$

$$= \underbrace{E\left\{\left[h(\hat{\boldsymbol{\theta}}_n, \boldsymbol{x}) - E\left(h(\hat{\boldsymbol{\theta}}_n, \boldsymbol{x}) | \boldsymbol{x}\right)\right]^2 \middle| \boldsymbol{x}\right\}}_{\text{variance at } \boldsymbol{x}} + \underbrace{\left[E\left(h(\hat{\boldsymbol{\theta}}_n, \boldsymbol{x}) | \boldsymbol{x}\right) - E(z | \boldsymbol{x})\right]^2}_{(\text{bias at } \boldsymbol{x})^2}. \tag{13.2}$$

An unbiased estimator is one with $E\left(h(\hat{\boldsymbol{\theta}}_n, \boldsymbol{x}) | \boldsymbol{x}\right) = E(z | \boldsymbol{x})$ (implying that the second term on the right-hand side of (13.2) is zero).

As a final *overall* assessment of contributions toward the model MSE, one can average the squared bias and variance in (13.2) over all possible values of $\boldsymbol{x}$, yielding mean values, say $\overline{\text{bias}^2}$ and $\overline{\text{variance}}$. If $\boldsymbol{x}$ is generated randomly, then, of course, the averaging is with respect to the probability measure for $\boldsymbol{x}$. If $\boldsymbol{x}$ is chosen deterministically, then the averaging is with respect to plausible future values of $\boldsymbol{x}$ and their expected frequency. In either case, averaging the variance and squared bias terms in (13.2) leads to a global measure of the contributions to the MSE that does not depend on a specific value of $\boldsymbol{x}$:

$$\begin{aligned} \text{MSE}_{\text{overall}} &= E_{\boldsymbol{x}}\left[E\left\{\left[h(\hat{\boldsymbol{\theta}}_n, \boldsymbol{x}) - E(z | \boldsymbol{x})\right]^2 \middle| \boldsymbol{x}\right\}\right] \\ &= E_{\boldsymbol{x}}\left[\text{variance at } \boldsymbol{x} + (\text{bias at } \boldsymbol{x})^2\right] \\ &= \overline{\text{variance}} + \overline{\text{bias}^2}, \end{aligned} \tag{13.3}$$

where $E_x[\cdot]$ denotes the appropriate stochastic or deterministic average over possible values of $\boldsymbol{x}$. (Exercise 13.2 shows the importance of proper averaging with respect to the bias contribution. If one replaces the mean of bias^2 in (13.3) with the square of the mean of the bias, it is possible to obtain nonsensical values of overall MSE.)

13.1.2 Interpretation of the Bias–Variance Tradeoff

The subsection is devoted to interpreting the bias–variance tradeoff as it relates to picking the best model. Although unbiasedness is generally considered a desirable property, (13.2) and (13.3) show that an unbiased estimator can have a large MSE when the variance is large. In particular, a biased estimator may have a lower MSE than an unbiased estimator, even better than unbiased estimators that are best by some criterion (such as an estimator with the lowest variance among unbiased estimators that are linear combinations of the measurements). We begin with a simple example illustrating this point.

Example 13.1—Biased estimator with lower MSE than obvious unbiased estimator. Consider a sequence of n scalar independent, identically distributed (i.i.d.) measurements $\{z_1, z_2, \ldots, z_n\}$ having mean μ and variance $\sigma^2 > 0$. We can write $z_i = \mu + v_i$, where the v_i are mean-zero noises. The aim is to find a good estimator for the unknown μ. This is a simple case for the function $h(\cdot)$ since there is no input $\boldsymbol{x}$ (i.e., $h(\theta, \boldsymbol{x}) = \theta$, where $\theta = \mu$). Let $\hat{\theta}_n$ represent the estimate of μ (so $h(\hat{\theta}_n, \boldsymbol{x}) = \hat{\theta}_n$). The obvious unbiased estimator of μ is the sample mean, say $\hat{\theta}_n = \bar{z}$ (which may be computed in either recursive or batch form). Among all unbiased estimators that are linear combinations of the data, this estimator is minimum variance (and hence minimum MSE since the bias is zero; Wilks, 1962, pp. 279–280; Bickel and Doksum, 1977, p. 143).[2]

An alternative biased estimator is $\hat{\theta}_n = r\bar{z}$, where $0 < r < 1$. The bias and variance of the alternative estimator are

$$\text{bias} = E(r\bar{z}) - \mu = (r-1)\mu,$$

$$\text{variance} = \text{var}(r\bar{z}) = r^2 \frac{\sigma^2}{n}.$$

Expression (13.2) or, equivalently in this case of no dependence on $\boldsymbol{x}$, (13.3) then lead to an MSE of

---

[2] The properties of the sample mean are slightly different if *both* μ and σ^2 are being estimated (versus only μ). If σ is unknown and if the data are normally distributed, then the sample mean is the minimum MSE estimator of μ among *all* (not just linear) *unbiased* estimators (e.g., Bickel and Doksum, 1977, pp. 123–124). This is a consequence of the notion of completeness and sufficiency for estimators, a topic not considered here.

$$\text{MSE} = r^2 \frac{\sigma^2}{n} + (r-1)^2 \mu^2 .$$

If σ^2/n is sufficiently large relative to μ^2, then for a given r the alternative estimator has a lower MSE than $\overline{z}$ (which has an MSE of σ^2/n). For example, if $\sigma^2/n = 0.1$, $\mu^2 = 0.1$, and $r = 0.5$, then the MSE for $r\overline{z}$ is $0.25(0.1 + 0.1) = 0.05$, while the MSE for $\overline{z}$ is 0.1.

Unfortunately, this result is not particularly useful in practice, since the question of whether $r\overline{z}$ or $\overline{z}$ is a better estimator for a specific value of r can only be answered if one knows μ, the very quantity being estimated! Nevertheless, it does show that a biased estimator can yield a lower MSE than an unbiased estimator (see also Exercise 13.3). ❑

There is generally a tradeoff between the variance and bias contributions to the overall MSE. Regression functions $h(\cdot)$ with high variance tend to have low bias and vice versa. One can see this in Example 13.1 by letting r range from near 0 (high bias/low variance) to near 1 (low bias/high variance). More generally, when $h(\cdot)$ depends on an input $\boldsymbol{x}$, there is a relationship between the complexity of the model and the relative bias and variance. In particular, the following relationships typically hold:

Simple model	⇔	**High bias/low variance**
Complex model	⇔	**Low bias/high variance**

The bias–variance tradeoff provides a framework for choosing among candidate models. Of course, in practice, many factors other than bias, variance, and the resulting MSE may be relevant. These include cost, development time, historical precedent for particular model forms, desires of organizational leadership, and so on. Nevertheless, all other factors being equal, one would wish to pick the function $h(\cdot)$ with a balanced bias and variance. This balance in bias and variance results in the minimum MSE according to (13.2) or (13.3).

Figure 13.1 presents three plots to illustrate the above relationships on a simple problem with scalar input x. Each plot uses the same two sets of five data points (with both sets being at the same input values $x_1, x_2, \ldots, x_5$). One of the data sets is used to fit the curve and the other independent set is used to test the fit. Each of the three individual figures shows the result of a model fit based on a polynomial model with specified p (the polynomial has order $p - 1$ to allow for the additive constant term in the polynomial).

Figure 13.1(a) shows a case where the model is too complex relative to the data. Here the curve perfectly matches the five data points available for

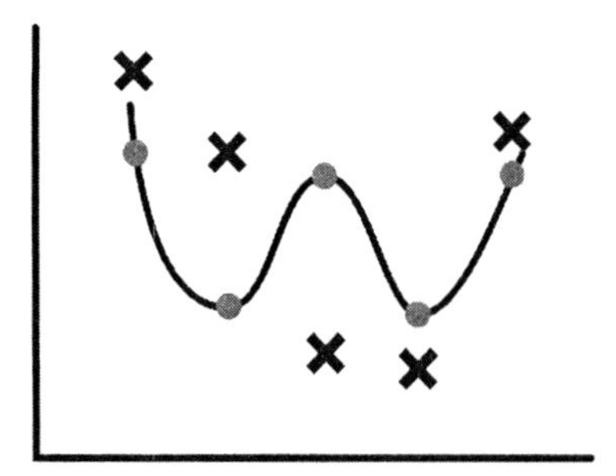

(a) High variance/low bias.
4th-order polynomial (p = 5).

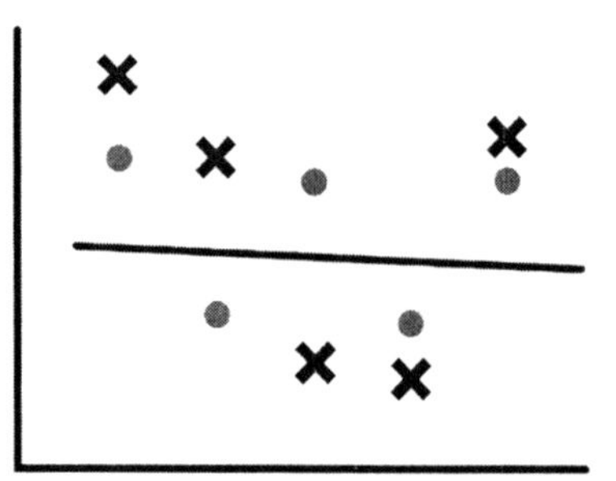

(b) Low variance/high bias.
1st-order polynomial (p = 2).

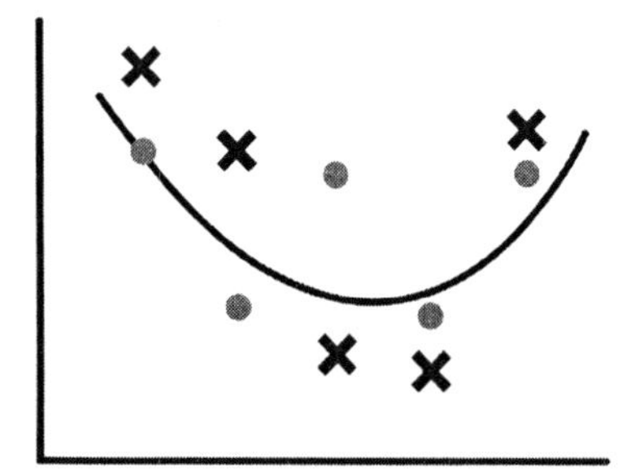

(c) Balanced variance & bias.
Minimum MSE.
2nd-order polynomial (p = 3).

● Data points for fitting

✖ Typical new data points

Figure 13.1. Illustration of the bias–variance tradeoff in model selection in a simple problem. Model in part (c), which has minimum MSE relative to the fitting data, is also most consistent with the new data points not used in fitting model.

fitting the model. Hence the variability of $h(\cdot)$ at each x will be identical to that of the data (z) itself. That is, at n = 5, $E\{[h(\hat{\theta}_n, x) - E(h(\hat{\theta}_n, x) \mid x)]^2 \mid x\}$ = $E[(z - E(z \mid x))^2 \mid x]$, for any $x \in \{x_1, x_2, \ldots, x_5\}$ and generic (batch or nonbatch) estimate $\hat{\theta}_n$. Note that the bias of $h(\cdot)$ relative to the fitting data set is zero since $z = z(x) = h(\hat{\theta}_n, x)$ trivially implies that $E(z|x) = E[h(\hat{\theta}_n, x) \mid x]$. Figure 13.1(b) shows a case with nearly the opposite character. This model is too simple relative to the data. Given the limited flexibility in the curve, there will be relatively little variation in $h(\cdot)$ at each x as new fitting data sets are collected. This provides for a small variance, $E\{[h(\hat{\theta}_n, x) - E(h(\hat{\theta}_n, x) \mid x)]^2 \mid x\}$, but a large bias since $h(\cdot)$ will not track z very well due to the rigidity in $h(\cdot)$. Figure 13.1(c) shows a curve that balances the bias–variance tradeoff with a curve of "reasonable" flexibility. A visual examination of the three curves relative to the new testing data (the **x** points) reveals that the balanced p = 3 (quadratic) model of Figure 13.1(c) seems to best match the new data.

Unfortunately, the bias–variance tradeoff is largely limited to gaining a *conceptual* understanding for comparing different models. Because the probability distributions for the input–output data, together with the resulting distribution of $\hat{\boldsymbol{\theta}}_n$, are not known (they depend on knowing the unknown true model), the values of the bias and variance will generally be unknown. It is, however, clear from the bias–variance tradeoff that there can be no universal best model form. One model form may provide nicely balanced bias and variance on one class of problems, but be too rigid (high bias) or flexible (high variance) in another example. For a *fixed* model form, the variance contribution to MSE tends to decrease when the sample size used in fitting the model is increased. Intuitively, this follows since the model quality improves from the greater information available for fitting the model. The variance contribution to MSE decreases with a greater amount of data since there is a reduced tendency to fit to the individual data points.

13.1.3 Bias–Variance Analysis for Linear Models

For the important special case of linear models, let us present an explicit form for the bias and variance. Although the bias and variance provide useful insight (see Example 13.2), they are generally not computable in practice because they depend on quantities that are unknown. Suppose that the *true process* generates data according to $z = f(\boldsymbol{x}) + \eta$, where $f(\boldsymbol{x})$ is an unknown (possibly nonlinear) function and η is an independent error having mean zero and variance σ^2. Suppose that the classical linear regression model (Section 3.1) is used to describe the process,

$$z_k = \boldsymbol{h}_k^T \boldsymbol{\theta} + v_k ,$$

where $\boldsymbol{h}_k$ is the design vector (dependent on input $\boldsymbol{x} = \boldsymbol{x}_k$) and v_k is a noise term having common variance across k. Suppose further that classical (batch) least-squares (Subsection 3.1.2) is used to estimate $\boldsymbol{\theta}$ (producing $\hat{\boldsymbol{\theta}}^{(n)}$). Then, the prediction for a *future* output $z = z(\boldsymbol{x})$ based on n input–output pairs $\boldsymbol{h}_k$, z_k being used for estimating $\boldsymbol{\theta}$ is given by

$$\begin{aligned}\hat{z}(\boldsymbol{x}) &= \boldsymbol{\ell}(\boldsymbol{x})^T \hat{\boldsymbol{\theta}}^{(n)} \\ &= \boldsymbol{\ell}(\boldsymbol{x})^T (\boldsymbol{H}_n^T \boldsymbol{H}_n)^{-1} \boldsymbol{H}_n^T \boldsymbol{Z}_n ,\end{aligned}$$

where $\boldsymbol{\ell}(\boldsymbol{x})$ is a $p \times 1$ vector dependent on the input $\boldsymbol{x}$, $\boldsymbol{Z}_n = [z_1, z_2, \ldots, z_n]^T$, and $\boldsymbol{H}_n$ is the $n \times p$ concatenated matrix of $\boldsymbol{h}_k^T$ row vectors.

In computing the average bias and variance according to (13.3), suppose that there are m future $\boldsymbol{x}$ values of interest, $\chi_1, \chi_2, \ldots, \chi_m$. We are interested in the bias and variance of $\hat{z}(\boldsymbol{x})$ averaged over these m input values (this is the averaging that is being used in concert with (13.3)). Given the estimate of $\boldsymbol{\theta}$

based on the original n input measurements $\boldsymbol{x}_1, \boldsymbol{x}_2, \ldots, \boldsymbol{x}_n$ (as reflected in $\boldsymbol{H}_n$) together with the corresponding output measurements $\boldsymbol{Z}_n$, the stacked value of predictions for the m new inputs is $[\hat{z}(\chi_1), \hat{z}(\chi_2), \ldots, \hat{z}(\chi_m)]^T$, where the $\hat{z}(\cdot)$ function is given above. The corresponding matrix of the m new inputs as reflected in $\mathcal{h}(\cdot)$ is $\mathcal{H}_m \equiv [\mathcal{h}(\chi_1), \mathcal{h}(\chi_2), \ldots, \mathcal{h}(\chi_m)]^T$.

Then, the vector of predictions for the m new inputs is

$$\begin{bmatrix} \hat{z}(\chi_1) \\ \hat{z}(\chi_2) \\ \vdots \\ \hat{z}(\chi_m) \end{bmatrix} = \mathcal{H}_m \hat{\theta}^{(n)} = \mathcal{H}_m (\boldsymbol{H}_n^T \boldsymbol{H}_n)^{-1} \boldsymbol{H}_n^T \boldsymbol{Z}_n \equiv \boldsymbol{S}_{m|n} \boldsymbol{Z}_n .$$

From this prediction function, the average bias-squared and variance for the m predictions based on the n data for estimating $\boldsymbol{\theta}$, now follow:

$$\overline{\text{bias}^2} = \frac{1}{m} \sum_{k=1}^{m} \left\{ \boldsymbol{s}_k^T \begin{bmatrix} f(\boldsymbol{x}_1) \\ f(\boldsymbol{x}_2) \\ \vdots \\ f(\boldsymbol{x}_n) \end{bmatrix} - f(\chi_k) \right\}^2 , \tag{13.4}$$

$$\overline{\text{variance}} = \frac{\sigma^2}{m} \text{trace}(\boldsymbol{S}_{m|n} \boldsymbol{S}_{m|n}^T) , \tag{13.5}$$

where $\boldsymbol{s}_k^T$ is the kth row in $\boldsymbol{S}_{m|n}$ (Exercise 13.4). In practice, of course, the $f(\cdot)$ values and (probably) σ^2 will be unknown. Expressions (13.4) and (13.5) represent the $\overline{\text{bias}^2}$ and $\overline{\text{variance}}$ terms that appear in (13.3).

Although (13.4) and (13.5) cannot typically be used *directly* in practice to evaluate candidate models, they can be used to provide valuable insight. Example 13.2 considers a specific $f(\cdot)$ and σ^2 as a means of illustrating the bias–variance tradeoff. This example compares the bias and variance contributions to MSE for several candidate models in the curvilinear form (Subsection 3.1.1) when the true data-generating process is nonlinear. Example 13.3 considers the effect of increasing sample size on the MSE.

Example 13.2—Bias–variance tradeoff for curvilinear models. Consider a problem where the true process is $f(x) = (x + x^2)^{1.1}$, with x a scalar input. Suppose that the additive independent noise η for the true process has $\sigma^2 = 100$. Using the standard batch least-squares estimate, let us compare the bias and variance for three candidate curvilinear models of the form $\hat{z}(x) = \mathcal{h}(x)^T \hat{\theta}^{(n)}$:

Model 1 (linear): $p = 1$; $\mathcal{h}(x) = x$,

$$\textbf{Model 2 (quadratic): } p = 2;\ \boldsymbol{h}(x) = [x, x^2]^T,$$

$$\textbf{Model 3 (cubic): } p = 3;\ \boldsymbol{h}(x) = [x, x^2,\ x^3]^T.$$

(Unlike Figure 13.1, there is no additive constant in the models; hence, the value of p corresponds directly to the polynomial order.) Note that model 2 with $\hat{\boldsymbol{\theta}}^{(n)} \approx [1, 1]^T$ is almost the correct model (differing in form by only the exponent 1.1). Hence, it might be expected that model 2 will provide the optimal bias–variance tradeoff. Let us see if this is true.

In computing the average bias and variance according to (13.3), suppose that $m = n$ and that the future x values are the same as the n values of x_k that were used in estimating $\boldsymbol{\theta}$. In particular, suppose that $x_k = \chi_k = k$, $k = 1, 2,\ldots, m$, and that $m = n = 10$ (i.e., $\boldsymbol{\mathcal{H}}_m = \boldsymbol{H}_n$). For example, in the case of $p = 2$,

$$\boldsymbol{\mathcal{H}}_m = \boldsymbol{H}_n = \begin{bmatrix} 1 & 2 & 3 & 4 & 5 & 6 & 7 & 8 & 9 & 10 \\ 1 & 4 & 9 & 16 & 25 & 36 & 49 & 64 & 81 & 100 \end{bmatrix}^T.$$

Table 13.1 presents the bias–variance analysis.

As shown in Table 13.1, model 2 is the preferred model (the lowest MSE), although model 3 has an MSE that is similar (due to the value in using a cubic term to capture some of the nonpolynomial effects caused by the exponent 1.1). Note the overall pattern of decreasing bias and increasing variance as the model complexity increases, consistent with Figure 13.1. ❑

Example 13.3—Effect of sample size. Consistent with the discussion of the preceding subsection, let us show how an increasing sample size (n) can reduce MSE via a reduction in the variance contribution. With the exception of the change in n, the setting is identical to Example 13.2. Suppose that $n = 20$, with the additional 10 inputs being identical to the initial 10 inputs, $x_k = x_{k+10} = k$, $k = 1, 2,\ldots, 10$. Then, using (13.5), $\overline{\text{variance}}$ for models 1, 2, and 3 is 5.0, 10.0, and 15.0, respectively. Because of the doubling of the sample size for fitting, these

Table 13.1. Average bias and variance and overall MSE for candidate curvilinear models.

	$\overline{\text{bias}^2}$	$\overline{\text{variance}}$	Overall MSE
Model 1 (linear)	510.6	10.0	520.6
Model 2 (quadratic)	0.53	20.0	20.53
Model 3 (cubic)	0.005	30.0	30.005

variances are half the magnitude of the variances in Table 13.1. The bias is unchanged. Hence, as expected, the overall MSE is reduced with an increase in the amount of fitting data. ❑

Given the conceptual insight of the bias–variance tradeoff, we are now in a position to consider practical means of optimizing this tradeoff. The next section addresses this issue.

13.2 MODEL SELECTION: CROSS-VALIDATION

There are a great number of methods for approximately addressing the bias–variance tradeoff in a manner that is feasible for implementation. These methods are variations on the theme of balancing low- and high-order requirements to produce a model that (implicitly at least) balances the bias and variance in a manner similar to Figure 13.1. Some of the best-known methods include the information criterion (AIC) (Akaike, 1974), the principle of minimum description length (Rissanen 1978; Wei, 1992), bootstrap model selection (Efron and Tibshirani, 1997), Bayesian model selection (Akaike, 1977; Schwarz, 1978; George and McCulloch, 1997), V-C dimension (Vapnik and Chervonenkis, 1971; Cherkassky and Mulier, 1998, Chap. 4), the Fisher information criterion (Wei, 1992), and cross-validation (Allen, 1974; Stone, 1974; Geisser, 1975). These methods rely on approaches such as maximum likelihood, the Fisher information matrix (see Section 13.3), information theory, regression, computer-based resampling, risk minimization, and sample fitting.

The basic principle in model selection is to minimize, implicitly or explicitly, a criterion of the generic form

$$f_1(\text{fitting error from given data}) + f_2(\text{model complexity}), \qquad (13.6)$$

where $f_1(\cdot)$ and $f_2(\cdot)$ are increasing functions of, respectively, some measure of the error in the model predicting the fitting data (the fitting error) and some measure of the number of terms in the model (model complexity). Good general discussions of model selection methods appear in Linhart and Zucchini (1986), Shao (1997), McQuarrie and Tsai (1998), and Ljung (1999, Chap. 16). Applications of some of these methods to neural networks are discussed in Geman et al. (1992). For reasons of wide applicability and popularity, this section focuses on the cross-validation method of model selection. In some applications, however, one of the other methods may be more effective. The reader seriously interested in a broader review of the important subject of model selection is directed to the references above.

The cross-validation approach is perhaps the most straightforward formal model selection method to understand and to implement. It is based on manipulations of the fitting (training) data (i.e., the data assumed available for model estimation.) Cross-validation does not require additional data and/or detailed prior information or analytical analysis beyond sample model fits. Cross-validation also has the advantage of applying to candidate models of virtually any form, not being restricted to specific classes of candidate models (e.g., linear/curvilinear regression models), as are some of the approaches mentioned above. Also, unlike some other approaches, it does not require that the underlying data be normally distributed. On the other hand, cross-validation is not necessarily the most powerful or discerning method in any specific problem, nor is it the most computationally efficient (since it requires repeated "sample" model fits). Cross-validation is one of the model-selection methods that *implicitly* optimizes the tradeoff criterion (13.6), as there is no direct construction of a performance metric dependent on $f_1(\cdot)$ and $f_2(\cdot)$. (Cross-validation has other applications as well. For example, one other use is helping determine when to stop an algorithm's iteration process; see Amari et al., 1997.)

Let us sketch how cross-validation works in the context of selecting a model. A more formal step-by-step description is given later in this section. Suppose that two or more candidate model forms are to be evaluated. For example, an analyst may have small-, medium-, and large-scale neural networks as candidate model forms and wishes to know which neural network is likely to produce the best predictions. Cross-validation is a commonsense approach based on sequentially partitioning the full data set into fitting and test *subsets*. For each partition, estimates are produced for the candidate model forms from the fitting subset. Then, the performance of each candidate model is measured on the test subset. This procedure is repeated for all partitions of the full data set.

Let n_T denote the size of the test subsets, where, of course, $n_T < n$, with n the size of the full data set. A common strategy—called *leave-one-out*—is to pick $n_T = 1$ and cycle through all n possible combinations of fitting and test subsets (e.g., Stone, 1974; Allen, 1974). This approach produces n model fits from the n possible fitting subsets of size $n - 1$. Each of these model fits generates a prediction error on the one data point left out (i.e., the difference between the outcome of the point left out and the predicted value based on the model fit from the remaining $n - 1$ points). The best model form is the one for which the chosen type of average for these n prediction errors—say, the sample MSE or mean absolute deviation (MAD)—is lowest. (We do not include the qualifier "sample" below, but it should be clear from context that the MSE and other values are not analytically based, but are derived from the specific sample.)

There are often advantages, however, to choosing $n_T > 1$. The advantages include greater efficiency (i.e., fewer model fits) for *some* implementations with $n_T > 1$ (but definitely not all implementations, as we see shortly). There is also some theoretical and empirical evidence that this n_T-fold ($n_T > 1$) approach produces more accurate results than the leave-one-out strategy (Breiman and Spector, 1992; Shao, 1993; Breiman, 1996). In fact, for linear models, the leave-one-out strategy has been shown in Shao (1993) to be biased to picking models with excessive complexity (i.e., p too large). When $n_T > 1$, the test subsets may be chosen deterministically or randomly, with or without replacement. For test subsets chosen deterministically with replacement (i.e., all possible combinations of test subsets of size n_T are used), there are a potentially huge number of possible fitting/test subset combinations. In particular, the number of combinations is "n choose n_T" $\binom{n}{n_T}$. For example, with $n = 30$ and $n_T = 6$ (as in Example 13.4), cross-validation in this manner would require over 590,000 sample model fits!

One way of mitigating this explosion is to randomly select (usually with replacement) a relatively small number of test subsets of size n_T (e.g., Shao, 1993). Another approach is to choose n_T such that n is divisible by n_T and then choose the test subsets *without replacement* so that all of the data appear once and only once in a test subset. The allocation of the n data to the n/n_T test subsets may be done randomly or deterministically (e.g., Neter et al., 1996, p. 437; Cherkassky and Mulier, 1998). The "once and only once" aspect may be viewed as an extension of the leave-one-out strategy. This allocation reduces the number of modelfits from n in leave-one-out and from $\binom{n}{n_T}$ in deterministic replacement selection to n/n_T (e.g., from 30 and over 590,000, respectively, to only 5 in the illustration above).

Figure 13.2 presents a schematic of the process of partitioning the data into three combinations of fitting and test subsets when the test subsets are chosen deterministically or randomly so that all data serve once (and only once) in a test subset. Hence, the test subsets are disjoint. The deterministic version of this disjoint sampling procedure is used in Examples 13.4 and 13.5. Note that while the test subsets may be independent (given that the raw data are independent), the fitting subsets typically share data as in Figure 13.2. For example, each fitting subset in Figure 13.2 shares half of its data with each of the other fitting subsets. As described in the leave-one-out strategy above, the model form that produces the best performance across the sequence of test subsets with respect to a specified metric is chosen as the best model form. Typically, the performance metric is the MSE (equivalently, the root-mean-squared error, RMS) of the predictions over the multiple test subsets, although other approaches may be used as well (such as MAD in Example 13.5).

After determining the best model form via cross-validation, the *full* data set is used to produce the final estimates for the parameters of this model. Obviously, there are some choices to be made in implementing the cross-validation approach. In particular: the type of partitioning for the data, the metric by which the models will be compared on the sequence of test sets (MSE, MAD, etc.), whether to use random or deterministic sampling for the test subsets, and so on. Guidelines for these choices are discussed, for example, in Shao (1993), Neter et al. (1996, pp. 437–439), Cherkassky and Mulier, (1998, pp. 78–79), and McQuarrie and Tsai (1998, pp. 251–261). In practice, however, an intuitive "feel" is often the primary guide since the more formal guidelines are generally restricted to relatively narrow model classes.

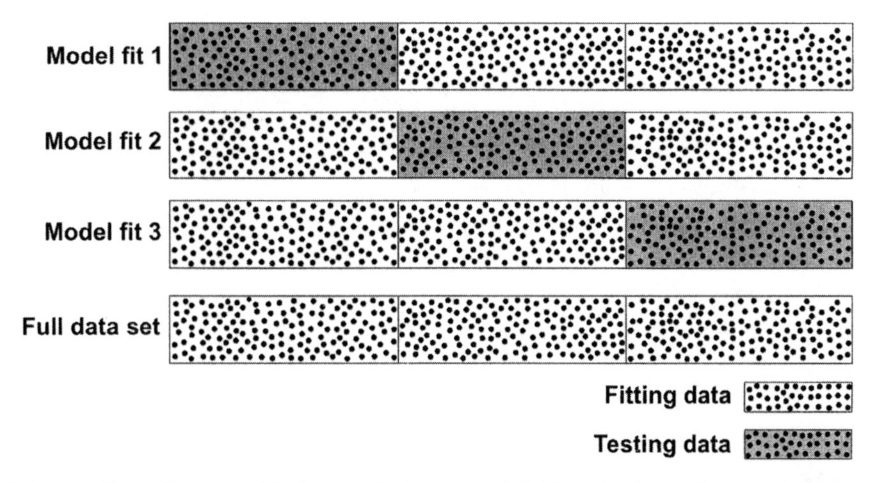

Figure 13.2. Cross-validation with three model fits based on three subsets of fitting data within the full data set. The data are illustrated schematically as random points. The test subsets are disjoint, with the union covering the full data set. The models are compared based on performance over three test subsets. The winning model form is fit with the full data set after the cross-validation is complete.

The steps given below are a formal summary of the procedure outlined above. These steps describe a typical implementation of cross-validation. The performance metric discussed below is MSE, but the overall approach is identical if a different metric is used. Unlike the bias–variance calculations of Section 13.1, the following steps are based solely on the *available* model $h(\cdot)$ and data. It is not necessary to know the unknown: the true data-generating mechanism or the distribution of the noise in the process.

Implementation of Cross-Validation with Disjoint Test Subsets

Step 0 **(Initialization)** Determine n_T and the strategy for choosing the disjoint test subsets; n is assumed divisible by n_T. Set $m = 1$ and $i = 1$, where m is the counter for the candidate model being considered and i is the counter for the test subset being used from the n/n_T possible test subsets.

Step 1 Consider the mth model (the mth candidate form for $h(\cdot)$). For the ith test subset, let the remaining $n - n_T$ elements be the ith fitting subset. Estimate $\boldsymbol{\theta}$ for the mth model from this fitting subset.

Step 2 Based on the estimate for $\boldsymbol{\theta}$ from step 1, compare the predictions of the mth model and the data in the ith test subset. Suppose that the MSE is being used for this comparison. Then let $\text{MSE}_i(m)$ denote the mean of the sum of squared errors over the n_T points in the ith test subset. (If another metric is used, simply replace MSE in the steps below with that metric.)

Step 3 Update i to $i+1$ and return to step 1. Terminate when all data have been included once and only once in a test subset (corresponding to a terminal value of $i = n/n_T$). Upon termination, let $\overline{\text{MSE}}(m)$ denote the mean of the n/n_T values for $\text{MSE}_i(m)$ (i.e., $\overline{\text{MSE}}(m)$ is the overall MSE for model m across the n/n_T test subsets).

Step 4 Repeat steps 1 to 3 for the next model by updating m to $m+1$ and resetting i to $i = 1$. Terminate when the cross-validation calculations have been performed for all models.

Step 5 Choose the model corresponding to the lowest $\overline{\text{MSE}}(m)$ as the best model.

One of the limitations in formal model selection in general—including cross-validation—is that statistical tests of whether one model is better than another "...are difficult to construct and one must then rely on a simple comparison between estimated expected discrepancies" (Linhart and Zucchini, 1986, p. 15). For cross-validation, this is associated with step 5 above. Although it is usually relatively straightforward to determine the model with the lowest $\overline{\text{MSE}}(m)$ (or sample mean of other metric if appropriate), it is difficult to know if the lowest value is statistically significantly lower than the value of other candidate models.[3] Might the choice of a particular winning model just be an anomaly of the particular data set?

Unfortunately, there is very little information available to formally determine the P-value (say) associated with the difference in $\overline{\text{MSE}}(m)$ values for the candidate models. This question of statistical significance appears to have no easy answer given that the n/n_T values of $\overline{\text{MSE}}(m)$ will be statistically dependent in a complicated way. That is, the $\overline{\text{MSE}}(m)$ contributions depend—through the parameter estimates and testing subsets—on the *same* full set of data. An additional complication follows from this being a multiple comparisons problem (Chapter 12) since, in general, there will be more than two candidate models.

The dependence and multiple comparisons aspects preclude the use of standard techniques for comparisons such as the matched or unmatched t-tests

---

[3]The lack of formal statistical justification for choosing one criterion value as being significantly better than another (in the sense of something like a P-value) is not unique to cross-validation. Other approaches, not surprisingly, suffer the same shortcoming, as they, too, depend on nontrivial transformations of the same data set, leading to a comparison based on highly dependent test measures. There are, however, special cases where it is possible to derive (at least approximate) distributions for test statistics associated with the difference in models and/or derive the expected value and variance associated with such differences (see, e.g., McQuarrie and Tsai, 1998, pp. 25–27, for the approximate mean and variance of the difference for AIC in linear regression models). Bayesian methods for model selection eliminate the need to work with test statistics per se, but introduce other complications associated with the choice of prior distribution and the need to carry out numerical integration (e.g., Schwarz, 1978; George and McCulloch, 1997).

(Appendix B). Hence, typical applications of cross-validation simply rely on the outcome in step 5 without further statistical testing and without assigning a *P*-value to the outcome. (*Limited* inference results, however, are available. For instance, Example 13.5 uses an independent data set to confirm the outcome of the cross-validation. McQuarrie and Tsai, 1998, pp. 254, 258–259, describe approximate tests for leave-one-out cross-validation that apply in comparing two linear models with normally distributed noise when one of the candidate models is the true model.)

Two demonstrations of cross-validation are given below. Example 13.4 is for an artificial data set, and Example 13.5 is for the oboe reed data of Section 3.4. Example 13.4 illustrates cross-validation for a relatively low noise level in the measurements, while Example 13.5 involves a larger noise contribution.

Example 13.4—Cross-validation on artificial data set. This example is similar to a problem in Cherkassky and Mulier (1998, pp. 85–88). Although the system and data are artificial, the general model selection process here is identical to what typically occurs with a real system. A simulated set of $n = 30$ data points is produced from the true system

$$z_k = \sin(2\pi x_k) + \eta_k, \tag{13.7}$$

where η_k is i.i.d. $N(0, 0.1^2)$ and the scalar inputs x_k are generated as i.i.d. $U(0, 1)$ random variables. (As needed in Exercise 13.6, the input–output data for this example are in the file **`sinedata`** at the book's Web site.) For a scalar input x, consider candidate regression models of the polynomial form

$$h(x, \boldsymbol{\theta}) = \sum_{i=0}^{p-1} t_{i+1} x^i, \tag{13.8}$$

where $\boldsymbol{\theta} = [t_1, t_2, \ldots, t_p]^T$, as usual. Note that this model is of the curvilinear form (Subsection 3.1.1) and hence specialized techniques for model determination in standard linear regression (as reviewed in Cherkassky and Mulier, 1998, p. 229, or Ljung, 1999, Chap. 16) can be used. However, we will use a generic cross-validation approach, which applies equally in linear and nonlinear models. All model fitting for cross-validation here is carried out with ordinary least-squares estimation (Section 3.1).

We choose $n_T = 6$ points for the $n/n_T = 5$ disjoint testing subsets. Hence, $i = 1, 2, 3, 4$, or 5 in the notation of the five-step procedure for cross-validation above. The five test subsets are chosen as elements 1 to 6, 7 to 12,..., and 25 to 30. Because the input data are generated uniformly, we know that each of these five test subsets is statistically representative of the full data set. For each division of the full data set, we fit the model with the subset of $n - n_T = 30 - 6 = 24$ data points left in the fitting set after removing the points for the testing subset (e.g., elements 7 to 30 are the fitting subset for the first test subset of elements 1 to 6). We then choose the model that provides the best overall fit over the multiple test subsets of data. The measure of fit reported here is the RMS error over all five test subsets, calculated by averaging the MSEs over the test subsets and taking the square root (i.e., $[\overline{\text{MSE}}(m)]^{1/2}$ in the notation of the cross-validation implementation steps above; RMS is reported here to maintain an error measure in the same units as the z_k values).

Let us consider three candidate polynomial models: a linear (affine) model ($p = 2$, including the additive constant), a third-order polynomial ($p = 4$), and a tenth-order polynomial ($p = 11$). Hence, the model counter $m = 1, 2$, or 3 in the notation of the five-step procedure above. Table 13.2 shows the RMS errors across the five test subsets for the three candidate models.[4]

Table 13.2. RMS errors in cross-validation for the three candidate models. RMS errors derived from MSE over all test subsets.

Linear model ($p = 2$)	Third-order polynomial ($p = 4$)	Tenth-order polynomial ($p = 11$)
0.610	0.129	3.78

As described in the five-step procedure above, we choose as the best model the one with the lowest overall RMS error (equivalent to the lowest MSE, of course). Hence, the third-order (cubic) polynomial yields the best fit according to the cross-validation principle. The linear function exhibited excessive bias with the inadequate flexibility of a straight line, while the tenth-order polynomial tended to fit too closely to the 24 data points of each fitting subset, causing some very large individual errors on the testing subsets (and hence contributing to the large overall RMS). The cubic polynomial did a nice job of balancing the bias and the variance. Figure 13.3 shows the 30 data points in **sinedata**, the true sine function, and the cubic polynomial with coefficients estimated from the full set of 30 points. Visually, the cubic polynomial provides a nice fit to the true sine function. ❑

[4]The RMS error for the $p = 11$ model in the table is only an approximation. The matrices corresponding to $(\boldsymbol{H}_{n-n_T}^T \boldsymbol{H}_{n-n_T})^{-1}$ in the basic least-squares formula in Subsection 3.1.2 are ill-conditioned, leading to some numerical instability (an example of multicollinearity in regression). A MATLAB-based calculation of the RMS error is 3.90 (versus the MS EXCEL result of 3.78 in the table). Despite the instability, it is clear that the RMS error for $p = 11$ is significantly larger than the error for $p = 4$.

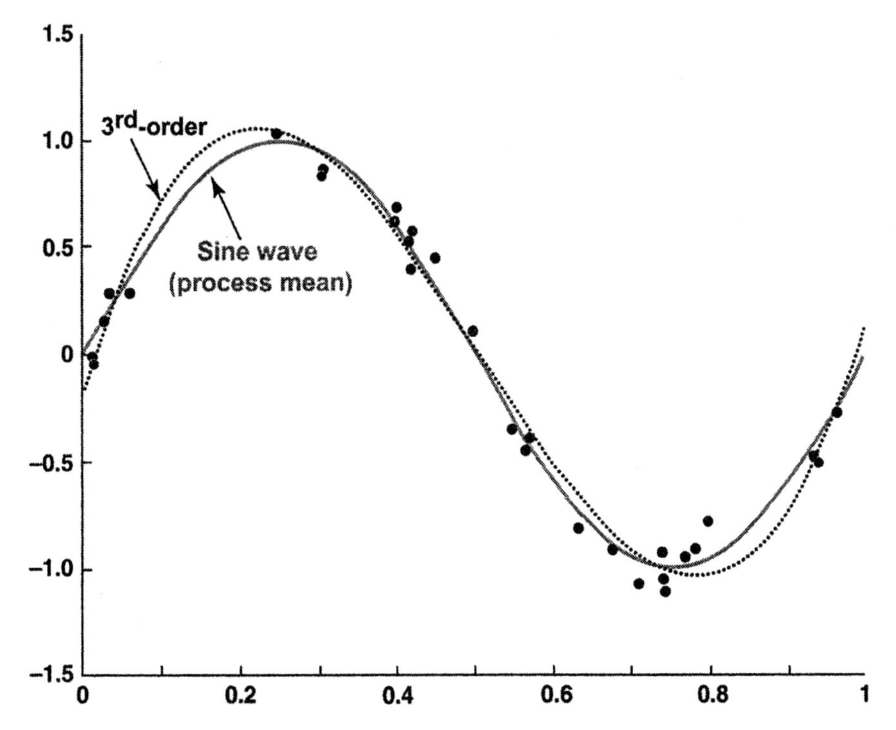

Figure 13.3. True mean and fitted model in a low-noise environment. The 30 data points are from the true process (13.7). The third-order polynomial model is chosen from cross-validation. The sine function corresponds to the mean of the data-generating process.

Example 13.5—Cross-validation with oboe reed data. Section 3.4 described the problem of predicting the final quality of an oboe reed from indicators that are available early in the reed-making process. These indicators represent the input variables $\boldsymbol{x}$. Based on the data set **reeddata-fit** (at the book's Web site), let us use cross-validation to compare two candidate models. The first model is the standard linear model given in Section 3.4, with $\boldsymbol{x}$ composed of all six input variables T, A, E, V, S, F:

$$z = \theta_{\text{const}} + \theta_T T + \theta_A A + \theta_E E + \theta_V V + \theta_S S + \theta_F F + v. \tag{13.9}$$

This model has $p = 7$ (including the additive constant). The second model is a simplified form with only first element of $\boldsymbol{x}$, "top close" (T), as an input:

$$z = \theta_{\text{const}} + \theta_T T + v, \tag{13.10}$$

implying that $p = 2$. As with Example 13.4, all model fitting here (for cross-validation and final estimation) is carried out with ordinary least-squares estimation.

The cross-validation process uses the 160 input–output measurements in **reeddata-fit**. As in Figure 13.2, we sequentially divide the full data set into pairs of fitting and disjoint test subsets, with each fitting subset containing 120 measurements and each test subset containing the remaining 40 measurements. These four pairs of fitting/testing subsets are chosen based on picking the four test subsets as disjoint measurements 1 to 40, 41 to 80, 81 to 120, and 121 to 160. The fitting data for each of the pairs are the remaining 120 elements (e.g., measurements 41 to 160 in the first pair). Relative to the five steps of cross-validation above, the index i runs over 1, 2, 3, and 4. Because the data exhibit trends in $\boldsymbol{x}$ due to certain similarities in the raw cane for measurements collected near the same time, it may have been useful to randomize the order of the measurements before constructing the disjoint test subsets. This randomization was not done in the study here.

Table 13.3 shows the RMS and MAD errors over the four test subsets for the two candidate models. The RMS estimates are the square roots of the mean of the four sample MSEs taken over the test subsets (analogous to Example 13.4). The MAD values are the sample means of the prediction errors $|\hat{z}_k - z_k| = |\boldsymbol{x}_k^T\hat{\boldsymbol{\theta}} - z_k|$, where $\boldsymbol{x}_k$ denotes the kth input vector (or input scalar for model (13.10)), z_k is the corresponding output, and $\hat{\boldsymbol{\theta}}$ denotes one of the four estimates for $\boldsymbol{\theta}$ formed from one of the fitting subsets containing 120 measurements. The MAD estimate is the mean over the four test subsets.

From Table 13.3, both the MAD and RMS errors suggest that the full linear model provides superior predictions. This is not surprising given that the reduced model omits the critical "first blow" (F) input variable. Note, however, that we have not provided a formal statistical justification for this choice. As discussed following the generic cross-validation steps above, the individual MAD or MSE contributions for the four subsets of test data are dependent in a complicated way, implying that there is no known method for calculating P-values or providing other formal justification for the choice of model (13.9) over model (13.10). In particular, the four MAD or MSE contributions depend—through the parameter estimates and testing subsets—on the *same* data in **reeddata-fit**.

For this problem, however, we have the luxury of a *separate* (independent) test set of data, **reeddata-test** (at the book's Web site). This is not typical in applications of cross-validation, where it is usually assumed that

Table 13.3. RMS and MAD errors from cross-validation for the full- and reduced-order linear models.

	Full linear model (13.9)	Reduced linear model (13.10)
RMS	0.327	0.386
MAD	0.266	0.306

all available data are used in the cross-validation. (Further, we have the luxury of only two candidate models, thus avoiding the need to appeal to multiple comparisons methods, as in Chapter 12.)

This separate test set can be used to provide an independent assessment of the models based on parameter estimates using the full set of data in **reeddata-fit**. In this way, we can create statistically independent predictions (conditional on the parameter estimates from **reeddata-fit**) and conduct a matched-pairs t-test. This provides an independent assessment of the value of cross-validation in this relatively high-noise problem (higher noise than the previous sine example, Example 13.4). Carrying out this test on the 80 measurements in **reeddata-test** gives a t-statistic of 1.68, yielding a one-sided P-value of 0.049. This provides substantial—but not overwhelming—support for the superiority of model (13.9) over (13.10), as predicted by cross-validation. ❑

13.3 THE INFORMATION MATRIX: APPLICATIONS AND RESAMPLING-BASED COMPUTATION

13.3.1 Introduction

The Fisher information matrix plays a central role in the practice and theory of estimation. This matrix provides a summary of the amount of information in the data relative to the quantities of interest. Some of the specific applications of the information matrix include confidence region calculation for parameter estimates, input determination in experimental design, performance-bound determination in an adaptive system (such as a control system), model selection via some of the methods *other* than cross-validation (as mentioned at the beginning of Section 13.2), and uncertainty-bound calculation for predictions (such as with a neural network). The information matrix has several connections to the theme of this book: (i) It is the limiting covariance matrix in the asymptotic distribution of root-finding stochastic approximation when the optimal gain sequence is used (e.g., Sections 4.4 and 7.8); (ii) it provides the basis for the important *D-optimal* criterion that will be seen in the optimal design discussion of Chapter 17; and (iii) it can be computed using Monte Carlo resampling together with a technique originally developed for search and optimization (the efficient technique for estimating the Hessian matrix discussed in Section 7.8).

Subsection 13.3.2 provides some formal background on the information matrix. Subsection 13.3.3 discusses two key properties that closely connect the information matrix to the covariance matrix of general parameter estimates. This connection provides the prime rationale for applications of the information matrix in the areas of uncertainty regions for parameter estimation, experimental design, and predictive inference, as summarized in Subsection 13.3.4. Finally, Subsection 13.3.5 describes the above-mentioned resampling-based approach to approximating the matrix. The definitions and key facts here are a prerequisite to most of Chapter 17.

13.3.2 Fisher Information Matrix: Definition and Two Equivalent Forms

Consider a sequence of random vectors $\{z_1, z_2, \ldots, z_n\}$. For example, z_k may be modeled to represent the output in a multivariate version of our traditional model of the process, $z_k = \boldsymbol{h}(\boldsymbol{\theta}, \boldsymbol{x}_k) + \boldsymbol{v}_k$ (i.e., z_k, $\boldsymbol{h}(\cdot)$, and $\boldsymbol{v}_k$ may be vectors). More generally, it may simply be assumed that the z_k are random vectors without such a model form. The Fisher information matrix can be defined once there is *some* additional structure. If inputs $\boldsymbol{x}_k$ are relevant to the problem, let us assume that they are chosen deterministically. Further, let us assume that the *general form* for the joint probability density or probability mass (or hybrid density/mass) function for the stacked vector of random output data $\boldsymbol{Z}_n = [z_1^T, z_2^T, \ldots, z_n^T]^T$ is known, but that this function depends on an unknown vector $\boldsymbol{\theta}$ and, if relevant, on the inputs $\boldsymbol{x}_k$. So, for example, it might be assumed that the z_k are i.i.d. normal with mean $\boldsymbol{\mu} = \boldsymbol{\mu}(\boldsymbol{\theta})$ and covariance matrix $\boldsymbol{\Sigma} = \boldsymbol{\Sigma}(\boldsymbol{\theta})$ dependent on the unknown parameters $\boldsymbol{\theta}$ (there are no inputs $\boldsymbol{x}_k$ in this i.i.d. case). Once $\boldsymbol{\theta}$ is specified, the distribution for the outputs z_k is known precisely.

To define the information matrix, we need to define the *likelihood function* based on the probability density/mass function for the outputs. The likelihood function is identical to the probability density/mass function with the exception that there is a reversal of the conditioning for the arguments. The probability density/mass function is a representation of the relative frequency with which a collection of outcomes $\boldsymbol{Z}_n$ will be observed conditioned on $\boldsymbol{\theta}$. The likelihood function, on the other hand, is the probability density/mass function considering $\boldsymbol{\theta}$ as a variable *conditioned* on the data $\boldsymbol{Z}_n$.

Let the probability density/mass function for $\boldsymbol{Z}_n$ be $p_Z(\boldsymbol{\zeta}|\boldsymbol{\theta})$, where $\boldsymbol{\zeta}$ (zeta) is a dummy vector representing the possible outcomes for the elements in $\boldsymbol{Z}_n$ (in $p_Z(\boldsymbol{\zeta}|\boldsymbol{\theta})$, the index n on $\boldsymbol{Z}_n$ is being suppressed for notational convenience). The corresponding likelihood function, say $\ell(\boldsymbol{\theta}|\boldsymbol{\zeta})$, satisfies

$$\ell(\boldsymbol{\theta}|\boldsymbol{\zeta}) \equiv p_Z(\boldsymbol{\zeta}|\boldsymbol{\theta}). \tag{13.11}$$

One important application of the likelihood interpretation of the probability density/mass function is in maximum likelihood estimation (as seen in the production function example of Section 4.2). In maximum likelihood, the estimate for $\boldsymbol{\theta}$ is the value maximizing the likelihood function (13.11) based on an observed collection of data $\boldsymbol{\zeta} = \boldsymbol{Z}_n$ (roughly speaking, the value of $\boldsymbol{\theta}$ "most likely" given the observed data). In cases where the z_k are independent, one usually maximizes $\log \ell(\boldsymbol{\theta}|\boldsymbol{\zeta})$ (rather than $\ell(\boldsymbol{\theta}|\boldsymbol{\zeta})$) because the criterion then simplifies to an additive form (why?).[5] Taking the logarithm, of course, does not alter the maximum likelihood estimate.

---

[5] The basic likelihood function is a product over $k = 1, 2, \ldots, n$ of the density function for each data point because of the independence. Taking the logarithm converts this product to a sum.

With the definition of the likelihood function in (13.11), we are now in a position to define the Fisher information matrix. For convenience in the derivatives and integrals below, we will generally suppress the arguments in $\ell(\cdot)$ since the interpretation should be clear from the context. When ℓ appears as part of an integrand, then $\ell = \ell(\boldsymbol{\theta}|\boldsymbol{\zeta})$, where $\boldsymbol{\zeta}$ is the dummy vector of integration. When ℓ appears in an expectation, then $\ell = \ell(\boldsymbol{\theta}|\boldsymbol{Z}_n)$ is the *random variable* dependent on $\boldsymbol{Z}_n$. All expectations below are with respect to the data $\boldsymbol{Z}_n$ (i.e., all randomness manifests itself in the output data; unlike some of the discussion in Sections 13.1 and 13.2, the inputs $\boldsymbol{x}_i$ are not random). For example,

$$E\left(\frac{\partial \log \ell}{\partial \boldsymbol{\theta}}\bigg|\boldsymbol{\theta}\right) = E\left(\frac{\partial \log \ell(\boldsymbol{\theta}|\boldsymbol{Z}_n)}{\partial \boldsymbol{\theta}}\bigg|\boldsymbol{\theta}\right)$$
$$= \int \frac{\partial \log \ell(\boldsymbol{\theta}|\boldsymbol{\zeta})}{\partial \boldsymbol{\theta}} p_Z(\boldsymbol{\zeta}|\boldsymbol{\theta})\, d\boldsymbol{\zeta} = \int \frac{\partial \log \ell}{\partial \boldsymbol{\theta}} \ell\, d\boldsymbol{\zeta},$$

where the integrals are over the domain for $\boldsymbol{Z}_n$. The conditioning on $\boldsymbol{\theta}$ here and elsewhere emphasizes that in some cases the value of $\boldsymbol{\theta}$ for which the information matrix is being computed will represent a random quantity (such as the estimate based on the n measurements).

With the above notational convention, the $p \times p$ Fisher information matrix $\boldsymbol{F}_n(\boldsymbol{\theta})$ for a differentiable log-likelihood function is given by

$$\boldsymbol{F}_n(\boldsymbol{\theta}) \equiv E\left(\frac{\partial \log \ell}{\partial \boldsymbol{\theta}} \cdot \frac{\partial \log \ell}{\partial \boldsymbol{\theta}^T}\bigg|\boldsymbol{\theta}\right)$$
$$= \int \frac{\partial \log \ell}{\partial \boldsymbol{\theta}} \cdot \frac{\partial \log \ell}{\partial \boldsymbol{\theta}^T} \ell\, d\boldsymbol{\zeta}. \qquad (13.12)$$

(Following the notational convention of Section A.1 in Appendix A, note that the argument in the expectation of (13.12) involves the product of $p \times 1$ and $1 \times p$ vectors.)

In the common case where the underlying data are assumed independent, the magnitude of $\boldsymbol{F}_n(\boldsymbol{\theta})$ grows at a rate proportional to n since $\log \ell$ represents a sum of n random terms. To see this, recall that when the z_k are independent, $\ell(\boldsymbol{\theta}|\boldsymbol{\zeta}) = p_z(\boldsymbol{\zeta}|\boldsymbol{\theta})$ is a product of the n density functions for each of the z_k. The log operator then converts this product into a sum of n terms.

The bounded quantity $\boldsymbol{F}_n(\boldsymbol{\theta})/n$ is employed as an average information matrix over all measurements. When the data depend on some inputs $\boldsymbol{x}_k$, then $\boldsymbol{F}_n(\boldsymbol{\theta})$ also depends on these (deterministic) inputs (i.e., $\boldsymbol{F}_n(\boldsymbol{\theta}) = \boldsymbol{F}_n(\boldsymbol{\theta}|\boldsymbol{x}_1, \boldsymbol{x}_2, \ldots, \boldsymbol{x}_n)$). For notational convenience—and since many applications depend on cases (such as i.i.d. data) where there are no varying inputs—we suppress this dependence and write $\boldsymbol{F}_n(\boldsymbol{\theta})$ for the information matrix. In experimental design, however, this dependence on the $\boldsymbol{x}_i$ is critical, as we see in Subsection 13.3.4 and Chapter 17.

Let us now derive some important relationships that are used in establishing properties of $\boldsymbol{F}_n(\boldsymbol{\theta})$. Assume that the likelihood function is regular in the sense that standard conditions such as in Wilks (1962, pp. 408–411 and 418–419) or Bickel and Doksum (1977, pp. 126–127) hold. One of these conditions is that the set $\{\zeta\text{: } \ell(\boldsymbol{\theta}|\zeta) > 0\}$ does not depend on $\boldsymbol{\theta}$. A fundamental implication of the regularity for the likelihood is that the necessary interchanges of differentiation and integration below are valid (see Theorem A.3 in Appendix A). Let us begin by noting that

$$\frac{\partial \log \ell}{\partial \boldsymbol{\theta}} = \frac{1}{\ell}\frac{\partial \ell}{\partial \boldsymbol{\theta}} \quad \text{if } \ell = \ell(\boldsymbol{\theta}|\zeta) \neq 0. \tag{13.13}$$

A key contributor to the important properties of $\boldsymbol{F}_n(\boldsymbol{\theta})$ is the fact the random vector $\partial \log \ell / \partial \boldsymbol{\theta}$ has mean zero. This follows from (13.13) and the above-mentioned interchange of derivative and integral:

$$E\left(\frac{\partial \log \ell}{\partial \boldsymbol{\theta}}\bigg|\,\boldsymbol{\theta}\right) = \int \frac{\partial \ell}{\partial \boldsymbol{\theta}}\, d\zeta = \frac{\partial}{\partial \boldsymbol{\theta}}\int \ell\, d\zeta = \frac{\partial}{\partial \boldsymbol{\theta}}\int p_Z(\zeta|\boldsymbol{\theta})\, d\zeta = \frac{\partial(\text{constant})}{\partial \boldsymbol{\theta}} = \boldsymbol{0}, \tag{13.14}$$

where the constant in the last derivative expression of (13.14) is unity. One interesting consequence of the mean-zero result in (13.14), as applied to (13.12), is that $\boldsymbol{F}_n(\boldsymbol{\theta})$ is the *covariance matrix* of the random vector $\partial \log \ell / \partial \boldsymbol{\theta}$. Except for relatively simple problems, however, the form in (13.12) is generally not useful in the practical calculation of the information matrix. Computing the expectation of the indicated vector product of multivariate nonlinear functions is usually a hopeless task.

Fortunately, there is an expression equivalent to (13.12) that is more amenable to computation. Suppose that $\log \ell$ is twice differentiable in $\boldsymbol{\theta}$. That is, the Hessian matrix

$$\boldsymbol{H}_{\log \ell}(\boldsymbol{\theta};\zeta) \equiv \frac{\partial^2 \log \ell(\boldsymbol{\theta}\,|\,\zeta)}{\partial \boldsymbol{\theta}\,\partial \boldsymbol{\theta}^T}$$

exists. (The subscript $\log \ell$ is included on $\boldsymbol{H}$ to distinguish this Hessian matrix from the generic Hessian of the loss function $L = L(\boldsymbol{\theta})$, denoted $\boldsymbol{H}(\boldsymbol{\theta}) = \partial^2 L / \partial \boldsymbol{\theta}\, \partial \boldsymbol{\theta}^T$, which appeared a number of times in Chapters 4–7.) Let us now derive the Hessian-based form. Differentiating (13.14), of course, yields zero. Hence,

$$\frac{\partial}{\partial \boldsymbol{\theta}} E\left(\frac{\partial \log \ell}{\partial \boldsymbol{\theta}^T}\middle| \boldsymbol{\theta}\right) = \frac{\partial}{\partial \boldsymbol{\theta}} \int \frac{\partial \log \ell}{\partial \boldsymbol{\theta}^T} \ell \, d\zeta$$

$$= \int \frac{\partial \log^2 \ell}{\partial \boldsymbol{\theta} \partial \boldsymbol{\theta}^T} \ell \, d\zeta + \int \frac{\partial \log \ell}{\partial \boldsymbol{\theta}} \cdot \frac{\partial \ell}{\partial \boldsymbol{\theta}^T} d\zeta$$

$$= \int \boldsymbol{H}_{\log \ell} \, \ell \, d\zeta + \boldsymbol{F}_n(\boldsymbol{\theta})$$

$$= \mathbf{0},$$

where the second line (once again) uses an interchange of derivative and integral and the third line follows by (13.13) applied to the definition of $\boldsymbol{F}_n(\boldsymbol{\theta})$ in (13.12). Solving for $\boldsymbol{F}_n(\boldsymbol{\theta})$, the information matrix is related to the Hessian matrix of $\log \ell$ through

$$\boldsymbol{F}_n(\boldsymbol{\theta}) = -E\left[\boldsymbol{H}_{\log \ell}(\boldsymbol{\theta}; \boldsymbol{Z}_n) \middle| \boldsymbol{\theta}\right]. \tag{13.15}$$

The form in (13.15) is usually more amenable to practical calculation than the product-based form in (13.12).

Let us present two examples of the computation of $\boldsymbol{F}_n(\boldsymbol{\theta})$. The first example pertains to a case where the scalar outcomes z_i are discrete. The second example is a signal-plus-noise problem.

Example 13.6—Information number with a Poisson distribution. Suppose that i.i.d. scalar measurements z_i are collected where the z_i can take on integer values $\{0, 1, 2, \ldots\}$. These data are modeled as coming from a Poisson distribution, a common assumption for the distribution of the number of events that occur in a given period of time (as heavily used, say, in queuing theory for transportation and communications systems). The joint probability mass function for the n measurements is

$$p_{\boldsymbol{Z}}(\boldsymbol{Z}_n \mid \theta) = \prod_{i=1}^{n} \frac{e^{-\theta} \theta^{z_i}}{z_i!}, \; \theta > 0,$$

where, for convenience, we are evaluating the mass function at the random measurements $\{z_i\}$ instead of the "dummy" vector $\boldsymbol{\zeta}$. It is known that $E(z_i) = \text{var}(z_i) = \theta$ (Bickel and Doksum, 1977, p. 456). From the mass function,

$$\log \ell(\theta \mid \boldsymbol{Z}_n) = -n\theta + \sum_{i=1}^{n} [z_i \log \theta - \log(z_i!)],$$

$$\frac{\partial \log \ell}{\partial \theta} = -n + \frac{1}{\theta} \sum_{i=1}^{n} z_i \, .$$

So, by (13.12) and (13.14),

$$F_n(\theta) = \operatorname{var}\left(\frac{\partial \log \ell}{\partial \theta}\right) = \operatorname{var}\left(\frac{1}{\theta}\sum_{i=1}^{n} z_i\right) = \frac{1}{\theta^2} n\theta = \frac{n}{\theta}.$$

The same result applies, of course, if the second-derivative-based form in (13.15) is used:

$$F_n(\theta) = -E\left(\frac{\partial^2 \log \ell}{\partial \theta^2}\right) = E\left(\frac{1}{\theta^2}\sum_{i=1}^{n} z_i\right) = \frac{1}{\theta^2} n\theta = \frac{n}{\theta}.$$

While there is little difference in the effort required to derive the product-based or second-derivative-based forms for the information number in this simple example, we will see in the next example a more typical case where the second-derivative (Hessian)-based form is easier to compute. ❑

Example 13.7—Information matrix in a signal-plus-noise problem. Suppose independent, nonidentically distributed scalar data z_i are collected with $z_i \sim N(\mu, \sigma^2 + q_i)$ where $\boldsymbol{\theta} = [\mu, \sigma^2]^T$ and the q_i are known parameters. Such a framework arises when each z_i represents an independent measurement of a signal (with distribution $N(\mu, \sigma^2)$) that is obscured by independent, nonidentically distributed noise (having distribution $N(0, q_i)$). The nonidentically distributed noise arises in settings where some measurements are of better quality than other measurements (due perhaps to the varying orientation of the measurement apparatus in a series of signal measurements). Examples of application for this statistical model include parameter estimation for the state-space model in Kalman filtering (Shumway et al., 1981; Sun, 1982) and dose response analysis (Hui and Berger, 1983). The probability density for the sequence of measurements is given by

$$p_Z(\boldsymbol{Z}_n \mid \boldsymbol{\theta}) = \left\{\prod_{i=1}^{n}\left[2\pi(\sigma^2 + q_i)\right]\right\}^{-1/2} \exp\left[-\frac{1}{2}\sum_{i=1}^{n}\frac{(z_i - \mu)^2}{\sigma^2 + q_i}\right],$$

where, as in Example 13.6, the random measurements $\{z_i\}$ are replacing the dummy variables (elements of $\boldsymbol{\zeta}$) in the density function.

The density above leads to the log-likelihood function

$$\log \ell(\boldsymbol{\theta} \mid \boldsymbol{Z}_n) = -\frac{n \log(2\pi)}{2} - \frac{1}{2}\sum_{i=1}^{n} \log(\sigma^2 + q_i) - \frac{1}{2}\sum_{i=1}^{n}\frac{(z_i - \mu)^2}{\sigma^2 + q_i}.$$

The derivatives with respect to the two elements of $\boldsymbol{\theta}$ are

$$\frac{\partial \log \ell}{\partial \mu} = \sum_{i=1}^{n} \frac{z_i - \mu}{\sigma^2 + q_i},$$

$$\frac{\partial \log \ell}{\partial (\sigma^2)} = -\frac{1}{2} \sum_{i=1}^{n} \frac{1}{\sigma^2 + q_i} + \frac{1}{2} \sum_{i=1}^{n} \frac{(z_i - \mu)^2}{(\sigma^2 + q_i)^2}.$$

The above lead to the following second derivatives appearing in the 2×2 Hessian matrix:

$$\frac{\partial^2 \log \ell}{\partial \mu^2} = -\sum_{i=1}^{n} \frac{1}{\sigma^2 + q_i},$$

$$\frac{\partial^2 \log \ell}{\partial \mu \partial (\sigma^2)} = \frac{\partial^2 \log \ell}{\partial (\sigma^2) \partial \mu} = -\sum_{i=1}^{n} \frac{z_i - \mu}{(\sigma^2 + q_i)^2},$$

$$\frac{\partial^2 \log \ell}{\partial (\sigma^2)^2} = \frac{1}{2} \sum_{i=1}^{n} \frac{1}{(\sigma^2 + q_i)^2} - \sum_{i=1}^{n} \frac{(z_i - \mu)^2}{(\sigma^2 + q_i)^3}.$$

Hence,

$$F_n(\theta) = -E\left[H_{\log \ell}(\theta; Z_n) \middle| \theta\right] = \begin{bmatrix} \sum_{i=1}^{n} (\sigma^2 + q_i)^{-1} & 0 \\ 0 & \frac{1}{2} \sum_{i=1}^{n} (\sigma^2 + q_i)^{-2} \end{bmatrix}. \quad (13.16)$$

Of course, the above result could have been obtained using the equivalent vector-product definition of the information matrix in (13.12). As suggested in the discussion following (13.14), however, the vector-product form is usually more cumbersome than the Hessian-based form used here. In this example, the vector-product definition requires some messy bookkeeping in computing the (2, 2) matrix component corresponding to the mean of the product of the $\partial \log \ell / \partial (\sigma^2)$ terms. (Equivalence for a simplified version of this problem is considered in Exercise 13.10.) ❑

Despite the advantages of the Hessian-based form, the analytical calculation of $F_n(\theta)$ is often difficult or impossible in many nonlinear problems. Obtaining the required first or second derivatives of $\log \ell$ may be a formidable task in some applications, and computing the required expectation of the generally nonlinear multivariate function is often impossible in problems of practical interest. The resampling approach described in Subsection 13.3.5 is oriented to such cases.

13.3.3 Two Key Properties of the Information Matrix: Connections to the Covariance Matrix of Parameter Estimates

The above discussion focused on the definition of the information matrix and the equivalence of two representations for the matrix (the gradient-product form and the Hessian-based form). We now discuss two of the most important analytical properties of the matrix. The primary rationale for $\boldsymbol{F}_n(\boldsymbol{\theta})$ as a measure of information about $\boldsymbol{\theta}$ within the data $\boldsymbol{Z}_n$ comes from its connection to the covariance matrix for the estimate of $\boldsymbol{\theta}$ constructed from $\boldsymbol{Z}_n$. As in Sections 13.1 and 13.2, let $\hat{\boldsymbol{\theta}}_n$ denote a generic estimate of $\boldsymbol{\theta}$ based on the n data points in $\boldsymbol{Z}_n$. That is, $\hat{\boldsymbol{\theta}}_n$ may be a batch estimate or a recursive estimate where the data pairs, $\boldsymbol{x}_i$, z_i, are processed one at a time.

The first of the key properties makes the connection to the covariance matrix of an estimate via an asymptotic normality result. For some common forms of estimates $\hat{\boldsymbol{\theta}}_n$ (e.g., maximum likelihood and Bayesian maximum a posteriori), it is known that, under modest conditions,

$$\sqrt{n}(\hat{\boldsymbol{\theta}}_n - \boldsymbol{\theta}^*) \xrightarrow{\text{dist}} N(\boldsymbol{0}, \bar{\boldsymbol{F}}^{-1}) \tag{13.17}$$

where $\xrightarrow{\text{dist}}$ denotes convergence in distribution (Appendix C, Subsection C.2.4) and

$$\bar{\boldsymbol{F}} \equiv \lim_{n\to\infty} \frac{\boldsymbol{F}_n(\boldsymbol{\theta}^*)}{n}$$

provided that the indicated limit exists and is invertible (e.g., Hoadley, 1971; Rao, 1973, pp. 415–417). Hence, in practice, for n reasonably large, $\hat{\boldsymbol{\theta}}_n$ is approximately $N(\boldsymbol{\theta}, \boldsymbol{F}_n(\boldsymbol{\theta})^{-1})$ distributed when $\boldsymbol{\theta}$ is chosen close to the unknown $\boldsymbol{\theta}^*$. Because $\hat{\boldsymbol{\theta}}_n$ is generally convergent to $\boldsymbol{\theta}^*$ in some stochastic sense under the conditions in which (13.17) holds, $\boldsymbol{\theta}$ is usually chosen to be $\hat{\boldsymbol{\theta}}_n$ for the evaluation of $\boldsymbol{F}_n(\boldsymbol{\theta})$.

Let us comment on the special case where $\hat{\boldsymbol{\theta}}_n$ is a recursive estimate, particularly the gradient-based stochastic approximation (SA) algorithm discussed in Chapter 5. Recall that SA includes popular algorithms such as least-mean-squares (LMS) (Sections 3.2, 3.3, and 5.1) and neural network backpropagation (Section 5.2) as special cases. From Section 5.1, the stochastic gradient interpretation of root-finding SA is

$$\hat{\boldsymbol{\theta}}_{k+1} = \hat{\boldsymbol{\theta}}_k - a_k \boldsymbol{Y}_k(\hat{\boldsymbol{\theta}}_k),\ k = 0, 1, \ldots, n-1,$$

where $\boldsymbol{Y}_k(\hat{\boldsymbol{\theta}}_k)$ is an unbiased measurement of the gradient $\boldsymbol{g}(\boldsymbol{\theta}) = \partial L/\partial\boldsymbol{\theta}$ evaluated at $\boldsymbol{\theta} = \hat{\boldsymbol{\theta}}_k$. Then, as shown, for example, in Kushner and Yang (1995) and Kushner and Yin (1997, pp. 332–333), (13.17) holds for the SA recursion above if one of the following hold: (i) The optimal matrix gain sequence $\boldsymbol{a}_k = \boldsymbol{H}(\boldsymbol{\theta}^*)^{-1}/(k+1)$ is used where $\boldsymbol{H}(\boldsymbol{\theta}) = \partial^2 L/\partial\boldsymbol{\theta}\partial\boldsymbol{\theta}^T$ (Section 4.4); (ii) an adaptive gain of the form $\boldsymbol{a}_k = \boldsymbol{H}_k^{-1}/(k+1)$ is used where $\boldsymbol{H}_k \to \boldsymbol{H}(\boldsymbol{\theta}^*)$ a.s. as $k \to \infty$ (Sections 4.5 and 7.8); or (iii) iterate averaging is used based on a scalar gain $a_k = a/(k+1)^{\alpha}$ for $\alpha < 1$ (Subsection 4.5.3). In practice, only options (ii) and (iii) are generally feasible since one will rarely know $\boldsymbol{H}(\boldsymbol{\theta}^*)$.

Returning to the general estimation context (not restricted to the recursive form in the preceding paragraph), let us present the second key property of the information matrix. This property applies in finite samples. If $\hat{\boldsymbol{\theta}}_n$ is *any unbiased* estimator of $\boldsymbol{\theta}$ (not just one for which (13.17) holds),

$$\text{cov}(\hat{\boldsymbol{\theta}}_n) \geq \boldsymbol{F}_n(\boldsymbol{\theta}^*)^{-1} \text{ for all } n, \tag{13.18}$$

where the inequality is in the matrix sense (Appendix A: Section A.2). There is also an expression analogous to (13.18) for biased estimators, but it is not especially useful in practice because it requires knowledge of the gradient of the bias with respect to $\boldsymbol{\theta}$ (Rao, 1973, pp. 323–327; Bickel and Doksum, 1977, pp. 127–128). Expression (13.18) is often referred to as the *Cramér–Rao inequality*, but priority of discovery is now given to French mathematician M. Frêchet (Bickel and Doksum, 1977, p. 142).

Expressions (13.17) and (13.18) point to the close connection between the inverse Fisher information matrix and the covariance matrix of the estimator. While (13.17) is an asymptotic result, (13.18) applies for all sample sizes subject to the unbiasedness requirement. It is also clear why the name *information matrix* is used for $\boldsymbol{F}_n(\boldsymbol{\theta})$: A larger $\boldsymbol{F}_n(\boldsymbol{\theta})$ (in the matrix sense) is associated with a smaller covariance matrix (i.e., more information), while a smaller $\boldsymbol{F}_n(\boldsymbol{\theta})$ is associated with a larger covariance matrix (i.e., less information). In particular, suppose that $\tilde{\boldsymbol{F}} \geq \tilde{\tilde{\boldsymbol{F}}} > \boldsymbol{0}$ for two information matrices $\tilde{\boldsymbol{F}}$ and $\tilde{\tilde{\boldsymbol{F}}}$ (e.g., perhaps $\tilde{\boldsymbol{F}}$ and $\tilde{\tilde{\boldsymbol{F}}}$ are associated with different sample sizes). Then by Appendix A (Section A.2), $\tilde{\boldsymbol{F}}^{-1} \leq \tilde{\tilde{\boldsymbol{F}}}^{-1}$, which by the covariance interpretation in (13.17) and (13.18) suggests less uncertainty in the parameter estimate for $\boldsymbol{\theta}$ under the scenario producing $\tilde{\boldsymbol{F}}$.

13.3.4 Selected Applications

This section is composed of several short discussions of applications of the information matrix. Some areas not discussed, where the information matrix has also been prominently used, include model selection via some of the techniques *other* than cross-validation (Section 13.2) and the determination of *noninformative* prior distributions for Bayesian analysis. Noninformative priors allow the use of the famous Bayes' rule in inference and estimation, but are intended to be "neutral" (i.e., noninformative) relative to the data.

Uncertainty Bounds and Hypothesis Tests

This application is based on the fundamental asymptotic normality result (13.17). The asymptotic normality provides an approximate distribution for $\hat{\boldsymbol{\theta}}_n$ given that n is sufficiently large. This distribution, in turn, can be used to characterize the uncertainty in $\hat{\boldsymbol{\theta}}_n$ and test hypotheses about specific values of $\boldsymbol{\theta}$.

For example, in a hypothesis testing context, suppose that we are testing a null hypothesis of $\boldsymbol{\theta} = \overline{\boldsymbol{\theta}}$ against the alternative hypothesis $\boldsymbol{\theta} \neq \overline{\boldsymbol{\theta}}$ for some nominal value $\overline{\boldsymbol{\theta}}$. In testing such a hypothesis, we treat $\overline{\boldsymbol{\theta}}$ as the true value of $\boldsymbol{\theta}$, and test whether the observed $\hat{\boldsymbol{\theta}}_n$ is in or out of some specified acceptance region. Using (13.17), the hypothesis test is based on the assumption that

$$\hat{\boldsymbol{\theta}}_n \overset{\text{approx.}}{\sim} N\left(\overline{\boldsymbol{\theta}}, \boldsymbol{F}_n(\overline{\boldsymbol{\theta}})^{-1}\right)$$

for the finite n of practical interest. Because of the difficulties in interpreting simultaneous intervals for the multiple components of $\boldsymbol{\theta}$, it is common to map the difference $\hat{\boldsymbol{\theta}}_n - \overline{\boldsymbol{\theta}}$ into a scalar via the following inner product form of test statistic:

$$(\hat{\boldsymbol{\theta}}_n - \overline{\boldsymbol{\theta}})^T \boldsymbol{F}_n(\overline{\boldsymbol{\theta}})(\hat{\boldsymbol{\theta}}_n - \overline{\boldsymbol{\theta}}).$$

Based on the approximate normality for $\hat{\boldsymbol{\theta}}_n$, the above test statistic is, under the null hypothesis, approximately chi-squared distributed with p degrees of freedom. (Note that the test statistic represents an approximate sum of p squared $N(0, 1)$ random variables; see Exercise 13.14.).

Hence, $\hat{\boldsymbol{\theta}}_n$ provides evidence to *reject* the null hypothesis that $\boldsymbol{\theta} = \overline{\boldsymbol{\theta}}$ if the P-value associated with the above test statistic is small. This P-value is small if $\hat{\boldsymbol{\theta}}_n - \overline{\boldsymbol{\theta}}$ is large in magnitude, where *large* here is relative to the approximate covariance matrix given by the inverse information matrix (so the normalization in the inner product of the test statistic is the inverse of the inverse information matrix—the matrix itself). The P-value is computed based on the above-mentioned chi-squared distribution.

Choice of Optimal Input Values

Recall that the input and output vectors may be modeled according to $z_k = \boldsymbol{h}(\boldsymbol{\theta}, \boldsymbol{x}_k) + \boldsymbol{v}_k$. A problem in experimental design for control and other applications is to choose the inputs $\boldsymbol{x}_1, \boldsymbol{x}_2, \ldots, \boldsymbol{x}_n$ so as to maximize the useful information in the data $z_1, z_2, \ldots, z_n$. Much can (and has!) been said about this problem when the underlying model is linear in $\boldsymbol{\theta}$. In the context of nonlinear models (such as neural networks), significantly less is known.

The field of optimal experimental design is devoted to the question of picking the best inputs. Obviously, for this question to make sense, we must formally define what is meant by *best*. Intuitively, the aim is to maximize the information in the data with respect to the estimation of $\boldsymbol{\theta}$ under the assumed model form. In this way, we use the resources devoted to data collection in the most efficient way. This aim, however, is still too vague. Fortunately, the information matrix provides a formal means of measuring information in the data relative to $\boldsymbol{\theta}$.

Recall that, for convenience, we suppressed the inputs in writing $\boldsymbol{F}_n(\boldsymbol{\theta})$ above. More completely, the information matrix is $\boldsymbol{F}_n(\boldsymbol{\theta}) = \boldsymbol{F}_n(\boldsymbol{\theta}|\boldsymbol{x}_1, \boldsymbol{x}_2,\ldots, \boldsymbol{x}_n)$. Hence, picking the n inputs to maximize $\boldsymbol{F}_n(\boldsymbol{\theta}|\boldsymbol{x}_1, \boldsymbol{x}_2,\ldots, \boldsymbol{x}_n)$ in some sense is a means of providing the data with the most information about $\boldsymbol{\theta}$. Because $\boldsymbol{F}_n(\boldsymbol{\theta}|\boldsymbol{x}_1, \boldsymbol{x}_2,\ldots, \boldsymbol{x}_n)$ is a *matrix*, there is no unique notion of maximum. For this reason, the most popular criterion in optimal design is the determinant of the information matrix, $\det[\boldsymbol{F}_n(\boldsymbol{\theta}|\boldsymbol{x}_1, \boldsymbol{x}_2,\ldots, \boldsymbol{x}_n)]$. The determinant is a standard measure of the size of a matrix and reduces the information matrix to a scalar criterion that can be uniquely maximized. In the field of experimental design, this is the famous *D-optimal criterion* (D for determinant).

There are other criteria based on the information matrix, but the D-optimal criterion is the most popular. Chapter 17 considers the subject of optimal experimental design—including D-optimality—in much greater detail.

Prediction Intervals for Neural Networks and Other Function Approximators

Related to the problem of putting uncertainty bounds on the estimate of $\boldsymbol{\theta}$ (as discussed above) is the problem of putting some type of probabilistic bounds on the accuracy of a prediction coming from nonlinear models such as neural networks (NNs). Suppose, for convenience, that z_k (and $h(\boldsymbol{\theta}, \boldsymbol{x}_k)$, of course) are scalar (the ideas here also apply in the multivariate case). For discussion purposes, suppose that $h(\boldsymbol{\theta}, \boldsymbol{x})$ represents a NN output based on weight parameters $\boldsymbol{\theta}$. The essential problem here is that one uses a set of fitting data to estimate the NN weights, and then one wants to use a trained NN to make predictions about the outcomes for new input values. However, since there is inevitable error in the weight estimates, the predictions will also be in error to some extent. As discussed in Chryssolouris et al. (1996) and Hwang and Ding (1997), the information matrix is valuable for finding the prediction bounds.

Suppose that the data vector $\boldsymbol{Z}_n$ has a known distributional form with unknown parameters. Commonly, $\boldsymbol{Z}_n$ is assumed multivariate normal, with the mean for each z_k being $h(\boldsymbol{\theta}, \boldsymbol{x}_k)$ (i.e., v_k has mean zero). When $h(\boldsymbol{\theta}, \boldsymbol{x})$ is differentiable in $\boldsymbol{\theta}$ for an $\boldsymbol{x}$ of interest and the asymptotic normality in (13.17) holds, then

$$\sqrt{n}\left[h(\hat{\boldsymbol{\theta}}_n, \boldsymbol{x}) - h(\boldsymbol{\theta}^*, \boldsymbol{x})\right] \xrightarrow{\text{dist}} N\left(0, h'(\boldsymbol{\theta}^*, \boldsymbol{x})^T \bar{\boldsymbol{F}}^{-1} h'(\boldsymbol{\theta}^*, \boldsymbol{x})\right),$$

where $h'(\cdot)$ denotes the gradient of $h(\cdot)$ with respect to $\boldsymbol{\theta}$. Hence, the NN prediction satisfies

$$h(\hat{\boldsymbol{\theta}}_n, \boldsymbol{x}) \overset{\text{approx.}}{\sim} N\left(h(\boldsymbol{\theta}, \boldsymbol{x}),\ h'(\boldsymbol{\theta}, \boldsymbol{x})^T \boldsymbol{F}_n(\boldsymbol{\theta})^{-1} h'(\boldsymbol{\theta}, \boldsymbol{x})\right) \tag{13.19}$$

for $\boldsymbol{\theta}$ close to $\boldsymbol{\theta}^*$ when n is reasonably large. In practice, $\boldsymbol{\theta}$ is often set to $\hat{\boldsymbol{\theta}}_n$ in the mean and variance expressions on the right-hand side of (13.19). Thus, the prediction $h(\hat{\boldsymbol{\theta}}_n, \boldsymbol{x})$ has an uncertainty given by a normal distribution with an approximate variance given by the variance in (13.19) evaluated at $\hat{\boldsymbol{\theta}}_n$. This uncertainty provides some sense of how much $h(\hat{\boldsymbol{\theta}}_n, \boldsymbol{x})$ is likely to differ from $E(z|\boldsymbol{x})$ when $h(\cdot)$ is such that $E(z|\boldsymbol{x}) \approx h(\boldsymbol{\theta}^*, \boldsymbol{x})$.

As opposed to the prediction error $E(z|\boldsymbol{x}) - h(\hat{\boldsymbol{\theta}}_n, \boldsymbol{x})$, the approach above can also be used to form an approximate distribution for the actual *observation* error, namely $z - h(\hat{\boldsymbol{\theta}}_n, \boldsymbol{x})$, where $z = z(\boldsymbol{x})$ is some future measurement. Suppose that the true process and model are both of the form $z_k = h(\boldsymbol{\theta}, \boldsymbol{x}_k) + v_k$, where v_k is i.i.d. $N(0, \sigma^2)$ with a reliable estimate of σ^2 available. Then, from (13.19), $z - h(\hat{\boldsymbol{\theta}}_n, \boldsymbol{x})$ is approximately normally distributed with mean zero and variance σ^2 plus the $O(1/n)$ variance appearing in (13.19).

13.3.5 Resampling-Based Calculation of the Information Matrix[6]

The calculation of $\boldsymbol{F}_n(\boldsymbol{\theta})$ is often difficult or impossible in many nonlinear problems. Obtaining the required first or second derivatives of the log-likelihood function may be a formidable task in some applications, and computing the required expectation of the generally nonlinear multivariate function is often impossible in problems of practical interest. To address this difficulty, this subsection outlines a computer resampling approach to estimating $\boldsymbol{F}_n(\boldsymbol{\theta})$. This approach is useful when analytical methods for computing $\boldsymbol{F}_n(\boldsymbol{\theta})$ are infeasible. The approach makes use of an idea introduced for optimization—the Hessian estimation for stochastic approximation introduced in Section 7.8—even though this problem is not directly one of optimization.

The basis for the technique below is to use computational horsepower in lieu of traditional detailed theoretical analysis to determine $\boldsymbol{F}_n(\boldsymbol{\theta})$. The method here is an example of a Monte Carlo-based method for producing an estimate. Such methods have become very popular as a means of handling problems that were formerly infeasible. Two other notable Monte Carlo techniques are the bootstrap method for determining statistical distributions of estimates (e.g., Efron and Tibshirani, 1986; Lunneborg, 2000) and the Markov chain Monte Carlo method for producing pseudorandom numbers and related quantities, considered in Chapter 16. Part of the appeal of the Monte Carlo method here for estimating $\boldsymbol{F}_n(\boldsymbol{\theta})$ is that it can be implemented with only evaluations of the log-likelihood (typically much easier to obtain than the customary gradient or second derivative information). Alternatively, if the gradient of the log-likelihood is available, that information can be used to enhance performance.

---

[6]This subsection may be omitted at first reading. It provides a Monte Carlo method for estimating the information matrix when it is not possible to obtain an analytical solution.

The essence of the method is to produce a large number of efficient "almost unbiased" estimates of the Hessian matrix of $\log \ell(\cdot)$ and then average the negative of these estimates to obtain an approximation to $\boldsymbol{F}_n(\boldsymbol{\theta})$. This approach is directly motivated by the definition of $\boldsymbol{F}_n(\boldsymbol{\theta})$ as the mean value of the negative Hessian matrix (eqn. (13.15)). To produce these estimates, we generate *pseudodata vectors* in a Monte Carlo manner analogous to the bootstrap method mentioned above. The pseudodata are generated according to a bootstrap resampling scheme treating the chosen $\boldsymbol{\theta}$ as "truth." The pseudodata are generated according to the probability model (13.11). So, for example, if the real data $\boldsymbol{Z}_n = [z_1^T, z_2^T, \ldots, z_n^T]^T$ are assumed to be jointly normally distributed, $N(\boldsymbol{\mu}(\boldsymbol{\theta}), \boldsymbol{\Sigma}(\boldsymbol{\theta}))$, then the pseudodata are generated by Monte Carlo according to a normal distribution based on a mean $\boldsymbol{\mu}$ and covariance matrix $\boldsymbol{\Sigma}$ evaluated at the chosen $\boldsymbol{\theta}$.

In particular, let the ith pseudodata vector be $\boldsymbol{Z}_{\text{pseudo}}(i)$, where $\dim(\boldsymbol{Z}_{\text{pseudo}}(i)) = \dim(\boldsymbol{Z}_n)$ (some multiple of n corresponding to the dimension of the z_i). This pseudodata vector represents a sample of size n (analogous to the real data $\boldsymbol{Z}_n$). The use of the notation $\boldsymbol{Z}_{\text{pseudo}}$ without the argument (i) is a generic reference to a pseudodata vector. The form of the distribution used to generate $\boldsymbol{Z}_{\text{pseudo}}$ is identical to the form represented by the likelihood function (13.11); the choice of $\boldsymbol{\theta}$ in the distribution depends on the application.

Given the aim to avoid the complex calculations usually needed to obtain second derivative information, the critical part of this conceptually simple scheme is the efficient Hessian estimation. Section 7.8 introduced an efficient scheme for estimating Hessian matrices in the context of optimization. Although there is no optimization here per se, we use the same formula for Hessian estimation. This formula is based on the simultaneous perturbation principle.

The approach given below can work with either $\log \ell(\boldsymbol{\theta}|\boldsymbol{Z}_{\text{pseudo}})$ values (alone) or with the gradient $\boldsymbol{g}(\boldsymbol{\theta}|\boldsymbol{Z}_{\text{pseudo}}) \equiv \partial \log \ell(\boldsymbol{\theta}|\boldsymbol{Z}_{\text{pseudo}})/\partial\boldsymbol{\theta}$ if that is available. The former usually corresponds to cases where the likelihood function and associated nonlinear process are so complex that no gradients are available. The latter allows for the fact that sometimes the gradient is available in even relatively complex problems and that such information should be used to enhance the estimation process if available. To highlight the fundamental commonality of approach, we let $\boldsymbol{G}(\boldsymbol{\theta}|\boldsymbol{Z}_{\text{pseudo}})$ represent either a gradient *approximation* (based on $\log \ell(\boldsymbol{\theta}|\boldsymbol{Z}_{\text{pseudo}})$ values) or the exact gradient $\boldsymbol{g}(\boldsymbol{\theta}|\boldsymbol{Z}_{\text{pseudo}})$. Because of its efficiency, the simultaneous perturbation gradient approximation is recommended in the case where only $\log \ell(\boldsymbol{\theta}|\boldsymbol{Z}_{\text{pseudo}})$ values are available (see Section 7.8).

We now present the Hessian estimate. Let $\hat{\boldsymbol{H}}_k$ denote the kth estimate of the Hessian $\boldsymbol{H}_{\log \ell}(\cdot)$. The estimate here differs slightly from that in Section 7.8 with the decaying c_k sequence because there is no iterating towards a solution in the optimization sense (hence no need for an explicit c_k sequence). The Hessian estimate is

$$\hat{\boldsymbol{H}}_k = \tfrac{1}{2}\left\{\frac{\delta \boldsymbol{G}_k}{2}\left[\Delta_{k1}^{-1}, \Delta_{k2}^{-1}, \ldots, \Delta_{kp}^{-1}\right] + \left(\frac{\delta \boldsymbol{G}_k}{2}\left[\Delta_{k1}^{-1}, \Delta_{k2}^{-1}, \ldots, \Delta_{kp}^{-1}\right]\right)^T\right\}, \quad (13.20)$$

where $\delta \boldsymbol{G}_k = \boldsymbol{G}(\boldsymbol{\theta} + \boldsymbol{\Delta}_k \mid \boldsymbol{Z}_{\text{pseudo}}) - \boldsymbol{G}(\boldsymbol{\theta} - \boldsymbol{\Delta}_k \mid \boldsymbol{Z}_{\text{pseudo}})$ and the vector $\boldsymbol{\Delta}_k \equiv [\Delta_{k1}, \Delta_{k2}, \ldots, \Delta_{kp}]^T$ is a mean-zero random perturbation such that the $\{\Delta_{kj}\}$ are "small" symmetrically distributed random variables that for all k, j are independent, identically distributed, uniformly bounded, and satisfy $E(|1/\Delta_{kj}|)$ $< \infty$ uniformly in k, j. The latter condition *excludes* such commonly used Monte Carlo distributions as uniform and Gaussian. Assume that $|\Delta_{kj}| \leq c$ for some small $c > 0$. Note that the user has full control over the choice of the Δ_{kj} distribution. A valid (and simple) choice is the Bernoulli $\pm c$ distribution (it is not known at this time if this is the "best" distribution to choose). To illustrate how the *individual* Hessian estimates may be quite poor, note that $\hat{\boldsymbol{H}}_k$ in (13.20) has (at most) rank two (and may not be positive semidefinite). This low quality, however, does not prevent the information matrix estimate from being accurate. The averaging process eliminates inadequacies in the Hessian estimates.

Given the form for the Hessian estimate in (13.20), it is now relatively straightforward to estimate $\boldsymbol{F}_n(\boldsymbol{\theta})$. From the results in Section 7.8, the Hessian estimate has an $O(c^2)$ bias. That is, $E(\hat{\boldsymbol{H}}_k \mid \boldsymbol{Z}_{\text{pseudo}}) = \boldsymbol{H}_{\log \ell}(\boldsymbol{\theta}; \boldsymbol{Z}_{\text{pseudo}}) + O(c^2)$. Hence, averaging many $\hat{\boldsymbol{H}}_k$ values yields an estimate of

$$E[\boldsymbol{H}_{\log \ell}(\boldsymbol{\theta}; \boldsymbol{Z}_{\text{pseudo}})] = -\boldsymbol{F}_n(\boldsymbol{\theta})$$

to within an $O(c^2)$ bias (the expectation in the left-hand side above is with respect to the pseudodata). The resulting estimate can be made as accurate as desired through reducing c and increasing the number of $\hat{\boldsymbol{H}}_k$ values being averaged. The averaging of the $\hat{\boldsymbol{H}}_k$ values may be done recursively (as in Section 7.8) to avoid having to store many matrices.

Let us now present a step-by-step summary of the above Monte Carlo resampling approach for estimating $\boldsymbol{F}_n(\boldsymbol{\theta})$. A numerical example is given in Spall (1998); this example is an extension of the signal-plus-noise problem in Example 13.7 to the setting where the data (z_i) are multivariate.

Monte Carlo Resampling Method for Estimating $\boldsymbol{F}_n(\boldsymbol{\theta})$

Step 0 **(Initialization)** Determine $\boldsymbol{\theta}$ and the number of pseudodata vectors that will be generated (N). Determine whether log-likelihood $\log \ell(\cdot)$ or gradient information $\boldsymbol{g}(\cdot)$ will be used to form the $\hat{\boldsymbol{H}}_k$ estimates. Pick the small number c in the Bernoulli $\pm c$ distribution used to generate the perturbations Δ_{kj}; $c = 0.0001$ has been effective in the author's experience (non-Bernoulli distributions may also be used subject to the conditions mentioned below (13.20)). Set $i = 1$.

Step 1 **(Generating pseudodata)** Based on $\boldsymbol{\theta}$ given in step 0, generate by Monte Carlo the ith pseudodata vector $\boldsymbol{Z}_{\text{pseudo}}(i)$.

Step 2 **(Hessian estimation)** With the pseudodata vector in step 1, use the Hessian estimation formula in (13.20) to determine one or more $\hat{\boldsymbol{H}}_k$. (Forming an average of more than one $\hat{\boldsymbol{H}}_k$ is useful at each pseudodata vector if the vectors are relatively expensive to generate.)

Step 3 **(Averaging Hessian estimates)** Repeat steps 1 and 2 a large number of times (N). Take the negative of the average of the N Hessian estimates produced in step 2; this is the estimate of $\boldsymbol{F}_n(\boldsymbol{\theta})$. It is usually convenient to use the standard recursive representation of a sample mean to avoid the storage of the N Hessian estimates.

13.4 CONCLUDING REMARKS

It was largely assumed in Chapters 1–12 that there was some given model, as reflected in the choice of the parameters to estimate ($\boldsymbol{\theta}$) and the associated loss or root-finding function. We considered a number of search and optimization methods for estimating $\boldsymbol{\theta}$. This chapter, on the other hand, has looked at some issues that are relevant *before* and *after* the estimation of $\boldsymbol{\theta}$.

In particular, this chapter discussed the bias–variance tradeoff in fitting a model to a set of data, the choice of the "best" structure for a model in light of the bias–variance tradeoff, and the Fisher information matrix as a summary description of the amount of information in a set of data relative to a given model form. Among many other areas for application, these topics arise in constructing simulation models or in building open-loop models for use with control systems. Some of these modeling issues are typically encountered prior to the application of a search and optimization method. In particular, one must select the form of the model before it is possible to estimate the parameters of the model!

The bias–variance tradeoff is enlightening largely for the conceptual understanding it provides. The tradeoff has little direct utility in assessing the adequacy of candidate models because it depends on information that is generally unavailable (such as the true probability distribution of the data). One consequence of the tradeoff is that no single type of model can be universally superior. Models that are very flexible (have many parameters) are valuable in certain complex problems but will hew too closely to the nuances of specific data sets in problems with simple input–output relationships. Application of a complex model to a fundamentally simple problem results in a model with a variance that is too high. On the other hand, a simple model without sufficient flexibility will suffer in problems with complex (highly nonlinear) input–output relationships. In particular, models that are too simple create excess bias, as the details of the input–output relationship are smoothed over.

As a practical means of capturing the essence of the bias–variance tradeoff, we discussed model selection methods, focusing on the popular cross-validation approach. Cross-validation is based on the commonsense idea of repeatedly partitioning an existing set of data into fitting and test subsets, choosing the best model as the one that provides the best average fit over all test subsets. This method has been used in a wide variety of applications, as it is simple to implement and imposes minimal assumptions on the problem structure. The method's generality, of course, means that in certain problems, other approaches that exploit specific problem structure may work better. Further, a downside of standard cross-validation (as discussed here) is that the complete set of data must be available in advance. Other approaches are better suited to (say) on-line applications where the model selection may be determined adaptively as data arrive.

Cross-validation is ideally suited to comparing candidate simulation models based on data from an actual physical process. Such models are often too complex for the use of methods that require simpler analytical structure (e.g., most of the other methods mentioned at the start of Section 13.2).

We concluded this chapter with a discussion of the Fisher information matrix. This matrix has wide application before *and* after the estimation of $\boldsymbol{\theta}$. Two important applications prior to estimating $\boldsymbol{\theta}$ are in model selection and experimental design. Although cross-validation as discussed in Section 13.2 is one model selection method that does not use the information matrix, some of the other methods that were mentioned *do* rely on the information matrix. In experimental design, the information matrix is used to pick the input ($\boldsymbol{x}$) values to provide the best estimate for $\boldsymbol{\theta}$. Chapter 17 considers the subject of experimental design further. One prominent application for the information matrix after $\boldsymbol{\theta}$ is estimated is approximate confidence region calculation.

While the information matrix is very useful in several applications, we also saw that it can be difficult or impossible to analytically compute in certain problems. To this end, we were able to draw on one of the methods introduced in the context of stochastic search and optimization: the simultaneous perturbation-based estimate of the Hessian matrix. By averaging a large number of these matrix estimates, we are able to generate an accurate estimate of the information matrix in problems for which no analytical solution is available. This is an example of a resampling-based estimate, where Monte Carlo sampling is used to estimate a quantity of physical interest. Another prominent example of estimation via Monte Carlo sampling is Markov chain Monte Carlo, as considered in Chapter 16.

EXERCISES

13.1 Prove the bias–variance decomposition in eqn. (13.2).

13.2 Provide a brief description or sketch of a case where the model provides a poor description of the process, but where a global MSE defined as $\overline{\text{variance}} + (\overline{\text{bias}})^2$ (instead of the form in (13.3)) is zero. This shows the importance of a proper definition of averaging with respect to the inputs $\boldsymbol{x}$.

13.3 **(a)** In the setting of Example 13.1, show that for any $0 < |\mu| < \infty$ and $0 < \sigma < \infty$, there exists an estimator of the form $r\bar{z}$, $0 < r < 1$, that produces a lower MSE than the unbiased estimator $\bar{z}$.

(b) For a fixed μ and σ satisfying the conditions of part (a), let r_n denote the value of r producing the estimator $r\bar{z}$ with the lowest MSE at a sample size n. Show that the difference $1 - r_n$ decays at rate $O(1/n)$ (i.e., $\bar{z}$ is nearly the lowest MSE estimator for large n).

13.4 Prove the bias and variance relationships in (13.4) and (13.5).

13.5 Consider the setting of Example 13.2 except that the scalar inputs $x_k = \chi_k = 2k$ (versus $x_k = \chi_k = k$), $k = 1, 2, \ldots, m$, $m = 10$. Produce a table analogous to Table 13.1 in Example 13.2 and comment on the differences in the results here and the results in Example 13.2.

13.6 With the same 30 data points (**`sinedata`** at the book's Web site) and cross-validation approach used in Example 13.4, compute the RMS error of a fifth-order polynomial. Show that this model is superior to the linear and tenth-order models, but slightly inferior to the cubic polynomial.

13.7 Consider the cross-validation/oboe reed problem in Example 13.5. Produce an additional column in Table 13.3 showing the RMS and MAD values for the five-input linear model

$$z = \theta_{\text{const}} + \theta_T T + \theta_A A + \theta_E E + \theta_V V + \theta_S S + v,$$

which is a model with all input variables except "first blow" (F). Among the three models being considered in this expanded Table 13.3, note that cross-validation continues to favor the full linear model (13.9).

13.8 There exists some useful special structure with leave-one-out cross-validation when applied to linear models. When the ith measurement $(\boldsymbol{h}_i, z_i)$ is excluded from the original n measurements, show that the ordinary least-squares estimate of $\boldsymbol{\theta}$ is:

$$\left[(\boldsymbol{H}_n^T\boldsymbol{H}_n)^{-1} + \frac{(\boldsymbol{H}_n^T\boldsymbol{H}_n)^{-1}\boldsymbol{h}_i\boldsymbol{h}_i^T(\boldsymbol{H}_n^T\boldsymbol{H}_n)^{-1}}{1-\boldsymbol{h}_i^T(\boldsymbol{H}_n^T\boldsymbol{H}_n)^{-1}\boldsymbol{h}_i}\right]\left(\boldsymbol{H}_n^T\boldsymbol{Z}_n - \boldsymbol{h}_i z_i\right).$$

(The value of this result will be apparent in the next exercise.) (Hint: Begin with the basic batch least-squares formula $\hat{\boldsymbol{\theta}}^{(n)} = (\boldsymbol{H}_n^T\boldsymbol{H}_n)^{-1}\boldsymbol{H}_n^T\boldsymbol{Z}_n$ and apply the matrix inversion lemma [matrix relationship (xxii) in Appendix A] as needed.)

13.9 Using the result in Exercise 13.8, show that the difference between the prediction and actual outcome for the ith measurement in leave-one-out cross-validation is $(\boldsymbol{h}_i^T\hat{\boldsymbol{\theta}}^{(n)} - z_i)/[1-\boldsymbol{h}_i^T(\boldsymbol{H}_n^T\boldsymbol{H}_n)^{-1}\boldsymbol{h}_i]$ for all i. (That is, the only regression estimate needed to form the overall MSE for a given model is the estimate $\hat{\boldsymbol{\theta}}^{(n)}$ from *all* the data.)

13.10 For the special case of the signal-plus-noise problem in Example 13.7 where $q_i = q$ for all i, compute $\boldsymbol{F}_n(\boldsymbol{\theta})$ using the vector-product form in (13.12).

Show the equality of this expression to the Hessian-based form in (13.16) when $q_i = q$. (Hint: If $X \sim N(0, a^2)$, then $E(X^3) = 0$ and $E(X^4) = 3a^4$.)

13.11 Suppose that i.i.d. binary data z_i satisfy $P(z_i = 0) = 1 - \theta$ and $P(z_i = 1) = \theta$. Using Fisher information, show that the sample mean $\bar{z}$ is a minimum-variance unbiased estimator for θ.

13.12 Consider data z_i that are i.i.d. according to the exponential distribution function $1 - e^{-\lambda z}$, where λ, $z > 0$. Using Fisher information, show that the sample mean $\bar{z}$ is a minimum-variance unbiased estimator of $\theta = 1/\lambda$. (Note that $E(z) = 1/\lambda$ and $\text{var}(z) = 1/\lambda^2$.)

13.13 Let $\phi = \phi(\theta)$ be a one-to-one, continuously differentiable transformation. Show that the information matrix for ϕ is $(\partial\theta^T/\partial\phi)\boldsymbol{F}_n(\theta)(\partial\theta/\partial\phi^T)$. (Note: The derivative $\partial\theta^T/\partial\phi$ is guaranteed to exist by the inverse function theorem; see, e.g., Apostol, 1974, p. 417.)

13.14 Provide justification for the statement in Subsection 13.3.4 that the test statistic $(\hat{\theta}_n - \bar{\theta})^T \boldsymbol{F}_n(\bar{\theta})(\hat{\theta}_n - \bar{\theta})$ represents an approximate sum of p squared $N(0, 1)$ random variables where $\bar{\theta}$ is some value being tested under the null hypothesis (i.e., the test statistic is approximately chi-squared distributed with p degrees of freedom).

13.15 For a statistical model of your choice, implement the Monte Carlo resampling scheme in Subsection 13.3.5 for determining the information matrix. For this model:

(a) Compute the true information matrix.

(b) Estimate the information matrix using only likelihood evaluations (no gradients).

(c) Estimate the information matrix using gradients of the log-likelihood function. Relative to truth as determined in part (a), comment on the relative accuracy of the results in parts (b) and (c).

CHAPTER 14

SIMULATION-BASED OPTIMIZATION I: REGENERATION, COMMON RANDOM NUMBERS, AND SELECTION METHODS

This and the next chapter consider the important case where Monte Carlo simulations are the primary source of input information during the search process. Search and optimization play a major role in two aspects of simulation analysis—building the simulation (parameter estimation) and simulation-based optimization (using the simulation as a proxy for the actual system in an optimization process). The focus here is on the latter problem of simulation-based optimization. In particular, we discuss some of the ways in which special properties associated with simulations—properties not generally available in other settings—can be used to enhance search and optimization. The resulting algorithms are generally special cases of some of the algorithms seen previously.

Section 14.1 provides general background. Section 14.2 introduces the important concept of regeneration, whereby a system periodically resets itself. Section 14.3 is a short summary of finite-difference-type methods for simulation-based optimization with emphasis on the standard finite-difference and simultaneous perturbation estimators that were encountered in general form in Chapters 6 and 7. Based on these gradient estimators for stochastic approximation, Section 14.4 discusses the important concept of common random numbers, one of the most useful properties of simulations and one that is not typically available in physical systems. Section 14.5 considers statistical methods for selecting the best option among several candidate solutions, including techniques based on common random numbers. This section connects to some of the discrete optimization methods of Chapter 12. Section 14.6 offers some concluding remarks, including a brief discussion of the limitations of simulation-based optimization.

14.1 BACKGROUND

14.1.1 Focus of Chapter and Roles of Search and Optimization in Simulation

Many real-world problems are too complex to be solved by analytical means and are therefore studied via computer simulation. Although simulation has traditionally been viewed as a method of last resort, recent advances in hardware, software, and user interfaces have made simulation more of a first-line means of attacking many problems. The focus in this and the next chapter will be on *stochastic simulations*, sometimes called *Monte Carlo simulations*. These rely on the internal generation of pseudorandom numbers (Appendix D) to represent the randomness in relevant events (although some of the methods apply as well in simulations that are purely deterministic). As we will see, stochastic simulations are intimately connected to the problem of *noisy* measurements for search and optimization as posed in Chapter 1 (Property A in Subsection 1.1.3) and seen repeatedly in other chapters of this book.

This and the next chapter discuss ways in which some of the unique properties associated with simulations can be used to enhance search and optimization. The specific algorithms here are generally special cases of some of the algorithms seen previously. This chapter focuses on methods where the simulation output is used *directly* in the optimization process. This is in contrast to stochastic gradient-based methods, which require extensive knowledge of the inner workings of the simulation. This chapter considers stochastic approximation algorithms with the finite-difference or simultaneous perturbation gradient estimation methods of Chapters 6 and 7 and statistical comparison methods similar to the material in Chapter 12. The exception to this focus on direct optimization is a brief discussion on gradient-based methods in the regeneration discussion (Section 14.2).

In contrast, in Chapter 15, knowledge of the internal structure of the simulation is used to construct sophisticated gradient estimators for use in optimization algorithms of the stochastic gradient form (Chapter 5) or the deterministic nonlinear programming form (via the sample path method). One special case of the results in Chapter 15 (the infinitesimal perturbation method) was summarized in Section 5.3 as an illustration of the stochastic gradient algorithm.

Some of the discussion in this and the next chapter pertains to queuing systems. A standard notation for such systems is *GI*/*G*/*c*, where *GI* stands for the distribution of the times between arrivals into the queuing network, *G* stands for the distribution of the service times to process an arrival, and *c* stands for the number of servers. The notation "*GI*" here means that the interarrivals have a general (*G*) distribution and are independent (*I*). An important special case is the *M*/*M*/1 queue, where *M* represents an exponential distribution for the interarrival and service time distributions and there is one server in the network. (The notation "*M*" comes from the fact that with an exponential distribution, the probability of a future occurrence—arrival or service completion, as appropriate—is independent of how long it has been since the last arrival or since the service began. This is the ***M***arkovian or ***M***emoryless property of the exponential distribution.)

A cautionary note: Serious, "industrial strength" uses of simulation in search and optimization (or other applications) generally involve many implementation details and an exploitation of structure unique to the particular application. It is obviously not possible in the two chapters here to discuss many of the clever strategies that are exploited in practice. For example, a large amount of the simulation-based literature is focused on *discrete-event* systems. Such systems are characterized by jumps in the state of a system as a result of the occurrence of an event (e.g., an arrival into a queuing network). A large number of modern systems in communications, manufacturing, transportation, healthcare, and other areas are discrete-event systems. Properties associated with discrete events can be heavily exploited in many serious simulation-based optimization problems to enhance the performance of the relevant algorithms.

The focus of this and the next chapter is on some important generic principles rather than the many specialized strategies associated with discrete-event or other systems. A user with a serious application is directed to the vast specialized literature in simulation, including the texts of Glasserman (1991a), Pflug (1996), Fu and Hu (1997), Banks (1998), Rubinstein and Melamed (1998), Cassandras and Lafortune (1999), and Law and Kelton (2000). A survey with accompanying discussion is given in Fu (2002).

Search and optimization algorithms play a critical role in at least two aspects of simulation analysis. One is in *building* the simulation model through the determination of the optimal values of the parameters internal to the simulation. The other role is in *using the simulation* to optimize the *real system* of interest after the simulation has been built (i.e., after the internal model parameters are determined). In the first role, $\boldsymbol{\theta}$ in the notation of this book represents the fundamental internal simulation model parameters that are assumed to hold across a variety of scenarios. In the second role for optimization involving simulations, $\boldsymbol{\theta}$ has a different meaning, representing parameters that may be varied in the real system and in running the simulation (i.e., $\boldsymbol{\theta}$ is associated with the simulation input).

For example, in a simulation of vehicle traffic in a network, the parameters in the first role (model building) might represent fundamental characteristics associated with the network structure (e.g., the mean arrival rates into the network and/or the topology of the network as reflected by the distances between intersections). These correspond to fixed quantities that the analyst cannot generally alter in the real system. Parameters in the second role (system optimization via simulation) might, for example, be the settings for the traffic signals in the real network. So, one might run the simulation for a variety of traffic signal settings, observing which setting produces the best performance for the *fixed* network characteristics represented by the simulation parameters found in the model-building phase. The critical process of building simulations is a large area that is closely tied to some of issues seen elsewhere in this book, especially parameter estimation for models (a.k.a. system identification) (Chapter 3; Sections 4.2, 5.1, and 5.2); the bias-variance tradeoff, model selection, and the information matrix (Chapter 13); and experimental design (Chapter 17).

Because many of the methods discussed elsewhere in this book are directly applicable to the process of building the simulation model, this and the next chapter focus on the *other* (second) role for simulation optimization, as discussed in the preceding paragraphs. So, $\boldsymbol{\theta}$ typically represents some system design or control parameters that are being determined with the aid of the simulation. It is assumed that the simulation has been properly constructed using the appropriate combination of prior knowledge of the process and estimation of the fundamental model parameters.

The use of simulations for optimizing real systems is sometimes controversial. The controversy usually centers on the question of whether the simulation is an adequate representation of the real process. We will not dwell on this debate. The answer to such a question centers on the specific goals of the analysis relative to the inherent strengths and weaknesses of the simulation software, the validity of the mathematical conditions embedded in the software (e.g., are the true system arrivals really Poisson distributed?), and the care with which the simulation is exercised by the users of the software. For motivation, we will refer several times to the following example in the sections below.

Example 14.1—Plant layout. Suppose that an industrial engineer is interested in finding the optimal placement of machines on a plant (factory) floor to maximize the product output in an assembly operation. Important constraints in this process include the types of material being assembled and the order of assembly, the kinds of labor skills that can reasonably be employed, workplace safety, product life, and so on. An obvious—but very costly—means for solving this problem is to simply try different machine placements in the real plant and conduct operational tests. This, clearly, is almost never feasible in practice.

Suppose, on the other hand, that there is a credible simulation of the plant output as a function of machine placement. This simulation must take into account the overall design of the plant, the movement of material, the role of people on the plant floor, and the other constraints mentioned above, all weighted by relative importance. (The parameters determined during the model-building phase would encompass these fixed aspects of the system, corresponding to the first role of search and optimization in simulation, as discussed above.) One can try different machine placements in the *simulation* of the system and evaluate the performance at each configuration. Here, $\boldsymbol{\theta}$ represents the vector of placement coordinates for the machines. This simulation-based approach may be used to provide the optimal solution if the simulation is a faithful replication of the actual system. A key issue here is the choice of intermediate test machine placements in the simulation for evaluation purposes en route to obtaining the best placement. Optimization methods provide the mechanism for determining these test machine placements via the rules that govern the iteration-by-iteration selection of $\boldsymbol{\theta}$ (machine location) values. ❑

14.1.2 Statement of the Optimization Problem

In simulation-based optimization, we are faced with solving a version of the problem introduced in Chapter 5, namely to minimize the loss function:

$$L(\boldsymbol{\theta}) = E[Q(\boldsymbol{\theta}, V)] \qquad (14.1)$$

over $\boldsymbol{\theta} \in \Theta$, where V represents the amalgamation of the (pseudo) random effects in the simulation (presumably manifesting themselves in a way indicative of the random effects in the actual system) and Q represents a *sample* realization of the loss function calculated from running the simulation. In general, the solution to $\min_{\boldsymbol{\theta}\in\Theta} L(\boldsymbol{\theta})$ is not unique. For ease of discussion, however, we will follow precedent and generally refer to *the* solution $\boldsymbol{\theta}^* = \arg\min_{\boldsymbol{\theta}\in\Theta} L(\boldsymbol{\theta})$. Most of the results here follow directly when there is a *set* of solutions Θ^*.

In the plant layout example above, Q might represent the (negative) output for one day given a particular set of random effects V (individual machine performance, labor productivity, material quality, arrival rates for needed parts from external suppliers or other areas of the plant, etc.), while $\boldsymbol{\theta}$ represents the vector of location coordinates for the machines being positioned. Hence one simulation run produces one value of Q. Note that the direct simulation output may or may not represent a value of Q (which is assumed to be a scalar number). For example, in the plant layout problem again, if the simulation output were a stream of hour-by-hour production rates broken down by different product categories, then a transformation to cumulative daily output weighted by the relative importance of different categories may be needed to obtain Q.

To focus the discussion in this and the next chapter, we suppose—unless noted otherwise—that V represents the collection of *direct* random processes in the simulation rather than the collection of underlying uniformly distributed $U(0, 1)$ (pseudo) random variables that may be generating the processes. Hence, for example, in an *M*/*M*/*c* queuing problem, V represents the Poisson-distributed arrivals and exponential-distributed service times versus the underlying uniform variables that are being used (say, with the inverse transform method—Appendix D) to generate the relevant Poisson and exponential distributed random variables. We denote the ith element of V by $\mathcal{V}_i$.

Although $Q = Q(\boldsymbol{\theta}, V)$ is shown in (14.1) as an explicit function of both $\boldsymbol{\theta}$ and V, there are many cases where the dependence on $\boldsymbol{\theta}$ is only implicit via the effect of $\boldsymbol{\theta}$ on the probability distribution used in generating V. For example, in a queuing system where $\mathcal{V}_i$ represents the processing time of customer i, there may be interest in minimizing the total processing time of all customers. Here $Q = Q(\boldsymbol{\theta}, V) = \sum_i \mathcal{V}_i$. The dependence on $\boldsymbol{\theta}$ is implicit via its effect on the probability distribution of the processing times. Suppose, in particular, that the distribution of V is governed by a probability density function $p_V(\cdot\,|\boldsymbol{\theta})$, where $\boldsymbol{\theta}$ represents parameters of the density that can be controlled by the system designer. The ultimate dependence on $\boldsymbol{\theta}$ (although not explicit in Q) appears in the loss function, as shown by writing $L(\boldsymbol{\theta}) = E(\sum_i \mathcal{V}_i)$ where the expectation is calculated based on $p_V(\cdot\,|\boldsymbol{\theta})$ over the domain for V.

The goal, therefore, is to find the $\boldsymbol{\theta}$ producing the best *mean* value of Q by using runs of the available simulation. Subject to the caveats above on the validity of the simulation as a representation of the real system, this is then tantamount to finding the best value of $\boldsymbol{\theta}$ for the *real* system. The remainder of this chapter discusses several approaches to simulation-based optimization.

14.2 REGENERATIVE SYSTEMS

14.2.1 Background and Definition of the Loss Function $L(\boldsymbol{\theta})$

Consider the common case where the system under study is a dynamic (time-varying) process. One of the significant issues in simulation-based optimization is the choice of how much time should be represented in the runs of the simulation. That is, in the process of simulation-based optimization, one chooses $\boldsymbol{\theta}$ values and runs simulations to produce $Q(\boldsymbol{\theta}, V)$ values (or some quantity related to $Q(\boldsymbol{\theta}, V)$ such as its derivative with respect to $\boldsymbol{\theta}$). In many cases, however, the amount of time to be *represented* by the simulation (not to be confused with the CPU or clock time involved in actually executing the simulation) may not be obvious. For example, in the plant layout problem above, an analyst might wonder if the loss function—and hence simulation-based loss *measurement*—should represent one hour of plant operations? One day? One year? The answer to this question clearly depends on the goals of the analysis. For example, if one is concerned with optimizing an inventory control system in a plant or retail establishment, it would be useful to run the simulation over *at least* one complete cycle of initial stocking, drawing the inventory down, and restocking.

A common means of answering the question above is by appealing to the concept of *regeneration.* A regenerative system has the property of returning periodically to some particular probabilistic state from which it effectively starts anew. That is, the system returns to conditions under which the probabilities of various future outcomes are the same as the a priori probabilities of achieving those outcomes at earlier incarnations of that state. An equivalent way of viewing regenerative processes is that at the particular regeneration times, the future behavior of the system is independent of the past behavior.

Some of the most common examples of regenerative processes are queuing networks where the system periodically "starts over." For example, daily traffic flow in a road network may be *probabilistically* approximately identical day-to-day over the course of the Monday–Friday workweek. Note that the regeneration times may be random for some systems and deterministic (i.e., knowable in advance) for others. The traffic flow example is a candidate for either of these settings. For example, key traffic indicators may reach a defined free-flow state immediately prior to the buildup of traffic congestion in the morning rush period at *randomly* varying times in the morning as a function of ambient traffic, weather conditions, accidents, and so on. In this setting the length of the regeneration period will vary about a mean of 24 hours. On the other hand, for some regions with very predictable conditions, the random effects may be negligible, indicating that the regeneration time associated with the defined pre-rush traffic conditions is known with nearly perfect certainty. Another common example of a regenerative process is an (s, S) inventory control system. When the inventory drops below s, an order is triggered to bring the inventory to level S, where the system regenerates (under the appropriate conditions on demand, backorders, etc.).

In general, the regeneration points are a set of deterministic or random times $\tau_0 < \tau_1 < \tau_2 < \ldots$ such that the system probabilistically restarts itself at each τ_i. Under modest conditions, the regenerative periods (or cycles) will be independent, identically distributed (i.i.d.) stochastic processes. If the regeneration points are also random, the periods have i.i.d. random lengths, $\tau_1 - \tau_0, \tau_2 - \tau_1, \ldots$. The example below illustrates this point.

Example 14.2—Queuing system. Consider a single-server queuing system $GI/G/1$ with a general (unknown) distribution for the interarrival and service times. Let W_k and S_k be the waiting and service time, respectively, for the kth customer into the system. Let $A_{k|k+1}$ be the interarrival time between the kth and $(k + 1)$st customer. Suppose that the system starts with zero wait ($W_0 = 0$) (so the first customer always encounters zero wait). Assume that $\{S_k\}$ and $\{A_{k|k+1}\}$ are i.i.d. sequences with means μ_S and μ_A, respectively. By this i.i.d. assumption, a customer arriving later in the process who encounters a zero wait ($W_k = 0$) will face the same probabilistic conditions as an earlier customer. If $\mu_S < \mu_A$, it is known that W_k will, in fact, return to the state of zero wait for an infinite subset of indices within $k \in [0, 1, 2,\ldots)$ (Rubinstein, 1981, p. 193). Hence this process is regenerative based on a return state of zero wait.

Figure 14.1 shows a plot of the waiting time encountered by 17 customers into a queuing system. We see that customers 1, 3, 4, 7, 11, and 16 face zero wait. Let us assume that the first regeneration period begins with the arrival of the first customer. So, the first regeneration period represents the time between the arrival of the first customer and the arrival of the next customer with zero wait (customer 3 here). This system is one where regeneration times are random.

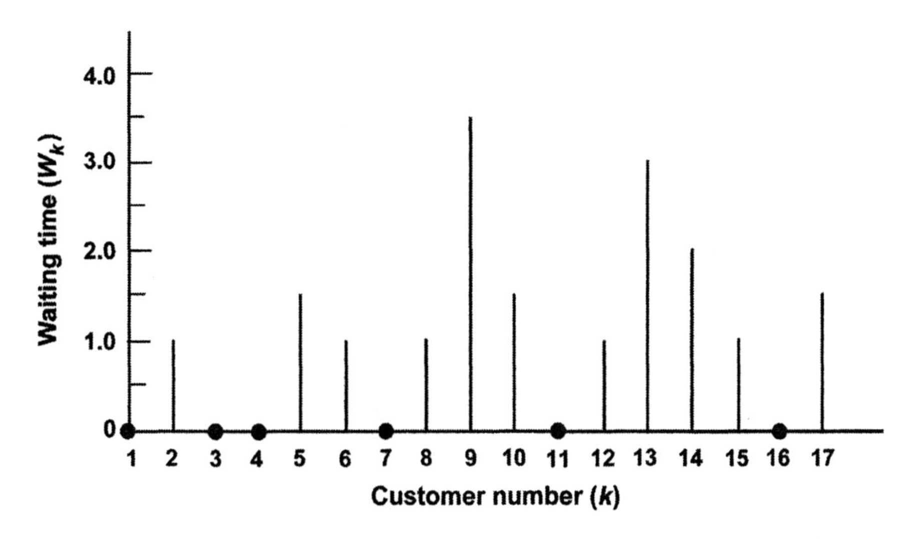

Figure 14.1. Zero and nonzero wait times depicting regeneration for $GI/G/1$ queue. Beginning with customer 1, the system regenerates at customers 3, 4, 7, 11, and 16.

Figure 14.1 depicts five complete regeneration periods, encompassing the system being turned on (say, time 0) and customers 1–15. The five periods correspond to the intervals with customers 1–2, 3, 4–6, 7–10, and 11–15. Also shown is the beginning of a sixth period with customer 16 (with the period having an unknown end). The length of the five random regeneration periods are $A_{1|2} + A_{2|3}$, $A_{3|4}$, $A_{4|5} + A_{5|6} + A_{6|7}$, $A_{7|8} + A_{8|9} + A_{9|10} + A_{10|11}$, and $A_{11|12} + A_{12|13} + A_{13|14} + A_{14|15} + A_{15|16}$. ❑

Simulation-based optimization when the system has a regenerative structure has been considered extensively (e.g., Rubinstein, 1981, Sects. 6.5 and 6.6; Fu, 1990; L'Ecuyer and Glynn, 1994; Fu and Hill, 1997; Rubinstein and Melamed, 1998, Sect. 3.7; Tang et al., 1999). Regenerative systems provide a logical basis for determining how long to run a simulation at each iteration of the optimization algorithm. In particular, one can consider the optimization of system performance using sets of data that completely cover one or more regeneration periods. This has the advantage of providing groups of data that are i.i.d., greatly easing analytical analysis of the system, such as required in some optimization approaches. This contrasts with nonregenerative simulations, where one may have no formal basis for determining how long to run a simulation to provide a representative sample of the process.

A typical loss function used in concert with regeneration is

$$L(\boldsymbol{\theta}) = \frac{E(C_{\boldsymbol{\theta}})}{E(\ell_{\boldsymbol{\theta}})}, \tag{14.2}$$

where $C_{\boldsymbol{\theta}}$ represents some random cost accrued during a regeneration period of length $\ell_{\boldsymbol{\theta}}$ with $E(\ell_{\boldsymbol{\theta}}) > 0$. In the $L(\boldsymbol{\theta}) = E[Q(\boldsymbol{\theta}, V)]$ notation in Section 14.1 and elsewhere, $Q(\boldsymbol{\theta}, V) = C_{\boldsymbol{\theta}}/E(\ell_{\boldsymbol{\theta}})$ (so V represents the random terms entering $C_{\boldsymbol{\theta}}$). Below, we sometimes add an index i to $C_{\boldsymbol{\theta}}$ and $\ell_{\boldsymbol{\theta}}$ to indicate the cost and length associated with a specific regeneration period—the ith. When no index is used, as in (14.2), the reference is to the generic random variable for an arbitrary regeneration period.

In (14.2), "length" $\ell_{\boldsymbol{\theta}}$ typically represents elapsed time or number of objects or individuals that are processed during a period. So, the loss function is the expected cost accrued during a regeneration period relative to the expected length of the period. Because the regeneration periods are i.i.d., there is no index on L associated with a particular period; all periods produce the same mean values in the numerator and denominator. While the example above is in terms of queuing problems, regeneration applies to arbitrary stochastic processes. Rubinstein and Melamed (1998, Examples 3.7.2 and 3.7.4), for instance, consider a problem in inventory control. The example below shows how the general structure above produces a specific loss function for use in an optimization problem.

Example 14.3—Loss function for a queuing system. Consider the *GI*/*G*/1 queuing problem of Example 14.2. Suppose that the i.i.d. service and interarrival times S_k and $A_{k|k+1}$ depend on $\boldsymbol{\theta}$. Then, we can define a cost $C_{\boldsymbol{\theta}}$ as the total time in the system for customers in a given regeneration period. The *k*th customer's time in the system is the sum of the wait time W_k and service time S_k. Further, let $\ell_{\boldsymbol{\theta}}$ denote the total number of customers processed during a period. Given the i.i.d. nature of the interarrivals and the service times, $E(C_{\boldsymbol{\theta}})$ and $E(\ell_{\boldsymbol{\theta}})$ do not depend on the specific period (i.e., the regeneration periods are i.i.d.). Hence, $L(\boldsymbol{\theta})$ according to (14.2) is an expression of the typical wait time for each customer. Following Fu (1990) and L'Ecuyer and Glynn (1994), this loss for an arbitrary regeneration period can be written as

$$L(\boldsymbol{\theta}) = \frac{E\left[\sum_{k\in\text{period}}(W_k + S_k)\right]}{E(\text{number of customers in period})},$$

where the period is arbitrary because the regeneration process is i.i.d. (the numerator sum is over a *random* number of customers in one period). A similar optimization criterion is considered in Rubinstein and Melamed (1998, Sect. 3.7), except that the problem is inverted from that above in that the one seeks to *maximize* the number of customers processed per unit time. ❑

14.2.2 Estimators of $L(\boldsymbol{\theta})$

Given the regenerative structure, there is a natural estimator of L at any value of $\boldsymbol{\theta}$. This estimator can form the basis for input into one of the search and optimization algorithms that depend only on noisy measurements of the loss function (such as the stochastic approximation [SA] algorithms, finite-difference SA [FDSA] and simultaneous perturbation SA [SPSA], introduced in Chapters 6 and 7). Let $C_{\boldsymbol{\theta},i}$ denote the total cost over the *i*th regeneration period, with $\ell_{\boldsymbol{\theta},i}$ the corresponding length. Then, directly motivated by the definition in (14.2), an estimator of $L = L(\boldsymbol{\theta})$ is

$$\hat{L}_N(\boldsymbol{\theta}) \equiv \frac{N^{-1}\sum_{i=1}^{N} C_{\boldsymbol{\theta},i}}{N^{-1}\sum_{i=1}^{N} \ell_{\boldsymbol{\theta},i}} = \frac{\sum_{i=1}^{N} C_{\boldsymbol{\theta},i}}{\sum_{i=1}^{N} \ell_{\boldsymbol{\theta},i}}, \tag{14.3}$$

where N is the number of regeneration periods. Suppose that the variances of $C_{\boldsymbol{\theta},i}$ and $\ell_{\boldsymbol{\theta},i}$ exist (i.e., are finite). Then, by the strong law of large numbers (Appendix C, including Exercise C.6),

$$\lim_{N\to\infty} \hat{L}_N(\boldsymbol{\theta}) = \lim_{N\to\infty} \frac{N^{-1}\sum_{i=1}^{N} C_{\boldsymbol{\theta},i}}{N^{-1}\sum_{i=1}^{N} \ell_{\boldsymbol{\theta},i}} = \frac{E(C_{\boldsymbol{\theta}})}{E(\ell_{\boldsymbol{\theta}})} = L(\boldsymbol{\theta}) \text{ a.s.}$$

However, $\hat{L}_N(\boldsymbol{\theta})$ is a *biased* estimator for any finite N because

$$E[\hat{L}_N(\boldsymbol{\theta})] = E\left(\frac{\sum_{i=1}^{N} C_{\boldsymbol{\theta},i}}{\sum_{i=1}^{N} \ell_{\boldsymbol{\theta},i}}\right) \neq \frac{E\left(N^{-1}\sum_{i=1}^{N} C_{\boldsymbol{\theta},i}\right)}{E\left(N^{-1}\sum_{i=1}^{N} \ell_{\boldsymbol{\theta},i}\right)} = L(\boldsymbol{\theta}).$$

This bias indicates that, in general, the use of $\hat{L}_N(\boldsymbol{\theta})$ as the input $y = y(\boldsymbol{\theta})$ in an optimization algorithm must be done with caution. Further, the bias generally depends on $\boldsymbol{\theta}$; so, in forming *differences* of loss measurements at two distinct values of $\boldsymbol{\theta}$, the biases generally do not cancel. The FDSA (Chapter 6) and SPSA (Chapter 7) algorithms have varying conditions on the bias of the input. If the input is unbiased, the relevant conditions for all of the algorithms are satisfied. Hence, it is desirable to seek an unbiased input. We analyze the bias in more detail after the following example illustrating (14.3).

Example 14.4—Estimate of $L(\boldsymbol{\theta})$ for a queuing system. Let us use the data depicted in Figure 14.1 to construct the estimator of $L(\boldsymbol{\theta})$ given in (14.3). As in Example 14.3, the cost $C_{\boldsymbol{\theta}}$ is the waiting time plus the service time. Further, let the length $\ell_{\boldsymbol{\theta}}$ be measured in terms of the number of customers processed in a regeneration period. For the $\boldsymbol{\theta}$ used in generating the data, the estimate of $L(\boldsymbol{\theta})$ based on (14.3) is

$$\frac{\sum_i \text{wait \& service times for customers in period } i}{\sum_i \text{number of customers in period } i} = \frac{\sum_i \sum_{k \in \text{period } i} (W_k + S_k)}{\text{total number of customers}},$$

where the sums are over all regeneration periods. Figure 14.1 shows five complete regeneration periods and the beginning of a sixth. Suppose that the cumulative service time over the five complete periods is 10.0. This leads to an estimate of

$$\frac{\overbrace{1.0 + 0 + 2.5 + 6.0 + 7.0}^{\text{wait times}} + \overbrace{10.0}^{\substack{\text{cum. service} \\ \text{time}}}}{2 + 1 + 3 + 4 + 5} = \frac{26.5}{15} = 1.77.$$

(Note that only 15 of the 17 depicted arrivals are used due to the lack of information on the sixth regeneration period, which is only known to contain *at least* two arrivals.) As discussed above, however, this estimate of $L(\boldsymbol{\theta})$ at the given $\boldsymbol{\theta}$ is biased. Modified conditions or estimators are considered below to mitigate this bias. ❑

Let us now discuss three cases where $\hat{L}_N(\boldsymbol{\theta})$ (or a closely related estimate) may be unbiased—or nearly unbiased—and may therefore be useful as an input to an optimization algorithm. The first case is where the number of regeneration periods N is large enough so that $\hat{L}_N(\boldsymbol{\theta})$ is reasonably close to $L(\boldsymbol{\theta})$ *and* where it is valid, correspondingly, to say that $E[\hat{L}_N(\boldsymbol{\theta})]$ is close to $L(\boldsymbol{\theta})$. (The latter result does not automatically follow since a.s. convergence does *not* generally imply convergence of the mean; see, e.g., Example C.4 in Appendix C.) The well-known dominated convergence theorem from analysis provides sufficient conditions for $E[\hat{L}_N(\boldsymbol{\theta})]$ to converge to $L(\boldsymbol{\theta})$ as $N \to \infty$; see Subsection C.2.3 in Appendix C. One specific form of the dominated convergence theorem, the bounded convergence theorem (Corollary 2 to Theorem C.1), holds if $\ell_{\boldsymbol{\theta}} \geq m > 0$ and $0 \leq C_{\boldsymbol{\theta}} \leq M < \infty$. Under these conditions on $\ell_{\boldsymbol{\theta}}$ and $C_{\boldsymbol{\theta}}$, it follows that $0 \leq \hat{L}_N(\boldsymbol{\theta}) \leq M/m < \infty$ for all N, satisfying the key condition of the bounded convergence theorem and guaranteeing that $E[\hat{L}_N(\boldsymbol{\theta})] \to L(\boldsymbol{\theta})$ as $N \to \infty$. Suppose that the dominated (or bounded) convergence theorem hold for all $\boldsymbol{\theta}$. Then, for each $\boldsymbol{\theta}$, $E[\hat{L}_N(\boldsymbol{\theta})]$ is close to $L(\boldsymbol{\theta})$ for N reasonably large. For many practical applications, the resulting small bias (i.e., $E[\hat{L}_N(\boldsymbol{\theta})] \approx L(\boldsymbol{\theta})$, not $E[\hat{L}_N(\boldsymbol{\theta})] = L(\boldsymbol{\theta})$) would be acceptable in an optimization algorithm that formally requires an unbiased loss estimate because it would not be expected to affect the iteration process in a significant way.

The second case associated with controlling the bias of $\hat{L}_N(\boldsymbol{\theta})$ is when $\ell_{\boldsymbol{\theta}}$ is not a random quantity. We mentioned above a couple of periodic systems where this might be true (e.g., daily operations in a factory, where $\ell_{\boldsymbol{\theta}}$ involves the processing of a repeatable number of parts or the elapsed time over which a certain number of operations are performed). Many other such systems exist. Here the denominator in (14.3) is replaced by the deterministic $\ell_{\boldsymbol{\theta}}$, leading to

$$\hat{L}_N(\boldsymbol{\theta}) \equiv \frac{N^{-1}\sum_{i=1}^{N} C_{\boldsymbol{\theta},i}}{\ell_{\boldsymbol{\theta}}}, \tag{14.4}$$

which is clearly unbiased as an estimator of $L(\boldsymbol{\theta})$. Note that, in general, $L(\boldsymbol{\theta})$ in this setting still depends on a *ratio* as in (14.2) (i.e., $\ell_{\boldsymbol{\theta}}$ continues to depend on $\boldsymbol{\theta}$). So it is not sufficient to simply minimize the numerator $E(C_{\boldsymbol{\theta}})$.

The third case associated with controlling the bias returns to the general definition of $L(\boldsymbol{\theta})$ involving random $C_{\boldsymbol{\theta}}$ and $\ell_{\boldsymbol{\theta}}$ values. The distinction here is that the estimate of $L(\boldsymbol{\theta})$ is based on the $C_{\boldsymbol{\theta}}$ values being statistically independent of the $\ell_{\boldsymbol{\theta}}$ values. This contrasts with the estimate in (14.3) where the $C_{\boldsymbol{\theta}}$ and $\ell_{\boldsymbol{\theta}}$ values come from the same regeneration periods and will likely be highly dependent. Suppose that there are two groups of mutually exclusive regeneration periods, one group generating the values for $C_{\boldsymbol{\theta}}$ and the other group generating the values for $\ell_{\boldsymbol{\theta}}$. Then, an obvious modification of the estimator in (14.3) is

$$\hat{L}_{N_1,N_2}(\boldsymbol{\theta}) \equiv \frac{N_1^{-1}\sum_{i\in\text{Group}(1)} C_{\boldsymbol{\theta},i}}{N_2^{-1}\sum_{i\in\text{Group}(2)} \ell_{\boldsymbol{\theta},i}}, \tag{14.5}$$

where Group(1) and Group(2) denote the indices for the two groups of mutually exclusive regeneration periods and N_1 and N_2 denote the number of regeneration periods in each of the two groups. So, the numerator sum represents the sample mean cost over the periods in the first group and the denominator sum is the sample mean length of all periods in the second group. For example, if there are a total of five regeneration periods at the indicated $\boldsymbol{\theta}$, the numerator sum might be based on data from $N_1 = 2$ of the periods while the denominator sum relies on data from the remaining $N_2 = 3$ periods. (This mutually exclusive idea is also used below in the context of gradient-based methods.)

In contrast to the estimate $\hat{L}_N(\boldsymbol{\theta})$, the independence of the regeneration periods allows the factoring into separate expectations associated with $C_{\boldsymbol{\theta}}$ and with $\ell_{\boldsymbol{\theta}}$:

$$
\begin{aligned}
E[\hat{L}_{N_1,N_2}(\boldsymbol{\theta})] &= E\left(N_1^{-1}\sum\nolimits_{i\in \text{Group}(1)} C_{\boldsymbol{\theta},i}\right)E\left(\frac{1}{N_2^{-1}\sum\nolimits_{i\in \text{Group}(2)} \ell_{\boldsymbol{\theta},i}}\right) \\
&= E(C_{\boldsymbol{\theta}})E\left(\frac{N_2}{\sum\nolimits_{i\in \text{Group}(2)} \ell_{\boldsymbol{\theta},i}}\right). \qquad (14.6)
\end{aligned}
$$

Nevertheless, like $\hat{L}_N(\boldsymbol{\theta})$, $\hat{L}_{N_1,N_2}(\boldsymbol{\theta})$ remains a biased estimator because the rightmost expectation in (14.6) is *not* equal to $1/E(\ell_{\boldsymbol{\theta}})$. Unlike $\hat{L}_N(\boldsymbol{\theta})$, however, $\hat{L}_{N_1,N_2}(\boldsymbol{\theta})$ in (14.5) is in a form for which the bias can be bounded using a powerful tool from probability theory—the *Kantorovich inequality* (e.g., Clausing, 1982).

The Kantorovich inequality relates $1/E(X)$ to $E(1/X)$ for a positive random variable X. In particular, if $0 < m \le X \le M < \infty$ (a.s.) for some positive constants $m \le M$, the Kantorovich inequality states that

$$
1 \le E(X)E\left(\frac{1}{X}\right) \le \frac{(m+M)^2}{4mM}. \qquad (14.7)
$$

(More precise Kantorovich-type bounds are available under stronger conditions on the distribution of the random variable X; see, e.g., Wilkins, 1955; Clausing, 1982; or Watson, 1987. The lower bound in (14.7) also follows by Jensen's inequality of probability theory [see, e.g., Laha and Rohatgi, 1979, p. 368].) Hence, $E(1/X) \ge 1/E(X)$ for such bounded positive random variables X. In our case, $\ell_{\boldsymbol{\theta}}$ plays the role of X. Let $0 < m(\boldsymbol{\theta}) \le \ell_{\boldsymbol{\theta}} \le M(\boldsymbol{\theta})$ (in general, it is possible that $C_{\boldsymbol{\theta}} < 0$, implying that, potentially, $L(\boldsymbol{\theta}) < 0$). Applying the Kantorovich inequality (14.7) to the estimate in (14.5) yields the following upper bound to the magnitude of the bias of $\hat{L}_{N_1,N_2}(\boldsymbol{\theta})$:

$$0 \leq |\text{bias}| = \left|E[\hat{L}_{N_1,N_2}(\boldsymbol{\theta})] - L(\boldsymbol{\theta})\right| \leq |L(\boldsymbol{\theta})| \frac{[M(\boldsymbol{\theta}) - m(\boldsymbol{\theta})]^2}{4m(\boldsymbol{\theta})M(\boldsymbol{\theta})} \tag{14.8}$$

(Exercise 14.1). Note that the absolute value signs in (14.8) are superfluous in the common case where $L(\boldsymbol{\theta}) \geq 0$.

Table 14.1 shows some values for the Kantorovich-based bound as a function of the ratio $M(\boldsymbol{\theta})/m(\boldsymbol{\theta})$. The biases shown are normalized by $L(\boldsymbol{\theta})$ (i.e., the right-hand side of (14.8) divided through by $|L(\boldsymbol{\theta})|$, leaving the unit-free ratio in terms of $M(\boldsymbol{\theta})$ and $m(\boldsymbol{\theta})$). Hence the values shown are the bias as a fraction of the loss value. The table shows, as expected, that as $M(\boldsymbol{\theta})$ and $m(\boldsymbol{\theta})$ get closer, the bias gets smaller. In the limit, where $M(\boldsymbol{\theta}) = m(\boldsymbol{\theta})$, we have the deterministic $\ell_{\boldsymbol{\theta}}$ case of (14.4), where the bias is zero. When the upper bound to $\ell_{\boldsymbol{\theta}}$ is twice the lower bound, we see that the bias can be no more that 12.5 percent of the loss value. When the upper bound to $\ell_{\boldsymbol{\theta}}$ is only 10 percent greater than the lower bound, the bias can be no more than 0.23 percent of the loss value. For most practical applications, such a small bias would be considered negligible, indicating that the estimate $\hat{L}_{N_1,N_2}(\boldsymbol{\theta})$ would be acceptable in an optimization algorithm that formally requires an unbiased loss estimate.

Applications such as the traffic and inventory examples mentioned at the beginning of this section, where *overall* patterns may exhibit little variation, would be the type of problems for which a fairly narrow range on $\ell_{\boldsymbol{\theta}}$ might be specified, leading to a negligible bias in the estimate of the loss function. Another class of problems would be queuing systems where the arrival times and service times are fairly predictable, leading to a relatively tight bound on $\ell_{\boldsymbol{\theta}}$.

Table 14.1. Kantorovich-based upper bound to the magnitude of the bias of loss estimate (14.5) as a fraction of $L(\boldsymbol{\theta})$.

$\dfrac{M(\boldsymbol{\theta})}{m(\boldsymbol{\theta})}$	Upper bound to $\left\lvert\dfrac{\text{bias}}{L(\boldsymbol{\theta})}\right\rvert$
1.10	0.0023
1.25	0.0125
1.50	0.0417
2.00	0.125

14.2.3 Estimates Related to the Gradient of $L(\boldsymbol{\theta})$

As introduced in the stochastic gradient discussion of Chapter 5, and as considered in detail for simulations in Chapter 15, powerful optimization methods are available if a direct unbiased estimate of the gradient $\boldsymbol{g}(\boldsymbol{\theta}) = \partial L/\partial\boldsymbol{\theta}$ can be determined. Such stochastic gradient methods represent an application of (Robbins–Monro) root-finding stochastic approximation. While the rest of this chapter emphasizes methods for optimization that do not rely on direct measurements of the gradient, this subsection shows how regeneration can be exploited to create a direct gradient estimator. This gradient-based structure motivates a nongradient (SPSA)-based estimator in Section 14.3.

Consider the basic loss estimate $\hat{L}_N(\boldsymbol{\theta})$ in (14.3). A direct estimate of the gradient is available by differentiating $\hat{L}_N(\boldsymbol{\theta})$ with respect to $\boldsymbol{\theta}$. Fu (1990) shows that this estimate is strongly convergent (i.e., a.s. convergent to $\boldsymbol{g}(\boldsymbol{\theta})$ as $N \to \infty$) if the following condition governing the interchange of differentiation and expectation (see Appendix A) holds:

$$E(\ell_{\boldsymbol{\theta}})E\left(\frac{\partial C_{\boldsymbol{\theta}}}{\partial\boldsymbol{\theta}}\right) - E(C_{\boldsymbol{\theta}})E\left(\frac{\partial \ell_{\boldsymbol{\theta}}}{\partial\boldsymbol{\theta}}\right) = E(\ell_{\boldsymbol{\theta}})\frac{\partial E(C_{\boldsymbol{\theta}})}{\partial\boldsymbol{\theta}} - E(C_{\boldsymbol{\theta}})\frac{\partial E(\ell_{\boldsymbol{\theta}})}{\partial\boldsymbol{\theta}}. \quad (14.9)$$

Analogous to $\hat{L}_N(\boldsymbol{\theta})$ itself, however, the estimate $\partial\hat{L}_N(\boldsymbol{\theta})/\partial\boldsymbol{\theta}$ is *biased* for finite N (i.e., $E[\partial\hat{L}_N(\boldsymbol{\theta})/\partial\boldsymbol{\theta}] \neq \boldsymbol{g}(\boldsymbol{\theta})$). So, in light of the regularity conditions for stochastic gradient methods requiring an unbiased gradient estimate (Section 5.1), this estimate is formally inappropriate.

By recasting the root-finding problem slightly, however, Fu (1990) and L'Ecuyer and Glynn (1994) introduce a means for coping with the bias in the estimate $\partial\hat{L}_N(\boldsymbol{\theta})/\partial\boldsymbol{\theta}$. In particular, under the assumptions stated for the definition of $L(\boldsymbol{\theta})$ in (14.2), finding the optimizing root $\boldsymbol{\theta}^*$ of $\boldsymbol{g}(\boldsymbol{\theta}) = \boldsymbol{0}$ (the root producing a global minimum of $L(\boldsymbol{\theta})$) is equivalent to finding the optimizing root of

$$\boldsymbol{g}_{\text{eqv}}(\boldsymbol{\theta}) \equiv E(\ell_{\boldsymbol{\theta}})\frac{\partial E(C_{\boldsymbol{\theta}})}{\partial\boldsymbol{\theta}} - E(C_{\boldsymbol{\theta}})\frac{\partial E(\ell_{\boldsymbol{\theta}})}{\partial\boldsymbol{\theta}} = \boldsymbol{0}, \quad (14.10)$$

where the notation $\boldsymbol{g}_{\text{eqv}}(\boldsymbol{\theta})$ is meant to evoke an expression *equivalent* to $\boldsymbol{g}(\boldsymbol{\theta})$ in the sense that both functions have the same zero (see Exercise 14.2). Now, we need to find an unbiased estimate of $\boldsymbol{g}_{\text{eqv}}(\boldsymbol{\theta})$. One way is to observe at least two regenerative periods at a specified $\boldsymbol{\theta}$, calculating $C_{\boldsymbol{\theta}}$ and $\partial C_{\boldsymbol{\theta}}/\partial\boldsymbol{\theta}$ from one or more regenerative periods and $\ell_{\boldsymbol{\theta}}$ and $\partial\ell_{\boldsymbol{\theta}}/\partial\boldsymbol{\theta}$ from the remaining regenerative periods (so $C_{\boldsymbol{\theta}}$ and $\partial C_{\boldsymbol{\theta}}/\partial\boldsymbol{\theta}$ are statistically independent of $\ell_{\boldsymbol{\theta}}$ and $\partial\ell_{\boldsymbol{\theta}}/\partial\boldsymbol{\theta}$). The same notion was used in the modified loss estimate in (14.5), where N_1 regenerative periods were used to estimate $C_{\boldsymbol{\theta}}$ while N_2 different periods were used to estimate $\ell_{\boldsymbol{\theta}}$.

Extending this idea to the gradient-based setting here, one can form the estimate of $\boldsymbol{g}_{\text{eqv}}(\boldsymbol{\theta})$:

$$\hat{\boldsymbol{g}}_{\text{eqv}}(\boldsymbol{\theta}) = \overline{\ell_{\boldsymbol{\theta}}}\,\overline{\frac{\partial C_{\boldsymbol{\theta}}}{\partial \boldsymbol{\theta}}} - \overline{C_{\boldsymbol{\theta}}}\,\overline{\frac{\partial \ell_{\boldsymbol{\theta}}}{\partial \boldsymbol{\theta}}}, \qquad (14.11)$$

where the overbars denote a sample mean of the indicated quantity over the relevant regeneration periods. For example, if there are five regeneration periods at the indicated $\boldsymbol{\theta}$, the sample means of $C_{\boldsymbol{\theta}}$ and $\partial C_{\boldsymbol{\theta}}/\partial \boldsymbol{\theta}$ might be based on data from $N_1 = 2$ of the periods while the sample means of $\ell_{\boldsymbol{\theta}}$ and $\partial \ell_{\boldsymbol{\theta}}/\partial \boldsymbol{\theta}$ rely on data from the remaining $N_2 = 3$ periods. With the minimum number of required regeneration periods—two—the sample means are the individual values of $C_{\boldsymbol{\theta}}$ and $\partial C_{\boldsymbol{\theta}}/\partial \boldsymbol{\theta}$ from one period and $\ell_{\boldsymbol{\theta}}$ and $\partial \ell_{\boldsymbol{\theta}}/\partial \boldsymbol{\theta}$ from the other period (i.e., equivalent to the right-hand side of (14.11) with no overbars). Fu (1990) establishes the following result:

Proposition 14.1. Suppose that the samples of $\ell_{\boldsymbol{\theta}}$ are independent of the samples of $\partial C_{\boldsymbol{\theta}}/\partial \boldsymbol{\theta}$ and that the samples of $\partial \ell_{\boldsymbol{\theta}}/\partial \boldsymbol{\theta}$ are independent of the samples of $C_{\boldsymbol{\theta}}$. Further, suppose that the interchange condition given in (14.9) holds. Then, $\hat{\boldsymbol{g}}_{\text{eqv}}(\boldsymbol{\theta})$ is an unbiased estimator of $\boldsymbol{g}_{\text{eqv}}(\boldsymbol{\theta})$.

Proof. We have

$$\begin{aligned} E[\hat{\boldsymbol{g}}_{\text{eqv}}(\boldsymbol{\theta})] &= E\left(\overline{\ell_{\boldsymbol{\theta}}}\,\overline{\frac{\partial C_{\boldsymbol{\theta}}}{\partial \boldsymbol{\theta}}} - \overline{C_{\boldsymbol{\theta}}}\,\overline{\frac{\partial \ell_{\boldsymbol{\theta}}}{\partial \boldsymbol{\theta}}}\right) \\ &= E(\overline{\ell_{\boldsymbol{\theta}}})E\left(\overline{\frac{\partial C_{\boldsymbol{\theta}}}{\partial \boldsymbol{\theta}}}\right) - E(\overline{C_{\boldsymbol{\theta}}})E\left(\overline{\frac{\partial \ell_{\boldsymbol{\theta}}}{\partial \boldsymbol{\theta}}}\right) \\ &= E(\ell_{\boldsymbol{\theta}})E\left(\frac{\partial C_{\boldsymbol{\theta}}}{\partial \boldsymbol{\theta}}\right) - E(C_{\boldsymbol{\theta}})E\left(\frac{\partial \ell_{\boldsymbol{\theta}}}{\partial \boldsymbol{\theta}}\right), \end{aligned}$$

where the second equality follows by the assumed independence. By (14.9) and the definition of $\boldsymbol{g}_{\text{eqv}}(\boldsymbol{\theta})$ in (14.10), the proof is complete. ❑

Given the estimate in (14.11), one can now use the root-finding version of the stochastic approximation algorithm (Chapter 4) to attempt to find the optimizing root $\boldsymbol{\theta}^*$ of $\boldsymbol{g}_{\text{eqv}}(\boldsymbol{\theta}) = \boldsymbol{0}$. In particular, the required input to the root-finding algorithm at a given value of $\boldsymbol{\theta}$ is $\hat{\boldsymbol{g}}_{\text{eqv}}(\boldsymbol{\theta})$. So, in the root-finding notation, in going from the estimate of $\boldsymbol{\theta}$ at iteration k to the estimate at iteration $k+1$, the input for unconstrained or constrained SA is $\boldsymbol{Y}_k(\hat{\boldsymbol{\theta}}_k) = \hat{\boldsymbol{g}}_{\text{eqv}}(\hat{\boldsymbol{\theta}}_k)$. From Proposition 14.1, an unbiasedness condition of root-finding SA (Section 4.3) is satisfied: $E[\boldsymbol{Y}_k(\hat{\boldsymbol{\theta}}_k) \mid \hat{\boldsymbol{\theta}}_0, \hat{\boldsymbol{\theta}}_1, \ldots, \hat{\boldsymbol{\theta}}_k] = \boldsymbol{g}_{\text{eqv}}(\hat{\boldsymbol{\theta}}_k)$ provided that the required regeneration periods at $\hat{\boldsymbol{\theta}}_k$ satisfy the independence conditions of Proposition 14.1 *and* are also independent from the periods generated at earlier values, $\hat{\boldsymbol{\theta}}_0$, $\hat{\boldsymbol{\theta}}_1,\ldots,\hat{\boldsymbol{\theta}}_{k-1}$. Hence, the Robbins–Monro root-finding SA algorithm can be used when at least two regeneration periods are observed in the simulation at each $\hat{\boldsymbol{\theta}}_k$ (in order to produce the estimate $\boldsymbol{Y}_k(\hat{\boldsymbol{\theta}}_k) = \hat{\boldsymbol{g}}_{\text{eqv}}(\hat{\boldsymbol{\theta}}_k)$) *and* when the regeneration periods are produced from an independent random number seed at each $\hat{\boldsymbol{\theta}}_k$. This unbiasedness holds at each $\hat{\boldsymbol{\theta}}_k$ for any *finite* number of regeneration periods satisfying Proposition 14.1.

For a *GI*/*G*/1 queue, Fu (1990) and L'Ecuyer and Glynn (1994) employ $\boldsymbol{Y}_k(\hat{\boldsymbol{\theta}}_k) = \hat{\boldsymbol{g}}_{\text{eqv}}(\hat{\boldsymbol{\theta}}_k)$ with root-finding SA to find the solution to $\boldsymbol{g}_{\text{eqv}}(\boldsymbol{\theta}) = \boldsymbol{0}$, where

$$\boldsymbol{g}_{\text{eqv}}(\boldsymbol{\theta}) = \boldsymbol{0} \Leftrightarrow \frac{\partial L(\boldsymbol{\theta})}{\partial \boldsymbol{\theta}} = \frac{\partial(\text{wait time/customer})}{\partial \boldsymbol{\theta}} = \boldsymbol{0}$$

($\Leftrightarrow$ is read as "if and only if"). However, the requirement for obtaining the gradients $\partial C_{\boldsymbol{\theta}}/\partial \boldsymbol{\theta}$ and $\partial \ell_{\boldsymbol{\theta}}/\partial \boldsymbol{\theta}$ needed to compute $\hat{\boldsymbol{g}}_{\text{eqv}}(\hat{\boldsymbol{\theta}}_k)$ is stringent, as it involves detailed knowledge of the formulas and probability distributions embedded in the simulation. Thus, Section 14.3 discusses an application of this general setting when it is not possible to directly calculate the gradients needed in $\hat{\boldsymbol{g}}_{\text{eqv}}(\hat{\boldsymbol{\theta}}_k)$. This approach uses the simultaneous perturbation gradient estimate (Chapter 7) to approximate the gradients.

14.3 OPTIMIZATION WITH FINITE-DIFFERENCE AND SIMULTANEOUS PERTURBATION GRADIENT ESTIMATORS

Chapters 4–7 discussed general methods and theory for stochastic approximation. SA methods appear to be the leading formal techniques for optimization in a simulation context. Among the reasons for the popularity are the algorithmic similarity to standard deterministic methods (i.e., the similarity in structure between SA and steepest descent-type algorithms) and the strong theoretical basis for accommodating noisy loss and/or gradient measurements. As discussed previously, the latter concern is inherent in the Monte Carlo simulation context. Central to the SA approach is some type of estimator for the gradient $\partial L/\partial \boldsymbol{\theta}$.

Chapters 6 and 7 discussed two gradient estimators that do not rely on direct gradient information (FDSA and SPSA). Both have been applied in simulation-based optimization, with the basic FDSA being the oldest and best-known approach. (Based on the root-finding SA algorithm, Chapter 15 focuses on direct stochastic gradient estimators of $\partial L/\partial \boldsymbol{\theta}$ in the simulation context, as discussed in general form in Chapter 5.) In general, FDSA and SPSA apply to arbitrary simulations and loss functions $L(\boldsymbol{\theta}) = E[Q(\boldsymbol{\theta}, V)]$, with Q representing the simulation output, as defined in Section 14.1. In particular, there is no need to assume a special form for the simulation, such as regenerative, to employ these gradient-free methods (although the end of this section discusses some unique structure that *is* available if the simulation is for a regenerative system).

In fact, these gradient-free algorithms are the most popular of the iterative optimization algorithms for simulation: "...this estimator [FDSA], or some variant of it, remains the method of choice for the majority of practitioners" (Fu and Hu, 1997, p. 5). This may be surprising given the much greater effort that has gone into the more sophisticated model-based approaches to be discussed in Chapter 15 (such as infinitesimal perturbation analysis and the likelihood ratio method), but this is a reflection of the (usually) much greater ease of implementation. The finite-difference and simultaneous perturbation gradient estimators are based on values of $Q(\boldsymbol{\theta}, V)$ (representing the noisy measurements of $L(\boldsymbol{\theta})$) at appropriately perturbed values of $\boldsymbol{\theta}$, as discussed in Chapters 6 and 7. There is no need to know the inner workings of the simulation, as required in the model-based approaches, which need gradients such as $\partial Q/\partial\boldsymbol{\theta}$. Rather, one simply specifies $\boldsymbol{\theta}$ and runs the simulation—precisely what the simulation was designed to do!—to produce outputs $Q(\boldsymbol{\theta}, V)$.

The FDSA and SPSA gradient estimates are as follows. The ith component of the two-sided finite-difference gradient estimate is

$$\hat{g}_{ki}(\hat{\boldsymbol{\theta}}_k) = \frac{Q(\hat{\boldsymbol{\theta}}_k + c_k\boldsymbol{\xi}_i, V_{ki}^{(+)}) - Q(\hat{\boldsymbol{\theta}}_k - c_k\boldsymbol{\xi}_i, V_{ki}^{(-)})}{2c_k},$$

where the $V_{ki}^{(\pm)}$ terms represent the Monte Carlo random sequences in the given simulations, $\boldsymbol{\xi}_i$ is the unit vector with a 1 in the ith place and a 0 elsewhere, and $c_k \to 0$ as $k \to \infty$. (Note that $V_{ki}^{(\pm)}$, in general, depends on the value of $\boldsymbol{\theta}$, either directly or indirectly through the dependence of the distribution generating the V values on $\boldsymbol{\theta}$.)

Analogously, the ith component of the simultaneous perturbation gradient estimate is

$$\hat{g}_{ki}(\hat{\boldsymbol{\theta}}_k) = \frac{Q(\hat{\boldsymbol{\theta}}_k + c_k\boldsymbol{\Delta}_k, V_k^{(+)}) - Q(\hat{\boldsymbol{\theta}}_k - c_k\boldsymbol{\Delta}_k, V_k^{(-)})}{2c_k\Delta_{ki}},$$

where the Monte Carlo generated p-dimensional random perturbation vector, $\boldsymbol{\Delta}_k = [\Delta_{k1}, \Delta_{k2},\ldots, \Delta_{kp}]^T$, has a user-specified distribution satisfying conditions discussed in Sections 7.2 and 7.3 (a valid distribution is Bernoulli ± 1 for the individual components Δ_{ki}). Unlike the FDSA gradient estimate, there is only a single subscript index on V, reflecting the simultaneous perturbation of all components in $\boldsymbol{\theta}$.

Given a gradient estimate above, the procedure for determining an estimate of the optimal parameters $\boldsymbol{\theta}^*$ follows exactly as in Chapters 6 and 7. Namely, an initial guess $\hat{\boldsymbol{\theta}}_0$ for the parameters is made and then simulations are run to form the gradient estimates, $\hat{\boldsymbol{g}}_0(\hat{\boldsymbol{\theta}}_0)$, $\hat{\boldsymbol{g}}_1(\hat{\boldsymbol{\theta}}_1),\ldots$, needed in updating the $\boldsymbol{\theta}$ estimate according to $\hat{\boldsymbol{\theta}}_{k+1} = \hat{\boldsymbol{\theta}}_k - a_k\hat{\boldsymbol{g}}_k(\hat{\boldsymbol{\theta}}_k)$ (or $\hat{\boldsymbol{\theta}}_{k+1} = \Psi_\Theta[\hat{\boldsymbol{\theta}}_k - a_k\hat{\boldsymbol{g}}_k(\hat{\boldsymbol{\theta}}_k)]$ when there is a constraint mapping Ψ_Θ, as discussed in Sections 4.1 and 7.7).

With the exception of certain properties associated with common random numbers discussed below, the relative performance of FDSA and SPSA in the simulation context will be as discussed in the generic description of Chapters 6 and 7. The relative performance follows the familiar fundamental relationship on the cost of the gradient approximations: Given that one simulation run produces one value of Q, then $2p$ runs of the simulation are required to produce the standard two-sided finite-difference gradient estimate while only two runs are needed for the simultaneous perturbation estimate. An example simulation-based implementation illustrating this relative cost is in Fu and Hill (1997).

There is also a convenient connection of the FDSA and SPSA approaches to the regenerative structure of Section 14.2. Suppose that the system under study is regenerative and that the loss function in (14.2), $L(\boldsymbol{\theta}) = E(C_{\boldsymbol{\theta}})/E(\ell_{\boldsymbol{\theta}})$, will be used. It was shown in Subsection 14.2.3 that finding the optimum $\boldsymbol{\theta}^*$ via $\boldsymbol{g}(\boldsymbol{\theta}) = \partial L/\partial\boldsymbol{\theta} = \boldsymbol{0}$ is equivalent to finding the optimizing root of $\boldsymbol{g}_{\text{eqv}}(\boldsymbol{\theta}) = \boldsymbol{0}$, where $\boldsymbol{g}_{\text{eqv}}(\boldsymbol{\theta})$ is defined in (14.10). An unbiased estimate $\hat{\boldsymbol{g}}_{\text{eqv}}(\boldsymbol{\theta})$ was described in (14.11), which is based on sample means of the gradients $\partial C_{\boldsymbol{\theta}}/\partial\boldsymbol{\theta}$ and $\partial\ell_{\boldsymbol{\theta}}/\partial\boldsymbol{\theta}$ across several simulations (with the simulations for $\partial C_{\boldsymbol{\theta}}/\partial\boldsymbol{\theta}$ being independent of the simulations for $\partial\ell_{\boldsymbol{\theta}}/\partial\boldsymbol{\theta}$). This unbiased estimate is appropriate for a root-finding SA algorithm, as discussed in Subsection 14.2.3.

Suppose that the simulation is sufficiently complex so that it is not possible to compute the gradients, $\partial C_{\boldsymbol{\theta}}/\partial\boldsymbol{\theta}$ and $\partial\ell_{\boldsymbol{\theta}}/\partial\boldsymbol{\theta}$. Finite differences (L'Ecuyer and Glynn, 1994) or simultaneous perturbation (Fu and Hill, 1997) could then be used to estimate the gradients. The SPSA-based approximations to the two required gradients lead to the following gradient-free estimate of the ith component of $\boldsymbol{g}_{\text{eqv}}(\boldsymbol{\theta})$ at the kth iteration:

$$\begin{aligned}\hat{g}_{\text{eqv};ki}(\hat{\boldsymbol{\theta}}_k) &= \frac{C_k^{(+)} - C_k^{(-)}}{2c_k\Delta_{ki}}\ell_k^{(-)} - C_k^{(-)}\frac{\ell_k^{(+)} - \ell_k^{(-)}}{2c_k\Delta_{ki}} \\ &= \frac{C_k^{(+)}\ell_k^{(-)} - C_k^{(-)}\ell_k^{(+)}}{2c_k\Delta_{ki}},\end{aligned}$$

where $C_k^{(\pm)} = C_{\boldsymbol{\theta}}(\hat{\boldsymbol{\theta}}_k \pm c_k\boldsymbol{\Delta}_k, \boldsymbol{V}_k^{(\pm)})$ and $\ell_k^{(\pm)} = \ell_{\boldsymbol{\theta}}(\hat{\boldsymbol{\theta}}_k \pm c_k\boldsymbol{\Delta}_k, \boldsymbol{V}_k^{(\pm)})$. (In the spirit of SPSA, the same two pairs of values, $C_k^{(+)}$, $\ell_k^{(+)}$ and $C_k^{(-)}$, $\ell_k^{(-)}$, representing two runs of the simulation, are used for all p components of

$\hat{g}_{\text{eqv};k}(\hat{\boldsymbol{\theta}}_k)$.) Fu and Hill (1997) discuss the conditions under which the above gradient estimate produces a convergent SPSA algorithm.

14.4 COMMON RANDOM NUMBERS

14.4.1 Introduction

Common random numbers (CRNs) provide one of the most useful tools in simulation-based optimization, being widely applied because of their effectiveness, ease of implementation, and intuitive appeal. CRNs provide—among other benefits—a means for reducing the variability of the gradient estimate that serves as the input to the SA iteration process. This, in turn, is able to reduce the error in the $\boldsymbol{\theta}$ estimation process. CRNs have been studied extensively (see, e.g., Glasserman and Yao, 1992; L'Ecuyer and Perron, 1994; Rubinstein and Melamed, 1998, pp. 90–94; and Kleinman et al., 1999). We restrict our attention here to their use in the gradient-free estimators for SA algorithms discussed in Section 14.3 (FDSA and SPSA). Most of the explicit technical points are illustrated with SPSA, but it should be clear how the points will translate to the FDSA approach. Some of the references above include discussion on applications of CRNs beyond their use in such SA gradient estimators.

The essence of CRNs in a simulation-based optimization context is to use the same random numbers in the simulations being differenced in forming the gradient approximations. That is, the two function measurements in the numerator of one component of the gradient approximation each use a sequence of Monte Carlo-generated random variables, V, to generate the various random events in the process. CRNs are based on using the *same* values of V in both simulations. In this way, the change information being provided by the differences in the numerators of the gradient approximations is the result of the changes in the $\boldsymbol{\theta}$ elements and not the changes due to different random variables in the Monte Carlo simulations. Consequently, CRNs can reduce the variability of the gradient approximations and thereby improve the performance of the optimization process.

One of the key conditions for CRNs to be effective is that the random numbers being generated in the simulations forming the difference should be used in comparable ways. For example, the random numbers generating the mth arrival into a queue in one simulation should correspond to the random numbers generating the mth arrival in the other simulation. This comparability manifests itself in requirements of monotonicity and synchronization, as described below. The following example is intended to help clarify the role of CRNs in practice.

Example 14.5—Plant layout revisited. Let us return to the plant layout problem (Example 14.1) for a *conceptual* example of CRNs. Suppose that we are using the SPSA algorithm to maximize the mean number of assembly operations completed on a daily basis and that each calculation of $-Q$ represents one day's worth of completed operations (the negative sign reflects our standard use of Q as a loss measurement for a *minimization* problem). Suppose further that the only randomness in the process is the arrival times of parts into the assembly operations area. In forming the gradient estimate at one iteration, we need to collect the two noisy loss measurements $Q(\hat{\boldsymbol{\theta}}_k \pm c_k \boldsymbol{\Delta}_k, \boldsymbol{V}_k^{(\pm)})$ where $\boldsymbol{V}_k^{(+)}$ and $\boldsymbol{V}_k^{(-)}$ represent the sequence of arrival times throughout the day of the parts under the machine locations $\hat{\boldsymbol{\theta}}_k + c_k \boldsymbol{\Delta}_k$ and $\hat{\boldsymbol{\theta}}_k - c_k \boldsymbol{\Delta}_k$.

If the calculations were to be done based on *real* experiments on the factory floor, then it is likely that the parts would arrive at different times in the two experiments needed (equivalent to $\boldsymbol{V}_k^{(+)} \neq \boldsymbol{V}_k^{(-)}$ in the simulation) since the experiments would be conducted on different days. Hence the difference in $Q(\hat{\boldsymbol{\theta}}_k + c_k \boldsymbol{\Delta}_k, \boldsymbol{V}_k^{(+)})$ and $Q(\hat{\boldsymbol{\theta}}_k - c_k \boldsymbol{\Delta}_k, \boldsymbol{V}_k^{(-)})$ would be due to differences in $\boldsymbol{\theta}$ *and* the difference in random effects (i.e., $\boldsymbol{V}_k^{(+)} \neq \boldsymbol{V}_k^{(-)}$). However, in simulation-based optimization, we may be able to impose the requirement that $\boldsymbol{V}_k = \boldsymbol{V}_k^{(+)} = \boldsymbol{V}_k^{(-)}$ through picking the same random number seed in generating the uniform random variables that are transformed to generate $\boldsymbol{V}_k^{(+)}$ and $\boldsymbol{V}_k^{(-)}$. Note that CRNs do not apply across iterations (i.e., $\boldsymbol{V}_k \neq \boldsymbol{V}_j$ for $k \neq j$) in order to ensure sufficient variability in the optimization process.

In determining the appropriateness of CRNs, consider two cases: (i) Parts can enter the assembly area at any time, irrespective of the state of the current assembly operations, and (ii) parts are not allowed to enter the assembly area until queues at certain critical machines are sufficiently cleared (i.e., there is limited buffer capacity). In (i), the random effects $\boldsymbol{V}$ are independent of the value of $\boldsymbol{\theta}$ (the machine locations). In (ii), the value of $\boldsymbol{\theta}$ *does* affect the arrival times because some arrival times will be delayed if the overall set of machine locations is causing excessive queue lengths at certain critical machines.

Under case (i) above, it is automatic to be able to impose CRNs by simply having the arrival times be the same in the two simulations. Under (ii), CRNs are feasible *provided* that the two sets of machine locations represented by $\hat{\boldsymbol{\theta}}_k \pm c_k \boldsymbol{\Delta}_k$ are sufficiently close so that parts can be accepted into the assembly area equally in both simulations (so the parts enter the assembly area at the same times for each pair of simulated days). Under CRNs in either case (i) or (ii), the variability of the gradient estimate will be isolated to the difference in production output due to the change in machine location $\boldsymbol{\theta}$. Of course, to ensure that an optimal solution to $\boldsymbol{\theta}$ is found, the CRNs $\boldsymbol{V}_k$ should change at every iteration to represent the range of reasonable part arrivals in practice. ❑

Slightly more formally, we can see the positive effect of CRNs by recalling the well-known formula governing the variance of the difference of two random variables X and Y:

$$\text{var}(X - Y) = \text{var}(X) + \text{var}(Y) - 2\,\text{cov}(X, Y), \tag{14.12}$$

where we may view X and Y as representing the two numerator terms in one of the gradient approximations. It is apparent that *increasing* the covariance (i.e., increasing the correlation) between X and Y *decreases* the overall variance. In the simulation-based context, we can usually increase the covariance by using the same random number stream in the underlying simulations generating the two parts of the differences. CRNs are implemented very simply by using the same random number seed(s) in the two simulations (although this alone may not guarantee that the benefits of CRNs apply; see the discussion on partial CRNs in Subsection 14.4.3).[1]

14.4.2 Theory and Examples for Common Random Numbers

The benefit of the CRN method comes from a reduction in the variance of the difference $y_k^{(+)} - y_k^{(-)} = Q(\hat{\boldsymbol{\theta}}_k + c_k\boldsymbol{\Delta}_k, \boldsymbol{V}_k^{(+)}) - Q(\hat{\boldsymbol{\theta}}_k - c_k\boldsymbol{\Delta}_k, \boldsymbol{V}_k^{(-)})$ appearing in the numerator of the SPSA gradient approximation, where the $\boldsymbol{V}_k^{(\pm)}$ terms represent the vector of Monte Carlo-generated random variables used in the two measurements. Intuitively, this variance reduction is achieved by CRNs when $Q(\hat{\boldsymbol{\theta}}_k + c_k\boldsymbol{\Delta}_k, \boldsymbol{V})$ and $Q(\hat{\boldsymbol{\theta}}_k - c_k\boldsymbol{\Delta}_k, \boldsymbol{V})$ tend to move in the same way for a change in $\boldsymbol{V}$. This is formalized in Proposition 14.2. Suppose (as is typical) that the $\boldsymbol{V}_k^{(\pm)}$ are generated as streams of random variables that have been formed from a transformation of a stream of $U(0, 1)$ random variables, say $\boldsymbol{U}_k^{(\pm)}$. Further, suppose that the $\boldsymbol{V}_k^{(\pm)}$ can be generated by the inverse transform method (Appendix D), with each component of $\boldsymbol{V}_k^{(\pm)}$ generated by a corresponding single component of $\boldsymbol{U}_k^{(\pm)}$. That is, write $V_{ki}^{(+)} = F_{ki}^{-1}(U_{ki}^{(+)})$ and $V_{ki}^{(-)} = G_{ki}^{-1}(U_{ki}^{(-)})$, where F_{ki} and G_{ki} represent the cumulative distribution functions for scalar components $V_{ki}^{(+)}$ and $V_{ki}^{(-)}$ (not necessarily the same distribution functions in general) and $U_{ki}^{(\pm)}$ represent the corresponding scalar components of $\boldsymbol{U}_k^{(\pm)}$.

Note that there are two sources of variability in $y_k^{(+)} - y_k^{(-)}$: that due to $\pm c_k\boldsymbol{\Delta}_k$ and that due to $\boldsymbol{V}_k^{(\pm)}$. The former source is desirable since it reveals information about the shape of the loss function L. The latter source, on the other hand, is undesirable since the $\boldsymbol{V}_k^{(\pm)}$ are the direct causes of the measurement noise (i.e., if there were no $\boldsymbol{V}_k^{(\pm)}$ contributions, then $y_k^{(+)} - y_k^{(-)} = L(\hat{\boldsymbol{\theta}}_k + c_k\boldsymbol{\Delta}_k) - L(\hat{\boldsymbol{\theta}}_k - c_k\boldsymbol{\Delta}_k)$).

With CRNs, one may reduce the conditional variance $\text{var}[(y_k^{(+)} - y_k^{(-)}) \,|\, \hat{\boldsymbol{\theta}}_k, \boldsymbol{\Delta}_k]$ by picking $\boldsymbol{V}_k^{(+)}$ to be as close as possible to $\boldsymbol{V}_k^{(-)}$ through generating $\boldsymbol{U}_k^{(+)}$ and $\boldsymbol{U}_k^{(-)}$ in a manner so that they are as close as

---

[1] A complementary form to CRNs, *antithetic random numbers*, uses the summation form, var($X+Y$), to reduce the variability for sums of simulation runs. Rubinstein and Melamed (1998, pp. 90–94) show how antithetic random variables can be used to reduce variability when considering averages of simulation outputs. Chapter 16 also mentions an application in averaging multiple runs of the Markov chain Monte Carlo algorithms.

possible to each other. (Note that the conditional variance expressions here depend only on the current value of the $\boldsymbol{\theta}$ estimate, $\hat{\boldsymbol{\theta}}_k$, not on previous $\boldsymbol{\theta}$ estimates. This contrasts with the conditioning in the general theoretical results of Chapter 7, which depend on the current and the previous $\boldsymbol{\theta}$ estimates. In the simulation-based context of the current chapter, it is unnecessary to condition on the previous $\boldsymbol{\theta}$ estimates because the V values generated at iteration k are assumed independent of the V values generated at previous iterations.) Reducing this conditional variance then translates into a corresponding reduction in the conditional and unconditional variance of the elements of the gradient estimate $\hat{g}_k(\hat{\boldsymbol{\theta}}_k)$. In the ideal case, $\boldsymbol{U}_k^{(+)} = \boldsymbol{U}_k^{(-)}$, implying that $\boldsymbol{V}_k^{(+)} = \boldsymbol{V}_k^{(-)}$ (not always feasible in practice, as discussed below).

Typically, the practical implementation of CRNs is achieved by simply using the same random number seed in generating the sequences $\boldsymbol{U}_k^{(+)}$ and $\boldsymbol{U}_k^{(-)}$ (i.e., using the same seed achieves the same result as storing the sequence of random numbers for reuse, but it is usually easier to implement). Minimizing the variance of $y_k^{(+)} - y_k^{(-)}$ contributes to reducing the overall variability of the gradient estimate $\hat{g}_k(\hat{\boldsymbol{\theta}}_k)$, which, in turn, reduces the variability of $\hat{\boldsymbol{\theta}}_k$ (the ultimate interest!). The following result from Rubinstein et al. (1985) provides useful sufficient conditions for determining when CRNs can minimize the variance in the gradient estimate. Note that CRNs can *reduce* the variance (as opposed to *minimize* the variance) under conditions much weaker than those below (see, e.g., Exercise 14.6(b)). All statements involving conditional expectations (e.g., $\text{var}[\hat{g}_{ki}(\hat{\boldsymbol{\theta}}_k)\,|\,\hat{\boldsymbol{\theta}}_k]$) apply a.s.

Proposition 14.2. Suppose that $\boldsymbol{V}_k^{(+)}$ and $\boldsymbol{V}_k^{(-)}$ are each composed of a common number of independent components and that dependence is allowed only between like components of $\boldsymbol{V}_k^{(+)}$ and $\boldsymbol{V}_k^{(-)}$ (i.e., $V_{ki}^{(+)}$ and $V_{ki}^{(-)}$ may be dependent for all i, but $V_{ki}^{(+)}$ and $V_{kj}^{(-)}$ must be independent for $i \neq j$). Assume that $\boldsymbol{V}_k^{(+)}$ and $\boldsymbol{V}_k^{(-)}$ can be generated by the inverse transform method applied to each component of $\boldsymbol{U}_k^{(+)}$ and $\boldsymbol{U}_k^{(-)}$ (Appendix D). For almost all values of $\boldsymbol{\Delta}_k$, suppose that $Q(\hat{\boldsymbol{\theta}}_k + c_k\boldsymbol{\Delta}_k, \boldsymbol{V}_k^{(+)})$ and $Q(\hat{\boldsymbol{\theta}}_k - c_k\boldsymbol{\Delta}_k, \boldsymbol{V}_k^{(-)})$ are monotonic (nonincreasing or nondecreasing) in the same direction in each component of $\boldsymbol{V}_k^{(+)}$ and $\boldsymbol{V}_k^{(-)}$ within the range of allowable values for the components. (The direction may be nonincreasing for some components, nondecreasing for others.) Then $\text{var}[\hat{g}_{ki}(\hat{\boldsymbol{\theta}}_k)\,|\,\hat{\boldsymbol{\theta}}_k]$ is minimized at each i when $\boldsymbol{U}_k^{(+)} = \boldsymbol{U}_k^{(-)} = \boldsymbol{U}_k$.

Sketch of Proof. Taking $y_k^{(\pm)} = Q(\hat{\boldsymbol{\theta}}_k \pm c_k\boldsymbol{\Delta}_k, \boldsymbol{V}_k^{(\pm)})$, Rubinstein et al. (1985) (see also Rubinstein and Melamed, 1998, pp. 90–92) show that $\text{var}[(y_k^{(+)} - y_k^{(-)})\,|\,\hat{\boldsymbol{\theta}}_k, \boldsymbol{\Delta}_k]$ is minimized when $\boldsymbol{U}_k^{(+)} = \boldsymbol{U}_k^{(-)} = \boldsymbol{U}_k$ under the stated assumptions. The result then follows by working with the definition $\hat{g}_{ki}(\hat{\boldsymbol{\theta}}_k) = (y_k^{(+)} - y_k^{(-)})/(2c_k\Delta_{ki})$ and converting to a conditioning solely on $\hat{\boldsymbol{\theta}}_k$ (see Exercise 14.5). ❑

The result in Proposition 14.2 is intuitively plausible since $Q(\hat{\boldsymbol{\theta}}_k + c_k\boldsymbol{\Delta}_k, V)$ and $Q(\hat{\boldsymbol{\theta}}_k - c_k\boldsymbol{\Delta}_k, V)$ tend to move together as V changes. The monotonicity assumption states that *both* $Q(\hat{\boldsymbol{\theta}}_k + c_k\boldsymbol{\Delta}_k, V)$ and $Q(\hat{\boldsymbol{\theta}}_k - c_k\boldsymbol{\Delta}_k, V)$ must increase (or decrease) as the ith component of V increases. The direction of change in Q may differ with increases in different components of V, but $Q(\hat{\boldsymbol{\theta}}_k + c_k\boldsymbol{\Delta}_k, V)$ and $Q(\hat{\boldsymbol{\theta}}_k - c_k\boldsymbol{\Delta}_k, V)$ must still change in the same direction.

The full result in Rubinstein et al. (1985) allows for more general dependence among the elements of $V_k^{(+)}$ and $V_k^{(-)}$ than the independence required in Proposition 14.2. We do not explore those extensions here, as they get into concepts (e.g., total positivity) that are beyond the scope of this book (the independence stated here is a special case). Glasserman and Yao (1992) provide a comprehensive discussion of the role of conditions such as monotonicity and the inverse transforms for generating the random variables (V). They note, for example, that although the inverse transform method is part of the proposition here (and similar results), it is not necessary to *use* the inversion method in actual implementation. Rather, it must be assumed only that the inversion method is feasible for this problem. Other methods are acceptable if they produce the correct distribution. Glasserman and Yao (1992) provide a detailed discussion on the relationship of monotonicity to the synchronization of the random variables in the simulations.

In practice, how does one know if the monotonicity assumption in Proposition 14.2 is satisfied? There is no easy answer to this in general, but let us sketch a scenario where the assumption will be satisfied. Suppose, for example, that $Q(\boldsymbol{\theta}, V)$ is a continuous function of $\boldsymbol{\theta}$ or, more generally, that small changes in $\boldsymbol{\theta}$ do not alter the fundamental ordering and number of events represented in the simulation output. Then, for sufficiently small c_k (large k) the functions $Q(\hat{\boldsymbol{\theta}}_k + c_k\boldsymbol{\Delta}_k, V)$ and $Q(\hat{\boldsymbol{\theta}}_k - c_k\boldsymbol{\Delta}_k, V)$ will behave nearly identically in V (since $\hat{\boldsymbol{\theta}}_k + c_k\boldsymbol{\Delta}_k$ will be nearly the same as $\hat{\boldsymbol{\theta}}_k - c_k\boldsymbol{\Delta}_k$). Suppose further that $Q(\hat{\boldsymbol{\theta}}_k + c_k\boldsymbol{\Delta}_k, V)$ and $Q(\hat{\boldsymbol{\theta}}_k - c_k\boldsymbol{\Delta}_k, V)$ are differentiable in V near some operating point V^* where the derivatives $\partial Q(\hat{\boldsymbol{\theta}}_k \pm c_k\boldsymbol{\Delta}_k, V)/\partial V$ are nonzero, and that the randomly generated elements of V deviate from V^* by only a small magnitude. Then, there is evidence that the monotonicity is satisfied because $Q(\boldsymbol{\theta}, V)$ is locally (near V^*) monotonic in V by the continuity, and $V_k^{(+)}$ and $V_k^{(-)}$ take on values near V^*.

The following examples illustrate the types of problems for which Proposition 14.2 applies.

Example 14.6—Network design. Consider a transportation network with multiple origins and destinations. Suppose that an "omniscient authority" wishes to optimize some network design parameters ($\boldsymbol{\theta}$) to minimize the mean overall network cost. Let $\mathcal{V}_i$ represent the (random) cost of travel over the ith link in the network, with the links indexed by i. Suppose that the design parameters $\boldsymbol{\theta}$

appear in the probability distribution for the random costs (e.g., $\boldsymbol{\theta}$ may represent aspects of the traffic signal operations). Let $OD(j)$ represent the collection of links associated with the jth customer's origin–destination pair, $j = 1, 2,\ldots, m$. For example, $OD(j) = \{2, 4, 7, 16\}$ indicates that the jth customer traveled over the links labeled 2, 4, 7, and 16 in going from origin to destination. With $\boldsymbol{V}$ representing the collection of travel times $\{\mathcal{V}_i\}$ the overall cost associated with travel over the network is

$$Q(\boldsymbol{\theta},\boldsymbol{V}) = \sum_{j=1}^{m} \sum_{i\in OD(j)} \mathcal{V}_i .$$

Note that $\boldsymbol{\theta}$ does not appear explicitly in Q; rather, it appears implicitly via its effect on the distribution of the travel costs $\mathcal{V}_i$ (this phenomenon is considered in more detail in Section 15.2 in the context of likelihood ratio methods for gradient estimation). It is clear that Q is nondecreasing in each component of $\boldsymbol{V}$. Hence, if the elements of $\boldsymbol{V}$ can be generated by the inverse transform method, Proposition 14.2 indicates that CRNs provide a minimum variance gradient estimate. ❑

Example 14.7—Linear model. Suppose that the simulation output can be represented as a simple linear function of $\boldsymbol{\theta}$, i.e., $\boldsymbol{Z} = \boldsymbol{B}\boldsymbol{\theta}$, where $\boldsymbol{B} = \boldsymbol{B}(\boldsymbol{V})$ ($\boldsymbol{B}$ may be a *nonlinear* function of the random inputs $\boldsymbol{V}$). Given that $L(\boldsymbol{\theta}) = E[Q(\boldsymbol{\theta}, \boldsymbol{V})] = E[(\boldsymbol{Z} - \boldsymbol{d})^T(\boldsymbol{Z} - \boldsymbol{d})]$, where $\boldsymbol{d}$ is some "desired" vector, let us consider the variability of the SPSA gradient estimator. Note that

$$\begin{aligned} y_k^{(\pm)} &= Q(\hat{\boldsymbol{\theta}}_k \pm c_k\boldsymbol{\Delta}_k, \boldsymbol{V}_k^{(\pm)}) = [(\boldsymbol{\theta}_k^{(\pm)})^T \boldsymbol{B}(\boldsymbol{V}_k^{(\pm)})^T - \boldsymbol{d}^T][\boldsymbol{B}(\boldsymbol{V}_k^{(\pm)})\boldsymbol{\theta}_k^{(\pm)} - \boldsymbol{d}] \\ &= (\boldsymbol{\theta}_k^{(\pm)})^T \boldsymbol{B}(\boldsymbol{V}_k^{(\pm)})^T \boldsymbol{B}(\boldsymbol{V}_k^{(\pm)})\boldsymbol{\theta}_k^{(\pm)} - 2\boldsymbol{d}^T \boldsymbol{B}(\boldsymbol{V}_k^{(\pm)})\boldsymbol{\theta}_k^{(\pm)} + \boldsymbol{d}^T\boldsymbol{d}, \end{aligned} \tag{14.13}$$

where $\boldsymbol{\theta}_k^{(\pm)} = \hat{\boldsymbol{\theta}}_k \pm c_k\boldsymbol{\Delta}_k$. Proposition 14.2 indicates that CRNs provide a minimum variance gradient estimate when a positive change in the ith element of $\boldsymbol{V}_k^{(+)}$ and $\boldsymbol{V}_k^{(-)}$, $i = 1, 2,\ldots,$ causes the same direction of change (either positive or negative) in $y_k^{(+)}$ and $y_k^{(-)}$ (i.e., monotonicity). Suppose that $\boldsymbol{V}$ can lie in a small domain surrounding some $\boldsymbol{V}^*$ and that $\boldsymbol{B}(\boldsymbol{V})$ is differentiable. Then if the corresponding components of $\partial y_k^{(+)}/\partial \boldsymbol{V}$ and $\partial y_k^{(-)}/\partial \boldsymbol{V}$ have the same sign at $\boldsymbol{V}^*$, the monotonicity assumption is satisfied.

As a specific example, suppose that $p = 2$, $\boldsymbol{V} = [\mathcal{V}_1, \mathcal{V}_2]^T$, and

$$\boldsymbol{B}(\boldsymbol{V}) = \begin{bmatrix} \mathcal{V}_1 & 0 \\ 0 & \mathcal{V}_2^2 \end{bmatrix}.$$

Then, from (14.13) and letting $\boldsymbol{d} = [d_1, d_2]^T$, the assumptions of Proposition 14.2 can be evaluated by considering the signs of

$$\frac{\partial y_k^{(\pm)}}{\partial \mathcal{V}_1} = 2(\theta_{k1}^{(\pm)})^2 \mathcal{V}_1 - 2\theta_{k1}^{(\pm)} d_1,$$

$$\frac{\partial y_k^{(\pm)}}{\partial \mathcal{V}_2} = 4(\theta_{k2}^{(\pm)})^2 \mathcal{V}_2^3 - 4\theta_{k2}^{(\pm)} d_2 \mathcal{V}_2,$$

over the range of possible values for $\boldsymbol{\theta}_k^{(\pm)}$ and $\boldsymbol{V}$. For example, if $\hat{\boldsymbol{\theta}}_k = [0,\ 0]^T$, the elements of $c_k\boldsymbol{\Delta}_k$ are i.i.d. Bernoulli ± 0.1, and $\boldsymbol{d} = [1,\ 1]^T$, then from the derivative expressions above, it is known that $y_k^{(+)}$ and $y_k^{(-)}$ satisfy Proposition 14.2 if the possible $\boldsymbol{V}$ values lie in the region $(10, \infty] \times (\sqrt{10}, \infty]$. Both $y_k^{(+)}$ and $y_k^{(-)}$ are monotonically increasing in both components of $\boldsymbol{V}$ for $\boldsymbol{V}$ restricted to this region.

The essential form of the numerator $y_k^{(+)} - y_k^{(-)}$ in the SPSA gradient approximation is

$$\hat{\boldsymbol{\theta}}_k^T [\boldsymbol{B}(\boldsymbol{V}_k^{(+)})^T \boldsymbol{B}(\boldsymbol{V}_k^{(+)}) - \boldsymbol{B}(\boldsymbol{V}_k^{(-)})^T \boldsymbol{B}(\boldsymbol{V}_k^{(-)})] \hat{\boldsymbol{\theta}}_k$$
$$- 2\boldsymbol{d}^T [\boldsymbol{B}(\boldsymbol{V}_k^{(+)}) - \boldsymbol{B}(\boldsymbol{V}_k^{(-)})] \hat{\boldsymbol{\theta}}_k + O_P(c_k), \qquad (14.14)$$

where the term $O_P(c_k)$ represents a collection of random terms that are multiplied by c_k (the subscript P denotes a *probabilistic* version of standard big-O). Under assumptions such as those discussed in the preceding paragraph, Proposition 14.2 indicates that the resulting gradient approximation has minimum variance under CRNs. From (14.14), we see that under CRNs, $\text{var}[(y_k^{(+)} - y_k^{(-)}) \,|\, \hat{\boldsymbol{\theta}}_k, \boldsymbol{\Delta}_k] = O(c_k^2)$ ($c_k \to 0$ as usual) since all terms but the $O_P(c_k)$ term on the right-hand side of (14.14) cancel each other. (Because the conditional variances here are random variables, all statements here involving a conditional variance are only guaranteed to hold on a set of probability 1 [i.e., in the a.s. sense]—see Appendix C.) Under non-CRNs, $\text{var}[(y_k^{(+)} - y_k^{(-)}) \,|\, \hat{\boldsymbol{\theta}}_k, \boldsymbol{\Delta}_k] \geq \text{constant} > 0$ for all k. This leads to $\text{var}[\hat{g}_{ki}(\hat{\boldsymbol{\theta}}_k) \,|\, \hat{\boldsymbol{\theta}}_k]$ being $O(1)$ under CRNs, but $O(c_k^{-2})$ under non-CRNs. (Because $a_k \to 0$ sufficiently fast, the SPSA iterate $\hat{\boldsymbol{\theta}}_k$ converges under non-CRNs as $k \to \infty$ despite this diverging $O(c_k^{-2})$ variance.) This difference is indicative of the benefits of CRNs in forming the gradient estimate, and, in fact, is one example of a more general property of gradient estimates under CRNs (see the discussion on asymptotics below). ❑

The above results focus on finite-sample analysis of CRNs. If one considers *asymptotic* performance, the benefits of CRNs occur without the need to invoke the monotonicity and inverse transform restrictions. These benefits follow from $c_k \to 0$, and the resulting negligible distinction between the two inputs $Q(\hat{\boldsymbol{\theta}}_k + c_k\boldsymbol{\Delta}_k, \boldsymbol{V})$ and $Q(\hat{\boldsymbol{\theta}}_k - c_k\boldsymbol{\Delta}_k, \boldsymbol{V})$. There exists theory to formalize these benefits. The theory pertains to the asymptotic distribution of the iterate $\hat{\boldsymbol{\theta}}_k$, as we now discuss.

The convergence conditions for FDSA and SPSA with a gradient estimate based on CRNs are, not surprisingly, very similar to those in Sections 6.4–6.5 and 7.3–7.4 (see L'Ecuyer and Yin, 1998; Kleinman et al., 1999). Suppose that each iteration uses a common value for the V input in the Q values forming the gradient estimate. For example, with SPSA, one has $V_k = V_k^{(+)} = V_k^{(-)}$ by picking the same random number seed in generating the uniform random variables that are transformed to generate $V_k^{(+)}$ and $V_k^{(-)}$ (subject to $V_k^{(+)}$ and $V_k^{(-)}$ being used in the same way—the comparability condition—as discussed at the beginning of the section). L'Ecuyer and Yin (1998) showed that CRNs increased the best asymptotic rate of stochastic convergence for FDSA from the standard $O(1/k^{1/3})$ rate discussed in Section 6.5 to the same $O(1/\sqrt{k})$ rate achieved in the gradient-based version of the root-finding SA algorithm (Section 4.4). Hence the use of CRNs compensates in some sense for the fact that only function measurements (versus gradients) are being used.

Kleinman et al. (1999) show that SPSA can also achieve the $O(1/\sqrt{k})$ rate with CRNs. This is achieved via an asymptotic normality result similar to Section 7.4 for basic SPSA. Recall the standard gain forms: $a/(k+1+A)^{\alpha}$ and $c_k = c/(k+1)^{\gamma}$ with a, c, α, and γ all strictly positive and $A \geq 0$. Under conditions identical to those in Section 7.4, except that $4\gamma - \alpha > 0$ (versus $3\gamma - \alpha/2 \geq 0$ and $\alpha - 2\gamma > 0$ in Section 7.4), SPSA with CRNs satisfies

$$k^{\alpha/2}(\hat{\boldsymbol{\theta}}_k - \boldsymbol{\theta}^*) \xrightarrow{\text{dist}} N(\mathbf{0}, \boldsymbol{\Sigma}_{\text{CRN}}), \tag{14.15}$$

where $\boldsymbol{\Sigma}_{\text{CRN}}$ is a covariance matrix that resembles $\boldsymbol{\Sigma}_{\text{SP}}$ as cited in Section 7.4. Because $\alpha \leq 1$, the maximum rate of convergence is $O(1/\sqrt{k})$. Unlike the SPSA asymptotic normality in Section 7.4—but like the root-finding SA algorithm result in Section 4.4—the mean of the asymptotic distribution is $\mathbf{0}$. The formerly optimal gain values ($\alpha = 1$, $\gamma = 1/6$) are not allowed here by the condition $4\gamma - \alpha > 0$; rather, $\gamma > 1/4$ with the asymptotically optimal $\alpha = 1$. The increase in rate of convergence is driven by a reduction of the variance of the elements in $\hat{\boldsymbol{g}}_k(\hat{\boldsymbol{\theta}}_k)$ from $O(c_k^{-2})$ to $O(1)$. Examples 14.6 and 14.7 conceptually or analytically illustrated this variance reduction in two problems. The two-part study in Examples 14.8 and 14.9 shows specific *numerical* improvement in a particular problem.

Example 14.8—Numerical illustration of CRNs versus non-CRNs for SPSA. Consider the loss function

$$L(\boldsymbol{\theta}) = \boldsymbol{\theta}^T\boldsymbol{\theta} + E\left[\sum_{i=1}^{p} \exp(-\mathcal{V}_i t_i)\right], \tag{14.16}$$

where $\boldsymbol{\theta} = [t_1, t_2, \ldots, t_p]^T$, the components satisfy $t_i \geq 0$ for all $i = 1, 2, \ldots, p$, and the $\mathcal{V}_i$ are independently generated, exponentially distributed random variables with means $1/\lambda_i$, $i = 1, 2, \ldots, p$. This loss function is used in an example in Kleinman et al. (1999). In SPSA, we constrain the elements of $\hat{\boldsymbol{\theta}}_k$ and $\hat{\boldsymbol{\theta}}_k \pm c_k\boldsymbol{\Delta}_k$ to be nonnegative. (Because constraints on $\hat{\boldsymbol{\theta}}_k \pm c_k\boldsymbol{\Delta}_k$ interfere with the "natural" simultaneous perturbation process, the gradient estimate may not have the standard $O(c_k^2)$ bias. This only affects the early iterations because once the elements of $\hat{\boldsymbol{\theta}}_k$ are safely away from zero and c_k is sufficiently small, elements of $\hat{\boldsymbol{\theta}}_k \pm c_k\boldsymbol{\Delta}_k$ will also automatically be nonnegative.) Using the relatively simple loss function above in lieu of a complex loss based on a full Monte Carlo simulation allows for a computationally efficient study and complete control over the implementation of CRNs. Hence, the measured loss $Q(\boldsymbol{\theta}, V) = \boldsymbol{\theta}^T\boldsymbol{\theta} + \sum_{i=1}^{p}\exp(-\mathcal{V}_i t_i)$ may be viewed as a proxy for a large simulation output where $V = [\mathcal{V}_1, \mathcal{V}_2, \ldots, \mathcal{V}_p]^T$.

Suppose that $p = 10$ and that the exponentially distributed $\mathcal{V}_i$ are generated according to the inverse-transform method (Appendix D). The distributional parameters λ_i and the corresponding elements within $\boldsymbol{\theta}^*$ are given in Table 14.2 (the λ_i were chosen according to a uniform distribution between 0.5 and 2.0; these values are taken from Kleinman et al., 1999). The $\boldsymbol{\theta}^*$ values are found by use of a deterministic optimization method applied to the equivalent analytical form for the loss in (14.16): $L(\boldsymbol{\theta}) = \boldsymbol{\theta}^T\boldsymbol{\theta} + \sum_{i=1}^{10}\lambda_i/(\lambda_i + t_i)$ (Exercise 14.8). Of course, $\boldsymbol{\theta}^*$ is only known because of the simplicity of the "simulation" represented by Q; in practice, $\boldsymbol{\theta}^*$ is not known prior to the stochastic search process. Note that the $\mathcal{V}_i$ are independent and that Q is monotonically

Table 14.2. Values of distribution parameters λ_i and elements of $\boldsymbol{\theta}^*$.

i	λ_i	ith element of $\boldsymbol{\theta}^*$
1	1.1025	0.286
2	1.6945	0.229
3	1.4789	0.248
4	1.9262	0.211
5	0.7505	0.325
6	1.3267	0.263
7	0.8428	0.315
8	0.7247	0.327
9	0.7693	0.323
10	1.3986	0.256

nonincreasing for an increasing $\mathcal{V}_i$ for any value of $t_i \geq 0$. Because of the constraints on the elements of $\hat{\theta}_k$ and $\hat{\theta}_k \pm c_k \Delta_k$ to be nonnegative, Proposition 14.2 indicates that CRNs produce the minimum variance for the components of $\hat{g}_k(\hat{\theta}_k)$.

Let us use the above values to compare the CRN and non-CRN results with SPSA based on a total of n iterations (i.e., the terminal value of the iterate is $\hat{\theta}_n$). Consider the standard gain sequence forms, $a_k = a/(k+1)^{\alpha}$ and $c_k = c/(k+1)^{\gamma}$, $0 \leq k \leq n-1$. For both CRN and non-CRN, let $a = 0.7$, $c = 0.5$. For the CRN case, $\alpha = 1$ and any $1/4 < \gamma < 1/2$ are asymptotically optimal (Kleinman et al. 1999; see also the above-mentioned condition $4\gamma - \alpha > 0$); we use $\alpha = 1$ and $\gamma = 0.49$. For the non-CRN case, the asymptotically optimal $\alpha = 1$ and $\gamma = 1/6$ are used (see Section 7.4). An initial condition $\hat{\theta}_0 = [1, 1, \ldots, 1]^T$ is employed for all runs. Table 14.3 presents the results of the study; the loss values and normalized parameter estimation errors represent the sample mean of the 50 terminal values based on 50 replications at the indicated number of iterations. The values of $L(\hat{\theta}_n)$ in the table were computed based on the above-mentioned analytical form for the loss function.

We can also compute P-values associated with the comparison of the CRN and non-CRN results. Consider the terminal loss values $L(\hat{\theta}_n)$ as the basis for the comparisons. Consider a statistical test with a null hypothesis that the mean terminal L value with CRNs is no different than the mean terminal L value for non-CRNs. The terminal losses for the two samples (CRN and non-CRN) were generated independently. Hence, we use an unmatched pairs t-test with nonidentical variance, as discussed in Appendix B. (While the $L(\hat{\theta}_n)$ values from the 50 replications are not likely to represent a sample from a normal

Table 14.3. Comparison of CRN and non-CRN terminal loss values and parameter estimate errors with SPSA. Values shown are the sample means of the loss and the sample means of the norm of the error in $\boldsymbol{\theta}$ estimate from 50 replications ($L(\hat{\theta}_0) = 15.3025$; $L(\theta^*) = 8.72266$).

Total iterations n	CRN		Non-CRN	
	$L(\hat{\theta}_n)$	$\frac{\lVert\hat{\theta}_n - \theta^*\rVert}{\lVert\hat{\theta}_0 - \theta^*\rVert}$	$L(\hat{\theta}_n)$	$\frac{\lVert\hat{\theta}_n - \theta^*\rVert}{\lVert\hat{\theta}_0 - \theta^*\rVert}$
100	9.5222	0.2581	9.7000	0.2931
1000	8.7268	0.02195	8.7354	0.04103
10,000	8.7230	0.00658	8.7253	0.01845
100,000	8.7227	0.00207	8.7232	0.00819

distribution, the reasonably large sample size of 50 for each of the two samples provides some justification for the t-test, as discussed in Appendix B.) For n = 1000, 10,000, and 100,000, the P-values for the null hypothesis of equality between CRN and non-CRN are less than 10^{-10}, indicating strong rejection of the null hypothesis in favor of the alternative hypothesis that CRN produces lower terminal loss values. For $n = 100$, the mean loss for CRN is lower than for non-CRN, but the large two-sided P-value of 0.63 does not indicate statistical significance of this advantage. The power of CRN strongly manifests itself *after* the algorithm reaches 100 iterations. ❑

Example 14.9—Rate of convergence analysis for CRN and non-CRN results in Example 14.8. Figure 14.2 plots the means of the 50 values of $\|n^{\beta/2}(\hat{\boldsymbol{\theta}}_n - \boldsymbol{\theta}^*)\|$ for $\beta = 1$ or $\beta = 2/3$, as a function of n = 100, 1000, 10,000, and 100,000. In the CRN case, the magnitudes of the norms stabilize nicely at $\beta = 1$, consistent with the asymptotic normality result in (14.15) that shows a stochastic rate of convergence for the $\boldsymbol{\theta}$ estimate that is $O(1/\sqrt{n})$ at the selected values of α and γ. In the non-CRN case, the magnitudes *grow* with increasing n at $\beta = 1$ (beginning with n = 1000), but are nicely bounded at $\beta = 2/3$; this is consistent with the slower rate of convergence of $O(1/n^{1/3})$ for basic SPSA, as given in Section 7.4. ❑

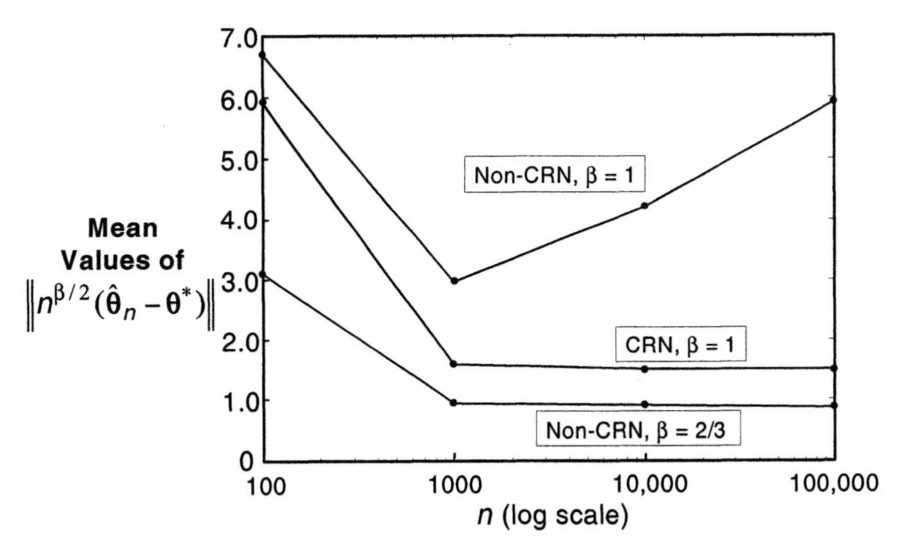

Figure 14.2. Sample means of $\|n^{\beta/2}(\hat{\boldsymbol{\theta}}_n - \boldsymbol{\theta}^*)\|$ for $\beta = 1$ or 2/3. Stochastic rate of convergence for CRN is $O(1/\sqrt{n})$ and for non-CRN is $O(1/n^{1/3})$, consistent with the indicated two curves that are essentially flat for large n. The remaining growing curve shows that non-CRN values are diverging when multiplied by scale factor that is too large ($\sqrt{n}$). (No curve is shown for CRN with $\beta = 2/3$; such a curve would decay to zero.)

14.4.3 Partial Common Random Numbers for Finite Samples

A difficulty in applying CRNs is that even when initializing both simulations with the same seed, the difference in the two $\boldsymbol{\theta}$ values used, $\hat{\boldsymbol{\theta}}_k \pm c_k \boldsymbol{\Delta}_k$, may cause the simulations to process the sequences of Monte Carlo variables $V_k^{(\pm)}$ in different ways. In particular, the use of the random numbers in the two simulations may not be synchronized. This issue is especially relevant in the finite-sample ($k << \infty$) setting, where c_k may be relatively large (so the two design values $\hat{\boldsymbol{\theta}}_k \pm c_k \boldsymbol{\Delta}_k$ may be relatively far apart). This lack of synchronization was suggested in the context of case (ii) of the plant layout problem in Example 14.5, where the dependence of part arrivals on $\boldsymbol{\theta}$ may alter the number of parts in the system.

More generally, it is easy to see how the simulations may become unsynchronized by considering a queuing system. For example, suppose that $\boldsymbol{\theta}$ represents parameters associated with service rates in a network and V represents the (random) length of service times for the arrivals into the system. Then, if the simulation associated with $\hat{\boldsymbol{\theta}}_k + c_k \boldsymbol{\Delta}_k$ allows for faster service than the simulation associated with $\hat{\boldsymbol{\theta}}_k - c_k \boldsymbol{\Delta}_k$, it is possible that more arrivals will be processed in the $\hat{\boldsymbol{\theta}}_k + c_k \boldsymbol{\Delta}_k$ simulation. Hence, $V_k^{(+)}$ contains elements (service times) not present in $V_k^{(-)}$. More generally, if V represents many random effects (arrivals, services, etc.), then the processing of extra customers in one simulation will not only make the number of elements in the two V vectors different, but will generally make the definitions of ordinally corresponding elements different. In particular, the mth element of $V_k^{(+)}$ generally has a different physical meaning than the mth element of $V_k^{(-)}$.

Kleinman et al. (1999) discuss the implications of the above based on the notion of *partial CRNs*. In this discussion of partial CRNs, we focus on the SPSA formulation, but analogous ideas apply in the FDSA approach. This approach is built from the observation that in the early iterations, when c_k is relatively large, the $\boldsymbol{\theta}$ values used in the two simulations (i.e., $\hat{\boldsymbol{\theta}}_k \pm c_k \boldsymbol{\Delta}_k$) will differ by a greater amount, increasing the likelihood that the two random streams $V_k^{(\pm)}$ will not be synchronized. However, in the later iterations, when c_k is small, there is little difference in the two $\boldsymbol{\theta}$ values. Hence, it is more likely that the sequences $V_k^{(\pm)}$ will be synchronized. In fact, since $c_k \to 0$, the sequences are guaranteed in the limit to be fully synchronized if the simulation output is a continuous function of $\boldsymbol{\theta}$ (e.g., if $Q(\boldsymbol{\theta}, V)$ is continuous in $\boldsymbol{\theta}$ and V and its distribution function do not depend on $\boldsymbol{\theta}$). This reminds us that the *asymptotic* distribution theory mentioned above for CRNs, expression (14.15), is applicable even in cases where the synchronization is not automatically true at two given parameter values $\hat{\boldsymbol{\theta}}_k \pm c_k \boldsymbol{\Delta}_k$ for a specific k.

What can one expect when the simulation budget does not allow k to get large enough so that c_k is small enough to have full synchronization of $V_k^{(\pm)}$?

Kleinman et al. (1999) consider this situation. The analysis of the partial CRN setting proceeds by decomposing $Q(\boldsymbol{\theta}, V)$ into two parts, the first corresponding to the sequence of events where the random elements in $V_k^{(\pm)}$ are synchronized and the second where they are unsynchronized. This decomposition applies when $Q(\boldsymbol{\theta}, V)$ is naturally defined in terms of a sum of events occurring over a period (e.g., the sum of the wait times for all customers in a queuing system over a specified period). The decomposition is not typically applicable when the simulation output is only available in toto at the end of the simulation run (say, a war gaming situation where the criterion is whether a battle is won or lost, which is known only after a simulation is completed).

As an example of applicability of partial CRNs, suppose in the queuing problem that the period being simulated represents two hours of a network's operations. In the first hour, the difference in the service rates governed by $\hat{\boldsymbol{\theta}}_k \pm c_k\boldsymbol{\Delta}_k$ does not cause any of the random elements in the two sequences $V_k^{(\pm)}$ to be used differently. At the beginning of the second hour, an extra arrival is processed in the $\hat{\boldsymbol{\theta}}_k + c_k\boldsymbol{\Delta}_k$ simulation (relative to the $\hat{\boldsymbol{\theta}}_k - c_k\boldsymbol{\Delta}_k$ simulation), causing all subsequent uses of $V_k^{(+)}$ to be unsynchronized with $V_k^{(-)}$. Intuitively, one might expect some benefits since there should be less variability for at least some of the input into the difference $y_k^{(+)} - y_k^{(-)} = Q(\hat{\boldsymbol{\theta}}_k + c_k\boldsymbol{\Delta}_k, V_k^{(+)}) - Q(\hat{\boldsymbol{\theta}}_k - c_k\boldsymbol{\Delta}_k, V_k^{(-)})$. This can be formalized for general problems by writing

$$\begin{aligned} y_k^{(+)} - y_k^{(-)} &= Q_1(\hat{\boldsymbol{\theta}}_k + c_k\boldsymbol{\Delta}_k, V_{k_1}^{(+)}) + Q_2(\hat{\boldsymbol{\theta}}_k + c_k\boldsymbol{\Delta}_k, V_{k_2}^{(+)}) \\ &\quad - Q_1(\hat{\boldsymbol{\theta}}_k - c_k\boldsymbol{\Delta}_k, V_{k_1}^{(-)}) - Q_2(\hat{\boldsymbol{\theta}}_k - c_k\boldsymbol{\Delta}_k, V_{k_2}^{(-)}) \\ &= \underbrace{Q_1(\hat{\boldsymbol{\theta}}_k + c_k\boldsymbol{\Delta}_k, V_{k_1}) - Q_1(\hat{\boldsymbol{\theta}}_k - c_k\boldsymbol{\Delta}_k, V_{k_1})}_{\text{synchronized part}} \\ &\quad + \underbrace{Q_2(\hat{\boldsymbol{\theta}}_k + c_k\boldsymbol{\Delta}_k, V_{k_2}^{(+)}) - Q_2(\hat{\boldsymbol{\theta}}_k - c_k\boldsymbol{\Delta}_k, V_{k_2}^{(-)})}_{\text{unsynchronized part}}, \end{aligned} \tag{14.17}$$

where the subscripts 1 or 2 denote the first (synchronized) or second (unsynchronized) part as defined above and $V_{k_1} = V_{k_1}^{(+)} = V_{k_1}^{(-)}$. With such a decomposition, it is possible to show that the variance of the gradient estimate $\hat{g}_k(\hat{\boldsymbol{\theta}}_k)$ is reduced over that from a standard (independent random number) implementation. As in the full CRNs case above, this guarantee follows when $Q_1(\hat{\boldsymbol{\theta}}_k + c_k\boldsymbol{\Delta}_k, V)$ and $Q_1(\hat{\boldsymbol{\theta}}_k - c_k\boldsymbol{\Delta}_k, V)$ are both monotonic in the same direction in each component of V (Exercise 14.10). Hence, there is some benefit in attempting to synchronize the simulation input as much as possible, which is usually achieved via the simple method of using the same random number seed for the two simulations entering the gradient approximation.

Kleinman et al. (1999) show some numerical results associated with partial CRNs for the relatively simple $p = 10$ exponential problem in Example 14.8. This problem is designed so that it is possible to implement full CRNs (not always possible in practice, of course!). As expected, the error improves as we move from the independent random number case to the partial case and then to the full CRNs case. Kleinman et al. (1999) use the initial condition and SPSA gain settings of Example 14.8. In the study of this reference, the sample mean (100 replications) of the normalized distance between $\boldsymbol{\theta}^*$ and the terminal $\hat{\boldsymbol{\theta}}_k$ at $k = 10{,}000$ reduces from 0.019 in the independent random number case (no CRNs or partial CRNs) to 0.0071 in the partial CRNs setting and to 0.0065 in the full CRNs setting.

14.5 SELECTION METHODS FOR OPTIMIZATION WITH DISCRETE-VALUED $\boldsymbol{\theta}$

Frequently, simulations are used to choose the best system design among several candidate designs. This can be cast into a problem of simulation-based optimization over a discrete search space. For example, in the plant layout example of Section 14.1, it may be that each of the machines can only be placed at a small set of specified locations, perhaps constrained to this set of locations due to the need to meet safety regulations and other restrictions on the flow of people and material on the factory floor. This section is a summary of certain techniques that have been used when the elements of $\boldsymbol{\theta}$ can take on only discrete values. If, say, each of p machines can be placed at one of M locations, then the search domain Θ has M^p values to be considered.

The simplest discrete setting, of course, is when Θ contains only a small number of elements. Here, relatively simple statistical methods can be used to estimate the best option. Even in this relatively simple case, however, it may require significant effort to reliably determine $\boldsymbol{\theta}^*$, the fundamental reason being that we do not (typically) observe L at each candidate solution. Rather, a noisy measurement of L is observed as a consequence of the Monte Carlo aspect of the simulation. So it is usually necessary to use some type of averaging procedure and associated statistical tests to identify an optimal value of $\boldsymbol{\theta}$.

In cases where the search space Θ contains many options, the relatively simple averaging ideas above are usually not feasible. Here, we are driven toward other methods that offer the prospect of at least an approximately optimal solution in a reasonable time. Section 2.3 on random search discussed some methods for coping with noise; Andradóttir (1998) discusses some others. All of these methods are based on the principle of moving iteratively from one feasible solution to a neighboring feasible solution in the discrete space. The difference in the algorithms rests on the mechanics of moving from one point to the next and the means by which the value of a new point is compared with that of the old point.

Aside from the basic random search algorithms above, Chapters 8–10 show that simulated annealing (SAN) and evolutionary algorithms can be used for discrete optimization. Recall, however, that standard SAN or evolutionary algorithms are not designed to cope with *noisy* loss measurements. Some work has been done for the noisy case, especially with a discrete search space (e.g., Gelfand and Mitter, 1989; Fox and Heine, 1995). Andradóttir (1998) also presents a variant of SAN for the simulation-based discrete optimization case. Section 7.7 mentioned that SPSA—which *is* explicitly designed for the case of noisy function measurements—can sometimes be used for discrete optimization. Ahmed and Alkhamis (2002) embed a selection method of the type in this section *within* SAN to cope with the randomness.

To address the problem of choosing the best option from among a relatively small number of possibilities, we return to the theme of Chapter 12 on statistical comparisons. *Relatively small* typically means up to 20 options (Goldsman and Nelson, 1998), although there is no hard upper bound to the number of options (see Nelson et al., 2001, for extensions to a greater number of options). In particular, we summarize a statistical comparisons approach that provides statistical guarantees on the likelihood of choosing the best option or an option that is close to the best option. The approach here, which is based on the notion of an indifference zone, is certainly not the only useful approach in simulation-based optimization via statistical comparisons. The reader is directed to Goldsman and Nelson (1998) or Swisher and Jacobson (1999) for a broader survey of related techniques. Although this section is largely self-contained, the reader might find it useful to skim Chapter 12, especially Section 12.1, for some motivational material. The general approach described below is one of the most popular means for selecting among competing alternatives. The approach is also well suited to exploit common random numbers.

Suppose that the aim is to find a unique $\boldsymbol{\theta}^* = \arg\min_{\boldsymbol{\theta}\in\Theta} L(\boldsymbol{\theta})$, where $\Theta = \{\boldsymbol{\theta}_1, \boldsymbol{\theta}_2, \ldots, \boldsymbol{\theta}_K\}$ represents the K possible values for $\boldsymbol{\theta}$. Hence, Θ is a set of discrete options. As in Chapter 12, let $\overline{L}_i$ represent a sample mean of measured values of L at $\boldsymbol{\theta} = \boldsymbol{\theta}_i$. In particular, let $y_k(\boldsymbol{\theta}_i) = L(\boldsymbol{\theta}_i) + \varepsilon_{ki}$ represent the kth measurement of L at the given $\boldsymbol{\theta}_i$, $k = 1, 2, \ldots, n_i$ (say), and ε_{ki} represent the mean-zero noise term. For our purposes here, each $y_k(\boldsymbol{\theta}_i)$ represents the output $Q(\cdot)$ of one simulation run where the simulation parameter vector $\boldsymbol{\theta}$ has been set to $\boldsymbol{\theta}_i$. Then, $\overline{L}_i = n_i^{-1}\sum_{k=1}^{n_i} y_k(\boldsymbol{\theta}_i)$. Note that when $K = 2$, the standard two-sample t-tests from basic statistics (Appendix B) can be used to statistically determine whether $\boldsymbol{\theta}_1$ or $\boldsymbol{\theta}_2$ is preferred via whether $\overline{L}_1$ is lower than $\overline{L}_2$—or vice versa—in a statistically significant sense. The main interest here is the more difficult *multiple comparisons* case, where $K \geq 3$.

Recall from Section 12.5 that the indifference zone approaches are guaranteed with probability at least $1 - \alpha$ to provide a solution equal to $\boldsymbol{\theta}^*$ when the true loss value at $\boldsymbol{\theta}_i \neq \boldsymbol{\theta}^*$ is at least δ units greater than $L(\boldsymbol{\theta}^*)$. The term *indifference* arises because the user is willing to tolerate any solution $\boldsymbol{\theta}_i$ such that the loss $L(\boldsymbol{\theta}_i)$ is in the range $[L(\boldsymbol{\theta}^*), L(\boldsymbol{\theta}^*)+\delta)$, called the *indifference zone*. As

δ gets smaller, the user has more stringent requirements, requiring a greater amount of sampling to form the sample means $\bar{L}_i$ as a way of distinguishing the difference between the candidate loss values $L(\boldsymbol{\theta}_i)$.

The selection process is based on picking the $\boldsymbol{\theta}_i$ associated with the lowest $\bar{L}_i$ as best. Let the random event "correct selection" correspond to correctly selecting the unique value of $\boldsymbol{\theta}_i$ equal to $\boldsymbol{\theta}^*$ based on a set of data (simulation runs). Formally, an indifference zone estimate satisfies the following relationship:

$$P(\text{correct selection}) \geq 1 - \alpha$$
$$\text{whenever } L(\boldsymbol{\theta}_i) - L(\boldsymbol{\theta}^*) \geq \delta \text{ for all } \boldsymbol{\theta}_i \neq \boldsymbol{\theta}^*$$

(Rinott, 1978). Although it is always possible that the analyst will make an incorrect selection, the expression above provides a guarantee that a correct selection is highly likely when $L(\boldsymbol{\theta}_i)$ is at least δ units larger than $L(\boldsymbol{\theta}^*)$ for $\boldsymbol{\theta}_i \neq \boldsymbol{\theta}^*$. There are no such probabilistic guarantees if there are suboptimal $\boldsymbol{\theta}_i$ having a loss value within δ of $L(\boldsymbol{\theta}^*)$ (hence the term *indifference*!).

As with most other techniques we have seen, there are variations in implementation. Initial results for indifference zone selection assumed *known* variances for the measurements (e.g., Bechhofer, 1954). Later results—as presented here—consider the more practical setting where the variances are unknown and must be estimated (e.g., Rinott, 1978). We present below two versions of a general two-stage approach that seems to be the most useful in typical simulation applications. In both versions, the first stage is directed at estimating the variances of the measurements $y_k(\boldsymbol{\theta}_i)$ for all i. These variances are used to determine the sample size needed to meet the indifference zone requirement. The second stage is used to complete the analysis based on collecting additional measurements. In general, the approach does not require that the (unknown) variances be identical across candidate solutions $\boldsymbol{\theta}_i$.

The two step-by-step techniques below represent the two versions of the general two-stage indifference zone approach. The first technique assumes independent simulations, while the second uses common random numbers. Both procedures assume that the $y_k(\boldsymbol{\theta}_i)$ (equivalently, the ε_{ki}) are normally distributed. At first glance, this normality assumption may seem restrictive in simulation problems. However, to the extent that the simulation output is the average of many "small" random effects, central limit theory (Subsection C.2.4 in Appendix C) suggests that the output will be nearly normally distributed. Likewise, if each $y_k(\boldsymbol{\theta}_i)$ represents the average of several independent replications of a simulation, then $y_k(\boldsymbol{\theta}_i)$ will be approximately normally distributed. Nelson and Goldsman (2001) discuss this further.

The independent sampling technique is introduced in Rinott (1978). The CRN-based technique is described in Nelson and Matejcik (1995).

Two-Stage Indifference Zone Selection with Independent Sampling

Step 0 **(Initialization)** Choose δ, α, and the sample size for the first stage n_0.

Step 1 **(Beginning of first stage)** At each $\boldsymbol{\theta}_i$, run the simulation n_0 times, collecting an i.i.d. sample of size n_0 for each of the candidate $\boldsymbol{\theta}_i$. The simulations should also be independent across the $\boldsymbol{\theta}_i$ (so all Kn_0 required simulation runs should be independent).

Step 2 For each $\boldsymbol{\theta}_i$, compute the sample variance from the appropriate n_0 measurements in step 1 (using the standard s^2 form in Appendix B). Denote the sample variances by s_i^2, $i = 1, 2,\ldots, K$.

Step 3 Find the value h from Table 8.3 in Goldsman and Nelson (1998) (or equivalent table in, say, Wilcox, 1984). Compute the total sample sizes n_i for each $\boldsymbol{\theta}_i$, including both the first and second stages of testing:

$$n_i = \max\left\{n_0, \left\lceil \frac{h^2 s_i^2}{\delta^2} \right\rceil\right\}, \tag{14.18}$$

where $\lceil \cdot \rceil$ means to round up to the next integer.

Step 4 **(Beginning of second stage)** At each $\boldsymbol{\theta}_i$, run the simulation $n_i - n_0$ times, collecting an i.i.d. sample. This sample should be independent of the initial n_0 measurements and independent of measurements for other values of $\boldsymbol{\theta}$.

Step 5 Use the combined measurements collected in steps 1 and 4 to form the sample means $\bar{L}_i = n_i^{-1}\sum_{k=1}^{n_i} y_k(\boldsymbol{\theta}_i)$ for $i = 1, 2,\ldots, K$.

Step 6 Select the $\boldsymbol{\theta}_i$ corresponding to the lowest $\bar{L}_i$ as the best. It is known that the probability of this $\boldsymbol{\theta}_i$ corresponding to $\boldsymbol{\theta}^*$ is at least $1 - \alpha$ whenever $L(\boldsymbol{\theta}_j) - L(\boldsymbol{\theta}^*) \geq \delta$ for all $\boldsymbol{\theta}_j \neq \boldsymbol{\theta}^*$.

We now present the related indifference zone technique based on CRNs. The advantage of this technique is that for a given indifference zone (δ) and probability of selection ($1 - \alpha$), smaller sample sizes (i.e., fewer simulation runs) can provide the same performance guarantees as the independent sampling above. The smaller required sample sizes follow from the reduced variability provided by CRNs in comparing different scenarios. The CRN procedure is implemented at each k by having the simulation runs that produce $y_k(\boldsymbol{\theta}_1)$, $y_k(\boldsymbol{\theta}_2),\ldots, y_k(\boldsymbol{\theta}_K)$ use the same random number seed and satisfy the synchronicity requirements discussed in Section 14.4. The common seed is changed at each k.

It was noted above that the indifference zone procedures are based on an assumption that the $y_k(\boldsymbol{\theta}_i)$ are normally distributed. To ensure the validity of the CRN procedure, Nelson and Matejcik (1995) introduce the additional assumption that the covariance matrix of the vector of correlated measurements $[y_k(\boldsymbol{\theta}_1), y_k(\boldsymbol{\theta}_2),\ldots, y_k(\boldsymbol{\theta}_K)]^T$ has a structure called *sphericity*. We will not consider the full generality of sphericity, but note that one important special case is that at each k,

$\text{var}[y_k(\boldsymbol{\theta}_i)]$ is a constant independent of i and $\text{cov}[y_k(\boldsymbol{\theta}_i), y_k(\boldsymbol{\theta}_j)]$ is a constant independent of i and j. Nelson and Matejcik (1995) demonstrate that the CRN procedure is robust to departures from these assumptions under the usual CRN condition that $\text{cov}[y_k(\boldsymbol{\theta}_i), y_k(\boldsymbol{\theta}_j)] > 0$ for all k, i, j. Nelson and Matejcik (1995) also discuss a method based on the Bonferroni inequality (Section 12.3) that does not require the sphericity condition, but this method requires larger sample sizes to compensate for the relative lack of structure.

Two-Stage Indifference Zone Selection with CRN (Dependent) Sampling

Step 0 **(Initialization)** Choose δ, α, and the sample size for the first stage n_0.

Step 1 **(Beginning of first stage)** At each $\boldsymbol{\theta}_i$, run the simulation n_0 times, collecting an i.i.d. sample of size n_0 for each of the candidate $\boldsymbol{\theta}_i$. Use CRNs across the $\boldsymbol{\theta}_i$. That is, at each k, the simulation runs producing $y_k(\boldsymbol{\theta}_1)$, $y_k(\boldsymbol{\theta}_2)$,..., $y_k(\boldsymbol{\theta}_K)$ use the same random number seed and satisfy the synchronicity requirements; the seed is changed at each k.

Step 2 Let $M_1(i)$, $M_2(k)$, and M_3 represent the following three sample means:

$$M_1(i) = \frac{1}{n_0}\sum_{k=1}^{n_0} y_k(\boldsymbol{\theta}_i),$$

$$M_2(k) = \frac{1}{K}\sum_{i=1}^{K} y_k(\boldsymbol{\theta}_i),$$

$$M_3 = \frac{1}{Kn_0}\sum_{i=1}^{K}\sum_{k=1}^{n_0} y_k(\boldsymbol{\theta}_i) = \frac{1}{K}\sum_{i=1}^{K} M_1(i) = \frac{1}{n_0}\sum_{k=1}^{n_0} M_2(k).$$

Then compute the overall sample variance according to

$$s^2 = \frac{2\sum_{i=1}^{K}\sum_{k=1}^{n_0}\left[y_k(\boldsymbol{\theta}_i) - M_1(i) - M_2(k) + M_3\right]^2}{(K-1)(n_0-1)}. \tag{14.19}$$

Step 3 Find the value $t = t^{(\alpha)}_{K-1,(K-1)(n_0-1)}$, where t is a critical value from the multivariate t-distribution in, say, Table 8.2 in Goldsman and Nelson (1998) (or equivalent table in, say, Hochberg and Tamhane, 1987, App. 3, Table 4). Compute the common total sample size n for each $\boldsymbol{\theta}_i$, which includes both the first and second stages of testing:

$$n = \max\left\{n_0, \left\lceil \frac{t^2 s^2}{\delta^2} \right\rceil\right\}. \tag{14.20}$$

Step 4 **(Beginning of second stage)** At each $\boldsymbol{\theta}_i$, run the simulation $n - n_0$ times to collect an i.i.d. sample (independent of the initial n_0 measurements). Use CRNs across the $\boldsymbol{\theta}_i$.

Step 5 Use the combined measurements collected in steps 1 and 4 to form the sample means $\bar{L}_i = n^{-1}\sum_{k=1}^{n} y_k(\boldsymbol{\theta}_i)$ for $i = 1, 2, \ldots, K$.

Step 6 Select the $\boldsymbol{\theta}_i$ corresponding to the lowest $\bar{L}_i$ as the best. It is known that the probability of this $\boldsymbol{\theta}_i$ corresponding to $\boldsymbol{\theta}^*$ is at least $1 - \alpha$ whenever $L(\boldsymbol{\theta}_j) - L(\boldsymbol{\theta}^*) \geq \delta$ for all $\boldsymbol{\theta}_j \neq \boldsymbol{\theta}^*$.

Let us now present two examples of the above indifference zone procedures. The first example considers the procedure based on independent sampling while the second uses CRNs. The second example illustrates the reduction in number of simulation runs made possible by use of CRNs.

Example 14.10—Selection for $K = 4$ problem with independent sampling. Let us consider an example using the indifference zone method with independent sampling. Example 14.11 considers the same general setting with CRNs. Suppose that an analyst is evaluating four options via simulation and that the output is at least approximately normally distributed. Unknown to the analyst, suppose that the four options produce output having the distributions:

For all k: $y_k(\boldsymbol{\theta}_1) \sim N(0, 1)$, $y_k(\boldsymbol{\theta}_2) \sim N(0.25, 1)$, $y_k(\boldsymbol{\theta}_3) \sim N(0.25, 1)$, and $y_k(\boldsymbol{\theta}_4) \sim N(0.4, 1)$.

Because $y_k(\boldsymbol{\theta}_1)$ has the lowest mean (i.e., the lowest $L(\boldsymbol{\theta}_i)$), $\boldsymbol{\theta}^* = \boldsymbol{\theta}_1$. Suppose that $n_0 = 15$, $\delta = 0.25$, and $\alpha = 0.05$. Hence, only $L(\boldsymbol{\theta}_1)$ lies in the indifference zone $[L(\boldsymbol{\theta}^*), L(\boldsymbol{\theta}^*)+\delta)$. From Goldsman and Nelson (1998, Table 8.3), the appropriate critical value is $h = 3.285$. Using simulated data generated via the MATLAB **`randn`** generator, we obtain the samples of size n_0 and the associated sample variances s_i^2. The sample variances and the resulting total sample sizes n_i using (14.18) are as follows:

$$
\begin{aligned}
s_1^2 &= 0.912 &\Rightarrow&\quad n_1 = 158,\\
s_2^2 &= 1.042 &\Rightarrow&\quad n_2 = 180,\\
s_3^2 &= 0.873 &\Rightarrow&\quad n_3 = 151,\\
s_4^2 &= 0.708 &\Rightarrow&\quad n_4 = 123.
\end{aligned}
$$

This completes the first stage of the procedure.

Based on collecting the additional $n_i - n_0$ measurements for each of the four options, the analyst obtains the data needed for the second stage. The resulting sample means based on the combined data from the first and second stages are $\bar{L}_1 = -0.095$, $\bar{L}_2 = 0.238$, $\bar{L}_3 = 0.142$, and $\bar{L}_4 = 0.422$. Because $\bar{L}_1$ is the lowest sample mean and the true loss functions $L(\boldsymbol{\theta}_j)$ at $\boldsymbol{\theta}_j \neq \boldsymbol{\theta}^*$ are at least $\delta = 0.25$ unit greater than $L(\boldsymbol{\theta}^*)$, the analyst knows with probability at least $1 - \alpha = 0.95$ that $\boldsymbol{\theta}_1$ corresponds to $\boldsymbol{\theta}^*$. Of course, because truth is available here, we know that $\boldsymbol{\theta}_1$ is, in fact, equal to $\boldsymbol{\theta}^*$. ❑

Example 14.11—Selection for $K = 4$ problem with CRN sampling. Suppose that the overall problem setting and n_0, δ, and α are the same as in Example 14.10. In the approach here, however, suppose that the analyst can implement the CRN-based indifference zone approach. This example illustrates that more efficient analysis is possible through a reduction in the sample size needed to achieve equivalent accuracy. Suppose that the analyst knows that the simulation outputs $y_k(\boldsymbol{\theta}_i)$ have the same variance for all k, i. Further, suppose that the analyst knows that with CRNs for the simulations run at $\boldsymbol{\theta}_1$, $\boldsymbol{\theta}_2$, $\boldsymbol{\theta}_3$, and $\boldsymbol{\theta}_4$, $\text{cov}(y_k(\boldsymbol{\theta}_i), y_k(\boldsymbol{\theta}_j))$ does not depend on k, i, or j. Because the data are assumed normally distributed, the above-mentioned conditions for the CRN-based indifference zone method are met.

Unknown to the analyst, suppose that the four options produce output having the joint distribution

$$\begin{bmatrix} y_k(\boldsymbol{\theta}_1) \\ y_k(\boldsymbol{\theta}_2) \\ y_k(\boldsymbol{\theta}_3) \\ y_k(\boldsymbol{\theta}_4) \end{bmatrix} \sim N\left(\begin{bmatrix} 0 \\ 0.25 \\ 0.25 \\ 0.4 \end{bmatrix}, \begin{bmatrix} 1 & 0.5 & 0.5 & 0.5 \\ 0.5 & 1 & 0.5 & 0.5 \\ 0.5 & 0.5 & 1 & 0.5 \\ 0.5 & 0.5 & 0.5 & 1 \end{bmatrix} \right)$$

for all k. Because the variances are all unity, the off-diagonal covariances of 0.5 are equal to the correlations. Note that the means (i.e., the $L(\boldsymbol{\theta}_i)$) are identical to the means in Example 14.10, indicating that $\boldsymbol{\theta}^* = \boldsymbol{\theta}_1$. From Goldsman and Nelson (1998, Table 8.2), the appropriate critical value is $t^{(\alpha)}_{K-1,(K-1)(n_0-1)} = t^{(0.05)}_{3,42} = 2.127$. To generate the jointly distributed data, we applied the MATLAB **sqrtm** function to the above covariance matrix, producing a matrix square root (see Appendix A). This square root is used to multiply the output from the **randn** generator. Then, the appropriate mean components are added. From the resulting data and the MATLAB M-file **var_CRNindiffzone** at the book's Web site (which implements (14.19)), it is found that $s^2 = 0.8066$. This leads to a sample size of $n = 59$ by (14.20). As a reflection of the value of CRNs, this sample size is much smaller than the sample sizes of 123 to 180 needed in the independent sampling case of Example 14.10.

The above completes the first stage of the procedure. Based on collecting the additional $n - n_0 = 44$ measurement for each of the four options, the analyst obtains the data needed for the second stage. The resulting sample means from the data in the first and second stages are $\bar{L}_1 = -0.126$, $\bar{L}_2 = 0.015$, $\bar{L}_3 = 0.329$, and $\bar{L}_4 = 0.407$. Hence, as in Example 14.10, the analyst selects $\boldsymbol{\theta}_1$ as the best option. The analyst would know with probability at least $1 - \alpha = 0.95$ that this selection correctly corresponds to θ^*. ❑

14.6 CONCLUDING REMARKS

This chapter considered several issues that are important in using Monte Carlo simulations to optimize a process. We did not address the issue of building or validating the simulation model, as it was assumed that the simulation is a faithful representation of the process of interest. (Chapter 13 considered issues relevant to the construction of a simulation—or other—model of a process.) The use of simulations for optimizing (or at least improving) real processes is one of the most powerful applications of Monte Carlo methods. Using simulations can provide substantial cost savings over attempting to optimize the real process directly. In fact, in many applications it would be impossible to directly optimize a system by making changes to the process. Think, for example, of the problem of optimizing the operations of an airport by varying the terminal, taxiway, and runway locations. It is obviously not possible to carry out that optimization on the actual airport. There are, of course, countless other such systems.

Certainly, simulations are not a panacea for system design and optimization. There are many opportunities for abuse of the general approach if one is not aware of fundamental limitations. Many of the general issues and limitations on modeling discussed in Chapter 13 apply to simulation models. Misusing a "good" simulation or using a fundamentally flawed simulation may lead to solutions that are far from optimal and perhaps even dangerous to the operation of the associated physical system.

As abstractions of the real process, all simulations have limits of credibility. One must be aware of the limits in using the simulation for optimization and design. Such limits may result from simplifications in the mathematical representations in the simulation, restrictions following from regularity conditions for the validity of the embedded mathematical procedures, potential flaws in the pseudorandom number generator(s) (Appendix D), neglect of subtle interactions between a system and its environment, and the omission of critical inputs that perhaps are only relevant in some settings. For the validity of a simulation-based optimization of a real system, the search space (Θ) should correspond to the domain over which the simulation produces credible output.

An analyst should be aware that the slick graphics, animation, and data presentation of many modern commercial simulation packages do not automatically confer greater credibility to the output. Further, in many commercial packages, the underlying mathematical assumptions may be hidden from the user. In using a simulation to solve an optimization or other problem, there is simply no shortcut to having a sufficiently deep understanding of the real system and its relationship to the simulation.

With the limitations above in mind, the methods of this chapter are powerful means of improving processes. We discussed regenerative systems and how that structure can be exploited in simulations. Further, we discussed the use of gradient-free methods of optimization (FDSA and SPSA) that are designed to cope with noisy measurements of the loss function. The principle of common random numbers is a powerful means for variance reduction that can be used in concert with optimization to increase the effectiveness of the simulation runs. Finally, we discussed methods for selecting the best option from several candidates based on the notion of an indifference zone. These selection methods provide statistical guarantees on the quality of the choice.

The next chapter continues the theme of simulation-based optimization, but with an emphasis on gradient-based methods. These methods are important special cases of root-finding stochastic approximation as introduced in Chapter 4 and the general stochastic gradient methods of Chapter 5.

EXERCISES

14.1 Derive the bound in (14.8) to the bias for the regenerative estimate $\hat{L}_{N_1,N_2}(\boldsymbol{\theta})$ of the loss function (14.2).

14.2 For the regenerative-process-based loss function in (14.2), show that finding $\boldsymbol{\theta}$ such that $\boldsymbol{g}_{\text{eqv}}(\boldsymbol{\theta}) = \boldsymbol{0}$ in (14.10) yields the same solution as $\boldsymbol{g}(\boldsymbol{\theta}) = \boldsymbol{0}$.

14.3 Let C_θ and ℓ_θ be independent random variables with respective density functions $p_C(c) = e^{-c}$, $c > 0$, and $p_\ell(l)$ equal to a uniform density over the interval (m, M) (i.e., $U(m, M)$), $0 < m < M$. Consider the problem of estimating $L = E(C_\theta)/E(\ell_\theta)$ using (14.5) with $N_1 = N_2 = 1$. Compute the exact bias in $\hat{L}_{N_1,N_2} = \hat{L}_{1,1}$ (normalized by the value of L) for $m = 1$, $M = 2$ and $m = 1$, $M = 3$. Contrast the two biases with the corresponding bounds found by the Kantorovich inequality in (14.7) and (14.8).

14.4 Let X and Y be scalar random variables with known distribution functions F_X and F_Y. Assume that both X and Y can be generated in Monte Carlo fashion by the inverse transform method based on an underlying uniform random variable (Appendix D). Suppose that one is going to estimate $E(X - Y)$ by repeated generation of values of X and Y.

(a) Using the inverse transform method, describe a CRN and a non-CRN means by which this estimation can be carried out.

(b) Prove using (14.12) that the CRN method will yield a variance for the estimate that is less than or equal to the variance for the non-CRN estimate. (Hint: If $f_1(\cdot)$ and $f_2(\cdot)$ are two nondecreasing functions, $E[f_1(Z)f_2(Z)] \geq E[f_1(Z)]E[f_2(Z)]$ for any random variable Z taking values in the domains of $f_1(\cdot)$ and $f_2(\cdot)$ provided that the indicated expectations exist [Tong, 1980, p. 13].)

14.5 Complete the proof of Proposition 14.2. In particular, take the stated result of Rubinstein et al. (1985) as a given and expand on the second sentence of the current sketch of the proof.

14.6 Suppose that the simulation output can be represented as a linear regression $\boldsymbol{Z} = \boldsymbol{B}\boldsymbol{\theta} + \boldsymbol{V}$, where $\boldsymbol{V}$ is a noise vector with finite covariance matrix ($\boldsymbol{B}$ is not dependent on $\boldsymbol{V}$). Given that $L(\boldsymbol{\theta}) = E[Q(\boldsymbol{\theta}, \boldsymbol{V})] = E[(\boldsymbol{Z} - \boldsymbol{d})^T(\boldsymbol{Z} - \boldsymbol{d})]$, where $\boldsymbol{d}$

is some "desired" vector, analyze the variability of $y_k^{(+)} - y_k^{(-)}$ under CRNs in a manner similar to that of Example 14.7. In particular:

(a) Write down the expression for the measurements $y_k^{(\pm)}$ (in terms of the given regression model) and isolate the specific part of $y_k^{(+)}$ and $y_k^{(-)}$ that have the opposite sign under CRNs. In practice, what tends to mitigate the effects of this opposite sign?

(b) Compare in general terms (detailed calculation not required) the value of $\text{var}[(y_k^{(+)} - y_k^{(-)}) \mid \hat{\boldsymbol{\theta}}_k, \boldsymbol{\Delta}_k]$ under CRNs and under non-CRNs (with independent noise terms $\boldsymbol{V}_k^{(+)}$ and $\boldsymbol{V}_k^{(-)}$). Show that the variance under CRNs is reduced even if the monotonicity assumption of Proposition 14.2 is not satisfied.

14.7 Consider the setting of Example 14.7 with $\dim(\boldsymbol{Z}) = 2$, $\boldsymbol{V} = [\mathcal{V}_1, \mathcal{V}_2, \mathcal{V}_3]^T \sim N(\boldsymbol{0}, \boldsymbol{I}_3)$, $\boldsymbol{d} = \boldsymbol{0}$, and $\boldsymbol{B} = \begin{bmatrix} \mathcal{V}_1^2 & \mathcal{V}_2 \\ 0 & \mathcal{V}_3 \end{bmatrix}$. Suppose that the SPSA gradient approximation will be used conditioned on $\boldsymbol{\theta} = \hat{\boldsymbol{\theta}}_k = [1, 1]^T$, where the Δ_{ki} are symmetric Bernoulli ± 1 distributed.

(a) Let $c_k = 1$. For both the CRN and non-CRN case, determine by any convenient means (analytically or Monte Carlo) the covariance matrix of the gradient estimate conditional on $\boldsymbol{\theta} = \hat{\boldsymbol{\theta}}_k$ (this conditional covariance takes account of the randomness in the perturbations and $\boldsymbol{V}$). For the non-CRN case, the two $\boldsymbol{V}$ processes (in the two required values of $\boldsymbol{Z}$) should be independent of each other. Contrast the two covariance matrices; is one of the covariance matrices smaller in the matrix sense (see Appendix A)?

(b) Repeat part (a) except for setting $c_k = 0.1$.

(c) Comment on the relative performance of CRN and non-CRN for the two c_k values in parts (a) and (b). Include comments about the implications for CRNs in optimization.

14.8 In Example 14.8, show that $L(\boldsymbol{\theta}) = \boldsymbol{\theta}^T\boldsymbol{\theta} + \sum_{i=1}^{10} \lambda_i/(\lambda_i + t_i)$. (Hint: It is useful in this derivation to recall that the inverse transform method is used to generate the exponentially distributed $\mathcal{V}_i$.)

14.9 For $n = 1500$ iterations in the setting of Example 14.8, compare CRN and non-CRN when they are run with the *same* gain sequence values (versus the different γ in the example). Set $a = 0.7$, $c = 0.5$, $\alpha = 1$, and $\gamma = 0.49$, as in the CRN case of Example 14.8.

(a) Establish that these are valid gain values for non-CRN SPSA (i.e., condition B.1″ and other gain conditions in Theorem 7.2 in Section 7.4 hold).

(b) From 50 replications, calculate the mean terminal loss values and the normalized $\boldsymbol{\theta}$ estimates (as in Table 14.3) for the CRN and non-CRN cases. Compare the terminal loss values with the unmatched pairs two-sample t-test.

14.10 Consider the partial CRN setting of (14.17). Assume that $\boldsymbol{V}_{k_1}$ is independent of $\{\boldsymbol{V}_{k_2}^{(+)}, \boldsymbol{V}_{k_2}^{(-)}\}$ and that the conditions of Proposition 14.2 apply to the synchronized input $Q_1(\hat{\theta}_k + c_k\Delta_k, V)$ and $Q_1(\hat{\theta}_k - c_k\Delta_k, V)$. Show that the partial CRN form leads to a variance reduction in the SPSA gradient approximation (versus the case with no CRNs). (Hint: Some of the same arguments in Exercise 14.5 apply here; it is sufficient to simply reference those arguments if you completed Exercise 14.5.)

14.11 For the indifference zone approach with K options, identify a bound to α such that it never makes sense to choose α larger than this bound. This bound should be less than 1.

14.12 Carry out a simulated indifference zone selection process under independent sampling and the following problem settings: $K = 5$, $\alpha = 0.05$, $\delta = 0.2$, $n_0 = 40$, and $h = 3.264$. For all k, suppose that the simulation outputs are distributed according to: $y_k(\theta_1) \sim N(0, 1)$, $y_k(\theta_2) \sim N(0.2, 1)$, $y_k(\theta_3) \sim N(0.2, 1)$, $y_k(\theta_4) \sim N(0.2, 1)$, and $y_k(\theta_5) \sim N(0.4, 1)$.

14.13 Consider the setting and goals of Exercise 14.12 with the exception of using CRN-based sampling. In the simulated indifference zone selection process, assume that $\text{cov}[y_k(\theta_i), y_k(\theta_j)] = 0.5$ for all k, i, and j. The required critical value is $t^{(\alpha)}_{K-1,(K-1)(n_0-1)} = t^{(0.05)}_{4,156} = 2.17$. If you also did Exercise 14.12, contrast the sample sizes needed for the independent and CRN sampling implementations.

CHAPTER 15

SIMULATION-BASED OPTIMIZATION II: STOCHASTIC GRADIENT AND SAMPLE PATH METHODS

The gradient-free methods that were emphasized in Chapter 14 essentially treat the simulation as a black box, assuming minimal knowledge of the inner workings of the simulation. In some cases, however, one *does* have knowledge of how the simulation is constructed. This chapter describes how that knowledge can be exploited to enhance simulation-based optimization. Calculation of the stochastic gradient (as in Chapter 5) provides the mechanism through which the knowledge is used. Over the past several decades, a large amount of effort has gone into developing stochastic gradient methods for simulation-based optimization. This chapter is a sampling of available methods.

Section 15.1 covers some general issues in gradient estimation, building on some of the results in the stochastic gradient analysis of Chapter 5. This section includes a discussion on the role of interchange of derivative and integral. Section 15.2 considers two important cases of gradient estimation methods in a simulation-based context—the likelihood ratio (score function) method and the infinitesimal perturbation method. The resulting search algorithms have an interpretation as a root-finding (Robbins–Monro) stochastic approximation algorithm (Chapter 4). Section 15.3 considers the implementation of the stochastic gradient-based optimization methods, including a numerical example. Section 15.4 presents the sample path method of simulation-based optimization. Here, the *stochastic* simulation-based problem is converted to a *deterministic* problem that provides a solution closely approximating the stochastic problem. Section 15.5 closes the chapter with some concluding remarks and a mention of some other methods for simulation-based gradient estimation.

15.1 FRAMEWORK FOR GRADIENT ESTIMATION

15.1.1 Some Issues in Gradient Estimation

Let us consider some of the main gradient-based methods for optimization in a simulation-based context. Such methods have been a major focus of research in simulation modeling for many years. The primary motivation for use of such methods is their potential for faster convergence (i.e., fewer simulation runs) than the nongradient methods discussed in Chapter 14. The focus continues to be on the optimization problem introduced in Section 14.1, namely, to minimize

$$L(\boldsymbol{\theta}) = E[Q(\boldsymbol{\theta}, V)],$$

where V represents the amalgamation of the (pseudo) random effects in the simulation (presumably manifesting themselves in a way indicative of the random effects in the actual system) and Q represents a *sample* realization of the loss function calculated from running the simulation.

As discussed in Chapter 1 and elsewhere, the gradient $\boldsymbol{g}(\boldsymbol{\theta}) = \partial L/\partial\boldsymbol{\theta}$—or a noisy estimate of the gradient—is an important tool in optimization of differentiable loss functions. A number of methods exist for estimating the gradient as it arises in the simulation-based context. We discuss methods here that use the structure of the root-finding stochastic approximation (SA) algorithm (Chapters 4 and 5) and the structure of deterministic nonlinear programming (briefly introduced in Section 1.4). While the emphasis in most stochastic or deterministic gradient-based methods is local optimization, recall from Section 8.4 that gradient-based methods can also be used in *global* optimization via the principle of injected noise with annealing.

Within the root-finding context, Section 5.3 was a brief introduction to one of the more popular approaches—the infinitesimal perturbation analysis (IPA) method. This chapter will broaden our horizons, considering a general setting that includes IPA as a special case. Another method that has received considerable attention is the *likelihood ratio*—sometimes called *score function*—approach (LR/SF). LR/SF for simulation-based optimization is discussed, for example, in Rubinstein (1986, Sect. 3.6), Glynn (1987), L'Ecuyer (1990a), Fu (1994), and Rubinstein and Melamed (1998, Part II). The aim of both IPA and LR/SF is to produce an estimate of $\boldsymbol{g}(\boldsymbol{\theta}) = \partial L/\partial\boldsymbol{\theta}$. While this estimate has several applications, one of the most important is in the context of a root-finding SA-based optimization method. Recall from Chapter 5 that root-finding SA with a noisy measurement of the gradient is called a *stochastic gradient method*. For purposes of efficiency, an aim in simulation-based optimization using the stochastic gradient formulation is to produce the gradient measurement with only one simulation run (versus, say, the $2p$ runs of FDSA or the two runs of SPSA, as discussed in Section 14.3[1]). Another aim in the gradient estimation is to produce *unbiased* estimates of $\boldsymbol{g}(\boldsymbol{\theta})$.

---

[1]Recall from Section 7.7 that there is also a one-measurement version of SPSA (versus the standard two-measurement form). This yields an SPSA gradient estimate from one simulation run without the need for the modeling information required in the gradient-based SA methods. However, as discussed in Section 7.7, this one-run SPSA is generally inferior to the standard two-run form. Further, it has a slower asymptotic rate of convergence than the gradient-based methods.

This chapter also considers a method that is based on converting the inherently stochastic simulation-based problem to a potentially easier-to-solve deterministic problem. The solution via this *sample path method* can be made arbitrarily close to the true optimal solution $\boldsymbol{\theta}^*$ by increasing the number of simulation runs in an averaging process. One can use deterministic nonlinear programming methods (e.g., Bazaraa et al., 1993) to find the solution in the sample path method. Because of the relatively well-developed methods for constrained optimization in nonlinear programming, this approach has particular appeal when there are difficult constraints to be satisfied.

In discussing the methods for gradient estimation, it is useful to distinguish between two types of parameters:

> ***Distributional parameters*** are those elements of $\boldsymbol{\theta}$ that enter via their effect on the probability distribution of the random terms V in the system. So, for example, if a process involves $N(\mu, \sigma^2)$-generated inputs, where μ and σ^2 are subject to design considerations (i.e., are a part of $\boldsymbol{\theta}$), then μ and σ^2 would be distributional parameters. These parameters enter the probability density or mass function for the random effects V, $p_V(\cdot|\boldsymbol{\theta})$, discussed below.
>
> ***Structural parameters*** in $\boldsymbol{\theta}$ have their effect directly on the output Q, typically involving an input–output relationship due, say, to the physics of a process. The plant layout problem in Chapter 14 (Examples 14.1 and 14.5) depicts a case with structural parameters when the probability distributions for the random effects in the process (e.g., the distribution of the parts arrivals in the plant layout example) do not depend on $\boldsymbol{\theta}$. The machine locations ($\boldsymbol{\theta}$) have a direct effect on the product output Q by governing the flow of material and people that have an impact on operations. For general problems, structural parameters are those entering the observed cost $Q(\boldsymbol{\theta}, V)$.

It is easy to envision cases where the division between distributional and structural parameters is not obvious. See, for example, L'Ecuyer (1990a), Fu and Hu (1997, pp. 3–4 and 11–12), Rubinstein and Melamed (1998, pp. 115–117), and Subsection 5.1.1 (Example 5.1), Subsection 15.1.2, and Section 15.2 here. In fact, the division is often somewhat arbitrary in the sense that V could represent the raw uniform random variables generated for the simulation. Then all the distributional transformation operations are embedded in Q, implying that all parameters $\boldsymbol{\theta}$ are structural parameters. In a mathematical sense, therefore, the distinction may not always be critical. But, for *applications*, the distinction may be important since the parameters may be allocated to the two categories in a physically meaningful way. As in Chapter 14, we have adopted the common approach here of having V be the "natural" random variables of the process (rather than the underlying $U(0, 1)$ random variables) because of the easier physical interpretability of the analysis. Sections 15.2 and 15.3 show that the long-standing division between two of the most popular gradient estimators

(LR/SF and IPA) rests critically on the distinction between whether $\boldsymbol{\theta}$ represents distributional or structural parameters.

Initially, the LR/SF and IPA approaches developed largely independent of each other. However, it has long been recognized that there is a common basis for the approaches (e.g., L'Ecuyer, 1990a). We adopt this more general framework here, but first let us make a few comments about the original versions of LR/SF and IPA. A critical assumption of the original LR/SF formulation (e.g., Aleksandrov et al., 1968; Glynn, 1987) is that the elements of $\boldsymbol{\theta}$ enter only as *distributional* parameters rather than through some explicit structure in the problem affecting Q. So, the original LR/SF would not apply in the plant layout problem of Chapter 14 (Examples 14.1 and 14.5) when $\boldsymbol{\theta}$ corresponds to the machine locations as they appear in the sample loss Q. LR/SF would apply in the case where $\boldsymbol{\theta}$ appears in the distribution function for $\boldsymbol{V}$, representing the means of the service times for a set of machines as derived from prior knowledge of how location and service time are related (so that optimizing the service times is tantamount to optimizing the machine locations).

In a complementary manner, the original IPA approach (e.g., Ho, 1987) applies to problems where the $\boldsymbol{\theta}$ dependence appears only as structural parameters in the sample loss function Q, not in the distribution function generating the $\boldsymbol{V}$. Of course, much research has occurred since the original work in extending the LR/SF and IPA approaches to broader settings; a summary is in Rubinstein and Melamed (1998, pp. 112–118). We will adopt the broader setting where $\boldsymbol{\theta}$ can enter both (or either) of Q or the distribution function.

15.1.2 Gradient Estimation and the Interchange of Derivative and Integral

Consistent with previous usage, suppose that there exists a probability density or mass function for the cumulative random effects $\boldsymbol{V}$ in the simulation, $p_V(\boldsymbol{\upsilon}|\boldsymbol{\theta})$, where $\boldsymbol{\upsilon}$ represents the dummy variable for use in integrals associated with expected value calculations. In general, this density or mass function depends on $\boldsymbol{\theta}$, although there are special cases where it does not (i.e., $p_V(\boldsymbol{\upsilon}|\boldsymbol{\theta}) = p_V(\boldsymbol{\upsilon})$). For the discussion below, let us assume that $p_V = p_V(\boldsymbol{\upsilon}|\boldsymbol{\theta})$ is a *density* function. When p_V is a mass function, we simply replace the indicated integrals with sums. The "bottom line" result below, estimate (15.4), applies for either the density or mass case. Recall the familiar formulation

$$L(\boldsymbol{\theta}) = E[Q(\boldsymbol{\theta}, \boldsymbol{V})] = \int_\Lambda Q(\boldsymbol{\theta}, \boldsymbol{\upsilon}) p_V(\boldsymbol{\upsilon} \,|\, \boldsymbol{\theta})\, d\boldsymbol{\upsilon}, \tag{15.1}$$

where Λ is the domain for $\boldsymbol{V}$ (i.e., $p_V(\boldsymbol{\upsilon}|\boldsymbol{\theta}) = 0$ outside of Λ).

Assume that Q and p_V are differentiable with respect to $\boldsymbol{\theta}$ (this assumption is also needed when p_V is a mass function). A central issue in the necessary gradient calculations of this section is the question of validity of the following interchange of the order of integration and differentiation:

$$\frac{\partial L}{\partial \boldsymbol{\theta}} = \frac{\partial}{\partial \boldsymbol{\theta}}\int_{\Lambda} Q(\boldsymbol{\theta},\boldsymbol{\upsilon})p_V(\boldsymbol{\upsilon}\,|\,\boldsymbol{\theta})\,d\boldsymbol{\upsilon} \stackrel{?}{=} \int_{\Lambda}\frac{\partial[Q(\boldsymbol{\theta},\boldsymbol{\upsilon})p_V(\boldsymbol{\upsilon}\,|\,\boldsymbol{\theta})]}{\partial \boldsymbol{\theta}}\,d\boldsymbol{\upsilon}.$$

The following theorem provides sufficient conditions for this interchange to hold. Recall that $\Theta \times \Lambda = \{(\boldsymbol{\theta}, \boldsymbol{\upsilon}): \boldsymbol{\theta} \in \Theta, \boldsymbol{\upsilon} \in \Lambda\}$ (the Θ here is the constraint domain for $\boldsymbol{\theta}$ introduced in Chapter 1). The result is a direct application of Theorem A.3 in Appendix A. Similar theorems are given in L'Ecuyer (1990a), Glasserman (1991a, pp. 49–53, 102–104; 1991b), and Rubinstein and Melamed (1998, pp. 123–124).

Theorem 15.1. Suppose that $\Theta \subset \mathbb{R}^p$ is an open set. Let $Q\times p_V$ and $\partial(Q\times p_V)/\partial\boldsymbol{\theta}$ be continuous on $\Theta\times\Lambda$. Suppose that there exist nonnegative functions $q_0(\boldsymbol{\upsilon})$ and $q_1(\boldsymbol{\upsilon})$ such that

$$\left|Q(\boldsymbol{\theta},\boldsymbol{\upsilon})p_V(\boldsymbol{\upsilon}\,|\,\boldsymbol{\theta})\right| \le q_0(\boldsymbol{\upsilon}); \quad \left\|\frac{\partial[Q(\boldsymbol{\theta},\boldsymbol{\upsilon})p_V(\boldsymbol{\upsilon}\,|\,\boldsymbol{\theta})]}{\partial\boldsymbol{\theta}}\right\| \le q_1(\boldsymbol{\upsilon}) \text{ for all } (\boldsymbol{\theta},\boldsymbol{\upsilon}) \in \Theta\times\Lambda,$$

where $\int_{\Lambda} q_0(\boldsymbol{\upsilon})\,d\boldsymbol{\upsilon} < \infty$ and $\int_{\Lambda} q_1(\boldsymbol{\upsilon})\,d\boldsymbol{\upsilon} < \infty$. Then

$$\frac{\partial}{\partial \boldsymbol{\theta}}\int_{\Lambda} Q(\boldsymbol{\theta},\boldsymbol{\upsilon})p_V(\boldsymbol{\upsilon}\,|\,\boldsymbol{\theta})\,d\boldsymbol{\upsilon} = \int_{\Lambda}\frac{\partial[Q(\boldsymbol{\theta},\boldsymbol{\upsilon})p_V(\boldsymbol{\upsilon}\,|\,\boldsymbol{\theta})]}{\partial \boldsymbol{\theta}}\,d\boldsymbol{\upsilon}.$$

Although the conditions of Theorem 15.1 may appear rather abstract, there is a strong association with practical concerns in simulations. In fact, it is not automatic that the conditions of Theorem 15.1 will hold in many practical simulations. For example, in the case where $\boldsymbol{\theta}$ represents only structural parameters (so $\boldsymbol{\theta}$ does not enter p_V), the continuity conditions on Q and $\partial Q/\partial\boldsymbol{\theta}$ require that small changes in $\boldsymbol{\theta}$ (the infinitesimal perturbations of IPA) correspond to small changes in the simulation output that is relevant to the computation of Q. In a discrete-event simulation, this is equivalent to assuming that small perturbations in $\boldsymbol{\theta}$ do not change the sequence of events but rather change in a small way only the occurrence times or possibly the magnitude of the events.

As discussed in several contexts in Chapter 14 (especially in the discussion of common random numbers—Section 14.4), small changes in $\boldsymbol{\theta}$ may sometimes lead to large changes in the output via a change in the way the random effects V are processed. If, for instance, the aim is to maximize the mean number of times that a queue is clear when a customer arrives (e.g., maximize the number of regeneration cycles in a setting such as Figure 14.1), then Theorem 15.1 will not apply when a small change in $\boldsymbol{\theta}$ causes one of the customers currently facing a zero wait to face a nonzero wait (e.g., a loss of one of the regeneration cycles in Figure 14.1). In general, because of the flexibility one may have in defining p_V vis-à-vis Q, there may be cases where one definition of p_V and Q may lead to a violation of the theorem conditions while another equivalent definition (in the sense that L is the same for any $\boldsymbol{\theta}$) will satisfy the conditions (recall that the conditions are *sufficient*, but not necessary). Example 5.1 (Subsection 5.1.1) illustrated the flexibility in a particular stochastic gradient problem; the same principles apply in the simulation-based context.

The nonuniqueness in defining distributional and structural parameters discussed above is one way in which equivalent definitions may lead to different conclusions regarding the applicability of Theorem 15.1. Example 15.1 and Exercise 15.1 illustrate this principle in a reliability problem. Similarly, sometimes a small modification in the goals of the optimization may alter the applicability of Theorem 15.1. For example, in the queuing problem of the preceding paragraph, if the aim is changed from maximizing the number of zero waits to minimizing mean wait time, then a small change in $\boldsymbol{\theta}$ may have only a small effect on Q, enhancing the applicability of Theorem 15.1.

Before presenting successful examples of the interchange of derivative and integral in Sections 15.2 and 15.3, let us present a conceptual example where the interchange is *not* allowed by Theorem 15.1. This example is a realistic problem in reliability with multiple components. Following some subsequent discussion on gradient estimates for implementation we will present a second example of the failure of interchange. Suri (1989), L'Ecuyer (1990a), and Fu and Hu (1997, pp. 146–147) present a number of other examples where the interchange of derivative and integral is not allowed (see also Example A.3 in Appendix A here). These include communications network simulations, repair/reliability models different from those below, and general queuing networks. The fundamental principle in these examples is that small changes in $\boldsymbol{\theta}$ can have a dramatic effect on the simulation output (e.g., number of customers processed in a queue).

Example 15.1—Reliability system (L'Ecuyer, 1990a). This example shows how a small change in $\boldsymbol{\theta}$ can sometimes lead to a large change in Q due to a different ordering of events. Such a discontinuous Q violates one of the conditions of Theorem 15.1. Consider a system with multiple identical components, each operating independently with a random failure time given by a density function that is strictly positive on the interval $(0, T)$, T being the maximum time of interest. The aim is to develop a component replacement strategy that minimizes the total costs over time $(0, T)$ due to component failures and replacements. The costs include a failure cost every time a component fails, a fixed cost for having to intervene in the system to replace one or more components, and a replacement cost that is directly proportional to the number of components that are to be replaced. Preventive replacements are used to reduce the number of component failures and to reduce the number of times that system intervention (with its associated fixed cost) is required.

Suppose that $\boldsymbol{\theta}$ contains two parameters: $t_1 > t_2 > 0$. Whenever a component fails or reaches age t_1 that component is replaced together with all other components having an age at least t_2. Suppose that V represents the collection of natural component failure times (i.e., the failure times that would occur if $t_1 = t_2 = \infty$). Note that $p_V(\boldsymbol{\upsilon}|\boldsymbol{\theta}) = p_V(\boldsymbol{\upsilon})$.

The aim is to optimize the mean cost of system operations, where the total cost reflects the sum of failure, fixed, and replacement cost; a single experiment produces an observed cost Q. Note, however, that $Q = Q(\boldsymbol{\theta}, \boldsymbol{\upsilon})$ is discontinuous in $\boldsymbol{\theta}$ whenever t_1 or t_2 is close enough to one of the failure times in the fixed value for $\boldsymbol{\upsilon}$ (the dummy variable for V). So a small change in $\boldsymbol{\theta}$ causes a change in the sequence of failures (i.e., a step change in the observed total cost). So, for example, if an element of V has a value 100, then $\lim_{\eta\to 0}\{Q([100+\eta, t_2]^T, V) - Q([100-\eta, t_2]^T, V)\} \neq 0$ due to the inclusion of the failure cost in $Q([100+\eta, t_2]^T, V)$ but not in $Q([100-\eta, t_2]^T, V)$ for any $\eta > 0$ (the indicated limit is as η *decreases* to 0). Hence, Theorem 15.1 does not apply here. As shown in Exercise 15.1, however, a redefinition of V can lead to the continuity condition being satisfied. ❑

We now show how the derivative–integral interchange is used to construct an estimate of $\boldsymbol{g}(\boldsymbol{\theta}) = \partial L/\partial\boldsymbol{\theta}$. Under the conditions of Theorem 15.1 (or one of the equivalent results mentioned above),

$$\begin{aligned} \boldsymbol{g}(\boldsymbol{\theta}) &= \frac{\partial}{\partial\boldsymbol{\theta}}\int_\Lambda Q(\boldsymbol{\theta},\boldsymbol{\upsilon})p_V(\boldsymbol{\upsilon}|\boldsymbol{\theta})\,d\boldsymbol{\upsilon} \\ &= \int_\Lambda\left[Q(\boldsymbol{\theta},\boldsymbol{\upsilon})\frac{\partial p_V(\boldsymbol{\upsilon}|\boldsymbol{\theta})}{\partial\boldsymbol{\theta}} + \frac{\partial Q(\boldsymbol{\theta},\boldsymbol{\upsilon})}{\partial\boldsymbol{\theta}}p_V(\boldsymbol{\upsilon}|\boldsymbol{\theta})\right]d\boldsymbol{\upsilon}, \end{aligned} \tag{15.2}$$

where the last equality follows by the well-known product rule of calculus applied to $\partial[Q(\boldsymbol{\theta},\boldsymbol{\upsilon})p_V(\boldsymbol{\upsilon}|\boldsymbol{\theta})]/\partial\boldsymbol{\theta}$. The expression above does not lend itself to direct representation as an expectation since the first summand in the integral after the second equality does not represent an integral against a density function. Fortunately, it is easy to convert this part of the integral to an expected value. Assuming that $p_V(\boldsymbol{\upsilon}|\boldsymbol{\theta}) > 0$, simply multiply by $p_V(\boldsymbol{\upsilon}|\boldsymbol{\theta})/p_V(\boldsymbol{\upsilon}|\boldsymbol{\theta})$ to obtain

$$\begin{aligned} \boldsymbol{g}(\boldsymbol{\theta}) &= \int_\Lambda\left[Q(\boldsymbol{\theta},\boldsymbol{\upsilon})p_V(\boldsymbol{\upsilon}|\boldsymbol{\theta})^{-1}\frac{\partial p_V(\boldsymbol{\upsilon}|\boldsymbol{\theta})}{\partial\boldsymbol{\theta}} + \frac{\partial Q(\boldsymbol{\theta},\boldsymbol{\upsilon})}{\partial\boldsymbol{\theta}}\right]p_V(\boldsymbol{\upsilon}|\boldsymbol{\theta})\,d\boldsymbol{\upsilon} \\ &= E\left[Q(\boldsymbol{\theta},V)\frac{\partial\log p_V(V|\boldsymbol{\theta})}{\partial\boldsymbol{\theta}} + \frac{\partial Q(\boldsymbol{\theta},V)}{\partial\boldsymbol{\theta}}\right], \end{aligned} \tag{15.3}$$

where the second line follows by the fundamental formula $\partial\log f(\boldsymbol{x})/\partial\boldsymbol{x} = f(\boldsymbol{x})^{-1}\partial f(\boldsymbol{x})/\partial\boldsymbol{x}$ for a differentiable function $f(\boldsymbol{x}) > 0$.

From (15.3), there is a ready way to produce unbiased estimates of $\boldsymbol{g}(\boldsymbol{\theta})$ at any $\boldsymbol{\theta}$. In particular, for a sample V, one unbiased gradient estimate is

$$Q(\boldsymbol{\theta},V)\frac{\partial\log p_V(V|\boldsymbol{\theta})}{\partial\boldsymbol{\theta}} + \frac{\partial Q(\boldsymbol{\theta},V)}{\partial\boldsymbol{\theta}}. \tag{15.4}$$

In practice, the V in (15.4) is associated with one simulation run. In contrast to the gradient-free methods of Sections 14.3 and 14.4 (FDSA and SPSA), detailed knowledge of the inner working of the simulation is required to form the estimate in (15.4). In particular, it must be possible to compute the two indicated gradients. To smooth out noise effects, an obvious variation on (15.4) is to average independent samples (independent V values), with each sample associated with one simulation run. Because this changes nothing fundamental in the discussion below, we focus on the one-simulation gradient estimate in (15.4).

In the spirit of Example 15.1, the example below is an invalid interchange of derivative and expectation, leading to a misapplication of the result (15.4). This is presented as a cautionary illustration on the importance of the regularity conditions for the interchange (such as in Theorem 15.1).

Example 15.2—Incorrect gradient estimate. This example demonstrates how a naïve approach to interchanging the derivative and integral can lead to an inappropriate gradient estimate. Suppose that θ and V are scalars with $Q(\theta, V) = (1-V)^2$ and $\theta > 0$ a distributional parameter arising via the assumption that $V \sim U(0, \theta)$ (so $\Lambda = (0, \theta)$). Straightforward calculations show that $L(\theta) = 1 - \theta + \theta^2/3$, implying that $g(\theta) = -1 + 2\theta/3$ and $\theta^* = 3/2$.

Let us now show how the conditions of Theorem 15.1 are violated. The density function is $p_V(\upsilon|\theta) = \theta^{-1} I_{\{0<\upsilon<\theta\}}$, where $I_{\{\cdot\}}$ is the indicator function (defined in the section "Frequently Used Notation" near the back of the book). Then one needs to compute $\partial(Q \times p_V)/\partial\theta = Q\partial p_V/\partial\theta$. Following Fu and Hu (1997, pp. 15–16), the product rule for derivatives yields

$$\frac{\partial p_V(\upsilon \mid \theta)}{\partial \theta} = \theta^{-1}\delta(\theta - \upsilon) - \theta^{-2} I_{\{0<\upsilon<\theta\}}, \tag{15.5}$$

where the Dirac δ-function is defined in relation to the indicator (step) function by

$$I_{\{\upsilon \geq \theta\}} = \int_{-\infty}^{\upsilon} \delta(y - \theta)\, dy, \tag{15.6}$$

or, equivalently,

$$I_{\{\upsilon<\theta\}} = 1 - I_{\{\upsilon \geq \theta\}} = \int_{\upsilon}^{\infty} \delta(y - \theta)\, dy. \tag{15.7}$$

Note that the δ-function appearing in (15.5) can be thought of, informally, as the derivative resulting from the antiderivative property of calculus applied to the integral in (15.7) (also noting the equivalence of $I_{\{\upsilon<\theta\}}$ and $I_{\{0<\upsilon<\theta\}}$ given the definitions of Λ). In this sense, the δ-function can be thought of as a type of derivative of the indicator function. This is an intuitive—not mathematically rigorous—statement.

Observe that the δ-function in (15.5) implies that the gradient $\partial p_V/\partial\theta$ is unbounded, indicating that $\partial(Q\times p_V)/\partial\theta$ is unbounded. Hence, there exists no $q_1(\upsilon)$ with finite integral that can bound this gradient, a violation of a condition in Theorem 15.1.

Suppose that one ignores the violation of conditions for Theorem 15.1 and plows ahead with interchanging the derivative and integral via (15.2), leading to an estimator of the form (15.4). Because $\partial Q/\partial\theta = 0$, the invalid estimator of $g(\theta)$ at an observed V is then $Q\partial\log p_V/\partial\theta = (Q/p_V)\partial p_V/\partial\theta = (1-V)^2[\theta^{-1}\delta(\theta-V)-\theta^{-2}I_{\{0<V<\theta\}}]/\theta^{-1}$ on Λ. Given that $\delta(\theta - V) = 0$ on Λ, this estimator can be simplified to $-(1-V)^2/\theta = -Q(\theta,V)/\theta$. (This type of result also appears in Fu and Hu, 1997, p. 147, but in a different problem context.) The mean of this invalid gradient estimator is

$$E\left[-\frac{Q(\theta,V)}{\theta}\right] = -\int_\Lambda (1-\upsilon)^2\theta^{-2}d\upsilon = \frac{-3+3\theta-\theta^2}{3\theta}. \tag{15.8}$$

Because we are seeking an unbiased gradient estimator, the mean in (15.8) *should* equal $g(\theta) = -1 + 2\theta/3$. But it does not.

As an indication of the severity of the bias in the invalid estimator, let us compare values of $g(\theta)$ with the means from (15.8) at $\theta = 1, 3/2$ (the value θ^*), and 10. We find, respectively, that $g(\theta) = -0.33$, 0, and 5.67 while the corresponding means from (15.8) are −0.33, −0.17, and −2.43. These discrepancies illustrate the danger of a cavalier interchange of derivative and integral. ❑

The setting above is a broad basis for gradient estimation without regard to the specific search algorithm being used. The estimate in (15.4) can be used in general algorithms requiring an unbiased estimate of the gradient. The remainder of this chapter focuses on important search approaches that make use of the above gradient estimation framework.

15.2 PURE LIKELIHOOD RATIO/SCORE FUNCTION AND PURE INFINITESIMAL PERTURBATION ANALYSIS

As indicated in Section 15.1, there has been a traditional segregation of the LR/SF and IPA approaches. Although we have adopted a general framework that encompasses these two approaches—as manifested in the general form of gradient estimate in (15.4)—let us briefly comment on the relative properties of the traditional "pure" LR/SF and IPA approaches for simulation-based optimization. Recall from Section 15.1 that pure LR/SF corresponds to the setting where $Q(\theta, V) = Q(V)$, and pure IPA is where $p_V(\upsilon|\theta) = p_V(\upsilon)$. So, in the former, the parameter effects must all be captured in the distribution function for V, while in the latter they must all be captured in the observed loss (i.e., the distinction between distributional and structural parameters discussed in Section 15.1). This section discusses gradient estimation per se; the next section discusses the application to search and optimization.

For pure LR/SF, only a simulation output $Q(V)$ and density derivative $\partial \log p_V / \partial \boldsymbol{\theta}$ (derived directly from $\partial p_V / \partial \boldsymbol{\theta}$) are required, while with pure IPA, only the performance derivative $\partial Q / \partial \boldsymbol{\theta}$ is required. The labels LR and SF come from two related concepts. LR comes from the ratio $p_V(\boldsymbol{\upsilon} | \boldsymbol{\theta}) / p_V(\boldsymbol{\upsilon} | \boldsymbol{\theta}')$ that appears in the sample path version of the method that is discussed in Section 15.5, with the ratio measuring the relative likelihood of the outcome $V = \boldsymbol{\upsilon}$ under the two values $\boldsymbol{\theta}$ and $\boldsymbol{\theta}'$. SF comes from standard statistics terminology, where $\partial \log p_V / \partial \boldsymbol{\theta}$ is the score function set to zero for finding a maximum likelihood estimate. In particular, from (15.4), the LR/SF and IPA gradient estimates at a given $\boldsymbol{\theta}$ are

$$\textbf{LR/SF:} \quad Q(V) \frac{\partial \log p_V(V | \boldsymbol{\theta})}{\partial \boldsymbol{\theta}} = \frac{Q(V)}{p_V(V | \boldsymbol{\theta})} \frac{\partial p_V(V | \boldsymbol{\theta})}{\partial \boldsymbol{\theta}}, \quad p_V(V|\boldsymbol{\theta}) > 0, \tag{15.9}$$

$$\textbf{IPA:} \quad \frac{\partial Q(\boldsymbol{\theta}, V)}{\partial \boldsymbol{\theta}}. \tag{15.10}$$

In practice, neither LR/SF nor IPA may provide a framework for reasonable implementation. Pure LR/SF may impose excessive demands in deriving a distribution function that encompasses all of the $\boldsymbol{\theta}$ effects, beginning with the $U(0, 1)$ random variables. On the other hand, pure IPA requires that $Q(\boldsymbol{\theta}, V)$ absorb all of the $\boldsymbol{\theta}$ effects, possibly making the computation of $\partial Q / \partial \boldsymbol{\theta}$ intractable or invalidating the validity of interchange of derivative and integral due to nondifferentiability of Q. Between these extremes, the analyst may have many choices, all resulting in a gradient form of the type in (15.4), with *all* terms present. For example, if the underlying $U(0, 1)$ random variables go through many transformations in creating the ultimate random effects of interest in the simulation, then the analyst may define V as the random variables at any of the levels of transformation. So, given the choice in where the $\boldsymbol{\theta}$ effects are allocated for purposes of the optimization, it is worth considering an allocation that produces a lower variance in the gradient estimate.

Unfortunately, there are no easy general rules pointing toward the lowest variance gradient estimate that has reasonable implementation demands. At the extremes, with the pure LR/SF estimate it may be easier to satisfy the interchange conditions of Theorem 15.1 since the density (or mass) functions tend to be smoother (as a function of $\boldsymbol{\theta}$) than the measured performance Q when Q must capture all of the $\boldsymbol{\theta}$ effects (as we saw in Example 15.1 and Exercise 15.1; see, e.g., Fu and Hu, 1997, p. 146, and Rubinstein and Melamed, 1998, pp. 218–221). LR/SF, however, has the undesirable property that the variance of the gradient estimate generally grows with the number of elements in V (often corresponding to simulation length).

To see that the variance grows in the LR/SF estimate, suppose that V is composed of M independent random effects $\{\mathcal{V}_j\}$ with associated density/mass functions $p_{\mathcal{V}_j}(\upsilon \mid \boldsymbol{\theta}), j = 1, 2, \ldots, M$ (υ the dummy variable for $\mathcal{V}_j$). Because of the independence among the elements of V, the number of summands in $\log p_V(V|\boldsymbol{\theta})$ is the same as the number of elements (i.e., M). From the fundamental LR/SF assumption that $\partial Q/\partial\boldsymbol{\theta} = \mathbf{0}$ (i.e., Q does not depend on $\boldsymbol{\theta}$), the gradient estimate (15.9) is

$$Q(V)\frac{\partial \log p_V(V \mid \boldsymbol{\theta})}{\partial \boldsymbol{\theta}} = Q(V)\sum_{j=1}^{M}\frac{\partial \log p_{\mathcal{V}_j}(\mathcal{V}_j \mid \boldsymbol{\theta})}{\partial \boldsymbol{\theta}}. \tag{15.11}$$

Hence, the gradient estimate tends to have a variance of magnitude $O(M)$ (the "tends to" qualifier is due to the dependence introduced in the M-term sum through the $Q(V)$ multiplier).

A number of techniques have been introduced to reduce the variance of pure LR/SF. One such technique exploits regenerative structure (Section 14.2). With regeneration, one can average over the regenerative cycles as a means of variance reduction (e.g., Glynn, 1987, 1990; L'Ecuyer et al., 1994). In particular, the above sum containing M components can be reduced to a sum of a smaller number of components, each representing the result of one regenerative cycle. So, the sum in (15.11) is replaced by an analogous averaging over the number of regenerative cycles.

At the other extreme, pure IPA does not generally suffer from the increasing variance problem mentioned above for pure LR/SF because $\partial Q/\partial\boldsymbol{\theta}$ does not generally grow with the number of components in V and because $\partial \log p_V/\partial\boldsymbol{\theta} = \mathbf{0}$. However, the complexity of Q as a function of $\boldsymbol{\theta}$ is a frequent barrier to implementation. Sometimes, a straightforward definition of a gradient estimate may yield a biased estimate, as we saw in Example 15.2 (other examples are in Cao, 1985; L'Ecuyer and Glynn, 1994; and Fu and Hu, 1997, Chap. 4). However, special structure can sometimes be exploited to create an unbiased estimate.

A large literature has evolved for the development of techniques for computing $\partial Q/\partial\boldsymbol{\theta}$ in a simulation-based optimization context. The books by Glasserman (1991a), Fu and Hu (1997), and Cassandras and Lafortune (1999, especially Chap. 11) and survey papers by Suri (1989) and Ho and Cassandras (1997) are some of the references that summarize many of the available techniques. These techniques are designed to exploit special structure in the simulation to allow for the practical calculation of the gradient $\partial Q/\partial\boldsymbol{\theta}$ and to provide the necessary smoothness in Q to allow for the required interchange of derivative and integral (i.e., to yield an unbiased estimator). In cases where such techniques are feasible, IPA—or close cousins termed *smoothed PA*, *rare PA*, and so on—frequently provide the lowest variance gradient estimate (e.g., L'Ecuyer et al., 1994; Rubinstein and Melamed, 1998, pp. 221–223).

The example below contrasts pure LR/SF and pure IPA on a simple problem of gradient estimation (not necessarily in an optimization context). This example shows how different gradient estimators—with different properties—can result from the seemingly arbitrary decision to allocate $\boldsymbol{\theta}$ totally to the distribution function or totally to the performance measure.

Example 15.3—IPA and LR/SF estimates for exponentially distributed random variable. Let Z be an exponential random variable with mean θ. That is, $p_Z(z\,|\,\theta) = e^{-z/\theta}/\theta$ for $z \geq 0$. Define $L = E(Z) = \theta$. We can derive both LR/SF and IPA estimates of the gradient of L for illustration. (Note that in this trivial case, the gradient of L can be computed as $\partial L/\partial\theta = 1$.)

To compute the LR/SF gradient estimate, we define $V = Z$ and Q as the identity function, that is, $Q(\theta, V) = V$. Because θ does not enter Q, θ is only a distributional parameter. Therefore, only the first term in equation (15.4) contributes to the derivative estimate (as in (15.9)). Using $p_Z(Z|\theta) = p_V(V|\theta)$, the LR/SF gradient estimate can be derived as

$$V\frac{\partial \log p_V(V\,|\,\theta)}{\partial\theta} = \frac{V}{\theta}\left(\frac{V}{\theta}-1\right).$$

To apply the IPA technique, we change the definition of V. Let $V \sim U(0, 1)$ and $Z = -\theta \log V$. As above, Z is exponentially distributed with mean θ. Then, $L(\theta) = E(Z) = E[Q(\theta, V)] = -E[\theta \log V]$. With these new definitions, θ only appears as a structural parameter (i.e., the distribution of V is independent of θ). Therefore, the IPA gradient estimate can be derived from the second term in (15.4) (as in (15.10)):

$$\frac{\partial Q(\theta, V)}{\partial\theta} = -\log V\,.$$

We can easily check that both LR/SF and IPA gradient estimates are unbiased. It can also be shown that the IPA estimate has smaller variance than the LR/SF estimate (Exercise 15.2). ❑

15.3 GRADIENT ESTIMATION METHODS IN ROOT-FINDING STOCHASTIC APPROXIMATION: THE HYBRID LR/SF AND IPA SETTING

Let us return to the general setting of Section 15.1, where the gradient estimation form is generally a hybrid LR/SF and IPA form. Recall the basic root-finding SA recursions in Section 4.1:

$$\hat{\boldsymbol{\theta}}_{k+1} = \hat{\boldsymbol{\theta}}_k - a_k \boldsymbol{Y}_k(\hat{\boldsymbol{\theta}}_k) \quad \text{(unconstrained)},$$

$$\hat{\boldsymbol{\theta}}_{k+1} = \Psi_{\Theta}[\hat{\boldsymbol{\theta}}_k - a_k \boldsymbol{Y}_k(\hat{\boldsymbol{\theta}}_k)] \quad \text{(constrained)},$$

where $\boldsymbol{Y}_k(\boldsymbol{\theta})$ represents a (noisy) measurement of the function for which a zero is to be found. Under appropriate conditions (e.g., Section 4.3), it is known that $\hat{\boldsymbol{\theta}}_k$ converges to an optimum $\boldsymbol{\theta}^* = \arg\min_{\boldsymbol{\theta}\in\Theta} L(\boldsymbol{\theta})$ as $k \to \infty$. The point $\boldsymbol{\theta}^*$ may be unique or may be one of multiple points in a *set* of equivalent minima Θ^*, although, as we have done previously for ease of discussion, we usually consider $\boldsymbol{\theta}^*$ as a unique minimum. For simulation-based optimization, the stochastic gradient setting of Chapter 5 applies. Hence, $\boldsymbol{Y}_k(\boldsymbol{\theta})$ represents a measurement of $\boldsymbol{g}(\boldsymbol{\theta}) = \partial L/\partial\boldsymbol{\theta}$. (As discussed in Section 8.4, if there are multiple local minima to $L(\boldsymbol{\theta})$, the SA recursions above produce a *global* optimum in the limit if a Monte Carlo noise term is injected on the right-hand side.)

Recall that a basic philosophy in SA is to perform a type of *across-iteration* averaging to smooth out noise effects versus expending a large amount of resources in getting accurate estimates for $\boldsymbol{g}(\boldsymbol{\theta})$ at each iteration. The implication relative to (15.4) is that the amount of *per-iteration* averaging of $\boldsymbol{Y}_k(\boldsymbol{\theta})$ values (corresponding to multiple samples of $\boldsymbol{V}$) should be small. Clearly, in a dynamic system where the simulation environment is changing rapidly (possibly affecting the optimal value $\boldsymbol{\theta}^*$), there are advantages to avoiding a large amount of averaging at a given $\boldsymbol{\theta}$ since the simulations being averaged may not represent identically distributed samples.

The bottom line relative to a root-finding SA implementation is that (15.4) is usually implemented directly as it is presented, that is, without averaging from multiple values of $\boldsymbol{V}$ at each value of $\boldsymbol{\theta}$. So the unbiased estimate of $\boldsymbol{g}(\boldsymbol{\theta})$ for use in the SA recursions above is

$$\boldsymbol{Y}_k(\hat{\boldsymbol{\theta}}_k) = Q(\hat{\boldsymbol{\theta}}_k, \boldsymbol{V}_k)\left.\frac{\partial \log p_{\boldsymbol{V}}(\boldsymbol{V}_k \mid \boldsymbol{\theta})}{\partial \boldsymbol{\theta}}\right|_{\boldsymbol{\theta}=\hat{\boldsymbol{\theta}}_k} + \left.\frac{\partial Q(\boldsymbol{\theta}, \boldsymbol{V}_k)}{\partial \boldsymbol{\theta}}\right|_{\boldsymbol{\theta}=\hat{\boldsymbol{\theta}}_k}, \tag{15.12}$$

where $\boldsymbol{V}_k$ represents the realization of $\boldsymbol{V}$ for use in the kth iteration, generated from $p_{\boldsymbol{V}}(\boldsymbol{\upsilon} \mid \boldsymbol{\theta} = \hat{\boldsymbol{\theta}}_k)$. Note that *one* simulation run—corresponding to one realization of $\boldsymbol{V}$—can be used to produce this estimate at each value of $\boldsymbol{\theta}$. (A closely related approach is summarized in Section 15.4, where one realization of $\boldsymbol{V}$ can be repeatedly used to produce unbiased gradient estimates for *multiple* values of $\boldsymbol{\theta}$.)

Unlike the FDSA and SPSA methods, one must be able to compute the gradients $\partial \log p_{\boldsymbol{V}}(\boldsymbol{V}_k \mid \boldsymbol{\theta})/\partial\boldsymbol{\theta}$ and $\partial Q(\boldsymbol{\theta}, \boldsymbol{V}_k)/\partial\boldsymbol{\theta}$ to obtain this estimate (with FDSA and SPSA, only values of $Q(\boldsymbol{\theta}, \boldsymbol{V})$ are needed). As discussed in Sections 6.5 and 7.4, however, the stochastic gradient algorithm buys an increased rate of convergence relative to FDSA or SPSA, with the optimal stochastic rate increasing from $O(1/k^{1/3})$ to $O(1/\sqrt{k})$. We also saw in Section 14.4 that common random numbers in FDSA and SPSA can increase the rate from

$O(1/k^{1/3})$ to $O(1/\sqrt{k})$, although the relative mean-squared errors of root-finding SA and CRN-based FDSA or SPSA at the common rate of convergence do not appear to have been analyzed.

The two-part example below illustrates the method above for computing the gradient estimate together with the application in an SA-based optimization process.

Example 15.4—Laboratory experimental response (part 1—general setting). Consider a laboratory setting where a specimen is exposed to a sequence $\{V_k\}$ of randomly generated independent, identically distributed (i.i.d.) "on–off" stimuli with "on" probability λ (and, of course, "off" probability of $1-\lambda$). After each exposure, the specimen responds in some way to reveal important information to an experimenter. The aim is to design an experiment such that the specimen response is maximized. Suppose that the function measuring the negative of this response is $Q(\boldsymbol{\theta}, V_k) = Q(\lambda, \beta, V_k)$, where β is a design parameter (negative response is used here so that we have a minimization problem). Further, suppose that $Q(\lambda, \beta, V_k)$ is differentiable in $\boldsymbol{\theta} = [\lambda, \beta]^T$ and that the derivative can be computed. The aim is to pick $\boldsymbol{\theta}$ in a way to maximize the expected response. Note that this problem does not have a clear demarcation between distributional parameters (λ) and structural parameters (λ, β) since Q depends on both λ and β.

Each stimulus will be associated with one iteration of the stochastic gradient algorithm. The Bernoulli probability mass function can be expressed as $p_V(\upsilon|\boldsymbol{\theta}) = \lambda^{\upsilon}(1-\lambda)^{1-\upsilon}$, where $\upsilon = 0$ or 1. This function is clearly differentiable in λ as required in Theorem 15.1, yielding the following contribution to the unbiased gradient estimate in (15.12):

$$\frac{\partial \log p_V(\upsilon\,|\,\boldsymbol{\theta})}{\partial \boldsymbol{\theta}} = \begin{bmatrix} \dfrac{\lambda-\upsilon}{(\lambda-1)\lambda} \\ 0 \end{bmatrix}.$$

Assume that Q is such that the conditions of Theorem 15.1 are satisfied (see Example 15.5 below). Then, from (15.12), a valid input to the root-finding algorithm at the kth iteration is

$$\boldsymbol{Y}_k(\hat{\boldsymbol{\theta}}_k) = \begin{bmatrix} Q(\hat{\boldsymbol{\theta}}_k, V_k)\dfrac{\hat{\lambda}_k - V_k}{(\hat{\lambda}_k - 1)\hat{\lambda}_k} + Q'_{\lambda}(\hat{\boldsymbol{\theta}}_k, V_k) \\ Q'_{\beta}(\hat{\boldsymbol{\theta}}_k, V_k) \end{bmatrix}, \qquad (15.13)$$

where $\hat{\boldsymbol{\theta}}_k = [\hat{\lambda}_k, \hat{\beta}_k]^T$ and $Q'_x(\boldsymbol{\theta}, V) = \partial Q(\boldsymbol{\theta}, V)/\partial x$, $x = \lambda$ or β. (See also Exercise 15.3.) ❑

Example 15.5—Laboratory experimental response (part 2—numerical results). We continue the experimental response example with a simple case where an analytical solution is available. This example is intended to convey concepts and provide a known solution to compare to the stochastic gradient (hybrid LR/SF and IPA) solution. In practice, of course, simulation-based optimization is oriented to cases where such simple solutions do not exist. Suppose that the simulation output produces a (negative) response function as follows:

$$Q(\boldsymbol{\theta}, V) = \beta^2 + (1-\lambda)(\beta - V),$$

where $V = 1$ represents the "on" event (and, of course, $V = 0$ represents the "off" event) and $-1 < \beta < 1$.

Given that this response function depends on both λ and β (i.e., there is no strict segregation of the distributional and structural parameters), this is a setting where the response of the system to one stimulus is dependent on the likelihood of other stimuli (i.e., dependent on λ). So there will be some cases where the system tends to respond strongly to a stimulus $V = 1$ when the probability (λ) of other stimuli is high (say, a drug response increases with the likelihood of greater past or future dosages, perhaps through a placebo effect) and other cases where the system responds weakly in the presence of a high probability of other stimuli (say, as when a drug loses effectiveness with repeated usage).

With the above response function, we obtain the following closed-form expression for L:

$$L(\boldsymbol{\theta}) = \sum_{\upsilon=0}^{1} Q(\boldsymbol{\theta}, \upsilon)\lambda^{\upsilon}(1-\lambda)^{1-\upsilon} = -\lambda + \lambda^2 + \beta - \lambda\beta + \beta^2.$$

Setting the gradient $\boldsymbol{g}(\boldsymbol{\theta}) = \boldsymbol{0}$ and examining the Hessian matrix yields a unique minimum at $\boldsymbol{\theta}^* = [1/3, -1/3]^T$, having $L(\boldsymbol{\theta}^*) = -1/3$.

Before proceeding with the implementation for optimization, let us note that the full power of Theorem 15.1 is unnecessary in this problem. The expectation "integral" in this case is simply a summation over the two possible outcomes for V. Hence, the interchange of derivative and "integral" (i.e., summation) is trivially true because the required derivatives exist.

Figure 15.1 shows the result of one typical realization of 1000 iterations of the stochastic gradient algorithm as applied with the Q above. Each iteration of the algorithm depends on one simulation producing an output Q. Based on the principles of Section 4.4, the gain sequence $a_k = 0.1/(100+k)^{0.501}$ is used. From an initial condition of $\hat{\boldsymbol{\theta}}_0 = [1/2, 1/2]^T$, the algorithm moves to $\hat{\boldsymbol{\theta}}_{1000} = [0.33028, -0.32361]^T$. This yielded a final loss value of -0.33320 and a final $\boldsymbol{\theta}$ estimate having Euclidean distance of 0.0102 to the solution. As an indication of the "typical" nature of this run, a separate Monte Carlo study based on 30

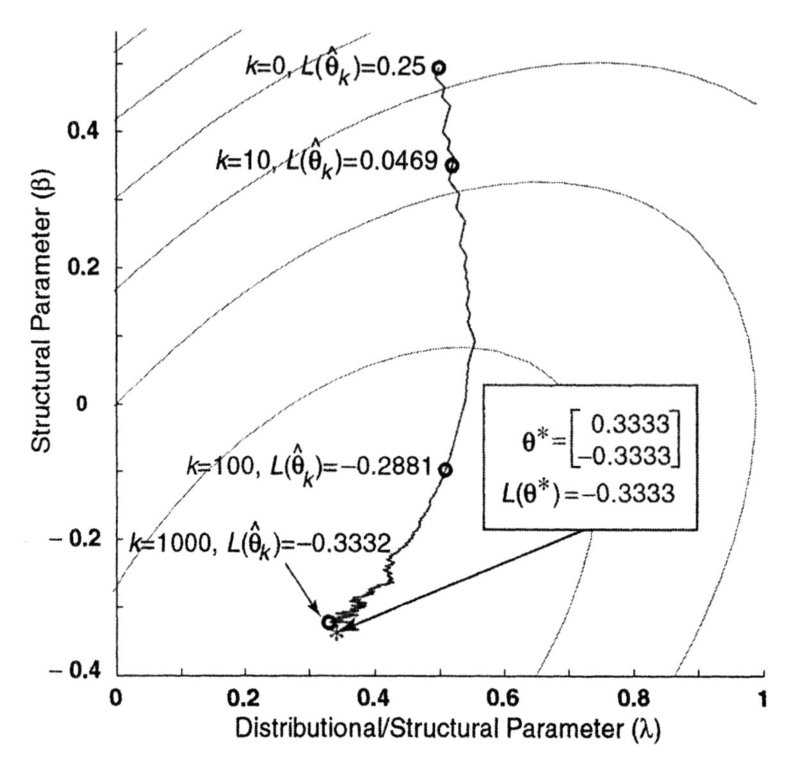

Figure 15.1. Search path for typical realization of the stochastic gradient algorithm in experimental response model.

independent replications of the root-finding algorithm was conducted, yielding final loss values ranging from −0.33069 to −0.33333 and final distances to the solution of 0.0612 to 0.0058.

Figure 15.1 shows that the distance from the final estimate to the optimal point $\boldsymbol{\theta}^*$ is visually small and that the loss value has effectively converged (note that the noise $\boldsymbol{e}$ in the gradient measurements is relatively small in this problem). The $\boldsymbol{\theta}$ estimate being relatively farther from $\boldsymbol{\theta}^*$ than the corresponding $L(\boldsymbol{\theta})$ is from $L(\boldsymbol{\theta}^*)$ is a typical pattern in simulation-based (and other) optimization given the flatness of loss functions near $\boldsymbol{\theta}^*$.

As an illustration of the degrading effects of the noise in the measurements of $\boldsymbol{g}(\boldsymbol{\theta})$, let us consider a noise-free steepest descent algorithm based on direct values of $\boldsymbol{g}(\boldsymbol{\theta})$ (as in Section 1.4). Using the same a_k as above, the steepest descent algorithm was able to get within a distance of 0.0069 of the solution, which is significantly better than the 0.0102 distance in the typical run of Figure 15.1, but not as good as the minimum distance of 0.0058 in the 30 noisy test cases mentioned above. (However, an a_k tuned to the noise-free case leads to better results for steepest descent, significantly better than the best noisy run.) Of course, this noise-free solution is not available in practical simulation-based optimization unless one does a very large (and costly!) amount of averaging of the noisy measurements at each iteration. (A different form of noise-free optimization is considered in the next section.) ❑

15.4 SAMPLE PATH OPTIMIZATION

In general, the main cost of each iteration of the stochastic gradient algorithm is the simulation needed to produce the gradient estimate. Hence, n iterations require n simulations. The *sample path* (sometimes called *stochastic counterpart*) method of optimization is based on a different principle. Here, structure in the problem is exploited to allow for the *multiple* gradient estimates (say, n) needed in the search process to be computed from a relatively small *single* set of simulation runs. That is, a surrogate loss function is created from a set of V values, say V_0, $V_1, \ldots, V_{N-1}$, representing N simulation runs. (We are using indices 0, 1,..., $N-1$, rather than 1, 2,..., N, to be consistent with the indexing in the stochastic gradient form of Section 15.3, where the first V value is V_0.) These N runs are then averaged in an appropriate way, leading to an approximation to the true loss function. The n gradients needed in the search process can be computed based on this approximate loss function serving as a proxy for the unavailable true loss $L(\boldsymbol{\theta})$. This is tantamount to reusing the same set of N realizations of V (the sample path) for multiple gradient estimates. Note that, in general, there is no relationship between n and N.

Because the sample path method is based on reusing the fixed set of simulation runs, one may effectively treat the resulting optimization problem as a *deterministic* optimization problem, using standard deterministic nonlinear programming techniques to solve for the optimal $\boldsymbol{\theta}$. This reuse of the same sequence of V values for multiple gradient calculations is a means for squeezing more information from the limited number of simulation runs. The nonlinear programming aspect is especially useful in problems with complicated constraints because there is much machinery in deterministic search and optimization for handling constraints that is not readily available in stochastic optimization (e.g., Bazaraa et al., 1993; Polak, 1997).

An additional significant benefit of the sample path method follows if the simulation is very costly to run. Because the number of simulations is fixed (N), a large number of iterations of the deterministic optimization method can be conducted at nominal cost beyond the cost of the N simulations. The nominal cost here reflects the cost of the gradient calculation based on the simulation output and possible other costs of the algorithm if an approach more complex than steepest descent (Section 1.4) is being used.

In the sample path method, rather than optimize $L(\boldsymbol{\theta})$ in (15.1), one optimizes a function $\overline{L}_N(\boldsymbol{\theta})$, representing an average of N independent simulation runs. (In particular, it is shown in (15.15) and (15.16) below how knowledge of the functional relationships, $Q(\boldsymbol{\theta}, V)$ and $p_V(\upsilon|\boldsymbol{\theta})$, is used to form $\overline{L}_N(\boldsymbol{\theta})$.) So, although there are the potential implementation and computational benefits mentioned above, a price to be paid in the sample path method is that the limiting point for the $\boldsymbol{\theta}$ estimate depends on the specific sequence of N random processes, $V_0, V_1, \ldots, V_{N-1}$. That is, under appropriate conditions, $\hat{\boldsymbol{\theta}}_n$ computed in a sample path implementation of a search algorithm converges as $n \to \infty$ to

$$\boldsymbol{\theta}_N^* \equiv \arg \min_{\boldsymbol{\theta} \in \Theta} \bar{L}_N(\boldsymbol{\theta})$$

where $\boldsymbol{\theta}_N^* = \boldsymbol{\theta}_N^*(V_0, V_1, \ldots, V_{N-1})$ and $\boldsymbol{\theta}_N^* \neq \boldsymbol{\theta}^*$ in general. (For the discussion here, $\boldsymbol{\theta}_N^*$ is assumed to be unique.) So, a central point is to establish conditions such that $\boldsymbol{\theta}_N^*$ is very close to $\boldsymbol{\theta}^*$ $(= \arg\min_{\boldsymbol{\theta} \in \Theta} L(\boldsymbol{\theta})$, of course). This implies that for sufficiently large N, the optimization of $\bar{L}_N(\boldsymbol{\theta})$ yields a solution that is close to the optimization of the true loss function $L(\boldsymbol{\theta})$.

One might suspect that $\boldsymbol{\theta}_N^*$ and $\boldsymbol{\theta}^*$ will be close for large N since, by the laws of large numbers, $\bar{L}_N(\boldsymbol{\theta}) \to L(\boldsymbol{\theta})$ for each $\boldsymbol{\theta}$ in some stochastic sense when $N \to \infty$. But this is not automatic. For example, well-known counterexamples exist where having a function converge does not imply that the derivative of that function will converge (see, e.g., Rudin, 1976, pp. 152–154). Specifically, $\bar{L}_N(\boldsymbol{\theta}) \to L(\boldsymbol{\theta})$ does not imply $\partial \bar{L}_N(\boldsymbol{\theta})/\partial \boldsymbol{\theta} \to \partial L(\boldsymbol{\theta})/\partial \boldsymbol{\theta}$; so, a zero of $\partial \bar{L}_N(\boldsymbol{\theta})/\partial \boldsymbol{\theta}$ may not be close to a zero of $\partial L(\boldsymbol{\theta})/\partial \boldsymbol{\theta}$. Hence, in general, one cannot assume that $\boldsymbol{\theta}_N^*$ and $\boldsymbol{\theta}^*$ are approximately equal, even for large N. Shapiro (1991, 1996), Robinson (1996), and Rubinstein and Melamed (1998, p. 152), among others, discuss conditions under which $\boldsymbol{\theta}_N^*$ *does* converge to $\boldsymbol{\theta}^*$. Let us present one of those results here. As in Theorem 15.1, $p_V(\boldsymbol{\upsilon}|\boldsymbol{\theta})$ may represent either a probability density or a mass function; integrals are to be interpreted as sums in the discrete (mass function) case. The conditions here bear some similarity to the conditions of Theorem 15.1, but there are some nontrivial differences as well.

Theorem 15.2 (Rubinstein and Melamed, 1998, p. 152). Suppose that the search space $\Theta \subset \mathbb{R}^p$ is a compact (i.e., closed and bounded) set known to contain $\boldsymbol{\theta}^*$. Assume that for almost all V (recall definition of "almost all" in Subsection C.2.1, Appendix C), $p_V(V|\boldsymbol{\theta})$ is continuous on Θ and that there exists a nonnegative function $q_0(\boldsymbol{\upsilon})$ such that

$$\left|Q(\boldsymbol{\theta}, \boldsymbol{\upsilon}) p_V(\boldsymbol{\upsilon} \,|\, \boldsymbol{\theta})\right| \leq q_0(\boldsymbol{\upsilon}) \text{ for all } (\boldsymbol{\theta}, \boldsymbol{\upsilon}) \in \Theta \times \Lambda$$

where $\int_\Lambda q_0(\boldsymbol{\upsilon})\, d\boldsymbol{\upsilon} < \infty$. Then $\boldsymbol{\theta}_N^* \to \boldsymbol{\theta}^*$ a.s. as $N \to \infty$.

Shapiro (1991) and Rubinstein and Melamed (1998, p. 152) also present asymptotic distribution results for $\boldsymbol{\theta}_N^*$, showing that $\sqrt{N}(\boldsymbol{\theta}_N^* - \boldsymbol{\theta}^*)$ is asymptotically normally distributed under conditions stronger than those in

Theorem 15.2. These results can be used, as in other applications of asymptotic distributions in this book, to assess the approximate error of an estimator ($\boldsymbol{\theta}_N^*$ in this case). The error is proportional in a stochastic sense to $1/\sqrt{N}$. Based on a large number of studies with typical discrete-event simulations, Rubinstein and Melamed (1998, pp. 151–152) state that when $\bar{L}_N(\boldsymbol{\theta})$ is a convex function (Appendix A), values of N on the order of 1000 or more are typically needed to ensure that $\boldsymbol{\theta}_N^*$ is an acceptable approximation of $\boldsymbol{\theta}^*$.

More discussion on Theorem 15.2 and the associated asymptotic normality, including more detail on the implications of the constraints embodied in Θ, are given in Shapiro (1991, 1996). Shapiro (1996) includes the study of regenerative processes (Section 14.2) as a special case. He also shows how asymptotic normality results pertaining to the error in using $\bar{L}_N(\boldsymbol{\theta})$ as an approximation to $L(\boldsymbol{\theta})$ can be applied in making statistical inference about the solution $\boldsymbol{\theta}_N^*$. Analogous to what has been done in other contexts in the preceding chapters, Shapiro (1991, 1996) uses the asymptotic normality to assess the efficiency of the sample path method relative to SA-based stochastic gradient method.

Although the sample path method converts the problem to a deterministic process, the LR/SF and IPA forms in Sections 15.2 and 15.3 are still relevant for gradient calculation. In particular, one must still make a choice about the allocation of the elements in $\boldsymbol{\theta}$ to being distributional parameters, structural parameters, or a hybrid form. That choice affects whether (15.4) or one of the "pure" forms discussed in Section 15.2 may be used to construct the gradient estimate.

In some sense, however, Theorem 15.2 begs the fundamental question: How close will the estimate $\hat{\boldsymbol{\theta}}_n$ be to $\boldsymbol{\theta}^*$ when a sample path method is used? This question cannot be answered without knowledge of the convergence properties of the specific nonlinear programming algorithm being used to minimize $\bar{L}_N(\boldsymbol{\theta})$. Section 1.4 discussed the convergence properties of the steepest descent and Newton–Raphson algorithms. Bazaraa et al. (1993), Polak (1997), and many other texts discuss these properties in more detail, including the convergence properties of other popular deterministic search algorithms not discussed in Section 1.4. To assess the closeness of $\hat{\boldsymbol{\theta}}_n$ to $\boldsymbol{\theta}^*$, one also needs insight into the error in using $\boldsymbol{\theta}_N^*$ to approximate $\boldsymbol{\theta}^*$. As mentioned above, an asymptotic normality result exists that may provide an approximate error bound. Combining the two sources of error, the total error can then be bounded by the triangle inequality,

$$\left\|\hat{\boldsymbol{\theta}}_n - \boldsymbol{\theta}^*\right\| \le \left\|\hat{\boldsymbol{\theta}}_n - \boldsymbol{\theta}_N^*\right\| + \left\|\boldsymbol{\theta}_N^* - \boldsymbol{\theta}^*\right\|, \tag{15.14}$$

where the first error on the right-hand side reflects the inaccuracy in the nonlinear program and the second error is due to substitution of the sample path minimum for the true minimum. Let us present an application of this result.

Example 15.6—Error analysis for sample path optimization. Suppose that $p = 5$. From the properties of the deterministic search algorithm, suppose that each of the elements of $\hat{\boldsymbol{\theta}}_n$ is known to be within 0.2 unit of the corresponding elements in the optimum $\boldsymbol{\theta}_N^*$. Further, assume that from asymptotic theory, the approximate distribution of $\boldsymbol{\theta}_N^*$ is $N(\boldsymbol{\theta}^*, 0.05\boldsymbol{I}_5)$ (see the discussion above following Theorem 15.2). Suppose that we wish to identify a distance that it is known with 90 percent certainty to be at least as large as the true distance between $\hat{\boldsymbol{\theta}}_n$ and $\boldsymbol{\theta}^*$.

From (15.14), one knows that the true distance can be bounded by the sum of the distance between $\hat{\boldsymbol{\theta}}_n$ and $\boldsymbol{\theta}_N^*$ and the distance between $\boldsymbol{\theta}_N^*$ and $\boldsymbol{\theta}^*$. By the above-mentioned assumption on the deterministic search algorithm, the distance between $\hat{\boldsymbol{\theta}}_n$ and $\boldsymbol{\theta}_N^*$ is known to be no greater than $\sqrt{5\times 0.2^2} = 0.45$. From the asymptotic normality for $\boldsymbol{\theta}_N^*$, it is known that $(\boldsymbol{\theta}_N^* - \boldsymbol{\theta}^*)^T(\boldsymbol{\theta}_N^* - \boldsymbol{\theta}^*)/0.05$ is approximately chi-squared distributed with five degrees of freedom (the 0.05 term is a *variance*, not a standard deviation). The 90 percent critical value from a chi-squared distribution table is 9.25, indicating that the distance between $\boldsymbol{\theta}_N^*$ and $\boldsymbol{\theta}^*$ is known with 90 percent certainty to be no greater than $\sqrt{0.05\times 9.25} = 0.68$. Hence, from (15.14), and subject to the asymptotic distribution of $\boldsymbol{\theta}_N^*$ being a credible representation of the actual finite sample distribution, one can be 90 percent certain that the distance between $\hat{\boldsymbol{\theta}}_n$ and $\boldsymbol{\theta}^*$ is no more than $0.45 + 0.68 = 1.13$. ❑

A fundamental issue in sample path optimization is the determination of the distribution for generating the fixed sample $V_0, V_1, \ldots, V_{N-1}$ that is used throughout the search process. Clearly, the user must be concrete and specific in this choice since Monte Carlo random variables cannot be generated without an exact distribution, including full specification of the distribution's parameters. In a simulation setting, this is equivalent to choosing the distributional form and associated parameter values for all of the random effects being generated inside the simulation. It seems, however, that a potential problem arises because the distribution for the V_i typically depends on the very parameters $\boldsymbol{\theta}$ being estimated during the search process. That is, it seems that a new set of V_i must be generated for each new value of $\boldsymbol{\theta}$ during the search process. This would defeat the very principle of sample path optimization! Fortunately, there is a ready way around this problem, as we now describe.

Let us now show how knowledge of the functional relationships, $Q(\boldsymbol{\theta}, V)$ and $p_V(\upsilon|\boldsymbol{\theta})$, is used to form $\bar{L}_N(\boldsymbol{\theta})$ via an averaging scheme. Suppose that there exists a density function $q_V(\upsilon)$ such that for every υ where $q_V(\upsilon) = 0$, then $p_V(\upsilon|\boldsymbol{\theta}) = 0$ for all possible values of $\boldsymbol{\theta} \in \Theta$. It is sometimes said that $q_V(\upsilon)$ *dominates* $p_V(\upsilon|\boldsymbol{\theta})$ (more formally, that the associated *distribution*—rather than density/mass—function dominates the distribution function associated with $p_V(\upsilon|\boldsymbol{\theta})$). Then, the fundamental definition for the loss function, (15.1), can be expressed as

$$
\begin{aligned}
L(\boldsymbol{\theta}) &= \int_{\Lambda} Q(\boldsymbol{\theta}, \upsilon) p_V(\upsilon \mid \boldsymbol{\theta})\, d\upsilon \\
&= \int_{\Lambda} Q(\boldsymbol{\theta}, \upsilon) \frac{p_V(\upsilon \mid \boldsymbol{\theta})}{q_V(\upsilon)} q_V(\upsilon)\, d\upsilon \\
&= E_{q_V}\left[Q(\boldsymbol{\theta}, V) \frac{p_V(V \mid \boldsymbol{\theta})}{q_V(V)} \right],
\end{aligned} \tag{15.15}
$$

where $E_{q_V}[\cdot]$ denotes the expectation based on $q_V(\upsilon)$.[2] An important special case is where $q_V(\upsilon) = p_V(\upsilon|\boldsymbol{\theta}')$ for some fixed value $\boldsymbol{\theta}'$. The significance of (15.15) is quite profound. Namely, when $q_V(\upsilon) = p_V(\upsilon|\boldsymbol{\theta}')$, one can generate sample values of V from the distribution based on the *fixed* $\boldsymbol{\theta}'$ to obtain an unbiased estimate of the loss function at *any* value of $\boldsymbol{\theta}$. We refer to $\boldsymbol{\theta}'$ as the *reference value* of $\boldsymbol{\theta}$.

In particular, letting $V_0, V_1, \ldots, V_{N-1}$ be a sequence of N values of V generated according to $q_V(\upsilon) = p_V(\upsilon|\boldsymbol{\theta}')$, then

$$
\overline{L}_N(\boldsymbol{\theta}) \equiv \frac{1}{N} \sum_{i=0}^{N-1} Q(\boldsymbol{\theta}, V_i) \frac{p_V(V_i|\boldsymbol{\theta})}{p_V(V_i|\boldsymbol{\theta}')} \tag{15.16}
$$

is an unbiased estimate of $L(\boldsymbol{\theta})$. For a fixed reference value $\boldsymbol{\theta}'$, (15.16) is viewed solely as a function of $\boldsymbol{\theta}$, as is the true loss function $L(\boldsymbol{\theta})$. Further, from (15.15), the famous weak or strong laws of large numbers (Section C.2 in Appendix C here, or Laha and Rohatgi, 1979, Chap. 2) state that $\overline{L}_N(\boldsymbol{\theta})$ converges in probability (pr.) or almost surely (a.s.) to $L(\boldsymbol{\theta})$ for all $\boldsymbol{\theta}$. Let us illustrate the sample path method in a simple problem involving scalar Bernoulli-distributed V. Although much simpler than most realistic simulation-based problems, this example depicts the essential properties of the sample path method.

Example 15.7—Sample path gradient estimate for binary-outcome problem. Suppose V is a binary random variable such that $V = 1$ with probability $0 < \theta < 1$ and $V = 0$ with probability $1 - \theta$. Let $Q(\theta, V) = V + b/\theta$ where $0 < b < 1$. This leads to $L(\theta) = \theta + b/\theta$, from which it is easily seen that the unique minimizing

---

[2]The discussion here represents a special case of a broader measure-theoretic framework of probability, where instead of a density or mass function for V, a general probability measure is used (L'Ecuyer, 1990a). (These results require a background beyond the prerequisites for this book.) Here the primary measure of interest for V (based on $\boldsymbol{\theta}$) is said to be *absolutely continuous* with respect to a dominating measure based on $q_V(\upsilon)$. Further, the ratio $p_V(\upsilon \mid \boldsymbol{\theta})/q_V(\upsilon)$ appearing in (15.15) represents the *Radon–Nikodym derivative* of the primary measure with respect to the dominating measure. This more general framework allows for some elegant convergence results and is potentially useful if $p_V(\upsilon|\boldsymbol{\theta})$ and $q_V(\upsilon)$ do not represent "nice" density or mass functions.

point is $\theta^* = \sqrt{b}$. Let us now construct a sample path gradient estimate and compare the zero of that function with the known value of θ^*. The mass function is $p_V(\upsilon|\theta) = \theta^\upsilon(1-\theta)^{1-\upsilon}$ where $\upsilon = 0$ or 1. From (15.16),

$$\frac{\partial \bar{L}_N(\theta)}{\partial\theta} = \frac{1}{N}\sum_{i=0}^{N-1}\left[V_i \frac{p_V(V_i\,|\,\theta)}{p_V(V_i\,|\,\theta')}\times\frac{V_i-\theta}{\theta(1-\theta)}\right]-\frac{b}{\theta^2}. \tag{15.17}$$

When $V_i = 0$, then $[\cdot] = 0$ in (15.17); when $V_i = 1$, then $[\cdot] = 1/\theta'$. Let us take $\theta' = 1/2$ and $b = 0.01$.

For $N = 5$ and 25, we generate the following random sequences for $\{V_i\}$ from the same seed:

$$N = 5:\ \{0, 0, 0, 1, 0\},$$

$$N = 25:\ \{0, 0, 0, 1, 0;\ 1, 1, 0, 1, 0;\ 1, 0, 1, 0, 0;\ 0, 1, 0, 1, 1;\ 0, 1, 0, 0, 1\}.$$

From (15.17), the calculations for $N = 5$ and 25, respectively, are

$$\frac{\partial \bar{L}_N(\theta)}{\partial\theta} = \frac{1}{5}\left[4\times 0 + 1\times\left(\tfrac{1}{2}\right)^{-1}\right]-\frac{0.01}{\theta^2} = 0 \Rightarrow \theta_N^* = \sqrt{\frac{0.01}{0.400}} = 0.158,$$

$$\frac{\partial \bar{L}_N(\theta)}{\partial\theta} = \frac{1}{25}\left[14\times 0 + 11\times\left(\tfrac{1}{2}\right)^{-1}\right]-\frac{0.01}{\theta^2} = 0 \Rightarrow \theta_N^* = \sqrt{\frac{0.01}{0.880}} = 0.107.$$

(These solutions are global minimums of $\bar{L}_N(\theta)$ since $\partial^2\bar{L}_N(\theta)/\partial\theta^2 > 0$ for all θ.) The above solutions compare with $\theta^* = \sqrt{0.01} = 0.10$. As expected, the difference between θ_N^* and θ^* decreases when N is increased from 5 to 25. ❑

Let us return to an issue mentioned at the beginning of this section: How does the efficiency of the sample path method compare with the classical stochastic gradient (i.e., root-finding SA) approach? In seeking a *general* answer to this question, we are, of course, forced into a theoretical (versus numerical) framework. But as we have seen repeatedly in this book in similar contexts, this question is unanswerable in finite samples due to a lack of finite-sample theory. However, asymptotically it is possible to partially answer this question. Suppose that the true sample path minimum $\boldsymbol{\theta}_N^*$ is available (i.e., we are assuming that the deterministic search algorithm has converged perfectly to the minimum). How does the accuracy of $\boldsymbol{\theta}_N^*$ compare to the solution resulting from the N simulations runs used in N iterations of the stochastic gradient algorithm?

Shapiro (1996) shows that, asymptotically, both the sample path and stochastic gradient (root-finding SA) with optimal gains yield statistically equivalent solutions for the same amount of input information in the case of *unconstrained* optimization ($\Theta = \mathbb{R}^p$). As discussed in Subsection 4.5.2, the optimal gain for root-finding SA is the matrix gain $\boldsymbol{a}_k = \boldsymbol{H}(\boldsymbol{\theta}^*)^{-1}/(k+1)$, $k \geq 0$, which, of course, is unavailable in practice ($\boldsymbol{H}$ is the Hessian matrix of the loss function). Practical methods of achieving the asymptotic rate of convergence associated with the optimal gain are the iterate averaging method (Subsection 4.5.3) and the second-order stochastic gradient (2SG) implementation of the adaptive SPSA approach (Section 7.8). (Recall, however, that the iterate averaging method may not yield improved performance in finite samples and that the 2SG implementation requires *three* simulations per iteration as long as the Hessian matrix is being estimated.) Using a simple quadratic loss function, the example below shows that sometimes the sample path and stochastic gradient methods are equivalent in finite-sample problems.

Example 15.8—Equivalence between sample path and stochastic gradient methods. Consider $Q(\boldsymbol{\theta}, V) = \boldsymbol{\theta}^T\boldsymbol{B}\boldsymbol{\theta}/2 + \boldsymbol{\theta}^T V$, where $\boldsymbol{B}$ is positive definite and V is p-dimensional. Then the sample path loss is $\bar{L}_N(\boldsymbol{\theta}) = \boldsymbol{\theta}^T\boldsymbol{B}\boldsymbol{\theta}/2 + \boldsymbol{\theta}^T\bar{V}_N$, where $\bar{V}_N = N^{-1}\sum_{i=0}^{N-1} V_i$. It follows easily that $\boldsymbol{\theta}_N^* = -\boldsymbol{B}^{-1}\bar{V}_N$. Relative to the root-finding SA approach, $\boldsymbol{H}(\boldsymbol{\theta}^*) = \boldsymbol{B}$, indicating that $\boldsymbol{a}_k = \boldsymbol{B}^{-1}/(k+1)$. The root-finding algorithm then has the form

$$\hat{\boldsymbol{\theta}}_{k+1} = \hat{\boldsymbol{\theta}}_k - \frac{\boldsymbol{B}^{-1}}{k+1}(\boldsymbol{B}\hat{\boldsymbol{\theta}}_k + V_k),\ k \geq 0,$$

which can be solved in closed form as $\hat{\boldsymbol{\theta}}_N = -\boldsymbol{B}^{-1}\bar{V}_N$ (Exercise 15.11). Hence $\hat{\boldsymbol{\theta}}_N = \boldsymbol{\theta}_N^*$ (both using N simulations). ❑

In constrained cases ($\Theta \neq \mathbb{R}^p$), it is no longer true that the stochastic gradient and sample path methods are asymptotically equivalent. Shapiro (1996) presents examples showing that the sample path method has a faster rate of convergence in at least some constrained problems, but it appears to be unknown if this is a general result for constrained problems (it likely depends on the specific form of the constraints). The examples presented in Shapiro (1996) pertain to constrained problems where the solution lies on the boundary of the constraint set.

One obvious question for the sample path method is: What is a good choice for the reference value $\boldsymbol{\theta}'$ in the usual case where $q_V(\upsilon) = p_V(\upsilon|\boldsymbol{\theta}')$? Choosing $\boldsymbol{\theta}'$ wisely can have a large impact on the performance of the sample path method. A good choice of $\boldsymbol{\theta}'$ can reduce the variance of the loss and gradient estimators, $\bar{L}_N(\boldsymbol{\theta})$ and $\partial\bar{L}_N(\boldsymbol{\theta})/\partial\boldsymbol{\theta}$, and contribute to a better estimate for $\boldsymbol{\theta}$. One formulation for picking $\boldsymbol{\theta}'$ is to solve the auxiliary optimization problem, $\min_{\boldsymbol{\theta}'} \text{var}[\bar{L}_N(\boldsymbol{\theta})]$, prior to solving the main problem of interest (recall that $\bar{L}_N(\boldsymbol{\theta})$ depends on $\boldsymbol{\theta}'$ as shown in (15.16)). Unfortunately, this auxiliary problem is unsolvable in most practical applications. Special cases of this auxiliary problem have been considered in a number of references, including Dussault et al. (1997). For instance, in the context of queuing networks, it is beneficial to choose $\boldsymbol{\theta}'$ such that the amount of traffic in the network *exceeds* that found under more typical values of $\boldsymbol{\theta}$. Nevertheless, despite the lack of an easy general result, we saw in Example 15.7 above that even if $\boldsymbol{\theta}'$ is not chosen optimally, the sample path method may still yield reasonable results.

15.5 CONCLUDING REMARKS

This chapter continued in the spirit of Chapter 14 in showing how simulations can be used for optimizing physical processes. Of course, as in Chapter 14, the results here are critically dependent on the simulation being a valid representation of the actual process. In contrast to Chapter 14, however, the focus here has been on methods involving the direct measurement of the gradient of the loss function.

The gradient-based methods require much more information than the nongradient methods of Chapter 14, which use only simulation inputs and outputs. In particular, the analyst must know the inner workings of the simulation and/or the complete probability distribution(s) of the random variables being generated as part of the Monte Carlo aspect of the simulation. If such information is available, the gradient-based methods of this chapter usually result in improved performance (e.g., increased rate of convergence) relative to the methods of Chapter 14. Nevertheless, while the bulk of published research in simulation-based optimization for discrete-event systems has focused on gradient-based methods, most *applications* appear to rely on the simpler finite-difference and related search methods (Fu and Hu, 1997, p. 5).

Section 15.1 discussed conditions under which a derivative and integral can be interchanged (see also Appendix A). Such an interchange is critical to the validity of the gradient-based optimization methods. Sections 15.2 and 15.3 focused on the LR/SF and IPA forms of gradient estimators, and, through (15.4) in Section 15.1, on a general form including LR/SF and IPA as limiting special cases. The LR/SF and IPA algorithms are special cases of the stochastic gradient algorithm of Chapter 5, which itself is a special case of root-finding stochastic approximation (Chapter 4). Section 15.4 considered a sample path version of the gradient-based algorithms. Unlike the SA-based methods of Sections 15.2 and 15.3, the sample path method is based on a gradient from a fixed set of data resulting from a fixed set of simulation runs (not directly connected to the number of iterations of the algorithm). In this way, the *stochastic* optimization problem associated with Monte Carlo simulations is converted to a *deterministic* problem for which classical deterministic search methods can be used.

Arguably, among many possible unbiased direct gradient measurements, the basic gradient estimators of Sections 15.2–15.4 have attracted the most attention in the simulation-based optimization community. For the sake of thoroughness, let us briefly discuss some other approaches.

There is a whole family of *perturbation analysis methods* besides the IPA approach. One such relative of IPA is the smoothed perturbation analysis (SPA) method. Here, the observed loss $Q(\boldsymbol{\theta}, V)$ is replaced by a conditional expectation of Q, with the conditioning on a part of V. The book by Fu and Hu (1997) is devoted largely to this SPA approach. Such conditioning can often cope with cases where discontinuities cause a violation of the conditions for the interchange of derivative and integral. Another gradient estimation method is rare perturbation analysis (see, e.g., Vázquez-Abad, 1999, and references therein), which is based on deleting or adding events with very small probabilities of occurrence. Like SPA, this has the effect of smoothing out the measured loss function Q.

In a different spirit from the PA family of estimation methods is the frequency domain (harmonic analysis) method. This method is based on building an approximation to the simulation (the approximation is sometimes called a metamodel) and then estimating the gradient directly from the approximation (Mitra and Park, 1991; Jacobson, 1994). The frequency domain approach is based on oscillating the components of $\boldsymbol{\theta}$ in a sinusoidal fashion during a single simulation run. This provides sensitivity information that in turn leads to a gradient estimate through the metamodel. A key aspect of this approach is the determination of the time index, the frequency, and the amplitude of the sinusoidal input. However, there appear to be no recent results that alter the conclusion of L'Ecuyer (1991): "Whether or not (and in what situations) frequency domain estimation would be competitive in practice for derivative estimation is not clear at this point."

Other gradient estimation methods are discussed in L'Ecuyer (1991), Pflug (1996), and Arsham (1998), among other references. These include the weak derivative method, the gradient surface method, and higher-order (Newton–Raphson-like) methods. All of these methods are based on exploiting structure internal to the simulation to improve the performance of the search and optimization process. A good recent discussion of other issues associated with simulation-based optimization, including a list of some commercial packages, is Fu (2002).

EXERCISES

15.1 Consider the reliability problem in Example 15.1. Assume that instead of defining V as the "natural" component lifetimes, it is defined as having three elements, the elements being the number of times that the system incurs a fixed, failure, and replacement cost over the interval $[0, T]$. Derive the expression for $Q(\boldsymbol{\theta}, \boldsymbol{\upsilon})$ with the new definition of V. In contrast to Example 15.1, sketch an argument to show that $Q(\boldsymbol{\theta}, \boldsymbol{\upsilon})p_V(\boldsymbol{\upsilon}|\boldsymbol{\theta})$ is now a continuous function in $\boldsymbol{\theta}$. This may be done by considering the special case of two components in the system and deriving the probability of observing one failure and two replacements over $[0, T]$. (Note that the full power of Theorem 15.1 is unnecessary because the expectation integral is replaced by a triple summation.)

15.2 Verify that both the LR/SF and IPA gradient estimates in Example 15.3 are unbiased and show that the IPA estimate has smaller variance than the LR/SF estimate. (Hint: for the exponential random variable Z of Example 15.3, $E(Z^3) = 6\theta^3$ and $E(Z^4) = 24\theta^4$.)

15.3 Fill in the missing details in going from the definition of $p_V(\upsilon|\boldsymbol{\theta})$ to the stochastic gradient input $\boldsymbol{Y}_k(\hat{\boldsymbol{\theta}}_k)$ in Example 15.4.

15.4 Consider the setting of Example 15.4. With the same probability mass function, but with the negative response function Q changed to $Q(\boldsymbol{\theta}, V) = (\beta - 10)^2 + 2V\lambda - V$, do the following:

(a) Determine $L(\boldsymbol{\theta})$.

(b) Analytically derive the value for $\boldsymbol{\theta}^*$ and $L(\boldsymbol{\theta}^*)$.

(c) Derive the stochastic gradient input $\boldsymbol{Y}_k(\hat{\boldsymbol{\theta}}_k)$.

15.5 For the problem of Exercise 15.4, do the following:

(a) Using an initial $\boldsymbol{\theta}$ of $[0.5, 10.5]^T$ and $a_k = 0.1/(50+k)^{0.501}$, run five independent replications of the stochastic gradient algorithm, each for 1000 iterations. Show that the distance from $\boldsymbol{\theta}^*$ to the final $\boldsymbol{\theta}$ estimate (i.e., $\hat{\boldsymbol{\theta}}_{1000}$) is negligibly different for each of the five runs. For one of the runs, produce a table showing the $\boldsymbol{\theta}$ estimates and loss values at 0, 10, 100, and 1000 iterations (analogous to the numbers shown in Figure 15.1)

(b) Interpret the results here vis-à-vis the results of Example 15.5. Provide some conceptual discussion on the reasons for the difference in performance.

15.6 Consider the setting of Example 15.4. With the same probability mass function, but with the negative response function Q changed to $Q(\boldsymbol{\theta}, V) = (\beta - 1)^2 + 2V(\lambda - 1) + \beta(V + 1)$, do the following three tasks:

(a) Determine $L(\boldsymbol{\theta})$.

(b) Analytically derive the value for $\boldsymbol{\theta}^*$ and $L(\boldsymbol{\theta}^*)$.

(c) Derive the stochastic gradient input $\boldsymbol{Y}_k(\hat{\boldsymbol{\theta}}_k)$.

15.7 For the problem of Exercise 15.6 with an initial $\boldsymbol{\theta}$ of $[0.75, -0.75]^T$ and $a_k = 0.025/(100+k)^{0.501}$, run five independent replications of the stochastic gradient algorithm, each for 10,000 iterations (not 1000 as in Example 15.4). Report the distance from $\boldsymbol{\theta}^*$ to the final $\boldsymbol{\theta}$ estimate (i.e., $\hat{\boldsymbol{\theta}}_{10,000}$) for each of the five runs. For one of the runs, produce a table showing the $\boldsymbol{\theta}$ estimate and loss values at 0, 10, 100, 1000, and 10,000 iterations (analogous to the numbers shown in Figure 15.1).

15.8 Consider the sample path method. Suppose that $p = 4$ and that each of the elements of $\hat{\boldsymbol{\theta}}_n$ are known to be within 0.1 unit of the corresponding elements in the optimum $\boldsymbol{\theta}_N^*$. Let the approximate distribution of $\boldsymbol{\theta}_N^*$ be $N(\boldsymbol{\theta}^*, \boldsymbol{C})$, $\boldsymbol{C} = [c_{ij}]$, $c_{ii} = 0.02$ for all i; $c_{ij} = 0.015$ for all $i \neq j$.

(a) Show that $\boldsymbol{C}$ is a valid covariance matrix.

(b) Using simulation or analytical analysis, provide a distance that is known with 99 percent certainty to be at least as large as the true distance between $\hat{\boldsymbol{\theta}}_n$ and $\boldsymbol{\theta}^*$.

(c) Contrast the solution in part (b) with an erroneous solution resulting from ignoring the correlations in $\boldsymbol{C}$ (i.e., a calculation based on incorrectly assuming that $\boldsymbol{C} = 0.02\boldsymbol{I}_4$).

15.9 Suppose that a scalar V has an exponential distribution with mean θ. Let $L(\theta) = \theta$. Derive the sample path gradient estimate using the generic LR/SF form and a reference parameter value θ'.

15.10 Consider the binary-outcome setting of Example 15.7. Generate one realization of $N = 20$ measurements based on taking $\theta' = 1/2$. Letting $b = 1/16$, contrast $\boldsymbol{\theta}_N^*$ and $\boldsymbol{\theta}^*$.

15.11 Letting $\hat{\boldsymbol{\theta}}_0 = \mathbf{0}$, show by the method of induction that $\hat{\boldsymbol{\theta}}_N = -\boldsymbol{B}^{-1}\bar{\boldsymbol{V}}_N$, as stated in Example 15.8.

CHAPTER 16

MARKOV CHAIN MONTE CARLO

The preceding two chapters considered the interface of simulation and optimization. This chapter on Markov chain Monte Carlo (MCMC) continues the study of simulation-related methods, but with a different focus. MCMC is a powerful means for generating random samples that can be used in computing statistical estimates, numerical integrals, and marginal and joint probabilities. The approach is especially useful in statistical applications where one is forming an estimate based on a multivariate probability distribution or density function that is only available to within an unknown constant. MCMC provides a means for generating samples from *joint* distributions based on easier sampling from *conditional* distributions. The approach has had a large impact on the theory and practice of statistical modeling. In fact, MCMC sometimes applies in problems where it is hard to imagine any other approach working.

The two most popular specific implementations of MCMC are the Metropolis–Hastings (M-H) algorithm and Gibbs sampling. Over the past 10–15 years, these two implementations of MCMC have revolutionized aspects of statistical modeling and data analysis by providing a practical general framework for many problems that formerly would have required significant application-specific methodology. Gibbs sampling, in particular, has had an especially significant impact in Bayesian problems, but Gibbs sampling applies to non-Bayesian problems as well.

Section 16.1 provides some general background related to random sampling and ergodic averaging. Section 16.2 discusses the M-H algorithm and Section 16.3 discusses Gibbs sampling. Section 16.4 presents some theoretical justification for Gibbs sampling and Section 16.5 presents several numerical examples. Section 16.6 discusses the important special case of Bayesian analysis. Finally, Section 16.7 offers some summary remarks on the relative properties of M-H and Gibbs sampling.

16.1 BACKGROUND

Markov chain Monte Carlo (MCMC) is a powerful means for generating random samples that can be used in computing statistical estimates and in computing marginal and conditional probabilities. MCMC methods rely on a *dependent* (Markov) sequence with a limiting distribution corresponding to a distribution of interest.[1] This contrasts with many classical Monte Carlo methods, which are based on *independent* samples. MCMC methods apply to a broader class of multivariate problems and are frequently easier to implement.

Although MCMC has general applicability, one area where MCMC has had a revolutionary impact is Bayesian analysis. MCMC has greatly expanded the range of problems for which Bayesian methods can be applied. Metropolis et al. (1953) introduced MCMC-type methods in the area of physics. Following key papers by Hastings (1970), Geman and Geman (1984), and Tanner and Wong (1987), the paper of Gelfand and Smith (1990) is largely credited with introducing the application of MCMC methods in modern statistics, specifically in Bayesian modeling.

Over the last decade, many papers and books have been published displaying the power of MCMC in dealing with realistic problems in a wide variety of areas. Among the excellent review papers in this area are the survey by Besag et al. (1995) and the vignettes by Cappé and Robert (2000) and Gelfand (2000). The survey by Evans and Swartz (1995) puts MCMC in perspective relative to other powerful methods for numerical integration in a statistical setting. Among the many books devoted to MCMC and its close algorithmic relatives are Gilks et al. (1996), Robert and Casella (1999), Chen et al. (2000), and Liu (2001). The February 2002 issue of the *IEEE Transactions on Signal Processing* is devoted to MCMC and closely related methods, as used in signal processing and tracking applications.

The treatment here is merely a glance at this booming area of research and practice. The aim is to provide the reader with some of the central motivation and the rudiments needed for a straightforward application. Obviously, the books and other references provide extensive detail not included here. Following convention, we use the terms *density* and *distribution* interchangeably when referring to the mechanism for generating a random process; this does not imply, of course, that a density is the same as a distribution function.

The prototype problem is as follows. Suppose that there is a process generating a random vector $\boldsymbol{X}$, and we wish to compute $E[f(\boldsymbol{X})]$ for some function $f(\cdot)$. We are interested in the case where this expectation is not readily available via standard analytical means. For the moment, let us suppose that $\boldsymbol{X}$ is a continuous random vector with an associated probability density function $p(\boldsymbol{x}) = p_{\boldsymbol{X}}(\boldsymbol{x})$. The methods to be described are, in fact, quite general and are not inherently tied to this assumption of continuity. We usually adopt this restriction for ease of presentation and because continuous random processes are very common. Hence, we may write

[1]As discussed in Appendix E, the term *Markov chain* is sometimes reserved for use with processes having discrete outcomes. In general applications of MCMC, the relevant processes may have discrete, continuous, or hybrid outcomes. For consistency with standard terminology in the MCMC area, we follow suit in this chapter in using the term *Markov chain* under the more general application to discrete or continuous outcomes.

$$E[f(\boldsymbol{X})] = \int f(\boldsymbol{x})p(\boldsymbol{x})\,d\boldsymbol{x}\,, \tag{16.1}$$

where the integral is over the domain for $\boldsymbol{X}$. The density $p(\boldsymbol{x})$ is sometimes called the *target density*. More generally, the target density (distribution) represents the distribution for the random variables of interest for the analysis. In some cases, for example, the target will pertain to a subset of the elements in $\boldsymbol{X}$ (e.g., it may represent the marginal distribution for only the first component of $\boldsymbol{X}$).

A standard Monte Carlo method for approximating the integral in (16.1) is to draw n independent, identically distributed (i.i.d.) samples of $\boldsymbol{X}$, say $\boldsymbol{X}_k$, $k = 1, 2,\ldots, n$, from the density function $p(\boldsymbol{x})$. We then form an average based on these independent samples, leading to

$$E[f(\boldsymbol{X})] \approx \frac{1}{n}\sum_{k=1}^{n} f(\boldsymbol{X}_k)\,.$$

Because the samples $\boldsymbol{X}_k$ are i.i.d., the strong and weak laws of large numbers (Appendix C) ensure that the approximation can be made as accurate as desired with increasing n.

Often, however, drawing samples from the density $p(\boldsymbol{x})$ is not feasible. The density may be very complicated and perhaps not even available analytically. Standard random number generation methods, such as in Appendix D, are generally quite limited in the type of randomness that can be produced. In practice, many distributions are not of the restricted "named" type for which most standard methods apply. The accept–reject method, which is much more general than the inverse transform method, is usually only computationally practical if the structure of $p(\boldsymbol{x})$ is exploited to find a "tight" majorizing function (typically requiring an optimization process). A related general method for random number generation is a version of the accept–reject method called adaptive rejection sampling (Robert and Casella, 1999, pp. 56–59 and 232). This method is restricted to densities that are log-concave (i.e., the logarithm of the density is a concave function) and is known to be very inefficient in high-dimensional problems (computations proportional to the *fifth power* of dim($\boldsymbol{X}$); see, e.g., Gilks et al., 1996, p. 84). Moreover, this sampling method may not be as efficient as the Markov chain-based methods described below (Robert and Casella, 1999, p. 241).

Fortunately, the integral approximation above—with its i.i.d. summands and requirement for direct sampling from $p(\boldsymbol{x})$—is more restrictive than necessary to form an estimate of $E[f(\boldsymbol{X})]$. An integral can, in fact, be approximated by a possibly *dependent* sample $\{\boldsymbol{X}_k\}$ that properly reflects the proportions associated with $p(\boldsymbol{x})$. This less stringent requirement opens up the possibility of efficient Markov chain-based schemes that avoid the need to directly sample from $p(\boldsymbol{x})$. In particular, one can use Markov chain-based Monte Carlo methods to efficiently produce dependent sequences having $p(\boldsymbol{x})$ as a limiting distribution *without the difficult or impossible task of sampling directly* from $p(\boldsymbol{x})$. An additional important benefit of MCMC methods is that $p(\boldsymbol{x})$ need only be known to within a scale factor. This is especially relevant in Bayesian applications, as discussed in Sections 16.2 and 16.6.

Consider a sequence $\boldsymbol{X}_0, \boldsymbol{X}_1, \boldsymbol{X}_2, \ldots$ such that $\boldsymbol{X}_{k+1}$ is generated from the (conditional) distribution for $\{\boldsymbol{X}_{k+1} | \boldsymbol{X}_k\}$ and $\boldsymbol{X}_0$ represents some initial condition (not needed in the i.i.d. case above). By the form of the conditional distribution, knowledge of $\boldsymbol{X}_k$ provides the information required to probabilistically characterize the behavior of the state $\boldsymbol{X}_{k+1}$. That is, the distribution for $\boldsymbol{X}_{k+1}$ depends only on the current state $\boldsymbol{X}_k$, not on the earlier states $\boldsymbol{X}_0, \boldsymbol{X}_1, \ldots, \boldsymbol{X}_{k-1}$. Hence, $\boldsymbol{X}_0, \boldsymbol{X}_1, \boldsymbol{X}_2, \ldots$ is a Markov chain, as discussed in Appendix E (but see footnote 1 in the current chapter on the use of the term *Markov chain*).

Under standard conditions for Markov chains, the dependence of $\boldsymbol{X}_k$ on any fixed number of early states, say $\boldsymbol{X}_0, \boldsymbol{X}_1, \ldots, \boldsymbol{X}_M$, $M < \infty$, disappears as $k \to \infty$. Hence, the density (distribution) of $\boldsymbol{X}_k$ approaches a stationary form, say $p^*(\cdot)$. That is, as k gets large, the (dependent) random vectors in the Markov sequence have a common distribution with, say, density $p^*(\cdot)$. Ignoring the first M iterations in the chain (called the *burn-in period*), we can form an *ergodic average*

$$\frac{1}{n-M} \sum_{k=M+1}^{n} f(\boldsymbol{X}_k), \tag{16.2}$$

so called because it is a practical realization of the famous ergodic theorem of stochastic processes.

The ergodic theorem guarantees that the normalized sum in (16.2) will approach the mean of $f(\boldsymbol{X})$ (usually in the mean square [m.s.] or almost sure [a.s.] sense; see Appendix C) as $n \to \infty$ for any fixed M, where this mean is computed with respect to $p^*(\cdot)$. For the ergodic theorem for m.s. convergence to hold, it is necessary and sufficient that: (i) $\text{cov}[f(\boldsymbol{X}_j), f(\boldsymbol{X}_k)]$ is uniformly bounded in magnitude for all j, k and (ii) the correlation between $f(\boldsymbol{X}_n)$ and the sample mean in (16.2) goes to zero as $n \to \infty$ (Parzen, 1962, pp. 72–75). In other words, the process is ergodic in the m.s. sense if there is less correlation between the terminal observation $f(\boldsymbol{X}_n)$ and the sample mean in (16.2) as n is increased. In the MCMC methods, if the Markov chain is generated properly, then $p^*(\cdot)$ equals the target density $p(\cdot)$, as desired. That is, the limit of the ergodic mean in (16.2) corresponds to the desired value $E[f(\boldsymbol{X})]$ computed with respect to $p(\cdot)$.

We outlined above a general formulation for generating random samples via the output of a Markov chain. There are, of course, some fundamental elements that need to be specified to justify the approach and to provide the details required for implementation. These elements form the basis for the MCMC approach.

In the remainder of this chapter devoted to MCMC, we present an overview of the Metropolis–Hastings (M-H) and Gibbs sampling implementations of MCMC, discuss applications in Bayesian modeling, sketch the theoretical foundations, and present some examples. There are also some important aspects of MCMC that we do not cover in this relatively brief review. For example, we do not discuss formal methods for diagnosing convergence. This is an active area of current research, with a large number of specialized techniques. Robert and Casella (1999, Chap. 8) is one of many references on this subject. As with general stochastic search and optimization methods, there is no universal method for knowing when to stop a chain. We also do not review the many software packages that are available for both simple and sophisticated implementations of MCMC.[2]

16.2 METROPOLIS–HASTINGS ALGORITHM

As discussed above, the sum of dependent random variables in (16.2) converges to the mean of $f(\boldsymbol{X})$ under appropriate conditions. Although this is hopeful, we must show that this is the "right" mean, that is, the mean computed with respect to $p(\boldsymbol{x})$. Fortunately, it is surprisingly easy to produce such a Markov process via a variant of the Metropolis sampling seen in the simulated annealing algorithm of Chapter 8. The form given here, introduced by Hastings (1970), builds on the criterion in Metropolis et al. (1953).

Given an initial condition $\boldsymbol{X}_0$, the M-H algorithm is a mechanism for producing the Markov process $\boldsymbol{X}_1, \boldsymbol{X}_2, \ldots$ for use in (16.2). From a state $\boldsymbol{X}_k = \boldsymbol{x}$, the next state is chosen by generating a candidate point $\boldsymbol{W}$ from a *proposal distribution* (sometimes called an *instrumental distribution* or a *candidate-generating distribution*), $q(\cdot|\boldsymbol{x})$. In principle, the proposal distribution may be chosen arbitrarily, although there may be efficiency advantages to one form over another in some applications. The proposal distribution satisfies the key property for density functions, namely

$$\int q(\boldsymbol{w} \,|\, \boldsymbol{X} = \boldsymbol{x})\, d\boldsymbol{w}$$

for any $\boldsymbol{X} = \boldsymbol{x}$, as appropriate. There are also some very modest conditions for the proposal distribution, as discussed in Robert and Casella (1999, pp. 233–235). For example, the set of points $\boldsymbol{w}$ where $q(\boldsymbol{w}|\boldsymbol{x}) > 0$, as $\boldsymbol{X} = \boldsymbol{x}$ ranges over the set of points where $f(\boldsymbol{x}) \neq \boldsymbol{0}$, should be a *superset* of the set of points $\boldsymbol{x}$ where $f(\boldsymbol{x}) \neq \boldsymbol{0}$. A common example for $q(\cdot|\boldsymbol{x})$ is a uniform distribution centered around $\boldsymbol{x}$. This example satisfies the superset condition. One implication of the superset and other conditions is that it is possible to generate candidate points that "fill up" the support of the target density (i.e., provide an adequate number of points throughout the region where $p(\cdot) > 0$).

[2] Let us mention one prominent package, however. BUGS (*B*ayesian inference *U*sing *G*ibbs *S*ampling) is available free on the Web and is one of the most popular standard packages. BUGS is available at *www.mrc-bsu.cam.ac.uk/bugs*.

Analogous to step 2 of the simulated annealing algorithm in Section 8.2, the candidate point $\boldsymbol{W}$ is accepted with probability $\rho(\boldsymbol{X}_k, \boldsymbol{W})$, where

$$\rho(\boldsymbol{x}, \boldsymbol{w}) = \min\left\{\frac{p(\boldsymbol{w})}{p(\boldsymbol{x})}\frac{q(\boldsymbol{x}\,|\,\boldsymbol{w})}{q(\boldsymbol{w}\,|\,\boldsymbol{x})},\ 1\right\}. \tag{16.3}$$

If the candidate point is accepted, then $\boldsymbol{X}_{k+1} = \boldsymbol{W}$; otherwise, $\boldsymbol{X}_{k+1} = \boldsymbol{X}_k$. Note that $q(\cdot|\cdot)$, appearing in both the numerator and denominator of (16.3), is the *same function* with only the conditioning interchanged ($\boldsymbol{x}$ and $\boldsymbol{w}$ have the same dimension). (In general, of course, the functional *form* for the distribution of one random variable conditioned on another depends on the order of the conditioning.) Likewise, $p(\cdot)$ in the numerator and denominator is the same function with only the arguments changed. (The connection to simulated annealing is discussed further in Robert and Casella, 1999, p. 281.)

Because of the ratio form, an important implication of (16.3) is that one only needs to know $p(\cdot)$ to within a constant because the constant cancels out. Eliminating the need to determine the constant has significant practical advantages by removing the need for a formidable numerical integration. For example, in Bayesian applications, as considered in Section 16.6, the target density represents a posterior density that is conditioned on some set of data (the data conditioning does not affect the mechanics of the M-H algorithm). It is notoriously difficult to obtain the constant associated with the posterior density function because the constant is the marginal density for the data appearing in the denominator of Bayes' rule. This marginal density requires difficult numerical integration. We discuss this further in Section 16.6.

Table 16.1 summarizes two common proposal distributions. Suppose that $m = \dim(\boldsymbol{X})$ (so $m = \dim(\boldsymbol{W})$). Let us denote an m-fold uniform distribution by $U_m(\boldsymbol{a}, \boldsymbol{b})$, where $\boldsymbol{a}$ and $\boldsymbol{b}$ are m-dimensional vectors. This distribution is such that each component of the m-dimensional random vector has an independent uniform distribution with lower and upper endpoints given by the corresponding component of $\boldsymbol{a}$ and $\boldsymbol{b}$ (so the probability is uniform over the hypercube defined

Table 16.1. Examples of two popular general forms for proposal distributions.

General form of proposal distribution	$q(\boldsymbol{w}\|\boldsymbol{x})$	$q(\boldsymbol{x}\|\boldsymbol{w})$
Normal with covariance matrix $\boldsymbol{\Sigma}$	$N(\boldsymbol{x}, \boldsymbol{\Sigma})$	$N(\boldsymbol{w}, \boldsymbol{\Sigma})$
Uniform of width 2δ for each component	$U_m(\boldsymbol{x}-\delta\boldsymbol{1}_m,\ \boldsymbol{x}+\delta\boldsymbol{1}_m)$	$U_m(\boldsymbol{w}-\delta\boldsymbol{1}_m,\ \boldsymbol{w}+\delta\boldsymbol{1}_m)$

by $\boldsymbol{a}$ and $\boldsymbol{b}$). These two candidate-generating processes are examples of *random walk processes*. That is, the candidate point may be written as the current point plus noise: $\boldsymbol{W} = \boldsymbol{X}$ + noise, where the noise is a mean-zero normal or uniform distribution. Note further that $q(\boldsymbol{w}|\boldsymbol{x}) = q(\boldsymbol{x}|\boldsymbol{w})$, an example of the important special case where the proposal distribution is symmetric. An implication of the symmetric proposal is that criterion (16.3) simplifies to

$$\rho(\boldsymbol{x}, \boldsymbol{w}) = \min\left\{\frac{p(\boldsymbol{w})}{p(\boldsymbol{x})},\ 1\right\}$$

as a result of $q(\boldsymbol{x}\,|\,\boldsymbol{w})/q(\boldsymbol{w}\,|\,\boldsymbol{x}) = 1$. For the m-fold uniform distribution in Table 16.1, let $\mathbf{1}_m$ denote an m-dimensional vector of 1's and δ be a positive constant.

A remarkable result associated with (16.3) is that although the proposal distribution $q(\cdot\,|\,\cdot)$ may have almost any form (subject to the modest conditions mentioned above), the stationary distribution of the chain satisfies $p^*(\cdot) = p(\cdot)$ (Hastings, 1970). Chib and Greenberg (1995) and Robert and Casella (1999, pp. 235–238) elaborate on some of the arguments in Hastings (1970), establishing an appropriate form of convergence in distribution. Below is a summary of the steps for the M-H algorithm. The first several steps pertain to the burn-in period and the remaining steps are used to form the ergodic average in (16.2).

M-H Algorithm for Estimating $E[f(\boldsymbol{X})]$

Step 0 **(Initialization)** Choose the length of the burn-in period M and an initial state $\boldsymbol{X}_0$. Set $k = 0$.

Step 1 Generate a candidate point $\boldsymbol{W}$ according to the proposal distribution $q(\cdot\,|\boldsymbol{X}_k)$.

Step 2 Generate a point U from a $U(0, 1)$ distribution. Set $\boldsymbol{X}_{k+1} = \boldsymbol{W}$ if $U \leq \rho(\boldsymbol{X}_k, \boldsymbol{W})$ from (16.3). Otherwise, set $\boldsymbol{X}_{k+1} = \boldsymbol{X}_k$.

Step 3 Repeat steps 1 and 2 until $\boldsymbol{X}_M$ is available. Terminate the burn-in process and proceed to step 4 with $\boldsymbol{X}_k = \boldsymbol{X}_M$.

Step 4 Carry out step 1.

Step 5 Carry out step 2.

Step 6 Repeat steps 4 and 5 until it is possible to compute the ergodic average of $n - M$ evaluations in (16.2). (Of course, if desired, this average can be computed recursively without storing all of $f(\boldsymbol{X}_{M+1}), f(\boldsymbol{X}_{M+2}),\ldots, f(\boldsymbol{X}_n)$.) This ergodic average is the estimate of $E[f(\boldsymbol{X})]$ under the target density $p(\cdot)$.

There are a number of specific ways in which the overall M-H algorithm can be implemented. The most obvious variation in implementation is in the choice of the proposal distribution $q(\cdot\,|\,\cdot)$. Although almost any choice of $q(\cdot\,|\,\cdot)$ works in the sense that the ergodic average in (16.2) converges to $E[f(\boldsymbol{X})]$, there are clear differences in the rate of convergence depending on the nature of the problem. There are also forms of averaging that differ from the standard ergodic averaging. One variation is to run many independent chains, with each chain terminating at X_{M+1}. In this way, $E[f(X)]$ is estimated by forming a sample mean of *independent* values $f(X_{M+1})$. Regeneration, as discussed in Section 14.2, may also be used to improve the performance of M-H (Gilks et al., 1998). This creates independent blocks of iterations, allowing for the proposal distribution to be adapted at each block to improve the sampling.

The rate at which candidate values $\boldsymbol{W}$ are accepted affects M-H performance. This rate should be neither too small nor too large to allow for an adequate exploration of the space. As summarized in Roberts et al. (1997) and Roberts and Rosenthal (1998), if the proposal distribution is normal, the approximate optimal acceptance rate is 23 percent as $\dim(\boldsymbol{X}) \to \infty$ (i.e., 23 percent of the time, $\boldsymbol{X}_{k+1} \neq \boldsymbol{X}_k$ in the M-H steps). In addition, if the target distribution is also normal, the approximate optimal acceptance rate is 45 percent when $\dim(\boldsymbol{X}) = 1$ (Chen et al., 2000, p. 23). These results pertain to the *random walk* proposal distribution, implying that the target and proposal distributions are not identical to each other (if they were identical, a 100 percent acceptance rate is optimal!). Roberts and Rosenthal (2001) point out that the 23 percent rate is, in fact, quite robust to deviations from the assumptions above (e.g., the optimal acceptance rate for $\dim(\boldsymbol{X}) = 10$ is negligibly different from the asymptotic rate of 23 percent; further, the results apply for certain types of nonnormal proposal distributions). This robustness can be seen both theoretically and empirically. As we have seen with other aspects of stochastic search and optimization, tuning is necessary to achieve approximately optimal performance. In practice, this can be achieved by running test iterations of the algorithm, adjusting parameters in the proposal distribution until the acceptance rate is near the optimal rate.

Let us present a simple example where the target density is bivariate normal. This example demonstrates the performance of M-H in a setting that is easy to understand. In practice, there are other (likely more) efficient methods of generating samples from a multivariate normal distribution (see Appendix D).

Example 16.1—Simulating a bivariate normal distribution. Consider a problem where the target density is bivariate normal with the two variables highly correlated. In particular, suppose that

$$\boldsymbol{X} \sim N\left(\begin{bmatrix} 0 \\ 0 \end{bmatrix}, \begin{bmatrix} 1 & 0.9 \\ 0.9 & 1 \end{bmatrix}\right).$$

Further, suppose that the proposal distribution $q(\boldsymbol{w}|\boldsymbol{x})$ is a shifted uniform distribution as in Table 16.1. That is, $\boldsymbol{W}$ (conditioned on $\boldsymbol{X} = \boldsymbol{x}$) is generated according to $U_2(\boldsymbol{x} - 0.5\mathbf{1}_2, \boldsymbol{x} + 0.5\mathbf{1}_2)$. The distribution for each of the two elements in $\boldsymbol{W}$ has the standard unit length (centered around the elements of the moving point $\boldsymbol{x}$). Because $q(\boldsymbol{w}|\boldsymbol{x}) = q(\boldsymbol{x}|\boldsymbol{w})$, the form of the normal target density function implies that

$$\rho(\boldsymbol{x}, \boldsymbol{w}) = \min\left\{ \frac{\exp\left(-\frac{1}{2}\boldsymbol{w}^T \begin{bmatrix} 1 & 0.9 \\ 0.9 & 1 \end{bmatrix}^{-1} \boldsymbol{w}\right)}{\exp\left(-\frac{1}{2}\boldsymbol{x}^T \begin{bmatrix} 1 & 0.9 \\ 0.9 & 1 \end{bmatrix}^{-1} \boldsymbol{x}\right)}, 1 \right\}.$$

As discussed above, note that the constant terms in the bivariate normal density functions are not needed in constructing $\rho(\cdot)$ (a trivial advantage here, but a major advantage in applications such as Bayesian analysis).

Figure 16.1 shows the results of a study where M-H was used to estimate $E[f(\boldsymbol{X})]$, where $f(\boldsymbol{X}) = [1, 1]\boldsymbol{X}$. Thus, we are estimating the sum of the means for the two elements of $\boldsymbol{X}$. The figure shows the evolution of the ergodic averages for three independent runs. The $M = 500$ burn-in period for each run is initialized at $\boldsymbol{X}_0 = [-1, 1]^T$. It is clear that the three runs are all settling down near the true value $E[f(\boldsymbol{X})] = 0$. Nevertheless, some improvement in the estimates is possible, especially for the run represented by the light-gray line.

Tuning of the proposal distribution $U_2(\cdot)$ from the current naïve unit-length intervals provides better results. For each of the runs, about 70 percent of the candidate points $\boldsymbol{W}$ are accepted. This rate is higher than the above-mentioned optimal rate of 23 percent under a normal proposal and large dim($\boldsymbol{X}$)

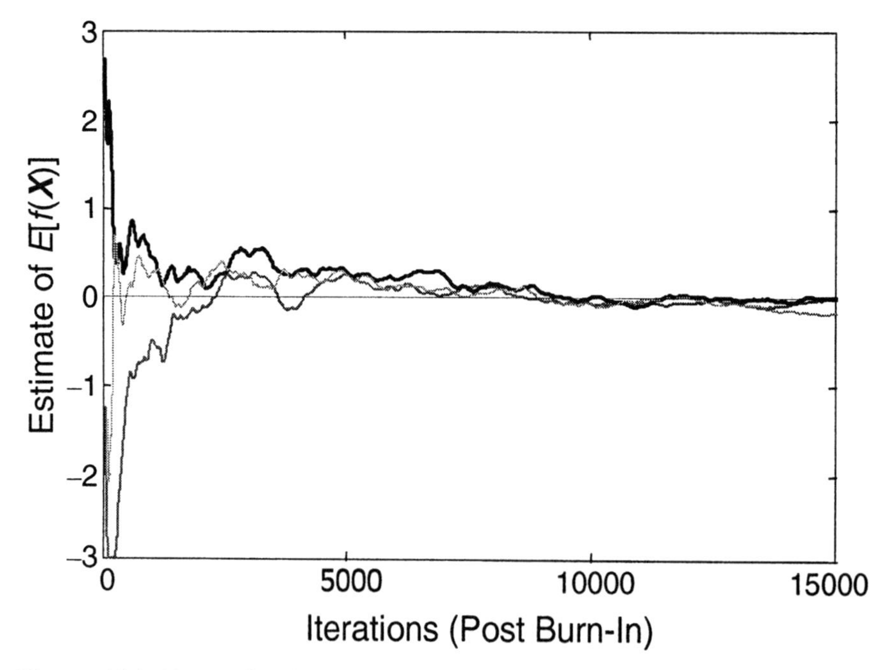

Figure 16.1. Traces for three independent runs of M-H sampler in estimating $E[f(\boldsymbol{X})]$ (i.e., estimating the sum of the two components in $E(\boldsymbol{X})$). Target value is 0. Burn-in period (M) is 500 for each run. Standard (unit length) uniform proposal distribution used here. Improved performance is possible by optimizing the proposal distribution.

(neither of which are true here). Based on numerical experimentation, a wider support for the uniform distribution decreases the acceptance rate in this case. In fact, as a testament to the robustness of the 23 percent guideline (see the discussion above), a decreased acceptance rate *did* improve the performance in this case even though the proposal distribution is not normal and dim(X) is not large. Changing the proposal distribution to $U_2(x - 2 1_2, x + 2 1_2)$ caused the acceptance rate to drop to approximately 24 percent and decreased the magnitude of the maximum terminal error in 50 independent runs of M-H from 0.33 to 0.20. Other *forms* for the proposal distribution may also be valuable; as mentioned earlier, the normal distribution is a frequent proposal form. (Exercise 16.2 considers the normal proposal form and the wider $U_2(x - 2 1_2, x + 2 1_2)$ form.)

While the M-H algorithm has wide applicability, there is some loss of information in the ergodic sampling (with its serial dependence) relative to independent sampling. For comparison purposes, suppose that one could directly obtain independent samples of X (an idealized case since one would then not need M-H!). An estimate of $E[f(X)]$ based on the mean of 15,000 independent samples of $[1, 1]X$ would have a standard deviation of 0.0159 (Exercise 16.1). This contrasts with an approximate standard deviation of 0.138 for the terminal estimate from 15,000 samples from the M-H algorithm as represented in Figure 16.1. This approximation is the sample standard deviation of the terminal estimate from 50 independent runs of M-H. Hence, for the same number of samples (ignoring the burn-in samples), the M-H algorithm produces an estimate over eight times more variable than direct sampling ($0.138/0.0159 \approx 8.7$). The increased variation in the M-H estimate is due to the positive correlation in the iterates (Exercise 16.1). Of course, M-H is predicated on direct sampling not being available, so this type of comparison is only useful to establish a bound on behavior. ❑

We next consider what is likely the most popular specific implementation of the M-H algorithm.

16.3 GIBBS SAMPLING

Gibbs sampling represents an implementation of the M-H algorithm on an element-by-element basis for the components in X. The term *Gibbs sampling* was introduced by Geman and Geman (1984) in a specific implementation of a Gibbs distribution for sampling on lattices.[3] The term is now used more generally (and casually) to refer to the special case where the proposal distribution is built directly from the density of interest $p(\cdot)$. Because of this restriction, the method is more limited than M-H. The restriction sometimes leads to advantages in efficiency and ease of implementation via the elimination of the tuning typically needed in M-H. Gibbs sampling is especially important in

---

[3]Gibbs sampling derives its name from the physicist Josiah W. Gibbs, 1839–1903, based on the connection to Gibbs random fields identified in Geman and Geman (1984).

Bayesian implementations. Gibbs sampling is uniquely designed for multivariate problems. In fact, "...the crucial issue is replacement of the sampling of a high-dimensional vector with sampling of lower-dimensional component blocks, thus breaking the so-called curse of dimensionality" (Gelfand, 2000).

Gibbs sampling may be interpreted as a concatenation of m M-H algorithms, one for each variable in the random vector of interest, $\boldsymbol{X}$. (Because this version of M-H does not generate a new vector in toto, it is not precisely the same as the multivariate form in Section 16.2.) This concatenation has m target distributions, each representing a conditional distribution for one variable given values for all other variables (called the *full conditional distribution*, as discussed below). In contrast to the M-H algorithm, where the proposal distributions may be chosen almost arbitrarily, the proposals here have a required form. Namely, the proposal distribution for the ith element of $\boldsymbol{X}$ is the conditional distribution of that variable given the most recent values for all other variables during the iterative process. Thus, the target and proposal distributions are the same.

To make the above concepts more concrete, consider a trivariate ($m = 3$) problem based on density functions, the three variables being X, Y, and Z. The three target densities are $p_{X|Y,Z}(\cdot\,|\,\cdot)$, $p_{Y|X,Z}(\cdot\,|\,\cdot)$, and $p_{Z|X,Y}(\cdot\,|\,\cdot)$. As with generic M-H, let W represent a candidate random variable generated according to the candidate density. For the first element, X, the candidate-generating density is then

$$q(w\,|\,y,z) = p_{X|Y,Z}(w\,|\,y,z). \tag{16.4}$$

With the target density $p(\cdot) = p_{X|Y,Z}(\cdot\,|\,\cdot)$, the substitution of (16.4) into the probability of acceptance for the M-H algorithm as given in (16.3) yields

$$\rho(x,w) = \min\left\{\frac{p_{X|Y,Z}(w\,|\,y,z)}{p_{X|Y,Z}(x\,|\,y,z)}\frac{q(x\,|\,y,z)}{q(w\,|\,y,z)},\ 1\right\} = \min\{1,\ 1\} = 1. \tag{16.5}$$

Unlike the general M-H algorithm, relationship (16.5) implies that the new point W is always accepted as a representation for X. Hence, from (16.5), given any previous values for Y and Z, the candidate value W generated from $q(w\,|\,y,z) = p_{X|Y,Z}(w\,|\,y,z)$ is guaranteed to be the new value for X. Identical arguments apply for the other two variables, Y and Z. In fact, because of the automatic acceptance of a candidate point, there is no need for a separate (W) candidate process for each variable; one simply generates a new X, Y, or Z as appropriate. The general multivariate relationship between Gibbs sampling and M-H is discussed in Robert and Casella (1999, pp. 296–297) and Gilks et al. (1996, pp. 10–12). This relationship is an obvious extension of the trivariate setting above.

As a consequence of the theory of Markov processes, the iterative values X_k, Y_k, Z_k generated via the Gibbs sampler represent, in the limit, observations from the joint density $p_{X,Y,Z}(x,y,z)$. Further, each of X_k, Y_k, and Z_k has a distribution that approaches its respective marginal, $p_X(x)$, $p_Y(y)$, and $p_Z(z)$.

The implementation of the Gibbs sampler for the trivariate problem of generating samples from $p_{X,Y,Z}(\cdot)$ is as follows. Suppose that the sampler begins with an initial guess at Y and Z, say Y_0 and Z_0. Using the full conditional $p_{X|Y,Z}(x|Y_0, Z_0)$, we can then generate a sample point X_1 by Monte Carlo. We next use the full conditional $p_{Y|X,Z}(y|X_1, Z_0)$ to generate a sample point Y_1. Likewise, $p_{Z|X,Y}(z|X_1, Y_1)$ is used to generate Z_1. At this point, the Gibbs sampler has completed one iteration, producing X_1, Y_1, and Z_1. We now repeat the process, using Y_1 and Z_1 to initiate the conditioning and producing a sample X_2, Y_2, and Z_2. This Monte Carlo sampling forms a sequence

$$Y_0, Z_0; X_1, Y_1, Z_1; X_2, Y_2, Z_2; \ldots; X_n, Y_n, Z_n. \tag{16.6}$$

The sequence in (16.6) is called the *Gibbs sequence*. The sequential sampling approach above extends in an obvious way to the general m-dimensional case.

The Gibbs sequence above can be used in several ways to estimate $E[f(X,Y,Z)]$. For example, let M denote the burn-in period, as with the M-H algorithm. Under modest conditions, the later observations, $X_{M+1}, Y_{M+1}, Z_{M+1}; \ldots; X_n, Y_n, Z_n$, in the Gibbs sequence represent measurements from $p_{X,Y,Z}(\cdot)$. These can then be substituted in the averaging of (16.2) to produce an estimate of $E[f(X,Y,Z)]$. A variation on standard ergodic averaging is to pick off every (say) ℓth value in the chain, and average only these values. If ℓ is reasonably large, this is roughly equivalent to averaging independent samples. Yet another way of estimating $E[f(X,Y,Z)]$ is to generate N independent Gibbs sequences, using only the final output in each sequence. Now, $E[f(X,Y,Z)]$ is estimated by an average of N samples, where the summands are i.i.d.

The example below shows the derivation of the required conditional distributions for generating the Gibbs sequence in a bivariate (m = 2) setting. This example pertains to a problem with discrete outcomes.

Example 16.2—Gibbs sampling with a Bernoulli distribution. The sampling need not be performed according to continuous random variables and associated probability *density* functions. Consider the following example associated with a Bernoulli distribution, as given in Casella and George (1992). Let the bivariate random vector $[X, Y]^T$ have a joint probability (mass) function

$$\begin{bmatrix} P(X=0,Y=0) & P(X=0,Y=1) \\ P(X=1,Y=0) & P(X=1,Y=1) \end{bmatrix} = \begin{bmatrix} p_{00} & p_{01} \\ p_{10} & p_{11} \end{bmatrix},$$

where $p_{ij} \geq 0$ and $p_{00} + p_{01} + p_{10} + p_{11} = 1$. That is, each variable, X or Y, is a Bernoulli-distributed random variable that can take on a value 0 or 1. The two variables are dependent. So, for example, the likelihood of Y = 1 depends on whether X = 0 or 1.

Using basic rules of probability, it is simple to compute the *conditional* probabilities $P(Y = a|X = b)$ and $P(X = a|Y = b)$, where a and b are 0 or 1. These probabilities are given in the two matrices below:

$$\begin{bmatrix} P(Y=0|X=0) & P(Y=1|X=0) \\ P(Y=0|X=1) & P(Y=1|X=1) \end{bmatrix} = \begin{bmatrix} \dfrac{p_{00}}{p_{00}+p_{01}} & \dfrac{p_{01}}{p_{00}+p_{01}} \\ \dfrac{p_{10}}{p_{10}+p_{11}} & \dfrac{p_{11}}{p_{10}+p_{11}} \end{bmatrix}$$

and

$$\begin{bmatrix} P(X=0|Y=0) & P(X=1|Y=0) \\ P(X=0|Y=1) & P(X=1|Y=1) \end{bmatrix} = \begin{bmatrix} \dfrac{p_{00}}{p_{00}+p_{10}} & \dfrac{p_{10}}{p_{00}+p_{10}} \\ \dfrac{p_{01}}{p_{01}+p_{11}} & \dfrac{p_{11}}{p_{01}+p_{11}} \end{bmatrix}.$$

The Gibbs sampling procedure applied with these two sets of conditional probabilities yields a sequence of 0's and 1's. We can then use this sequence to estimate $E[f(X, Y)]$. ❑

The above two- and three-variable discussions serve to motivate the general multivariate setting. In this more general setting, we continue to be interested in estimating quantities of the form $E[f(\boldsymbol{X})] = \int f(\boldsymbol{x})p(\boldsymbol{x})d\boldsymbol{x}$ (or its discrete analogue), where $\boldsymbol{X}$ is a collection of m (say) univariate or multivariate components. In the common case where the m components are univariate, the kth sample $\boldsymbol{X}$ from the Gibbs sampling algorithm is:

$$\boldsymbol{X}_k = \begin{bmatrix} X_{k1} \\ X_{k2} \\ \vdots \\ X_{km} \end{bmatrix},$$

where X_{ki} denotes the ith component for the kth replicate of $\boldsymbol{X}$ generated via the sampling algorithm. As we saw above, in Gibbs sampling, it is not necessary to introduce a separate candidate process $\boldsymbol{W}$ (as in the general M-H approach) because the candidate point is always accepted. While each X_{ki} is usually a scalar element, there are problems where it is useful to have at least some X_{ki} be multivariate (see Example 16.7).

The idea of the proposal distribution for M-H can be used to update each component of $\boldsymbol{X}$. This updating is done on a component-by-component basis. Central to the updating is the conditional random variable $\{X_{k+1,i}|\boldsymbol{X}_{k\backslash i}\}$, where

$$\mathbf{X}_{k\backslash i} \equiv \{X_{k+1,1}, X_{k+1,2}, \ldots, X_{k+1,i-1}, X_{k,i+1}, \ldots, X_{km}\},$$

$i = 1, 2, \ldots, m$. To avoid very cumbersome subscript notation associated with the relevant random variables and the associated conditioning, let $p_i(\cdot)$ represent the sampling density for the conditional random variable:

$$\{X_{k+1,i} | \mathbf{X}_{k\backslash i}\} \sim p_i(x | \mathbf{X}_{k\backslash i}), \; i = 1, 2, \ldots, m.$$

That is, $p_i(\cdot)$ represents the sampling density for the random variable $X_{k+1,i}$ conditioned on $\mathbf{X}_{k\backslash i}$, where the first $i-1$ elements of $\mathbf{X}_{k\backslash i}$ represent sample points at the same $(k+1)$st iteration while the remaining $m-i$ elements are points available from the kth iteration. This strange-looking conditioning follows naturally from the sequential component-wise processing in the Gibbs sampling procedure, as given below. The conditioning represents the most recent information available when generating the ith component of $\mathbf{X}$.

The density $p_i(\cdot)$ is the generalization of the full conditional densities that were given above for the $m = 3$ problem. Note that the full conditional is a *univariate* sampling density in the common case where each X_{ki} is univariate. Thus, even when $\mathbf{X}$ is high dimensional, the sampling is univariate. This has significant potential advantages. The derivation of the full conditional follows from basic laws of probability:

$$p_i(x | \mathbf{X}_{k\backslash i}) = \frac{p_{\mathbf{X}}(x, \mathbf{X}_{k\backslash i})}{\int p_{\mathbf{X}}(x, \mathbf{X}_{k\backslash i})\,dx}, \qquad (16.7)$$

where the denominator integral is over the domain for $X_{k+1,i} = x$, and $p_{\mathbf{X}}(\cdot)$ appearing on the right-hand side represents (as before) the density for $\mathbf{X}$. In some practical applications, this definition can be used directly to obtain the full conditionals required for the steps of the Gibbs sampling procedure. In other cases, standard numerical methods for random number generation can be used. A good discussion of methods for obtaining samples from full conditionals appears in Gilks et al. (1996, Chap. 5). There are some applications for which sampling from full conditionals is difficult. One popular method for coping with such a difficulty is the *Metropolis within Gibbs* approach. This method involves the use of M-H within the Gibbs sampling steps below to produce samples from the full conditionals (see, e.g., Gilks et al., 1996, pp. 84–85; Robert and Casella, 1999, pp. 322–326).

Let us now present a standard implementation of the Gibbs sampling algorithm for estimating $E[f(\mathbf{X})]$ in (16.1). After presenting these steps, we comment on several variations of the standard algorithm.

Gibbs Sampling Algorithm for Estimating $E[f(X)]$

Step 0 **(Initialization)** Choose the length of the burn-in period M and an arbitrary initial state $\boldsymbol{X}_0$. Set $k = 0$.

Step 1 Generate $\boldsymbol{X}_{k+1}$ according to the following m steps:
1. Generate $X_{k+1,1} \sim p_1(x \mid \boldsymbol{X}_{k\backslash 1})$.
2. Generate $X_{k+1,2} \sim p_2(x \mid \boldsymbol{X}_{k\backslash 2})$.
⋮
***m*.** Generate $X_{k+1,m} \sim p_m(x \mid \boldsymbol{X}_{k\backslash m})$.

Step 2 Repeat step 1 until $\boldsymbol{X}_M$ is available. Terminate the burn-in process and proceed to step 3 with $\boldsymbol{X}_k = \boldsymbol{X}_M$.

Step 3 Carry out step 1.

Step 4 Repeat step 3 until it is possible to compute the ergodic average of $n - M$ evaluations in (16.2). (Of course, if desired, this average can be computed recursively without storing all of $f(\boldsymbol{X}_{M+1}), f(\boldsymbol{X}_{M+2}), \ldots, f(\boldsymbol{X}_n)$.) This ergodic average is the estimate of $E[f(\boldsymbol{X})]$ under the target distribution $p(\cdot)$.

As with M-H, there are implementations of Gibbs sampling for estimating $E[f(\boldsymbol{X})]$ other than the standard ergodic averaging of (16.2). One variation is to run many independent chains, each chain terminating at $\boldsymbol{X}_{M+1}$. In this way, $E[f(\boldsymbol{X})]$ is estimated by forming a sample mean of *independent* values $f(\boldsymbol{X}_{M+1})$. Another variation is to average two dependent chains, where the dependence is introduced via the notion of antithetic random variables (an analogue of common random numbers discussed in Chapter 14). Such antithetic averaging based on pairs of dependent chains can lead to significant reduction of the variance of the estimate for $E[f(\boldsymbol{X})]$ (Frigessi et al., 2000).

16.4 SKETCH OF THEORETICAL FOUNDATION FOR GIBBS SAMPLING

It is not immediately obvious why the above Markov sampling procedure works. How can sequential sampling from *conditional* distributions produce samples from a *joint* distribution? Let us sketch the rationale. Given the close connection of Gibbs sampling to the M-H algorithm introduced earlier, the arguments here also provide some flavor of the basis for M-H, although the details are somewhat different. This relatively informal discussion is a simplified version of the discussion in Gelfand and Smith (1990) and Robert and Casella (1999, Sect. 7.1.3).

Let us consider the three-variable case, X, Y, and Z, and sketch how the sampling from the full conditionals as above provides the information necessary to obtain samples from the density $p_{X,Y,Z}(x,y,z)$ (or from the marginals for any of the three variables). The ideas for three variables below extend immediately to an arbitrary number of variables (m as above). From basic rules of conditional probability,

$$p_X(x) = \int p_{X|Y,Z}(x\,|\,y,z)p_{Y,Z}(y,z)\,dy\,dz\,,$$

$$p_Y(y) = \int p_{Y|X,Z}(y\,|\,x,z)p_{X,Z}(x,z)\,dx\,dz\,,$$

$$p_Z(z) = \int p_{Z|X,Y}(z\,|\,x,y)p_{X,Y}(x,y)\,dx\,dy\,,$$

where the integrals are over the relevant domains in $\mathbb{R}^2$. Note the presence of a full conditional in each of the integrands above. The full conditionals form the basis for the Markov aspect of the sampling because the next random variate is generated based on only the most recent conditioning.

The expressions above provide the basis for the Gibbs sampling in (16.6). Suppose that we begin the sampler with the top expression above and make an initial guess at Y and Z, say Y_0 and Z_0. Using the full conditional $p_{X|Y,Z}(x|Y_0, Z_0)$, we can then generate a sample point X_1 by Monte Carlo. Proceeding downward through the expressions above, we next use the full conditionals $p_{Y|X,Z}(y|X_1, Z_0)$ and $p_{Z|X,Y}(z|X_1, Y_1)$ to generate sample points Y_1 and Z_1, respectively. At this point, we have completed one iteration of the Gibbs sampler, producing a sample X_1, Y_1, and Z_1. The process is then repeated, using Y_1 and Z_1 to initiate the conditioning and producing a sample X_2, Y_2, and Z_2. Continuing this process, it can be shown under modest conditions that X_k, Y_k, and Z_k jointly converge in distribution as $k \to \infty$ to the distribution associated with $p_{X,Y,Z}(x,y,z)$. Likewise, X_k, Y_k, and Z_k individually converge in distribution to the respective marginal distributions associated with $p_X(x)$, $p_Y(y)$, and $p_Z(z)$ (see Robert and Casella, 1999, Sect. 7.1.3).

The formal basis for convergence follows from the *fixed-point integral equation* that establishes a relationship between marginal and conditional distributions. For instance, if the marginal $p_X(x)$ in the three-variable problem above is of interest, then the Gibbs sampling routine provides a sample having limiting density $p_X(x)$, where the *function* $p_X(\cdot)$ is the unique solution (i.e., $\phi(\cdot) = p_X(\cdot)$) to the integral equation:

$$\begin{aligned}\phi(\cdot) &= \int\left[\int p_{X|Y,Z}(\cdot\,|\,y,z)p_{Y,Z|X}(y,z\,|\,\tau)\,dy\,dz\right]\phi(\tau)\,d\tau\\ &\equiv \int K(\cdot\,|\,\tau)\phi(\tau)\,d\tau,\end{aligned}$$

and where the integral inside the [ ] in the top expression, represented by $K(\,\cdot\,|\tau)$, is over the appropriate subspace of $\mathbb{R}^2$ (Tanner and Wong, 1987; Gelfand and Smith, 1990). The term $K(\cdot)$ is often called the *transition kernel*. From the expression above, it can be shown (Tanner and Wong, 1987; Gelfand and Smith, 1990) that the Gibbs recursion at a specific x can be written in Markov transition form as

$$\phi_{k+1}(x) = \int K(x \mid \tau)\phi_k(\tau)\,d\tau,$$

where $\phi_k(\cdot)$ denotes the true density for X_k. The fundamental result in Gibbs sampling is that $\phi_k(\cdot)$ converges to $p_X(\cdot)$ as $k \to \infty$. Similar ideas apply—with analogous transition kernels—for other target densities and dimensions $m \neq 3$.

Analogous situations apply when the random variables have a discrete distribution. Here the kernel-based form above is replaced with a Markov transition matrix (Appendix E). Consider, for example, the scalar element X in the trivariate illustration above. Let $\boldsymbol{p}_k$ represent the vector of probabilities associated with the possible outcomes for X_k (so, e.g., if X is a random variable having 10 possible outcomes, there are 10 nonnegative elements in $\boldsymbol{p}_k$ for all k, with the elements summing to 1). Let $\boldsymbol{P}$ represent the transition matrix governing the probability of going from X_k to X_{k+1} (so $\boldsymbol{P}$ has dimension $\dim(\boldsymbol{p}_k)\times\dim(\boldsymbol{p}_k)$). The elements of this transition matrix are directly available through the conditional probabilities of the variables in the problem. Thus, in the trivariate setting above, an individual element of $\boldsymbol{P}$ is available by applying the total probability theorem to first determine the probabilities of going from X_k to Y_{k+1}, then from Y_{k+1} to Z_{k+1}, and finally, from Z_{k+1} to X_{k+1}.

By standard Markov chain theory,

$$\boldsymbol{p}_{k+1}^T = \boldsymbol{p}_k^T \boldsymbol{P} = \boldsymbol{p}_0^T \boldsymbol{P}^{k+1}.$$

Then, if all elements of $\boldsymbol{P}$ are strictly positive, $\boldsymbol{p}_k$ converges to the limiting $\bar{\boldsymbol{p}}$ that is the solution to the balance equation

$$\bar{\boldsymbol{p}}^T = \bar{\boldsymbol{p}}^T \boldsymbol{P} \tag{16.8}$$

(Theorem E.1 in Appendix E). In particular, the $\bar{\boldsymbol{p}}$ that satisfies this stationarity condition must be the marginal distribution for X. Thus, the Gibbs sampler converges to the marginal distribution, as desired. Building on Example 16.2, let us now illustrate this balance equation.

Example 16.3—Balance equation for discrete bivariate problem. In the context of Example 16.2, suppose that the matrix of joint probabilities is

$$\begin{bmatrix} P(X=0,Y=0) & P(X=0,Y=1) \\ P(X=1,Y=0) & P(X=1,Y=1) \end{bmatrix} = \begin{bmatrix} 0.1 & 0.2 \\ 0.5 & 0.2 \end{bmatrix}.$$

The vector of marginal probabilities associated with X is

$$\begin{bmatrix} P(X=0) \\ P(X=1) \end{bmatrix} = \begin{bmatrix} 0.3 \\ 0.7 \end{bmatrix}.$$

We are interested in the 2×2 transition matrix $\boldsymbol{P}$ governing the probability of going from X_k to X_{k+1}. From basic laws of conditional probability, the $(1,1)$ component of $\boldsymbol{P}$ is given by

$$\begin{aligned} P(X_{k+1}=0 \mid X_k=0) &= P(X_{k+1}=0 \mid Y_{k+1}=0)P(Y_{k+1}=0 \mid X_k=0) \\ &\quad + P(X_{k+1}=0 \mid Y_{k+1}=1)P(Y_{k+1}=1 \mid X_k=0). \end{aligned}$$

The other three components of $\boldsymbol{P}$ are found in a like manner. The X to Y and Y to X transition matrices in Example 16.2 provide the required conditional probabilities (Exercise 16.5). It is found that

$$\boldsymbol{P} = \begin{bmatrix} 0.3889 & 0.6111 \\ 0.2619 & 0.7381 \end{bmatrix}.$$

One may verify that $\overline{\boldsymbol{p}}^T = \overline{\boldsymbol{p}}^T \boldsymbol{P}$, where $\overline{\boldsymbol{p}}^T = [P(X=0), P(X=1)] = [0.3, 0.7]$, indicating convergence of the Gibbs sampler. ❑

16.5 SOME EXAMPLES OF GIBBS SAMPLING

This section presents three examples of Gibbs sampling. The first is for a (multivariate) normal target distribution, which leads to conditional distributions that are also normal. The second is for a truncated exponential distribution. One point illustrated in the second example is that the target $p(\boldsymbol{x}) = p_{\boldsymbol{X}}(\boldsymbol{x})$ does not automatically exist even when the full conditionals do exist. The Gibbs sampler only produces meaningful results (of course!) if the target distribution exists. General regularity conditions for the existence of the target distribution (which is often the joint density for the elements in $\boldsymbol{X}$) are beyond the scope of the treatment here, but may, for example, be found in Robert and Casella (1999, Sect. 7.1.5). In the special case of two variables, Exercise 16.8 provides a key regularity condition. More generally, one should be aware that the relative ease of implementing the Gibbs sampler in some problems does not obviate the need for careful mathematical analysis of the problem structure and the results. The last of the three examples here illustrates the derivation of the full conditional for a particular trivariate model used in spatial modeling.

Example 16.4—Gibbs sampling for a normal distribution. Suppose that there is interest in generating samples $\boldsymbol{X} \sim N(\boldsymbol{\mu}, \boldsymbol{\Sigma})$ for some mean vector $\boldsymbol{\mu}$ and covariance matrix $\boldsymbol{\Sigma}$. As mentioned above for M-H, the Gibbs sampler may not be the most efficient method of generating samples from a multivariate normal distribution. Nevertheless, this example is useful as an illustration of the process of constructing the full conditionals.

A standard result from multivariate normality is that the distribution of any selection of components within $\boldsymbol{X}$ conditioned on the remaining components is also normal (e.g., Mardia et al., 1979, pp. 62–63). Specifically, the distribution of the ith component conditioned on the remaining components provides the sampling distribution:

$$\{X_{ki} | \boldsymbol{X}_{k\backslash i}\} \sim N\left(\mu_i + \boldsymbol{\Sigma}_{i,\backslash i}^T \boldsymbol{\Sigma}_{\backslash i,\backslash i}^{-1}(\boldsymbol{X}_{k\backslash i} - \boldsymbol{\mu}_{\backslash i}),\ \sigma_i^2 - \boldsymbol{\Sigma}_{i,\backslash i}^T \boldsymbol{\Sigma}_{\backslash i,\backslash i}^{-1} \boldsymbol{\Sigma}_{i,\backslash i}\right), \tag{16.9}$$

where μ_i and σ_i^2 are, respectively, the ith component of $\boldsymbol{\mu}$ and ith diagonal component of $\boldsymbol{\Sigma}$, $\boldsymbol{\mu}_{\backslash i}$ is the vector containing all components of $\boldsymbol{\mu}$ except μ_i, $\boldsymbol{\Sigma}_{i,\backslash i}$ is the column vector of the elements of $\boldsymbol{\Sigma}$ corresponding to the covariances between the ith component of $\boldsymbol{X}$ and all *other* components of $\boldsymbol{X}$, and $\boldsymbol{\Sigma}_{\backslash i,\backslash i}$ contains the elements of $\boldsymbol{\Sigma}$ with the row and column corresponding to the ith component of $\boldsymbol{X}$ removed. (Note that $\boldsymbol{\mu}_{\backslash i}$ and $\boldsymbol{\Sigma}_{i,\backslash i}$ are $(m-1)$-dimensional and $\boldsymbol{\Sigma}_{\backslash i,\backslash i}$ is $(m-1)\times(m-1)$-dimensional.) Hence, to generate sample values of $\boldsymbol{X}$ via the Gibbs sampler, the densities $p_i(x|\boldsymbol{X}_{k\backslash i})$, $i = 1, 2,\ldots, m$, in steps 1 and 3 of the procedure previously outlined are equal to the right-hand side of (16.9). ❑

Example 16.5—Gibbs sampling for truncated exponential distributions. Following Casella and George (1992), let X and Y have conditional exponential probability density functions on an interval $(0, B)$:

$$p_{X|Y}(x\,|\,y) = \frac{ye^{-yx}}{1-e^{-By}}, \quad 0 < x < B,$$

$$p_{Y|X}(y\,|\,x) = \frac{xe^{-xy}}{1-e^{-Bx}}, \quad 0 < y < B.$$

With the sampling densities $p_1(\cdot) = p_{X|Y}(\cdot)$ and $p_2(\cdot) = p_{Y|X}(\cdot)$, the Gibbs algorithm can be used to produce samples from the joint density $p_{X,Y}(x,y)$. It is easy to generate from these sampling densities using the inverse-transform method (Section D.2 of Appendix D). Suppose for the application here that the interest is in the marginal density $p_X(x)$ rather than the joint density.

As noted in Casella and George (1992), the density $p_X(x)$ does not exist for $B = \infty$ (see also Exercise 16.8). That is, at $B = \infty$, the marginal "density" that results is improper in the sense that $\int_0^\infty \left[\int_0^\infty p_{X,Y}(x,y)dy\right]dx = \infty$ even though both conditional densities above exist and are proper. From Robert and Casella (1999, Sect. 7.1.5), the existence of the joint density in this problem (from which the marginal for X can be determined) requires that $\int_0^B p_{Y|X}(y\,|\,x)/p_{X|Y}(x\,|\,y)dy < \infty$. As seen in Exercise 16.8, this condition is violated for $B = \infty$.

Nontrivial calculations show that for the truncated $B < \infty$ case, the marginal density exists and satisfies

$$p_X(x) = c\frac{1-e^{-Bx}}{x},$$

where c is the normalizing constant (Casella and George, 1992). Using the property $\int_0^B p_X(x)dx = 1$ and letting $B = 5$, it is found that $c \approx 0.2634$. This known marginal density can be used for comparisons with the output of the Gibbs sampler. (In practice, of course, the marginal or joint densities are usually unknown.)

Figure 16.2 shows a histogram of output for a Gibbs sampler based on $n = 40$. The histogram is constructed from the terminal output of the chain using 5000 independent replications, with each replication being initialized at $Y_0 = 2.5$ (from which X_1, Y_1, X_2, Y_2, etc. are generated). The histogram closely matches the marginal density, indicating that the chain output has a distribution close to the desired distribution.

We also tested the ability of the chain to provide an ergodic average close to the true value. Based on n above, let the burn-in period be $M = 10$. The chain is used to estimate the mean $E(X)$; that is, $f(X, Y) = X$ as in step 4 of the Gibbs sampling procedure above. For comparison purposes, the true mean is given by

$$E(X) = 0.2634\int_0^5 x\frac{1-e^{-5x}}{x}dx = 1.264.$$

Let us compare results based on the ergodic average without and with burn-in ($n = 40$, $M = 0$ and $n = 40$, $M = 10$, respectively). Based on 120 independent replications (using the M-file **Gibbs** available at the book's Web site), it is found that the mean of the 120 estimates for $E(X)$ is 1.296 without burn-in and 1.254 with burn-in. The corresponding sample standard deviations of the estimates are 0.259 and 0.294, respectively (so the sample standard deviations of the *means* of the 120 estimates are 0.0236 and 0.0268). In comparing these estimates relative to the true $E(X) = 1.264$, the resulting one-sample t-statistics are 1.356 for the no-burn-in implementation and 0.373 for the burn-in implementation. These values indicate that there is no evidence that either of the no-burn-in or burn-in implementations provides an estimate of $E(X)$ significantly different from the true value of 1.264 (i.e., the P-values are larger than 0.10). More detailed analysis would be required to determine if there is a significant difference between the no-burn-in and burn-in implementation. Note that the burn-in period (M) and overall run length (n) for this problem are shorter than needed in many other problems, where these numbers may easily run into the 100s or 1000s. ❑

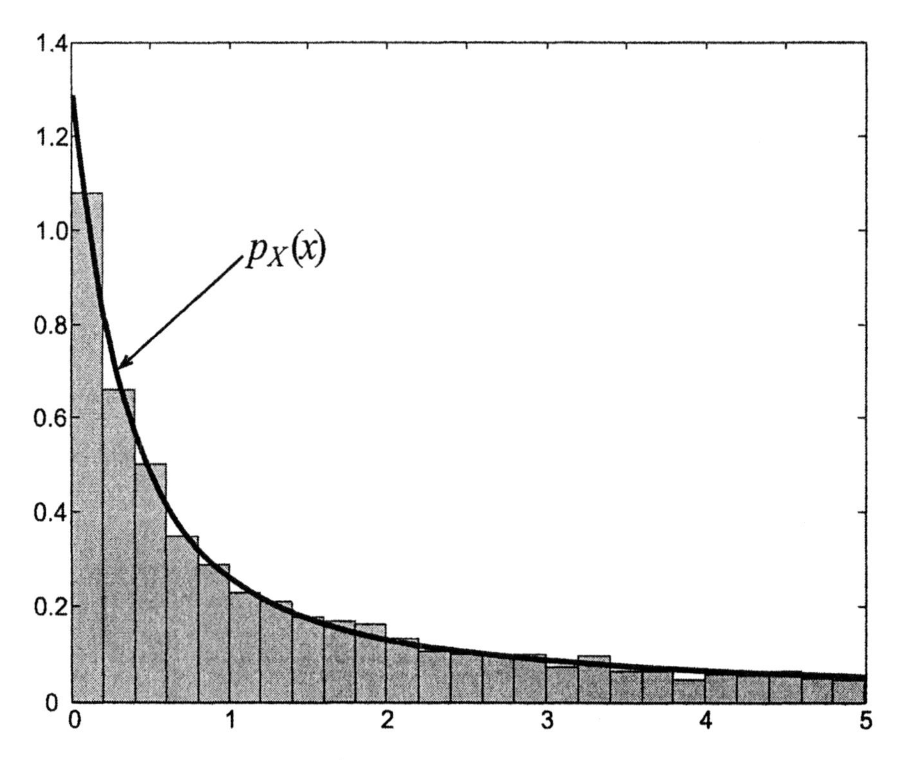

Figure 16.2. Histogram for terminal X from Gibbs sampling process with truncated exponential sampling. Histogram constructed from 5000 independent replications. The desired marginal density is shown by the solid line.

Example 16.6—Autoexponential model. The trivariate autoexponential model in Besag et al. (1995) has applications in some aspects of spatial modeling. The relevant random variables are represented by X, Y, Z, each defined on the interval $(0, \infty)$. The trivariate autoexponential density has the form

$$p_{X,Y,Z}(x, y, z) = c\exp[-(x + y + z + \theta_{xy}xy + \theta_{yz}yz + \theta_{xz}xz)],$$

where c is the normalizing constant and θ_{xy}, θ_{yz}, and θ_{xz} are known parameters. The density $p_{X,Y,Z}(x, y, z)$ represents the target density. There is no easy way to directly sample from $p_{X,Y,Z}(x, y, z)$. We are interested in constructing the full conditionals to be used in generating samples from this target density.

The Gibbs sampling procedure can be easily implemented since the three full conditional densities are simple scalar exponential densities (Appendix D shows how exponential random variables can be easily generated). For example, the full conditional for the random variable $\{Z|X=x, Y=y\}$ is an exponential density function with parameter $1+\theta_{yz}y+\theta_{xz}x$ (i.e., the mean of $\{Z|X=x, Y=y\}$ is $(1+\theta_{yz}y+\theta_{xz}x)^{-1}$; see Exercise 16.10). The two other full conditionals have analogous convenient exponential forms. Note that the derivation for these full conditionals does not depend on the normalizing constant c (due to cancellation via (16.7)). This is a critical advantage to sampling from the full conditionals. Note that *other* conditional and unconditional densities deriving from the autoexponential density above are not so convenient. For example, Robert and Casella (1999, p. 287) present the densities for $\{Y|X=x\}$ and X, both of which are nonexponential and cannot be used to easily generate samples. ❑

16.6 APPLICATIONS IN BAYESIAN ANALYSIS

The above discussion of MCMC has been for general problems where the desire is to obtain quantities related to a target distribution for a random vector $\boldsymbol{X}$. More specifically, MCMC—particularly Gibbs sampling—has had an especially profound impact on *Bayesian* methods of analysis. There seem to be at least two reasons for this: (i) The structure of Bayes' rule is well matched to the requirements of MCMC for drawing samples from appropriate conditional densities, and (ii) MCMC fills a long-standing need for a general-purpose method for constructing quantities related to posterior distributions that does not require cumbersome numerical integration. The seminal paper related to Bayesian applications of MCMC is Gelfand and Smith (1990).

Let us review the Bayesian framework. Suppose that $\boldsymbol{\Pi}$ represents a vector of terms important to the analysis of some system and that data $\boldsymbol{Z}$ can be collected on that system. For example, $\boldsymbol{\Pi}$ might be the parameters $\boldsymbol{\theta}$ of a model that are to be estimated; in an example below, $\boldsymbol{\Pi}$ represents both model parameters and the response to a varying input factor. In the Bayesian approach, $\boldsymbol{\Pi}$ is treated as a random vector instead of a fixed constant. Let us suppose that $\boldsymbol{\Pi}$ has a probability density, say $p_{\boldsymbol{\Pi}}(\boldsymbol{\pi})$. (The general Bayesian approach can work with discrete or hybrid $\boldsymbol{\Pi}$ as well.) The density $p_{\boldsymbol{\Pi}}(\boldsymbol{\pi})$ is referred to as the *prior density*, as it reflects a priori information available about $\boldsymbol{\Pi}$. There are many issues associated with the prior—philosophical and mathematical—but we will not delve into those here.

Bayes' rule takes the prior information on $\boldsymbol{\Pi}$, expressed via the prior density function, combines it with the conditional density function for the data $p_{\boldsymbol{Z}|\boldsymbol{\Pi}}(\boldsymbol{z}|\boldsymbol{\pi})$ (sometimes called the *likelihood function*), and forms the *posterior density function* $p_{\boldsymbol{\Pi}|\boldsymbol{Z}}(\boldsymbol{\pi}|\boldsymbol{z})$ according to

$$p_{\boldsymbol{\Pi}|\boldsymbol{Z}}(\boldsymbol{\pi}\,|\,\boldsymbol{z}) = \frac{p_{\boldsymbol{Z}|\boldsymbol{\Pi}}(\boldsymbol{z}\,|\,\boldsymbol{\pi})p_{\boldsymbol{\Pi}}(\boldsymbol{\pi})}{\int p_{\boldsymbol{Z}|\boldsymbol{\Pi}}(\boldsymbol{z}\,|\,\boldsymbol{\pi})p_{\boldsymbol{\Pi}}(\boldsymbol{\pi})d\boldsymbol{\pi}} = \frac{p_{\boldsymbol{\Pi},\boldsymbol{Z}}(\boldsymbol{\pi},\boldsymbol{z})}{p_{\boldsymbol{Z}}(\boldsymbol{z})}, \tag{16.10}$$

where the integral is over the domain for $\boldsymbol{\Pi}$. This simple-looking formula has profound implications and applications relative to modern learning and statistical analysis. The posterior density provides a fundamental means of characterizing the system parameters (or other quantities) $\boldsymbol{\Pi}$ by combining data ($\boldsymbol{Z}$) with prior information.

In almost all practical applications, numerical methods will be needed to calculate the integrals required for forming and using the posterior density. Rarely will the integration be feasible in closed form. One area requiring integration is the computation of the conditional expectation, $E[f(\boldsymbol{\Pi})|\boldsymbol{Z}] = \int f(\boldsymbol{\pi}) p_{\Pi|Z}(\boldsymbol{\pi}|\boldsymbol{Z})\, d\boldsymbol{\pi}$, particularly the special case of the conditional mean $E[\boldsymbol{\Pi}|\boldsymbol{Z}]$. Further, in using the posterior density function in (16.10) to construct posterior *probabilities* associated with $\boldsymbol{\Pi}$, multifold integration of the left side of (16.10) is required. In particular, there might be interest in probabilities of the form $P(\boldsymbol{\Pi} \in S|\boldsymbol{Z}) = \int_S p_{\Pi|Z}(\boldsymbol{\pi}|\boldsymbol{Z})\, d\boldsymbol{\pi}$, where S is some subset of the domain for $\boldsymbol{\Pi}$. (This type of integration arises, for instance, in computing credible regions for $\boldsymbol{\Pi}$, the Bayesian analogue of confidence regions.) A final aspect involving multivariate integration is the computation of the marginal density for $\boldsymbol{Z}$ (i.e., the denominator of (16.10)). This marginal is needed in some applications (e.g., Chib, 1995), but (happily!) not in the Gibbs sampling for producing samples from the posterior, as we discuss below.

As mentioned at the beginning of this section, MCMC—particularly Gibbs sampling—is especially well suited to Bayesian analysis. Recall that the Gibbs sampling procedure can be implemented if one can construct full conditional densities for each element of the vector of interest and samples can be generated from the full conditionals. Following the discussion of Section 16.3, $\{\Pi_{k+1,i}|\boldsymbol{\Pi}_{k\backslash i}, \boldsymbol{Z}\} \sim p_i(\pi|\boldsymbol{\Pi}_{k\backslash i}, \boldsymbol{Z})$, $i = 1, 2, \ldots, m$, where the notation is analogous to the generic notation of Section 16.3. In particular, $\boldsymbol{\Pi}_k \equiv [\Pi_{k1}, \Pi_{k2}, \ldots, \Pi_{km}]^T$ and the set of $m-1$ components without the ith component is $\boldsymbol{\Pi}_{k\backslash i} \equiv \{\Pi_{k+1,1}, \Pi_{k+1,2}, \ldots, \Pi_{k+1,i-1}, \Pi_{k,i+1}, \ldots, \Pi_{km}\}$. In the special case where $\boldsymbol{\Pi} = \boldsymbol{\theta}$, then $m = \dim(\boldsymbol{\theta})$ ($=p$, following previous notation) and the ith component of $\boldsymbol{\Pi}$ corresponds to the ith element of $\boldsymbol{\theta}$. As mentioned earlier, the individual components Π_{ki} may be scalar or multivariate. Note the extra conditioning ($\boldsymbol{Z}$) due to the data, reflecting the posterior aspect of the Bayesian formulation. This conditioning—although critical to the Bayesian analysis—may be treated as a constant relative to the Gibbs sampling process. We emphasize this treatment as a constant by writing $\boldsymbol{Z} = \boldsymbol{z}$ in the conditioning arguments below.

Expression (16.7) provides the fundamental form for the full conditionals. Substituting the right-hand side of (16.10) into (16.7) yields the full conditionals for the random variables,

$$p_i(\pi|\boldsymbol{\Pi}_{k\backslash i}, \boldsymbol{Z} = \boldsymbol{z}) = \frac{p_{\Pi|Z}(\pi, \boldsymbol{\Pi}_{k\backslash i}|\boldsymbol{z})}{\int p_{\Pi|Z}(\pi, \boldsymbol{\Pi}_{k\backslash i}|\boldsymbol{z})\, d\pi} = \frac{p_{\Pi,Z}(\pi, \boldsymbol{\Pi}_{k\backslash i}, \boldsymbol{z})}{\int p_{\Pi,Z}(\pi, \boldsymbol{\Pi}_{k\backslash i}, \boldsymbol{z})\, d\pi}, \tag{16.11}$$

where the second equality follows by the cancellation of the marginal density $p_Z(\boldsymbol{z})$ in the numerator and denominator. Hence, to create samples from the posterior density function via the Gibbs sampler it is *not* necessary to compute the denominator integral in Bayes' rule (16.10). In another context, we saw this desirable property in Example 16.6.

Example 16.7 sets up a Gibbs sampling implementation for the popular *variance-components model* (sometimes called the *random-effects model*). This example is considered in Gelfand and Smith (1990). The variance-components model is popular in engineering, clinical trials, and survey design as a means of studying systems where multiple input factors may have multiple levels.

Example 16.7—Variance-components model. Suppose that data arrive according to $Z_{ij} = \beta_i + \varepsilon_{ij}$, $i = 1, 2, \ldots, N_I$, $j = 1, 2, \ldots, N_J$, where $\{\beta_i \,|\, \mu, \sigma_\beta^2\} \sim N(\mu, \sigma_\beta^2)$ and $\{\varepsilon_{ij} \,|\, \sigma_\varepsilon^2\} \sim N(0, \sigma_\varepsilon^2)$. Here, Z_{ij} represents the measured response of the jth replication of the ith level of the factor. The term β_i represents the underlying random response to the ith level of the factor, and the noise ε_{ij} represents the measurement error for the jth replication of the ith level of the factor. Given that $\{\beta_i \,|\, \mu, \sigma_\beta^2\}$ and $\{\varepsilon_{ij} \,|\, \sigma_\varepsilon^2\}$ are independent, it follows that $\{Z_{ij} \,|\, \beta_i, \sigma_\varepsilon^2\} \sim N(\beta_i, \sigma_\varepsilon^2)$. Let $\boldsymbol{Z} = [Z_{11}, \ldots, Z_{1,N_J}; Z_{21}, \ldots; Z_{N_I,1}, \ldots, Z_{N_I N_J}]^T$ and $\boldsymbol{\beta} = [\beta_1, \beta_1, \ldots, \beta_{N_I}]^T$.

In the Bayesian context, we take the parameters μ, σ_β^2, and σ_ε^2 as random (and independent). The prior distributions for μ, σ_β^2, and σ_ε^2 are $N(\mu_0, \sigma_0^2)$, $\mathrm{IG}(a_\beta, b_\beta)$, and $\mathrm{IG}(a_\varepsilon, b_\varepsilon)$, respectively, where μ_0, σ_0^2, a_β, b_β, a_ε, and b_ε are known parameters of the prior densities and $\mathrm{IG}(\cdot)$ denotes an inverse gamma distribution. (For $\mathrm{IG}(a, b)$, the density with argument σ^2 is proportional to $(\sigma^2)^{-a-1}\exp(-b/\sigma^2)$; the constant of proportionality depends on a, b and the gamma function, as shown in Gelfand and Smith, 1990, and Chen et al., 2000, p. 244.)

Putting the above pieces together, the joint density for $[\boldsymbol{\Pi}; \boldsymbol{Z}] = [\boldsymbol{\beta}, \mu, \sigma_\beta^2, \sigma_\varepsilon^2; \boldsymbol{Z}]$ is

$$\{\boldsymbol{Z} \,|\, \boldsymbol{\beta}, \sigma_\varepsilon^2\} * \{\boldsymbol{\beta} \,|\, \mu, \sigma_\beta^2\} * \{\mu\} * \{\sigma_\beta^2\} * \{\sigma_\varepsilon^2\}$$

$$= N(\boldsymbol{\beta}, \sigma_\varepsilon^2 \boldsymbol{I}_{N_I N_J}) * N(\mu \mathbf{1}_{N_I}, \sigma_\beta^2 \boldsymbol{I}_{N_I}) * N(\mu_0, \sigma_0^2) * \mathrm{IG}(a_\beta, b_\beta) * \mathrm{IG}(a_\varepsilon, b_\varepsilon), \tag{16.12}$$

where the $*$ operator denotes the multiplication of the density functions associated with (as appropriate) the indicated random variables or the indicated distributions, and $\mathbf{1}_{N_I}$ denotes an N_I-dimensional vector of 1's. The density in (16.12) is the numerator in Bayes' rule (16.10). For our purposes, the four terms in $\boldsymbol{\Pi} = [\boldsymbol{\beta}, \mu, \sigma_\beta^2, \sigma_\varepsilon^2]$ are of interest. Note that the first component of $\boldsymbol{\Pi}$ (i.e., $\boldsymbol{\beta}$) is multivariate, illustrating the point above that for some applications it is beneficial to have multivariate components in forming full conditionals.

We can use the following four full conditionals to construct samples from the posterior $p_{\Pi|Z}(\pi|z)$. As given in Gelfand and Smith (1990), the full conditionals are built from the joint density in (16.12) according to (16.11):

$$\{\boldsymbol{\beta} \mid \mu, \sigma_\beta^2, \sigma_\varepsilon^2, \boldsymbol{Z}\}$$
$$\sim N\left(\frac{N_J\sigma_\beta^2}{N_J\sigma_\beta^2+\sigma_\varepsilon^2}\bar{\boldsymbol{Z}}+\frac{\mu\sigma_\varepsilon^2}{N_J\sigma_\beta^2+\sigma_\varepsilon^2}\mathbf{1}_{N_I},\ \frac{\sigma_\beta^2\sigma_\varepsilon^2}{N_J\sigma_\beta^2+\sigma_\varepsilon^2}\boldsymbol{I}_{N_I}\right),$$

$$\{\mu \mid \boldsymbol{\beta}, \sigma_\beta^2, \sigma_\varepsilon^2, \boldsymbol{Z}\} = \{\mu \mid \boldsymbol{\beta}, \sigma_\beta^2\}$$
$$\sim N\left(\frac{\sigma_\beta^2\mu_0+\sigma_\beta^2\sum_i\beta_i}{N_I\sigma_\beta^2+\sigma_\beta^2},\ \frac{\sigma_\beta^2\sigma_\varepsilon^2}{N_I\sigma_\beta^2+\sigma_\beta^2}\right),$$

$$\{\sigma_\beta^2 \mid \boldsymbol{\beta}, \mu, \sigma_\varepsilon^2, \boldsymbol{Z}\} = \{\sigma_\beta^2 \mid \boldsymbol{\beta}, \mu\}$$
$$\sim \mathrm{IG}\left(a_\beta+\tfrac{1}{2}N_I,\ b_\beta+\tfrac{1}{2}\sum_i(\beta_i-\mu)^2\right),$$

$$\{\sigma_\varepsilon^2 \mid \boldsymbol{\beta}, \mu, \sigma_\beta^2, \boldsymbol{Z}\} = \{\sigma_\varepsilon^2 \mid \boldsymbol{\beta}, \boldsymbol{Z}\}$$
$$\sim \mathrm{IG}\left(a_\varepsilon+\tfrac{1}{2}N_IN_J,\ b_\varepsilon+\tfrac{1}{2}\sum_i\sum_j(Z_{ij}-\mu)^2\right),$$

where $\bar{\boldsymbol{Z}} = N_J^{-1}[\sum_j Z_{1j},\ldots,\sum_j Z_{N_Ij}]^T$. It is relatively easy to sample from the above four distributions in the cyclic manner of the Gibbs algorithm. In particular, the sampling densities, $p_i(\pi \mid \boldsymbol{\Pi}_{k\backslash i}, \boldsymbol{Z})$, $i = 1, 2, 3, 4$, correspond to the four full conditionals above. Hence, provided that the sampler has run long enough and/or that burn-in effects are removed, the easy sampling of the full conditionals provides a means for estimating quantities from the otherwise formidable $p_{\Pi|Z}(\boldsymbol{\pi}|\boldsymbol{z})$. ❑

As we have seen with so many other aspects of stochastic search, there are connections between root-finding stochastic approximation (SA) (Chapter 4) and MCMC as well. Gu and Kong (1998) and Gu and Zhu (2001) show how maximum likelihood estimates can be computed via SA, where MCMC methods are used to approximate the gradient and Hessian matrix of the log-likelihood function. Hence, the MCMC-based calculations provide the noisy input that fits into the classical SA framework (analogous to $\boldsymbol{Y}(\boldsymbol{\theta})$ in Chapter 4 and elsewhere). Chen et al. (2000, pp. 323–325) summarizes an application of SA in finding the mode of unimodal posterior distributions for a subset of parameters of interest. In this application, one is interested in solving the root-finding problem:

$$\frac{\partial \log p_{\Pi|Z}}{\partial \boldsymbol{\theta}_{(1)}} = \mathbf{0},$$

where $\boldsymbol{\Pi} = [\boldsymbol{\theta}_{(1)}, \boldsymbol{\theta}_{(2)}]$ represents a collection of model parameters with $\boldsymbol{\theta}_{(1)}$ representing the parameters of interest and $\boldsymbol{\theta}_{(2)}$ representing other parameters. A standard root-finding SA recursion may be applied to estimate $\boldsymbol{\theta}_{(1)}$; the values for $\boldsymbol{\theta}_{(2)}$ that are required to get the noisy measurement of the gradient above (for use in the SA algorithm) are generated via MCMC. This is a convenient way of getting the required noisy gradient estimate $\boldsymbol{Y}(\boldsymbol{\theta})$.

One of the areas of application for Bayesian-based Gibbs sampling is state and parameter estimation in dynamic (state-space) models (see Subsection 3.3.4 for a summary of such models). The Gibbs sampler allows for the treatment of nonstandard conditions, such as nonlinearity for the system dynamics and/or nonnormality for the noise terms. In such conditions, the Kalman filter may yield poor results or may be inapplicable. A summary of the overall approach and some pointers to the broader literature is given in Spall (2003).

16.7 CONCLUDING REMARKS

The discussion above summarizes important aspects of Markov chain Monte Carlo, including the motivation, theory, implementation, and connection to Bayesian analysis. The focus is on the Metropolis–Hastings and Gibbs sampling versions of MCMC. Many large-scale practical implementations of MCMC borrow aspects from both M-H and Gibbs sampling.

Although Gibbs may be considered a component-wise M-H, the techniques have developed largely independent of each other. Recognizing this, we discuss M-H and Gibbs as separate approaches. As with other stochastic search methods, no one approach is to be universally preferred. One strong aspect of both M-H and Gibbs is the theory supporting the methods and guaranteeing convergence under modest conditions.

The M-H method is more general than Gibbs sampling. The Gibbs sampler has the relatively strong requirement that the full conditional distributions be available. M-H has no such requirement and in fact provides almost complete flexibility in the choice of the distribution from which to simulate. Assuming that the full conditional distributions are available for simulation, the Gibbs sampler has more intuitive appeal than the M-H algorithm. Building up the joint distribution from the set of full conditionals does not seem to require the "leap of logic" that is needed with the M-H algorithm and its arbitrary choice of sampling (proposal) distribution (although M-H *is* on a fully sound theoretical footing). Bayesian problems provide a framework especially well suited to Gibbs sampling as a result of the frequent availability of the full conditionals.

Because the structure of Gibbs sampling is more restrictive than M-H, Gibbs has the advantage of not needing the tuning that is required in M-H. A serious application of M-H will usually require some experimentation with the proposal distribution, not only in the specific parameters of the distribution, but perhaps in the general *form* of the distribution (e.g., gamma or uniform?). The restrictions in Gibbs sampling are analogous to certain search methods (e.g., Newton–Raphson in Section 1.4 or random search algorithm A in Section 2.2), where the tuning is eliminated because the structure of the algorithm is sufficiently proscribed.

It is not possible to say which of M-H or Gibbs is computationally more efficient in general. Much depends on the specifics of the implementation. Examples are available in the literature illustrating both efficient and inefficient results for either or both approaches.

We saw, for instance, a highly efficient Gibbs implementation in the truncated exponential problem of Example 16.5 (i.e., very few iterations required to obtain samples having the required distribution). This is consistent with the general principle that Gibbs—like other methods in other settings—may benefit by taking into account the structure of the problem. Gibbs also has the advantage of low-dimensional—usually *scalar*—random number generation for arbitrarily high-dimensional problems. On the other hand, the component-wise generation in Gibbs sampling may produce a phenomenon analogous to multivariate optimization with only one component at a time. In optimization, this may lead to convergence to a saddlepoint or local optimum; in Gibbs sampling, this may lead to slow exploration of the space of possible outcomes for $f(X)$, leading to a poor estimate for $E[f(X)]$. Robert and Casella (1999, pp. 318–319) illustrate slow convergence via this phenomenon in a problem involving a mixture distribution (a mixture of normals). In M-H, slow convergence is generally associated with a poor choice of the proposal distribution relative to the target distribution. This may cause M-H to miss some of the finer structure of the target distribution. There are also highly efficient M-H implementations, especially when the algorithm benefits by some tuning (e.g., Chib and Greenberg, 1995; Robert and Casella, 1999, Chap. 6).

The ultimate value of M-H or Gibbs sampling, of course, is their usefulness in solving practical problems. Both approaches have proved to be important tools in the modern analyst's toolbox.

EXERCISES

16.1 Discuss why the ergodic average of a given number of samples in typical applications of the M-H (and other) algorithms will have greater variability than a corresponding average of the same number of independent samples of $f(X)$. In the demonstration of this point in Example 16.1, verify analytically that the standard deviation for the *independent samples* case is 0.0159.

16.2 **(a)** Based on 50 independent replications of the M-H algorithm in Example 16.1 with the uniform proposal distribution used in Figure 16.1, test whether the mean of the terminal estimate is statistically indistinguishable from the true value of zero.

(b) Repeat the test with a normal proposal distribution but other aspects of Example 16.1 unchanged. In particular, assume that $\{W|X=x\} \sim N(x, I_2/12)$ (note that this proposal distribution has the same mean and variance as the original uniform distribution).

(c) Finally, do the same test above, but with a $U_2(\boldsymbol{x} - 2\mathbf{1}_2, \boldsymbol{x} + 2\mathbf{1}_2)$ proposal distribution (see Table 16.1). Comment on the observed differences in the performance for the three proposal distributions.

16.3 Consider the setting of Example 16.1 except that the off-diagonal (covariance) term in the covariance matrix for $\boldsymbol{X}$ changes from 0.9 to 0.5. Based on a $U_2(\boldsymbol{x} - 3\mathbf{1}_2, \boldsymbol{x} + 3\mathbf{1}_2)$ proposal distribution, produce a plot of four independent replications of estimates for $E[f(\boldsymbol{X})]$ over the range of 0 to 10,000 post-burn-in iterations (analogous to Figure 16.1) and determine the approximate acceptance rate for the candidate points $\boldsymbol{W}$. Use the burn-in period and $\boldsymbol{X}_0$ of Example 16.1.

16.4 Consider a special case of M-H where $q(\boldsymbol{w}|\boldsymbol{x}) = q(\boldsymbol{w})$. Suppose that there exists a $C \geq 1$ such that $p(\boldsymbol{x}) \leq Cq(\boldsymbol{x})$ for all $\boldsymbol{x}$ in the support of $p(\boldsymbol{x})$. Let $p(\boldsymbol{x})$ and $q(\boldsymbol{x})$ be continuous functions. Prove that the expected acceptance probability at any iteration (i.e., $E[\rho(\boldsymbol{X}, \boldsymbol{W})]$) is at least $1/C$. (Hint: The "Answers to Selected Exercises" section near the end of the book provides an outline of the proof.) How does this bound relate to the acceptance probability for the accept–reject method (Appendix D)?

16.5 In Example 16.3, derive the values for the transition matrix $\boldsymbol{P}$ and verify that the balance equation $\overline{\boldsymbol{p}}^T = \overline{\boldsymbol{p}}^T \boldsymbol{P}$ holds.

16.6 Suppose that $\boldsymbol{X}$ is bivariate normally distributed where the marginal distribution for the two components is $N(0, 1)$ and the correlation is ρ. Present the two sampling distributions, $p_1(x|X_{k2})$ and $p_2(x|X_{k1})$, for use in step 1 (and 3) of the Gibbs sampling algorithm.

16.7 Consider the setting of Example 16.5 with the exception that the Gibbs sampler is now used to estimate the *second moment* of X (relative to $p_X(x)$) rather than the mean (first moment) of X. Assume the same burn-in period, number of replications, and initial condition for Y. Compute the true second moment and test for a bias in the estimate of the second moment when using an ergodic average with no burn-in ($M = 0$) and when using the specified burn-in ($M = 10$).

16.8 As discussed in Example 16.5, the existence of the full conditional densities does not guarantee the existence of the joint density $p_{X,Y}(x, y)$. In this two-variable problem, the existence of the target density requires that

$$\int \frac{p_{Y|X}(y\,|\,x)}{p_{X|Y}(x\,|\,y)}\,dy < \infty,$$

where the integral is over the domain for Y. Show that this condition is violated for Example 16.5 when $B = \infty$.

16.9 Consider the setting of Example 16.5 except that $B = 4$. Determine the normalizing constant c and produce a plot of the true density for X and corresponding histogram (analogous to Figure 16.2) based on 2000 independent replications and the terminal output at $n = 500$.

16.10 For the setting of Example 16.6, establish that the full conditional density function for $\{Z\,|\,X = x, Y = y\}$ has the exponential form given in the example.

CHAPTER 17

OPTIMAL DESIGN FOR EXPERIMENTAL INPUTS

This chapter wades into the vast subject of experimental design for establishing user-determined input values. We are concerned here with using experimental resources in some intelligent way to obtain the most information possible about a process. While the general subject of experimental design is large, encompassing both qualitative and quantitative aspects, our focus is much narrower (although still quite broad). We discuss *optimal design*, an area within experimental design dealing with the formulation and solution to the problem of picking input values according to a formal criterion.

Section 17.1 provides some general background about optimal experimental design, including a description of the main (D-optimal) criterion of interest. Section 17.2 considers optimal design in the context of linear models, while Section 17.3 discusses some specialized methods for response surface methods. Section 17.4 considers the more difficult nonlinear problem. After some concluding remarks in Section 17.5, an appendix (Section 17.6) summarizes some results for optimal inputs in dynamic systems. Stochastic methods of search and optimization play a role in recursively determining the optimal design.

17.1 INTRODUCTION

17.1.1 Motivation

Experimental design is an important subject in areas such as statistics, engineering design and quality control, social science, agriculture, process control, and medical clinical trials and pharmaceuticals. The subject involves both qualitative and quantitative aspects. In particular, experimental design deals with issues such as:

- **Comparison of treatments.** A key aspect of experimental design is to determine the best of several candidate inputs. For example, in the operations of a bank, one might be interested in comparing the profit for various combinations of administrative structure, number of tellers, interest rates paid on bank deposits, and so on. The testing might involve computer simulation and/or tests in the actual bank.

- **Variable screening.** When there are a large number of possible variables, screening experiments can be used for a sensitivity analysis that helps identify the input and output variables most relevant to the application.
- **Response surface exploration.** After the key inputs have been identified, it is often of interest to get detailed insight into the effect of these inputs on the output. This is usually carried out via a low-order (first- or second-order) polynomial in a local region of the input space.
- **Enhanced estimation.** Building mathematical models usually requires the collection of data from the real system. Among other applications, these data are used to estimate unknown parameters for the model. It is desirable to pick the experimental inputs so that the resulting output data can be used to estimate model parameters as accurately as possible.
- **Model validation.** Suppose that an analyst wishes to evaluate whether an existing model (simulation or otherwise) is an accurate representation of the process under study. If the response surface analysis above (or the closely related metamodel discussed later in this section) indicates that one or more factors have an effect very different from that based on predictions from an existing model, the model may be flawed.[1]
- **Choice of inputs in dynamic systems.** An important issue in building control systems and related models of dynamic systems is the choice of the inputs while the dynamic system is operating. The interest here is to excite the physical system in a way that is useful for generating information that will subsequently be used in building a model and controlling the system (for linear time series models, the term *persistency of excitation* is used to refer to a choice of inputs guaranteeing that all model parameters can be estimated).

In the confines of one chapter, we are not able to treat all of the above subjects in any detail. Textbooks such as Box and Draper (1987), Khuri and Cornell (1987), Atkinson and Donev (1992), Wu and Hamada (2000), and Montgomery (2001), and review papers such as Nair (1992), Rosenberger (1996), Miller and Wu (1996), and Cox and Reid (2000), provide a much fuller discussion of some or all of the aspects above. Reviews from the control systems perspective are given in Mehra (1974), Goodwin and Payne (1977), Walter and Pronzato (1990), and Ljung (1999, Chap. 13).

Consider the familiar input–output process relating a scalar output z to an input vector $\boldsymbol{x}$. Suppose, as in previous chapters, that we model this input–output process with a regression function $h(\boldsymbol{\theta}, \boldsymbol{x})$, where $\boldsymbol{\theta}$ is the usual p-dimensional parameter vector that needs to be determined. In particular, the model for the output z is

---

[1]Kleijnen (1998, p. 210) mentions a validation study where the design of experiments together with a regression metamodel exposed a serious flaw in a simulation model of the environment related to the greenhouse gas problem.

$$z = h(\boldsymbol{\theta}, \boldsymbol{x}) + v, \tag{17.1}$$

where v is a noise term possibly dependent on $\boldsymbol{x}$ or $\boldsymbol{\theta}$. Suppose that the dimension of $\boldsymbol{x}$ is r.

In the language of experimental design, $\boldsymbol{x}$ is composed of factors, each factor representing one of the quantities that is going to be studied in the experiment. Each component of $\boldsymbol{x}$ represents one factor. The range of possible physical meanings for $\boldsymbol{x}$ is endless, but might include terms related to factory or machine settings, drug dosages, the amount of studying needed for an exam, and so on. It is assumed that the analyst has full control over the levels of the factors in producing a set of data. The fundamental problem here is to choose n (say) input vectors, $\boldsymbol{x}_1, \boldsymbol{x}_2, \ldots, \boldsymbol{x}_n$, such that when experiments are run at these n values, the corresponding output values, $z_1, z_2, \ldots, z_n$, are as informative as possible with respect to building the model in (17.1). Let

$$\boldsymbol{X}_n \equiv [\boldsymbol{x}_1, \boldsymbol{x}_2, \ldots, \boldsymbol{x}_n]^T$$

be the $n \times r$ matrix of n input vectors that will be used to estimate $\boldsymbol{\theta}$. Of course, finding inputs $\boldsymbol{X}_n$ that provide data that are "informative as possible" is a nebulous and perhaps hopeless goal. The optimal experimental design framework provides enough structure to make this goal attainable in a restricted sense.

The focus here will be on choosing the input levels in some optimal way to enhance the process of estimating $\boldsymbol{\theta}$. This represents only one aspect of the more general goals of experimental design that were listed above. Nevertheless, it is an important aspect. It is assumed that the input and output variables have been defined and that there exists a known structure for the mathematical (regression) model $h(\cdot)$ describing the process. The interest is in determining the best values of the various factors in the vector $\boldsymbol{x}$ to use when running experiments and collecting data. In most input–output processes, the choice of input levels will have a significant effect on the observed outputs.

Optimal design is the rubric used for the aspect of experimental design dealing with optimization of the n design vectors contained in $\boldsymbol{X}_n$ based on a formal optimization criterion. Unlike the prior discussion in this book, the search and optimization focus here is not on $\boldsymbol{\theta}$. Rather, $\boldsymbol{\theta}$ will *ultimately* be estimated based on some optimization criterion after the design is implemented or will sequentially be estimated while the design is being established. Because of the inherent uncertainty in the information collected via experimentation and the possibly difficult form for the optimization criterion for $\boldsymbol{X}_n$, both of the defining characteristics of stochastic search—noisy loss measurements and injected algorithmic randomness, as discussed in Subsection 1.1.3—may be relevant in finding the optimal design factors.

The majority of traditional experimental design methods—including most of the methods treated in the textbooks and review articles mentioned above—

are motivated in a relatively informal way and not derived as formal optimal designs. These include popular approaches such as factorial and fractional factorial designs, composite designs, block designs (e.g., Latin squares or randomized blocks), and so on. Such "classical" methods have a long history of successful application in problems related to estimating $\boldsymbol{\theta}$ in linear models (linear in $\boldsymbol{\theta}$) as well as to other problems such as those listed at the beginning of this section.

In light of the long history of success in classical experimental design, one might wonder about the need for optimal design. Two important contributions of optimal design stand out. One is to add rigor and clarify the properties for some of the classical designs. As we will see, some classical designs are also optimal designs under certain conditions. The second important contribution is to provide an approach for constructing designs in nonstandard situations, including the very important cases where the model is nonlinear in $\boldsymbol{\theta}$ and/or the domain for the input factors is not a "nice" region such as a hypercube or hypersphere (i.e., r-dimensional cube or sphere). Atkinson (1996) discusses some other applications for optimal design. The following summary captures some of the benefits of optimal design.

Example 17.1—Contrast of classical and optimal designs. As an illustration of the benefits of optimal design, Montgomery (2001, p. 469) discusses a problem where an investigator is studying the properties of an adhesive material based on two input factors in $\boldsymbol{x}$ (i.e., $r = 2$). Because of constraints in the experiment, this problem yields a domain for $\boldsymbol{x}$ that is not of the square or circular form typically assumed in classical design when $r = 2$ (although the underlying regression model is linear). The domain is square with two opposite corners diagonally sheared off. On the other hand, the optimal design *is* readily able to cope with the irregular domain. Montgomery compares the accuracy of the estimate for $\boldsymbol{\theta}$ resulting from a classical design (the central composite design, which is a combination of the cube and star designs discussed in Subsection 17.2.3) and from an optimal design (based on D-optimality, as introduced in Subsection 17.1.3). Based on a sum of the variances of the estimates for the elements of $\boldsymbol{\theta}$ as the measure of accuracy, it is found that approximately twice as many experiments are needed with the classical design as with the optimal design to achieve the same accuracy in estimating $\boldsymbol{\theta}$. ❑

While the benefits from optimal design may be significant, one limitation of the approach is its close connection to assumptions in the model.[2] If the underlying model structure is imperfectly known, then the resulting "optimal"

[2]Note that the strong connection to the model here differs from some of the other approaches that we have seen in this text in other (nonexperimental design) contexts. In particular, methods such as random search, FDSA, SPSA, simulated annealing, and evolutionary computation in Chapters 2 and 6–10 require minimal information about underlying models.

design may, in fact, be suboptimal. For example, if the assumed model is linear in $\boldsymbol{x}$ and the true system is nonlinear in $\boldsymbol{x}$, the choice of inputs under an optimal design may be suboptimal. Further, optimal designs are inherently tied to the choice of an optimality criterion. In practice, the choice of criterion may not be obvious, although an equivalence theorem introduced in Section 17.2 provides a powerful rationale for one popular criterion (the *D*-optimal criterion). The use of optimal design is sometimes advocated for follow-up studies. Here, information from initial standard designs can be used to produce preliminary information about the model, which then forms the basis for the subsequent optimal design. Among the texts and monographs that emphasize the optimal design aspect of experimental design are Fedorov (1972), Silvey (1980), Atkinson and Donev (1992), and Pukelsheim (1993).

To close this subsection, let us illustrate the different characteristics of optimal design for linear or nonlinear models in the context of Monte Carlo simulation. Recall from Chapters 14 and 15 that the aim is to use a simulation to optimize some aspect of a real system. It is assumed that the simulation itself is fully trained as an accurate representation of the actual system within the domain of interest. Now suppose that we step back and concern ourselves with building either an efficient approximation to an existing simulation or with building the original simulation itself.

When building an approximation of a large-scale simulation, one is concerned with building a *metamodel* of the simulation. Metamodels are typically curve fits of the actual simulation that can be used in cases where the simulation would otherwise have to be run a prohibitive number of times. Simulation runs provide the data for building the metamodel. The advantage of the metamodel is that, once constructed, it can be executed in a small fraction of the time required for the full simulation. Among other uses, metamodels can be applied for sensitivity analysis and model validation. Typical metamodels are low-order polynomials of the input variables $\boldsymbol{x}$ weighted by elements of a parameter vector $\boldsymbol{\theta}$. This corresponds to a curvilinear model (as discussed in Section 3.1) because the model is linear in $\boldsymbol{\theta}$. Hence, design methods for linear models can be used to determine the inputs for the simulation runs required to build the metamodel. Optimal design provides the means for squeezing the most information out of a set of n simulation runs. Response surface methods (Section 17.3) are often used to build metamodels.

The second, more challenging, case for optimal design pertains to the building of the simulation itself. Here, there are parameters $\boldsymbol{\theta}$ inside the simulation that need to be estimated to produce reliable simulation output. This is quite different from the metamodel approximation for an *existing* simulation. For example, in a simulation of traffic flow in a network, $\boldsymbol{\theta}$ might include terms related to the mean arrival rates into the network, the capacity of various streets inside the network, the mean times for daily traffic surges into various parts of the network, and so on. Hence, the design involves choosing aspects of the real traffic system, corresponding to the inputs $\boldsymbol{x}$, that enhance the ability to estimate $\boldsymbol{\theta}$. This might include the choice of the timing strategies for the traffic signals in the network, the placement of variable message signs notifying drivers of congestion, and so on. For the very reason that a simulation is being used to represent the system (versus a simple analytical model), it is expected that $\boldsymbol{\theta}$ enters in a complex nonlinear way. Hence, nonlinear design methods (Section 17.4) are appropriate. In contrast to the metamodel problem above, optimal design provides guidance on how best to conduct the n experiments on the *physical system*, as needed to build the simulation.

In summary, for the simulation context, *linear* design is appropriate in the building of simulation metamodel if the metamodel is a polynomial form in the inputs (a nonlinear metamodel such as a neural network requires nonlinear design). On the other hand, *nonlinear* design is usually required for parameter estimation inside the simulation.

17.1.2 Finite-Sample and Asymptotic (Continuous) Designs

The above discussion motivates some of the interest in experimental design and the need to pick input values intelligently. Let us now introduce more formally the concept of an optimal design for finding the "best" n input vectors $\boldsymbol{X}_n$. In particular, as a step toward developing practical methods for determining an optimal design, this subsection discusses the distinction between finite-sample and limiting (asymptotic) designs.

In finite-sample designs, the n inputs are allocated to achieve an optimal value of some criterion, such as the D-optimal criterion discussed in the next subsection. In an asymptotic (continuous) design, inputs are allocated according to proportions that would apply *as if* there were an infinite amount of data available. The proportions in an optimal asymptotic continuous design may be any real number in the interval [0, 1]. Hence, in finite-sample practice, it may not be possible to find an exact design that corresponds to the optimal continuous design. If n is sufficiently large, however, one can find an approximation to the optimum that is likely to be satisfactory. The example below illustrates the contrast between finite-sample and asymptotic designs.

Example 17.2—Finite-sample and asymptotic designs. Suppose that $n = 10$, and the input x is a scalar such that $0 \le x \le 100$. Based on optimizing an appropriate criterion for the given sample size n, it might be found that four measurements should be taken at $x = 0$, two at $x = 50$, and four at $x = 100$ (i.e., the optimal $\boldsymbol{X}_n = [0, 0, 0, 0, 50, 50, 100, 100, 100, 100]^T$). Suppose, however, that the best value of the criterion is possible if we place 42 percent of the measurements at the endpoints ($x = 0$ or 100) and the remaining 16 percent at the center point ($x = 50$). This exact proportion is only possible if we have n as some multiple of 50 (e.g., at $n = 50$, 21 measurements are placed at the endpoints and 8 measurements are placed at the center point). The allocation chosen above for $n = 10$ is only an approximation of the asymptotic—or continuous—design. This example is one where it is possible to achieve an exact asymptotic design for at least some values of n (i.e., multiples of 50). In other problems, at least some allocations may be according to irrational numbers, in which case no finite-sample design can ever achieve an exact asymptotic design. ❑

Mathematical tractability provides the prime rationale for seeking optimal asymptotic designs. In a manner analogous to the use of asymptotic theory in many other aspects of stochastic algorithms, the mathematical problem of finding an optimal design is often simplified by considering asymptotic designs. Hence, in finding an asymptotic design, we ignore the constraint that the number of measurements at any design value ($\boldsymbol{x}$) must be an integer. In *implementing* the asymptotic solution, we approximate the design by allocating the measurements in proportions as close as possible to the asymptotic design.

The experimental designs of interest here are represented by a set of distinct input support points ($\boldsymbol{x} = \boldsymbol{\chi}_i$) and a corresponding set of weights (w_i) representing the allocations to the support points:

$$\xi \equiv \begin{Bmatrix} \boldsymbol{\chi}_1 & \boldsymbol{\chi}_2 & \cdots & \boldsymbol{\chi}_N \\ w_1 & w_2 & \cdots & w_N \end{Bmatrix}. \tag{17.2}$$

The first row gives the values of the input factors and the second row gives the associated weights. Note that $N \leq n$ and $\sum_{i=1}^{N} w_i = 1$ with $0 \leq w_i \leq 1$. Because the total number of inputs is n, the number of inputs taking value $\boldsymbol{\chi}_i$ is exactly or approximately nw_i (*exactly* if the weights are associated with a finite-sample design; *approximately* if the weights are for an asymptotic design that does not allow a strictly proportional allocation over the n measurements).

So, determining a design involves choosing the number of support points N and associated values $\boldsymbol{\chi}_i$, together with the proportions of measurements assigned to those support points (the w_i). Let

$$\begin{aligned} \xi_n^* &\equiv \text{optimal finite-sample } (n) \text{ design}, \\ \xi_\infty^* &\equiv \text{optimal asymptotic (continuous) design.} \end{aligned} \tag{17.3}$$

An optimal design ξ^*, where ξ^* represents either ξ_n^* or ξ_∞^* as appropriate, is the combination of inputs and weights that optimizes the chosen criterion. Hence, for Example 17.2 discussed above, (17.3) implies that the optimal finite-sample ($n = 10$) and asymptotic designs would be written as

$$\xi_{10}^* = \begin{Bmatrix} 0 & 50 & 100 \\ 0.40 & 0.20 & 0.40 \end{Bmatrix} \text{ and } \xi_\infty^* = \begin{Bmatrix} 0 & 50 & 100 \\ 0.42 & 0.16 & 0.42 \end{Bmatrix}.$$

Often, for simple models with $\boldsymbol{\theta}$, as usual, having p elements, the optimal asymptotic design assigns a weight $w_i = 1/p$ to p different input points. So, if $n = p$ (or n is some integer multiple of p), it is possible to implement the optimal asymptotic design exactly. In more complex models, the allocations are usually less tidy, including cases with weights that are not rational numbers (e.g., $w_i = 1/\sqrt{5}$ for some i). As n gets larger, the allocation of measurements can be made to approach any optimal asymptotic design arbitrarily closely.

The values for $\boldsymbol{x}_k$ are drawn from the $N \le n$ inputs $\boldsymbol{\chi}_i$ in the design ξ from (17.2). The allocation of the N support points to the n inputs applied in estimating $\boldsymbol{\theta}$ is done according to the proportions reflected in the weights w_i, as illustrated in the example below.

Example 17.3—Mapping from design inputs (support points) to applied inputs. Suppose that $n = 20$ measurements will be used to estimate $\boldsymbol{\theta}$. Based on an analysis for optimal designs, suppose that the optimal finite-sample design is

$$\xi_{20}^{*} = \begin{Bmatrix} \boldsymbol{\chi}_1 & \boldsymbol{\chi}_2 & \boldsymbol{\chi}_3 & \boldsymbol{\chi}_4 & \boldsymbol{\chi}_5 & \boldsymbol{\chi}_6 \\ 0.05 & 0.15 & 0.05 & 0.30 & 0.25 & 0.20 \end{Bmatrix},$$

where the $\boldsymbol{\chi}_i$ are some specified inputs. Based on allocating the $\boldsymbol{x}_k$ sequentially according to the order of the $\boldsymbol{\chi}_i$, we have the mapping of $\boldsymbol{\chi}_i$ to $\boldsymbol{x}_k$ shown in Table 17.1. For example, from the table, $\boldsymbol{\chi}_2 = \boldsymbol{x}_2 = \boldsymbol{x}_3 = \boldsymbol{x}_4$. ❑

17.1.3 Precision Matrix and *D*-Optimality

Recall that the fundamental goal is to pick a design ξ such that $\boldsymbol{\theta}$ is estimated with as much precision as possible. An important quantity in optimal design is the $p \times p$ precision matrix, say $\boldsymbol{M}(\boldsymbol{\theta}, \xi)$, which is an expression of the accuracy of

Table 17.1. Design inputs $\boldsymbol{\chi}_i$ allocated to the input vectors $\boldsymbol{x}_k$ applied in estimating $\boldsymbol{\theta}$ in Example 17.3. For each row, the value in the left column is assigned to the inputs in the right column.

Inputs from ξ_{20}^{*}	Input vector applied in model (17.1)
$\boldsymbol{\chi}_1$	$\boldsymbol{x}_1$
$\boldsymbol{\chi}_2$	$\boldsymbol{x}_2, \boldsymbol{x}_3, \boldsymbol{x}_4$
$\boldsymbol{\chi}_3$	$\boldsymbol{x}_5$
$\boldsymbol{\chi}_4$	$\boldsymbol{x}_6, \boldsymbol{x}_7, \boldsymbol{x}_8, \boldsymbol{x}_9, \boldsymbol{x}_{10}, \boldsymbol{x}_{11}$
$\boldsymbol{\chi}_5$	$\boldsymbol{x}_{12}, \boldsymbol{x}_{13}, \boldsymbol{x}_{14}, \boldsymbol{x}_{15}, \boldsymbol{x}_{16}$
$\boldsymbol{\chi}_6$	$\boldsymbol{x}_{17}, \boldsymbol{x}_{18}, \boldsymbol{x}_{19}, \boldsymbol{x}_{20}$

the $\boldsymbol{\theta}$ estimate based on the n inputs X_n. In general, this matrix depends on $\boldsymbol{\theta}$, as indicated. This matrix reflects the variability in the estimate that is induced from the stochastic variability of the outputs $z_1, z_2, \ldots, z_n$. A "larger" value of $\boldsymbol{M}$ reflects more precision—lower variability—in the estimate. We are most interested in two specific representations of $\boldsymbol{M}$: (i) the case where $\boldsymbol{M}$ is proportional to the inverse of the covariance matrix of the $\boldsymbol{\theta}$ estimate, and (ii) the case where $\boldsymbol{M}$ is equal to the (Fisher) information matrix. The quantity $\boldsymbol{M}$ is frequently called the information matrix in the optimal design literature, but we avoid that terminology because in case (i) above, $\boldsymbol{M}$ does not strictly equal the information matrix (see the definition in Section 13.3). In fact, the covariance matrix used in case (i) can be defined without the distributional assumptions needed to define the information matrix. Case (i) is appropriate for linear models while case (ii) applies to more general nonlinear models. Case (ii) relies on the close connection between the inverse information matrix and the covariance matrix of the $\boldsymbol{\theta}$ estimate (discussed in detail in Section 13.3).

Given the close connection of the inverse of the covariance matrix to the precision matrix $\boldsymbol{M}$, a natural goal in picking the design ξ is to find the design that "maximizes" the matrix $\boldsymbol{M}$. We might first ask if it is possible, in general, to find such a design. That is, can we find a design ξ^* such that $\boldsymbol{M}(\boldsymbol{\theta}, \xi^*) > \boldsymbol{M}(\boldsymbol{\theta}, \xi)$ in the matrix sense[3] for all $\xi \neq \xi^*$? The answer is no.

Because no design exists with a uniformly larger precision matrix, we can work with a weaker metric that is similar in spirit, the *D-optimal criterion*. The D-optimal criterion is the determinant of the precision matrix (D for determinant). In particular, the optimal design is found according to

$$\xi^* = \arg\max_{\xi} \{\det[\boldsymbol{M}(\boldsymbol{\theta}, \xi)]\}. \tag{17.4}$$

The D-optimal criterion $\det[\boldsymbol{M}(\boldsymbol{\theta}, \xi)]$ is the most popular formal means of determining an optimal design ξ^*. This criterion expresses a goal *similar* to the goal of a larger precision matrix given the interpretation of the determinant as some measure of the size of the matrix. This criterion has several desirable properties, as we discuss below. It is, however, easy to illustrate that a precision matrix based on ξ^* as defined in (17.4) is not uniformly larger than all other precision matrices, as we now show.

Example 17.4—Cautionary illustration related to the *D*-optimal criterion. Let us demonstrate that while the D-optimal measure provides a useful basis for comparing the size of matrices, caution is necessary regarding the strength of the conclusion. Suppose that $\boldsymbol{M}^*$ represents the optimal precision matrix based on

---

[3]The inequality here means that $\boldsymbol{M}(\boldsymbol{\theta}, \xi^*) - \boldsymbol{M}(\boldsymbol{\theta}, \xi)$ is a positive definite matrix—see Appendix A (Section A.2).

$\boldsymbol{\xi}^*$ and that $\boldsymbol{M}$ represents the precision matrix based on some $\boldsymbol{\xi} \neq \boldsymbol{\xi}^*$. Suppose that

$$\boldsymbol{M}^* = \begin{bmatrix} 3 & 2 \\ 2 & 2 \end{bmatrix} \text{ and } \boldsymbol{M} = \begin{bmatrix} 2 & 0.5 \\ 0.5 & 1 \end{bmatrix}.$$

Both of $\boldsymbol{M}^*$ and $\boldsymbol{M}$ are positive definite and $\det(\boldsymbol{M}^*) = 2 \geq \det(\boldsymbol{M}) = 1.75$. However, $\boldsymbol{M}^* - \boldsymbol{M} = \begin{bmatrix} 1 & 1.5 \\ 1.5 & 1 \end{bmatrix}$ is an indefinite (nonpositive definite) matrix, having one positive and one negative eigenvalue. So, $\boldsymbol{M}^* \not\geq \boldsymbol{M}$. (See Exercise 17.1.) ❑

There has been much further discussion of the interpretation of the *D*-optimal criterion in the experimental design literature together with discussion of alternative criteria (one of which turns out to be equivalent to (17.4) under certain circumstances, as shown in Theorem 17.1). We will not reiterate the debates and discussion here (see, e.g., Silvey, 1980, Chaps. 1 and 2; Atkinson and Donev, 1992, Chap. 10; and Montgomery, 2001, pp. 466–472, for statistical perspectives on these issues; Ljung, 1999, Chap. 13, discusses some criteria popular in control systems engineering, both in open- and closed-loop problems). Rather, we accept this widely used criterion and proceed to make connections to the fundamental themes of this text.

One desirable property of *D*-optimality is *transform invariance*, such as what we saw with the Newton–Raphson and second-order stochastic approximation algorithms (Sections 1.4, 4.5, and 7.8). It is obviously desirable that an optimal design be independent of whether (say) we choose to measure elements of $\boldsymbol{\theta}$ in meters or centimeters. In particular, invertible linear transformations of $\boldsymbol{\theta}$ do not affect $\boldsymbol{\xi}^*$ (Exercise 17.2). Linear transformations *do* affect the solution in nondeterminant-based criteria (Pukelsheim, 1993, pp. 137, 344–345).

Directly solving for $\boldsymbol{\xi}^*$ in most practical problems is difficult. Rather, for linear models, one often "guesses" at a design based on intuitive considerations and then verifies it by an important theorem on equivalence of certain solutions. This theorem is presented in Section 17.2. In nonlinear models, alternative methods are needed.

17.2 LINEAR MODELS

17.2.1 Background and Connections to *D*-Optimality

The subject of experimental design for linear models is a large subject unto itself. For example, almost all of the well-known text by Montgomery (2001–5th ed.) is devoted to classical experimental design for linear models. Nevertheless, some of the issues for linear models are generic and apply in nonlinear problems as well. Recall that a model is linear if it is linear in $\boldsymbol{\theta}$; nonlinear functions of $\boldsymbol{x} = \boldsymbol{x}_k$ are allowed within the linear framework.

Suppose that the classical linear regression model (eqn. (3.1)) is used to describe the process,

$$z_k = \boldsymbol{h}_k^T\boldsymbol{\theta} + v_k, \quad k = 1, 2, \ldots, n, \tag{17.5}$$

where $\boldsymbol{h}_k = \boldsymbol{h}(\boldsymbol{x}_k)$ is the design vector of dimension p dependent on input $\boldsymbol{x} = \boldsymbol{x}_k$, $\boldsymbol{h}(\boldsymbol{x})$ is a function mapping the inputs to the design vector (as in Chapter 13), and v_k is a mean-zero noise term uncorrelated with v_j for all $j \neq k$ and having common variance σ^2 across k. The values for $\boldsymbol{x}_k$ are drawn from the $N \leq n$ inputs $\boldsymbol{\chi}_i$ in the design ξ from (17.2). The allocation of the n inputs is done according to the proportions reflected in the weights w_i, as illustrated in Example 17.3.

As in Subsections 3.1.2 and 13.1.3, let $\boldsymbol{Z}_n = [z_1, z_2, \ldots, z_n]^T$ and $\boldsymbol{H}_n$ be the $n \times p$ stacked matrix of $\boldsymbol{h}_k^T$ row vectors. The leads to the (batch) ordinary least-squares estimate of $\boldsymbol{\theta}$ based on the n input–output pairs

$$\hat{\boldsymbol{\theta}}^{(n)} = (\boldsymbol{H}_n^T\boldsymbol{H}_n)^{-1}\boldsymbol{H}_n^T\boldsymbol{Z}_n \tag{17.6}$$

(assuming that the indicated inverse exists, of course). In the linear formulation of (17.6), $\boldsymbol{H}_n$ is a general (possibly nonlinear) function of the collection of inputs $\boldsymbol{X}_n$. In the common case where the regression model includes an additive constant and where the elements of $\boldsymbol{h}_k$ correspond directly to the elements of $\boldsymbol{x}_k$, then

$$\begin{aligned} z_k &= \boldsymbol{h}_k^T\boldsymbol{\theta} + v_k \\ &= [1, x_{k1}, x_{k2}, \ldots, x_{k,p-1}]\boldsymbol{\theta} + v_k \\ &= [1, \boldsymbol{x}_k^T]\boldsymbol{\theta} + v_k, \end{aligned} \tag{17.7}$$

where x_{ki} is the ith element of $\boldsymbol{x}_k$. Here, the first element of $\boldsymbol{\theta}$ is the intercept term (the above-mentioned additive constant), and the remaining $p - 1$ elements are the slope terms. The p elements of $\boldsymbol{\theta}$ jointly represent the linear regression parameters to be estimated and $\boldsymbol{x}_k = [x_{k1}, x_{k2}, \ldots, x_{k,p-1}]^T$ represents the $p - 1$ input factors that are to be determined in the experimental design at each k. (In the nonlinear problems of Section 17.4, there is generally no such "easy" relationship between the dimension p and the number of terms to be determined in the design.) Note that $n \geq p$ to ensure that the p elements of $\boldsymbol{\theta}$ can be uniquely estimated. That is, we need at least as many data points as there are parameters to be estimated.

In the general linear case of (17.5) and (17.6), there is a ready expression for the dispersion of $\hat{\boldsymbol{\theta}}^{(n)}$ via the covariance matrix for the estimate. This provides an expression directly usable in the *D*-optimal criterion. Specifically, the covariance matrix for the standard least-squares regression estimate $\hat{\boldsymbol{\theta}}^{(n)}$ appearing in (17.6) has a particularly simple form:

$$\mathrm{cov}\left(\hat{\boldsymbol{\theta}}^{(n)}\right) = (\boldsymbol{H}_n^T \boldsymbol{H}_n)^{-1}\sigma^2 \tag{17.8}$$

(Subsection 3.1.2; Box and Draper, 1987, p. 74). Note that this covariance matrix is not dependent on $\boldsymbol{\theta}$ or on the probability distribution for the noise (beyond knowledge of the variance σ^2), a property rarely true of the covariance matrix for the estimate of $\boldsymbol{\theta}$ in nonlinear models.

Because the inverse of the covariance matrix is a representation of the precision of the estimate, (17.8) provides the basis for the precision matrix used in the *D*-optimal criterion. In particular, the precision matrix is proportional to the inverse of the covariance matrix. So, the *D*-optimal solution of maximizing the determinant of the precision matrix is equivalent to minimizing the determinant of the covariance matrix. Because σ^2 does not affect the minimization of $\det[\mathrm{cov}(\hat{\boldsymbol{\theta}}_n)]$, the *D*-optimal solution is found according to

$$\begin{aligned}\boldsymbol{\xi}^* &= \arg\min_{\boldsymbol{\xi}}\left\{\det\left[(\boldsymbol{H}_n^T \boldsymbol{H}_n)^{-1}\right]\right\} = \arg\min_{\boldsymbol{\xi}}\left\{\left[\det(\boldsymbol{H}_n^T \boldsymbol{H}_n)\right]^{-1}\right\} \\ &= \arg\max_{\boldsymbol{\xi}}\left\{\det(\boldsymbol{H}_n^T \boldsymbol{H}_n)\right\},\end{aligned} \tag{17.9}$$

where $\boldsymbol{\xi}^* = \boldsymbol{\xi}_\infty^*$ if one is determining the optimal asymptotic design and $\boldsymbol{\xi}^* = \boldsymbol{\xi}_n^*$ if the optimization is based directly on only n measurements being available (so $w_i n$ is an integer for all i). Note that for any selected $\boldsymbol{\xi}$,

$$\boldsymbol{H}_n^T \boldsymbol{H}_n = \sum_{k=1}^{n} \boldsymbol{h}_k \boldsymbol{h}_k^T = n\sum_{i=1}^{N} w_i \boldsymbol{h}(\boldsymbol{\chi}_i)\boldsymbol{h}(\boldsymbol{\chi}_i)^T \tag{17.10}$$

(Exercise 17.3).

For convenience and consistency with most literature in optimal design, the precision matrix for use in calculating a *D*-optimal solution is usually scaled so that it does not grow with n. Then, the maximization of the criterion may be interpreted as maximizing the information per sample. Hence, in the notation of the *D*-optimal criterion (17.4), let the precision matrix $\boldsymbol{M}(\boldsymbol{\xi}) = \boldsymbol{M}(\boldsymbol{\theta}, \boldsymbol{\xi})$ be

$$\boldsymbol{M}(\boldsymbol{\xi}) = \sum_{i=1}^{N} w_i \boldsymbol{h}(\boldsymbol{\chi}_i)\boldsymbol{h}(\boldsymbol{\chi}_i)^T, \tag{17.11}$$

where $M(\xi)$ is written for $M(\theta, \xi)$ because the right-hand side of (17.11) does not depend on θ (this contrasts with nonlinear models in Section 17.4, where the precision matrix *does* depend on θ). Note that $M(\xi)$ differs from $H_n^T H_n$ by only the multiple n in (17.10) (i.e., $nM(\xi) = H_n^T H_n$). From (17.8)–(17.10), maximizing the determinant of $M(\xi)$ in (17.11) is equivalent to minimizing $\det[\text{cov}(\hat{\theta}^{(n)})]$. Hence, we let the expression in (17.11) be the precision matrix for use with D-optimality in (17.4) for the linear model. Although not shown in (17.11), another form of weighting (to complement the w_i) is to weight the summands by an "efficiency function" (e.g., Fedorov, 1972, p. 88). This weighting can be used, say, to compensate for greater variance in the outputs at particular inputs. We do not pursue that extension here.

There is a useful geometric interpretation of the problem in (17.9) in terms of confidence regions for θ. Suppose that the errors v_k are normally distributed. Then, the set

$$\left\{\theta : \left(\hat{\theta}^{(n)} - \theta\right)^T H_n^T H_n \left(\hat{\theta}^{(n)} - \theta\right) \leq \text{constant}\right\} \tag{17.12}$$

is the collection of θ values that would be in a confidence ellipsoid about $\hat{\theta}^{(n)}$ at a specified probability level. That is, the null hypothesis that θ is the true parameter vector would not be rejected for any θ in the above set. The indicated constant depends on σ^2 and the specified probability level (e.g., 0.95). Regardless of the specific value of the constant, the volume of the set above is proportional to $[\det(H_n^T H_n)]^{-1/2}$ (Silvey, 1980, p. 10; Box and Draper, 1987, pp. 490–491). It is natural to want to make this confidence region as "tight" (small) as possible. Because maximizing $\det[M(\xi)]$ according to D-optimality is equivalent to minimizing $[\det(H_n^T H_n)]^{-1}$, the D-optimal solution minimizes the size of the confidence region.

Let us now illustrate D-optimality based on a simple example. This example illustrates the common phenomenon of multiple (equivalent) D-optimal solutions. This is also one of the relatively rare problems that can be solved by simple analytical means. In most practical cases, a numerical search procedure is required to find a solution.

Example 17.5—*D*-optimality for simple regression model. Consider the linear model

$$z = [1, x]\theta + v,$$

where $-1 \leq x \leq 1$ is a scalar input. Hence, θ is composed of two terms, an intercept and a slope. Suppose that $n = 3$ measurements will be used to estimate θ. The aim is to determine the three input values x_1, x_2, and x_3. The measurement and precision matrices are

$$\boldsymbol{H}_3 = \begin{bmatrix} 1 & x_1 \\ 1 & x_2 \\ 1 & x_3 \end{bmatrix} \text{ and } \boldsymbol{M}(\xi) = \tfrac{1}{3}\boldsymbol{H}_3^T\boldsymbol{H}_3 = \tfrac{1}{3}\begin{bmatrix} 3 & x_1+x_2+x_3 \\ x_1+x_2+x_3 & x_1^2+x_2^2+x_3^2 \end{bmatrix}.$$

Hence,

$$\det[3\boldsymbol{M}(\xi)] = \det(\boldsymbol{H}_3^T\boldsymbol{H}_3) = 3(x_1^2+x_2^2+x_3^2) - (x_1+x_2+x_3)^2.$$

By inspection, there are six combinations of input values that yield the maximum value $\det(\boldsymbol{H}_3^T\boldsymbol{H}_3) = 8$. Two of the six solutions are

$$x_1 = x_2 = 1, x_3 = -1 \text{ and } x_1 = x_2 = -1, x_3 = 1. \quad (17.13)$$

More generally, all six solutions fit one of the two following general patterns for the optimal finite-sample design ξ_3^*:

$$\xi_3^* = \begin{Bmatrix} \chi_1 = 1 & \chi_2 = -1 \\ \tfrac{2}{3} & \tfrac{1}{3} \end{Bmatrix} \text{ or } \xi_3^* = \begin{Bmatrix} \chi_1 = -1 & \chi_2 = 1 \\ \tfrac{2}{3} & \tfrac{1}{3} \end{Bmatrix}.$$

So, for the two sample solutions in (17.13) above, $\chi_1 = x_1 = x_2$ and $\chi_2 = x_3$. ❑

17.2.2 Some Properties of Asymptotic Designs

Asymptotic designs hold a special place in the theory and practice of optimal designs. They provide a useful approximation to the often-intractable optimal finite-sample design. Listed below are a few key properties of *D*-optimal asymptotic designs as applied to linear models (e.g., Atkinson and Donev, 1992, pp. 116–117; Pukelsheim, 1993, Chap. 8).

(i) **Nonuniqueness of design.** For a given problem, a *D*-optimal design ξ_∞^* may not be unique. If $\xi_{\infty;1}^*$ and $\xi_{\infty;2}^*$ are two asymptotically optimal designs, $\lambda\xi_{\infty;1}^*$ + $(1-\lambda)\xi_{\infty;2}^*$ is also an asymptotically optimal design for any $0 \le \lambda \le 1$.

(ii) **Bounds on number of support points.** There exists a *D*-optimal design ξ_∞^* where the number of support points *N* satisfies $p \le N \le p(p+1)/2$. (Property (i) above also indicates that there may be optimal designs with $N > p(p+1)/2$. The number of points *N* cannot be less than *p* because that would cause $\boldsymbol{H}_n^T\boldsymbol{H}_n$, which is inverted to form the least-squares estimate in (17.6), to be singular.)

(iii) **Bounds on optimality criteria.** The *D*-optimal criteria for the asymptotic and finite-sample designs are related by

$$1 \le \frac{\det \boldsymbol{M}(\xi_\infty^*)}{\det \boldsymbol{M}(\xi_n^*)} \le \frac{n^p}{n(n-1)\cdots(n+1-p)}.$$

A famous result in experimental design, the equivalence theorem of Kiefer and Wolfowitz (1960), provides the foundation for several methods of solving for *D*-optimal designs.[4] This result shows that the asymptotic *D*-optimal design ξ_∞^* is the same as a design from a minimax strategy producing the minimum variance for model predictions. Aside from this useful alternative interpretation of the *D*-optimal solution, the necessary and sufficient conditions of this theorem provide a practical structure for developing numerical algorithms to obtain solutions in nontrivial problems.

Theorem 17.1 (Kiefer–Wolfowitz equivalence theorem). Consider the linear model in (17.5) and let $\mathcal{X}$ be a compact (closed and bounded) set representing the domain of allowable values for inputs $\boldsymbol{x}$. The following statements are equivalent:

(a) The asymptotic design ξ_∞^* is *D*-optimal.

(b) $\boldsymbol{h}(\boldsymbol{x})^T \boldsymbol{M}(\xi_\infty^*)^{-1} \boldsymbol{h}(\boldsymbol{x}) \leq p$ for all $\boldsymbol{x} \in \mathcal{X}$.

(c) ξ_∞^* minimizes $\max_{\boldsymbol{x}\in\mathcal{X}} \boldsymbol{h}(\boldsymbol{x})^T \boldsymbol{M}(\xi)^{-1} \boldsymbol{h}(\boldsymbol{x})$ with this minimax solution implying that $\boldsymbol{h}_i^T \boldsymbol{M}(\xi_\infty^*)^{-1} \boldsymbol{h}_i = p$ at all points $\chi_i \in \mathcal{X}$ of the design ξ_∞^*, where $\boldsymbol{h}_i = \boldsymbol{h}(\chi_i)$.

Aside from Kiefer and Wolfowitz (1960), Theorem 17.1 is proved in Silvey (1980, pp. 19–23) and Pukelsheim (1993, p. 212). Because $\hat{z}(\boldsymbol{x}) = \boldsymbol{h}(\boldsymbol{x})^T \hat{\boldsymbol{\theta}}^{(n)}$ and $\text{var}[\hat{z}(\boldsymbol{x})]$ is proportional to $\boldsymbol{h}(\boldsymbol{x})^T \boldsymbol{M}(\xi_\infty^*)^{-1} \boldsymbol{h}(\boldsymbol{x})$ (from (17.8), (17.10) and (17.11)), (b) and (c) are statements regarding the variance of the predictions $\hat{z}(\boldsymbol{x})$. The maximum variance (over $\mathcal{X}$) of the predictions $\hat{z}(\boldsymbol{x})$ is minimized when an optimal design ξ_∞^* is used to form $\hat{\boldsymbol{\theta}}^{(n)}$. Further, this maximum variance is realized at the design points $\boldsymbol{h}_i = \boldsymbol{h}(\chi_i)$. The example below shows how the equivalence theorem can be used to obtain a practical solution to a design problem. In the example, the theorem is used to confirm a hunch regarding a likely optimal design.

Example 17.6—Application of equivalence theorem to simple curvilinear model. Consider a $p = 3$ model involving a second-order polynomial in a scalar x: $z = [1, x, x^2]\boldsymbol{\theta} + v$, where it is known that $-1 \leq x \leq 1$. Based on intuition, it seems that an $N = 3$ design is a candidate for the optimal, where we take $n/N = n/3$ observations at each of the design points $\chi_1 = -1$, $\chi_2 = 0$, and $\chi_3 = 1$. So, the same three input values are repeated until n measurements are reached. For small n, this design may not be achievable (since n may not be divisible by 3), but for

[4] This is the Kiefer–Wolfowitz of finite-difference stochastic approximation (FDSA) fame (Chapter 6).

large n, this design is achievable to within a negligible error even if n is not divisible by 3. By (17.2), this candidate design may be written as

$$\xi = \begin{Bmatrix} \chi_1 & \chi_2 & \chi_3 \\ w_1 & w_2 & w_3 \end{Bmatrix} = \begin{Bmatrix} -1 & 0 & 1 \\ 1/3 & 1/3 & 1/3 \end{Bmatrix}. \tag{17.14}$$

Because $\boldsymbol{h}(x) = [1, x, x^2]^T$, this design leads to the three input vectors:

$$\boldsymbol{h}(\chi_1) = [1 \quad -1 \quad 1]^T,$$

$$\boldsymbol{h}(\chi_2) = [1 \quad 0 \quad 0]^T,$$

$$\boldsymbol{h}(\chi_3) = [1 \quad 1 \quad 1]^T.$$

From (17.11), a simple calculation shows that

$$\boldsymbol{M}(\xi) = \sum_{i=1}^{3} w_i \boldsymbol{h}(\chi_i)\boldsymbol{h}(\chi_i)^T = \frac{1}{3}\begin{bmatrix} 3 & 0 & 2 \\ 0 & 2 & 0 \\ 2 & 0 & 2 \end{bmatrix}.$$

With the design in (17.14), simple calculations show that

$$\boldsymbol{h}(x)^T \boldsymbol{M}(\xi)^{-1}\boldsymbol{h}(x) = \frac{3}{4}\left[4 - 6x^2(1-x^2)\right]$$
$$\leq 3 \quad \text{for all } x \in \mathcal{X} = [-1, 1].$$

Because the upper bound above is the same as p, it is known from (b) of the equivalence theorem that the solution in (17.14) is, in fact, the asymptotically optimal solution ξ_∞^* (see also Exercise 17.5). Figure 17.1 depicts the variance function $\boldsymbol{h}(x)^T \boldsymbol{M}(\xi_\infty^*)^{-1}\boldsymbol{h}(x)$, showing that it is maximized at the three design points, as indicated in part (c) of Theorem 17.1. ❑

Let us emphasize that the results of Theorem 17.1 pertain to the optimal *asymptotic* design, not to the optimal finite-sample design. This can be illustrated using the problem of Example 17.6. The optimal asymptotic design requires that n be divisible by 3. If, say, $n = 4$, numerical analysis may be used to show that the optimal finite-sample design places the "extra" measurement at any one of the three design points, −1, 0, or 1, above (Atkinson and Donev, 1992, p. 99). The D-optimal criterion is identical for any such allocation of the extra measurement. So, if the center point is the one that is replicated, the optimal finite-sample design is

$$\xi_4^* = \begin{Bmatrix} -1 & 0 & 1 \\ 1/4 & 1/2 & 1/4 \end{Bmatrix}. \tag{17.15}$$

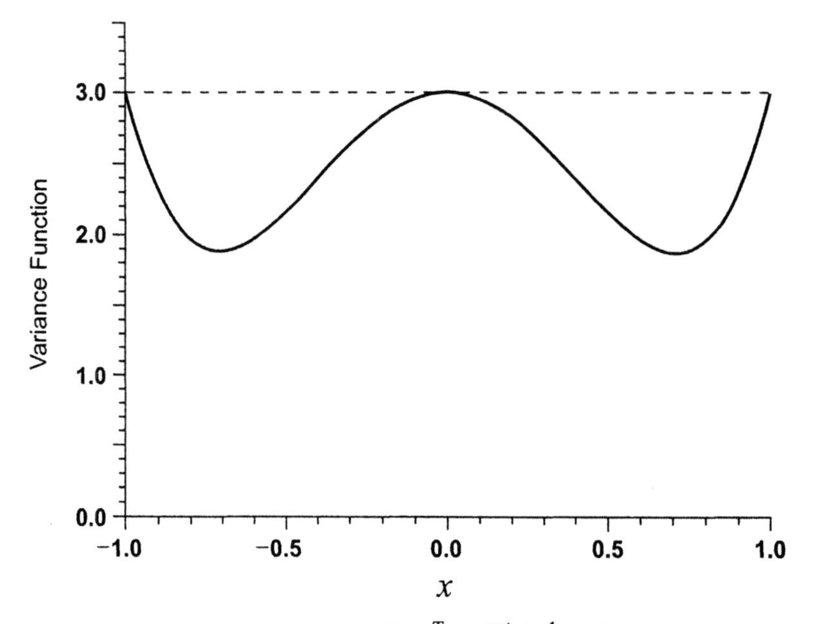

Figure 17.1. Plot of variance function $\boldsymbol{h}(x)^T \boldsymbol{M}(\xi_\infty^*)^{-1}\boldsymbol{h}(x)$ for varying input x in Example 17.6. This plot illustrates a consequence of Theorem 17.1: the variance function is bounded above by p and is maximized at the design points $x = \chi_1, \chi_2$, and χ_3.

Exercise 17.8 asks the reader to produce a plot of the variance function $\boldsymbol{h}(x)^T \boldsymbol{M}(\xi_4^*)^{-1}\boldsymbol{h}(x)$. In contrast to Figure 17.1, this plot illustrates that for this finite-sample design the variance function is *not* maximized at the design points.

The example below illustrates one of the standard "tricks" in developing an optimal design. In particular, it is shown that a change of variables may sometimes transform a messy or intractable problem to one that is quite straightforward. Furthermore, this problem illustrates a case where the emphasis is on maximizing the accuracy of only a subset of the parameters being estimated.

Example 17.7—Change of variables. It is convenient in many problems to redefine the variables. Suppose that $z = [e^{-\lambda x/2}, e^{-\lambda x}]\boldsymbol{\theta} + v$, where $\lambda > 0$ is some known parameter, $0 \leq x < \infty$, and v has mean zero. Directly solving for the optimal design in such a model is cumbersome. Consider the change of variable, $u = e^{-\lambda x/2}$. Then, the regression model has the relatively simple curvilinear form $z = [u, u^2]\boldsymbol{\theta} + v$, $0 < u \leq 1$.

Let $\boldsymbol{\theta} = [\beta_1, \beta_2]^T$ and $\hat{\boldsymbol{\theta}}^{(n)} = [\hat{\beta}_1, \hat{\beta}_2]^T$. Suppose that there is some question about whether the $e^{-\lambda x}$ contribution (corresponding to u^2) should be included in the model. So, the analyst intends to run a hypothesis test on whether

$\beta_2 = 0$. To strengthen the conclusions of the test, the aim is to choose the inputs (x) to minimize the variance of $\hat{\beta}_2$.

A candidate design for the variable u is

$$\xi_u \equiv \begin{Bmatrix} \mu & 1 \\ w & 1-w \end{Bmatrix}, \ 0 \le w \le 1,$$

where $0 < \mu \le 1$ is a value to be determined and the other value of the design measure is known to be $u = 1$ because any other design can be scaled up, with a resulting increase in all elements of $\boldsymbol{M}(\xi_u)$ (Atkinson and Donev, 1992, p. 102). Using (17.11),

$$\boldsymbol{M}(\xi_u) = w\begin{bmatrix} \mu \\ \mu^2 \end{bmatrix}\begin{bmatrix} \mu & \mu^2 \end{bmatrix} + (1-w)\begin{bmatrix} 1 \\ 1 \end{bmatrix}[1 \ \ 1] = \begin{bmatrix} 1-w+w\mu^2 & 1-w+w\mu^3 \\ 1-w+w\mu^3 & 1-w+w\mu^4 \end{bmatrix}.$$

Hence,

$$\begin{aligned} \det[\boldsymbol{M}(\xi_u)] &= (1-w+w\mu^2)(1-w+w\mu^4) - (1-w+w\mu^3)^2 \\ &= w(1-w)\mu^2(1-\mu)^2. \end{aligned}$$

Using the standard formula for the inverse of a 2×2 matrix, the (2, 2) element of $\boldsymbol{M}(\xi_u)^{-1}$ is

$$\frac{1-w+w\mu^2}{w(1-w)\mu^2(1-\mu)^2}.$$

Minimizing the above expression yields $\mu = \sqrt{2}-1$ and $w = 1/\sqrt{2}$. Converting back to the original variable (x) gives the following asymptotically optimal design for minimizing $\text{var}(\hat{\beta}_2)$:

$$\begin{Bmatrix} \chi_1 & \chi_2 \\ w_1 & w_2 \end{Bmatrix} = \begin{Bmatrix} -2\log(\sqrt{2}-1)/\lambda & 0 \\ 1/\sqrt{2} & 1-1/\sqrt{2} \end{Bmatrix} \approx \begin{Bmatrix} 1.763/\lambda & 0 \\ 0.707 & 0.293 \end{Bmatrix}.$$

(Exercise 17.9 asks you to verify via Theorem 17.1 that this solution is *not* the D-optimal design for the full vector $\boldsymbol{\theta}$.) ❑

17.2.3 Orthogonal Designs

The optimization problem for design with linear models, as presented in (17.9), is quite general. In fact, as we have seen, more than one design ξ may be a D-optimal solution. One of the ways in which the number of potential solutions can be reduced while introducing other desirable properties is to impose an orthogonality constraint. Intuitively, orthogonality tends to separate the effects of the input factors as much as possible. This is especially useful when effects may otherwise be *aliased* (so one effect masks another effect), as discussed in Wu and Hamada (2000, Sect. 4.4). Note that not all orthogonal designs are *D*-optimal. One advantage of orthogonal designs is that the estimates for the elements within $\boldsymbol{\theta}$ are uncorrelated. Hence, hypothesis tests under the normality assumption (where uncorrelatedness is equivalent to independence) may be carried out for each element separately without having to account for dependence of the estimates.

Recall that the model for the stacked vector of n measurements is $\boldsymbol{Z}_n = \boldsymbol{H}_n\boldsymbol{\theta} + [v_1, v_2, \ldots, v_n]^T$. Suppose that (17.7) applies (i.e., $z_k = [1, \boldsymbol{x}_k^T]\boldsymbol{\theta} + v_k$). Then,

$$\boldsymbol{H}_n = \begin{bmatrix} 1 & \\ 1 & \\ \vdots & , \boldsymbol{X}_n \\ 1 & \end{bmatrix},$$

where $\boldsymbol{X}_n = [\boldsymbol{x}_1, \boldsymbol{x}_2, \ldots, \boldsymbol{x}_n]^T$. The interest here is in designs where the column vectors within $\boldsymbol{H}_n$ are at right angles to one another (each column contains the inputs multiplying a specific element of $\boldsymbol{\theta}$). This achieves orthogonality and makes the estimates for all pairs of elements of $\boldsymbol{\theta}$ uncorrelated, as we now discuss.

Recall that $r = \dim(\boldsymbol{x}_k)$. The linear model (17.7) implies that $r = p - 1$ (for other model forms—e.g., nonlinear—r as a representation of the dimension of the input vector may not equal $p - 1$). Let the ith column of $\boldsymbol{X}_n$ be denoted by $\boldsymbol{x}_n^{(i)}$ (not to be confused with the ith row, $\boldsymbol{x}_i$) and let $\mathbf{1}_n$ denote an n-dimensional vector of 1's. There are $p - 1$ columns in $\boldsymbol{X}_n$. The orthogonality requirement is satisfied if $\left(\boldsymbol{x}_n^{(i)}\right)^T \boldsymbol{x}_n^{(j)} = 0$ for all $i \neq j$ and $\mathbf{1}_n^T \boldsymbol{x}_n^{(i)} = 0$ for all i. That is, the inner product of the n-dimensional vector corresponding to the contribution of one factor times the vector of contributions of a different factor is zero. In matrix notation, $\boldsymbol{X}_n$ should be chosen so that

$$\boldsymbol{H}_n^T \boldsymbol{H}_n = \text{diagonal matrix}. \qquad (17.16)$$

Designs satisfying (17.16) are called *fully orthogonal designs*. (Note that each $\boldsymbol{x}_n^{(i)}$ is a special case of the column $\boldsymbol{h}_{\bullet i+1}$, according to the general definition of columns in the $\boldsymbol{H} = \boldsymbol{H}_n$ matrix in Subsection 3.1.2.)

To see how (17.16) implies that the estimates within $\hat{\boldsymbol{\theta}}^{(n)}$ are uncorrelated, recall from (17.8) that $\text{cov}(\hat{\boldsymbol{\theta}}^{(n)}) = (\boldsymbol{H}_n^T \boldsymbol{H}_n)^{-1}\sigma^2$. In particular, with $\boldsymbol{\theta} = [\beta_0, \beta_1, \ldots, \beta_{p-1}]^T$ and $\hat{\boldsymbol{\theta}}^{(n)} = [\hat{\beta}_0, \hat{\beta}_1, \ldots, \hat{\beta}_{p-1}]^T$, then under the orthogonality requirement (17.16), $\text{cov}(\hat{\beta}_i, \hat{\beta}_j) = 0$ for all $i, j \geq 0$ and $i \neq j$ (following convention, the "0" subscript denotes the additive constant; i.e., β_0 is the additive constant).

Choices such as 2^r factorial (cubic) designs for experiments with r factors taking on one of two levels (e.g., Draper, 1988; Montgomery, 2001, Chaps. 6 and 7) satisfy the above orthogonality condition. This factorial design requires measurements at $n = 2^r$ levels of the factors during the experimentation process. Factorial designs are those such that factors in $\boldsymbol{x}$ are varied together rather than one at a time. As discussed in Montgomery (2001, Chap. 5), these designs are more efficient at exposing input–output relationships than varying only one input variable at a time. A simplex design (analogous to the simplex patterns of the search algorithm in Section 2.4) also satisfies this orthogonality condition while only requiring $n = r + 1$ levels of the factors. Other examples of acceptable orthogonal designs are given in Draper (1988), among many other references. Illustrations of two orthogonal designs for $r = 2$ and $r = 3$ are given in Figures 17.2 and 17.3, where $\boldsymbol{x}_k = [x_{k1}, x_{k2}]^T$ or $\boldsymbol{x}_k = [x_{k1}, x_{k2}, x_{k3}]^T$, as appropriate.

The above discussion pertains to fully orthogonal designs. In some cases, orthogonality may only apply to a subset of the columns within $\boldsymbol{H}_n$. This is especially the case in the curvilinear setting where elements within $\boldsymbol{X}_n$ may represent related inputs (e.g., two of the columns may represent x and x^2 effects). In such cases, only some of the estimates within $\hat{\boldsymbol{\theta}}^{(n)}$ will generally be uncorrelated. For example, Box and Draper (1987, p. 483) show a curvilinear case, where $\hat{\beta}_0$ is correlated with certain other elements in $\hat{\boldsymbol{\theta}}^{(n)}$, but where $\text{cov}(\hat{\beta}_i, \hat{\beta}_j) = 0$ for all $i, j \geq 1$ and $i \neq j$. Example 17.8 is another case of such a partially orthogonal design. This model has $r = p - 1 = 2$.

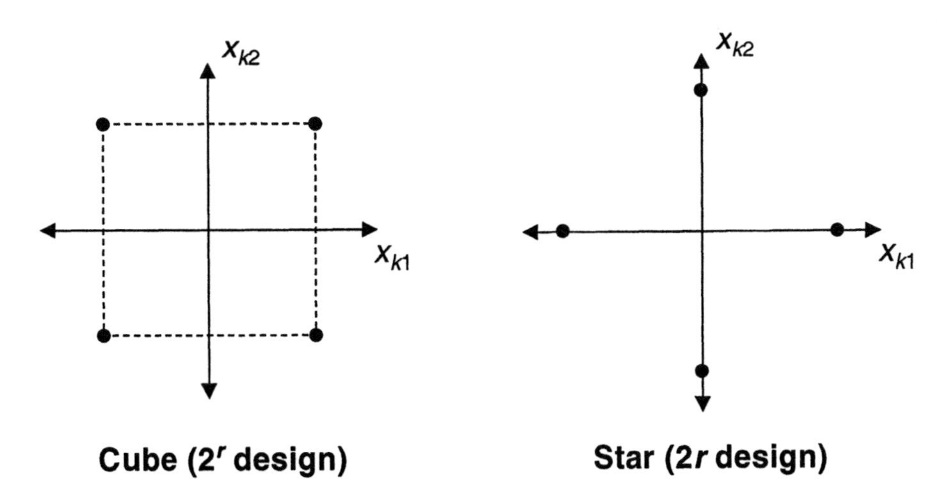

Figure 17.2. Two possible orthogonal designs for $r = 2$; points are symmetric with respect to origin.

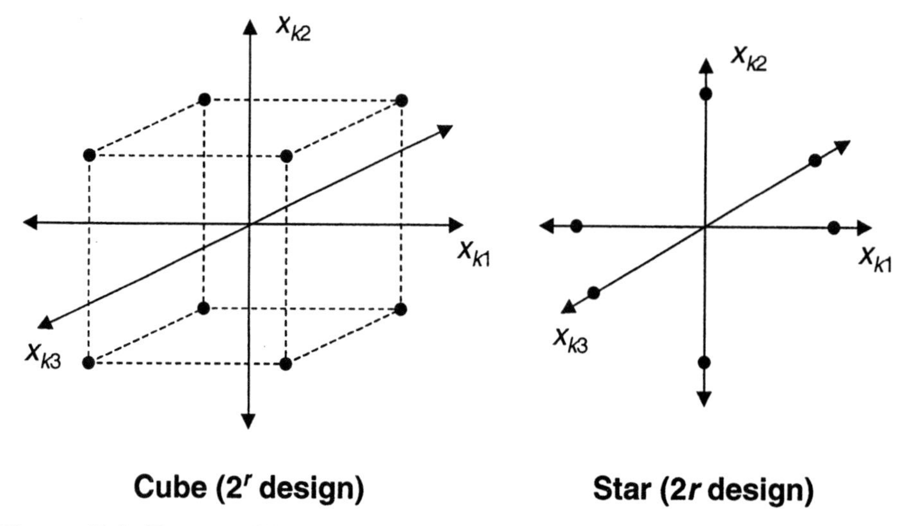

Figure 17.3. Two possible orthogonal designs for $r = 3$; points are symmetric with respect to origin.

Example 17.8—Partially orthogonal design for a simple curvilinear model. Consider again the $r = 2$ curvilinear model of Example 17.6: $z = [1, x, x^2]\boldsymbol{\theta} + v$, where it is known that $-1 \le x \le 1$. The aim is to pick a design based on three levels of the input x given that n is divisible by 3. So, the same three input values are repeated until n measurements are reached. An intuitively appealing candidate design—and one shown to be D-optimal in Example 17.6—is the design where $x = -1$, 0, and 1 with equal frequency. The resulting $n \times p$ design matrix $\boldsymbol{H}_n$ is composed of a vertical stack of $n/3$ blocks of the form

$$\boldsymbol{H}_3 = \begin{bmatrix} 1 & \\ 1, \boldsymbol{X} & \\ 1 & \end{bmatrix} = \begin{bmatrix} 1 & -1 & 1 \\ 1 & 0 & 0 \\ 1 & 1 & 1 \end{bmatrix}.$$

Therefore,

$$\boldsymbol{H}_3^T \boldsymbol{H}_3 = \begin{bmatrix} 1 & 1 & 1 \\ -1 & 0 & 1 \\ 1 & 0 & 1 \end{bmatrix} \begin{bmatrix} 1 & -1 & 1 \\ 1 & 0 & 0 \\ 1 & 1 & 1 \end{bmatrix} = \begin{bmatrix} 3 & 0 & 2 \\ 0 & 2 & 0 \\ 2 & 0 & 2 \end{bmatrix}.$$

Hence,

$$(\boldsymbol{H}_n^T \boldsymbol{H}_n)^{-1} = \frac{3}{n} \begin{bmatrix} 1 & 0 & -1 \\ 0 & \frac{1}{2} & 0 \\ -1 & 0 & \frac{3}{2} \end{bmatrix},$$

indicating that $\hat{\beta}_0$ is negatively correlated with $\hat{\beta}_2$, but that the two elements associated with the input x ($\hat{\beta}_1$ and $\hat{\beta}_2$) are uncorrelated with each other. ❑

17.2.4 Sketch of Algorithms for Finding Optimal Designs

One fundamental issue has been omitted in the discussion above: How can a *D*-optimal design be constructed in arbitrary linear models of the form (17.5)? In the examples above—and in many applied problems and examples elsewhere—one "guesses" at a design based on intuitive considerations and then verifies it by the equivalence theorem, Theorem 17.1. In some problems, however, an intuitive solution may be elusive. Fedorov (1972, Sects. 2.5 and 2.6), Silvey (1980, Chap. 4), Rustagi (1994, Sect. 2.5), Müller and Parmigiani (1995), and Angelis et al. (2001) are among the many references that discuss search algorithms for solving this nontrivial problem. Some of these algorithms are variations of the types of stochastic algorithms seen in earlier chapters. These algorithms are directed to finite-sample or asymptotic designs. Three of the recommended methods involve random search, stochastic approximation, and simulated annealing, core stochastic search methods that were considered in Chapters 2 and 4–8.

There are also some popular *exchange algorithms* for determining optimal finite-sample designs. Important variations on this general approach were introduced in Fedorov (1972, Chaps. 3 and 4), Mitchell (1974), and Cook and Nachtsheim (1980). These and other variations are widely available in commercial software (e.g., the OPTEX package in the SAS system of statistical software). Although we do not describe these deterministic algorithms in detail here, let us provide a sketch of the basic idea. Consider a candidate set of design points $\{\chi_i\}$. The list might include the points in a grid of the search space $\mathcal{X}$. Another possibility might be all or some of the points in an orthogonal design (such as the points in a 2^r or 3^r factorial design). The list of points should be rich enough to contain an adequate approximation of the optimal design ξ_n^*.

Based on the candidate set above, there are two broad steps in finding an optimal design by an exchange algorithm. First, one makes an initial guess by choosing the n inputs to occupy a subset of points in the candidate set. Depending on the value of n and the size of the candidate set, there may be points in the candidate set that are assigned multiple inputs or no inputs. This initial design produces an associated value of the *D*-optimal criterion $\det[\boldsymbol{M}(\xi)]$. In the second broad step, an iterative process is implemented to shuffle inputs among the elements in the candidate set with the aim of increasing the criterion. For example, an input that has been assigned to one of the candidates may be shifted to a previously unoccupied point in the candidate set. Various rules exist that govern such exchanges, but the overall theme is that at each iteration one seeks exchanges that maximize the improvement in the criterion. A good review of exchange and other procedures is in Atkinson and Donev (1992, Chap. 15).

An important property for optimization is that $\log\det[\boldsymbol{M}]$ is a strictly concave function on the set of positive definite matrices $\boldsymbol{M}$ (Exercise 17.11).

Such concavity is useful as it guarantees that the optimal $\boldsymbol{M}$ is unique. This follows because maximizing det[$\boldsymbol{M}$] is the same as maximizing $\log\det[\boldsymbol{M}]$ by the monotonicity of the log operation (i.e., if $a > b > 0$, then $\log(a) > \log(b)$). Unfortunately, this concavity in $\boldsymbol{M}$ does not tell us that $\log\det[\boldsymbol{M}] = \log\det[\boldsymbol{M}(\boldsymbol{\xi})]$ is concave in $\boldsymbol{\xi}$. Hence, there is no guarantee that $\boldsymbol{\xi}^*$ is unique. In particular, multiple global solutions $\boldsymbol{\xi}^*$ may yield the same maximizing value of $\boldsymbol{M}$ (such as in Example 17.5). Further, there may be multiple *local* maxima to $\log\det[\boldsymbol{M}(\boldsymbol{\xi})]$, complicating the optimization process.

17.3 RESPONSE SURFACE METHODOLOGY

Response surface methodology (RSM) is a sequential experimentation strategy for building an understanding of the relationship between input and output variables in a complex process. This section is a synopsis of RSM, including some discussion of its relationship to other aspects of experimental design.

Suppose that the *true process* (not model) for the output z satisfies

$$z = f(\boldsymbol{x}) + \text{noise}$$

for some unknown (usually nonlinear) function $f(\boldsymbol{x})$ and inputs $\boldsymbol{x}$. For simplicity and consistency with previous analysis, we consider a scalar output z and (generally) vector input $\boldsymbol{x}$. RSM is based on sequentially approximating $f(\boldsymbol{x})$ via low-order regression models of the form $\boldsymbol{h}(\boldsymbol{x})^T\boldsymbol{\theta}$, as shown in (17.5). Historically, RSM began as a tool in the chemical process industry (Box and Wilson, 1951), but it has evolved to be a general tool for the sequential analysis of complex processes (e.g., Khuri and Cornell, 1987; Wu and Hamada, 2000, Chap. 9). This section includes some pointers toward more complete discussions elsewhere.

RSM may be used to build an understanding of the input–output relationship, or, more specifically, to determine inputs that optimize the process. RSM is a sequential design strategy in that it is based on a recursive set of localized experiments aimed at converging to global optimal design. RSM applies to nonlinear problems by localizing the function surface to where a linear-type (pure linear or curvilinear) model can be used.

Let us focus on the use of RSM for finding an optimal value of $\boldsymbol{x}$ with the aim of minimizing the mean response z (hence, $E(z)$ is the loss function to be minimized). The basic idea is to first perform several experiments at $\boldsymbol{x}$ values near the currently estimated optimal $\boldsymbol{x}$ value. From these experiments, one builds up a "response surface" describing the local behavior of the process under study. This response surface is typically—but not necessarily—a first- or second-order regression polynomial in the vector of factors $\boldsymbol{x}$. The response surface is then used to move toward a better value of $\boldsymbol{x}$, and the process repeats itself. The success of RSM is based on the validity of repeated localized curve fitting and subsequent optimization. The methodology can be summarized in the following steps.

Outline of RSM for Optimizing $\boldsymbol{x}$

Step 0 **(Initialization)** Make an initial guess at the optimal value of $\boldsymbol{x}$.

Step 1 Collect responses z from several $\boldsymbol{x}$ values in the neighborhood of the current best estimate of the best $\boldsymbol{x}$ value. To enhance the fitting process, use experimental design to determine the values of $\boldsymbol{x}$ with which to collect measurements z.

Step 2 From the $\boldsymbol{x}$, z pairs in step 1, fit a regression (response surface) model over the region around the current best estimate of the optimal $\boldsymbol{x}$ levels.

Step 3 Based on the response surface in step 2, estimate a path of steepest descent in factor space.

Step 4 Perform a series of experiments at $\boldsymbol{x}$ values along the path of steepest descent until no additional improvement in the z response is obtained. This $\boldsymbol{x}$ value represents the new estimate of the best vector of factor levels.

Step 5 Go to step 1 and repeat the process until a final best factor level is obtained. Second-order (Hessian) analysis may be employed to ensure that the final solution is a minimum (given the aim to minimize $E(z)$).

A critical part of the RSM steps above is the choice of the experimental points in step 1. Optimal design for linear models, as described in Section 17.2, may be used to choose these points. The number and value of such points depend on the choice of the regression model in step 2. RSM is almost always based on the pure linear regression model or the second-order curvilinear form:

$$z_k = \beta_0 + \boldsymbol{\beta}^T \boldsymbol{x}_k + \boldsymbol{x}_k^T \boldsymbol{B} \boldsymbol{x}_k + v_k, \tag{17.17}$$

where β_0, $\boldsymbol{\beta}$, and $\boldsymbol{B}$ represent a scalar, vector, and symmetric matrix to be estimated (if the quadratic model is being used). Model (17.17) can be equivalently written in the linear form of (17.5), $z_k = \boldsymbol{h}_k^T \boldsymbol{\theta} + v_k$, where $\boldsymbol{h}_k = \boldsymbol{h}(\boldsymbol{x}_k)$ is the design vector dependent on input $\boldsymbol{x} = \boldsymbol{x}_k$. Hence, $\boldsymbol{\theta}$ represents the unique parameters in β_0, $\boldsymbol{\beta}$, and $\boldsymbol{B}$ that are to be estimated. It is often the case that the first few steps of RSM are based on a simple linear model $z_k = \beta_0 + \boldsymbol{\beta}^T \boldsymbol{x}_k + v_k$. A full factorial ($2^r$) design may be useful; this design is D-optimal for such simple linear models (Montgomery, 2001, p. 469). Later steps, requiring greater accuracy, employ the full quadratic-input model as in (17.17).

In the case where the quadratic (curvilinear) model (17.17) is applied, a large number of parameters may need to be estimated. Hence, some of the classical orthogonal designs discussed in Section 17.2 may no longer be appropriate because they do not provide the information needed to estimate the curvature effects. There are several popular ways to augment classical designs (such as the standard full factorial [cubic] designs illustrated in Figures 17.2 and 17.3). One is to take additional measurements at the center of the cube. Others, such as 3^r factorials, are discussed in Montgomery (2001, Chap. 9).

Given that $\dim(\boldsymbol{x}_k) = r$, the model in (17.17) has one constant, r linear terms, r squared terms, and $r(r-1)/2$ mixed second-order terms. This leads to a total of $p = r^2/2 + 3r/2 + 1$ terms in $\boldsymbol{\theta}$. To estimate the parameters in (17.17), the design must provide at least this number of data points. In cases where r is large, *fractional factorial* designs are sometimes used (e.g., Montgomery, 2001, Chaps. 8 and 9). These designs capture many of the benefits of full factorial designs, but with only a fraction of the number of measurements.

Lucas (1976) reports on the performance of several popular response surface designs using fewer measurements than the 2^r required in a full factorial design, including particular fractional factorial designs. In particular, certain "standard" designs requiring no more than $p + r$ measurements achieve a value of the D-optimal criterion that is at least 90 percent of the best possible value on problems in a given test set.

Given that one has settled on a design strategy, then step 3 of the RSM search proceeds along the path of steepest descent (as in Section 1.4). For the pure linear model, this is simply implemented by changing $\boldsymbol{x}$ by an amount proportional to the estimate of $-\boldsymbol{\beta}$. Responses are observed along this path until there is no significant improvement in the response (step 4). Of course, we do not know $\boldsymbol{\beta}$ (having only an estimate), so the gradient used in the steepest descent is only an estimate of the true gradient. Therefore, this represents a version of *stochastic* steepest descent as in the stochastic gradient form (Chapter 5) of the root-finding (Robbins–Monro) stochastic approximation algorithm. (Section 17.4 also discusses stochastic approximation for sequential design in nonlinear models.)

Alternatively, if the second-order curvilinear model (17.17) is used, then the stationary (zero-gradient) point of the second-order surface can be found and further analysis can be performed to determine if this stationary point is a minimum. Bisgaard and Ankenman (1996) report on a method for establishing confidence intervals for the eigenvalues of the Hessian matrix for the second-order model. This is useful in step 5 of the RSM process in order to statistically quantify (accounting for noise in the z values) whether the algorithm is converging to a minimum, maximum, or saddlepoint of the criterion $E(z)$. As above, stochastic approximation is relevant here since this response surface model (and associated gradient/Hessian) are only estimated (representing a noisy measurement of $E(z)$ and its derivatives).

While RSM has a long history of success, there is no guarantee of greater efficiency (i.e., fewer experiments) in reaching a solution than other stochastic optimization methods. In particular, stochastic gradient (Chapter 5) or gradient-free stochastic approximation (FDSA and SPSA, Chapters 6 and 7) may be useful in minimizing $E(z)$ without the process of sequential design. As with RSM, these methods are based on noisy measurements of the criterion (z values as noisy measurements of $E(z)$). There appears to be no formal study comparing RSM with such methods.

17.4 NONLINEAR MODELS

17.4.1 Introduction

Although significant challenges in experimental design arise with models $h(\boldsymbol{\theta}, \boldsymbol{x})$ linear in the parameters $\boldsymbol{\theta}$, much can (and has!) been said about the problem. This was discussed in Sections 17.2 and 17.3. For models that are nonlinear in $\boldsymbol{\theta}$ (such as neural networks, where $\boldsymbol{\theta}$ represent the connection weights being estimated), much less is known. A major challenge in design for nonlinear models is that the optimal design depends on $\boldsymbol{\theta}$. In contrast, in linear models of the form (17.5), the design does not depend on knowing $\boldsymbol{\theta}$. This dependence in nonlinear models leads to a conundrum. One is picking the design $\boldsymbol{\xi}$ with the aim of estimating the unknown $\boldsymbol{\theta}$, and yet one has to know $\boldsymbol{\theta}$ to pick the best $\boldsymbol{\xi}$!

This conundrum leads to various means of coping with the dependence on $\boldsymbol{\theta}$. These include simply producing a design based on one's best guess at $\boldsymbol{\theta}$, determining a design in a sequential manner by alternating between forming estimates of $\boldsymbol{\theta}$ and choosing a design, and augmenting the criterion in a Bayesian manner to reflect a prior distribution on $\boldsymbol{\theta}$. We discuss these three general approaches below.

As with the linear case, there are a number of design criteria. The dominant design criterion continues to be D-optimality (expression (17.4)), although several closely related performance measures are also used. As we have seen, D-optimality is directly applicable to both linear and nonlinear models through the Fisher information matrix and is oriented toward maximizing the precision of the parameter estimates. Following the previous sections, we continue to focus on D-optimality here, although the basic principles for search and optimization for an optimal design apply as well with other criteria. Recall that the D-optimal criterion depends on the precision matrix $\boldsymbol{M}(\boldsymbol{\theta}, \boldsymbol{\xi})$. The precision matrix here is the Fisher information matrix. In contrast to the batch estimator emphasized in the linear models of Sections 17.2 and 17.3, this section considers more general batch or recursive estimators. Hence, we use the generic notation $\hat{\boldsymbol{\theta}}_n$ to refer to an arbitrary estimator of $\boldsymbol{\theta}$ based on n data points (rather than the batch-specific notation $\hat{\boldsymbol{\theta}}^{(n)}$ used in Sections 17.2 and 17.3).

From Section 13.3, the information matrix $\boldsymbol{F}_n = \boldsymbol{F}_n(\boldsymbol{\theta}, \boldsymbol{X}_n)$ is

$$\begin{aligned} \boldsymbol{F}_n(\boldsymbol{\theta}, \boldsymbol{X}_n) &= E\left(\frac{\partial \log \ell}{\partial \boldsymbol{\theta}} \cdot \frac{\partial \log \ell}{\partial \boldsymbol{\theta}^T} \middle| \boldsymbol{\theta}, \boldsymbol{X}_n \right) \\ &= -E\left[\boldsymbol{H}_{\log \ell}(\boldsymbol{\theta}, \boldsymbol{Z}_n \,|\, \boldsymbol{X}_n) \middle| \boldsymbol{\theta}, \boldsymbol{X}_n \right], \end{aligned}$$

where ℓ is the likelihood function (the probability density or mass function in most typical cases), $\boldsymbol{H}_{\log \ell}$ is the Hessian matrix of the log-likelihood function, and the expectations are with respect to the randomness in the outputs $\boldsymbol{Z}_n = [z_1, z_2, \ldots, z_n]^T$ conditioned on $\boldsymbol{\theta}$ and the n input vectors represented by $\boldsymbol{X}_n$. We saw in Subsection 13.3.3 that there is a close connection between $\boldsymbol{F}_n^{-1}$ and the covariance matrix of general parameter estimates $\hat{\boldsymbol{\theta}}_n$. Loosely speaking, $\boldsymbol{F}_n^{-1}$ is an expression of uncertainty in $\hat{\boldsymbol{\theta}}_n$ for a large class of estimators based on a *general* (nonlinear) regression model of the form $z_k = h(\boldsymbol{\theta}, \boldsymbol{x}_k) + v_k$. In particular, $\boldsymbol{F}_n^{-1}$ evaluated at $\boldsymbol{\theta}$ close to the optimum $\boldsymbol{\theta}^*$ is approximately equal to $\text{cov}(\hat{\boldsymbol{\theta}}_n)$ for large n.

The Fisher information matrix depends on $\boldsymbol{\theta}$ in nonlinear models. In contrast, in linear models of the form (17.5), the precision matrix $\boldsymbol{M}(\boldsymbol{\theta}, \xi) = \boldsymbol{M}(\xi)$ $= \boldsymbol{H}_n^T \boldsymbol{H}_n / n$, which is a scaled form of the inverse covariance matrix, does not depend on $\boldsymbol{\theta}$. In Section 13.3 we suppressed the dependence of the information matrix on inputs, $\boldsymbol{X}_n$, simply writing $\boldsymbol{F}_n = \boldsymbol{F}_n(\boldsymbol{\theta})$. Here, of course, that dependence is critical, as we are fundamentally concerned with choosing the inputs, recalling that ξ is a convenient form for summarizing the allocation in $\boldsymbol{X}_n$. So, we will restore the dependence in the expressions for the information matrix here, writing $\boldsymbol{F}_n = \boldsymbol{F}_n(\boldsymbol{\theta}, \boldsymbol{X}_n)$.

Recall the fundamental goal of picking the design ξ that best helps in estimating $\boldsymbol{\theta}$. Given the close connection of $\boldsymbol{F}_n^{-1}$ to $\text{cov}(\hat{\boldsymbol{\theta}}_n)$, it is natural to have the precision matrix of the D-optimal criterion (17.4) satisfy $\boldsymbol{M}(\boldsymbol{\theta}, \xi) = \boldsymbol{F}_n(\boldsymbol{\theta}, \boldsymbol{X}_n)$. Hence, the aim in general nonlinear problems is to find a design satisfying

$$\arg\max_{\xi} \left\{ \det\left[\boldsymbol{M}(\boldsymbol{\theta}, \xi) \right] \right\} = \arg\max_{\boldsymbol{X}_n} \left\{ \det\left[\boldsymbol{F}_n(\boldsymbol{\theta}, \boldsymbol{X}_n) \right] \right\}. \qquad (17.18)$$

The two related examples below consider a relatively simple scalar problem to illustrate how, in general, the information matrix and resulting optimal design depend on $\boldsymbol{\theta}$ in a nonlinear problem.

Example 17.9—Nonlinear scalar problem in estimation of mean number of organisms via dilution. Consider a problem involving the estimation of the mean number of organisms per unit volume in some substance. Suppose that the unknown mean to be estimated is θ. If the original substance is diluted with pure water by a multiplicative factor $x \geq 1$, the mean number of organisms per unit volume is θ/x. The interest here is to enhance the estimation of θ by determining the best dilution factor x (the user-specified input here). Suppose the laboratory equipment can only check for the absence or presence of organisms rather than count the number of organisms (e.g., Cochran, 1973). Let $z_k \in \{0, 1\}$ represent a binary random variable denoting the absence or presence of at least one organism in a sample (0 if not present; 1 if present). If it is assumed that the number of organisms per unit volume is random according to a Poisson distribution (as in Cochran, 1973; Cox and Reid, 2000, pp. 178–179), the binary output z_k then has a Bernoulli (binary) probability distribution defined by

$$P(z_k = 0) = \exp(-\theta/x_k) \text{ and } P(z_k = 1) = 1 - \exp(-\theta/x_k).$$

(Of course, more information would be available for estimating θ if it were possible to count the number of organisms in a sample, but it is assumed here that the laboratory equipment provides only the above binary information.)

Assume that the experimenter obtains n unit samples, each diluted according to x_k. The aim in the design problem is to determine the sequence of x_k's to enhance the estimation of θ. From the Bernoulli probability distribution above, the log-likelihood function for θ is

$$\log \ell(\theta|\boldsymbol{Z}_n) = \sum_{k=1}^{n}\left\{\theta\frac{z_k - 1}{x_k} + z_k \log[1-\exp(-\theta/x_k)]\right\} \tag{17.19}$$

(see Exercise 17.13). Applying the basic formula for the information "matrix" (number) above yields

$$F_n(\theta, \boldsymbol{X}_n) = \sum_{k=1}^{n}\frac{\exp(-\theta/x_k)}{x_k^2[1-\exp(-\theta/x_k)]}$$

(see Exercise 17.13). Note that the information number depends on θ; hence the optimal design for the x_k's also depend on θ. Example 17.10 produces the D-optimal solution. ❑

Example 17.10—Optimal design for dilution problem. Consider the problem of Example 17.9. Suppose that the experimenter wants to determine the optimal finite-sample design. Because θ is a scalar, the D-optimal design is the design that maximizes $F_n(\theta, \boldsymbol{X}_n)$ (i.e., no determinant necessary). Note that maximizing one summand in $F_n(\theta, \boldsymbol{X}_n)$ is sufficient because each of the summands in $F_n(\theta, \boldsymbol{X}_n)$ has an identical form. Hence, it is sufficient to find an $x \geq 1$ that maximizes

$$\frac{e^{-\theta/x}}{x^2(1-e^{-\theta/x})}$$

(so that all n measurements have the same input x). Intuitively, it seems undesirable to have $x >> 1$ because then θ/x is near zero (i.e., in n unit samples, z_k would almost always indicate nonpresence). In the notation of design, it can be shown using the expression above that the optimal design satisfies

$$\xi_n^* = \begin{Bmatrix} 0.63\theta \\ 1.0 \end{Bmatrix}$$

for any n (Exercise 17.14). That is, the inputs are set to $x_k = \chi_1 = 0.63\theta$ for all k (i.e., $w_1 = 1.0$ as shown in ξ_n^*). Given the requirement to have $x \geq 1$, this solution is only valid if $\theta \geq 1/0.63 \approx 1.59$. This solution depends, of course, on the binary aspect of the data collection process. If, instead, each z_k represented a *count* of the number of organisms, the optimal design would be different. A practical implementation of the solution above would be an example of a "local design," local because it is valid in practice for θ near the value specified to obtain a realizable solution. ❑

17.4.2 Methods for Coping with Dependence on θ

There are three well-established means for coping with the dependence of the optimization criterion on $\boldsymbol{\theta}$ in nonlinear problems:

1. Simply assume a nominal value of $\boldsymbol{\theta}$ and develop an optimal design based on this fixed value during the optimization process. It is usually hoped that this nominal value is close to the unknown true value. This is called a *local design*, as illustrated in Example 17.10.
2. Use a sequential design strategy based on an iterated design and model fitting process. This strategy begins with an initial guess at $\boldsymbol{\theta}$, leading to an optimal design for a small number of inputs. Data are collected from the inputs and the value for $\boldsymbol{\theta}$ is updated, leading to a new design for another group of inputs. This process is repeated until a satisfactory estimate for $\boldsymbol{\theta}$ is obtained. Root-finding and stochastic gradient versions of stochastic approximation (SA) (Chapters 4 and 5) often play a role in sequential design.
3. Use a Bayesian strategy where a prior distribution is assigned to $\boldsymbol{\theta}$, reflecting uncertainty in the knowledge of the true value of $\boldsymbol{\theta}$. In this strategy, each value of the D-optimal (or other) criterion is evaluated at a design $\boldsymbol{\xi}$ after the $\boldsymbol{\theta}$ dependence has been integrated out according to the prior distribution.

While the local design approach 1 above is clearly the simplest, it has serious flaws in many applications. If, as usual, the dependence of the criterion on $\boldsymbol{\theta}$ is significant, the solution will be highly sensitive to the choice of the nominal value. For example, in the dilution problem above (Examples 17.9 and 17.10), the error in design is directly proportional to the error in the choice of the scalar θ. Nonetheless, it is sometimes useful to seek a locally optimal design in cases where one is confident in the knowledge of $\boldsymbol{\theta}$. It may also be useful in hypothesis testing where one is testing a null hypothesis of a specific value of $\boldsymbol{\theta}$; that value may be treated as truth in determining the optimal design. Further, local designs are used to provide the intermediate solutions in the sequential design approach.

We now give a more detailed discussion of approaches 2 and 3, sequential design and Bayesian design. Although there has been a fair amount of work published on these nonlinear design strategies, the number of practical applications has lagged. This is expected to change as commercial software is developed for particular approaches to nonlinear optimal design.

Sequential Design

Perhaps the most widely used approach to solving the problem posed in (17.18) is the sequential design method. The main idea is that the total resources available for the experiment are divided and the entire experiment is performed in a number of steps. Each step of the experiment uses a portion of the resources. At each step, analysis and parameter estimation are performed based on the step's available resources. Then a decision is made as to how to best use the next step's resources. The central benefit of sequential design is that one can take advantage of the natural learning process associated with experimentation. This allows the analyst to best focus resources in the latter stages of the design as more information becomes available about the nature of the process. For example, in medical studies, this may have ethical implications via allowing an adjustment of a treatment to avoid harm to patients.

The implementation of sequential design typically begins with an initial guess at a design. Then, using a small subset of the budget of input–output measurements, an estimate of $\boldsymbol{\theta}$ is formed from this initial design. This value of $\boldsymbol{\theta}$ is then used in the optimality criterion (17.18) to pick input levels for the next one or several inputs, leading to a new value of $\boldsymbol{\theta}$. This process is repeated until there is adequate convergence of the $\boldsymbol{\theta}$ estimate or until the full budget of measurements has been collected.

The steps below summarize the general sequential design strategy. A specific implementation requires details not specified in the steps below. We comment on some of these details in the discussion and example following the steps below.

Outline of Sequential Approach for Parameter Estimation and Optimal Design

Step 0 **(Initialization)** Make an initial guess at $\boldsymbol{\theta}$, say $\hat{\boldsymbol{\theta}}_0$. Allocate n_0 measurements to the initial design. Set $k = 0$, $n = 0$, and $\boldsymbol{F}_0(\hat{\boldsymbol{\theta}}_0, \boldsymbol{X}_0) = \boldsymbol{0}$ ($\boldsymbol{X}_0$ is only a notational placeholder here; it need not be explicitly defined in the initialization step).

Step 1 Given $\boldsymbol{X}_n$, choose the n_k inputs in $\boldsymbol{X} = \boldsymbol{X}_{n_k}$ to maximize $\det[\boldsymbol{F}_n(\hat{\boldsymbol{\theta}}_n, \boldsymbol{X}_n) + \boldsymbol{F}_{n_k}(\hat{\boldsymbol{\theta}}_n, \boldsymbol{X})]$.

Step 2 Collect n_k output measurements based on the n_k optimally chosen inputs from step 1. Use the n_k measurements to update the value of $\boldsymbol{\theta}$ from $\hat{\boldsymbol{\theta}}_n$ to $\hat{\boldsymbol{\theta}}_{n+n_k}$.

Step 3 Stop if the value of $\boldsymbol{\theta} = \hat{\boldsymbol{\theta}}_{n+n_k}$ is satisfactory or if the budget of measurements has been expended. Else return to step 1 with the new k set to the former $k + 1$ and the new n set to the former $n + n_k$ (so the updated $\boldsymbol{X}_n$ now includes the inputs $\boldsymbol{X}_{n_k}$ from step 1).

Let us offer a few comments about the steps above. From step 1, the optimization to find the new input points $\boldsymbol{X}_{n_k}$ takes account of the previously determined inputs ($\boldsymbol{X}_n$). If certain areas of the input space are well represented in the previous data, it is more likely that new inputs will be placed in "underrepresented" areas of the input space. Hence, the solution in step 1 is generally going to differ from a corresponding problem where one simply maximizes $\det[\boldsymbol{F}_{n_k}(\hat{\boldsymbol{\theta}}_n, \boldsymbol{X})]$ with respect to $\boldsymbol{X}$.

One of the distinctions in implementation is between a batch-sequential and a full-sequential design. In the batch-sequential implementation, $n_k > 1$ data points are collected at each iteration. The full-sequential implementation relies on $n_k = 1$ for all k.

There are two potentially difficult optimization problems embedded in the sequential design iteration process. One is in step 1, where the new design points $\boldsymbol{X}_{n_k}$ are to be determined; the other is in step 2, where the updated value of $\boldsymbol{\theta}$ is calculated. There are no universal methods for carrying out either of these optimization tasks.

A number of techniques have been examined for the optimization in step 1. Methods used here can be the same as methods used in local design because one is working with a fixed $\boldsymbol{\theta}$. In some fortunate cases, relatively simple analytical or numerical schemes may be used, such as in Examples 17.8 and 17.9. Atkinson and Donev (1992, Chap. 18) discuss an approach where the model $h(\boldsymbol{\theta}, \boldsymbol{x})$ is linearized about the current $\boldsymbol{\theta} = \hat{\boldsymbol{\theta}}_n$. Then, the methods of Section 17.2 for linear models can be used to determine an approximate design at each iteration. A demonstration of simulated annealing for nonlinear local design is given in Rustagi (1994, p. 294).

The optimization problem associated with estimating $\boldsymbol{\theta}$ (step 2) may be carried out in batch or recursive form. In the batch form, the estimate is simply computed "from scratch" at each iteration based on the cumulative $n + n_k$ data points that are available. Little or no use is made of the previous $\boldsymbol{\theta}$ estimate ($\hat{\boldsymbol{\theta}}_n$). This may be acceptable in simple problems. More commonly, the process of estimating $\boldsymbol{\theta}$ in nonlinear problems entails some difficult numerical optimization where one wants to exploit the knowledge of $\boldsymbol{\theta}$ that has been acquired through the previous experiments. In such cases, recursive methods are preferred, where the estimate $\hat{\boldsymbol{\theta}}_{n+n_k}$ is directly updated from the previous estimate $\hat{\boldsymbol{\theta}}_n$. Such recursive estimation is naturally connected to SA with a stochastic gradient input (Chapter 5), as we now discuss.

Consider recursive estimation in the context of the full-sequential version of the design strategy. Here $n_k = 1$ for all k. Hence, the process of choosing $\boldsymbol{X} = \boldsymbol{X}_{n_k}$ in step 1 above reduces to the problem of choosing the single input $\boldsymbol{x} = \boldsymbol{x}_{n+1}$ in the model $h(\hat{\boldsymbol{\theta}}_n, \boldsymbol{x})$. Let $\boldsymbol{Y}_n(\boldsymbol{\theta} | z_{n+1}, \boldsymbol{x}_{n+1})$ represent the instantaneous score vector when updating $\hat{\boldsymbol{\theta}}_n$ to $\hat{\boldsymbol{\theta}}_{n+1}$ in step 2. This corresponds to the gradient with respect to $\boldsymbol{\theta}$ of the part of the estimation criterion that depends on the current input. In a maximum likelihood setting, this typically involves the negative gradient of the logarithm of the probability density function for the $(n+1)$st measurement ("negative" so that the underlying root-finding problem corresponds to a minimization problem). In a least-squares setting, this will involve the gradient of the current squared error (i.e., $\boldsymbol{Y}_n(\boldsymbol{\theta}|z_{n+1}, \boldsymbol{x}_{n+1}) = \frac{1}{2}\partial[z_{n+1} - h(\boldsymbol{\theta}, \boldsymbol{x}_{n+1})]^2/\partial\boldsymbol{\theta}$).

The standard (unconstrained) SA algorithm in Chapters 4 and 5 in this design context is

$$\hat{\boldsymbol{\theta}}_{n+1} = \hat{\boldsymbol{\theta}}_n - a_n \boldsymbol{Y}_n(\hat{\boldsymbol{\theta}}_n \,|\, z_{n+1}, \boldsymbol{x}_{n+1}), \tag{17.20}$$

where a_n is a decaying, nonnegative gain sequence satisfying the usual conditions (Sections 4.3 and 4.4). Recall that Section 4.3 presents conditions under which the SA algorithm converges to the optimum $\boldsymbol{\theta}^*$ as the amount of data increases. (Because of the emphasis of this chapter on input design, the notation for $\boldsymbol{Y}_n$ above explicitly includes the dependence on z_{n+1} and $\boldsymbol{x}_{n+1}$. This contrasts with the notation in the SA algorithms of Chapters 4 and 5, where any such dependence is suppressed within the notation $\boldsymbol{Y}_n(\hat{\boldsymbol{\theta}}_n)$. The reader should be aware that the fundamental algorithm is the same, despite the notational difference.) The SA form in (17.20) is convenient in that the new estimate for $\boldsymbol{\theta}$ is directly derived from the previous estimate with an adjustment that tends to get smaller as more data are collected (due to $a_n \to 0$). Ford et al. (1989) discuss the use of a second-order SA algorithm (as in Subsection 4.5.2) in sequential design, where a_n is replaced by the scaled inverse of the Hessian matrix of the performance measure.

An important cautionary note on statistical inference is in order. Because each input $\boldsymbol{x}_{n+1}$ depends on prior data $z_1, \boldsymbol{x}_1, z_2, \boldsymbol{x}_2, \ldots, z_n, \boldsymbol{x}_n$ through step 1 above, the data are not independent. This contrasts with the classical assumptions associated with a fixed design, where the $\boldsymbol{x}_i$ are predetermined and the z_i are independent. Most of the theory associated with using the asymptotic distribution and Fisher information matrix for inference (Section 13.3) assumes independent data. Recall that under regularity conditions, a scaled estimate $\hat{\boldsymbol{\theta}}_n$ is asymptotically normally distributed with covariance matrix given by the inverse of the information matrix (Subsection 13.3.3). This result is very useful in statistical inference. However, because the data are not independent in sequential design, one should be cautious about a straightforward application of the cumulative information matrix $\boldsymbol{F}_n$ resulting from the steps above. Nonetheless, theoretical and numerical support exists for treating this matrix in the same way as the traditional information matrix based on independent data (e.g., Ford and Silvey, 1980; Chaudhuri and Mykland, 1995). Further, if the SA recursion (17.20), or its second-order form, is used to update the parameter estimates, theory exists under which the covariance matrix in the asymptotic distribution is the inverse information matrix (see Subsection 13.3.3). That is, the SA theory directly accommodates the dependence on the inputs and the measurements.

Let us now present an example of sequential design using (17.20). This example builds on the dilution problem of Examples 17.9 and 17.10.

Example 17.11—Sequential design and parameter estimation for dilution problem. Consider the sequential design approach in the context of the scalar problem of Examples 17.9 and 17.10. Suppose that a sequential maximum likelihood method is being used to estimate the mean θ. Hence, each input in the SA recursion (17.20) is the negative derivative of the most recent summand in the log-likelihood function (17.19). Taking the negative of the derivative of the last summand in $\log\ell(\theta|\boldsymbol{Z}_n)$ (see (17.19)), the input $\boldsymbol{Y}_n(\theta|z_{n+1}, x_{n+1})$ for use in (17.20) is

$$\boldsymbol{Y}_n(\theta|z_{n+1}, x_{n+1}) = \frac{1-z_{n+1}}{x_{n+1}} - \frac{z_{n+1}\exp(-\theta/x_{n+1})}{x_{n+1}[1-\exp(-\theta/x_{n+1})]}.$$

Consider a setting where $\theta^* = 4$ is used to generate the data (note that θ^* is not generally the same as the batch maximum likelihood estimate that would result from maximizing $\log\ell(\theta|\boldsymbol{Z}_{n+1})$ for a finite n). Let us use the SA recursion (17.20) with gain value a_n in the standard form, $a_n = a/(n+1+A)^\alpha$, $n \geq 0$, a and A nonnegative, and $1/2 < \alpha \leq 1$, as given in Section 4.4. Let the initial condition be $\hat{\theta}_0 = 2$. Based on some preliminary numerical experimentation and the theoretical principles of Sections 4.3 and 4.4, we chose $a = 25$, $A = 50$, and $\alpha = 1$. The local design for this model has only one support point $x = \chi_1 = 0.63\theta$ (Example 17.10). Hence, the optimal input in step 1 of the sequential design process is $x_{n+1} = 0.63\hat{\theta}_n$.

Table 17.2 shows the normalized estimation error, $|\hat{\theta}_n - \theta^*|/|\hat{\theta}_0 - \theta^*|$, based on the terminal estimate $\hat{\theta}_n$ for each of three sample sizes (100, 1000, or 10,000). Each table value represents the sample mean of 60 normalized errors resulting from 60 independent replications. The θ estimate throughout the iteration process is constrained to be in $[1.59, \infty)$ as discussed in Example 17.10. The table compares three design strategies. The sequential design method is as described in the steps above with $n_k = 1$ (corresponding to a full sequential design). The local design based on $\theta = \hat{\theta}_0$ (so $x_n = 0.63\hat{\theta}_0$ for all n) is meant to depict the results one would see using a fixed design based on a best guess at θ (corresponding to the initial guess at θ). The other local design based on $\theta = \theta^*$ (i.e., $x_n = 0.63\theta^*$) depicts an idealized case based on the true (unknown in practice) θ. Only the sequential design and the suboptimal local design based on $\theta = \hat{\theta}_0$ would be implementable in practice.

Table 17.2 shows that the sequential design and the idealized local design based on $\theta = \theta^*$ perform comparably. On the other hand, the local design based on $\theta = \hat{\theta}_0$ performs noticeably worse. The table illustrates the value in using a

Table 17.2. Normalized estimation errors for sequential design and for two strategies with local designs. Indicated sample mean is based on 60 realizations.

	Sample mean for $\|\hat{\theta}_n - \theta^*\|/\|\hat{\theta}_0 - \theta^*\|$		
Sample size per realization	Sequential design	Local design $\theta = \theta^*$	Local design $\theta = \hat{\theta}_0$
100	0.25	0.23	0.31
1000	0.065	0.063	0.096
10,000	0.021	0.022	0.032

near-optimal design via the sequential strategy. With the sample sizes of 1000 and 10,000, the sequential design produces mean parameter estimation errors that are about 2/3 the size of the errors under the $\hat{\theta}_0$-based local design. This equivalently leads to a significant reduction in the sample size required to obtain a given level of accuracy. In particular, the sequential strategy requires about *half* the measurements that are needed in the suboptimal ($\hat{\theta}_0$-based) local design to achieve the same level of accuracy (Exercises 17.15 and 17.16). ❑

Bayesian Design

In the Bayesian design approach, one copes with the uncertainty in $\boldsymbol{\theta}$ by assuming a prior distribution for $\boldsymbol{\theta}$. This prior distribution is used in averaging the design criterion over reasonable possible values. The prior is a reflection of one's knowledge of likely values for $\boldsymbol{\theta}$. The prior may be specified on a purely subjective basis or on the basis of information from previous estimation of $\boldsymbol{\theta}$ (e.g., using the asymptotic distribution for a maximum likelihood estimate, as discussed in Section 13.3, based on data collected previously). This approach does not require any other aspect of the analysis to be carried out in a Bayesian manner. That is, no posterior distribution is involved in obtaining the design, and the prior distribution need only be used in determining the design, not for the subsequent analysis. (Section 16.6 summarized some generic aspects of the Bayesian approach for parameter estimation.) Hence, one can implement a Bayesian design approach, as described here, without being a "card-carrying" Bayesian in the other aspects of the estimation and inference.

As with the non-Bayesian approach, there are competing lettered criteria (*A*-optimality, *D*-optimality, etc.). We will continue to focus on *D*-optimality. Unlike the non-Bayesian case, however, there is not a unique *D*-optimal criterion. In fact, *five* *D*-optimal criteria are given in Atkinson and Donev (1992, p. 214), and even more such criteria appear in other literature (e.g., Walter and Pronzato, 1990). Among other variations, the differences follow from whether the averaging for $\boldsymbol{\theta}$ appears inside or outside certain function evaluations. One D-optimal criterion that is advocated in Chaloner and Verdinelli (1995) is

$$E_{\boldsymbol{\theta}}\left\{\log\det\left[\boldsymbol{M}(\boldsymbol{\theta}, \boldsymbol{\xi})\right]\right\} = \int_{\Theta} \log\det\left[\boldsymbol{M}(\boldsymbol{\theta}, \boldsymbol{\xi})\right] p(\boldsymbol{\theta})\, d\boldsymbol{\theta}, \tag{17.21}$$

where $E_{\boldsymbol{\theta}}$ denotes expectation with respect to $\boldsymbol{\theta}$, $p(\boldsymbol{\theta})$ is the prior expressed as a density function, and Θ is, as in earlier chapters, the domain for $\boldsymbol{\theta}$. This criterion uses the log transformation as discussed at the end of Section 17.2. To illustrate the variations possible in D-optimality, an alternative criterion to the one in (17.21) is $\log\det\left\{E_{\boldsymbol{\theta}}\left[\boldsymbol{M}(\boldsymbol{\theta}, \boldsymbol{\xi})\right]\right\}$. Another alternative is considered in Exercise 17.17. One of the desirable properties of the criterion in (17.21) is that it is an approximation to a generally infeasible criterion that when optimized produces an experiment that maximizes the increase in Shannon information (we will not delve further into the deep subject of Shannon information).

As one might expect, finding the design that maximizes (17.21) (or other criteria) is usually a significant challenge. Among other issues, evaluating the integral in (17.21) may be numerically cumbersome (or worse!) at each candidate design $\boldsymbol{\xi}$. Deterministic and stochastic methods (such as genetic algorithms) are discussed in Chaloner and Verdinelli (1995), Müller and Parmigiani (1995), and Hamada et al. (2001). A practical method for avoiding the numerical integral in (17.21) is to choose a prior distribution that is discrete with only a small number of support points (the values of $\boldsymbol{\theta}$ having nonzero probability).

One of the major issues in determining a design $\boldsymbol{\xi}$ is the choice of the number of support points N (not to be confused with the support points for a discrete prior distribution as mentioned in the preceding paragraph). A convenient bound on N was presented in Subsection 17.2.2 for *linear* models. In contrast, for nonlinear models, there is no such convenient bound. Chaloner and Verdinelli (1995) discuss this issue with pointers to related literature. They note that the number of support points grows with the dispersion of the prior distribution. They also note that for tight priors (little dispersion in $\boldsymbol{\theta}$), an optimal design often exists with $N = p$, identical to a result for locally optimal solutions based on a fixed $\boldsymbol{\theta}$ (approach 1 at the beginning of this subsection).

The fact that N must be greater than p in many practical problems is sometimes beneficial. The points beyond p provide data for checking the model with other candidate model forms having more than p parameters (as discussed in Section 13.2). This can be done to ensure that the chosen model is, in fact, the "best" representation of the process. Such model checking is useful to address the general criticism of optimal design cited in Section 17.1 that the design is directly dependent on the chosen model form. The extra support points for the design help an analyst check the validity of alternative models.

Let us now present a simple example illustrating the process for finding a D-optimal design based on a prior distribution for a scalar θ.

Example 17.12—Optimal design with discrete prior for θ. Consider a nonlinear model of the form $z = e^{-\theta x} + v$, where $v \sim N(0, 1)$ and $\theta > 0$ and $x \geq 0$ are scalars. Suppose that for $k = 1, 2, \ldots, n$, the noises v_k are independent, identically distributed (i.i.d.), implying that the measurements z_k are independent. The information number for n measurements is

$$F_n(\theta, X_n) = \sum_{k=1}^{n} x_k^2 \exp(-2\theta x_k) .$$

As in Example 17.10 on local design, it is sufficient to work with only one summand in $F_n(\theta, X_n)$ because each of the summands has an identical form. Hence, based on criterion (17.21), the aim is to find a design that maximizes

$$E_\theta[\log(x^2 e^{-2\theta x})] = E_\theta(2\log x - 2\theta x) .$$

Suppose that the prior on θ is a Bernoulli (binary) distribution with probability ρ placed on value θ_1 and probability $1 - \rho$ placed on θ_2. Then, from the expression above, the aim is to find x that maximizes

$$\rho(\log x - \theta_1 x) + (1 - \rho)(\log x - \theta_2 x) \tag{17.22}$$

subject to the constraints on θ and x. For convenience, assume that $\theta_2 > \theta_1$. As shown in Haines (1995), standard calculus-based methods for optimization yield a unique (single support point) design when $\theta_2/\theta_1 \leq 2 + \sqrt{3}$. In particular, differentiating (17.22) with respect to x, setting the derivative to 0, and checking the second-order conditions to ensure that the solution is a maximum yields

$$\xi_n^* = \begin{Bmatrix} [\rho\theta_1 + (1-\rho)\theta_2]^{-1} \\ 1.0 \end{Bmatrix}.$$

When $\theta_2/\theta_1 > 2 + \sqrt{3}$, more complicated arguments are needed. The difficulties follow from the need to optimize along the boundary of the constraint set. In such a setting, the solution is either identical to the single-point result above or is a two-point solution of the generic form

$$\xi_n^* = \begin{Bmatrix} \chi_1^* & \chi_2^* \\ w_1^* & w_2^* \end{Bmatrix},$$

where χ_i^* and w_i^* are solutions to equations related to the curvature of the constraint set (see Haines, 1995). Whether the design has one or two support points when $\theta_2/\theta_1 > 2 + \sqrt{3}$ depends on whether a particular inequality is satisfied (see Haines, 1995, expression (3.7)).

Note that these results are consistent with the above discussion relating the number of support points in the design to the spread of the prior distribution. When the candidate θ values are close in the sense that $1 < \theta_2/\theta_1 \leq 2 + \sqrt{3}$, there is only one support point (corresponding to $p = 1$); when the θ values have greater spread in the sense that $\theta_2/\theta_1 > 2 + \sqrt{3}$, the number of support points may increase to two. ❑

17.5 CONCLUDING REMARKS

There are often many goals for an experiment. Some may conflict with each other. As a consequence, there is no "cut and dried" strategy to the problem of determining a good choice for the user-specified inputs. As discussed in this chapter, optimal design is a powerful general strategy. Applications arise in both static regression-type models and in dynamic models that are used, for example, in control systems. Another popular general strategy is classical experimental design (e.g., fractional factorial, Latin squares, etc.), which forms the basis for many popular books in the field. Classical design is based strongly on geometric and intuitive considerations.

The solution to a specific problem can sometimes be shown to be both an optimal design (say, a *D*-optimal design) and a classical design. Unfortunately, in the literature, the dichotomy between general optimal design and classical design often illustrates the adage "never the twain shall meet." For instance, there is only a brief discussion of optimal design in significant books on experimental design such as Wu and Hamada (2000) and Montgomery (2001). In fact, with its letter-based labeling of the design criteria (*A*-optimal, *D*-optimal, etc.), optimal design is sometimes derisively referred to as the "alphabet soup" approach.

Optimal design is often criticized on the basis of its close connection to the assumed model form. That is, a criterion such as the *D*-optimal measure emphasized here can only be computed with an exact specification of the model. (Classical designs also depend on model assumptions, although to a lesser extent than optimal designs. For example, prior knowledge of interactions that are likely to be negligible is essential in creating a practical factorial design by "confounding" [Montgomery, 2001, Chap. 7].) A criticism related to the model dependence is that it is difficult to capture the many goals of an experiment—as discussed in Section 17.1—in one mathematical criterion, as needed in optimal design. In practice, supplemental runs may be carried out to evaluate the model validity or address other goals for the experiment.

On the other hand, the relatively formal structure of optimal design provides a means of coping with the many nonstandard situations that arise in practical applications. The geometric and intuitive basis for classical design frequently breaks down when departing from the standard linear (including curvilinear) model over "nice" input domains. As noted in Cook and Nachtsheim (1989, p. 345), "...the situations in which classical designs work well are rather confining, occasionally leading to unfortunate results....Experimenters often have an overwhelming urge to tailor the scientific question or trim the experimental material to allow application of a particular classical design."

One important application of optimal design is in determining inputs in models that are nonlinear in the parameters $\boldsymbol{\theta}$ being estimated. This was demonstrated in Section 17.4. Classical design provides little or no guidance in the selection of inputs for nonlinear models. Further, there is an inherent difficulty in nonlinear problems because of the dependence of the criterion on the parameters being estimated. One is picking a design in order to estimate $\boldsymbol{\theta}$, but one has to know $\boldsymbol{\theta}$ to pick a design!

There are several approaches to dealing with the dependence of the criterion on $\boldsymbol{\theta}$ in nonlinear models, foremost among them sequential and Bayesian strategies. Another challenge in nonlinear design is the difficulty of calculating the criterion and carrying out the related optimization process. The information matrix forming the basis for the leading D-optimal criterion is not readily available in most practical nonlinear problems. (The Monte Carlo resampling method of Subsection 13.3.5 provides a means for determining the information matrix in difficult problems, but this is likely to be computationally burdensome in the repeated evaluations required during an optimization process.)

Finally (!), we come to the end of this book on stochastic search and optimization. Those who have read extensive parts of the book have been exposed to many powerful methods for use in estimation, Monte Carlo simulation, and control. It is also clear that there are many exciting areas—in both algorithm development and applications—that have yet to be fully explored. In using the methods here (or anywhere!), one should keep in mind the aphorism mentioned at the end of Chapter 1: "Better a rough answer to the right question than an exact answer to the wrong one." Whether rough or exact, the methods here include many tools useful in addressing many "right questions."

17.6 APPENDIX: OPTIMAL DESIGN IN DYNAMIC MODELS

The main body of this chapter explores optimal design in static input–output systems. There is also a large literature in dynamic systems, including control systems. See, for example, Mehra (1974), Goodwin and Payne (1977), Walter and Pronzato (1990), or Ljung (1999, Chap. 13). Of main interest here is the choice of inputs to enhance the estimation of the parameters of the dynamic model. For control systems, the inputs may be provided in closed-loop mode. One distinction with static models is that even for linear models, the D-optimal criterion (and resulting optimal inputs) may depend on the parameters $\boldsymbol{\theta}$ being estimated (see, e.g., Mehra, 1974, Example VIII.C).

The example sketched below provides a flavor of optimal designs for such systems. This example is extracted from Levadi (1966) and Mehra (1974).

Example 17.13—Optimal input for first-order model. Consider a scalar process $q(\tau)$ that is modeled by the first-order differential equation

$$\frac{dq(\tau)}{d\tau} = -q(\tau) + \theta x(\tau),$$

where τ is the time variable, $x(\tau)$ is a user-specified input (a time-varying analogue of the static $\boldsymbol{x}$ input in the main part of this chapter), and θ is a scale parameter to be estimated. The analyst collects data on a continuous basis over time $\tau \in [0, T]$ according to

$$z(\tau) = q(\tau) + v(\tau),$$

where $v(\tau)$ is a noise term. The aim is to determine $x(\tau)$ so that θ is estimated as accurately as possible. It is assumed that the input is bounded in the sense that $\int_0^T x(\tau)^2 d\tau = 1$. Suppose that the noise has mean zero and covariance function $E[v(\tau_1)v(\tau_2)] = c_0 \exp(-c_1|\tau_1 - \tau_2|)$, where c_0 and c_1 are two constants and τ_1 and τ_2 are two time instants. Hence, the correlation between noises at two different times decreases exponentially as the difference in the times gets larger.

The scalar precision quantity M here for the D-optimal solution is a dynamic analogue of the inverse variance quantity used with linear models, as described in Section 17.2. The inverse quantity $1/M$ reflects the variance in the estimate for θ based on data $z(\tau)$ over the interval $[0, T]$; the estimate for θ is constructed via a dynamic model analogue of the least-squares solution in Section 17.2.

As shown in Levadi (1966) and Mehra (1974), the time-varying D-optimal input $x^*(\tau)$ is

$$x^*(\tau) = c\sin(\omega\tau + \kappa),$$

where c and κ are constants dependent on T and the frequency ω. The value for ω depends on the degree of correlation in the noise as expressed through the magnitude of c_1 (the "bandwidth" of the noise). Unlike some solutions for linear dynamic systems, the optimal input $x^*(\tau)$ above does not depend on θ. ❑

EXERCISES

17.1 As a reminder of the limitations of the D-optimal criterion, produce two positive definite 3×3 matrices $\boldsymbol{M}^*$ and $\boldsymbol{M}$ such that $\det(\boldsymbol{M}^*) > \det(\boldsymbol{M})$, but $\boldsymbol{M}^* \not\geq \boldsymbol{M}$.

17.2 Suppose that $\boldsymbol{\theta}' \equiv A\boldsymbol{\theta}$ for some invertible matrix A. With $\boldsymbol{M}(\boldsymbol{\theta}, \xi)$ proportional to $\boldsymbol{F}_n(\boldsymbol{\theta}, \boldsymbol{X}_n)$, show that the D-optimal design ξ_n^* is the same for $\boldsymbol{\theta}'$ and $\boldsymbol{\theta}$ ("transform invariance").

17.3 Prove the equality in (17.10).

17.4 For the standard linear model (17.5), suppose that each successive block of N measurements relies on the same inputs $\boldsymbol{x}_1, \boldsymbol{x}_2, \ldots, \boldsymbol{x}_N$, where $n \geq N$ and n is divisible by N. Show using (17.9) that it is sufficient to consider $n = N$ to prove D-optimality.

17.5 For the problem setting of Example 17.6 assume that the solution is unknown, but that one knows to use three levels of x with equal weight assigned to the three levels ($w_1 = w_2 = w_3$). Based on the D-optimal criterion, find by one of the direct search methods in Chapter 2 a close approximation to the solution, $[\chi_1, \chi_2, \chi_3] = [-1, 0, 1]$ given in the example (or one of the permutations of this solution). Use an initial guess of $[0, 0, 0]$ in the search.

17.6 Suppose that the error v_k in the linear regression model (17.5) is normally distributed. Show that the covariance matrix in (17.8) is equal to the inverse Fisher information matrix as defined in Section 13.3.

17.7 Consider a problem with a linear model having $p = 6$ and $n = 50$. Suppose that the optimal asymptotic design has been found to satisfy $\det \boldsymbol{M}(\xi_\infty^*) = 1.5$. Provide upper and lower bounds to $\det \boldsymbol{M}(\xi_n^*)$.

17.8 For the optimal finite-sample design in (17.15), plot the variance function $\boldsymbol{h}(x)^T \boldsymbol{M}(\xi_4^*)^{-1} \boldsymbol{h}(x)$. In contrast to Figure 17.1, note how the variance function is *not* maximized at the design points.

17.9 Given the problem in Example 17.7:

(a) Use Theorem 17.1 to show that the solution in Example 17.7 is not D-optimal for the full vector $\boldsymbol{\theta}$.

(b) Determine the D-optimal solution for $\boldsymbol{\theta}$.

17.10 Verify that the two $r = 3$ designs in Figure 17.3 satisfy the orthogonality condition in (17.16). Sketch and verify the same for at least one *other* design with $r = 3$.

17.11 Prove that $\log\det(\boldsymbol{M})$ is a strictly concave function on the domain of matrices $\boldsymbol{M} > \boldsymbol{0}$. (Hint: Recall from Appendix A that the basic definitions of convexity/concavity apply for matrix arguments in a scalar function; use matrix relationship (xxiv) in Appendix A.)

17.12 Provide several strategies for determining when to stop the search in step 4 of the RSM algorithm in Section 17.3 given that random noise is present in the response measurements (as usual).

17.13 For the dilution problem in Example 17.9:

(a) Derive the log-likelihood function $\log \ell(\theta | \boldsymbol{Z}_n)$.

(b) Derive the Fisher information $F_n(\theta, X_n)$. (Hint: A convenient form for the Bernoulli probability mass function is given in Example 15.4.)

17.14 Verify the optimal design in Example 17.10.

17.15 With the same problem structure, initial condition, gain sequence, and θ^* as Example 17.11, compare in the manner of Table 17.2 the sequential design and optimal ($\theta = \theta^*$) and suboptimal ($\theta = \hat{\theta}_0$) local designs for sample sizes of 200, 2000, and 20,000. Use an average of 60 replications as in Table 17.2. Show that the results here and in Table 17.2, taken together, indicate that the sequential and optimal local designs need only about half the number of measurements of the suboptimal design to achieve the same level of accuracy.

17.16 Given the finding in Table 17.2 that the mean estimation errors with the sequential design are about 2/3 the size of the errors under the suboptimal local design, present an informal analytical argument explaining why the sequential design needs only half the number of measurements of the suboptimal design to achieve the same level of accuracy. (Hint: Use the asymptotic normality of SA as given in Section 4.4.)

17.17 An alternative Bayesian *D*-optimal criterion to the one in (17.21) is $E_\theta\{\det[\boldsymbol{M}(\theta, \xi)]\}$ (see Walter and Pronzato, 1990; Haines, 1995). Consider the setting of Example 17.12 with the exception of changing the form of the *D*-optimal criterion from (17.21) to this alternative criterion. Suppose that $1 < \theta_2/\theta_1 \leq 2 + \sqrt{3}$, implying that there is only one support point for the design (as in Example 17.12).

(a) Derive the transcendental equation that produces the optimal design using standard arguments as in Example 17.12.

(b) For $\theta_1 = 1$, $\theta_2 = 2$, and $\rho = 0.3$, determine the support point numerically and verify analytically or numerically that this point is at least a local maximum of the appropriate *D*-optimal criterion.

(c) Contrast the solution in part (b) with the corresponding solution using criterion (17.21), as given in Example 17.12.

APPENDIX A

SELECTED RESULTS FROM MULTIVARIATE ANALYSIS

This appendix presents some important results from multivariate analysis that are used throughout the book. It is expected that readers have seen at least some of this material in one or more contexts. The references cited here cover the results in more detail, together with other relevant material.

A.1 MULTIVARIATE CALCULUS AND ANALYSIS

This section reviews some important results in multivariate calculus and basic real analysis. The results here are used in various guises throughout the book. The real number line is denoted $\mathbb{R}^1$. More generally, $\mathbb{R}^m$ denotes Euclidean space of dimension m (m a positive integer). Elements of $\mathbb{R}^m$ are m-dimensional vectors. For example, an element of $\mathbb{R}^2$ is a vector lying in the real number plane.

For a vector $\boldsymbol{x} = [x_1, x_2,\ldots, x_m]^T$ and differentiable function $f(\boldsymbol{x})$, the gradient $\partial f(\boldsymbol{x})/\partial \boldsymbol{x}$ is the column vector of elements $\partial f(\boldsymbol{x})/\partial x_i$, $i = 1, 2,\ldots, m$. The row-vector version of the gradient is $\partial f(\boldsymbol{x})/\partial \boldsymbol{x}^T = \partial f(\boldsymbol{x})/\partial(\boldsymbol{x}^T) = [\partial f(\boldsymbol{x})/\partial \boldsymbol{x}]^T$. The Hessian matrix $\partial^2 f(\boldsymbol{x})/\partial \boldsymbol{x}\partial \boldsymbol{x}^T$ is the $m \times m$ matrix of second derivatives $\partial^2 f(\boldsymbol{x})/\partial x_i \partial x_j$.

We sometime use "big-O" or "little-o" order notation. A function $f(x) = O(h(x))$ is such that $f(x)/h(x)$ is bounded in magnitude as $x \to x_0$, where x_0 is some relevant limiting value and $h(x)$ is some function. A function $f(x) = o(h(x))$ is such that $f(x)/h(x) \to 0$ as $x \to x_0$. In practice, x_0 is usually 0 or ∞.

An important result from multivariate calculus for the analysis and derivation of optimization algorithms is *Taylor's theorem* (e.g., Fleming, 1977, pp. 94–97). This can be presented in many essentially equivalent ways. Two are given below in terms of a loss function $L(\boldsymbol{\theta})$ and associated domain Θ. The concept of a *line segment* is used in the theorem statements below. A line segment connecting two points, say $\boldsymbol{\theta}$ and $\boldsymbol{\theta}'$, is the set of points satisfying $\lambda\boldsymbol{\theta} + (1 - \lambda)\boldsymbol{\theta}'$ as λ ranges over $[0, 1]$. Before presenting the results related to Taylor's theorem, recall the definition of the gradient and Hessian matrix: $\boldsymbol{g}(\boldsymbol{\theta}) = \partial L/\partial\boldsymbol{\theta}$ and $\boldsymbol{H}(\boldsymbol{\theta}) = \partial^2 L/\partial\boldsymbol{\theta}\partial\boldsymbol{\theta}^T$, as given in Section 1.3.

Theorem A.1 (Taylor's theorem; first-order mean-value theorem). Suppose that $\boldsymbol{g}(\boldsymbol{\theta})$ exists and is continuous in $\Theta \subseteq \mathbb{R}^p$. Let $\boldsymbol{\theta}'$ be a point in Θ. Then for all $\boldsymbol{\theta} \in \Theta$ such that all the points on the line segment connecting $\boldsymbol{\theta}'$ and $\boldsymbol{\theta}$ lie in Θ, we have

$$L(\boldsymbol{\theta}) = L(\boldsymbol{\theta}') + \boldsymbol{g}(\bar{\boldsymbol{\theta}})^T(\boldsymbol{\theta} - \boldsymbol{\theta}'),$$

where $\bar{\boldsymbol{\theta}} = \lambda\boldsymbol{\theta} + (1 - \lambda)\boldsymbol{\theta}'$ for some $\lambda \in [0, 1]$.

Theorem A.2 (Taylor's theorem; second-order form). Suppose that $\boldsymbol{H}(\boldsymbol{\theta})$ exists and is continuous in some region $\Theta \subseteq \mathbb{R}^p$. Let $\boldsymbol{\theta}'$ be a point in Θ. Then for all $\boldsymbol{\theta} \in \Theta$ such that all the points on the line segment connecting $\boldsymbol{\theta}'$ and $\boldsymbol{\theta}$ lie in Θ, we have

$$L(\boldsymbol{\theta}) = L(\boldsymbol{\theta}') + \boldsymbol{g}(\boldsymbol{\theta}')^T(\boldsymbol{\theta} - \boldsymbol{\theta}') + \tfrac{1}{2}(\boldsymbol{\theta} - \boldsymbol{\theta}')^T \boldsymbol{H}(\bar{\boldsymbol{\theta}})(\boldsymbol{\theta} - \boldsymbol{\theta}')$$

where $\bar{\boldsymbol{\theta}} = \lambda\boldsymbol{\theta} + (1 - \lambda)\boldsymbol{\theta}'$ for some $\lambda \in [0, 1]$.

The above pattern can be continued to create higher-order forms of Taylor's theorem (i.e., expansions where the last term in the expansion is in terms of the third-, fourth-, or higher-order derivatives of L). In the multivariate case of interest here, however, the notation for these higher-order expansions can get messy, although the concept is not in any essential way much more difficult than above (see, e.g., Fleming, 1977, p. 94).

A key reason for the importance of Taylor expansions in optimization is that for $\boldsymbol{\theta}$ near $\boldsymbol{\theta}'$ one can convert the possibly nasty nonlinear function L into a relatively simple polynomial function in $\boldsymbol{\theta}$. For example, if the algorithm search is at a point near an optimal value $\boldsymbol{\theta}^*$, then by taking the expansion point $\boldsymbol{\theta}'$ as equal to the current point in the search process, the loss function behaves essentially as a polynomial between the current point and the desired point $\boldsymbol{\theta}^*$. Theoretical analysis of algorithm behavior is much easier when considering low (say, first or second)-order polynomial loss functions than when considering arbitrary nonlinear loss functions. From the above theorems, the first- and second-order approximations to $L(\boldsymbol{\theta})$ relative to an expansion point $\boldsymbol{\theta}'$ are

$$L(\boldsymbol{\theta}) \approx L(\boldsymbol{\theta}') + \boldsymbol{g}(\boldsymbol{\theta}')^T(\boldsymbol{\theta} - \boldsymbol{\theta}'), \tag{A.1}$$

$$L(\boldsymbol{\theta}) \approx L(\boldsymbol{\theta}') + \boldsymbol{g}(\boldsymbol{\theta}')^T(\boldsymbol{\theta} - \boldsymbol{\theta}') + \tfrac{1}{2}(\boldsymbol{\theta} - \boldsymbol{\theta}')^T \boldsymbol{H}(\boldsymbol{\theta}')(\boldsymbol{\theta} - \boldsymbol{\theta}'). \tag{A.2}$$

Because of the inherent error in the approximations, the above are not generally recommended for direct application in the numerical process of finding a solution to the minimization problem unless $\boldsymbol{\theta}'$ is allowed to evolve in some appropriate way during the course of the iteration process. (For example, the evolution of $\boldsymbol{\theta}'$ is essential to the response surface method of Section 17.3.)

The errors in the polynomial approximations (A.1) and (A.2) are related to the last term in the expansion. So, for example, from Taylor's theorem A.2 above, we know that the error in approximation (A.2) follows from the difference in $\boldsymbol{H}(\boldsymbol{\theta}')$ and $\boldsymbol{H}(\bar{\boldsymbol{\theta}})$. This difference will be small when $\boldsymbol{\theta}$ is near $\boldsymbol{\theta}'$ (as assumed!) since $\boldsymbol{H}$ is assumed to be continuous in the region around $\boldsymbol{\theta}'$. Another way to look at this (which is equivalent under the assumption of continuous third derivatives existing) is by going to a third-order form where $\boldsymbol{H}$ is simply evaluated at the known point $\boldsymbol{\theta}'$ (instead of the unknown point $\bar{\boldsymbol{\theta}}$). In this case, the relevant *third derivative* terms are evaluated at an intermediate point (on the line segment between $\boldsymbol{\theta}$ and $\boldsymbol{\theta}'$). Then the error in approximating L as a quadratic function is of order $\|\boldsymbol{\theta}-\boldsymbol{\theta}'\|^3$ (i.e., $O(\|\boldsymbol{\theta}-\boldsymbol{\theta}'\|^3)$), where the multiplier on this normed error is derived from the third derivative terms and $\|\boldsymbol{\theta}-\boldsymbol{\theta}'\|$ is small by assumption.

Example A.1—First- and second-order Taylor approximation. Suppose that $\boldsymbol{\theta} = [t_1, t_2]^T$ and

$$L(\boldsymbol{\theta}) = t_1^2 + t_2^2 + \exp(-2t_1 - t_2).$$

Then,

$$\boldsymbol{g}(\boldsymbol{\theta}) = \begin{bmatrix} 2t_1 - 2\exp(-2t_1 - t_2) \\ 2t_2 - \exp(-2t_1 - t_2) \end{bmatrix},$$

$$\boldsymbol{H}(\boldsymbol{\theta}) = \begin{bmatrix} 2 + 4\exp(-2t_1 - t_2) & 2\exp(-2t_1 - t_2) \\ 2\exp(-2t_1 - t_2) & 2 + \exp(-2t_1 - t_2) \end{bmatrix}.$$

Relative to an expansion point $\boldsymbol{\theta}' = [0, 0]^T$, Table A.1 compares the first- and second-order approximations in (A.1) and (A.2) with the exact value of L for a selected set of $\boldsymbol{\theta}$ values.

As expected, the second-order approximation does a better job of approximating the function than the first-order approximation over the selected values of $\boldsymbol{\theta}$. The first-order approximation is only reasonable when $\boldsymbol{\theta}$ is very close to $\boldsymbol{\theta}'$. Both approximations are relatively better when $\boldsymbol{\theta}$ is closer to $\boldsymbol{\theta}'$. ❑

Convex sets and convex functions are important notions in classical optimization theory. The set Θ is convex if for any two points $\boldsymbol{\theta}_1, \boldsymbol{\theta}_2 \in \Theta$, all points on the line segment between $\boldsymbol{\theta}_1$ and $\boldsymbol{\theta}_2$ lie in Θ. A loss function L is convex on the convex set Θ if $L(\lambda\boldsymbol{\theta}_1 + (1-\lambda)\boldsymbol{\theta}_2) \le \lambda L(\boldsymbol{\theta}_1) + (1-\lambda)L(\boldsymbol{\theta}_2)$ for $0 \le$

Table A.1. Comparisons of first- and second-order Taylor approximations relative to expansion point $[0, 0]^T$.

$\boldsymbol{\theta} = [t_1, t_2]^T$	First-order (A.1)	Second-order (A.2)	$L(\boldsymbol{\theta})$
[0.1, 0.1]	0.700	0.765	0.761
[−0.1,− 0.1]	1.300	1.365	1.370
[0.5, 0.5]	−0.500	1.125	0.723
[−0.5, −0.5]	2.500	4.125	4.982
[0.5, 0]	0	0.750	0.618
[0, 0.5]	0.500	0.875	0.857

$\lambda \leq 1$. For smooth functions, one may think of convex functions as being multivariate "bowl-shaped." More formally, a differentiable L is convex on Θ if and only if

$$L(\boldsymbol{\theta}) \geq L(\boldsymbol{\theta}') + \boldsymbol{g}(\boldsymbol{\theta}')^T(\boldsymbol{\theta} - \boldsymbol{\theta}') \tag{A.3}$$

subject to $\boldsymbol{\theta}, \boldsymbol{\theta}' \in \Theta$. *Strict* convexity replaces "≥" with ">" in (A.3) when $\boldsymbol{\theta} \neq \boldsymbol{\theta}'$. The function L is *concave* (or strictly concave) if $-L$ is convex (strictly convex). While the above discussion is in terms of loss functions, the basic definitions apply for an arbitrary scalar-valued function of a scalar, vector, or matrix argument.

If L is continuously twice differentiable, then convexity (strict convexity) is equivalent to the Hessian $\boldsymbol{H}$ being positive semidefinite (positive definite) on Θ (see Section A.2). This connection to convexity is the basis for the important role that $\boldsymbol{H}$ plays in making the distinction between local minima and maxima for loss functions that are twice continuously differentiable, as discussed in Section 1.3. Figure A.1 depicts convex and nonconvex loss functions for a scalar θ.

Another important topic in this review of multivariate calculus and analysis is the interchange of derivative and integral. Consider an integral of the form

$$\int_{\Lambda} f(\boldsymbol{\theta}, \boldsymbol{x})\, d\boldsymbol{x}, \tag{A.4}$$

where Λ is the region over which the integral is to be computed and f is a function mapping the vectors $\boldsymbol{\theta}$ and $\boldsymbol{x}$ into $\mathbb{R}^1$ (so f is a scalar). Assume that f is differentiable with respect to $\boldsymbol{\theta}$ for any $\boldsymbol{x} \in \Lambda$. A central issue in several aspects of stochastic search and optimization (e.g., Chapters 5 and 15) is the question of

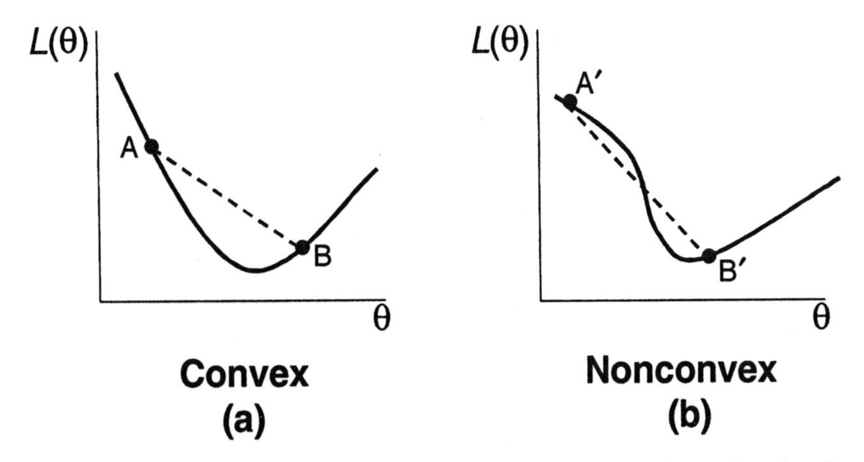

Figure A.1. Convex and nonconvex loss functions. In part (a), the function lies below any line segment connecting two points on the curve, such as shown in segment A−B. In part (b), the function lies above at least some points on some line segments, such as shown in segment A′−B′.

validity of the interchange of the order of integration and differentiation for the integral in (A.4):

$$\frac{\partial}{\partial \boldsymbol{\theta}} \int_{\Lambda} f(\boldsymbol{\theta}, \boldsymbol{x})\, d\boldsymbol{x} \stackrel{?}{=} \int_{\Lambda} \frac{\partial f(\boldsymbol{\theta}, \boldsymbol{x})}{\partial \boldsymbol{\theta}}\, d\boldsymbol{x}\,.$$

The following theorem provides sufficient conditions for this interchange to hold (i.e., for the above equality to hold). Let Θ be the domain for $\boldsymbol{\theta}$ (as introduced in Section 1.1). We use the fact that $\Theta \times \Lambda = \{(\boldsymbol{\theta}, \boldsymbol{x}): \boldsymbol{\theta} \in \Theta, \boldsymbol{x} \in \Lambda\}$. This theorem is a manifestation of the famous Lebesgue dominated convergence theorem, which provides conditions under which a limit may be brought through an integral (e.g., Fleming, 1977, p. 232); Appendix C, Subsection C.2.3, discusses this theorem in a different context. The limit in this case is associated with the definition of a gradient as a limit of finite differences for each element of $\boldsymbol{\theta}$. The proof of Theorem A.3 is given in Fleming (1977, pp. 237–238).

Theorem A.3 (Interchange of derivative and integral). Suppose that Θ is an open set. Let f and $\partial f/\partial \boldsymbol{\theta}$ be continuous on $\Theta \times \Lambda$. Suppose that there exist nonnegative functions $q_0(\boldsymbol{x})$ and $q_1(\boldsymbol{x})$ such that

$$\left| f(\boldsymbol{\theta}, \boldsymbol{x}) \right| \le q_0(\boldsymbol{x}),\ \left\| \frac{\partial f(\boldsymbol{\theta}, \boldsymbol{x})}{\partial \boldsymbol{\theta}} \right\| \le q_1(\boldsymbol{x}) \text{ for all } (\boldsymbol{\theta}, \boldsymbol{x}) \in \Theta \times \Lambda,$$

where $\int_\Lambda q_0(\boldsymbol{x})\,d\boldsymbol{x} < \infty$ and $\int_\Lambda q_1(\boldsymbol{x})\,d\boldsymbol{x} < \infty$.[1] Then

$$\frac{\partial}{\partial \boldsymbol{\theta}}\int_\Lambda f(\boldsymbol{\theta},\boldsymbol{x})\,d\boldsymbol{x} = \int_\Lambda \frac{\partial f(\boldsymbol{\theta},\boldsymbol{x})}{\partial \boldsymbol{\theta}}\,d\boldsymbol{x}.$$

Although the conditions of Theorem A.3 may appear rather abstract, there is a strong association with practical concerns in applications to stochastic search and optimization. In fact, it is not automatic that the conditions of Theorem A.3 hold in many practical problems. Chapters 5 and 15 consider this further. Examples A.2 and A.3 give simple illustrations where the interchange is valid and invalid, respectively.

Example A.2—Valid interchange of derivative and integral. Let $f(\theta, x) = \exp(-\theta x)/x$ for some scalar θ, x. Take the domains for θ and x to be $\Theta = (c, \infty)$, $c > 0$, and $\Lambda = [1, \infty)$, respectively. Note that $\partial f/\partial\theta = -\exp(-\theta x)$. The hypotheses of Theorem A.3 regarding boundedness of f and $\partial f/\partial\theta$ are satisfied with $q_0(x) = q_1(x) = \exp(-cx)$. Hence, Theorem A.3 implies that

$$\frac{\partial}{\partial\theta}\int_1^\infty \frac{e^{-\theta x}}{x}\,dx = -\int_1^\infty e^{-\theta x}\,dx = -\frac{e^{-\theta}}{\theta}.$$

Because $c > 0$ may be made arbitrarily small, the above formula holds for every $\theta > 0$. ❑

Example A.3—Invalid interchange of derivative and integral. The following example from Rudin (1976, pp. 242–243) shows that the interchange of derivative and integral may be invalid for relatively benign functions if at least one of the conditions in Theorem A.3 is violated. Let x and θ be scalars with

$$f(\theta, x) = \begin{cases} x, & 0 \le x \le \sqrt{\theta} \\ -x + 2\sqrt{\theta}, & \sqrt{\theta} \le x \le 2\sqrt{\theta} \\ 0, & \text{otherwise} \end{cases}$$

for $\theta \ge 0$. Let $f(\theta, x) = -f(|\theta|, x)$ for $\theta < 0$. Take the domains for θ and x to be $\Theta = (-1/4, 1/4)$ and $\Lambda = [-1, 1]$, respectively. While it can be shown by standard arguments that $f(\theta, x)$ is continuous on $\Theta \times \Lambda$ (see Exercise A.3), $\partial f(\theta,x)/\partial\theta$ is not continuous for all points in $\Theta \times \Lambda$. Hence, a condition of Theorem A.3 is violated.

---

[1] Strictly speaking, the functions $q_0(\boldsymbol{x})$ and $q_1(\boldsymbol{x})$ are required to be Lebesgue integrable, a subject we will not delve into. In only pathological cases (e.g., non-Lebesgue measurable functions) will this distinction be required relative to the bounded integral conditions in the theorem statement.

Let us now demonstrate that the derivative and integral are not interchangeable. It can be shown that $\int_{\Lambda} f(\theta, x)\,dx = \theta,\ \theta \in \Theta$. Further,

$$\left.\frac{\partial f(\theta, x)}{\partial \theta}\right|_{\theta=0} = 0$$

for all $x \in \Lambda$. (This is seen by noting that at $x = 0$ and $x \neq 0$, $[f(0+\delta\theta, x) - f(0, x)]/\delta\theta = 0$ for all $\delta\theta > 0$ sufficiently small.) Hence, at $\theta = 0$,

$$\frac{\partial}{\partial \theta}\int_{\Lambda} f(\theta, x)\,dx = 1 \neq \int_{\Lambda} \frac{\partial f(\theta, x)}{\partial \theta}\,dx = 0,$$

indicating that the derivative and integral cannot be interchanged. ❑

A final concept in this review of analysis methods is that of *infimum*. The infimum of a scalar-valued function $f(\boldsymbol{x})$ on a domain Λ is the *greatest lower bound* to the function on the domain. This is written as $\inf_{\boldsymbol{x}\in\Lambda} f(\boldsymbol{x})$. If a minimum on Λ is defined, the infimum is the same as the minimum. An infimum, however, is more general than a minimum. For example, with a scalar x, the function $f(x) = x$ has a well-defined minimum of 0 over the domain $\Lambda = [0, 1]$. If the domain is changed to $\Lambda = (0, 1]$, the function has no minimum. In the latter case, however, the greatest lower bound *is* defined, corresponding to the infimum (i.e., $\inf_{x\in(0,1]} f(x) = 0$). The *least upper bound*, the *supremum*, abbreviated $\sup_{\boldsymbol{x}\in\Lambda} f(\boldsymbol{x})$, is defined in an analogous way as a generalization of the concept of a maximum.

A.2 SOME USEFUL RESULTS IN MATRIX THEORY

This section summarizes a number of matrix properties that will be used (sometimes repeatedly) in this book. These are drawn from Bellman (1960), Mardia et al. (1979, App. A), Graybill (1983), Golub and Van Loan (1989), and Harville (2001). It is expected that the reader has previously seen most (but perhaps not all) of these relationships. Let $\boldsymbol{A}$ and $\boldsymbol{B}$ be general matrices, conformable as needed (e.g., dimensionally consistent so that, as needed, $\boldsymbol{AB}$ or $\boldsymbol{BA}$ are properly defined). Let $\boldsymbol{I}_p$ be the $p \times p$ identity matrix; $\boldsymbol{I}$ without a subscript is an identity matrix of unspecified dimension. All matrices and vectors in this section (and, more generally, in this book) will contain only real number entries; in particular, we do not consider matrices and vectors containing entries with complex numbers. Let $E(\cdot)$ and $\mathrm{cov}(\cdot)$ denote the expectation and covariance matrix of the argument random vector (formally defined in Section C.1 of Appendix C). Further, let $\boldsymbol{R}$ and $\boldsymbol{S}$ be *square* $p \times p$ matrices (not necessarily symmetric unless noted).

(i) $(\boldsymbol{AB})^T = \boldsymbol{B}^T\boldsymbol{A}^T$.

(ii) In general, $\boldsymbol{RS} \neq \boldsymbol{SR}$.

(iii) Let det(·) denote determinant. Then, $\det(\boldsymbol{RS}) = \det(\boldsymbol{R})\det(\boldsymbol{S})$ and $\det(\boldsymbol{S}^T) = \det(\boldsymbol{S})$.

(iv) The rank of $\boldsymbol{A}$ is the size of the largest nonsingular square submatrix of $\boldsymbol{A}$.

(v) $\text{rank}(\boldsymbol{AB}) \leq \min\{\text{rank}(\boldsymbol{A}), \text{rank}(\boldsymbol{B})\}$.

(vi) If $\boldsymbol{S}$ is full rank, the inverse $\boldsymbol{S}^{-1}$ exists and is the unique matrix satisfying $\boldsymbol{S}^{-1}\boldsymbol{S} = \boldsymbol{S}\boldsymbol{S}^{-1} = \boldsymbol{I}_p$. The converse also holds (invertibility implies full rank).

(vii) If $\boldsymbol{S}$ and $\boldsymbol{R}$ are full rank, $(\boldsymbol{RS})^{-1} = \boldsymbol{S}^{-1}\boldsymbol{R}^{-1}$.

(viii) The trace of $\boldsymbol{S}$, denoted trace($\boldsymbol{S}$), is the sum of the diagonal elements of $\boldsymbol{S}$.

(ix) $\det(\boldsymbol{S} - \lambda\boldsymbol{I}_p)$ represents a pth-order polynomial in the scalar λ. The p roots of $\det(\boldsymbol{S} - \lambda\boldsymbol{I}_p) = 0$, denoted $\lambda_1, \lambda_2, \ldots, \lambda_p$, are the possibly complex eigenvalues of $\boldsymbol{S}$.

(x) At least one eigenvalue of $\boldsymbol{S}$ is equal to zero if and only if $\boldsymbol{S}$ is singular (i.e., $\boldsymbol{S}^{-1}$ does not exist).

(xi) $\boldsymbol{S}$ is symmetric if $\boldsymbol{S} = \boldsymbol{S}^T$.

(xii) The eigenvalues of a symmetric matrix $\boldsymbol{S}$ are real numbers; the rank of $\boldsymbol{S}$ is the number of nonzero eigenvalues.

(xiii) A symmetric matrix $\boldsymbol{S}$ is positive definite (positive semidefinite) if $\boldsymbol{x}^T\boldsymbol{S}\boldsymbol{x} > 0$ ($\boldsymbol{x}^T\boldsymbol{S}\boldsymbol{x} \geq 0$) for all conformable $\boldsymbol{x} \neq \boldsymbol{0}$. We write $\boldsymbol{S} > \boldsymbol{0}$ or $\boldsymbol{S} \geq \boldsymbol{0}$ for (symmetric) $\boldsymbol{S}$ positive definite or positive semidefinite. Likewise, $\boldsymbol{S} > \boldsymbol{R}$ or $\boldsymbol{S} \geq \boldsymbol{R}$ if $\boldsymbol{S} - \boldsymbol{R}$ is positive definite or positive semidefinite.

(xiv) A necessary and sufficient condition for $\boldsymbol{S} > \boldsymbol{0}$ (or $\boldsymbol{S} \geq \boldsymbol{0}$) is that the eigenvalues $\lambda_i > 0$ (or $\lambda_i \geq 0$) for all i.

(xv) If $\boldsymbol{S} > \boldsymbol{0}$ (or $\boldsymbol{S} \geq \boldsymbol{0}$), then the diagonal elements of $\boldsymbol{S}$ are strictly positive (or nonnegative).

(xvi) If $\boldsymbol{S} > \boldsymbol{R} > \boldsymbol{0}$ (or $\boldsymbol{S} \geq \boldsymbol{R} > \boldsymbol{0}$), then $\boldsymbol{S}^{-1} < \boldsymbol{R}^{-1}$ (or $\boldsymbol{S}^{-1} \leq \boldsymbol{R}^{-1}$).

(xvii) If $\boldsymbol{S} > \boldsymbol{0}$ and $\boldsymbol{R} \geq \boldsymbol{0}$, then $\boldsymbol{S} + \boldsymbol{R} > \boldsymbol{0}$.

(xviii) If $\boldsymbol{S} \geq \boldsymbol{0}$, there is a *unique* positive semidefinite symmetric square root $\boldsymbol{S}^{1/2}$ such that $\boldsymbol{S} = \boldsymbol{S}^{1/2}\boldsymbol{S}^{1/2}$. (Without the stated restrictions on the square root, there are, in general, many matrices $\boldsymbol{R}$ such that $\boldsymbol{S} = \boldsymbol{R}^2$.)

(xix) If $\boldsymbol{S} \geq \boldsymbol{0}$, then there is at least one matrix $\boldsymbol{R}$ such that $\boldsymbol{S} = \boldsymbol{R}^T\boldsymbol{R}$. Three common examples of such $\boldsymbol{R}$ are the symmetric square root (see (xviii) above), the singular value decomposition, and the Cholesky factorization. (These are computed in MATLAB by the **`sqrtm`**, **`svd`**, and **`chol`** functions, respectively.)

(xx) For a random vector $\boldsymbol{x}$, $E(\boldsymbol{Ax}) = \boldsymbol{A}E(\boldsymbol{x})$. Let $\boldsymbol{S} \geq \boldsymbol{0}$ be the covariance matrix of $\boldsymbol{x}$. Then $\text{cov}(\boldsymbol{Ax}) = \boldsymbol{ASA}^T$.

(xxi) Let $\boldsymbol{A} = [a_{ij}]$. A Kronecker product of a $p \times q$ matrix $\boldsymbol{A}$ and $r \times s$ matrix $\boldsymbol{B}$ is the $pr \times qs$ matrix $\boldsymbol{A}\otimes\boldsymbol{B} = [a_{ij}\boldsymbol{B}]$, $i = 1, 2, \ldots, p$ and $j = 1, 2, \ldots, q$. See Example A.5 for an illustration of the increase in dimension.

(xxii) $(\boldsymbol{R}+\boldsymbol{ASB})^{-1} = \boldsymbol{R}^{-1} - \boldsymbol{R}^{-1}\boldsymbol{A}(\boldsymbol{BR}^{-1}\boldsymbol{A}+\boldsymbol{S}^{-1})^{-1}\boldsymbol{BR}^{-1}$, given that the indicated inverses are assumed to exist (sometimes called the *matrix inversion lemma* or the *Sherman–Morrison–Woodbury formula*).

(xxiii) Let $\boldsymbol{S}$ be symmetric. Then, $\partial(\boldsymbol{x}^T\boldsymbol{Sx})/\partial\boldsymbol{x} = 2\boldsymbol{Sx}$ given that $\boldsymbol{S}$ does not depend on $\boldsymbol{x}$.

(xxiv) If $\boldsymbol{S} > \boldsymbol{0}$ and $\boldsymbol{R} > \boldsymbol{0}$, then $\det[\alpha\boldsymbol{S} + (1-\alpha)\boldsymbol{R}] \geq (\det[\boldsymbol{S}])^{\alpha}(\det[\boldsymbol{R}])^{1-\alpha}$ for $0 < \alpha < 1$. Equality holds only if $\boldsymbol{S} = \boldsymbol{R}$.

(xxv) Let $\max_i|\lambda_i| < 1$, where λ_i is the ith eigenvalue of $\boldsymbol{S}$ (λ_i may not be a real number). Then, $(\boldsymbol{I}_p - \boldsymbol{S})^{-1}$ exists and $(\boldsymbol{I}_p - \boldsymbol{S})^{-1} = \boldsymbol{I}_p + \sum_{i=1}^{\infty}\boldsymbol{S}^i$.

Let us offer a few comments about the above. From relationship (v), any matrix factor $\boldsymbol{R}$ (as in (xix)) is of full rank if $\boldsymbol{S} > \boldsymbol{0}$. The singular value decomposition is useful in general matrix factorization, even for matrices that are not square. For positive definite matrices, this decomposition yields a matrix factor of the form $\boldsymbol{R}$ in property (xix). While the matrix inverse in (vi) is a useful theoretical concept, it is generally not recommended (or needed) in practical numerical analysis. Computations involving an inverse are generally much slower and more numerically unstable than other equivalent means. For example, in solving linear equations, methods based on matrix factorization (e.g., the *singular value decomposition* or the *Q-R factorization*) are usually recommended. So, for example, although one may *write* $\boldsymbol{S}^{-1}\boldsymbol{R}$ as representative of a term needed in some analysis, it is not necessary to explicitly compute $\boldsymbol{S}^{-1}$ in calculating $\boldsymbol{S}^{-1}\boldsymbol{R}$. Relationship (xxii) has major applications in control, signal processing, and linear systems analysis (such as in the derivation of the Kalman filter). It is used in several instances in Chapter 3.

Example A.4—Matrix square root. One useful special case of the covariance matrix relationship (xx) is when $\boldsymbol{S}$ is full rank (i.e., $\boldsymbol{S} > \boldsymbol{0}$) and $\boldsymbol{B}$ is square and satisfies $\boldsymbol{B} = (\boldsymbol{R}^T)^{-1}$ for any valid matrix factor satisfying $\boldsymbol{R}^T\boldsymbol{R} = \boldsymbol{S}$ (relationship (xix)). Then, cov($\boldsymbol{Bx}$) $= \boldsymbol{BSB}^T = (\boldsymbol{R}^T)^{-1}\boldsymbol{R}^T\boldsymbol{R}\boldsymbol{R}^{-1} = \boldsymbol{I}_p$. Hence, choosing $\boldsymbol{B} = (\boldsymbol{R}^T)^{-1}$ for any factor $\boldsymbol{R}$ allows one to standardize a random vector to have an identity covariance matrix, analogous to the well-known process of dividing a scalar random variable by the standard deviation to create a normalized random variable having variance of one. ❑

Example A.5—Kronecker product. Suppose that

$$\boldsymbol{A} = \begin{bmatrix} 0 \\ 1 \end{bmatrix} \text{ and } \boldsymbol{B} = \begin{bmatrix} 1 & 2 & 3 & 4 \\ 5 & 6 & 7 & 8 \end{bmatrix}.$$

Then,

$$A \otimes B = \begin{bmatrix} 0 & 0 & 0 & 0 \\ 0 & 0 & 0 & 0 \\ 1 & 2 & 3 & 4 \\ 5 & 6 & 7 & 8 \end{bmatrix}.$$

Unlike standard matrix multiplication, there is no requirement that the matrix dimensions be conformable. The Kronecker product is defined for any two matrices. ❑

EXERCISES (matrix properties (i) – (xxv) may be used as needed)

A.1 For the loss function in Example A.1, verify the calculations for $\boldsymbol{g}(\boldsymbol{\theta})$ and $\boldsymbol{H}(\boldsymbol{\theta})$ and assess (in the manner of Table A.1) the accuracy of the first- and second-order approximations for $\boldsymbol{\theta} = [0.05, -0.05]^T$ and $\boldsymbol{\theta} = [0.5, -0.5]^T$ (use $\boldsymbol{\theta}' = [0, 0]^T$, as in Example A.1).

A.2 Show that the loss function in Example A.1 is strictly convex over the domain $\Theta = \mathbb{R}^2$.

A.3 **(a)** Prove that $f(\theta, x)$ in Example A.3 is continuous on $\Theta \times \Lambda$. (Hint: Recall that a function is continuous at a point if and only if the limit of the function as the argument approaches the point is the same as the value of the function at the point.)[2]

(b) Plot $f(\theta, x)$ on $\Theta \times \Lambda$ and comment on how this illustrates the reason that the derivative and integral cannot be interchanged.

A.4 Let $A = \begin{bmatrix} 0 & 1 & 2 \\ 3 & 4 & 5 \end{bmatrix}$ and $B = [1 \;\; 2 \;\; 3]^T$. Compute "by hand" (i.e., not via MATLAB or other package) AB, $A \otimes B$, $B \otimes A$, and $A^T \otimes B^T$. Note that $(A \otimes B)^T = A^T \otimes B^T$ (this relationship holds more generally for any A and B).

A.5 As an illustration of matrix property (v), give examples of two matrices A and B such that:

(a) $\text{rank}(AB) < \min\{\text{rank}(A), \text{rank}(B)\}$.

(b) $\text{rank}(AB) = \min\{\text{rank}(A), \text{rank}(B)\}$.

A.6 For symmetric matrices, Q, R, and S, show that if $Q \geq R > 0$ and $R \geq S > 0$, then $Q^{-1} \leq S^{-1}$.

A.7 Consider the vector $t = (A^T A)^{-1} A^T x$, where x is a random vector with $\text{cov}(x) = \sigma^2 I$ (where σ is a scalar) and A is a nonrandom matrix such that the indicated inverse exists. Show that $\text{cov}(t) = (A^T A)^{-1}\sigma^2$. (With the appropriate interpretation of A and x, the vector t is the ordinary least-squares estimate of the parameters in a linear regression model. Chapters 3, 13, and 17 discuss properties and applications of this estimate and the associated covariance matrix.)

---

[2] We saw, however, that this continuity is not sufficient to guarantee the interchange of derivative and integral.

APPENDIX B

SOME BASIC TESTS IN STATISTICS

In many places in this book—and more broadly in the analysis of general stochastic systems—one is interested in determining whether a large or small value of a random quantity points toward something significant in the system under study. Some of the formal tests in statistics can play a useful role in carrying out such analysis. This appendix reviews several of the classical tests from statistics that will be most useful to us in the implementation and analysis of algorithms for stochastic search and optimization. It is assumed that the reader has previously been exposed to basic statistical testing; this appendix is not intended to be an introduction to statistics.

We begin in Section B.1 with a review of basic one-sample tests. The main emphasis in this book, however, is on two-sample tests. Section B.2 summarizes some of the most popular two-sample tests, which are versions of the conventional t-test, and presents several examples of their application. Section B.3 provides a brief discussion of tests other than the conventional t-tests.

B.1 STANDARD ONE-SAMPLE TEST

This short section summarizes the classical z- and t-tests from statistics. Suppose that there is a sequence of n independent, identically distributed (i.i.d.) scalar measurements, say $\{X_1, X_2, \ldots, X_n\}$. Further, suppose that there is a desire to use the measurements to perform tests on the process mean $\mu \equiv E(X_i)$. For example, an analyst may be interested in a one-sided test of the null hypothesis $\mu \leq 0$ versus the alternative hypothesis $\mu > 0$ or interested in a two-sided test of $\mu = 0$ versus $\mu \neq 0$. The classical z- and t-statistics for testing hypotheses about μ are

$$z = \frac{\overline{X} - \mu}{\sigma/\sqrt{n}} \text{ and } t = \frac{\overline{X} - \mu}{s/\sqrt{n}}, \tag{B.1}$$

where $\overline{X}$ is the sample mean of the n measurements, σ is the standard deviation of the X_i, and s is the sample standard deviation taken as the square root of

$$s^2 \equiv \frac{1}{n-1}\sum_{i=1}^{n}(X_i - \bar{X})^2.$$

The above estimate (with division by one less than the sample size) is called the *sample variance estimate* in the statistics literature. When the underlying data are normally distributed, the z- and t-statistics will, respectively, have a $N(0, 1)$ and Student-t distribution (or, simply, t-distribution) with $n - 1$ degrees of freedom.

The classical hypothesis testing formulation is based on whether z or t in (B.1), as appropriate, is small or large in magnitude relative to what is expected based on the $N(0, 1)$ or t-distribution. Large magnitudes in z or t support the *rejection* of a null hypothesis that $\mu = 0$. We illustrate some tests in Section B.2 for the important two-sample problem that arises frequently in, among other problems, the comparison of stochastic algorithms. These tests represent implementations of the basic t-statistic above when the data represent samples from two populations. Chapter 12 also focuses on extensions of the standard one- and two-sample tests, considering general *multiple* comparisons, as arise in optimization over a set of discrete outcomes.

The z-statistic in (B.1) (with its $N(0, 1)$ distribution) is used when σ^2 is known or when the sample size n is sufficiently large so that, for all practical purposes, the difference between the estimate s^2 and the true variance σ^2 is small enough so that the distributions of z and t are effectively identical. If the data are normally distributed, this number is often taken as $n = 60$, but lower or higher sample sizes are sometimes appropriate, depending on the requirements for accuracy in the statistical test.

Although the z- and t-test statistics in (B.1) are formally based on the X_i being normally distributed, both test statistics are widely used in nonnormal situations (e.g., Mardia et al., 1979, p. 147; Montgomery and Runger, 1999, p. 388). This will be the usual case in applications in this book, as there will rarely be evidence that the data entering the test statistics are normally distributed. In large samples, central limit theory (Appendix C) implies that $\sqrt{n}(\bar{X} - \mu)$ will be close to normally distributed, even when the X_i are far from normally distributed. This provides the key rationale for using z- and t-tests with nonnormal data (in the t-test, the fact that $s^2 \to \sigma^2$ almost surely [a.s.; see Appendix C] as $n \to \infty$ is also used). Nevertheless, if the true distribution of the data is seriously nonnormal and the sample size n is relatively small, one should exercise caution in the interpretation of the results, especially if the inference leads to some "close calls."

One of the most important concepts in classical (non-Bayesian) statistical testing is the *P-value* (probability value) associated with an experiment. The P-value is the probability that a future experiment (if conducted) would have a value at least as extreme as that observed in the current experiment. An equivalent definition, in the language of hypothesis testing, is that the P-value is the smallest level of significance that an experimenter would be able to use in rejecting a null hypothesis on the basis of the observed data.

So, for example, if the z-statistic from (B.1) yielded the value 1.96, the P-value for a two-sided test (null hypothesis of $\mu = 0$ versus an alternative of $\mu \neq 0$) would be 0.05, as available in any table of the normal distribution. That is, $z = 1.96$ is not large enough to reject the null hypothesis in a level 0.04 test, but is large enough to cause rejection at probability levels of at least 0.05. More generally, the P-value provides information beyond the simple binary "in/out" statement of whether the z- or t-test statistic lies in or out of some specified acceptance region for a hypothesis test. The P-value provides a measure of the *extent* to which the statistic appears inconsistent with the null hypothesis.

The following example illustrates the computation of a P-value for a t-test.

Example B.1—*P*-value in a *t*-test. Suppose that an experiment is conducted with i.i.d. data $\{X_i\}$ where $n = 40$ and σ^2 is unknown. The null hypothesis is $\mu \leq 0$; hence there is a *one-sided* alternative hypothesis $\mu > 0$. Using the formula in (B.1), a test statistic value $t = 2.20$ is calculated based on the 40 measurements and the most positive value of μ under the null hypothesis (i.e., $\mu = 0$). From the TDIST function in the MS EXCEL spreadsheet, the probability of a future value of t at least as extreme as the observed value (i.e., $t \geq 2.20$ based on $n - 1 = 39$ degrees of freedom) under the *additional* assumption that the X_i are normally distributed is the P-value of 0.017.

This P-value is much more informative than saying simply that the null hypothesis is rejected at one of the typical probability levels of 0.05 or 0.10. For instance, at a test level of 0.05, the P-value is only about 1/3 of the allowable probability that would result in rejection of a null hypothesis, providing stronger evidence against the null hypothesis. The P-value is also useful in settings where the data may not be normally distributed. A very small P-value provides evidence against the null hypothesis that compensates for doubts about the true distribution(s) for the $\{X_i\}$. ❑

A popular method for expressing uncertainty in an estimate is through the use of a *confidence interval*. A 95 percent confidence interval, for example, is known to contain the true value of quantity being estimated with probability 0.95. Note that the interval is a *random* quantity, while the true value is a constant. Although the concept is quite general, our interests focus on intervals for means for i.i.d. data, as derived from the t-distribution. Let $t_\nu^{(\alpha/2)} > 0$ denote the *critical value* for a t-distribution based on ν degrees of freedom and probability value α. That is, $t_\nu^{(\alpha/2)}$ denotes the value such that the probability of a t-distributed random variable being greater than $t_\nu^{(\alpha/2)}$ is $\alpha/2$. By the symmetry of the t-distribution, the probability of the *magnitude* of a t-distributed random variable being greater than the critical value $t_\nu^{(\alpha/2)}$ is α. From (B.1), the *random* interval $\left[\bar{X}-t_{n-1}^{(\alpha/2)}\, s/\sqrt{n}\,,\ \bar{X}+t_{n-1}^{(\alpha/2)}\, s/\sqrt{n}\right]$ is a level $1-\alpha$ confidence interval for the mean μ from i.i.d. data (consistent with the discussion above, this is only an *approximate* interval for nonnormally distributed data).

B.2 SOME BASIC TWO-SAMPLE TESTS

Is algorithm 1 better than algorithm 2 on problems of interest? Does one candidate simulation model provide more accurate predictions than another model? Will one set of input values for an experiment provide the same quality of estimate as another set of inputs? We will often be interested in answering such questions for the stochastic algorithms here, especially in comparing two search and optimization algorithms. In particular, despite the limitations of numerical testing discussed in Subsection 1.2.1, it is frequently convenient to attempt to answer such questions by running a computer experiment where both competitors are tested in a common setting.

With appropriate definitions and interpretation, the basic *t*-test outlined in Section B.1 (nominally for one population) provides a framework for testing for differences between *two* competing populations. In an experiment, suppose that we collect two sets of outcomes, say $\{X_i\}$ and $\{Y_i\}$, with each set representing a sequence of scalar i.i.d. measurements on one of the competitors. Although i.i.d. within each sequence, the two sequences may not be independent of each other (see the matched pairs test below). Our aim is to test whether $\mu_X \equiv E(X_i)$ differs significantly from $\mu_Y \equiv E(Y_i)$. This is a job for a two-sample test.

Suppose, for example, that we are interested in answering the above-mentioned question on whether algorithm 1 is better than algorithm 2. Then, μ_X and μ_Y will typically be the means of the loss function measurements L at the final parameter estimates $\boldsymbol{\theta} = \hat{\boldsymbol{\theta}}_k$ for the two algorithms. That is, the means will be $\mu_X = E[L(\hat{\boldsymbol{\theta}}_k)\,|\,\text{algorithm 1}]$ and $\mu_Y = E[L(\hat{\boldsymbol{\theta}}_k)\,|\,\text{algorithm 2}]$ (k might be different in algorithms 1 and 2, depending on how many iterations for each algorithm comprise a fair comparison).

Two-sample tests come in several varieties, and it is assumed that the reader has been exposed to such tests before. These are taught as part of most (all?) introductory statistics courses and are included as part of all modern spreadsheet software (e.g., in the data analysis tools of the MS EXCEL spreadsheet under *z*-test or *t*-test). We summarize these tests very briefly here because they will be used repeatedly in this book, not only in the evaluation of algorithms, but also as a basis for the multiple comparisons approach to optimization in Chapter 12. More thorough treatments are available in introductory textbooks, such as Anderson et al. (1991, Chap. 10) or Montgomery and Runger (1999, Chap. 9).

Consider the typical case where the variances of the scalar random variables X_i and Y_i are unknown, leading to tests based on the *t*-distribution. Although such tests are formally based on the X_i and Y_i being normally distributed, they are often successfully applied in settings where the population has a nonnormal distribution, as mentioned above for the one-sample test. Let σ_X^2 and σ_Y^2 be the unknown variances for the X_i and Y_i. There are three basic cases for the pairwise comparisons that are of interest to us:

(1) ***Matched pairs tests,*** where the data are collected in n (say) pairs (X_i, Y_i) and where each of the pairs share some underlying randomness. In our setting, this will usually be encountered when the *same* random numbers from the underlying random number generator are used in generating the two sequences $\{X_i\}$ and $\{Y_i\}$. Each of the sequences differ only as a result of two competing algorithms processing the common set of random numbers in a different way. The two sequences may be highly dependent. In attempting to understand inherent differences in the processes generating the data X_i and Y_i, it is usually advantageous to work with matched pairs instead of the independent pairs in cases (2) and (3) below. Having shared randomness reduces the variability due to effects *other* than those of inherent interest, allowing one to better focus on fundamental concerns, such as whether algorithm 1 is better than algorithm 2. (Incidentally, this notion of shared randomness is used to considerable advantage in simulation-based optimization, as discussed in Section 14.4 with common random numbers.)

(2) ***Unmatched pairs tests with common variance***, where the two sequences are independent, but where the unknown variances are equal, $\sigma_X^2 = \sigma_Y^2$. Such tests arise for us, for example, when it is not possible to "share" random numbers as in the matched pairs test above (e.g., as when each of the competing algorithms are tested by different individuals or organizations). The common variance is useful in allowing us to pool the two samples, leading to a larger effective sample for estimating the variance. Note that the two sample sizes, say n_X and n_Y, are not necessarily equal (e.g., when the cost of testing one algorithm exceeds the cost of testing the other algorithm, it may be possible to obtain a larger sample of the cheaper algorithm).

(3) ***Unmatched pairs tests with different variances,*** where the two sequences are as in case (2), except that $\sigma_X^2 \neq \sigma_Y^2$. This may arise when two candidate algorithms being tested differ in a fundamental way in their processing of the underlying randomness. Unlike case (2), it is not valid to pool the data for doing the variance estimation.

Variance estimates of some sort are fundamental to the t-tests mentioned above. For the matched pairs test, the variance estimate is on the differences $X_i - Y_i$ since the sequence of n differences will be i.i.d. For this case, the variance estimate is

$$s_{X-Y}^2 \equiv \frac{1}{n-1}\sum_{i=1}^{n}\left[X_i - Y_i - (\overline{X} - \overline{Y})\right]^2,$$

where $\overline{X}$ and $\overline{Y}$ denote the sample mean of the X_i and Y_i processes. For the unmatched pairs tests with or without a common variance, one can form the individual sample variance estimates

$$s_X^2 \equiv \frac{1}{n_X - 1}\sum_{i=1}^{n_X}(X_i - \overline{X})^2 \text{ and } s_Y^2 \equiv \frac{1}{n_Y - 1}\sum_{i=1}^{n_Y}(Y_i - \overline{Y})^2.$$

In the common variance case (case (2)), the two estimates are combined in a weighted (by relative sample size) fashion to form the following pooled variance estimate:

$$s_p^2 \equiv \frac{(n_X - 1)s_X^2 + (n_Y - 1)s_Y^2}{n_X + n_Y - 2}. \tag{B.2}$$

Given the variance estimates above and a specified false alarm rate α, we are now in a position to construct acceptance intervals that are guaranteed to contain the difference in means $\overline{X} - \overline{Y}$ with probability $1 - \alpha$ when there is no underlying difference in the true means μ_X and μ_Y. Hence, if an observed difference does not fall in the appropriate interval (depending on which one of the above three cases applies), there is evidence at level α that μ_X and μ_Y are *not* the same. As defined in Section B.1, $t_\nu^{(\alpha/2)} > 0$ denotes the critical value for a t-distribution based on ν degrees of freedom and probability value α. Then, for the three cases above, if there is no underlying difference in μ_X and μ_Y, the difference of sample means $\overline{X} - \overline{Y}$ should lie between the following upper and lower limits for symmetric two-sided acceptance intervals with probability $1 - \alpha$:

Case (1) (matched pairs): $\pm t_{n-1}^{(\alpha/2)}\left(\dfrac{s_{X-Y}^2}{n}\right)^{1/2}$. (B.3a)

Case (2) (identical variances): $\pm t_{n_X+n_Y-2}^{(\alpha/2)}\left(\dfrac{s_p^2}{n_X} + \dfrac{s_p^2}{n_Y}\right)^{1/2}$. (B.3b)

Case (3) (nonidentical variance): $\pm t_\nu^{(\alpha/2)}\left(\dfrac{s_X^2}{n_X} + \dfrac{s_Y^2}{n_Y}\right)^{1/2}$, (B.3c)

$$\text{where } \nu = \frac{\left(\dfrac{s_X^2}{n_X} + \dfrac{s_Y^2}{n_Y}\right)^2}{\dfrac{(s_X^2/n_X)^2}{n_X+1} + \dfrac{(s_Y^2/n_Y)^2}{n_Y+1}} - 2.$$

The bound in case (3)—with the rather ungainly expression for the degrees of freedom—is only an approximation. There is not an exact *t*-value available for the case of nonidentical variances because the distribution for $\overline{X} - \overline{Y}$ that is used in deriving the endpoints of the acceptance interval depends on the unknown ratio σ_X^2/σ_Y^2 (Bickel and Doksum, 1977, pp. 218–219). The form for the degrees of freedom in case (3) can be found, for example, in Montgomery and Runger (1999, pp. 392–393). Similar forms are available elsewhere; see, for example, Bickel and Doksum (1977, p. 219).

As discussed in Section B.1, one can also compute the *P*-value associated with an experiment, as a means of getting precise information on the strength at which the null hypothesis $\mu_X = \mu_Y$ can be rejected. The *P*-value provides information beyond whether the difference $\overline{X} - \overline{Y}$ lies in or out of an acceptance region by assigning a specific probability to the *extent* to which the difference is away from the nominal value of zero. The *P*-value is computed according to how far from zero the test statistic is relative to the *t*-distribution with the appropriate degrees of freedom as given in the bound above. In particular, the generic test statistic has the form

$$t = \frac{\overline{X} - \overline{Y}}{(\cdot)^{1/2}}, \tag{B.4}$$

where the $(\cdot)^{1/2}$ term represents the appropriate such term from one of the three bounds (B.3a, b, c) above (e.g., $(\cdot)^{1/2} = \left(s_{X-Y}^2/n\right)^{1/2}$ in case (1) associated with (B.3a)). The value of t in (B.4) has a *t*-distribution with the degrees of freedom indicated in (B.3a, b, c) as appropriate. Example B.2 illustrates this concept.

Example B.2—Implementation of *t*-test. Suppose that two optimization algorithms, one developed by the Hatfields and one developed by the McCoys, are being compared on a problem of the first type in Subsection 1.1.3 (Property A; i.e., a problem where only noisy measurements of L are available). The algorithms are run $n = 8$ times each, with all runs being statistically independent. It is assumed that the underlying (unknown) variance in the L measurements is the same for both algorithms (i.e., case (2) associated with (B.3b) applies).

Suppose further that the data are, at least approximately, normally distributed (so that the t-test formally applies). The terminal (noisy) loss values are as follows:

Run no.	Hatfield (H)	McCoy (M)
1	4.0	2.5
2	2.9	3.6
3	2.8	0.2
4	2.0	2.3
5	3.3	0.3
6	4.0	0.7
7	3.6	3.1
8	1.7	2.1

The above data yield the following sample means and standard deviations: $\overline{H} = 3.04$, $s_H = 0.860$, $\overline{M} = 1.85$, and $s_M = 1.296$. With $\alpha = 0.10$, $t_{n_H+n_M-2}^{(\alpha/2)} = t_{14}^{(0.05)} = 1.76$. The pooled standard deviation is $\sqrt{16.939/14} = 1.10$ (using (B.2)). Hence, the acceptance interval from (B.3b) is $[-0.97, 0.97]$. Because $\overline{H} - \overline{M} = 1.19 \notin [-0.97, 0.97]$, the null hypothesis of equality in means ($\mu_H = \mu_M$) can be rejected in favor of $\mu_H > \mu_M$ (i.e., the algorithm of the McCoys produces lower terminal loss values).

More precisely, the value of $t = 1.19/\sqrt{0.303} = 2.16$ from (B.4) has a two-sided P-value of 0.049 according to a t-distribution with 14 degrees of freedom (using the TDIST function in MS EXCEL). By traditional statistical measures this is only a moderately strong indication of rejection of the null hypothesis; tests that are more stringent usually require a P-value of less than 0.01 before a null hypothesis is rejected. Further, the data H and M may not, in fact, be normally distributed and the sample size is relatively small. So, the Hatfields may not yet cede to the McCoys! ❑

The next example illustrates the merits discussed above of using the matched pairs test in settings where it is appropriate. This is a harbinger of benefits to algorithm comparisons where it is possible to "match" randomness. A different example of the benefit of such matching is in common random numbers, as used in Monte Carlo simulations to reduce the variability of the output (Sections 14.4 and 14.5). The notation $N(\cdot,\cdot)$ represents a multivariate normal distribution with a specified mean vector and covariance matrix.

Example B.3—Implementation of matched pairs *t*-test. The following 10 data pairs were generated according to a $N\left(\begin{bmatrix} 2 \\ 2.5 \end{bmatrix}, \begin{bmatrix} 1 & 0.8 \\ 0.8 & 1 \end{bmatrix}\right)$ distribution:

i	X_i	Y_i
1	0.93	1.61
2	1.44	1.92
3	2.20	1.89
4	3.28	4.06
5	3.79	3.51
6	2.17	3.37
7	1.50	1.32
8	2.18	3.25
9	3.89	3.59
10	0.60	2.10

These correlated data represent a setting of matched pairs where the data tend to move together. For example, when X_i is above or below its mean, then Y_i is above or below its mean in the same way. (The correlation for the data equals $\text{cov}(X_i, Y_i)/(\sigma_X \sigma_Y) = 0.8/(1\times1) = 0.8$.)

Suppose that the process Y represents an existing (default) approach and there is interest in testing whether a new process X is superior in the sense that $\mu_X < \mu_Y$. This is a *one-sided* test where the null hypothesis (i.e., the nominal state to be *rejected*) is that $\mu_X \geq \mu_Y$. This is unlike the two-sided test in Example B.2, where there was not a default option to test against (so both options are a priori treated as equally likely to be the superior option). Let us compare the matched pairs (case (1)) and unmatched pairs (case (2)) tests. Of course, the correlated X_i and Y_i violate the independence condition of the unmatched pairs test; we are presenting results from this test here to illustrate what might happen if this test is *incorrectly* applied.

From the above data, we have $\overline{X} - \overline{Y} = -0.464$, $s_X = 1.145$, $s_Y = 0.986$, and $s_{X-Y} = 0.690$. From (B.4) this leads to *t*-statistics of -2.128 and -0.971 under case (1) and case (2), respectively. Then, using the TDIST function in MS EXCEL, this leads to corresponding *P*-values of 0.031 for the matched pairs case (1) (with 9 degrees of freedom) and 0.172 for the unmatched pairs case (2) (with 18 degrees of freedom). This clearly shows how the matched pairs test in case (1) leads to a sharper analysis when it is appropriate to use. Exercises B.4 and B.5 show, as expected, that the value of the matched pairs test is reduced when the data become more nearly independent. ❑

B.3 COMMENTS ON OTHER ASPECTS OF STATISTICAL TESTING

There are *many* commercial and noncommercial software packages available for implementing the procedures above (together with a vast number of other statistical procedures). A sampling of such packages is available by perusing the advertisements in any recent issue of *Amstat News* (the monthly publication of the American Statistical Association). MS EXCEL is one such package, notable for its popularity and relative ease of use. As discussed in the Preface, however, one should exercise caution in using EXCEL for certain statistical applications because of some documented failures (e.g., McCullough and Wilson, 2002). Although EXCEL is used for some of the numerical experiments in this book, the applications here are relatively benign and do not correspond to known weaknesses.

The philosophy and methods summarized in Sections B.1 and B.2 are from the *classical* (sometimes called *frequentist*) branch of statistics. An alternative means of performing analysis is the Bayesian approach. This has many advantages from the perspective of interpretation of the results. The Bayesian method, however, is not as commonly applied in practice. This appears to be largely for two reasons: (i) Difficulties in specifying the required probability distribution reflecting prior knowledge about a process (the prior distribution), and (ii) the computational demands, especially for problems with many parameters being estimated. The driving force behind the large computational demands is the need to carry out numerical integration in calculations related to the posterior distribution. (The Markov chain Monte Carlo sampling methods in Chapter 16 are sometimes effective in easing the computational burden.)

Despite the above difficulties, the Bayesian approach is ideally suited to many problems and has been widely used (although not as widely as the frequentist approach). A good introductory statistics textbook from the Bayesian view is Berry (1996).

Another general approach not considered here is *distribution-free inference*, sometimes called *nonparametric inference*. While the z- and t-tests discussed above are formally based on the underlying data being normally distributed, nonparametric statistical methods make almost no assumptions about the form of the distribution(s) for the underlying data. Despite their weak assumptions, some nonparametric methods (e.g., methods based on ranking the data from smallest to largest) can be surprisingly powerful in discerning deviations from the null hypothesis, with their efficiency being very close to tests based on parametric assumptions (such as normality).

For our purposes, however, the conventional t- (and z-) tests will usually be adequate, especially in light of the large-sample justification for these tests when the distributions of the underlying data are nonnormal. (Chapter 12 on multiple comparisons will, however, consider some distribution-free tests since the sample sizes in such problems are frequently small.) Lehmann (1975) and Hollander and Wolfe (1999) are good introductions to nonparametric methods.

EXERCISES

B.1 Suppose in a statistical experiment that $\overline{X} = 1.75$, $s^2 = 2.13$, and $n = 25$. Compute a symmetric 95 percent confidence interval using the standard t-distribution-based method. Without knowing more about the nature of the data, mention at least two reasons why this interval could be flawed.

B.2 Consider the formula for s^2 based on i.i.d. data, as given in Section B.1.

(a) Show that s^2 is an unbiased estimator of σ^2 (i.e., $E(s^2) = \sigma^2$).

(b) Show that s is a biased estimator of σ.

B.3 For the setting of Example B.2, suppose that it is *not* possible to assume that the underlying variances are equal. Calculate an approximate acceptance interval for the difference in sample means ($\alpha = 0.10$) and report a P-value for the observed $\overline{H} - \overline{M}$.

B.4 Transform the data in Example B.3 (i.e., do not generate a new independent data set) to represent a sample from a distribution with the same mean and variances as shown in the example, but with $\text{cov}(X_i, Y_i) = 0.2$ for all i (versus 0.8 in the example). In carrying out the transformation, use the *unique symmetric positive definite matrix square root* discussed in Appendix A (the matrix square root applies to the "before" and "after" covariance matrices of the data pairs). Compute P-values for the modified data under the matched and unmatched pairs settings. Show that the matched and unmatched results now differ little. (Hint: Using the matrix square root of the "before" covariance matrix, transform each of the 10 data vectors to represent a sample from a $N(\mathbf{0}, \boldsymbol{I}_2)$ distribution. Then, using the square root of the "after" covariance, transform the $N(\mathbf{0}, \boldsymbol{I}_2)$ data to have the required distribution. Example A.4 in Appendix A discusses such matrix transformations. Note that the mean should be subtracted before transforming the data.)

B.5 Carry out the tasks in Exercise B.4, except use the *Cholesky triangular factorization* (briefly discussed in Appendix A) as the needed matrix square root (versus the symmetric square root). In particular, compute P-values under the matched and unmatched pairs settings and show that the matched and unmatched results differ little.

B.6 In the two-sample problem, assume that $n_X = n_Y = n$ and suppose that the underlying variances satisfy $\sigma_X^2 >> \sigma_Y^2$.

(a) To the extent that the approximation in (B.3c) is correct, comment in qualitative terms on the error in the acceptance intervals for a two-sided test when one mistakenly assumes that the variances are identical, as in case (2) (i.e., (B.3b)). In particular, show that the intervals under (B.3b) will always be too narrow relative to the correct intervals under (B.3c).

(b) Provide quantification of the errors for $n = 10$ and $\alpha = 0.10, 0.05, 0.01, 0.005$, and 0.001.

APPENDIX C

PROBABILITY THEORY AND CONVERGENCE

...you can never know too much probability theory. If you are well grounded in probability theory, you will find it easy to integrate results from theoretical and applied statistics into the analysis of your applications.

—Daniel McFadden, recipient of 2000 Nobel Prize in Economics.

As a book devoted to stochastic algorithms, probability theory plays a central role. This appendix is a summary of some of the critical elements of probability theory, including the basics of measure-theoretic probability as applied to the convergence of random sequences. It is not possible to do formal convergence analysis of stochastic algorithms without such tools. Standard calculus arguments regarding the convergence of sequences and series are generally not sufficient in this stochastic environment. However, to be true to the intended level of the book, we will not delve into many of the subtleties of measure theory (fascinating though they are!).

Section C.1 provides some basic definitions related to random vectors and expectation. Section C.2 summarizes some key results in convergence.

C.1 BASIC PROPERTIES

As discussed in Chapter 1, to acquire objective insight into the relative behavior of stochastic algorithms, it is usually valuable to carry out both numerical (Monte Carlo) studies and theoretical analysis. Numerical experiments by Monte Carlo provide insight into practical finite-sample performance in *specific* problems while the theory provides general insight into *classes* of problems. The conditions of the theory can also point to restrictions in the range of problems for which a given algorithm should be used. This short section summarizes some of the essential elements of probability theory that are useful in the analysis of stochastic algorithms.

Consider an m-dimensional random vector $\boldsymbol{X}$. Formally, $\boldsymbol{X}$ may be viewed as a function of a point ω from an underlying sample space Ω. The point ω may be viewed as an experimental outcome in physical space and $X = X(\omega)$ may be viewed as the mapping of that experimental outcome into a space of mathematical convenience such as Euclidean space $\mathbb{R}^m$. For example, ω might represent the color of a liquid when a randomly chosen chemical is added, while X represents a particular *numerical* value for the color of the liquid on a chromatic scale. Hence, a probability statement such as $P(X \in \Lambda)$ for some set $\Lambda \subseteq \mathbb{R}^m$ may be written more formally as $P(\omega \in \Omega\colon X(\omega) \in \Lambda)$ (read as the probability of the subset of points ω in Ω such that the ω in that subset yield $X(\omega) \in \Lambda$). Note the extreme cases $P(\omega \in \Omega) = 1$ (the probability of *any* outcome) and $P(\omega \in \varnothing) = 0$ (the probability of *no* outcome).

Associated with any random vector X is a distribution function $0 \le F_X(x) \le 1$ representing the joint probability of the individual components in X being less than the corresponding components in the nonrandom vector x. Let the individual components of X and x be denoted by X_i and x_i, respectively. So, for example, if $X = [X_1, X_2]^T$ and $x = [x_1, x_2]^T$, then $F_X(x) = P(\omega \in \Omega\colon X_1(\omega) \le x_1, X_2(\omega) \le x_2)$. Note that $F_X(x)$ is a monotonically nondecreasing function as any component of x increases (or as multiple components increase).

An important special case of the distribution function $F_X(x)$ is when the underlying random vector X has a probability density function $p_X(x)$. Somewhat informally, X is often said to be continuously distributed in this case, in contrast to the discrete case, where elements of X can take only a countable number of values, such as the nonnegative integers. If a probability density exists and $X \in \mathbb{R}^m$, then

$$F_X(x_1, x_2, \ldots, x_m) = \int_{-\infty}^{x_1}\int_{-\infty}^{x_2}\cdots\int_{-\infty}^{x_m} p_X(q_1, q_2, \ldots, q_m)\, dq_1 dq_2 \cdots dq_m, \qquad \text{(C.1)}$$

where the components of x are $x_1, x_2, \ldots, x_m$ and the q_i are the corresponding dummy variables of integration.

Probably the most famous case of (C.1) is when X is (multivariate) normally distributed with mean vector μ and positive definite covariance matrix Σ (the distribution is denoted $N(\mu, \Sigma)$). Then the density function is

$$p_X(x) = \frac{1}{\sqrt{\det(2\pi\Sigma)}} \exp\left\{-\tfrac{1}{2}(x-\mu)^T \Sigma^{-1}(x-\mu)\right\}$$

for all $x \in \mathbb{R}^m$. A *key property* is that X is multivariate normally distributed if and only if all linear combinations $t^T X$ have a univariate normal distribution where t is a conformable deterministic vector. To allow for the case $t = \mathbf{0}$, we regard constants as degenerate forms of the normal distribution.

An interesting (and often forgotten) implication of this key property is that two (or more) normally distributed scalar random variables are not necessarily *jointly* (multivariate) normally distributed. Counterexamples illustrating this point are found by identifying at least one weighted sum of the variables under consideration that is not normally distributed (e.g., Romano and Siegel, 1986, p. 32–33; Stoyanov, 1997, pp. 87–100). Such counterexamples can occur for normal random variables that are dependent. Normal random variables that are mutually independent are, however, always jointly normally distributed by the well-known convolution property of the normal distribution (i.e., sums of mutually independent normal random variables are also normal). Another key property of joint normality is that for two random variables having a joint normal distribution, the random variables are independent if and only if they are uncorrelated (for other distributions, of course, uncorrelatedness does not generally imply independence).

The normal probability distribution is frequently seen as the limiting distribution for sequences that converge in distribution, as defined in Section C.2. Note, however, that there is no closed-form expression for the normal distribution function since the integral of the normal density $p_X(\boldsymbol{x})$ above (according to (C.1)) does not exist in closed form. An excellent compendium of additional facts about the multivariate normal distribution is provided in Mardia et al. (1979, Chap. 3). In our applications, we will usually drop the qualifier *multivariate* in referring to the distribution, simply calling it a normal distribution (the multivariate aspect being clear from the context).

A critical notion in probability is that of mathematical expectation. We will be most concerned with expectation for random variables that are purely discrete or purely continuous (in the sense that a probability density exists). Expectation can be defined much more generally than these two special cases, as discussed in Laha and Rohatgi (1979, Sect. 1.2) (e.g., for random variables that are hybrids of discrete and continuous). In defining the expectation of a discrete random vector, let p_i denote the probability of a discrete outcome $\boldsymbol{X} = \boldsymbol{x}(i)$. Of course, the discrete probabilities and the density function are normalized such that $\sum_i p_i = 1$ and $\int_{\mathbb{R}^m} p_X(\boldsymbol{x})d\boldsymbol{x} = 1$. For the discrete and continuous cases, the expectation of a possibly multivariate function $\boldsymbol{f}(\boldsymbol{X})$ of the random vector $\boldsymbol{X} \in \mathbb{R}^m$ is, respectively,

$$E[\boldsymbol{f}(\boldsymbol{X})] = \sum_i p_i \boldsymbol{f}(\boldsymbol{x}(i)), \tag{C.2}$$

$$E[\boldsymbol{f}(\boldsymbol{X})] = \int_{\mathbb{R}^m} \boldsymbol{f}(\boldsymbol{x}) p_X(\boldsymbol{x})d\boldsymbol{x}, \tag{C.3}$$

subject to the function $\boldsymbol{f}(\boldsymbol{x})$ being absolutely summable or integrable in the sense that $E[\|\boldsymbol{f}(\boldsymbol{X})\|] < \infty$ (to compute $E[\|\boldsymbol{f}(\boldsymbol{X})\|]$, $\|\boldsymbol{f}(\boldsymbol{X})\|$ replaces $\boldsymbol{f}(\cdot)$ in the sum or integral above). If the absolute summability or integrability condition is not satisfied, the expectation of $\boldsymbol{f}(\boldsymbol{X})$ does not exist (even if (C.2) or (C.3), as appropriate, is finite); see Example C.1 below. Perhaps the two most important functions $f(\cdot)$ are $f(X) = X \in \mathbb{R}^m$ and $f(X) = [X - E(X)][X - E(X)]^T \in \mathbb{R}^{m\times m}$, yielding the mean and covariance matrix of X, $E(X)$ and cov(X), respectively, when used in (C.2) or (C.3). Example C.1 presents a random variable for which the expected value of the variable (i.e., the mean) does not exist.

Example C.1—Random variable that does not have a mean. This example demonstrates that the computation of (C.2) or (C.3) alone is not sufficient to produce an expected value. Let the scalar discrete random variable X take on the values $x(i) = (-1)^i 2^i / i$, $i = 1, 2, \ldots$, with probabilities $p_i = 1/2^i$. Note that $\sum_{i=1}^{\infty} p_i = \frac{1}{2} + \frac{1}{4} + \frac{1}{8} + \ldots = 1$, as required. Then by (C.2),

$$\sum_{i=1}^{\infty} p_i x(i) = \sum_{i=1}^{\infty} \frac{(-1)^i}{i} = -\log 2$$

(here, and throughout the book, log($\cdot$) represents the *natural*, i.e., base e = 2.71828..., logarithm of x). However, this is not the expected value (the mean) of X since

$$\sum_{i=1}^{\infty} p_i |x(i)| = \sum_{i=1}^{\infty} \frac{1}{i} = \infty,$$

which violates the absolute summability condition. This random variable, therefore, does not have a mean (and consequently does not have any higher moments either; so the variance also does not exist). ❑

C.2 CONVERGENCE THEORY

C.2.1 Definitions of Convergence

It is usually hopeless to develop *finite-sample* theory related to the behavior of stochastic algorithms. Typically, the mathematics of finite-sample behavior is intractable. Hence, almost all of the relevant theory pertains to behavior *in the limit*, that is, as the amount of data and/or number of iterations goes to infinity. Many results related to the mathematical convergence properties of stochastic algorithms are presented in the main body of this book (Chapters 2–17), where convergence is in the sense of the input information (data and/or iterations) approaching infinity. Often, in concert with the implementation requirements and basic forms and assumptions for the algorithms, the convergence properties form the prime rationale for whether an algorithm should be applied to a specific problem. The limiting behavior of an algorithm can often shed critical light on the practical nonlimiting behavior in solving a real problem.

Because we are considering stochastic algorithms, the notion of convergence is *probabilistic*. This contrasts with the convergence arguments for deterministic techniques such as steepest descent and Newton–Raphson, as mentioned in Section 1.4. The probabilistic nature of the convergence indicates that measure-theoretic means are used to characterize the convergence. As suggested above, *the reader does not need to know measure theory beyond a few basic definitions and implications to understand most of the results in this book.* We give those here. In cases where more subtle implications of measure-theoretic probability might be useful, the reader will be directed to appropriate references. There are a large number of excellent textbooks that provide more extensive background on measure-theoretic probability (three "classics" are Chung, 1974; Laha and Rohatgi, 1979; and Chow and Teicher, 1997 [3rd ed.]).

The most common modes of probabilistic convergence are "almost surely (a.s.)," "in probability (pr.)," "in mean square (m.s.)," and "in distribution (dist.)." The first three of these pertain to a sequence of random vectors $\{X_k\}$ directly while the last one pertains to probability distributions associated with the sequence. The definitions of these modes of convergence are given below. When not specified, all limits are as $k \to \infty$. The indicated distance measure $\|\cdot\|$ denotes any valid norm (e.g., the classic Euclidean norm $\|x\| = \sqrt{\sum_i x_i^2}$, where x_i denotes the ith component of the vector x, as in Section C.1). The dependence of individual random vectors on ω as discussed in Section C.1, extends to sequences. Namely, an individual $\omega \in \Omega$ now generates the entire sequence (not just one random vector): $X_1(\omega), X_2(\omega), \ldots$. The limiting random variable (vector) $X = X(\omega)$ in the convergence below is often a constant in practice (e.g., with the constant being $\mathbf{0}$, a statement that $X_k \to X$ in some probabilistic sense reduces to $X_k \to \mathbf{0}$ in the same sense).

Convergence almost surely: $X_k \to X$ a.s. (or $X_k \to X$ with probability 1)

$$P\left(\omega \in \Omega: \lim_{k\to\infty} \|X_k(\omega) - X(\omega)\| = 0\right) = 1.$$

Convergence in probability: $X_k \to X$ in pr.

$$\lim_{k\to\infty} P\left(\omega \in \Omega: \|X_k(\omega) - X(\omega)\| \geq c\right) = 0 \text{ for any } c > 0.$$

Convergence in mean square: $X_k \to X$ in m.s.

$$\lim_{k\to\infty} E\left(\|X_k(\omega) - X(\omega)\|^2\right) = 0.$$

Convergence in distribution: $X_k \to X$ in dist.

$$\lim_{k\to\infty} F_{X_k}(x) = F_X(x) \text{ at every point } x \text{ where } F_X(x) \text{ is continuous.}$$

The interested reader may consult almost any graduate-level text on probability theory for a more complete discussion of these and related principles. From a practical point of view, the modes a.s., pr., and m.s. are expressing roughly the same idea: That the relevant process $X_k = X_k(\omega)$ closely approximates $X = X(\omega)$ as k gets large. The distinctions among these three modes, however, are not trivial. It is relatively easy to construct examples where a sequence converges in one mode but not another (see Example C.4). There are numerous practical problems in, for example, the field of stochastic processes where this convergence dichotomy occurs (e.g., Chung, 1974, pp. 68–69). Convergence in distribution is fundamentally different from the modes a.s., pr., and m.s., as it says nothing per se about X_k "settling down" near another random variable or constant. Rather it says that the *probability distributions* of X_k and X are close to each other for large k.

Figure C.1 shows the implications for convergence among the four modes above. Both of a.s. and m.s. convergence are stronger than pr. convergence; no general statement can be made about the relative strengths of m.s. and a.s convergence. If special conditions are introduced, additional implications are possible. (For example, with the additional requirements of the famous Lebesgue dominated convergence theorem, both a.s. and pr. convergence imply m.s. convergence; see Subsection C.2.3.)

We occasionally use the expression "for almost all" as in "for almost all sample points, such and such will happen." This standard expression in measure theory may apply to both finite-sample and asymptotic problems and implies that the "such and such" will happen for all ω except possibly those ω in a subset of Ω having probability zero. For asymptotic problems, this terminology corresponds to the a.s. convergence mode (where the limit exists for almost all ω).

When considering the convergence modes a.s., pr., and m.s. in the applications of this book, the iterate $\hat{\boldsymbol{\theta}}_k$—found as the estimate of $\boldsymbol{\theta}$ in the kth

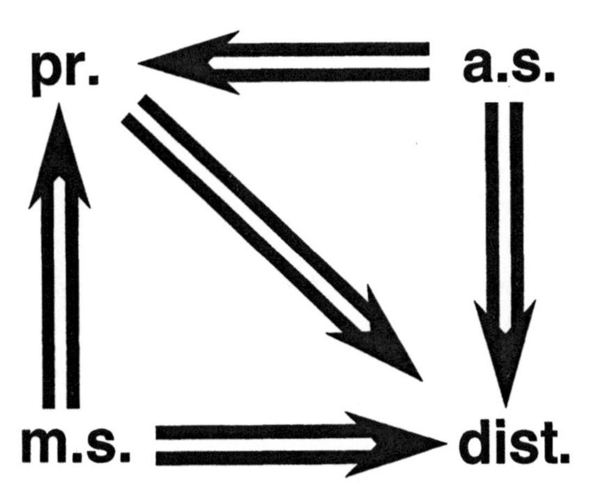

Figure C.1. Implications (indicated by $\Rightarrow$) among the four modes of convergence under general conditions. Stronger conditions allow for implications not shown.

iteration of a stochastic search or optimization algorithm—usually (but not always) plays the role of $\boldsymbol{X}_k$ above. Likewise, the unknown deterministic solution $\boldsymbol{\theta}^*$ to the search or optimization problem will replace the more general random limiting quantity $\boldsymbol{X}$. Hence, for instance, the general definition $\boldsymbol{X}_k \to \boldsymbol{X}$ a.s. above is reduced to the specific result $\hat{\boldsymbol{\theta}}_k \to \boldsymbol{\theta}^*$ a.s. in the search and optimization setting. A *single* sample point ω may be thought of as the sample value generating the *entire* set of individual random outcomes (the noise levels, the random search steps, etc.) that has led to the current iterate $\hat{\boldsymbol{\theta}}_k = \hat{\boldsymbol{\theta}}_k(\omega)$. Hence, if $\hat{\boldsymbol{\theta}}_k \to \boldsymbol{\theta}^*$ a.s., the probability is zero of encountering a combination of individual random outcomes such that $\hat{\boldsymbol{\theta}}_k$ does *not* get arbitrarily close to $\boldsymbol{\theta}^*$ as k gets large. An appealing aspect of a.s. convergence is that deterministic arguments (such as seen in basic calculus) may be applied for analyzing the convergence at each sample point. That is, for a fixed ω, the convergence of $\hat{\boldsymbol{\theta}}_k(\omega)$ can be analyzed deterministically.

C.2.2 Examples and Counterexamples Related to Convergence

Examples C.2, C.3, and C.4 demonstrate some of the convergence modes and illustrate—by counterexample—some of the relationships between the modes. Example C.2 concerns the *law of large numbers*, which is one of the most famous and useful results in probability theory and statistical inference. This law provides conditions for the convergence of the sample mean to the true mean. We show the *weak law* here, which corresponds to pr. convergence of the sample mean. Under more stringent conditions, there exists a *strong law* corresponding to a.s. convergence (see the end of Example C.2). Examples C.3 and C.4 are a microcosm of the possibilities for counterexamples.

Many other counterexamples related to convergence exist, suggesting that one should be careful about putting too much faith in intuition. The simplicity of many of these counterexamples also suggests that the points being illustrated are not drawn from only bizarre cases, but are indicative of problems that might be encountered in practical applications. Romano and Siegel (1986, Chap. 5) and Stoyanov (1997, Part 3) include many such counterexamples, including a sequence that converges in m.s. but not a.s., a sequence that converges in pr. but not a.s., and a sequence of discrete random variables that converges in dist. to a continuous random variable.

Example C.2—Weak law of large numbers (convergence in pr. of the sample mean). Let $\{X_i\}$, $i = 1, 2, \ldots, k$, be a sequence of uncorrelated scalar random variables with common mean μ and uniformly bounded, not necessarily identical variances $E[(X_i - \mu)^2] \leq c < \infty$. Hence, the $\{X_i\}$ need be neither independent (merely uncorrelated), nor identically distributed. Let $\overline{X}_k$ be the sample mean of $X_1, X_2, \ldots, X_k$. We have

$$E\left[(\bar{X}_k-\mu)^2\right]=\frac{1}{k^2}E\left[\left(\sum_{i=1}^{k}X_i-k\mu\right)^2\right]=\frac{1}{k^2}\sum_{i=1}^{k}E\left[(X_i-\mu)^2\right]\leq\frac{c}{k}\to 0$$

as $k\to\infty$, where the second equality follows by the uncorrelatedness. So, $\bar{X}_k$ is m.s. convergent to μ. By the implications in Figure C.1, $\bar{X}_k$ is also pr. convergent, which is the *weak law of large numbers* (e.g., Laha and Rohatgi, 1979, pp. 67–69). A special case of this law is that the sample mean of a sequence of independent, identically distributed (i.i.d.) random variables with finite variance converges in pr. to the true mean of the sequence. In fact, the sample mean for this special case will converge in the stronger a.s. sense (the strong law of large numbers), but the proof techniques for the strong law are more difficult than those for the weak law above. ❑

Example C.3—Convergence in dist. does not imply convergence in m.s. Consider a sequence of scalar i.i.d. data $\{X_k\}$ where $\mu=E(X_k)$ and $\sigma^2=\text{var}(X_k)>0$. This sequence is trivially convergent in dist. since $F_{X_k}(x)=F_X(x)$ for all x and k. However, X_k never "settles down," and hence does not converge in any of the senses, a.s., pr., or m.s.

More specifically, suppose, for example, that X is an independent random variable having distribution $F_X(x)$. Then, the lack of m.s. convergence of X_k to X is apparent from $E[(X_k-X)^2]=E[(X_k-\mu-X+\mu)^2]=E[(X_k-\mu)^2]+E[(X-\mu)^2]=2\sigma^2>0$ for all k, where the second equality follows by the independence of X_k and X. ❑

Example C.4—Convergence a.s. does not imply convergence in m.s. (or convergence of the mean). Suppose that sample points $\omega\in\Omega$ are generated with uniform probability on the sample space $\Omega=[0,1]$. Define the scalar random variable

$$X_k=X_k(\omega)=kI_{[0,\,1/k)}(\omega),$$

where $I_{[0,\,1/k)}(\omega)$ is the indicator function for the set $[0,1/k)$:

$$I_{[0,\,1/k)}(\omega)=\begin{cases}1 & \text{if } \omega\in[0,1/k)\\ 0 & \text{otherwise.}\end{cases}$$

Note that $X_k(\omega)\to 0$ for any $\omega\in\Omega$ except for ω in the "bad set" $\{0\}$. However, the measure of this bad set is 0 (i.e., $P(\omega=0)=0$). Hence,

$$P\left(\omega\in\Omega:\ \lim_{k\to\infty}\left|X_k(\omega)-0\right|=0\right)=1,$$

which corresponds to the definition of a.s. convergence above (i.e., $X_k \to 0$ a.s.). For m.s. convergence, note that

$$E[(X_k - 0)^2] = \int_0^1 X_k(\omega)^2 \, d\omega = k^2 \int_0^{1/k} d\omega = k \to \infty .$$

Hence, X_k is convergent to 0 a.s., but divergent m.s. In fact, through a substitution of any $\rho \geq 1$ for the exponent 2 in the expression $E[(X_k - 0)^2]$ above, this example illustrates a more general principle, since $E(X_k^\rho)$ does not converge to 0 as $k \to \infty$. Hence, in particular, the a.s. convergence of X_k to 0 does not imply convergence of the mean $E(X_k)$ to 0 (i.e., $\rho = 1$). ❑

C.2.3 Dominated Convergence Theorem

As suggested in Subsection C.2.1, it is sometimes possible to relate a.s. and pr. convergence to convergence of moments (such as the mean and variance) via the famous Lebesgue dominated convergence theorem. This theorem provides conditions under which a.s. or pr. convergence implies m.s. convergence. This implication of the dominated convergence theorem is of central importance in some aspects of convergence for stochastic algorithms, as the need frequently arises to show that "direct" convergence of some random variables (a.s. or pr.) implies convergence of the first or second moments (mean or variance). It is also of critical importance in justifying an interchange of derivative and integral for simulation-based optimization using gradients (Chapter 15). The proof of the dominated convergence theorem is found in many books on probability theory and real analysis, including Chung (1974, p. 67) or Chow and Teicher (1997, pp. 99–100). The theorem, together with two corollaries, is as follows.

Theorem C.1 (Dominated convergence). Suppose that $\boldsymbol{X}_k \to \boldsymbol{X}$ a.s. or pr. as $k \to \infty$ for some $\boldsymbol{X}$. Further, suppose there exists a scalar positive random variable Z such that $\|\boldsymbol{X}_k - \boldsymbol{X}\|^q \leq Z$ a.s. for all k and some $q > 0$, where $E(Z) < \infty$. Then $E\left(\|\boldsymbol{X}_k - \boldsymbol{X}\|^q\right) \to 0$ as $k \to \infty$. In particular, if $q = 2$, then a.s. or pr. convergence implies m.s. convergence.

Corollary 1 to Theorem C.1. Suppose that the conditions of the theorem hold with $q = 1$. Then $E(\boldsymbol{X}_k) \to E(\boldsymbol{X})$ as $k \to \infty$.

Corollary 2 to Theorem C.1. Suppose that $\|\boldsymbol{X}_k - \boldsymbol{X}\|^q \leq M$ a.s., where M is a constant. Then $E\left(\|\boldsymbol{X}_k - \boldsymbol{X}\|^q\right) \to 0$ as $k \to \infty$.

The first corollary follows from the fact that $\|E(\boldsymbol{X}_k) - E(\boldsymbol{X})\| \leq E(\|\boldsymbol{X}_k - \boldsymbol{X}\|) \to 0$ as $k \to \infty$. (The convergence $E(\|\boldsymbol{X}_k - \boldsymbol{X}\|) \to 0$ is sometimes called *convergence in mean*; on the other hand, $E(\boldsymbol{X}_k) \to E(\boldsymbol{X})$ is convergence *of* the mean.) The second corollary follows because a uniform nonrandom bound M automatically implies that the bounded integral condition $E(Z) = E(M) = M < \infty$

holds. The second corollary is sometimes referred to as the *bounded convergence theorem*. An important special case of the dominated convergence theorem and corollaries above is when the limiting random variable X is some constant. So, for instance, if it is known that a parameter estimate $\hat{\theta}_k \to \theta^*$ a.s. (or pr.) as $k \to \infty$, where θ^* represents some true value of θ, then under the conditions of the theorem, we also know that the mean $E(\hat{\theta}_k)$ converges to θ^*.

Note that Example C.4 shows a case where the dominated convergence theorem is *not* applicable. In particular, it is not possible to find the dominating Z (or M) because the magnitude of X_k grows without bound for "enough" sample points so that the relevant expected value (e.g., the mean or mean-squared error) is unbounded. On the other hand, in the law of large numbers setting (Example C.2), suppose that $|X_i| \le Z$ a.s. for all $i \le k$ where $E(Z) < \infty$. Then, $|\overline{X}_k| \le Z$ a.s. So by Theorem C.1, we have $E(\overline{X}_k) \to \mu$. Of course, in this specific example, the m.s. convergence of $\overline{X}_k$ to μ (and hence the convergence of $E(\overline{X}_k)$; see Exercise C.4(b)) is available via direct methods as shown in the middle of Example C.2. Nevertheless, this provides a simple application of the dominated convergence theorem.

C.2.4 Convergence in Distribution and Central Limit Theorem

Recall our interests in normalized versions of the search and optimization iterate $\hat{\theta}_k$ as a special case of X_k. For dist. convergence, X_k in the general definition of Subsection C.2.1 then plays the role of the normalized random vector, while X is a random vector having the limiting distribution. A typical normalization for dist. convergence is $X_k = k^{\eta}(\hat{\theta}_k - \theta^*)$, where $0 < \eta \le 1/2$ and X is a random vector with a (multivariate) normal distribution. So, for $k \to \infty$, the general dist. convergence definition above is reduced to

$$k^{\eta}(\hat{\theta}_k - \theta^*) \xrightarrow{\text{dist.}} X, \text{ where } X \sim N(\mu, \Sigma), \tag{C.4a}$$

which is also written as

$$k^{\eta}(\hat{\theta}_k - \theta^*) \xrightarrow{\text{dist.}} N(\mu, \Sigma), \tag{C.4b}$$

where $\xrightarrow{\text{dist.}}$ denotes *converges in distribution* and μ, Σ denote some mean vector and covariance matrix. Informally, (C.4a, b) imply that the probabilistic rate at which the iterate $\hat{\theta}_k$ approaches θ^* is $k^{-\eta}$. That is, $\hat{\theta}_k - \theta^*$ must decay at a rate $k^{-\eta}$ to balance the k^{η} "blow-up" factor on the left-hand side of (C.4a, b) and yield a well-behaved random vector X with the distribution $N(\mu, \Sigma)$ (i.e., well-behaved in the sense of being neither degenerate 0 nor ∞ in magnitude).

The central limit theorem, one of the most famous and useful results in probability and statistics, is an example of convergence in distribution. In fact, the central limit theorem is not one theorem, but is a collection of several related results governing the asymptotic distribution of normalized sums of independent random variables. The most common version of the normalized sum is a normalized sample mean of a random process. Central limit theory states that the sample mean (properly normalized) will be asymptotically normally distributed regardless of the distributions of the random summands in the sample mean.

In practice, countless systems have random effects that represent the amalgamation of many individual random contributions, with the contributions being approximately additive. For example, the externally measured vibrations of a machine are typically formed by the superposition of contributions (i.e., the weighted sum of contributions) from the numerous vibrations in the internal parts of the machine. To the extent that the internal vibrations may be treated as independent random effects, central limit theory says that the external vibrations will be approximately normally distributed regardless of the distribution of the internal vibrations. Example C.5 considers one important special case of central limit theory.

Example C.5—Special case of the central limit theorem. A specific central limit theorem pertains to the sample mean $\overline{\boldsymbol{X}}_k$ of k i.i.d. vectors, each having true mean $\boldsymbol{\mu}$ and nonsingular (bounded) covariance matrix $\boldsymbol{\Sigma}$. For this case, the central limit theorem implies that

$$\sqrt{k}(\overline{\boldsymbol{X}}_k - \boldsymbol{\mu}) \xrightarrow{\text{dist.}} N(\boldsymbol{0}, \boldsymbol{\Sigma}),\ k \to \infty. \tag{C.5}$$

Less formally, one can say from (C.5) that for large k the following approximately holds:

$$\overline{\boldsymbol{X}}_k \sim N\left(\boldsymbol{\mu}, \frac{\boldsymbol{\Sigma}}{k}\right).$$

Hence, by the discussion above, the probabilistic rate of convergence of $\overline{\boldsymbol{X}}_k$ to the true mean is $1/\sqrt{k}$. The asymptotic normality in (C.5) is a special case of the Lindeberg–Feller central limit theorem, which gives conditions such that a properly normalized sum of independent, but not necessarily identically distributed, random vectors is asymptotically normally distributed. The proof of this result—which is quite difficult—may be found in Laha and Rohatgi (1979, pp. 282–287) for the scalar case and Bhattacharya and Rao (1976, Sect. 3.18) for the multivariate case. ❑

A warning is in order, however: A normalized sum of independent random variables will not always be asymptotically normally distributed. Central limit effects may be destroyed if a small number of terms in the sum have undue influence. Lest one think that this negative result only happens in pathological cases, it is not difficult to find practical examples where central limit effects do *not* hold. For example, when the noise terms in the state-space model for a dynamical system are nonnormally distributed, Spall and Wall (1984) show that the popular Kalman filter estimate (see Subsection 3.3.4) is generally asymptotically *nonnormal* even though it is composed of a sum of independent random vectors. Other examples of the failure of the central limit theorem when applied to sums of independent random vectors are given in Romano and Siegel (1986, pp. 115–120). Nevertheless, when properly applied, the central limit theorem is one of the most useful results from probability theory.

EXERCISES

C.1 Suppose that $\boldsymbol{X} = [X_1, X_2]^T$ and that $p_{\boldsymbol{X}}(\boldsymbol{x}) = \{x_1 + x_2$ when $0 \leq x_1 \leq 1$ and $0 \leq x_2 \leq 1$; 0 otherwise$\}$. Compute $E(\boldsymbol{X})$ and $\text{cov}(\boldsymbol{X})$ (the mean vector and covariance matrix of $\boldsymbol{X}$).

C.2 Suppose that $\boldsymbol{X}$ has a multivariate normal distribution. Using the key property of joint normality, prove that $\boldsymbol{AX}$ will also have a multivariate normal distribution for any deterministic and conformable (not necessarily square) matrix $\boldsymbol{A}$.

C.3 Suppose that the scalar random variable X has a Cauchy distribution centered about zero[1]. The density function is $p_X(x) = 1/[\pi(1+x^2)]$, $-\infty < x < \infty$. For what values of $\rho > 0$ does $E(X^\rho)$ exist? In particular, does the mean of X exist?

C.4 Prove the following results in probability theory:

(a) Convergence in m.s. ⇒ convergence in pr.

(b) m.s. convergence is a special case ($q = 2$) of convergence in qth mean, $q > 0$, $E\left(\|\boldsymbol{X}_k - \boldsymbol{X}\|^q\right) \to 0$ as $k \to \infty$. Show that qth mean convergence ⇒ rth mean convergence for all $0 < r < q$.

(Hint: Two famous inequalities from probability theory may be useful. For two random vectors $\boldsymbol{Y}$ and $\boldsymbol{Z}$, one is the Chebyshev inequality, $P(\|\boldsymbol{Y}\| \geq c) \leq E\left(\|\boldsymbol{Y}\|^2\right)/c^2$ for any $c > 0$. The other is the Hölder inequality, $E\left(\|\boldsymbol{Y}\|\|\boldsymbol{Z}\|\right) \leq E\left(\|\boldsymbol{Y}\|^q\right)^{1/q} E\left(\|\boldsymbol{Z}\|^r\right)^{1/r}$, where $r, q > 0$ are such that $1/q + 1/r = 1$.)

C.5 Consider the setting of Example C.2 *except* that the variances are no longer uniformly bounded, but grow with k according to $E[(X_k - \mu)^2] = k^\rho c$ for some $\rho > 0$ and $c > 0$. What is the set of allowable values for ρ to guarantee m.s. convergence of $\overline{X}_k$?

C.6 Suppose that the scalar sequence $X_k \to X$ a.s. as $k \to \infty$. If $f(X)$ is a scalar-valued function that is continuous for almost all X, show that $f(X_k) \to f(X)$ a.s. (A direct consequence of this result is that if $X_k \to c$ a.s. for some constant c, then $f(X_k) \to f(c)$ a.s. if $f(\cdot)$ is continuous at c.)

---

[1]The Cauchy distribution is an example of a *heavy tailed distribution*, where there is a much greater probability of a large value of $|X|$ than there is with, say, a normal distribution. This distribution is sometimes used in practice to model systems having frequent outliers (outliers are points that are farther from the central tendency of the data than would be expected using a normal model).

APPENDIX D

RANDOM NUMBER GENERATION

This appendix gives a short survey of some issues in the generation of (pseudo) random numbers. Section D.1 discusses the popular linear congruential generator and Section D.2 discusses the transformation of uniform random numbers to other distributions.

D.1 BACKGROUND AND INTRODUCTION TO LINEAR CONGRUENTIAL GENERATORS

Random number generation plays a key role in many aspects of stochastic analysis, including search and optimization. For example, algorithms that make use of injected randomness (Property B in Subsection 1.1.3) almost all require the use of a computer-based Monte Carlo random number generator ("almost all" because there may be rare cases where random sequences are available by sampling independent random sources, such as internal system noise on a physical object). Further, for an evaluation of any stochastic search algorithm by simulation (even one with no injected randomness), one needs random numbers to represent the measurement noise in the loss function evaluations (Property A in Subsection 1.1.3). An additional critical application of random number generation is in various aspects of Monte Carlo simulation, including the random number generation internal to the simulation and the potential use of the simulation in a search and optimization context with injected algorithm randomness (Chapters 14–16).

With such computer-based number generators, one is, perhaps paradoxically, producing randomness from a precise sequence of steps. This sequence of steps is initiated with a *seed*, which is typically a user-specified number to start the generator. Having a seed available is often very useful in evaluating different optimization algorithms. Using the same seed eliminates or reduces the additional variability that would be caused by a new set of random variables, helping better expose the performance changes due to underlying variations in problem structure, algorithm implementation, or problem coefficient values.

Although the term *pseudo* is sometimes attached, as above, we will usually drop this distinction in referring to computer-generated (Monte Carlo) random numbers. From a practitioner's point of view, the (pseudo) random sequence should look as if were generated by a truly random process when subjected to any of the standard statistical tests for randomness.

In practice, the random numbers used in an implementation of an algorithm, an evaluation of an algorithm, or a simulation must have a specific distributional form (say, normal or Cauchy). Random numbers having specified distributions are built from the basic uniform (0, 1) (denoted $U(0, 1)$) numbers being produced in the random number generator according to standard transformation formulas as discussed in Section D.2. If the transformed *pseudorandom* variables that are actually used in the algorithm of interest behave according to the distribution expected by the algorithm, then the algorithm will have the statistical and convergence properties that have been established under the assumption of *actual* randomness in the random variables.

It is not the purpose here to thoroughly review the field of random number generation, as that is a large subject unto itself (some useful references are Rubinstein, 1981, Chaps. 2 and 3; Knuth, 1981; Marsaglia, 1985; Park and Miller, 1988; L'Ecuyer, 1990b; Press et al., 1992, Chap. 7; Neiderreiter, 1992; Rubinstein and Melamed, 1998, Chap. 2; and Robert and Casella, 1999, Chap. 2). Rather, we wish to provide some comments on the role and creation of random numbers in areas relevant to stochastic search and optimization.

Any Monte Carlo-type method or simulation is critically dependent on the nature of the randomness being employed, and may perform unreliably if the probability distribution of random variables assumed for the algorithms differs significantly from that actually being provided via the transformed uniform numbers coming out of the generator. Likewise, incorrect randomness generating the noise in the case of noisy loss measurements may lead to invalid statistical conclusions regarding the relative efficiency of an algorithm. Hence, the proper generation of the underlying $U(0, 1)$ random variables—from which the random variables of direct interest in the algorithm are generated—is critical. There are numerous statistical tests for establishing whether a random number generator is acceptable (see, e.g., Rubinstein, 1981, Chap. 2). Park and Miller (1988) is an especially good reference, including a description of the pitfalls of poor generators that have been applied in the past.

A core method for generating $U(0, 1)$ random numbers is the *linear congruential method* (e.g., the **rand** generator in earlier versions of MATLAB, say MATLAB version 4). These uniform random numbers may in turn be used to generate random variables from other distributions, as discussed in Section D.2. The linear congruential method is based on an integer recursion of the form

$$J_{k+1} = aJ_k + c \pmod{m}, \tag{D.1}$$

where J_{k+1} represents an integer, a is the multiplier, c is the increment, and m is the modulus. Here, mod m represents the operation of assigning to J_{k+1} the (integer) remainder in the division $(aJ_k + c)/m$. This remainder will be a value in $\{0, 1,\ldots, m-1\}$. The uniform random variable associated with J_k is simply the value $U_k \equiv J_k/m$, which will lie in [0, 1). In computer implementations, the modulus is taken as a large prime number (e.g., $2^{31} - 1$ on a 32-bit machine; even larger numbers on machines with greater accuracy).

Example D.1—Example of linear congruential generator. Let $m = 2^{31} - 1 = 2{,}147{,}483{,}647$, $c = 0$, and $a = 16{,}807$. These parameters were originally proposed in Lewis et al. (1969). Taking the seed $J_0 = 12{,}345$, by (D.1) we have

$$J_1 = 16{,}807 \times 12{,}345 \pmod{m} = 207{,}482{,}415$$

$$U_1 = \frac{207{,}482{,}415}{m} = 0.0966165285$$

$$J_2 = 16{,}807 \times 207{,}482{,}415 \pmod{m} = 1{,}790{,}989{,}824$$

$$U_2 = \frac{1{,}790{,}989{,}824}{m} = 0.8339946274$$

$$J_3 = 16{,}807 \times 1{,}790{,}989{,}824 \pmod{m} = 2{,}035{,}175{,}616$$

$$U_3 = \frac{2{,}035{,}175{,}616}{m} = 0.9477024977$$

and so on. Note that $16{,}807 \times 12{,}345 \pmod{m} = 16{,}807 \times 12{,}345 = 207{,}482{,}415$ in the calculation of J_1 because the indicated product is less than m. ❑

As stated in Press et al. (1992, p. 276): "Although this general framework is powerful enough to provide quite decent random numbers, its implementation in many, if not most, ANSI C libraries is quite flawed." L'Ecuyer (1998) also summarizes some shortcomings of the linear congruential algorithm, noting, in particular, that even the seemingly large modulus of $2^{31} - 1 \approx 2 \times 10^9$ is too small for many modern applications, especially in the area of Monte Carlo simulations (see Chapters 14–16).

One of the problems in linear generators of the form (D.1) pertains to cycling. That is, the pseudorandom sequence repeats itself within a given simulation run (from the structure of the linear congruential algorithm, an upper bound to this cycling time is given by the modulus). Further, there may be subtle statistical correlation in the successive random numbers. The effect is that vectors being generated to randomly distribute themselves in (say) p-dimensional space, may, in fact, lie in only a $(p - 1)$-dimensional subspace.

An example of deficient generators discussed in Law and Kelton (2000, p. 422) is the infamous RANDU generator of the 1960s, where $a = 65539$, $c = 0$, and the modulus is $2^{31} - 1$. This generator has the desirable property of being a full-period generator (i.e., only repeats itself after $2^{31} - 1$ random numbers have been generated), but for multivariate applications, the generated points clustered in a lower dimensional subspace. In particular, if the generator is being used to produce three-dimensional random vectors (each three values produce one vector), then the points being produced lie in the plane given by the equation $J_{k+1} = 6J_k - 9J_{k-1}$. This example reinforces the point that for any serious application requiring computer-generated random numbers, it is wise to carefully evaluate the underlying random number generator, perhaps performing some of the statistical tests for goodness of fit that are available.

Figure D.1 illustrates the performance of four simple linear generators with respect to the sample (empirical) mean of the random sequence being produced. We see that the two generators with very short periods ($m = 9$ and $m = 27$) produce sample means clearly away from the ideal mean of 0.5. The two generators with longer periods produce sample means closer to the ideal. Of course, the sample mean is only one measure of the desirability of the generator. A serious evaluation of these generators would require other tests (and by these other tests, the $m = 482$ generator would clearly be deficient because of relatively rapid cycling).

Although the linear congruential generators are very popular and are the best understood type of random number generators, there are many alternatives (see L'Ecuyer, 1998, for a survey). Most of these other generators have been developed in an attempt to achieve longer periods and better statistical properties. For example, the random number generator **rand** provided in version 6 of MATLAB uses a combination of a lagged Fibonacci generator combined with a shift register random integer generator, plus what is called an *ulp factor* based on the theory presented in Marsaglia (1991) (ulp = "units in the last

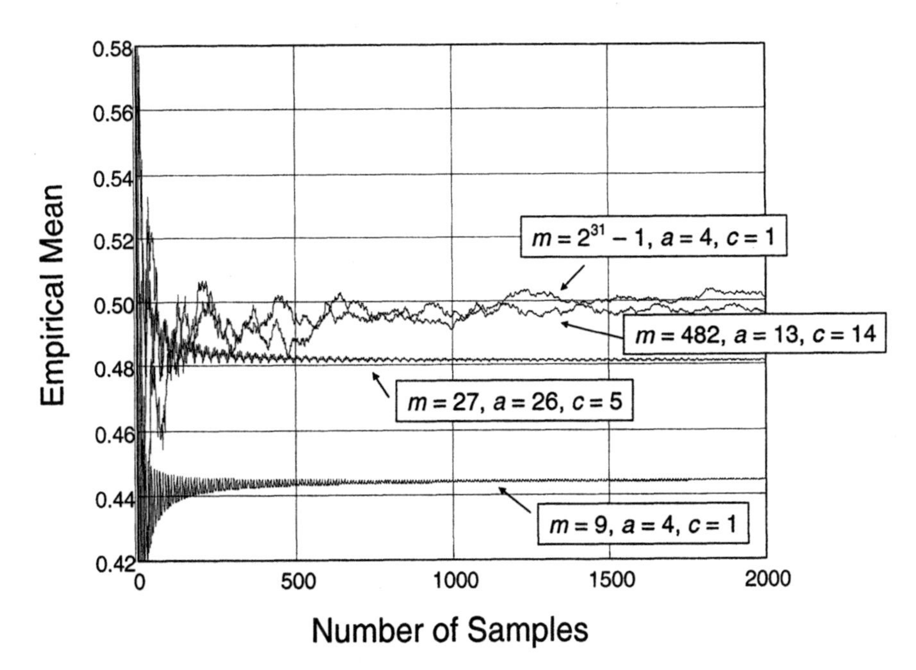

Figure D.1. The sample mean is *one* indication of the quality of pseudorandom number generators. The desired long-run mean is 0.50. The plot shows that generators with low modulus m are clearly deficient.

place"). Without the ulp factor, the numbers generated are not evenly distributed throughout the interval (0, 1). Their distribution is biased with fewer outcomes between [1/2, 1) than between [1/4, 1/2), and even more congregated in the interval [0, 1/4). The ulp factor corrects for this anomaly.

D.2 TRANSFORMATION OF UNIFORM RANDOM NUMBERS TO OTHER DISTRIBUTIONS

Random variables with distributions other than $U(0, 1)$ can be generated by proper transformation of uniformly distributed random numbers generated by the techniques described in the preceding section. Here we describe two common methods: the *inverse-transform method* and the *accept–reject method*. Among many other sources, descriptions of other transformation techniques can be found in Devroye (1986), Rubinstein and Melamed (1998, Chap. 2), or Robert and Casella (1999, Chap. 2).

The inverse-transform method is a direct method that requires derivation of the inverse function of the desired distribution function. Let X be a random variable with cumulative distribution function $F(x)$. Since F is a nondecreasing and right-continuous function, the inverse function $F^{-1}(U)$ exists and can be defined as

$$F^{-1}(y) = \inf\{x : F(x) \geq y\},\ 0 \leq y \leq 1.$$

Let U denote a random number with uniform distribution $U(0, 1)$. It is easy to show that the *random variable* defined by $F^{-1}(U)$ has the same distribution function as X. That is, $P(F^{-1}(U) \leq x) = F(x)$. Therefore, to generate a random variable X with distribution F, first collect a value U and then take X as the value $F^{-1}(U)$. The two examples below illustrate this method.

Example D.2—Creation of random numbers with exponential distribution. A random number X with exponential distribution has the distribution function $F(x) = 1 - e^{-\lambda x}$, where $\lambda, x > 0$. By the inverse-transform method, $X = F^{-1}(U) = -\log(1-U)/\lambda$. ❑

Example D.3—Creation of discrete random numbers. The inverse-transform method can also be applied to generate discrete random number. Let X be a discrete random variable with three possible outcomes according to

$$X = \begin{cases} x_1 \text{ with probablity } p_1 \\ x_2 \text{ with probablity } p_2 \\ x_3 \text{ with probablity } p_3 \end{cases}$$

with $p_1 + p_2 + p_3 = 1$. The distribution function $F(x)$ in this case is a piecewise constant function. By applying the inverse-transform method we can generate random samples of X from U by the following rules

$$X = \begin{cases} x_1 & \text{if } 0 \le U \le p_1 \\ x_2 & \text{if } p_1 < U \le p_1 + p_2 \\ x_3 & \text{if } p_1 + p_2 < U \le 1. \end{cases}$$

The inverse-transform method can also be applied to generate random numbers from an empirical distribution (i.e., histogram). An empirical distribution can be specified by a piecewise constant density function or a piecewise linear distribution function. ❑

One drawback of the inverse-transform technique is that it is often not possible to evaluate the inverse function $F^{-1}(U)$ in closed form (e.g., with the normal distribution). Let us briefly describe an indirect method, the accept–reject method, which does not require evaluations of the inverse function. The accept–reject method, however, calls for the existence of the density function $p_X(x)$ (or at least the density function to within an unknown scale factor) for the desired distribution and the specification of a function $\varphi(x)$ that *majorizes* $p_X(x)$. That is, $\varphi(x) \ge p_X(x)$ for all x.

To carry out the accept–reject method, one first identifies the majorizing function $\varphi(x)$. A popular way is to choose a density function $q(x)$ and a constant $C \ge 1$ such that $\varphi(x) = Cq(x) \ge p_X(x)$ for all x. Based on this choice, the steps of the accept–reject method below produce a random variable X, representing an outcome from the density $p_X(x)$.

Accept–Reject Method

Step 1 Generate U from $U(0, 1)$.

Step 2 Generate a random variable Y from the density function $q(\cdot)$, independent of U.

Step 3 If $U \le p_X(Y)/\varphi(Y)$, then set $X = Y$. Otherwise, go to step 1.

The density $q(x)$ is typically referred to as the *instrumental density* for the *target density* $p_X(x)$. Aside from its generality, an important advantage of the accept–reject method is that $p_X(x)$ is only required to be known up to a multiplicative scale factor. For example, consider a case where only $\pi_X(x) \equiv bp_X(x)$ is available with an unknown normalizing constant b. As long as a majorizing function $\varphi(x) \ge \pi_X(x)$ can be identified, the same procedure (using the ratio $\pi_X(x)/\varphi(x)$ in step 3) can be applied to generate variables for the target density $p_X(x)$ without the knowledge of the constant b. A demonstration is given in Robert and Casella (1999, pp. 50–51) with a scaled Cauchy distribution being used to generate normal random variables. This property is particularly important in Bayesian calculations where the normalizing constant for the posterior distribution (typically a marginal distribution) is generally difficult to compute. (We see in Chapter 16 that this desirable property is also partly responsible for the popularity of Markov chain Monte Carlo methods for generating pseudorandom numbers.)

The efficiency of the accept–reject method depends on the choice of the majorizing function $\varphi(x)$. Obviously, one wishes to identify a function $\varphi(x)$ that leads to an instrumental density $q(x)$ for which random variables are efficiently generated (e.g., the instrumental density beng $U(0, 1)$). However, this ease of generation for the instrumental density might lead to a significant probability of rejection in step 3 of the accept–reject procedure (i.e., a large percentage of samples generated from $q(x)$ being discarded) and result in inefficiency. The value of C in $\varphi(x) = Cq(x)$ has a strong impact on the performance of the algorithm. The probability of accepting the new point (accepting Y in step 3 above) is $1/C$ (Rubinstein and Melamed, 1998, p. 26; Robert and Casella, 1999, p. 50).

Hence, to achieve a high probability of acceptance, we will need to select a function $\varphi(x)$ that is close to the target density $p_X(x)$. Recall, however, that we are using the accept–reject method precisely because of the difficulty in generating random variables from $p_X(x)$! The ease of generation from $q(x)$ and a large probability of acceptance (a $C \approx 1$) are in conflict with each other in the designs of the accept–reject method. A general approach that does not require the specification of a majorizing function is the Markov chain Monte Carlo approach of Chapter 16.

Example D.4—Creation of $N(0, 1)$ random numbers. Note that the distribution function of X with distribution $N(0, 1)$ can be written as

$$\begin{aligned} F(x) &= P(X \le x \,|\, X \ge 0)P(X \ge 0) + P(X \le x \,|\, X < 0)P(X < 0) \\ &= 0.5P(X \le x \,|\, X \ge 0) + 0.5P(X \le x \,|\, X < 0). \end{aligned}$$

Therefore, to generate a random sample for X, it is sufficient to first generate a nonnegative random variable X with density function

$$\sqrt{\frac{2}{\pi}}\, e^{-x^2/2},\ x \ge 0,$$

and then assign to X a random sign (positive or negative with equal probability).

To generate a random variable with the above density function, let $q(x) = e^{-x}$, $x \ge 0$, an exponential distribution, and $C = \sqrt{2e/\pi}$. It is readily seen that $Cq(x)$ majorizes the density. Following the steps for the accept–reject method, we can generate a random variable having the above density. To produce a variable having the desired $N(0, 1)$ distribution, we then assign a random +/– sign, which can be done according to whether a separate $U(0, 1)$ process produces a value between 0 and 1/2 or between 1/2 and 1. ❑

In many simulations, the generation of random *vectors* with prescribed distribution is required. Both the inverse-transform and accept–reject methods can be extended for random vector generation (see, e.g., Rubinstein and Melamed, 1998, Sect. 2.3). A famous method for generating bivariate normal random vectors is the Box–Muller algorithm (Box and Muller, 1958). The current version of MATLAB (version 6) uses a sophisticated table look-up algorithm for generating multivariate normal vectors (the **randn** generator).

EXERCISES

D.1 Illustrate rapid cycling in the linear congruential random number generator with $J_0 = a = c = 3$ and $m = 5$. Note that the period is less than the modulus.

D.2 Illustrate the problem of random number generators projecting into lower-dimensional subspaces by considering a generator with $a = 23$, $c = 0$, and modulus = 97. Do two plots of J_k vs. J_{k-1}, first showing only 30 points and then showing 96 points. What do these two plots illustrate about the dangers of flawed generators? Do they suggest a possible way of using generators in practice?

D.3 Let $X_1, X_2,\ldots, X_n$ be independent, identically distributed (i.i.d.) random variables with exponential distribution $F(x) = 1-e^{-\lambda x}$. Consider the order statistics $Y = \min\{X_1, X_2,\ldots, X_n\}$. Use the inverse transform to generate Y from U. Compare the resulting formulas with the result in Example D.2. What does this comparison reveal about the distribution of Y?

D.4 Suppose that $X_1, X_2,\ldots, X_n$ is a random sample of size n from an unknown distribution. A piecewise linear empirical distribution can be constructed as an approximation to the true distribution as follows. First sort the sample X_i into ascending order: $a_1 \le a_2 \le \ldots \le a_n$, where we make the assignment $X_i = a_i$. The smoothed empirical distribution is given by

$$F(x) = \begin{cases} 0 \ \text{ if } x < a_1, \\ \dfrac{i-1}{n-1} + \dfrac{x-a_i}{(n-1)(a_{i+1}-a_i)} & \text{if } a_i \le x \le a_{i+1},\ i = 1,\ldots,n-1, \\ 1 \ \text{ if } x > a_n. \end{cases}$$

Use the inverse-transform method to derive an algorithm for generating random numbers with the empirical distribution defined above.

D.5 Consider the beta (4, 3) distribution with the density

$$p_X(x) = \begin{cases} 60x^3(1-x)^2 & \text{if } 0 \le x \le 1, \\ 0 & \text{otherwise.} \end{cases}$$

Use the accept–reject method to generate 50 samples for $p_X(x)$ using the following two majorizing functions: (i) $2.0736\,u[0, 1)$ and (ii) $1.875\,u[0, 0.5) + 2.0736\,u[0.5, 1)$, where $u[a, b)$ denotes the unit-height step function over $[a, b)$. Compare the probability of acceptance (the probability of accepting the new point Y given in step 2 of the algorithm) between these two choices of majorizing functions.

APPENDIX E

MARKOV PROCESSES

Markov processes are fundamental in the analysis of many stochastic algorithms, including random search, stochastic approximation, simulated annealing, evolutionary computation, reinforcement learning, and Markov chain Monte Carlo. This short appendix summarizes aspects of Markov processes that are used in the text. We restrict ourselves to processes that evolve in discrete time. Section E.1 provides some general remarks. Section E.2 provides a slightly more detailed treatment for the special case of Markov chains that take values on discrete state spaces.

E.1 BACKGROUND ON MARKOV PROCESSES

Markov processes have long held a central role in the theory and applications of stochastic processes. This appendix presents several results for Markov processes, as used in the text. Much more comprehensive general treatments are given in Parzen (1962, Chaps. 6 and 7), Meyn and Tweedie (1993), and Ross (1997, Chaps. 4 and 6), among many other references. We restrict ourselves to processes that evolve in discrete time.

Consider a stochastic process X_0, X_1, X_2,.... The sequence is a *Markov process* if the following relationship holds:

$$P(X_{k+1} \in \Lambda | X_0, X_1, \ldots, X_k) = P(X_{k+1} \in \Lambda | X_k) \tag{E.1}$$

for any dimensionally appropriate set Λ. That is, the conditional probability of the process depends only on the most recent value of the process, not on the full collection of all past values.

The following example depicts a popular Markov process.

Example E.1—Linear dynamical system. An important application of Markov processes is the first-order linear dynamical system. In particular, let

$$X_{k+1} = A_k X_k + W_k,$$

where $\{W_k\}$ is an independent noise sequence and A_k is a deterministic matrix. As discussed in Chapter 3, the above equation corresponds to a first-order vector autoregressive equation in time series analysis or a state equation in a linear state-space model (frequently associated with the Kalman filter). Because of the independence of W_k from past process information, X_0, $X_1,\ldots,$ X_{k-1}, the probabilistic characterization of the future state X_{k+1} depends only on the current state X_k and the probability distribution of the noise W_k. Hence, the fundamental Markov property (E.1) is satisfied. ❑

An especially important special case of the Markovian linear model in Example E.1 is the constant coefficient model having $A_k = A$. In such a model, stability may be analyzed by evaluating the eigenvalues of A. In particular, if the eigenvalues all have magnitude less than 1, and the noise W_k has mean $\mathbf{0}$ with constant covariance matrix, then the process will be (asymptotically) second-order stationary in the sense that $E(X_k)$ and $\text{cov}(X_k)$ will approach limiting values having finite magnitudes. The eigenvalue condition is equivalent to requiring that all eigenvalues lie inside the unit circle in the complex plane. Such stable processes have a fundamental role in control theory and time series analysis.

E.2 DISCRETE MARKOV CHAINS

An important special case of a Markov process is a discrete *Markov chain*, where the process X_k may take on only a discrete number of values. The number of possible values may be finite or infinite. Note that various authors use the terminology *Markov chain* differently. Some (e.g., Parzen, 1962; Ross, 1997) restrict the term to discrete-valued processes; others (e.g., Meyn and Tweedie, 1993) allow arbitrary (discrete and/or continuous) processes. (In the area of Markov chain Monte Carlo, considered in Chapter 16, the term *Markov chain* is used more generally to include continuous processes.) To avoid ambiguity, we refer to *discrete Markov chains* in this section.

Let $\{\chi_1, \chi_2,\ldots\}$ denote the possible values for the Markov chain (sometimes called the *state space* for the chain). That is, each X_k may take on one of the values $\{\chi_1, \chi_2,\ldots\}$. Therefore, from the right-hand side of the general expression (E.1), we can define the following transition probability:

$$P_{ij} \equiv P(X_{k+1} = \chi_j \,|\, X_k = \chi_i)\,,$$

where we have assumed that the transition probabilities are stationary across k (i.e., the P_{ij} do not depend on k). The transition matrix is the matrix of all possible transition probabilities. In the case where there is a finite number, say m, of possible states in the state space, the transition matrix $\boldsymbol{P}$ has the form:

$$\boldsymbol{P} \equiv \begin{bmatrix} P_{11} & P_{12} & \cdots & P_{1m} \\ P_{21} & P_{22} & \cdots & P_{2m} \\ \vdots & \vdots & \ddots & \vdots \\ P_{m1} & P_{m2} & \cdots & P_{mm} \end{bmatrix}.$$

Note that $P_{ij} \geq 0$ for all i, j and $\sum_{j=1}^{m} P_{ij} = 1$ for all i. Hence, every row in $\boldsymbol{P}$ sums to unity, which is equivalent to knowing that at every transition the process must move from the current state to one of the possible states in the state space.

The above transition matrix can be used to determine the marginal probability of reaching each state for one step of the process or across several steps. Let $\boldsymbol{p}_k$ be the vector of probabilities of being in each of the m states. Then

$$\boldsymbol{p}_{k+1}^T = \boldsymbol{p}_k^T \boldsymbol{P}.$$

More generally, $\boldsymbol{P}^k$ defines the transition matrix across k steps of the process. So

$$\boldsymbol{p}_k^T = \boldsymbol{p}_0^T \boldsymbol{P}^k.$$

In many cases, it is of interest to determine a stationary probability vector, say $\overline{\boldsymbol{p}}$. If a process is started with such a distribution, it will keep this distribution forever. That is, the following holds:

$$\overline{\boldsymbol{p}}^T = \overline{\boldsymbol{p}}^T \boldsymbol{P}. \tag{E.2}$$

This stationary distribution holds special significance in discrete Markov chain theory. To appreciate the significance, we first need to present some definitions. We then present a theorem that gives conditions under which $\overline{\boldsymbol{p}}$ in (E.2) is unique.

Let $P_{ij}^{(k)}$ be the ijth element of $\boldsymbol{P}^k$. State χ_j is *accessible* from state χ_i if $P_{ij}^{(k)} > 0$ for some $k \geq 1$. This implies that, starting in χ_i, it is possible that the process will eventually reach χ_j. States χ_i and χ_j *communicate* if each is accessible from the other. The chain is *irreducible* if all m states communicate. If there are two consecutive numbers s and $s+1$ such that the process can be in state χ_i at times s and $s+1$, then the state is called *aperiodic*. A state χ_i is *positive recurrent* if, starting in χ_i, the expected time for the process to reenter χ_i is finite. Positive recurrent states that are aperiodic are called *ergodic*. The chain is ergodic if this relationship holds for all states. Examples of discrete Markov chains satisfying (or not) these conditions are given in the references cited above.

Given the above definitions, we have the important result below, which is proved in Parzen (1962, Sect. 6.8), Meyn and Tweedie (1993, Chap. 15), and Robert and Casella (1999, Sect. 4.6). Theorem E.1 provides conditions under which $\boldsymbol{P}^k$ approaches a matrix with m identical rows.

Theorem E.1 (Convergence theorem for discrete Markov chains). Consider a discrete Markov chain with a finite state space. If the chain is irreducible and ergodic, then $\overline{\boldsymbol{p}}$ is the unique solution to (E.2) and

$$\lim_{k\to\infty} P_{ij}^{(k)} = \overline{p}_j$$

independent of i, where $\overline{p}_j$ is the jth element of $\overline{\boldsymbol{p}}$. Moreover, the convergence rate is geometrically fast in the sense that $\left|P_{ij}^{(k)} - \overline{p}_j\right| = O(c^k)$ for some $0 < c < 1$.

There exist algorithms for determining the unique solution $\overline{\boldsymbol{p}}$ to (E.2) (e.g., Iosifescu, 1980, pp. 123–124). Further, there exist sufficient conditions for the theorem conditions, as given, for example, in Parzen (1962, Chap. 6). The following simple three-state example illustrates Theorem E.1.

Example E.2—Convergence of discrete Markov chain. Suppose that

$$\boldsymbol{P} \equiv \begin{bmatrix} 0.4 & 0.3 & 0.3 \\ 0.2 & 0.3 & 0.5 \\ 0.3 & 0.2 & 0.5 \end{bmatrix}.$$

Then

$$\boldsymbol{P}^2 \equiv \begin{bmatrix} 0.310 & 0.270 & 0.420 \\ 0.290 & 0.250 & 0.460 \\ 0.310 & 0.250 & 0.440 \end{bmatrix} \text{ and } \boldsymbol{P}^4 \equiv \begin{bmatrix} 0.305 & 0.256 & 0.439 \\ 0.305 & 0.256 & 0.439 \\ 0.305 & 0.256 & 0.439 \end{bmatrix}.$$

Note that the elements of $\boldsymbol{P}^k$ rapidly converge to values that are independent of the row index (the i value in Theorem E.1), illustrating the geometric convergence rate. The three limiting values shown in the rows of $\boldsymbol{P}^4$ correspond to the three elements of $\overline{\boldsymbol{p}}$, as is easily verified (Exercise E.1). ❑

Relative to topics covered in this book, Theorem E.1 is related to the convergence of the simulated annealing algorithm (see Section 8.1), to the convergence or nonconvergence of genetic algorithms for discrete search (see Section 10.5), and to the convergence of estimates coming out of the Gibbs sampler in a Markov chain Monte Carlo sampling scheme (see Section 16.4).

EXERCISES

E.1 To within the indicated numerical accuracy, verify that the steady-state solution derived from the limiting transition matrix in Example E.2 yields a valid $\bar{\boldsymbol{p}}$.

E.2 Create a three-state transition matrix (other than $\boldsymbol{I}_3$) that violates the irreducibility requirement. Demonstrate numerically that (as expected) this matrix does *not* yield the convergence conclusion in Theorem E.1.

E.3 The following is known: State χ_i is positive recurrent if and only if $\sum_{k=1}^{\infty} P_{ii}^{(k)} = \infty$. Use this fact to show that if state χ_i is positive recurrent and χ_i communicates with χ_j, then χ_j is positive recurrent.

ANSWERS TO SELECTED EXERCISES

This section provides answers to selected exercises in the chapters and appendices. An asterisk (*) indicates that the solution here is incomplete relative to the information requested in the exercise.

CHAPTERS 1 – 17

1.2. On the domain [−1, 1], the maximum occurs at $\theta = 1/\sqrt{5}$ and the minimum occurs at $\theta = -1$.

1.10.* The table below provides the solution for the first two (of six) iterations:

k	$\hat{\theta}_k^T$	$L(\hat{\theta}_k)$
0	[0, 3.0]	52.0
1	[2.728, 1.512]	0.369
2	[2.500, 1.228]	0.0643

(Note: This problem corresponds to Example 8.6.2 in Bazaraa et al., 1993. The numbers here differ slightly from Bazaraa et al. This is apparently due to the limited accuracy used in Bazaraa et al.)

2.3.* A first-order Taylor approximation (Appendix A) yields

$$\log(1-P^*) \approx \log(1) - \left.\frac{1}{1-P^*}\right|_{P^*=0} \times P^*$$
$$= -P^*.$$

This implies from (2.4) that the number of loss measurements n satisfies $n = \log\rho/\log(1-P^*) \approx -\log\rho/P^*$ for small P^*, as was to be shown.

3.4. (b) With a = 0.07, the sample means of the terminal estimates for β and γ are 0.91 and 0.52, respectively (versus true values of 0.90 and 0.50).

4.3.* The table below shows the sample means of the terminal estimate from 20 replications. Under each sample mean is the approximate 95 percent confidence interval using the principles in Section B.1 (Appendix B).

	$n = 100$	$n = 10{,}000$	$n = 1{,}000{,}000$
Sample mean of terminal estimate (and 95% interval)	0.729 [0.700, 0.760]	0.920 [0.906, 0.934]	0.974 [0.970, 0.977]

5.10.* The table below presents the solution for the recursive processing with one pass through the data for one value of a. The mean absolute deviation (MAD) is computed with respect to the test data in **`reeddata-test`**.

$a =$ 0.02	$\hat{\beta}_0 = -0.183$, $\hat{\boldsymbol{\beta}} = \begin{bmatrix} -0.365 \\ -0.319 \\ -0.372 \\ -0.380 \\ -0.341 \\ -0.295 \end{bmatrix}, \hat{\boldsymbol{B}} = \begin{bmatrix} 0.330 & & & & & \\ & 0.423 & & & 0 & \\ & & 0.234 & & & \\ & & & 0.226 & & \\ & 0 & & & 0.329 & \\ & & & & & 0.525 \end{bmatrix}$	MAD= 0. 3647

6.4. For convenience, let $\boldsymbol{X}_k = k^{\beta/2}(\hat{\boldsymbol{\theta}}_k - \boldsymbol{\theta}^*)$. It is assumed that $\boldsymbol{X}_k$ converges in distribution to $N(\boldsymbol{\mu}_{\text{FD}}, \boldsymbol{\Sigma}_{\text{FD}})$. From Figure C.1 in Appendix C, pr. and a.s. convergence are both stronger than dist. convergence. However, even pr. and a.s. convergence, in general, are not strong enough to guarantee that the mean of the process converges (see Example C.4). Therefore, $E(\boldsymbol{X}_k) \nrightarrow \boldsymbol{\mu}_{\text{FD}}$ in general.

7.3. With valid (Bernoulli) perturbations, the mean terminal loss value $L(\hat{\boldsymbol{\theta}}_{1000})$ is approximately 0.03 over several independent replications. For the (invalid) uniform $[-\sqrt{3}, \sqrt{3}]$ and $N(0, 1)$ perturbation distributions, the $\boldsymbol{\theta}$ estimate quickly diverges. Based on several runs, the terminal loss value was typically between 10^{50} and 10^{100} for the uniform case and much greater than 10^{100} for the normal case. This illustrates the danger of using an invalid perturbation distribution, even one that has the same mean and variance as a valid distribution.

8.6. SAN is run with the coefficients of Example 8.2 on the indicated two-dimensional quartic loss function. The table below shows the sample-mean terminal loss values for initial temperatures of 0.001, 0.10, and 10.0 based on an average over 40 replications. The results for the lowest and highest initial temperature are poorer than the results in Table 8.2; the results for the intermediate initial temperature (0.10) are pulled directly from Table 8.2.

N	$T_{\text{init}}=$ 0.001	$T_{\text{init}}=$ 0.10	$T_{\text{init}}=$ 10.0
100	3.71	0.091	8.75
1000	2.12	0.067	6.11
10,000	0.429	0.0024	0.141

9.7.* The probability of a given (selected) binary-coded chromosome being changed from one generation to the next is bounded above by P(crossover $\cup$ mutation), where the event "mutation" refers to at least one bit being mutated. Note that

$$\begin{aligned}&P(\text{crossover} \cup \text{mutation})\\ &\quad = P(\text{crossover}) + P(\text{mutation}) - P(\text{crossover} \cap \text{mutation})\\ &\quad = P_c + (1-(1-P_m)^B) - P_c(1-(1-P_m)^B).\end{aligned}$$

The probability of passing intact is bounded below by 1 – P(crossover $\cup$ mutation).

10.8.* Applying Stirling's approximation for "large" N and B yields

$$N_P \approx \frac{1}{\sqrt{2\pi}}\left(1+\frac{2^B-1}{N}\right)^N\left(1+\frac{N}{2^B-1}\right)^{2^B-1}\left(\frac{1}{2^B-1}+\frac{1}{N}\right)^{1/2}.$$

11.3. (b) One aspect suggesting that the online form in (11.11) is superior is that it uses the most recent parameter estimate for all evaluations on the right-hand side of the parameter update recursion. The alternative form of TD in this exercise uses older parameter estimates for some of the evaluations on the right-hand side. Another potential shortcoming noted by Sutton (1988) is that the differences in the predictions will be due to changes in both $\boldsymbol{\theta}$ and the inputs (the $\boldsymbol{x}_\tau$), making it hard to isolate only the effects of interest. In contrast to the recursion of this exercise, the temporal difference in predictions on the right-hand side of the recursion in (11.11) is due solely to changes in the input values (since the parameter is the same in all predictions).

12.4. Under the null hypothesis, $L(\boldsymbol{\theta}_m) = L(\boldsymbol{\theta}_j)$ for all $j \neq m$. Further, $E(\delta_{mj}) = 0$ for all $j \neq m$ by the assumption of a common mean for the noise terms. Then

$$\begin{aligned}\text{cov}(\delta_{mj},\delta_{mi}) &= E[(\bar{L}_m - \bar{L}_j)(\bar{L}_m - \bar{L}_i)]\\ &= E(\bar{L}_m^2) - E(\bar{L}_j\bar{L}_m) - E(\bar{L}_i\bar{L}_m) + E(\bar{L}_i\bar{L}_j)\\ &= \text{var}(\bar{L}_m),\end{aligned}$$

where the last line follows by the mutual uncorrelatedness over all i, j, and m (so, e.g., $E(\bar{L}_j\bar{L}_m) = E(\bar{L}_j)E(\bar{L}_m) = [E(\bar{L}_m)]^2$ for $j \neq m$).

13.7. Using MS EXCEL and the data in **`reeddata-fit`**, we follow the approach outlined in Example 13.5 in terms of splitting the data into four pairs of fitting/testing subsets. The four MSE values are 0.092, 0.146, 0.157, and 0.181 and the four MAD values are 0.261, 0.328, 0.323, and 0.328. The modified Table 13.3 is given below; the values in italics are from Table 13.3 (so only the last column is new). The full linear model produces the lowest RMS and MAD values.

	Full linear model (13.9)	Reduced linear model (13.10)	Five-input linear model
RMS	*0.327*	*0.386*	0.379
MAD	*0.266*	*0.306*	0.310

14.7. (b) We approximate the CRN-based and non-CRN-based covariance matrices using the sample covariance matrix from 2×10^6 gradient approximations generated via Monte Carlo simulation (this large number is needed to get a stable estimate of the matrix). The approximate covariance matrices are as follows:

$$\text{CRN: } \operatorname{cov}[\hat{g}_k(\hat{\theta}_k)\,|\,\hat{\theta}_k] \approx \begin{bmatrix} 445.2 & 44.5 \\ 44.5 & 465.3 \end{bmatrix},$$

$$\text{Non-CRN: } \operatorname{cov}[\hat{g}_k(\hat{\theta}_k)\,|\,\hat{\theta}_k] \approx \begin{bmatrix} 5936.0 & 93.5 \\ 93.5 & 5956.4 \end{bmatrix}.$$

The above show that for $c_k = 0.1$, the CRN covariance matrix is much smaller (in the matrix sense) than the non-CRN covariance matrix.

15.3. We have $\log p_V(\upsilon|\theta) = \log p_V(\upsilon|\lambda) = \upsilon\log\lambda + (1-\upsilon)\log(1-\lambda)$. Hence, $\partial\log p_V(\upsilon\,|\,\lambda)/\partial\lambda = \upsilon/\lambda - (1-\upsilon)/(1-\lambda)$. This simplifies to $\lambda^{-1}(\lambda-\upsilon)/(\lambda-1)$, as shown in the first component of $\partial\log p_V(\upsilon\,|\,\theta)/\partial\theta$ in Example 15.4. The second component in $\partial\log p_V(\upsilon\,|\,\theta)/\partial\theta$ is zero since there is no dependence of $p_V(\upsilon|\theta)$ on β. The final form for $Y_k(\hat{\theta}_k)$ then follows immediately using expression (15.12).

16.4.* Let X be the current state value and W be the candidate point. The candidate point W is accepted with probability $\rho(X, W)$. This probability is random as it is depends on X and W. For convenience, let $\pi(x,w) = p(w)q(x)/[p(x)q(w)]$ (so $\rho(x,w) = \min\{\pi(x,w),\ 1\}$ according to (16.3)). The mean acceptance probability is given by

$$E[\rho(X,W)] = \int_{\pi(x,w)\geq 1} p(x)q(w)\,dx\,dw + \int_{\pi(x,w)<1} \pi(x,w)p(x)q(w)\,dx\,dw\,.$$

After several steps involving the re-expression of the integrals above (reader should show these steps), it is found that $E[\rho(X, W)] = 2\int_{\pi(x,w)\geq 1} p(x)q(w)\,dx\,dw$. The result to be proved then follows in several more steps (reader to show) by invoking the given inequality $p(\boldsymbol{x}) \leq Cq(\boldsymbol{x})$. (Incidentally, the form of M-H where $q(\boldsymbol{w}|\boldsymbol{x}) = q(\boldsymbol{w})$ is sometimes called the *independent M-H sampler.*)

16.6. For the bivariate setting here, the general expression in (16.9) simplifies considerably. In particular, $\mu_i = \mu_{\backslash i} = 0$, $\boldsymbol{\Sigma}_{i,\backslash i} = \rho$, and $\boldsymbol{\Sigma}_{\backslash i,\backslash i} = 1$. Hence, the bivariate sampling for the Gibbs sampling procedure is:

$$X_{k+1,1} \sim N(\rho X_{k2}, 1 - \rho^2) \text{ and } X_{k+1,2} \sim N(\rho X_{k+1,1}, 1 - \rho^2).$$

17.7. This problem is a direct application of the bound in property (iii) as shown in Subsection 17.2.2. The bound yields $1.098 \leq \det \boldsymbol{M}(\xi_n^*) \leq 1.5$.

17.17. (a) As in Example 17.12, consider the information number at each measurement (instead of F_n). This value is $x^2 e^{-2\theta x}$. Hence, the modified *D*-optimal criterion is $\rho x^2 \exp(-2\theta_1 x) + (1 - \rho)x^2 \exp(-2\theta_2 x)$. The derivative with respect to x yields

$$\rho[2x \exp(-2\theta_1 x) - 2\theta_1 x^2 \exp(-2\theta_1 x)] + (1 - \rho)[2x \exp(-2\theta_2 x) - 2\theta_2 x^2 \exp(-2\theta_2 x)].$$

Simplifying and setting to zero yields the following transcendental equation for x:

$$\rho \exp(-2\theta_1 x)(1 - \theta_1 x) + (1 - \rho) \exp(-2\theta_2 x)(1 - \theta_2 x) = 0.$$

This is a *starting point* for finding the support point for the design; we also need to ensure that the solution is a maximum (not a minimum or saddlepoint).

APPENDICES A – E

A.3. (b) A plot of $f(\theta, x)$ is given below. The plot depicts the continuity of $f(\theta, x)$ on $\Theta \times \Lambda$, but also illustrates that there are some "sharp edges" where the derivative will not be continuous. Hence, not all conditions of Theorem A.3 are satisfied.

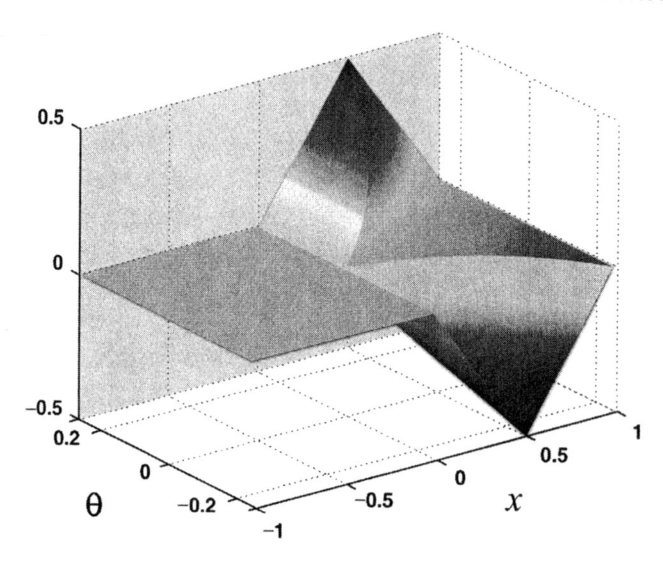

B.5.* With the exception of the choice of matrix square root, we follow the steps of Exercise B.4. Using the **chol** function in MATLAB, the upper triangular Cholesky factorization of the initial covariance matrix is $\begin{bmatrix} 1.0 & 0.8 \\ 0 & 0.6 \end{bmatrix}$. In transforming the data, note that $(\boldsymbol{\Sigma}^{1/2})^T\boldsymbol{\Sigma}^{1/2} = \boldsymbol{\Sigma}$ (but $\boldsymbol{\Sigma}^{1/2}(\boldsymbol{\Sigma}^{1/2})^T \neq \boldsymbol{\Sigma}$!). We find that the t-statistics are −0.742 and −0.734 for the matched and unmatched pairs, respectively. This leads to corresponding P-values (for a one-sided test) of 0.238 (9 degrees of freedom) and 0.236 (18 degrees of freedom), respectively. These values are close to one another. Hence, the merit of using the matched-pairs test has been almost totally lost as a result of the relatively low correlation here.

C.4. (b) Note that $E(\|\boldsymbol{X}_k - \boldsymbol{X}\|^r) = E(\|\boldsymbol{X}_k - \boldsymbol{X}\|^r \|\boldsymbol{X}_k - \boldsymbol{X}\|^0)$. Because $0 < r < q$, the Hölder inequality can be applied to the right-hand side of this expression:

$$E(\|\boldsymbol{X}_k - \boldsymbol{X}\|^r) \le \left\{E\left[\left(\|\boldsymbol{X}_k - \boldsymbol{X}\|^r\right)^{q/r}\right]\right\}^{r/q} \left\{E\left[\left(\|\boldsymbol{X}_k - \boldsymbol{X}\|^0\right)^{q/(q-r)}\right]\right\}^{(q-r)/q}$$

$$= \left\{E\left(\|\boldsymbol{X}_k - \boldsymbol{X}\|^q\right)\right\}^{r/q} \times 1.$$

But $E(\|\boldsymbol{X}_k - \boldsymbol{X}\|^q) \to 0$ as $k \to \infty$ by assumption, implying $E(\|\boldsymbol{X}_k - \boldsymbol{X}\|^r) \to 0$ from the bound above. This completes the proof.

D.1. Given $J_0 = a = c = 3$ and modulus = 5, we have $J_1 = aJ_0 + c \pmod{m} = (3\times3+3) \pmod{5} = 2$. Continuing the iterative process, it is found that $J_4 = J_0$. Therefore, this generator has period 4, which is smaller than the modulus.

E.1. From the matrix $\boldsymbol{P}^4$ in Example E.2, we hypothesize that $\bar{\boldsymbol{p}} = [0.305, 0.256, 0.439]^T$. As a first check on the validity of this solution, note that the elements sum to unity. Based on the given $\boldsymbol{P}$, it is then straightforward to check the defining relationship $\bar{\boldsymbol{p}}^T = \bar{\boldsymbol{p}}^T\boldsymbol{P}$ given in (E.2). This indicates that to within three digits, the hypothesized $\bar{\boldsymbol{p}}$ is, in fact, the actual $\bar{\boldsymbol{p}}$.

REFERENCES

Abreu, E., Lightstone, M., Mitra, S. K., and Arakawa, K. (1996), "A New Efficient Approach for the Removal of Impulse Noise from Highly Corrupted Images," *IEEE Transactions on Image Processing*, vol. 5, pp. 1012–1025.

Ahmed, M. A. and Alkhamis, T. M. (2002), "Simulation-Based Optimization Using Simulated Annealing with Ranking and Selection," *Computers and Operations Research*, vol. 29, pp. 387–402.

Akaike, H. (1974), "A New Look at the Statistical Model Identification," *IEEE Transactions on Automatic Control*, vol. AC-19, pp. 716–723.

Akaike, H. (1977), "An Objective Use of Bayesian Models," *Annals of the Institute of Statistical Mathematics*, vol. 29, pp. 9–20.

Albert, A. E. and Gardner, L. A. (1967), *Stochastic Approximation and Nonlinear Regression*, MIT Press, Cambridge, MA.

Aleksandrov, V. M., Sysoyev, V. I., and Shemeneva, V. V. (1968), "Stochastic Optimization," *Engineering Cybernetics*, vol. 5, pp. 11–16.

Allen, D. M. (1974), "The Relationship between Variable Selection and Data Augmentation and a Method for Prediction," *Technometrics*, vol. 16, pp. 125–127.

Alrefaei, M. H. and Andradóttir, S. (1999), "A Simulated Annealing Algorithm with Constant Temperature for Discrete Stochastic Optimization," *Management Science*, vol. 45, pp. 748–764.

Amari, S., Murata, N., Müller, K. R., Finke, M., and Yang, H. H. (1997), "Asymptotic Statistical Theory of Overtraining and Cross-Validation," *IEEE Transactions on Neural Networks*, vol. 8, pp. 985–996.

Anderson, B. D. O. and Moore, J. B. (1979), *Optimal Filtering*, Prentice Hall, Englewood Cliffs, NJ.

Anderson, D. R., Sweeney, D. J., and Williams, T. A. (1991), *Introduction to Statistics—Concepts and Applications* (2nd ed.), West Publishing, St. Paul, MN.

Anderssen, R. S. and Bloomfield, P. (1975), "Properties of the Random Search in Global Optimization," *Journal of Optimization Theory and Applications*, vol. 16, pp. 383–398.

Andradóttir, S. (1998), "Simulation Optimization," in *Handbook of Simulation: Principles, Methodology, Advances, Applications, and Practice* (J. Banks, ed.), Wiley, New York, Chapter 9.

Angelis, L, Bora-Senta, E., and Moyssiadis, C. (2001), "Optimal Exact Experimental Designs with Correlated Errors through a Simulated Annealing Algorithm," *Computational Statistics and Data Analysis*, vol. 37, pp. 275–296.

Apostol, T. M. (1974), *Mathematical Analysis* (2nd ed.), Addison-Wesley, Reading, MA.

Arsham, H. (1998), "Techniques for Monte Carlo Optimizing," *Monte Carlo Methods and Applications*, vol. 4, pp. 181–229.

Atkinson, A. C. (1996), "The Usefulness of Optimum Experimental Designs," *Journal of the Royal Statistical Society*, Series B, vol. 58, pp. 59–76 (see also discussion on pp. 95–111).

Atkinson, A. C. and Donev, A. N. (1992), *Optimum Experimental Designs*, Oxford University Press, Oxford.

Baba, N., Shoman, T., and Sawaragi, Y. (1977), "A Modified Convergence Theorem for a Random Optimization Method," *Information Sciences*, vol. 13, pp. 159–166.

Baker, J. E. (1985), "Adaptive Selection Methods for Genetic Algorithms," in *Proceedings of the First International Conference on Genetic Algorithms and Their Applications* (J. J. Grefenstette, ed.), L. Erlbaum Assoc., Hillsdale, NJ, pp. 101–111.

Banks, J. (ed.) (1998), *Handbook of Simulation: Principles, Methodology, Advances, Applications, and Practice*, Wiley, New York.

Barto, A. (1992), "Reinforcement Learning and Adaptive Critic Methods," in *Handbook of Intelligent Control* (D. White and S. Sofge, eds.), Van Nostrand Reinhold, New York, pp. 469–491.

Barton, R. R. and Ivey, J. S. (1996), "Nelder–Mead Simplex Modifications for Simulation Optimization," *Management Science*, vol. 42, pp. 954–973.

Baum, E. B., Boneh, D., and Garrett, C. (2001), "Where Genetic Algorithms Excel," *Evolutionary Computation*, vol. 9, pp. 93–124.

Bayard, D. S. (1991), "A Forward Method for Optimal Stochastic Nonlinear and Adaptive Control," *IEEE Transactions on Automatic Control*, vol. 36, pp. 1046–1053.

Bazaraa, M. S., Sherali, H. D., and Shetty, C. M. (1993), *Nonlinear Programming: Theory and Algorithms* (2nd ed.), Wiley, New York.

Bechhofer, R. E. (1954), "A Single-Sample Multiple Decision Procedure for Ranking Means of Normal Populations with Known Variances," *Annals of Mathematical Statistics*, vol. 25, pp. 16–39.

Bellman, R. (1960), *Introduction to Matrix Analysis*, McGraw-Hill, New York (reprinted by SIAM, 1995, ISBN 0-89871-346-3).

Benjamini, Y. and Yekutieli, D. (2001), "The Control of the False Discovery Rate in Multiple Testing Under Dependency," *Annals of Statistics*, vol. 29, pp. 1165–1188.

Benveniste, A., Metivier, M., and Priouret, P. (1990), *Adaptive Algorithms and Stochastic Approximations*, Springer-Verlag, New York (reprint of 1987 volume in French).

Berry, D. A. (1996), *Statistics: A Bayesian Perspective*, Wadsworth, Belmont, CA.

Bertoni, A. and Dorigo, M. (1993), "Implicit Parallelism in Genetic Algorithms," *Artificial Intelligence*, vol. 61, pp. 307–314.

Bertsekas, D. P. (1995a), *Dynamic Programming and Optimal Control* (vols. I and II), Athena Scientific, Belmont, MA.

Bertsekas, D. P. (1995b), "A Counterexample to Temporal Difference Learning," *Neural Computation*, vol. 7, pp. 270–279.

Besag, J., Green, P., Higdon, D., and Mengersen, K. (1995), "Bayesian Computation and Stochastic Systems" (with discussion), *Statistical Science*, vol. 10, pp. 3–66.

Beyer, H.-G. (1995), "Toward a Theory of Evolution Strategies: On the Benefits of Sex—the ($\mu/\mu,\lambda$) Theory," *Evolutionary Computation*, vol. 3, pp. 81–111.

Bhattacharya, R. N. and Rao, R. R. (1976), *Normal Approximations and Asymptotic Expansions*, Wiley, New York.

Bickel, P. J. and Doksum, K. (1977), *Mathematical Statistics: Basic Ideas and Selected Topics*, Holden-Day, San Francisco, CA.

Birge, J. R. and Louveaux, F. (1997), *Introduction to Stochastic Programming*, Springer-Verlag, New York.

Bisgaard, S. and Ankenman, B. (1996), "Standard Errors for the Eigenvalues in Second-Order Response Surface Models," *Technometrics*, vol. 38, pp. 238–246.

Bishop, C. (1992), "Exact Calculation of the Hessian Matrix for the Multilayer Perceptron," *Neural Computation*, vol. 4, pp. 494–501.

Blackmore, K. L., Williamson, R. C., Mareels, I. M. Y., and Sethares, W. A. (1997), "Online Learning via Congregational Descent," *Mathematics of Control, Signals, and Systems*, vol. 10, pp. 331–363.

Blum, J. R. (1954a), "Approximation Methods Which Converge with Probability One," *Annals of Mathematical Statistics*, vol. 25, pp. 382–386.

Blum, J. R. (1954b), "Multidimensional Stochastic Approximation Methods," *Annals of Mathematical Statistics*, vol. 25, pp. 737–744.

Bohachevsky, I. O., Johnson, M. E., and Stein, M. L. (1986), "Generalized Simulated Annealing for Function Optimization," *Technometrics*, vol. 28, pp. 209–217.

Borkar, V. S. and Meyn, S. P. (2000), "The O.D.E. Method for Convergence of Stochastic Approximation and Reinforcement Learning," *SIAM Journal on Control and Optimization*, vol. 38, pp. 447–469.

Box, G. E. P. and Draper, N. R. (1987), *Empirical Model-Building and Response Surfaces*, Wiley, New York.

Box, G. E. P. and Muller, M. E. (1958), "A Note on the Generation of Random Normal Variates," *Annals of Mathematical Statistics*, vol. 29, pp. 610–611.

Box, G. E. P. and Wilson, K. B. (1951), "On the Experimental Attainment of Optimum Conditions," *Journal of the Royal Statistical Society*, Series B, vol. 13, pp. 1–38.

Breiman, L. (1996), "Heuristics of Instability and Stabilization in Model Selection," *Annals of Statistics*, vol. 24, pp. 2350–2383.

Breiman, L. and Spector, P. (1992), "Submodel Selection and Evaluation in Regression—The Random *X* Case," *International Statistical Review*, vol. 60, pp. 291–319.

Brennan, R. W. and Rogers, P. (1995), "Stochastic Optimization Applied to a Manufacturing System Operation Problem," in *Proceedings of the Winter Simulation Conference* (C. Alexopoulos, K. Kang, W. Lilegdon, and D. Goldsman, eds.), pp. 857-864.

Brooks, S. P. and Morgan, B. J. T. (1995), "Optimization Using Simulated Annealing," *The Statistician*, vol. 44, pp. 241–257.

Cao, X.-R. (1985), "Convergence of Parameter Sensitivity Estimates in a Stochastic Experiment," *IEEE Transactions on Automatic Control*, vol. AC-30, pp. 845–853.

Cappé, O. and Robert, C. P. (2000), "Markov Chain Monte Carlo: 10 Years and Still Running!" *Journal of the American Statistical Association*, vol. 95, pp. 1282–1286.

Casella, G. and George, E. I. (1992), "Explaining the Gibbs Sampler," *The American Statistician*, vol. 46, pp. 167–174.

Cassandras, C. G. and Lafortune, S. (1999), *Introduction to Discrete Event Systems*, Kluwer Academic, Boston.

Ceasar-Spall, K. and Spall, J. C. (1997), "Regression Analysis as an Aid in Making Oboe Reeds," *Journal of Testing and Evaluation* (ASTM), vol. 25, pp. 439–444.

Cha, I. and Kassam, S. A. (1996), "RBFN Restoration of Nonlinearly Degraded Images," *IEEE Transactions on Image Processing*, vol. 5, pp. 964–975.

Chaloner, K. and Verdinelli, I. (1995), "Bayesian Experimental Design: A Review," *Statistical Science*, vol. 10, pp. 273–304.

Chaudhuri, P. and Mykland, P. A. (1995), "On Efficient Designing of Nonlinear Experiments," *Statistica Sinica*, vol. 5, pp. 421–440.

Chen, H.-F. (2002), *Stochastic Approximation and Its Application*, Kluwer Academic, Boston.

Chen, H.-F., Duncan, T. E., and Pasik-Duncan, B. (1999), "A Kiefer-Wolfowitz Algorithm with Random Differences," *IEEE Transactions on Automatic Control*, vol. 44, pp. 442–453.

Chen, M.-H, Shao, Q.-M., and Ibrahim, J. G. (2000), *Monte Carlo Methods in Bayesian Computation*, Springer-Verlag, New York.

Cherkassky, V. and Mulier, F. (1998), *Learning from Data—Concepts, Theory and Methods*, Wiley, New York.

Chib, S. (1995), "Marginal Likelihood from the Gibbs Sampler," *Journal of the American Statistical Association*, vol. 90, pp. 1313–1321.

Chib, S. and Greenberg, E. (1995), "Understanding the Metropolis–Hastings Algorithm," *The American Statistician*, vol. 49, pp. 327–335.

Chin, D. C. (1997), "Comparative Study of Stochastic Algorithms for System Optimization Based on Gradient Approximations," *IEEE Transactions on Systems, Man, and Cybernetics*, Part B, vol. 27, pp. 244–249.

Chow, Y. S. and Teicher, H. (1997), *Probability Theory: Independence, Interchangeability, and Martingales* (3rd ed.), Springer-Verlag, New York.

Chryssolouris, G., Lee, M., and Ramsey, A. (1996), "Confidence Interval Prediction for Neural Network Models," *IEEE Transactions on Neural Networks*, vol. 7, pp. 229–232.

Chung, K. L. (1954), "On a Stochastic Approximation Method," *Annals of Mathematical Statistics*, vol. 25, pp. 463–483.

Chung, K. L. (1974), *A Course in Probability Theory* (2nd ed.), Academic Press, New York.

Clausing, A. (1982), "Kantorovich-Type Inequalities," *American Mathematical Monthly*, vol. 89, pp. 314, 327–330.

Cochran, W. G. (1973), "Experiments for Nonlinear Functions," *Journal of the American Statistical Association*, vol. 68, pp. 771–781.

Cook, R. D. and Nachtsheim, C. J. (1980), "A Comparison of Algorithms for Constructing Exact *D*-Optimal Designs," *Technometrics*, vol. 22, pp. 315–324.

Cook, R. D. and Nachtsheim, C. J. (1989), "Computer-Aided Blocking of Factorial and Response Surface Designs," *Technometrics*, vol. 31, pp. 339–346.

Cox, D. R. and Reid, N. (2000), *The Theory of the Design of Experiments*, Chapman and Hall/CRC Press, Boca Raton, FL.

Culberson, J. C. (1998), "On the Futility of Blind Search: An Algorithmic View of 'No Free Lunch'," *Evolutionary Computation*, vol. 6, pp. 109–127.

Davis, L. (1991), "Bit Climbing, Representational Bias and Test Suite Design," in *Proceedings of the Fourth International Conference on Genetic Algorithms* (R. K. Belew and J. B. Booker, eds.), Morgan Kaufmann, San Francisco, CA, pp. 18–23.

Davis, L. (ed.) (1996), *Handbook of Genetic Algorithms*, International Thomson Computer Press, London (reprint of 1991 book of same title).

Dayan, P. and Sejnowski, T. J. (1994), "TD(λ) Converges with Probability 1," *Machine Learning*, vol. 14, pp. 295–301.

De Jong, K. A. (1975), "An Analysis of the Behavior of a Class of Genetic Adaptive Systems," Ph.D. dissertation, University of Michigan, Ann Arbor, MI (University Microfilms no. 76-9381).

De Jong, K. A. (1988), "Learning with Genetic Algorithms: An Overview," *Machine Learning*, vol. 3, pp. 121–137.

Delyon, B. and Juditsky, A. (1993), "Accelerated Stochastic Approximation," *SIAM Journal on Optimization*, vol. 3, pp. 868–881.

Delyon, B. and Juditsky, A. (1995), "Asymptotical Study of Parameter Tracking Algorithms," *SIAM Journal on Control and Optimization*, vol. 33, pp. 323–345.

Deming, S. N. and Parker, L. R. (1978), "A Review of Simplex Optimization of Variables in Analytical Chemistry," *CRC Critical Reviews in Analytical Chemistry*, vol. 45, pp. 278A–283A.

Dennis, J. E. and Schnabel, R. B. (1989), "A View of Unconstrained Optimization," in *Optimization* (Handbooks in OR and MS, vol. 1, G. L. Nemhauser et al., eds.), Elsevier, New York, pp. 1–72.

Dennis, J. E. and Woods, D. J. (1987), "Optimization on Microcomputers: The Nelder–Mead Simplex Algorithm," in *New Computing Environments: Microcomputers in Large Scale Computing* (A. Wouk, ed.), SIAM, Philadelphia, pp. 116–122.

Devroye, L. (1986), *Non-Uniform Random Variate Generation*, Springer-Verlag, New York.

Dippon, J. and Renz, J. (1997), "Weighted Means in Stochastic Approximation of Minima," *SIAM Journal of Control and Optimization*, vol. 35, pp. 1811–1827.

Dochain, D. and Bastin, G. (1984), "Adaptive Identification and Control Algorithms for Nonlinear Bacterial Growth Systems," *Automatica*, vol. 20, pp. 621–634.

Draper, N. R. (1988), "Response Surface Designs," in *Encyclopedia of Statistical Sciences*, vol. 8, Wiley, New York, pp. 107–119.

Dussault, J.-P., Labrecque, D., L'Ecuyer, P., and Rubinstein, R. Y. (1997), "Combining the Stochastic Counterpart and Stochastic Approximation Methods," *Discrete Event Dynamic Systems: Theory and Applications*, vol. 7, pp. 5–28.

Efron, B. and Tibshirani, R. (1986), "Bootstrap Methods for Standard Errors, Confidence Intervals, and Other Measures of Statistical Accuracy" (with discussion), *Statistical Science*, vol. 1, pp. 54-77.

Efron, B. and Tibshirani, R. (1997), "Improvements on Cross-Validation: The 0.632 + Bootstrap Method," *Journal of the American Statistical Association*, vol. 92, pp. 548–560.

Ermoliev, Y. (1969), "On the Method of Generalized Stochastic Gradients and Quasi-Fejer Sequences," *Cybernetics*, vol. 5, pp. 208–220.

Ernst, R. R. (1968), "Measurement and Control of Magnetic Field Homogeneity," *Review of Scientific Instruments*, vol. 39, pp. 998–1012.

Evans, M. and Swartz, T. (1995), "Methods for Approximating Integrals in Statistics with Special Emphasis on Bayesian Integration Problems," *Statistical Science*, vol. 10, pp. 254–272.

Evans, S. N. and Weber, N. C. (1986), "On the Almost Sure Convergence of a General Stochastic Approximation Procedure," *Bulletin of the Australian Mathematical Society*, vol. 34, pp. 335–342.

Fabian, V. (1967), "Stochastic Approximation of Minima with Improved Asymptotic Speed," *Annals of Mathematical Statistics*, vol. 38, pp. 191–200.

Fabian, V. (1968), "On Asymptotic Normality in Stochastic Approximation," *Annals of Mathematical Statistics*, vol. 39, pp. 1327–1332.

Fabian, V. (1971), "Stochastic Approximation," in *Optimizing Methods in Statistics* (J. S. Rustigi, ed.), Academic Press, New York, pp. 439–470.

Fang, H., Gong, G., and Qian, M. (1997), "Annealing of Iterative Stochastic Schemes," *SIAM Journal on Control and Optimization*, vol. 35, pp. 1886–1907.

Fedorov, V. V. (1972), *Theory of Optimal Experiments*, Academic Press, New York.

Fleming, W. (1977), *Functions of Several Variables*, Springer-Verlag, New York.

Fogel, D. B. (2000), *Evolutionary Computation: Toward a New Philosophy of Machine Intelligence* (2nd ed.), IEEE Press, Piscataway, NJ.

Fogel, L. J., Owens, A. J., and Walsh, M. J. (1966), *Artificial Intelligence through Simulated Evolution*, Wiley, New York.

Ford, I. and Silvey, S. D. (1980), "A Sequentially Constructed Design for Estimating a Nonlinear Parametric Function," *Biometrika*, vol. 67, pp. 381–388.

Ford, I., Titterington, D. M., and Kitsos, C. P. (1989), "Recent Advances in Nonlinear Experimental Design," *Technometrics*, vol. 31, pp. 49–60.

Forrest, S. and Mitchell, M. (1993), "What Makes a Problem Hard for a Genetic Algorithm? Some Anomalous Results and Their Explanation," *Machine Learning*, vol. 13, pp. 285–319.

Fox, B. L. and Heine, G. W. (1995), "Probabilistic Search with Overrides," *Annals of Applied Probability*, vol. 5, pp. 1087–1094.

Frigessi, A., Gasemyr, J., and Rue, H. (2000), "Antithetic Coupling of Two Gibbs Sampling Chains," *Annals of Statistics*, vol. 28, pp. 1128–1149.

Fu, M. C. (1990), "Convergence of a Stochastic Approximation Algorithm for the *GI*/*G*/1 Queue Using Infinitesimal Perturbation Analysis," *Journal of Optimization Theory and Applications*, vol. 65, pp. 149–160.

Fu, M. C. (1994), "Optimization via Simulation: A Review," *Annals of Operations Research*, vol. 53, pp. 199–248.

Fu, M. C. (2002), "Optimization for Simulation: Theory vs. Practice" (with discussion by S. Andradóttir, P. Glynn, and J. P. Kelly), *INFORMS Journal on Computing*, vol. 14, pp. 192–227.

Fu, M. C. and Hill, S. D. (1997), "Optimization of Discrete Event Systems via Simultaneous Perturbation Stochastic Approximation," *IIE Transactions*, vol. 29, pp. 233–243.

Fu, M. C. and Hu, J.-Q. (1997), *Conditional Monte Carlo: Gradient Estimation and Optimization Applications*, Kluwer Academic, Boston.

Geisser, S. (1975), "The Predictive Sample Reuse Method with Applications," *Journal of the American Statistical Association*, vol. 70, pp. 320–328.

Gelfand, A. E. (2000), "Gibbs Sampling," *Journal of the American Statistical Association*, vol. 95, pp. 1300–1304.

Gelfand, A. E. and Smith A. F. M. (1990), "Sampling-Based Approaches to Calculating Marginal Densities," *Journal of the American Statistical Association*, vol. 85, pp. 399–409.

Gelfand, S. and Mitter, S. K. (1989), "Simulated Annealing with Noisy or Imprecise Energy Measurements," *Journal of Optimization Theory and Applications*, vol. 62, pp. 49–62.

Gelfand, S. and Mitter, S. K. (1991), "Simulated Annealing Type Algorithms for Multivariate Optimization," *Algorithmica*, vol. 6, pp. 419–436.

Gelfand, S. and Mitter, S. K. (1993), "Metropolis-Type Annealing Algorithms for Global Optimization in $\mathbf{R}^d$," *SIAM Journal on Control and Optimization*, vol. 31, pp. 111–131.

Geman, S. and Geman, D. (1984), "Stochastic Relaxation, Gibbs Distributions, and the Bayesian Restoration of Images," *IEEE Transactions on Pattern Analysis and Machine Intelligence*, vol. PAMI-6, pp. 721–741.

Geman, S. and Hwang, C.-R. (1986), "Diffusions for Global Optimization," *SIAM Journal on Control and Optimization*, vol. 24, pp. 1031–1043.

Geman, S., Bienenstock, E., and Doursat, R. (1992), "Neural Networks and the Bias/Variance Dilemma," *Neural Computation*, vol. 4, pp. 1–58.

George, E. I. and McCulloch, R. E. (1997), "Approaches for Bayesian Variable Selection," *Statistica Sinica*, vol. 7, pp. 339–373.

Gerencsér, L. (1993), "Strong Approximation of the Recursive Prediction Error Estimator of the Parameters of an ARMA Process," *System and Control Letters*, vol. 21, pp. 347–351.

Gerencsér, L. (1995), "Rate of Convergence of the LMS Method," *System and Control Letters*, vol. 24, pp. 385–388.

Gerencsér, L. (1999), "Convergence Rate of Moments in Stochastic Approximation with Simultaneous Perturbation Gradient Approximation and Resetting," *IEEE Transactions on Automatic Control*, vol. 44, pp. 894–905.

Gerencsér, L. and Vágó, Z. (2001), "The Mathematics of Noise-Free SPSA," in *Proceedings of the IEEE Conference on Decision and Control*, 4–7 December 2001, Orlando, FL, pp. 4400–4405.

Gerencsér, L, Hill, S. D., and Vágó, Z. (1999), "Fixed Gain SPSA for Discrete Optimization," in *Proceedings of the IEEE Conference on Decision and Control*, 7–10 December 1999, Phoenix, AZ, pp. 1791–1795.

Gilks, W. R., Richardson, S., and Spiegelhalter, D. J. (eds.) (1996), *Markov Chain Monte Carlo in Practice*, Chapman and Hall, London.

Gilks, W. R., Roberts, G. O., and Sahu, S. K. (1998), "Adaptive Markov Chain Monte Carlo Through Regeneration," *Journal of the American Statistical Association*, vol. 93, pp. 1045–1054.

Glasserman, P. (1991a), *Gradient Estimation via Perturbation Analysis*, Kluwer Academic, Boston.

Glasserman, P. (1991b), "Derivative Estimates from Simulation of Continuous-Time Markov Chains," *Operations Research*, vol. 39, pp. 724–738.

Glasserman, P. and Yao, D. D. (1992), "Some Guidelines and Guarantees for Common Random Numbers," *Management Science*, vol. 38, pp. 884–908.

Glynn, P. (1987), "Likelihood Ratio Gradient Estimation: An Overview," in *Proceedings of the Winter Simulation Conference* (A. Thesen, H. Grant, W. D. Kelton, eds.), pp. 366–375.

Glynn, P. (1990), "Likelihood Ratio Gradient Estimation for Stochastic Systems," *Communications of the ACM*, vol. 33(10), pp. 75–84.

Goldberg, D. E. (1987), "Simple Genetic Algorithms and the Minimal Deceptive Problem," in *Genetic Algorithms and Simulated Annealing* (L. Davis, ed.), Morgan Kaufmann, San Francisco, CA, pp. 74–88.

Goldberg, D. E. (1989), *Genetic Algorithms in Search, Optimization, and Machine Learning*, Addison-Wesley, Reading, MA.

Goldberg, D. E. (2002), *The Design of Innovation: Lessons from and for Competent Genetic Algorithms*, Kluwer Academic, Boston.

Goldberg, D. E. and Deb, K. (1991), "A Comparative Analysis of Selection Schemes Used in Genetic Algorithms," in *Foundations of Genetic Algorithms* (G. Rawlins, ed.), Morgan Kaufmann, San Francisco, CA, pp. 69–93.

Goldberg, D. E., Deb, K., Kargupta, H., and Harik, G. (1993), "Rapid, Accurate Optimization of Difficult Problems Using Fast Messy Genetic Algorithms," in *Proceedings of the Fifth International Conference on Genetic Algorithms* (S. Forrest, ed.), Morgan Kaufmann, San Francisco, CA, pp. 56–64.

Goldsman, D. and Nelson, B. L. (1998), "Comparing Systems via Simulation," in *Handbook of Simulation: Principles, Methodology, Advances, Applications, and Practice* (J. Banks, ed.), Wiley, New York, Chapter 8.

Goldsman, D., Nelson, B. L., Opicka, T., and Pritsker, A. A. B. (1999), "A Ranking and Selection Project: Experiences from a University–Industry Collaboration," in *Proceedings of the Winter Simulation Conference* (P. Farrington, H. Nembhard, D. Sturrock, and G. Evans, eds.), 5–8 December 1999, Phoenix, AZ, pp. 83–92.

Golub, G. H. and Van Loan, C. E. (1989), *Matrix Computations* (2nd ed.), Johns Hopkins University Press, Baltimore, MD.

Gonzalez, R. C. and Woods, R. E. (1992), *Digital Image Processing*, Addison-Wesley, Reading, MA.

Goodsell, C. A. and Hanson, D. L. (1976), "Almost Sure Convergence for the Robbins-Monro Process," *Annals of Probability*, vol. 4, pp. 890–901.

Goodwin, G. C. and Payne, R. L. (1977), *Dynamic System Identification: Experiment Design and Data Analysis*, Academic Press, New York.

Graybill, F. A. (1983), *Matrices with Applications in Statistics* (2nd ed.), Wadsworth, Belmont, CA.

Griewank, A. and Corliss, G. F. (eds.) (1991), *Automatic Differentiation of Algorithms: Theory, Implementation, and Applications*, SIAM, Philadelphia.

Grubbs, F. E. (1964), "Approximate Circular and Noncircular Offset Probabilities of Hitting," *Operations Research*, vol. 12, pp. 51–62.

Gu, M. G. and Kong, F. H. (1998), "A Stochastic Approximation Algorithm with Markov Chain Monte Carlo Method for Incomplete Data Estimation Problems," *Proceedings of the National Academy of Sciences USA*, vol. 95, pp. 7270–7274.

Gu, M. G. and Zhu, H.-T. (2001), "Maximum Likelihood Estimation for Spatial Models by Markov Chain Monte Carlo Stochastic Approximation," *Journal of the Royal Statistical Society*, Series B, vol. 63, part 2, pp. 339–355.

Guo, L. and Ljung, L. (1995), "Performance Analysis of General Tracking Algorithms," *IEEE Transactions on Automatic Control*, vol. 40, pp. 1388–1402.

Gupta, S. S. (1965), "On Some Multiple Decision (Selection and Ranking) Rules," *Technometrics*, vol. 7, pp. 225–245.

Haines, L. M. (1995), "A Geometric Approach to Optimal Design for One-Parameter Nonlinear Models," *Journal of the Royal Statistical Society*, Series B, vol. 57, pp. 575–598.

Hajek, B. (1988), "Cooling Schedules for Optimal Annealing," *Mathematics of Operations Research*, vol. 13, pp. 311–329.

Hamada, M., Martz, H. F., Reese, C. S., and Wilson, A. G. (2001), "Finding Near Optimal Bayesian Experimental Designs via Genetic Algorithms," *American Statistician*, vol. 55, pp. 175–181.

Hart, W. E. (1997), "A Stationary Point Convergence Theory for Evolutionary Algorithms," in *Foundations of Genetic Algorithms 4* (R. K. Belew and M. D. Vose, eds.), Morgan Kaufmann, San Francisco, CA, pp. 325–342.

Harville, D. A. (2001), *Matrix Algebra: Exercises and Solutions*, Springer-Verlag, New York.

Hastings, W. K. (1970), "Monte Carlo Sampling Methods Using Markov Chains and their Applications," *Biometrika*, vol. 57, pp. 97–109.

Haykin, S. (1996), *Adaptive Filter Theory* (3rd ed.), Prentice Hall, Upper Saddle River, NJ.

Haykin, S. (1999), *Neural Networks: A Comprehensive Foundation* (2nd ed.), Prentice Hall, Upper Saddle River, NJ.

Hayter, A. J. (1984), "A Proof of the Conjecture that the Tukey–Kramer Procedure Multiple Comparisons Procedure is Conservative," *Annals of Statistics*, vol. 12, pp. 61–75.

Heckendorn, R. B. and Whitley, D. (1999), "Predicting Epistasis from Mathematical Models," *Evolutionary Computation*, vol. 7, pp. 69–101.

Hill, S. D. and Spall, J. C. (2000), "Inequality-Based Reliability Estimates for Complex Systems," in *Proceedings of the American Control Conference*, 28–30 June 2000, Chicago, IL, pp. 2704–2705.

Ho, Y.-C. (1987), Performance Evaluation and Perturbation Analysis of Discrete Event Dynamic Systems," *IEEE Transactions on Automatic Control*, vol. AC-32, pp. 563–572.

Ho, Y.-C. (1997), "On the Numerical Solutions of Stochastic Optimization Problems," *IEEE Transactions on Automatic Control*, vol. 42, pp. 727–729.

Ho, Y.-C. and Cassandras, C. G. (1997), "Perturbation Analysis for Control and Optimization of Queueing Systems: An Overview and State of the Art," in *Frontiers in Queueing* (J. Dshalalow, ed.), CRC Press, Boca Raton, FL, pp. 395–420.

Ho, Y.-C., Sreenivas, R., and Vakili, P. (1992), "Ordinal Optimization of Discrete Event Dynamic Systems," *Journal of Discrete Event Dynamic Systems*, vol. 2, pp. 61–88.

Hoadley, B. (1971), "Asymptotic Properties of Maximum Likelihood Estimates for the Independent Not Identically Distributed Case," *Annals of Mathematical Statistics*, vol. 42, pp. 1977–1991.

Hochberg, Y. and Tamhane, A. C. (1987), *Multiple Comparison Procedures*, Wiley, New York.

Holland, J. H. (1975), *Adaptation in Natural and Artificial Systems*, University of Michigan Press, Ann Arbor, MI.

Hollander, M. and Wolfe, D. A. (1999), *Nonparametric Statistical Methods* (2nd ed.), Wiley, New York.

Hromkovič, J. (2001), *Algorithmics for Hard Problems: Introduction to Combinatorial Optimization, Randomization, Approximation, and Heuristics*, Springer-Verlag, New York.

Hsu, J. C. (1996), *Multiple Comparisons Theory and Methods*, Chapman and Hall, London.

Hui, S. L. and Berger, J. O. (1983), "Empirical Bayes Estimation of Rates in Longitudinal Studies," *Journal of the American Statistical Association*, vol. 78, pp. 753–760.

Hwang, J. T. G. and Ding, A. A. (1997), "Prediction Intervals for Artificial Neural Networks," *Journal of the American Statistical Association*, vol. 92, pp. 748–757.

Iosifescu, M. (1980), *Finite Markov Processes and Their Applications*, Wiley, New York.

Jaakkola, T., Jordan, M. I., and Singh, S. P. (1994), "On the Convergence of Stochastic Iterative Dynamic Programming Algorithms," *Neural Computation*, vol. 6, pp. 1185–1201.

Jacobson, S. H. (1994), "Convergence Results for Harmonic Gradient Estimators," *ORSA Journal of Computing*, vol. 6, pp. 381–397.

Jang, J.-S. R., Sun, C.-T., and Mizutani, E. (1997), *Neuro-Fuzzy and Soft Computing*, Prentice Hall, Upper Saddle River, NJ.

Kallel, L., Naudts, B., and Rogers, A. (eds.) (2001), *Theoretical Aspects of Evolutionary Computing*, Springer-Verlag, New York.

Kalman, R. E. (1960), "A New Approach to Linear Filtering and Prediction Problems," *Transactions of the ASME, Journal of Basic Engineering*, Series D, vol. 82, pp. 35–45.

Kargupta, H. and Goldberg, D. E. (1997), "SEARCH, Blackbox Optimization, and Sample Complexity," in *Foundations of Genetic Algorithms 4* (R. K. Belew and M. D. Vose, eds.), Morgan Kaufmann, San Francisco, CA, pp. 291–324.

Karnopp, D. C. (1963), "Random Search Techniques for Optimization Problems," *Automatica*, vol. 1, pp. 111–121.

Kelly, C. T. (1999a), *Iterative Methods for Optimization*, SIAM, Philadelphia.

Kelly, C. T. (1999b), "Detection and Remediation of Stagnation in the Nelder–Mead Algorithm Using a Sufficient Decrease Condition," *SIAM Journal on Optimization*, vol. 10, pp. 43–55.

Kesten, H. (1958), "Accelerated Stochastic Approximation," *Annals of Mathematical Statistics*, vol. 29, pp. 41–59.

Khuri, A. L. and Cornell, J. A. (1987), *Response Surfaces: Design and Analysis*, Marcel Dekker, New York.

Kiefer, J. and Wolfowitz, J. (1952), "Stochastic Estimation of a Regression Function," *Annals of Mathematical Statistics*, vol. 23, pp. 462–466.

Kiefer, J. C. and Wolfowitz, J. (1960), "The Equivalence of Two Extremum Problems," *Canadian Journal of Mathematics*, vol. 12, pp. 363–366.

Kirkpatrick, S., Gelatt, C. D., and Vecchi, M. P. (1983), "Optimization by Simulated Annealing," *Science*, vol. 220, pp. 671–680.

Kleijnen, J. P. C. (1998), "Experimental Design for Sensitivity Analysis, Optimization, and Validation of Simulation Models," in *Handbook of Simulation: Principles, Methodology, Advances, Applications, and Practice* (J. Banks, ed.), Wiley, New York, Chapter 6.

Kleinman, N. L., Spall, J. C., and Naiman, D. Q. (1999), "Simulation-Based Optimization with Stochastic Approximation Using Common Random Numbers," *Management Science*, vol. 45, pp. 1570–1578.

Kmenta, J. (1997), *Elements of Econometrics* (2nd ed.), University of Michigan Press, Ann Arbor, MI.

Knuth, D. E. (1981), *Seminumerical Algorithms*, Addison-Wesley, Reading, MA.

Koehler, G. J., Bhattacharyya, S., and Vose, M. D. (1998), "General Cardinality Genetic Algorithms," *Evolutionary Computation*, vol. 5, pp. 439–459.

Koronacki, J. (1975), "Random-Seeking Methods for the Stochastic Unconstrained Optimization," *International Journal of Control*, vol. 21, pp. 517–527.

Kosko, B. (1992), *Neural Networks and Fuzzy Systems*, Prentice-Hall, Englewood Cliffs, NJ.

Koza, J. R. (1992), *Genetic Programming: On the Programming of Computers by Means of Natural Selection*, MIT Press, Cambridge, MA.

Kramer, C. Y. (1956), "Extensions of Multiple Range Tests to Group Means with Unequal Numbers of Replications," *Biometrics*, vol. 12, pp. 307–310.

Krishnakumar, K. and Goldberg, D. E. (1992), "Control System Optimization using Genetic Algorithms," *Journal of Guidance and Control* (AIAA), vol. 15, pp. 735–740.

Kuan, C.-M. and Hornik, K. (1991), "Convergence of Learning Algorithms with Constant Learning Rates," *IEEE Transactions on Neural Networks*, vol. 2, pp. 484–489.

Kushner, H. J. (1987), "Asymptotic Global Behavior for Stochastic Approximation and Diffusions with Slowly Decreasing Noise Effects: Global Minimization via Monte Carlo," *SIAM Journal on Applied Mathematics*, vol. 47, pp. 169–185.

Kushner, H. J. and Clark, D. S. (1978), *Stochastic Approximation Methods for Constrained and Unconstrained Systems*, Springer-Verlag, New York.

Kushner, H. J. and Huang. H. (1983), "Asymptotic Properties of Stochastic Approximations with Constant Coefficients," *SIAM Journal on Control and Optimization*, vol. 19, pp. 87–105.

Kushner, H. J. and Yang, J. (1993), "Stochastic Approximation with Averaging: Optimal Rates of Convergence for General Processes," *SIAM Journal on Control and Optimization*, vol. 31, pp. 1045–1062.

Kushner, H. J. and Yang, J. (1995), "Stochastic Approximation with Averaging and Feedback: Rapidly Convergent On-Line Algorithms," *IEEE Transactions on Automatic Control*, vol. 40, pp. 24–34.

Kushner, H. J. and Yin, G. G. (1997), *Stochastic Approximation Algorithms and Applications*, Springer-Verlag, New York.

Lagarias, J. C., Reeds, J. A., Wright, M. A., and Wright, P. E. (1998), "Convergence Properties of the Nelder–Mead Simplex Method in Low Dimensions," *SIAM Journal on Optimization*, vol. 9, pp. 112–147.

Laha, R. G. and Rohatgi, V. K. (1979), *Probability Theory*, Wiley, New York.

Lai, T. L. (1985), "Stochastic Approximation and Sequential Search for Optimum," in *Proceedings of the Berkeley Conference in Honor of Jerzy Neyman and Jack Kiefer*, vol. II (L. M. Le Cam and R. A. Olshen, eds.), Wadsworth, Belmont, CA, pp. 557–577.

Lai, T. L. (2001), "Sequential Analysis: Some Classical Problems and New Challenges" (with discussion), *Statistica Sinica*, vol. 11, pp. 303–408.

Landau, I. D., Lozano, R., and M'Saad, M. (1998), *Adaptive Control*, Springer-Verlag, New York.

Law, A. M. and Kelton, W. D. (2000), *Simulation Modeling and Analysis* (3rd ed.), McGraw-Hill, New York.

L'Ecuyer, P. (1990a), "A Unified View of the IPA, SF, and LR Gradient Estimation Techniques," *Management Science*, vol. 36, pp. 1364–1383.

L'Ecuyer, P. (1990b), "Random Numbers for Simulation," *Communications of the ACM*, vol. 33(10), pp. 85–97.

L'Ecuyer, P. (1991), "An Overview of Derivative Estimation," in *Proceedings of the Winter Simulation Conference* (B. Nelson, W. Kelton, and G. Clark, eds.), pp. 207–217.

L'Ecuyer, P. (1998), "Random Number Generation," in *Handbook of Simulation: Principles, Methodology, Advances, Applications, and Practice* (J. Banks, ed.), Wiley, New York, Chapter 4.

L'Ecuyer, P. and Glynn, P. W. (1994), "Stochastic Optimization by Simulation: Convergence Proofs for the *GI/G/*1 Queue in Steady State," *Management Science*, vol. 40, pp. 1562–1578.

L'Ecuyer, P. and Perron, G. (1994), "On the Convergence Rates of IPA and FDC Derivative Estimators for Finite-Horizon Stochastic Simulations," *Operations Research*, vol. 42, pp. 643–656.

L'Ecuyer, P. and Yin, G. (1998), "Budget-Dependent Convergence Rate of Stochastic Approximation," *SIAM Journal on Optimization*, vol. 8, pp. 217–247.

L'Ecuyer, P., Giroux, N., and Glynn, P. W. (1994), "Stochastic Optimization by Simulation: Numerical Experiments with the *M/M/*1 Queue in Steady State," *Management Science*, vol. 40, pp. 1245–1261.

Lehmann, E. L. (1975), *Nonparametrics: Statistical Methods Based on Ranks*, Holden-Day, San Francisco, CA (reprinted by Prentice Hall in paperback form in 1998, ISBN: 013997735X).

Lennon, W. K. and Passino, K. M. (1999), "Techniques for Genetic Adaptive Control," in *Soft Computing and Intelligent Systems: Theory and Applications* (M. Gupta and N. Sinha, eds.), Academic Press, New York, pp. 257–278.

Levadi, V. S. (1966), "Design of Input Signals for Parameter Estimation," *IEEE Transactions on Automatic Control*, vol. AC-11, pp. 205–211.

Lewis, P. A. W., Goodman, A. S., and Miller, J. M. (1969), "A Pseudo-Random Number Generator for the System/360," *IBM Systems Journal*, vol. 8, pp. 136–143.

Li, J. and Rhinehart, R. R. (1998), "Heuristic Random Optimization," *Computers and Chemical Engineering*, vol. 22, pp. 427–444.

Linhart, H. and Zucchini, W. (1986), *Model Selection*, Wiley, New York.

Liu, J. S. (2001), *Monte Carlo Strategies in Scientific Computing*, Springer-Verlag, New York.

Ljung, L. (1977), "Analysis of Recursive Stochastic Algorithms," *IEEE Transactions on Automatic Control*, vol. AC-22, pp. 551–575.

Ljung, L. (1984), "Analysis of Stochastic Gradient Algorithms for Linear Regression Problems," *IEEE Transactions on Information Theory*, vol. IT-30, pp. 151–160.

Ljung, L. (1999), *System Identification—Theory for the User* (2nd ed.), Prentice Hall PTR, Upper Saddle River, NJ.

Ljung, L. and Soderstrom, T. (1983), *Theory and Practice of Recursive Identification*, MIT Press, Cambridge, MA.

Ljung, L., Pflug, G., and Walk, H. (1992), *Stochastic Approximation and Optimization of Random Systems*, Birkhäuser, Basel.

Lucas, J. M. (1976), "Which Response Surface Design is Best," *Technometrics*, vol. 18, pp. 411–417.

Luman, R. R. (2000), "Upgrading Complex Systems of Systems: A CAIV Methodology for Warfare Area Requirements Analysis," *Military Operations Research*, vol. 5(2), pp. 53–75.

Lunneborg, C. E. (2000), *Data Analysis by Resampling: Concepts and Applications*, Duxbury Press, Pacific Grove, CA.

Macchi, O. and Eweda, E. (1983), "Second-Order Convergence Analysis of Stochastic Adaptive Linear Filtering," *IEEE Transactions on Automatic Control*, vol. AC-28, pp. 76–85.

Maeda, Y. and De Figueiredo, R. J. P. (1997), "Learning Rules for Neuro-Controller via Simultaneous Perturbation," *IEEE Transactions on Neural Networks*, vol. 8, pp. 1119–1130.

Mardia, K. V., Kent, J. T., and Bibby, J. M. (1979), *Multivariate Analysis*, Academic Press, New York.

Marsaglia, G. (1985), "A Current View of Random Number Generators," in *Computer Science and Statistics: The Interface* (L. Billard, ed.), Elsevier, New York, pp. 3–10.

Marsaglia, G. (1991), "A New Class of Random Number Generators," *Annals of Applied Probability*, vol. 1, pp. 462–480.

Maryak, J. L. and Chin, D. C. (2001), "Global Random Optimization by Simultaneous Perturbation Stochastic Approximation," in *Proceedings of the American Control Conference*, 25–27 June 2001, Arlington, VA, pp. 756–762.

Matyas, J. (1965), "Random Optimization," *Automation and Remote Control*, vol. 26, pp. 244–251.

McCullough, B. D. and Wilson, B. (2002), "On the Accuracy of Statistical Procedures in Microsoft Excel 2000 and Excel XP," *Computational Statistics and Data Analysis*, vol. 40, pp. 713–721.

McKinnon, K. I. M. (1998), "Convergence of the Nelder–Mead Simplex Algorithm to a Nonstationary Point," *SIAM Journal on Optimization*, vol. 9, pp. 148–158.

McQuarrie, A. D. R. and Tsai, C.-L. (1998), *Regression and Time Series Model Selection*, World Scientific, Singapore.

Mehra, R. K. (1974), "Optimal Input Signals for Parameter Estimation in Dynamic Systems—Survey and New Results," *IEEE Transactions on Automatic Control*, vol. AC-19, pp. 753–768.

Metivier, M. and Priouret, P. (1984), "Applications of a Kushner and Clark Lemma for General Classes of Stochastic Algorithms," *IEEE Transactions on Information Theory*, vol. IT-30, pp. 140–151.

Metropolis, N., Rosenbluth, A., Rosenbluth, M. Teller, A., and Teller, E. (1953), "Equation of State Calculations by Fast Computing Machines," *Journal of Chemical Physics*, vol. 21, pp. 1087–1092.

Meyn, S. P. and Tweedie, R. L. (1993), *Markov Chains and Stochastic Stability*, Springer-Verlag, New York.

Michalewicz, Z. (1996), *Genetic Algorithms + Data Structures = Evolution Programs* (3rd ed.), Springer-Verlag, New York.

Michalewicz, Z. and Fogel, D. B. (2000), *How to Solve It: Modern Heuristics*, Springer-Verlag, New York.

Michalski, R. S., Bratko, I., and Kubat, M. (eds.) (1998), *Machine Learning and Data Mining*, Wiley, New York.

Miller, A. and Wu, C. F. J. (1996), "Parameter Design for Signal Response Systems: A Different Look at Taguchi's Dynamic Parameter Design," *Statistical Science*, vol. 11, pp. 122–136.

Miller, R. G. (1981), *Simultaneous Statistical Inference* (2nd ed.), Springer-Verlag, New York.

Mitchell, M. (1996), *An Introduction to Genetic Algorithms*, MIT Press, Cambridge, MA.

Mitchell, T. J. (1974), "An Algorithm for the Construction of '*D*-Optimum' Experimental Designs," *Technometrics*, vol. 16, pp. 203–210.

Mitchell, T. M. (1997), *Machine Learning*, McGraw-Hill, New York.

Mitra, M. and Park, S. K. (1991), "Solution to the Indexing Problem of Frequency Domain Simulation Experiments," in *Proceedings of the Winter Simulation Conference* (B. Nelson, W. Kelton, and G. Clark, eds.), pp. 907–915.

Montgomery, D. C. (2001), *Design and Analysis of Experiments* (5th ed.), Wiley, New York.

Montgomery, D. C. and Runger, G. C. (1999), *Applied Statistics and Probability for Engineers* (2nd ed.), Wiley, New York.

Moon, T. K. and Stirling, W. C. (2000), *Mathematical Methods and Algorithms for Signal Processing*, Prentice Hall, Upper Saddle River, NJ.

Moré, J. J., Garbow, B. S., and Hillstrom, K. E. (1981), "Testing Unconstrained Optimization Software," *ACM Transactions on Mathematical Software*, vol. 7, pp. 17–41.

Mukherjee, S. and Fine, T. L. (1996), "Online Steepest Descent Yields Weights with Nonnormal Limiting Distributions," *Neural Computation*, vol. 8, pp. 1075–1084.

Müller, P. and Parmigiani, G. (1995), "Optimal Design via Curve Fitting of Monte Carlo Experiments," *Journal of the American Statistical Association*, vol. 90, pp. 1322–1330.

Naiman, D. Q. and Priebe, C. (2001), "Computing Scan Statistic *P*-Values Using Importance Sampling, with Applications to Genetics and Medical Image Analysis," *Journal of Computational and Graphical Statistics*, vol. 10, pp. 296–328.

Naiman, D. Q. and Wynn, H. P. (1992), "Inclusion-Exclusion Bonferroni Identities and Inequalities for Discrete Tube-Like Problems via Euler Characteristics," *Annals of Statistics*, vol. 20, pp. 43–76.

Nair, V. N. (ed.) (1992), "Taguchi's Parameter Design: A Panel Discussion," *Technometrics*, vol. 34, pp. 127–161.

Nakayama, M. K. (1997), "Multiple Comparison Procedures for Steady-State Simulations," *Annals of Statistics*, vol. 25, pp. 2433–2450.

Nandi, S., Ghosh, S., Tambe, S. S., and Kulkarni, B. D. (2001), "Artificial Neural-Network-Assisted Stochastic Process Optimization Strategies," *AIChE Journal*, vol. 47, pp. 126–141.

Naudts, B. and Kallel, L. (2000), "A Comparison of Predictive Measures for Problem Difficulty in Evolutionary Algorithms," *IEEE Transactions on Evolutionary Computation*, vol. 4, pp. 1–15.

Neiderreiter, H. (1992), *Random Number Generation and Quasi-Monte Carlo Methods*, SIAM, Philadelphia.

Nelder, J. A. and Mead, R. (1965), "A Simplex Method for Function Minimization," *The Computer Journal*, vol. 7, pp. 308–313.

Nelson, B. L. and Goldsman, D. (2001), "Comparisons with a Standard in Simulation Experiments," *Management Science*, vol. 47, pp. 449–463.

Nelson, B. L. and Matejcik, F. J. (1995), "Using Common Random Numbers for Indifference-Zone Selection and Multiple Comparisons in Simulation," *Management Science*, vol. 41, pp. 1935–1945.

Nelson, B. L., Swann, J., Goldsman, D., and Song, W. (2001), "Simple Procedures for Selecting the Best Simulated System when the Number of Alternatives is Large," *Operations Research*, vol. 49, pp. 950–963.

Neter, J., Kutner, M. H., Nachtsheim, C. J., and Wasserman, W. (1996), *Applied Linear Statistical Models* (4th ed.), Irwin, Chicago.

Nevel'son, M. B. and Has'minskii, R. Z. (1973), *Stochastic Approximation and Recursive Estimation*, American Mathematical Society, Providence, RI (publication year is 1973; copyright year is 1976).

Nevel'son, M. B. and Khas'minskii, R. Z. (1973), "An Adaptive Robbins-Monro Procedure," *Automation and Remote Control*, vol. 34, pp. 1594–1607. (Note: Second author is the same as in preceding reference; different translation from the Russian accounts for different spelling.)

Nissen, V. and Propach, J. (1998), "On the Robustness of Population-Based versus Point-Based Optimization in the Presence of Noise," *IEEE Transactions on Evolutionary Computation*, vol. 2, pp. 107–119.

Okamura, A., Kirimot, T., and Kondo, M. (1995), "A New Normalized Stochastic Approximation Algorithm Using a Time-Shift Parameter," *Electronics and Communications in Japan*, Part 3, vol. 78, pp. 41–51.

Olsson, C. K. (1993), "Simulated Annealing in Image Processing," in *Applied Simulated Annealing* (R. V. V. Vidal, ed.), Springer-Verlag, New York, pp. 313–334.

Park, S. K. and Miller, K. W. (1988), "Random Number Generators: Good Ones are Hard to Find," *Communications of the ACM*, vol. 31, pp. 1192–1201.

Parzen, E. (1962), *Stochastic Processes*, Holden-Day, New York.

Petridis, V., Kazarlis, S., and Bakirtzis, A. (1998), "Varying Fitness Functions in Genetic Algorithm Constrained Optimization: The Cutting Stock and Unit Commitment Problems," *IEEE Transactions on Systems, Man, and Cybernetics*, Part B, vol. 28, pp. 629–640.

Pflug, G. Ch. (1986), "Stochastic Minimization with Constant Step Size: Asymptotic Laws," *SIAM Journal on Control and Optimization*, vol. 24, pp. 655–666.

Pflug, G. Ch. (1996), *Optimization of Stochastic Models: The Interface Between Simulation and Optimization*, Kluwer Academic, Boston.

Pineda, F. (1997), "Mean-Field Theory for Batched TD(λ)," *Neural Computation*, vol. 9, pp. 1403–1419.

Polak, E. (1997), *Optimization: Algorithms and Consistent Approximations*, Springer-Verlag, New York.

Polyak, B. T. and Juditsky, A. B. (1992), "Acceleration of Stochastic Approximation by Averaging," *SIAM Journal on Control and Optimization*, vol. 30, pp. 838–855.

Polyak, B. T. and Tsypkin, Y. Z. (1973), "Pseudogradient Adaptation and Training Algorithms," *Automation and Remote Control*, vol. 34, pp. 377–397.

Press, W. H., Teukolsky, S. A., Vetterburg, W. T., and Flannery, B. R. (1992), *Numerical Recipes in C* (2nd ed.), Cambridge University Press, Cambridge.

Pukelsheim, F. (1993), *Optimal Design of Experiments*, Wiley, New York.

Qi, X. and Palmeiri, F. (1994), "Theoretical Analysis of Evolutionary Algorithms with Infinite Population Size in Continuous Space, Part I: Basic Properties," *IEEE Transactions on Neural Networks*, vol. 5, pp. 102–119.

Rana, S. B. and Whitley, L. D. (1997), "Bit Representations with a Twist," in *Proceedings of the Seventh International Conference on Genetic Algorithms* (T. Bäck, ed.), Morgan Kaufmann, San Francisco, CA, pp. 188–195.

Rana, S. B., Whitley, L. D., and Cogswell, R. (1996), "Searching in the Presence of Noise," in *Parallel Problem Solving from Nature—PPSN IV* (H. Voigt et al., eds.), Springer-Verlag, New York, pp. 198–207.

Rao, C. R. (1973), *Linear Statistical Inference and Its Applications* (2nd ed.), Wiley, New York.

Rechenberg, I. (1965), "Cybernetic Solution Path of an Experimental Problem," Royal Aircraft Translation 1122, Ministry of Aviation (U.K.).

Reeves, C. and Wright, C. (1994), "An Experimental Design Perspective on Genetic Algorithms," in *Foundations of Genetic Algorithms 3* (L. D. Whitley and M. Vose, eds.), Morgan Kaufmann, San Francisco, CA, pp. 7–22.

Reidys, C. M. and Stadler, P. F. (2002), "Combinatorial Landscapes," *SIAM Review*, vol. 44, pp. 3–54.

Rinott, Y. (1978), "On Two-Stage Selection Procedures and Related Probability Inequalities," *Communications in Statistics—Theory and Methods*, vol. 7, pp. 799–811.

Rissanen, J. (1978), "Modeling by Shortest Data Description," *Automatica*, vol. 14, pp. 465–471.

Robbins, H. and Monro, S. (1951), "A Stochastic Approximation Method," *Annals of Mathematical Statistics*, vol. 22, pp. 400–407.

Robert, C. P. and Casella, G. (1999), *Monte Carlo Statistical Methods*, Springer-Verlag, New York.

Roberts, G. O. and Rosenthal, J. S. (1998), "Markov Chain Monte Carlo: Some Practical Implications of Theoretical Results," *Canadian Journal of Statistics*, vol. 26, pp. 5–32.

Roberts, G. O. and Rosenthal, J. S. (2001), "Optimal Scaling for Various Metropolis–Hastings Algorithms," *Statistical Science*, vol. 16, pp. 351–367.

Roberts, G. O., Gelman, A., and Gilks, W. R. (1997), "Weak Convergence and Optimal Scaling of Random Walk Metropolis Algorithms," *Annals of Applied Probability*, vol. 7, pp. 110–120.

Robinson, S. M. (1996), "Analysis of Sample Path Optimization," *Mathematics of Operations Research*, vol. 21, pp. 513–528.

Romano, J. P. and Siegel, A. F. (1986), *Counterexamples in Probability and Statistics*, Wadsworth and Brooks, Monterey, CA.

Rosenberger, W. F. (1996), "New Directions in Adaptive Designs," *Statistical Science*, vol. 11, pp. 137–149.

Rosenbrock, H. H. (1960), "An Automatic Method for Finding the Greatest or Least Value of a Function," *Computer Journal*, vol. 3, pp. 175–184.

Ross, S. M. (1997), *Introduction to Probability Models* (6th ed.), Academic Press, New York.

Rubinstein, R. Y. (1981), *Simulation and the Monte Carlo Method*, Wiley, New York.

Rubinstein, R. Y. (1986), *Monte Carlo Optimization, Simulation and Sensitivity of Queuing Networks*, Wiley, New York.

Rubinstein, R. Y. and Melamed, B. (1998), *Modern Simulation and Modeling*, Wiley, New York.

Rubinstein, R. Y., Samorodnitsky, G., and Shaked, M. (1985), "Antithetic Variates, Multivariate Dependence, and Simulation of Complex Systems," *Management Science*, vol. 31, pp. 66–77.

Rudin, W. (1976), *Principles of Mathematical Analysis* (3rd ed.), McGraw-Hill, New York.

Rudolph, G. (1994), "Convergence Analysis of Canonical Genetic Algorithms," *IEEE Transactions on Neural Networks*, vol. 5, pp. 96–101.

Rudolph, G. (1997a), *Convergence Properties of Evolutionary Algorithms*, Verlag Kovac, Hamburg.

Rudolph, G. (1997b), "Convergence Rates of Evolutionary Algorithms for a Class of Convex Objective Functions," *Control and Cybernetics*, vol. 26, pp. 375–390.

Rudolph, G. (1998), "Finite Markov Chain Results in Evolutionary Computation: A Tour d'Horizon," *Fundamenta Informaticae*, vol. 34, pp. 1–22.

Ruppert, D. (1985), "A Newton–Raphson Version of the Multivariate Robbins–Monro Procedure," *Annals of Statistics*, vol. 13, pp. 236–245.

Ruppert, D. (1991), "Stochastic Approximation," in *Handbook of Sequential Analysis* (B. K. Ghosh and P. K. Sen, eds.), Marcel Dekker, New York, pp. 503–529.

Russell, S. and Norvig, P. (1995), *Artificial Intelligence—A Modern Approach*, Prentice Hall, Upper Saddle River, NJ.

Rustagi, J. S. (1994), *Optimization Techniques in Statistics*, Academic Press, New York.

Sacks, J. (1958), "Asymptotic Distribution of Stochastic Approximation Procedures," *Annals of Mathematical Statistics*, vol. 29, pp. 373–405.

Sadegh, P. (1997), "Constrained Optimization via Stochastic Approximation with a Simultaneous Perturbation Gradient Approximation," *Automatica*, vol. 33, pp. 889–892.

Sadegh, P. and Spall, J. C. (1998), "Optimal Random Perturbations for Stochastic Approximation with a Simultaneous Perturbation Gradient Approximation," *IEEE Transactions on Automatic Control*, vol. 43, pp. 1480–1484 (correction to references: vol. 44, p. 231).

Salomon, R. (1996), "The Influence of Different Coding Schemes on the Computational Complexity of Genetic Algorithms in Function Optimization," in *Parallel Problem Solving from Nature—PPSN IV* (H. Voigt et al., eds.), Springer-Verlag, New York, pp. 227–235.

Samuel, A. L. (1959), "Some Studies in Machine Learning Using the Game of Checkers," *IBM Journal of Research and Development*, vol. 3, pp. 210–229.

Saridis, G. N. (1977), *Self-Organizing Control of Stochastic Systems*, Marcel Dekker, New York.

Schwarz, G. (1978), "Estimating the Dimension of a Model," *Annals of Statistics*, vol. 6, pp. 461–464.

Schwefel, H.-P. (1977), *Numerische Optimierung von Computer-Modellen mittels der Evolutionsstrategie*, Birkhäuser, Basel (available in translation as *Numerical Optimization of Computer Models*, 1981, Wiley, Chichester).

Schwefel, H.-P. (1995), *Evolution and Optimum Seeking*, Wiley, New York.

Shao, J. (1993), "Linear Model Selection by Cross-Validation," *Journal of the American Statistical Association*, vol. 88, pp. 486–494.

Shao, J. (1997), "An Asymptotic Theory for Linear Model Selection," *Statistica Sinica*, vol. 7, pp. 221–264 (includes commentary by R. Beran, J. S. Rao, M. Stone, R. Tibshirani, and P. Zhang).

Shapiro, A. (1991), "Asymptotic Analysis of Stochastic Programs," *Annals of Operations Research*, vol. 30, pp. 169–186.

Shapiro, A. (1996), "Simulation-Based Optimization—Convergence Analysis and Statistical Inference," *Communications in Statistics—Stochastic Models*, vol. 12, pp. 425–454.

Shumway, R. H., Olsen, D. E., and Levy, L. J. (1981), "Estimation and Tests of Hypotheses for the Initial Mean and Covariance in the Kalman Filter Model," *Communications in Statistics—Theory and Methods*, vol. 10, pp. 1625–1641.

Silvey, S. D. (1980), *Optimal Design*, Chapman and Hall, London.

Solis, F. J. and Wets, J. B. (1981), "Minimization by Random Search Techniques," *Mathematics of Operations Research*, vol. 6, pp. 19–30.

Somerville, P. N. (1997), "Multiple Testing and Simultaneous Confidence Intervals: Calculation of Constants," *Computational Statistics and Data Analysis*, vol. 25, pp. 217–233.

Sorenson, H. W. (1980), *Parameter Estimation*, Marcel Dekker, New York.

Spall, J. C. (ed.) (1988a), *Bayesian Analysis of Time Series and Dynamic Models*, Marcel Dekker, New York.

Spall, J. C. (1988b), "A Stochastic Approximation Algorithm for Large-Dimensional Systems in the Kiefer–Wolfowitz Setting," in *Proceedings of the IEEE Conference on Decision and Control*, 7–9 December 1988, Austin, TX, pp. 1544–1548.

Spall, J. C. (1992), "Multivariate Stochastic Approximation Using a Simultaneous Perturbation Gradient Approximation," *IEEE Transactions on Automatic Control*, vol. 37, pp. 332–341.

Spall, J. C. (1997), "A One-Measurement Form of Simultaneous Perturbation Stochastic Approximation," *Automatica*, vol. 33, pp. 109–112.

Spall, J. C. (1998), "Resampling-Based Calculation of the Information Matrix for General Identification Problems," in *Proceedings of the American Control Conference*, 24–26 June 1998, Philadelphia, PA, pp. 3194–3198.

Spall, J. C. (2000), "Adaptive Stochastic Approximation by the Simultaneous Perturbation Method," *IEEE Transactions on Automatic Control*, vol. 45, pp. 1839–1853.

Spall, J. C. (2003), "Estimation via Markov Chain Monte Carlo," *IEEE Control Systems Magazine*, vol. 23(2), pp. 34–45.

Spall, J. C. and Chin, D. C. (1997), "Traffic-Responsive Signal Timing for System-Wide Traffic Control," *Transportation Research*, Part C, vol. 5, pp. 153–163.

Spall, J. C. and Cristion, J. A. (1994), "Nonlinear Adaptive Control Using Neural Networks: Estimation with a Smoothed Form of Simultaneous Perturbation Gradient Approximation," *Statistica Sinica*, vol. 4, pp. 1–27.

Spall, J. C. and Cristion, J. A. (1997), "A Neural Network Controller for Systems with Unmodeled Dynamics with Application to Wastewater Treatment," *IEEE Transactions on Systems, Man, and Cybernetics*, Part B, vol. 27, pp. 369–375.

Spall, J. C. and Cristion, J. A. (1998), "Model-Free Control of Nonlinear Stochastic Systems with Discrete-Time Measurements," *IEEE Transactions on Automatic Control*, vol. 43, pp. 1198–1210.

Spall, J. C. and Maryak, J. L. (1992), "A Feasible Bayesian Estimator of Quantiles for Projectile Accuracy from Non-i.i.d. Data," *Journal of the American Statistical Association*, vol. 87, pp. 676–681.

Spall, J. C. and Wall, K. D. (1984), "Asymptotic Distribution Theory for the Kalman Filter State Estimator," *Communications in Statistics—Theory and Methods*, vol. 13, pp. 1981–2003.

Spendley, W., Hext, G. R., and Himsworth, F. R. (1962), "Sequential Application of Simplex Design in Optimization and Evolutionary Operation," *Technometrics*, vol. 4, pp. 441–461.

Stark, D. R. and Spall, J. C. (2001), "Computable Bounds on the Rate of Convergence in Evolutionary Computation," in *Proceedings of the American Control Conference*, 25–27 June 2001, Arlington, VA, pp. 918–922.

Stephens, C. and Waelbroeck, H. (1999), "Schemata Evolution and Building Blocks," *Evolutionary Computation*, vol. 7, pp. 109–124.

Stone, M. (1974), "Cross-Validity Choice and Assessment of Statistical Predictors," *Journal of the Royal Statistical Society*, Series B, vol. 39, pp. 111–174.

Storn, R. and Price, K. (1997), "Differential Evolution—A Simple and Efficient Heuristic for Global Optimization Over Continuous Spaces," *Journal of Global Optimization*, vol. 11, pp. 341–359.

Stoyanov, J. M. (1997), *Counterexamples in Probability* (2nd ed.), Wiley, New York.

Stroud, P. D. (2001), "Kalman-Extended Genetic Algorithm for Search in Nonstationary Environments with Noisy Fitness Measurements," *IEEE Transactions on Evolutionary Computation*, vol. 5, pp. 66–77.

Styblinski, M. A. and Tang, T.-S. (1990), "Experiments in Nonconvex Optimization: Stochastic Approximation with Function Smoothing and Simulated Annealing," *Neural Networks*, vol. 3, pp. 467–483.

Sun, F. K. (1982), "A Maximum Likelihood Algorithm for the Mean and Covariance of Nonidentically Distributed Observations," *IEEE Transactions on Automatic Control*, vol. AC-27, pp. 245–247.

Suri, R. (1989), "Perturbation Analysis: The State of the Art and Research Issues Explained via the GI/G/1 Queue," *Proceedings of the IEEE*, vol. 77, pp. 114–137.

Sutton, R. S. (1988), "Learning to Predict by the Method of Temporal Differences," *Machine Learning*, vol. 3, pp. 9–44 (correction to Fig. 3 is in vol. 3, p. 377).

Sutton, R. S. and Barto, A. G. (1998), *Reinforcement Learning: An Introduction*, MIT Press, Cambridge, MA.

Sutton, R. S., Barto, A. G., and Williams, R. J. (1992), "Reinforcement Learning is Direct Adaptive Optimal Control," *IEEE Control Systems Magazine*, vol. 12(2), pp. 19–22.

Suzuki, J. (1995), "A Markov Chain Analysis on Simple Genetic Algorithms," *IEEE Transactions on Systems, Man, and Cybernetics*, vol. 25, pp. 655–659.

Swisher, J. R. and Jacobson, S. H. (1999), "A Survey of Ranking, Selection, and Multiple Comparisons Procedures for Discrete-Event Simulation," in *Proceedings of the Winter Simulation Conference* (P. Farrington, H. Nembhard, D. Sturrock, and G. Evans, eds.), pp. 492–501.

Szu, H. and Hartley, R. (1987), "Fast Simulated Annealing," *Physics Letters A*, vol. 122, pp. 157–162.

Tang, Q.-Y., L'Ecuyer, P., and Chen, H.-F. (1999), "Asymptotic Efficiency of Perturbation-Analysis-Based Stochastic Approximation with Averaging," *SIAM Journal on Control and Optimization*, vol. 37, pp. 1822–1847.

Tanner, M. and Wong, W. (1987), "The Calculation of Posterior Distributions by Data Augmentation (with discussion)," *Journal of the American Statistical Association*, vol. 82, pp. 528–550.

Tesauro, G. (1992), "Practical Issues in Temporal Difference Learning," *Machine Learning*, vol. 8, pp. 257–277.

Tesauro, G. (1995), "Temporal Difference Learning and TD-Gammon," *Communications of the ACM*, vol. 38(3), pp. 58–68.

Tomick, J. J., Arnold, S. F., and Barton, R. R. (1995), "Sample Size Selection for Improved Nelder–Mead Performance," in *Proceedings of the Winter Simulation Conference* (C. Alexopoulos et al., eds.), December 1995, Arlington, VA, pp. 341–345.

Tong, Y. L. (1980), *Probability Inequalities in Multivariate Distributions*, Academic Press, New York.

Tsitsiklis, J. N. and Van Roy, B. (1997), "An Analysis of Temporal-Difference Learning with Function Approximation," *IEEE Transactions on Automatic Control*, vol. 42, pp. 674–690.

Tsitsiklis, J. N. and Van Roy, B. (1999), "Average Cost Temporal-Difference Learning," *Automatica*, vol. 35, pp. 1799–1808.

Vapnik, V. and Chervonenkis, A. Y. (1971), "On the Uniform Convergence of Relative Frequencies of Events and Their Probabilities," *Theory of Probability and Its Applications*, vol. 16, pp. 264–280.

Vázquez-Abad, F. (1999), "Strong Points of Weak Convergence: A Study Using RPA Gradient Estimation for Automatic Learning," *Automatica*, vol. 35, pp. 1255–1274.

Vorontsov, M. A., Carhart, G. W., Cohen, M., and Cauwenberghs, G. (2000), "Adaptive Optics Based on Analog Parallel Stochastic Optimization: Analysis and Experimental Demonstration," *Journal of the Optical Society of America A*, vol. 17, pp. 1440–1453.

Vose, M. (1999), *The Simple Genetic Algorithm*, MIT Press, Cambridge, MA.

Walter, E. and Pronzato, L. (1990), "Qualitative and Quantitative Experiment Design for Phenomenological Models—A Survey," *Automatica*, vol. 26, pp. 195–213.

Wang, I.-J. (1996), "Analysis of Stochastic Approximation and Related Algorithms," Ph.D. dissertation, Purdue University, School of Electrical Engineering, West Lafayette, IN.

Wang, I.-J. and Chong, E. K. P. (1998), "A Deterministic Analysis of Stochastic Approximation with Randomized Differences," *IEEE Transactions on Automatic Control*, vol. 43, pp. 1745–1749.

Wang, I.-J. and Spall, J. C. (1999), "A Constrained Simultaneous Perturbation Stochastic Approximation Algorithm Based on Penalty Functions," in *Proceedings of the American Control Conference*, 2–4 June 1999, San Diego, CA, pp. 393–399.

Wang, Z.-Q., Manry, M. T., and Schiano, J. L. (2000), "LMS Learning Algorithms: Misconceptions and New Results on Convergence," *IEEE Transactions on Neural Networks*, vol. 11, pp. 47–56.

Watkins, C. and Dayan, P. (1992), "Q-Learning," *Machine Learning*, vol. 8, pp. 279–292.

Watson, G. S. (1987), "A Method for Discovering Kantorovich-Type Inequalities and a Probabilistic Interpretation," *Linear Algebra and Its Applications*, vol. 97, pp. 211–217.

Wei, C. Z. (1987), "Multivariate Adaptive Stochastic Approximation," *Annals of Statistics*, vol. 15, pp. 1115–1130.

Wei, C. Z. (1992), "On Predictive Least Squares Principles," *Annals of Statistics*, vol. 20, pp. 1–42.

Weile, D. S. and Michielssen, E. (1997), "Genetic Algorithm Optimization Applied to Electromagnetics: A Review," *IEEE Transactions on Antennas and Propagation*, vol. 45, pp. 343–353.

White, H. (1989), "Some Asymptotic Results for Learning in Single Hidden Layer Feedforward Neural Networks," *Journal of the American Statistical Association*, vol. 84, pp. 1003–1013.

Widrow, B. and Stearns, S. (1985), *Adaptive Signal Processing*, Prentice-Hall, Englewood Cliffs, NJ.

Wilcox, R. R. (1984), "A Table for Rinott's Selection Procedure," *Journal of Quality Technology*, vol. 16, pp. 97–100.

Wilkins, J. E. (1955), "The Average of the Reciprocal of a Function," *Proceedings of the American Mathematical Society*, vol. 6, pp. 806–815.

Wilks, S. S. (1962), *Mathematical Statistics*, Wiley, New York.

Wolpert, D. H. and Macready, W. G. (1997), "No Free Lunch Theorems for Optimization," *IEEE Transactions on Evolutionary Computation*, vol. 1, pp. 67–82.

Wu, C. F. J. and Hamada, M. (2000), *Experiments: Planning, Analysis, and Parameter Design Optimization*, Wiley, New York.

Yakowitz, S. J. and Fisher, L. (1973), "On Sequential Search for the Maximum of an Unknown Function," *Journal of Mathematical Analysis and Applications*, vol. 41, pp. 234–259.

Yin, G. (1999), "Rates of Convergence for a Class of Global Stochastic Optimization Algorithms," *SIAM Journal on Optimization*, vol. 10, pp. 99–120.

Yin, G. (2002), "Stochastic Approximation: Theory and Applications," in *Handbook of Stochastic Analysis and Applications* (D. Kannan and V. Lakshmikantham, eds.), Marcel Dekker, New York, Chapter 10.

Yin, G. and Yin, K. (1996), "Passive Stochastic Approximation with Constant Step Size and Window Width," *IEEE Transactions on Automatic Control*, vol. 41, pp. 90–106.

Yin, G. and Zhu, Y. (1992), "Averaging Procedures in Adaptive Filtering: An Efficient Approach," *IEEE Transactions on Automatic Control*, vol. 37, pp. 466–475.

Yin, G., Rudolph, G., and Schwefel, H.-P. (1996), "Analyzing the $(1, \lambda)$ Evolution Strategy via Stochastic Approximation Methods," *Evolutionary Computation*, vol. 3, pp. 473–489.

Young, P. C. (1984), *Recursive Estimation and Time-Series Analysis*, Springer-Verlag, New York.

Zhigljavsky, A. A. (1991), *Theory of Global Random Search*, Kluwer Academic, Boston.

Zhu, X. and Spall, J. C. (2002), "A Modified Second-Order SPSA Optimization Algorithm for Finite Samples," *International Journal of Adaptive Control and Signal Processing*, vol. 16, pp. 397–409.

FREQUENTLY USED NOTATION

Note: Inevitably, several terms are used in the text in more than one way, as indicated below. The intended meaning should always be clear from the context.

All vectors are treated as column vectors.

$\equiv$ denotes "defined as" (e.g., $a \equiv b$, a defined to be the same as b).

$\Rightarrow$ should be read as "implies" (as in $A \Rightarrow B$ means "A implies B").

❑ marks the end of a proof or example.

A dot $\cdot$ as an argument for a function or operator is a generic reference to that function or operator without regard to a specific argument.

A^T represents the transpose of a matrix or vector A.

S^c represents the complement of a set S.

$P(\cdot)$ denotes the probability of an event.

$E(\cdot)$ denotes the expected value of a random quantity.

$\text{cov}(\boldsymbol{x})$ denotes the covariance matrix of the random vector $\boldsymbol{x}$; $\text{cov}(\boldsymbol{x}) = E[(\boldsymbol{x} - E(\boldsymbol{x}))(\boldsymbol{x} - E(\boldsymbol{x}))^T]$.

$\log(x)$ or $\log x$ represents the *natural* (base $e = 2.71828\ldots$) logarithm of x.

$\exp(x) = e^x$ for $e = 2.71828\ldots$ (sometimes used if x is a complicated expression).

$\mathbb{R}^p$ represents Euclidean space of dimension p; hence $\mathbb{R}^1$ and $\mathbb{R}^2$ denote the real-number line and plane, respectively.

In discussions of stochastic convergence: a.s. denotes "almost surely"; pr. denotes "in probability"; m.s. denotes "in mean square"; and dist. denotes "in distribution."

$N(\boldsymbol{\mu}, \boldsymbol{\Sigma})$ denotes a (generally multivariate) normal distribution with mean vector $\boldsymbol{\mu}$ and covariance matrix $\boldsymbol{\Sigma}$. The terms *normal* and *Gaussian* are used interchangeably to describe this distribution.

$U(a, b)$ denotes a uniform distribution over the interval $(a, b) \subset \mathbb{R}^1$.

$\boldsymbol{\theta}$ is the vector of parameters being estimated in the search/optimization process.

$L(\boldsymbol{\theta})$ is a loss function being minimized.

$y(\boldsymbol{\theta})$ represents a noisy measurement of $L(\boldsymbol{\theta})$ ($y(\boldsymbol{\theta}) = L(\boldsymbol{\theta}) + \varepsilon$).

ε represents the noise in $y(\boldsymbol{\theta})$; $\varepsilon = \varepsilon(\boldsymbol{\theta})$ when necessary to emphasize dependence on $\boldsymbol{\theta}$.

$\boldsymbol{g}(\boldsymbol{\theta})$ is a function for which the root of $\boldsymbol{g}(\boldsymbol{\theta}) = \boldsymbol{0}$ is to be found.

$\boldsymbol{Y}(\boldsymbol{\theta})$ represents a noisy measurement of $\boldsymbol{g}(\boldsymbol{\theta})$ ($\boldsymbol{Y}(\boldsymbol{\theta}) = \boldsymbol{g}(\boldsymbol{\theta}) + \text{noise}$).

$\arg\min_{\boldsymbol{\theta}}\{L(\boldsymbol{\theta})\}$ denotes a value (or set of values) of the vector $\boldsymbol{\theta}$ (i.e., the argument) minimizing the criterion $L(\boldsymbol{\theta})$.

$\Theta \subseteq \mathbb{R}^p$ is the domain representing constraints on $\boldsymbol{\theta}$ in the search for a minimum to $L(\boldsymbol{\theta})$ or a root to $\boldsymbol{g}(\boldsymbol{\theta}) = \boldsymbol{0}$.

$\boldsymbol{\theta}^*$ is an optimal value of $\boldsymbol{\theta}$ (corresponding to $\arg\min_{\boldsymbol{\theta}}\{L(\boldsymbol{\theta})\}$) or a root of $\boldsymbol{g}(\boldsymbol{\theta}) = \boldsymbol{0}$; $\Theta^* \subseteq \Theta$ is a corresponding *set* of values when there is more than one $\boldsymbol{\theta}^*$.

e: (1) Base of natural logarithms ($e = 2.71828\ldots$).
(2) Error term in measurement of $\boldsymbol{g}(\boldsymbol{\theta})$ (bold $\boldsymbol{e}$ for $\boldsymbol{g}$ multivariate; $\boldsymbol{Y}(\boldsymbol{\theta}) = \boldsymbol{g}(\boldsymbol{\theta}) + \boldsymbol{e}(\boldsymbol{\theta})$).

$\boldsymbol{H}$: (1) Hessian matrix of $L(\boldsymbol{\theta})$ (or Jacobian matrix of $\boldsymbol{g}(\boldsymbol{\theta})$) (introduced in Chapter 1).
(2) Measurement matrix in a linear model (introduced in Chapter 3).

$\boldsymbol{I}_a$ denotes the $a \times a$ identity matrix ($\boldsymbol{I}$ without a subscript denotes an identity matrix without specified dimension).

In most cases $\hat{\boldsymbol{\theta}}_k$ is the estimate of $\boldsymbol{\theta}$ in the kth iteration of a search algorithm. In some cases $\hat{\boldsymbol{\theta}}_k$ refers to a generic (batch or nonbatch [i.e., recursive]) estimate of $\boldsymbol{\theta}$ based on k data points (this definition is used when describing methods that apply to general estimators, thereby avoiding the need to present separate cases for the batch and nonbatch settings). The use of $\hat{\boldsymbol{\theta}}_k$ should always be clear from the context.

$\hat{\boldsymbol{\theta}}^{(k)}$ is a batch (nonrecursive) estimate of $\boldsymbol{\theta}$ based on k data points. In most instances, $\hat{\boldsymbol{\theta}}^{(k)}$ is an ordinary least-squares estimate for $\boldsymbol{\theta}$ appearing in a linear regression model.

diag([a, b, c]) denotes a diagonal matrix with elements a, b, and c at the (1, 1), (2, 2), and (3, 3) locations, respectively (with obvious modifications for diagonal matrices that are not 3×3).

$[a, b]^i$, where i is a positive integer, denotes the i-fold Cartesian product of the interval $[a, b]$, where a and b are real numbers. Hence, $[a, b]^i = [a, b] \times [a, b] \times \ldots \times [a, b] \subseteq \mathbb{R}^i$.

$I_{\{\cdot\}}$ is an indicator function (i.e., $I_{\{\cdot\}} = \{1$ if the event $\{\cdot\}$ is true; 0 otherwise$\}$).

$O(\cdot)$ is the standard big-O order notation, implying that a function $f(x) = O(h(x))$ is such that $f(x)/h(x)$ remains bounded as $x \to x_0$, where x_0 is some limiting value and $h(x)$ is some function. In practice, x_0 is usually 0 or ∞.

Following the above, $o(\cdot)$ is the standard little-o order notation, implying that a function $f(x) = o(h(x))$ is such that $f(x)/h(x) \to 0$ as $x \to x_0$.

i.i.d. denotes "independent, identically distributed" in reference to a random sequence.

For two symmetric matrices $\boldsymbol{A}$ and $\boldsymbol{B}$ of identical dimension, $\boldsymbol{A} \geq (\leq)\ \boldsymbol{B}$ implies that $\boldsymbol{A} - \boldsymbol{B}$ is positive (negative) semidefinite; obvious analogous definition for $\boldsymbol{A} > (<)\ \boldsymbol{B}$.

det($\boldsymbol{A}$) denotes the determinant of a square matrix $\boldsymbol{A}$.

$\|\boldsymbol{x}\|$ denotes the standard Euclidean norm $\sqrt{\boldsymbol{x}^T\boldsymbol{x}}$ of a vector $\boldsymbol{x}$ unless noted otherwise.

MSE denotes mean-squared error.

INDEX

Note: For topics with multiple pages listed, the page numbers in bold face indicate the primary listing (when clearly identifiable).

WILEY-INTERSCIENCE SERIES IN DISCRETE MATHEMATICS AND OPTIMIZATION

AARTS AND KORST • Simulated Annealing and Boltzmann Machines: A Stochastic Approach to Combinatorial Optimization and Neural Computing
AARTS AND LENSTRA • Local Search in Combinatorial Optimization
ALON, SPENCER, AND ERDŐS • The Probabilistic Method, Second Edition
ANDERSON AND NASH • Linear Programming in Infinite-Dimensional Spaces: Theory and Application
ARLINGHAUS, ARLINGHAUS, AND HARARY • Graph Theory and Geography: An Interactive View E-Book
AZENCOTT • Simulated Annealing: Parallelization Techniques
BARTHÉLEMY AND GUÉNOCHE • Trees and Proximity Representations
BAZARRA, JARVIS, AND SHERALI • Linear Programming and Network Flows
CHANDRU AND HOOKER • Optimization Methods for Logical Inference
CHONG AND ŻAK • An Introduction to Optimization, Second Edition
COFFMAN AND LUEKER • Probabilistic Analysis of Packing and Partitioning Algorithms
COOK, CUNNINGHAM, PULLEYBLANK, AND SCHRIJVER • Combinatorial Optimization
DASKIN • Network and Discrete Location: Modes, Algorithms and Applications
DINITZ AND STINSON • Contemporary Design Theory: A Collection of Surveys
DU AND KO • Theory of Computational Complexity
ERICKSON • Introduction to Combinatorics
GLOVER, KLINGHAM, AND PHILLIPS • Network Models in Optimization and Their Practical Problems
GOLSHTEIN AND TRETYAKOV • Modified Lagrangians and Monotone Maps in Optimization
GONDRAN AND MINOUX • Graphs and Algorithms *(Translated by S. Vajdā)*
GRAHAM, ROTHSCHILD, AND SPENCER • Ramsey Theory, Second Edition
GROSS AND TUCKER • Topological Graph Theory
HALL • Combinatorial Theory, Second Edition
HOOKER • Logic-Based Methods for Optimization: Combining Optimization and Constraint Satisfaction
IMRICH AND KLAVŽAR • Product Graphs: Structure and Recognition
JANSON, LUCZAK, AND RUCINSKI • Random Graphs
JENSEN AND TOFT • Graph Coloring Problems
KAPLAN • Maxima and Minima with Applications: Practical Optimization and Duality
LAWLER, LENSTRA, RINNOOY KAN, AND SHMOYS, Editors • The Traveling Salesman Problem: A Guided Tour of Combinatorial Optimization
LAYWINE AND MULLEN • Discrete Mathematics Using Latin Squares
LEVITIN • Perturbation Theory in Mathematical Programming Applications
MAHMOUD • Evolution of Random Search Trees
MAHMOUD • Sorting: A Distribution Theory
MARTELLI • Introduction to Discrete Dynamical Systems and Chaos
MARTELLO AND TOTH • Knapsack Problems: Algorithms and Computer Implementations
McALOON AND TRETKOFF • Optimization and Computational Logic
MERRIS • Graph Theory

MINC • Nonnegative Matrices
MINOUX • Mathematical Programming: Theory and Algorithms *(Translated by S. Vajdā)*
MIRCHANDANI AND FRANCIS, Editors • Discrete Location Theory
NEMHAUSER AND WOLSEY • Integer and Combinatorial Optimization
NEMIROVSKY AND YUDIN • Problem Complexity and Method Efficiency in Optimization *(Translated bv E. R. Dawson)*
PACH AND AGARWAL • Combinatorial Geometry
PLESS • Introduction to the Theory of Error-Correcting Codes, Third Edition
ROOS AND VIAL • Ph. Theory and Algorithms for Linear Optimization: An Interior Point Approach
SCHEINERMAN AND ULLMAN • Fractional Graph Theory: A Rational Approach to the Theory of Graphs
SCHRIJVER • Theory of Linear and Integer Programming
SPALL • Introduction to Stochastic Search and Optimization: Estimation, Simulation, and Control
SZPANKOWSKI • Average Case Analysis of Algorithms on Sequences
TOMESCU • Problems in Combinatorics and Graph Theory *(Translated by R. A. Melter)*
TUCKER • Applied Combinatorics, Second Edition
WOLSEY • Integer Programming
YE • Interior Point Algorithms: Theory and Analysis

Printed in the United States
131144LV00001B/29-30/P

9 780471 330523